Student Solutions Manual

Biocalculus
Calculus, Probability, and Statistics for the Life Sciences

James Stewart
McMaster University and University of Toronto

Troy Day
Queen's University

Prepared by

Joshua D. Babbin

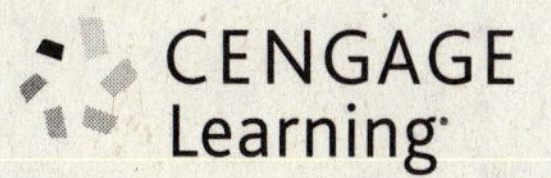

Australia • Brazil • Mexico • Singapore • United Kingdom • United States

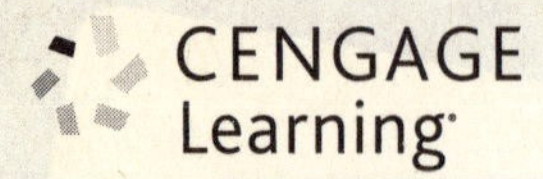

© 2016 Cengage Learning

WCN: 01-100-101

ALL RIGHTS RESERVED. No part of this work covered by the copyright herein may be reproduced, transmitted, stored, or used in any form or by any means graphic, electronic, or mechanical, including but not limited to photocopying, recording, scanning, digitizing, taping, Web distribution, information networks, or information storage and retrieval systems, except as permitted under Section 107 or 108 of the 1976 United States Copyright Act, without the prior written permission of the publisher.

For product information and technology assistance, contact us at
Cengage Learning Customer & Sales Support, 1-800-354-9706.

For permission to use material from this text or product, submit all requests online at
www.cengage.com/permissions
Further permissions questions can be emailed to
permissionrequest@cengage.com.

ISBN: 978-1-305-11406-7

Cengage Learning
20 Channel Center Street
Boston, MA 02210
USA

Cengage Learning is a leading provider of customized learning solutions with office locations around the globe, including Singapore, the United Kingdom, Australia, Mexico, Brazil, and Japan. Locate your local office at: **www.cengage.com/global**.

Cengage Learning products are represented in Canada by Nelson Education, Ltd.

To learn more about Cengage Learning Solutions, visit **www.cengage.com**.

Purchase any of our products at your local college store or at our preferred online store **www.cengagebrain.com**.

Printed in the United States of America
Print Number: 02 Print Year: 2015

Table of Contents

© 2016 Cengage Learning. All Rights Reserved. May not be scanned, copied or duplicated, or posted to a publicly accessible website, in whole or in part.

☐ DIAGNOSTIC TESTS

Test A Algebra

1. (a) $(-3)^4 = (-3)(-3)(-3)(-3) = 81$

(b) $-3^4 = -(3)(3)(3)(3) = -81$

(c) $3^{-4} = \dfrac{1}{3^4} = \dfrac{1}{81}$

(d) $\dfrac{5^{23}}{5^{21}} = 5^{23-21} = 5^2 = 25$

(e) $\left(\frac{2}{3}\right)^{-2} = \left(\frac{3}{2}\right)^2 = \frac{9}{4}$

(f) $16^{-3/4} = \dfrac{1}{16^{3/4}} = \dfrac{1}{\left(\sqrt[4]{16}\right)^3} = \dfrac{1}{2^3} = \dfrac{1}{8}$

2. (a) Note that $\sqrt{200} = \sqrt{100 \cdot 2} = 10\sqrt{2}$ and $\sqrt{32} = \sqrt{16 \cdot 2} = 4\sqrt{2}$. Thus $\sqrt{200} - \sqrt{32} = 10\sqrt{2} - 4\sqrt{2} = 6\sqrt{2}$.

(b) $(3a^3b^3)(4ab^2)^2 = 3a^3b^3 16a^2b^4 = 48a^5b^7$

(c) $\left(\dfrac{3x^{3/2}y^3}{x^2y^{-1/2}}\right)^{-2} = \left(\dfrac{x^2y^{-1/2}}{3x^{3/2}y^3}\right)^2 = \dfrac{(x^2y^{-1/2})^2}{(3x^{3/2}y^3)^2} = \dfrac{x^4y^{-1}}{9x^3y^6} = \dfrac{x^4}{9x^3y^6y} = \dfrac{x}{9y^7}$

3. (a) $3(x+6) + 4(2x-5) = 3x + 18 + 8x - 20 = 11x - 2$

(b) $(x+3)(4x-5) = 4x^2 - 5x + 12x - 15 = 4x^2 + 7x - 15$

(c) $\left(\sqrt{a}+\sqrt{b}\right)\left(\sqrt{a}-\sqrt{b}\right) = \left(\sqrt{a}\right)^2 - \sqrt{a}\sqrt{b} + \sqrt{a}\sqrt{b} - \left(\sqrt{b}\right)^2 = a - b$

Or: Use the formula for the difference of two squares to see that $\left(\sqrt{a}+\sqrt{b}\right)\left(\sqrt{a}-\sqrt{b}\right) = \left(\sqrt{a}\right)^2 - \left(\sqrt{b}\right)^2 = a - b$.

(d) $(2x+3)^2 = (2x+3)(2x+3) = 4x^2 + 6x + 6x + 9 = 4x^2 + 12x + 9$.

Note: A quicker way to expand this binomial is to use the formula $(a+b)^2 = a^2 + 2ab + b^2$ with $a = 2x$ and $b = 3$:

$(2x+3)^2 = (2x)^2 + 2(2x)(3) + 3^2 = 4x^2 + 12x + 9$

(e) See Reference Page 1 for the binomial formula $(a+b)^3 = a^3 + 3a^2b + 3ab^2 + b^3$. Using it, we get

$(x+2)^3 = x^3 + 3x^2(2) + 3x(2^2) + 2^3 = x^3 + 6x^2 + 12x + 8$.

4. (a) Using the difference of two squares formula, $a^2 - b^2 = (a+b)(a-b)$, we have

$4x^2 - 25 = (2x)^2 - 5^2 = (2x+5)(2x-5)$.

(b) Factoring by trial and error, we get $2x^2 + 5x - 12 = (2x-3)(x+4)$.

(c) Using factoring by grouping and the difference of two squares formula, we have

$x^3 - 3x^2 - 4x + 12 = x^2(x-3) - 4(x-3) = (x^2-4)(x-3) = (x-2)(x+2)(x-3)$.

(d) $x^4 + 27x = x(x^3 + 27) = x(x+3)(x^2 - 3x + 9)$

This last expression was obtained using the sum of two cubes formula, $a^3 + b^3 = (a+b)(a^2 - ab + b^2)$ with $a = x$ and $b = 3$. [See Reference Page 1 in the textbook.]

(e) The smallest exponent on x is $-\frac{1}{2}$, so we will factor out $x^{-1/2}$.

$3x^{3/2} - 9x^{1/2} + 6x^{-1/2} = 3x^{-1/2}(x^2 - 3x + 2) = 3x^{-1/2}(x-1)(x-2)$

(f) $x^3y - 4xy = xy(x^2 - 4) = xy(x-2)(x+2)$

© 2016 Cengage Learning. All Rights Reserved. May not be scanned, copied or duplicated, or posted to a publicly accessible website, in whole or in part.

5. (a) $\dfrac{x^2+3x+2}{x^2-x-2} = \dfrac{(x+1)(x+2)}{(x+1)(x-2)} = \dfrac{x+2}{x-2}$

(b) $\dfrac{2x^2-x-1}{x^2-9} \cdot \dfrac{x+3}{2x+1} = \dfrac{(2x+1)(x-1)}{(x-3)(x+3)} \cdot \dfrac{x+3}{2x+1} = \dfrac{x-1}{x-3}$

(c) $\dfrac{x^2}{x^2-4} - \dfrac{x+1}{x+2} = \dfrac{x^2}{(x-2)(x+2)} - \dfrac{x+1}{x+2} = \dfrac{x^2}{(x-2)(x+2)} - \dfrac{x+1}{x+2} \cdot \dfrac{x-2}{x-2} = \dfrac{x^2-(x+1)(x-2)}{(x-2)(x+2)}$

$$= \frac{x^2-(x^2-x-2)}{(x+2)(x-2)} = \frac{x+2}{(x+2)(x-2)} = \frac{1}{x-2}$$

(d) $\dfrac{\dfrac{y}{x}-\dfrac{x}{y}}{\dfrac{1}{y}-\dfrac{1}{x}} = \dfrac{\dfrac{y}{x}-\dfrac{x}{y}}{\dfrac{1}{y}-\dfrac{1}{x}} \cdot \dfrac{xy}{xy} = \dfrac{y^2-x^2}{x-y} = \dfrac{(y-x)(y+x)}{-(y-x)} = \dfrac{y+x}{-1} = -(x+y)$

6. (a) $\dfrac{\sqrt{10}}{\sqrt{5}-2} = \dfrac{\sqrt{10}}{\sqrt{5}-2} \cdot \dfrac{\sqrt{5}+2}{\sqrt{5}+2} = \dfrac{\sqrt{50}+2\sqrt{10}}{\left(\sqrt{5}\right)^2-2^2} = \dfrac{5\sqrt{2}+2\sqrt{10}}{5-4} = 5\sqrt{2}+2\sqrt{10}$

(b) $\dfrac{\sqrt{4+h}-2}{h} = \dfrac{\sqrt{4+h}-2}{h} \cdot \dfrac{\sqrt{4+h}+2}{\sqrt{4+h}+2} = \dfrac{4+h-4}{h\left(\sqrt{4+h}+2\right)} = \dfrac{h}{h\left(\sqrt{4+h}+2\right)} = \dfrac{1}{\sqrt{4+h}+2}$

7. (a) $x^2+x+1 = \left(x^2+x+\frac{1}{4}\right)+1-\frac{1}{4} = \left(x+\frac{1}{2}\right)^2+\frac{3}{4}$

(b) $2x^2-12x+11 = 2(x^2-6x)+11 = 2(x^2-6x+9-9)+11 = 2(x^2-6x+9)-18+11 = 2(x-3)^2-7$

8. (a) $x+5 = 14-\frac{1}{2}x \;\Leftrightarrow\; x+\frac{1}{2}x = 14-5 \;\Leftrightarrow\; \frac{3}{2}x = 9 \;\Leftrightarrow\; x = \frac{2}{3}\cdot 9 \;\Leftrightarrow\; x = 6$

(b) $\dfrac{2x}{x+1} = \dfrac{2x-1}{x} \;\Rightarrow\; 2x^2 = (2x-1)(x+1) \;\Leftrightarrow\; 2x^2 = 2x^2+x-1 \;\Leftrightarrow\; x = 1$

(c) $x^2-x-12 = 0 \;\Leftrightarrow\; (x+3)(x-4) = 0 \;\Leftrightarrow\; x+3 = 0$ or $x-4 = 0 \;\Leftrightarrow\; x = -3$ or $x = 4$

(d) By the quadratic formula, $2x^2+4x+1 = 0 \;\Leftrightarrow$

$$x = \frac{-4 \pm \sqrt{4^2-4(2)(1)}}{2(2)} = \frac{-4 \pm \sqrt{8}}{4} = \frac{-4 \pm 2\sqrt{2}}{4} = \frac{2\left(-2 \pm \sqrt{2}\right)}{4} = \frac{-2 \pm \sqrt{2}}{2} = -1 \pm \tfrac{1}{2}\sqrt{2}.$$

(e) $x^4-3x^2+2 = 0 \;\Leftrightarrow\; (x^2-1)(x^2-2) = 0 \;\Leftrightarrow\; x^2-1 = 0$ or $x^2-2 = 0 \;\Leftrightarrow\; x^2 = 1$ or $x^2 = 2 \;\Leftrightarrow$

$x = \pm 1$ or $x = \pm\sqrt{2}$

(f) $3\,|x-4| = 10 \;\Leftrightarrow\; |x-4| = \frac{10}{3} \;\Leftrightarrow\; x-4 = -\frac{10}{3}$ or $x-4 = \frac{10}{3} \;\Leftrightarrow\; x = \frac{2}{3}$ or $x = \frac{22}{3}$

(g) Multiplying through $2x(4-x)^{-1/2} - 3\sqrt{4-x} = 0$ by $(4-x)^{1/2}$ gives $2x-3(4-x) = 0 \;\Leftrightarrow$

$2x-12+3x = 0 \;\Leftrightarrow\; 5x-12 = 0 \;\Leftrightarrow\; 5x = 12 \;\Leftrightarrow\; x = \frac{12}{5}$.

9. (a) $-4 < 5-3x \leq 17 \;\Leftrightarrow\; -9 < -3x \leq 12 \;\Leftrightarrow\; 3 > x \geq -4$ or $-4 \leq x < 3$.
In interval notation, the answer is $[-4, 3)$.

(b) $x^2 < 2x+8 \;\Leftrightarrow\; x^2-2x-8 < 0 \;\Leftrightarrow\; (x+2)(x-4) < 0$. Now, $(x+2)(x-4)$ will change sign at the critical values $x = -2$ and $x = 4$. Thus the possible intervals of solution are $(-\infty, -2)$, $(-2, 4)$, and $(4, \infty)$. By choosing a single test value from each interval, we see that $(-2, 4)$ is the only interval that satisfies the inequality.

© 2016 Cengage Learning. All Rights Reserved. May not be scanned, copied or duplicated, or posted to a publicly accessible website, in whole or in part.

(c) The inequality $x(x-1)(x+2) > 0$ has critical values of $-2, 0$, and 1. The corresponding possible intervals of solution are $(-\infty, -2)$, $(-2, 0)$, $(0, 1)$ and $(1, \infty)$. By choosing a single test value from each interval, we see that both intervals $(-2, 0)$ and $(1, \infty)$ satisfy the inequality. Thus, the solution is the union of these two intervals: $(-2, 0) \cup (1, \infty)$.

(d) $|x-4| < 3 \quad \Leftrightarrow \quad -3 < x-4 < 3 \quad \Leftrightarrow \quad 1 < x < 7$. In interval notation, the answer is $(1, 7)$.

(e) $\dfrac{2x-3}{x+1} \leq 1 \quad \Leftrightarrow \quad \dfrac{2x-3}{x+1} - 1 \leq 0 \quad \Leftrightarrow \quad \dfrac{2x-3}{x+1} - \dfrac{x+1}{x+1} \leq 0 \quad \Leftrightarrow \quad \dfrac{2x-3-x-1}{x+1} \leq 0 \quad \Leftrightarrow \quad \dfrac{x-4}{x+1} \leq 0.$

Now, the expression $\dfrac{x-4}{x+1}$ may change signs at the critical values $x = -1$ and $x = 4$, so the possible intervals of solution are $(-\infty, -1)$, $(-1, 4]$, and $[4, \infty)$. By choosing a single test value from each interval, we see that $(-1, 4]$ is the only interval that satisfies the inequality.

10. (a) False. In order for the statement to be true, it must hold for all real numbers, so, to show that the statement is false, pick $p = 1$ and $q = 2$ and observe that $(1+2)^2 \neq 1^2 + 2^2$. In general, $(p+q)^2 = p^2 + 2pq + q^2$.

(b) True as long as a and b are nonnegative real numbers. To see this, think in terms of the laws of exponents: $\sqrt{ab} = (ab)^{1/2} = a^{1/2}b^{1/2} = \sqrt{a}\,\sqrt{b}$.

(c) False. To see this, let $p = 1$ and $q = 2$, then $\sqrt{1^2 + 2^2} \neq 1 + 2$.

(d) False. To see this, let $T = 1$ and $C = 2$, then $\dfrac{1 + 1(2)}{2} \neq 1 + 1$.

(e) False. To see this, let $x = 2$ and $y = 3$, then $\dfrac{1}{2-3} \neq \dfrac{1}{2} - \dfrac{1}{3}$.

(f) True since $\dfrac{1/x}{a/x - b/x} \cdot \dfrac{x}{x} = \dfrac{1}{a-b}$, as long as $x \neq 0$ and $a - b \neq 0$.

Test B Analytic Geometry

1. (a) Using the point $(2, -5)$ and $m = -3$ in the point-slope equation of a line, $y - y_1 = m(x - x_1)$, we get $y - (-5) = -3(x-2) \quad \Rightarrow \quad y + 5 = -3x + 6 \quad \Rightarrow \quad y = -3x + 1$.

(b) A line parallel to the x-axis must be horizontal and thus have a slope of 0. Since the line passes through the point $(2, -5)$, the y-coordinate of every point on the line is -5, so the equation is $y = -5$.

(c) A line parallel to the y-axis is vertical with undefined slope. So the x-coordinate of every point on the line is 2 and so the equation is $x = 2$.

(d) Note that $2x - 4y = 3 \quad \Rightarrow \quad -4y = -2x + 3 \quad \Rightarrow \quad y = \frac{1}{2}x - \frac{3}{4}$. Thus the slope of the given line is $m = \frac{1}{2}$. Hence, the slope of the line we're looking for is also $\frac{1}{2}$ (since the line we're looking for is required to be parallel to the given line). So the equation of the line is $y - (-5) = \frac{1}{2}(x-2) \quad \Rightarrow \quad y + 5 = \frac{1}{2}x - 1 \quad \Rightarrow \quad y = \frac{1}{2}x - 6$.

2. First we'll find the distance between the two given points in order to obtain the radius, r, of the circle: $r = \sqrt{[3-(-1)]^2 + (-2-4)^2} = \sqrt{4^2 + (-6)^2} = \sqrt{52}$. Next use the standard equation of a circle, $(x-h)^2 + (y-k)^2 = r^2$, where (h, k) is the center, to get $(x+1)^2 + (y-4)^2 = 52$.

© 2016 Cengage Learning. All Rights Reserved. May not be scanned, copied or duplicated, or posted to a publicly accessible website, in whole or in part.

3. We must rewrite the equation in standard form in order to identify the center and radius. Note that $x^2+y^2-6x+10y+9=0 \;\Rightarrow\; x^2-6x+9+y^2+10y=0$. For the left-hand side of the latter equation, we factor the first three terms and complete the square on the last two terms as follows: $x^2-6x+9+y^2+10y=0 \;\Rightarrow$ $(x-3)^2+y^2+10y+25=25 \;\Rightarrow\; (x-3)^2+(y+5)^2=25$. Thus, the center of the circle is $(3,-5)$ and the radius is 5.

4. (a) $A(-7,4)$ and $B(5,-12) \;\Rightarrow\; m_{AB}=\dfrac{-12-4}{5-(-7)}=\dfrac{-16}{12}=-\dfrac{4}{3}$

(b) $y-4=-\frac{4}{3}[x-(-7)] \;\Rightarrow\; y-4=-\frac{4}{3}x-\frac{28}{3} \;\Rightarrow\; 3y-12=-4x-28 \;\Rightarrow\; 4x+3y+16=0$. Putting $y=0$, we get $4x+16=0$, so the x-intercept is -4, and substituting 0 for x results in a y-intercept of $-\frac{16}{3}$.

(c) The midpoint is obtained by averaging the corresponding coordinates of both points: $\left(\frac{-7+5}{2},\frac{4+(-12)}{2}\right)=(-1,-4)$.

(d) $d=\sqrt{[5-(-7)]^2+(-12-4)^2}=\sqrt{12^2+(-16)^2}=\sqrt{144+256}=\sqrt{400}=20$

(e) The perpendicular bisector is the line that intersects the line segment $\overline{AB}$ at a right angle through its midpoint. Thus the perpendicular bisector passes through $(-1,-4)$ and has slope $\frac{3}{4}$ [the slope is obtained by taking the negative reciprocal of the answer from part (a)]. So the perpendicular bisector is given by $y+4=\frac{3}{4}[x-(-1)]$ or $3x-4y=13$.

(f) The center of the required circle is the midpoint of $\overline{AB}$, and the radius is half the length of $\overline{AB}$, which is 10. Thus, the equation is $(x+1)^2+(y+4)^2=100$.

5. (a) Graph the corresponding horizontal lines (given by the equations $y=-1$ and $y=3$) as solid lines. The inequality $y\ge -1$ describes the points (x,y) that lie on or *above* the line $y=-1$. The inequality $y\le 3$ describes the points (x,y) that lie on or *below* the line $y=3$. So the pair of inequalities $-1\le y\le 3$ describes the points that lie on or *between* the lines $y=-1$ and $y=3$.

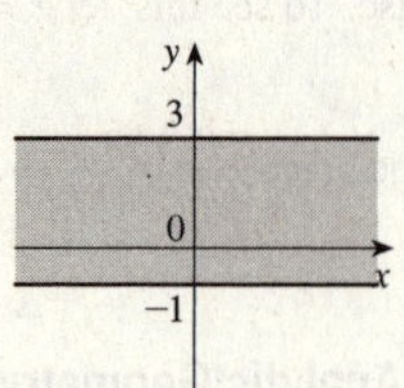

(b) Note that the given inequalities can be written as $-4<x<4$ and $-2<y<2$, respectively. So the region lies between the vertical lines $x=-4$ and $x=4$ and between the horizontal lines $y=-2$ and $y=2$. As shown in the graph, the region common to both graphs is a rectangle (minus its edges) centered at the origin.

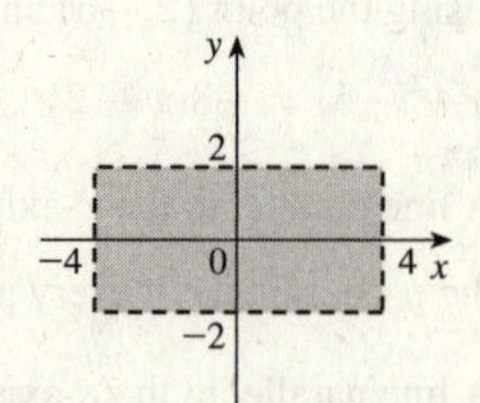

(c) We first graph $y=1-\frac{1}{2}x$ as a dotted line. Since $y<1-\frac{1}{2}x$, the points in the region lie *below* this line.

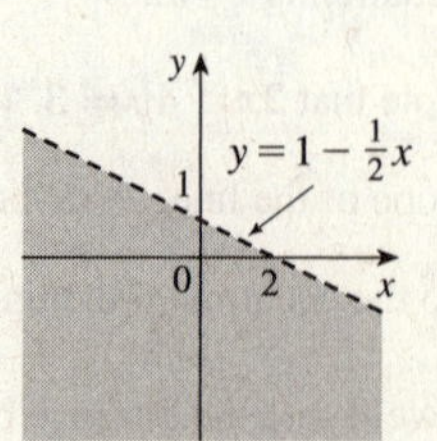

© 2016 Cengage Learning. All Rights Reserved. May not be scanned, copied or duplicated, or posted to a publicly accessible website, in whole or in part.

(d) We first graph the parabola $y = x^2 - 1$ using a solid curve. Since $y \geq x^2 - 1$, the points in the region lie on or *above* the parabola.

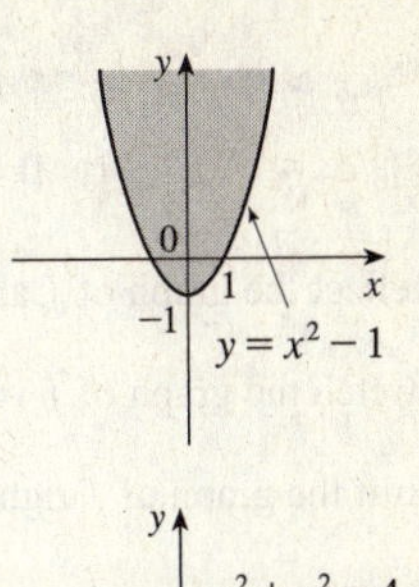

(e) We graph the circle $x^2 + y^2 = 4$ using a dotted curve. Since $\sqrt{x^2 + y^2} < 2$, the region consists of points whose distance from the origin is less than 2, that is, the points that lie *inside* the circle.

(f) The equation $9x^2 + 16y^2 = 144$ is an ellipse centered at $(0, 0)$. We put it in standard form by dividing by 144 and get $\dfrac{x^2}{16} + \dfrac{y^2}{9} = 1$. The x-intercepts are located at a distance of $\sqrt{16} = 4$ from the center while the y-intercepts are a distance of $\sqrt{9} = 3$ from the center (see the graph).

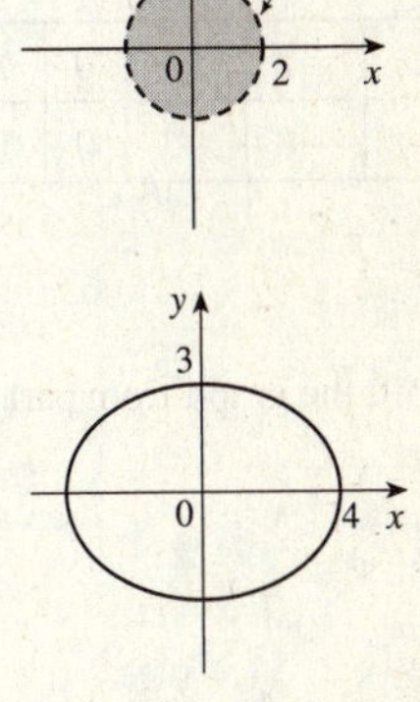

Test C Functions

1. (a) Locate -1 on the x-axis and then go down to the point on the graph with an x-coordinate of -1. The corresponding y-coordinate is the value of the function at $x = -1$, which is -2. So, $f(-1) = -2$.

(b) Using the same technique as in part (a), we get $f(2) \approx 2.8$.

(c) Locate 2 on the y-axis and then go left and right to find all points on the graph with a y-coordinate of 2. The corresponding x-coordinates are the x-values we are searching for. So $x = -3$ and $x = 1$.

(d) Using the same technique as in part (c), we get $x \approx -2.5$ and $x \approx 0.3$.

(e) The domain is all the x-values for which the graph exists, and the range is all the y-values for which the graph exists. Thus, the domain is $[-3, 3]$, and the range is $[-2, 3]$.

2. Note that $f(2 + h) = (2 + h)^3$ and $f(2) = 2^3 = 8$. So the difference quotient becomes

$$\frac{f(2+h) - f(2)}{h} = \frac{(2+h)^3 - 8}{h} = \frac{8 + 12h + 6h^2 + h^3 - 8}{h} = \frac{12h + 6h^2 + h^3}{h} = \frac{h(12 + 6h + h^2)}{h} = 12 + 6h + h^2.$$

3. (a) Set the denominator equal to 0 and solve to find restrictions on the domain: $x^2 + x - 2 = 0 \;\Rightarrow\; (x - 1)(x + 2) = 0 \;\Rightarrow\; x = 1$ or $x = -2$. Thus, the domain is all real numbers except 1 or -2 or, in interval notation, $(-\infty, -2) \cup (-2, 1) \cup (1, \infty)$.

(b) Note that the denominator is always greater than or equal to 1, and the numerator is defined for all real numbers. Thus, the domain is $(-\infty, \infty)$.

(c) Note that the function h is the sum of two root functions. So h is defined on the intersection of the domains of these two root functions. The domain of a square root function is found by setting its radicand greater than or equal to 0. Now,

© 2016 Cengage Learning. All Rights Reserved. May not be scanned, copied or duplicated, or posted to a publicly accessible website, in whole or in part.

$4 - x \geq 0 \;\Rightarrow\; x \leq 4$ and $x^2 - 1 \geq 0 \;\Rightarrow\; (x-1)(x+1) \geq 0 \;\Rightarrow\; x \leq -1$ or $x \geq 1$. Thus, the domain of h is $(-\infty, -1] \cup [1, 4]$.

4. (a) Reflect the graph of f about the x-axis.

(b) Stretch the graph of f vertically by a factor of 2, then shift 1 unit downward.

(c) Shift the graph of f right 3 units, then up 2 units.

5. (a) Make a table and then connect the points with a smooth curve:

x	-2	-1	0	1	2
y	-8	-1	0	1	8

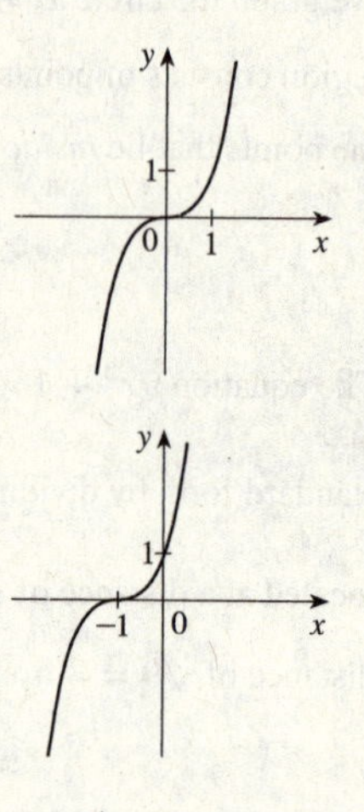

(b) Shift the graph from part (a) left 1 unit.

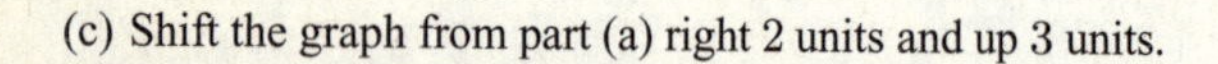
(c) Shift the graph from part (a) right 2 units and up 3 units.

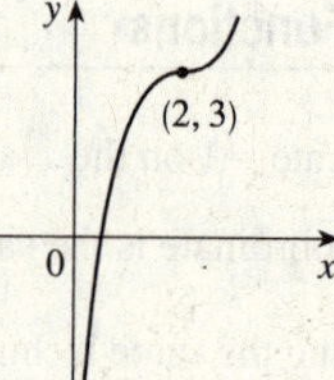

(d) First plot $y = x^2$. Next, to get the graph of $f(x) = 4 - x^2$,

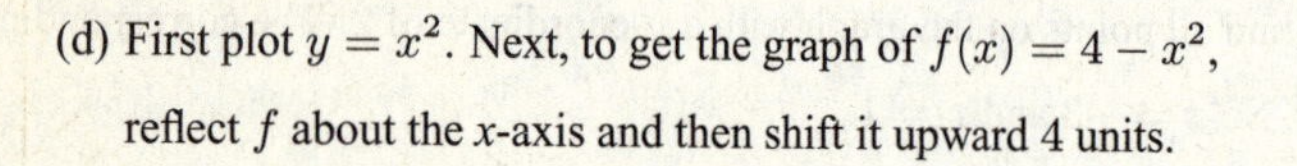
reflect f about the x-axis and then shift it upward 4 units.

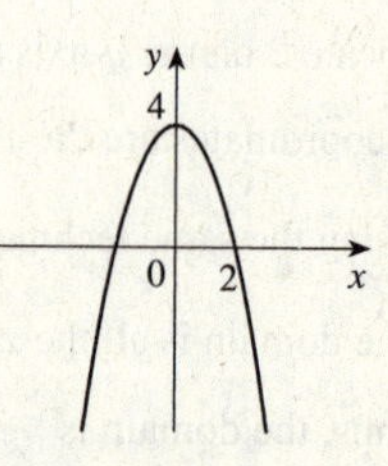

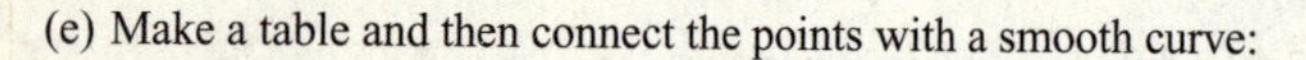
(e) Make a table and then connect the points with a smooth curve:

x	0	1	4	9
y	0	1	2	3

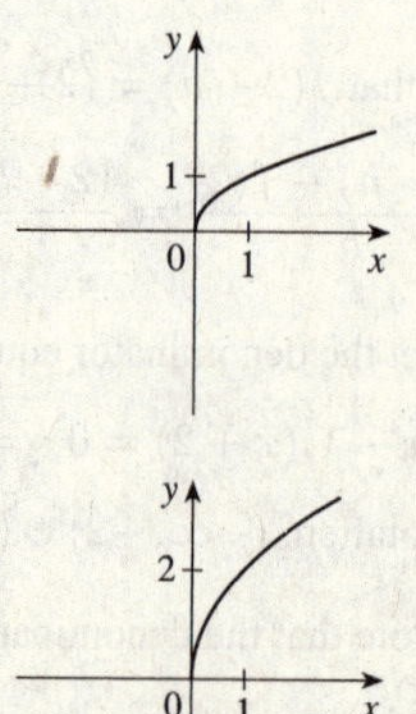

(f) Stretch the graph from part (e) vertically by a factor of two.

© 2016 Cengage Learning. All Rights Reserved. May not be scanned, copied or duplicated, or posted to a publicly accessible website, in whole or in part.

(g) First plot $y = 2^x$. Next, get the graph of $y = -2^x$ by reflecting the graph of $y = 2^x$ about the x-axis.

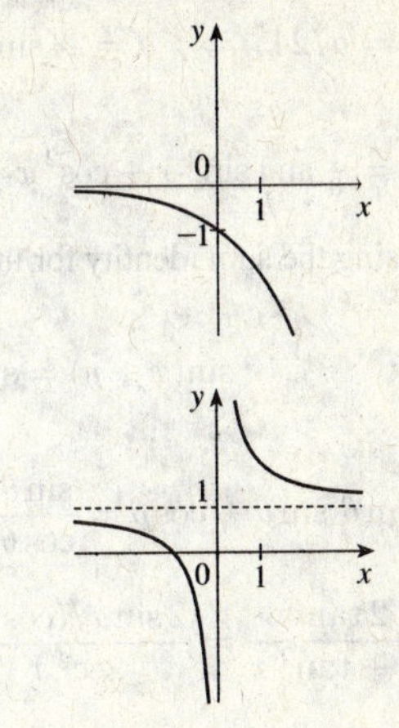

(h) Note that $y = 1 + x^{-1} = 1 + 1/x$. So first plot $y = 1/x$ and then shift it upward 1 unit.

6. (a) $f(-2) = 1 - (-2)^2 = -3$ and $f(1) = 2(1) + 1 = 3$

(b) For $x \leq 0$ plot $f(x) = 1 - x^2$ and, on the same plane, for $x > 0$ plot the graph of $f(x) = 2x + 1$.

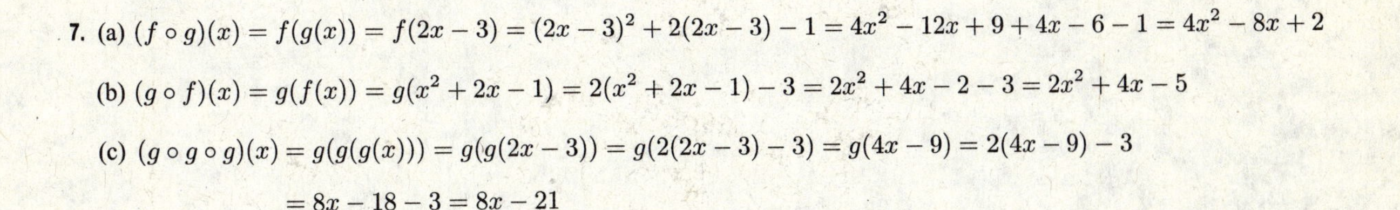

7. (a) $(f \circ g)(x) = f(g(x)) = f(2x-3) = (2x-3)^2 + 2(2x-3) - 1 = 4x^2 - 12x + 9 + 4x - 6 - 1 = 4x^2 - 8x + 2$

(b) $(g \circ f)(x) = g(f(x)) = g(x^2 + 2x - 1) = 2(x^2 + 2x - 1) - 3 = 2x^2 + 4x - 2 - 3 = 2x^2 + 4x - 5$

(c) $(g \circ g \circ g)(x) = g(g(g(x))) = g(g(2x-3)) = g(2(2x-3) - 3) = g(4x - 9) = 2(4x-9) - 3$
$= 8x - 18 - 3 = 8x - 21$

Test D Trigonometry

1. (a) $300° = 300°\left(\dfrac{\pi}{180°}\right) = \dfrac{300\pi}{180} = \dfrac{5\pi}{3}$ (b) $-18° = -18°\left(\dfrac{\pi}{180°}\right) = -\dfrac{18\pi}{180} = -\dfrac{\pi}{10}$

2. (a) $\dfrac{5\pi}{6} = \dfrac{5\pi}{6}\left(\dfrac{180}{\pi}\right)^{\circ} = 150°$ (b) $2 = 2\left(\dfrac{180}{\pi}\right)^{\circ} = \left(\dfrac{360}{\pi}\right)^{\circ} \approx 114.6°$

3. We will use the arc length formula, $s = r\theta$, where s is arc length, r is the radius of the circle, and θ is the measure of the central angle in radians. First, note that $30° = 30°\left(\dfrac{\pi}{180°}\right) = \dfrac{\pi}{6}$. So $s = (12)\left(\dfrac{\pi}{6}\right) = 2\pi$ cm.

4. (a) $\tan(\pi/3) = \sqrt{3}$ [You can read the value from a right triangle with sides 1, 2, and $\sqrt{3}$.]

(b) Note that $7\pi/6$ can be thought of as an angle in the third quadrant with reference angle $\pi/6$. Thus, $\sin(7\pi/6) = -\frac{1}{2}$, since the sine function is negative in the third quadrant.

(c) Note that $5\pi/3$ can be thought of as an angle in the fourth quadrant with reference angle $\pi/3$. Thus,

$\sec(5\pi/3) = \dfrac{1}{\cos(5\pi/3)} = \dfrac{1}{1/2} = 2$, since the cosine function is positive in the fourth quadrant.

© 2016 Cengage Learning. All Rights Reserved. May not be scanned, copied or duplicated, or posted to a publicly accessible website, in whole or in part.

5. $\sin\theta = a/24 \;\Rightarrow\; a = 24\sin\theta$ and $\cos\theta = b/24 \;\Rightarrow\; b = 24\cos\theta$

6. $\sin x = \frac{1}{3}$ and $\sin^2 x + \cos^2 x = 1 \;\Rightarrow\; \cos x = \sqrt{1 - \frac{1}{9}} = \dfrac{2\sqrt{2}}{3}$. Also, $\cos y = \frac{4}{5} \;\Rightarrow\; \sin y = \sqrt{1 - \frac{16}{25}} = \frac{3}{5}$.

So, using the sum identity for the sine, we have

$$\sin(x+y) = \sin x\,\cos y + \cos x\,\sin y = \frac{1}{3}\cdot\frac{4}{5} + \frac{2\sqrt{2}}{3}\cdot\frac{3}{5} = \frac{4+6\sqrt{2}}{15} = \frac{1}{15}(4+6\sqrt{2})$$

7. (a) $\tan\theta\,\sin\theta + \cos\theta = \dfrac{\sin\theta}{\cos\theta}\sin\theta + \cos\theta = \dfrac{\sin^2\theta}{\cos\theta} + \dfrac{\cos^2\theta}{\cos\theta} = \dfrac{1}{\cos\theta} = \sec\theta$

(b) $\dfrac{2\tan x}{1+\tan^2 x} = \dfrac{2\sin x/(\cos x)}{\sec^2 x} = 2\,\dfrac{\sin x}{\cos x}\,\cos^2 x = 2\sin x\,\cos x = \sin 2x$

8. $\sin 2x = \sin x \;\Leftrightarrow\; 2\sin x\,\cos x = \sin x \;\Leftrightarrow\; 2\sin x\,\cos x - \sin x = 0 \;\Leftrightarrow\; \sin x\,(2\cos x - 1) = 0 \;\Leftrightarrow$

$\sin x = 0$ or $\cos x = \frac{1}{2} \;\Rightarrow\; x = 0, \frac{\pi}{3}, \pi, \frac{5\pi}{3}, 2\pi.$

9. We first graph $y = \sin 2x$ (by compressing the graph of $\sin x$ by a factor of 2) and then shift it upward 1 unit.

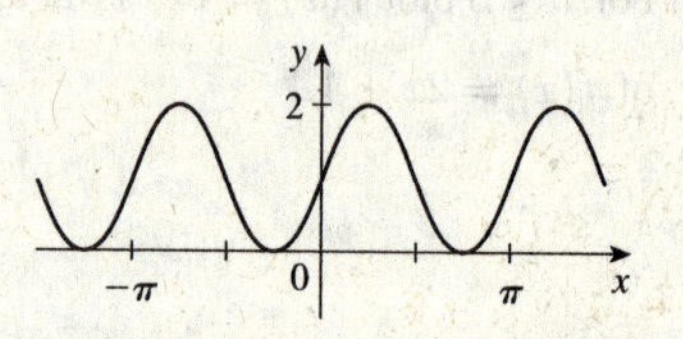

© 2016 Cengage Learning. All Rights Reserved. May not be scanned, copied or duplicated, or posted to a publicly accessible website, in whole or in part.

1 □ FUNCTIONS AND SEQUENCES

1.1 Four Ways to Represent a Function

1. The functions $f(x) = x + \sqrt{2-x}$ and $g(u) = u + \sqrt{2-u}$ give exactly the same output values for every input value, so f and g are equal.

3. (a) The point $(1, 3)$ is on the graph of f, so $f(1) = 3$.

(b) When $x = -1$, y is about -0.2, so $f(-1) \approx -0.2$.

(c) $f(x) = 1$ is equivalent to $y = 1$. When $y = 1$, we have $x = 0$ and $x = 3$.

(d) A reasonable estimate for x when $y = 0$ is $x = -0.8$.

(e) The domain of f consists of all x-values on the graph of f. For this function, the domain is $-2 \le x \le 4$, or $[-2, 4]$. The range of f consists of all y-values on the graph of f. For this function, the range is $-1 \le y \le 3$, or $[-1, 3]$.

(f) As x increases from -2 to 1, y increases from -1 to 3. Thus, f is increasing on the interval $[-2, 1]$.

5. No, the curve is not the graph of a function because a vertical line intersects the curve more than once. Hence, the curve fails the Vertical Line Test.

7. Yes, the curve is the graph of a function because it passes the Vertical Line Test. The domain is $[-3, 2]$ and the range is $[-3, -2) \cup [-1, 3]$.

9. (a) The graph shows that the global average temperature in 1950 was $T(1950) \approx 13.8\,°\text{C}$.

(b) By drawing the horizontal line $T = 14.2$ to the curve and then drawing the vertical line down to the horizontal axis, we see that $t \approx 1992$.

(c) The temperature was smallest in 1910 and largest in 2006.

(d) The range is $\{T \mid 13.5 \le T \le 14.5\} = [13.5, 14.5]$

11. If we draw the horizontal line pH $= 4.0$, we can see that the pH curve is less than 4.0 between 12:23AM and 12:52AM. Therefore, a clinical acid reflux episode occurred approximately between 12:23AM and 12:52AM at which time the esophageal pH was less than 4.0.

13. (a) At 30 °S and 20 °N, we expect approximately 100 and 134 ant species respectively.

(b) By drawing the horizontal line at a species richness of 100, we see there are two points of intersection with the curve, each having latitude values of roughly 30 °N and 30 °S.

(c) The function is even since its graph is symmetric with respect to the y-axis.

15. The person's weight increased to about 160 pounds at age 20 and stayed fairly steady for 10 years. The person's weight dropped to about 120 pounds for the next 5 years, then increased rapidly to about 170 pounds. The next 30 years saw a gradual increase to 190 pounds. Possible reasons for the drop in weight at 30 years of age: diet, exercise, health problems.

© 2016 Cengage Learning. All Rights Reserved. May not be scanned, copied or duplicated, or posted to a publicly accessible website, in whole or in part.

17. The water will cool down almost to freezing as the ice melts. Then, when the ice has melted, the water will slowly warm up to room temperature.

19. Initially, the bacteria population size remains constant during which nutrients are consumed in preparation for reproduction. In the second phase, the population size increases rapidly as the bacteria replicate. The population size plateaus in phase three at which point the "carrying capacity" has been reached and the available resources and space cannot support a larger population. Finally, the bacteria die due to starvation and waste toxicity and the population declines.

21. Of course, this graph depends strongly on the geographical location!

23 As the price increases, the amount sold decreases.

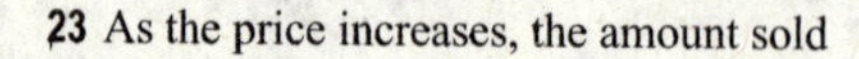

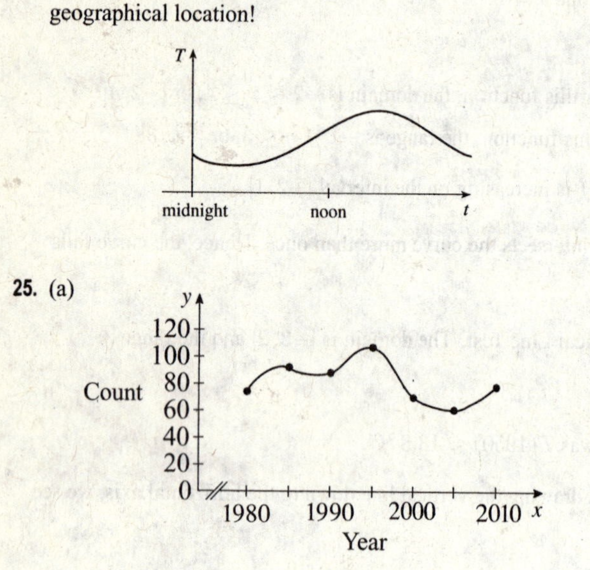

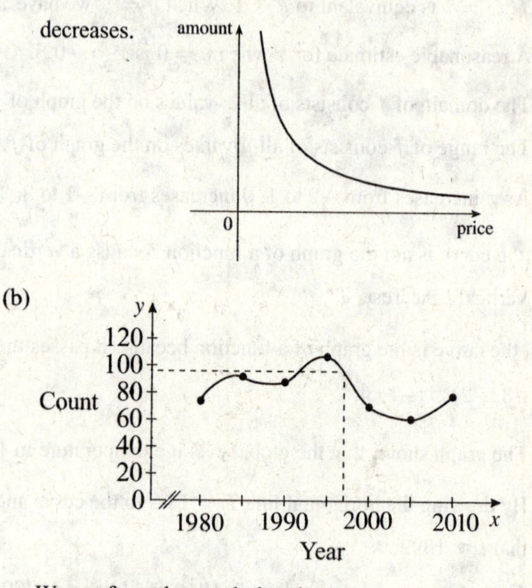

We see from the graph that there were approximately 92,000 birds in 1997.

27. $f(x) = 3x^2 - x + 2.$

$f(2) = 3(2)^2 - 2 + 2 = 12 - 2 + 2 = 12.$

$f(-2) = 3(-2)^2 - (-2) + 2 = 12 + 2 + 2 = 16.$

$f(a) = 3a^2 - a + 2.$

$f(-a) = 3(-a)^2 - (-a) + 2 = 3a^2 + a + 2.$

$f(a+1) = 3(a+1)^2 - (a+1) + 2 = 3(a^2 + 2a + 1) - a - 1 + 2 = 3a^2 + 6a + 3 - a + 1 = 3a^2 + 5a + 4.$

$2f(a) = 2 \cdot f(a) = 2(3a^2 - a + 2) = 6a^2 - 2a + 4.$

$f(2a) = 3(2a)^2 - (2a) + 2 = 3(4a^2) - 2a + 2 = 12a^2 - 2a + 2.$

$f(a^2) = 3(a^2)^2 - (a^2) + 2 = 3(a^4) - a^2 + 2 = 3a^4 - a^2 + 2.$

$$\begin{aligned}[f(a)]^2 &= \left[3a^2 - a + 2\right]^2 = \left(3a^2 - a + 2\right)\left(3a^2 - a + 2\right) \\ &= 9a^4 - 3a^3 + 6a^2 - 3a^3 + a^2 - 2a + 6a^2 - 2a + 4 = 9a^4 - 6a^3 + 13a^2 - 4a + 4.\end{aligned}$$

$f(a+h) = 3(a+h)^2 - (a+h) + 2 = 3(a^2 + 2ah + h^2) - a - h + 2 = 3a^2 + 6ah + 3h^2 - a - h + 2.$

© 2016 Cengage Learning. All Rights Reserved. May not be scanned, copied or duplicated, or posted to a publicly accessible website, in whole or in part.

29. $f(x) = 4 + 3x - x^2$, so $f(3+h) = 4 + 3(3+h) - (3+h)^2 = 4 + 9 + 3h - (9 + 6h + h^2) = 4 - 3h - h^2$,

and $\dfrac{f(3+h) - f(3)}{h} = \dfrac{(4 - 3h - h^2) - 4}{h} = \dfrac{h(-3-h)}{h} = -3 - h$.

31. $\dfrac{f(x) - f(a)}{x - a} = \dfrac{\dfrac{1}{x} - \dfrac{1}{a}}{x - a} = \dfrac{\dfrac{a - x}{xa}}{x - a} = \dfrac{a - x}{xa(x - a)} = \dfrac{-1(x - a)}{xa(x - a)} = -\dfrac{1}{ax}$

33. $f(x) = (x+4)/(x^2 - 9)$ is defined for all x except when $0 = x^2 - 9 \;\Leftrightarrow\; 0 = (x+3)(x-3) \;\Leftrightarrow\; x = -3$ or 3, so the domain is $\{x \in \mathbb{R} \mid x \neq -3, 3\} = (-\infty, -3) \cup (-3, 3) \cup (3, \infty)$.

35. $f(t) = \sqrt[3]{2t - 1}$ is defined for all real numbers. In fact $\sqrt[3]{p(t)}$, where $p(t)$ is a polynomial, is defined for all real numbers. Thus, the domain is $\mathbb{R}$, or $(-\infty, \infty)$.

37. $h(x) = 1 / \sqrt[4]{x^2 - 5x}$ is defined when $x^2 - 5x > 0 \;\Leftrightarrow\; x(x-5) > 0$. Note that $x^2 - 5x \neq 0$ since that would result in division by zero. The expression $x(x-5)$ is positive if $x < 0$ or $x > 5$. (See Appendix A for methods for solving inequalities.) Thus, the domain is $(-\infty, 0) \cup (5, \infty)$.

39. $F(p) = \sqrt{2 - \sqrt{p}}$ is defined when $p \geq 0$ and $2 - \sqrt{p} \geq 0$. Since $2 - \sqrt{p} \geq 0 \;\Rightarrow\; 2 \geq \sqrt{p} \;\Rightarrow\; \sqrt{p} \leq 2 \;\Rightarrow\; 0 \leq p \leq 4$, the domain is $[0, 4]$.

41. $f(x) = 2 - 0.4x$ is defined for all real numbers, so the domain is $\mathbb{R}$, or $(-\infty, \infty)$. The graph of f is a line with slope -0.4 and y-intercept 2.

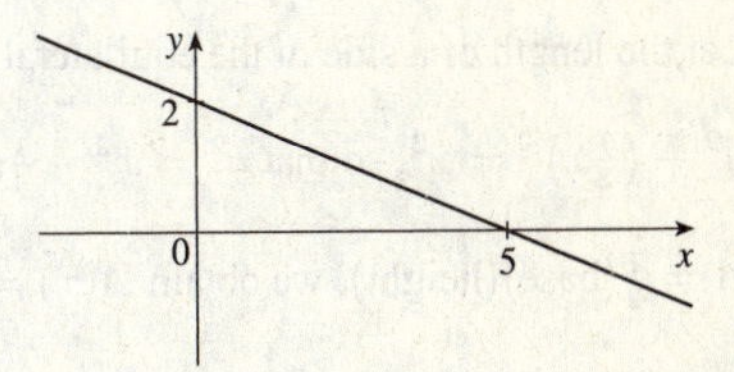

43. $f(t) = 2t + t^2$ is defined for all real numbers, so the domain is $\mathbb{R}$, or $(-\infty, \infty)$. The graph of f is a parabola opening upward since the coefficient of t^2 is positive. To find the t-intercepts, let $y = 0$ and solve for t. $0 = 2t + t^2 = t(2 + t) \;\Rightarrow\; t = 0$ or $t = -2$. The t-coordinate of the vertex is halfway between the t-intercepts, that is, at $t = -1$. Since $f(-1) = 2(-1) + (-1)^2 = -2 + 1 = -1$, the vertex is $(-1, -1)$.

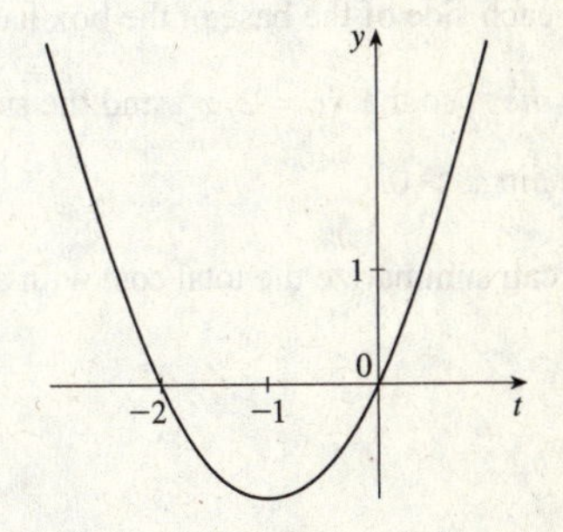

45. $g(x) = \sqrt{x - 5}$ is defined when $x - 5 \geq 0$ or $x \geq 5$, so the domain is $[5, \infty)$. Since $y = \sqrt{x - 5} \;\Rightarrow\; y^2 = x - 5 \;\Rightarrow\; x = y^2 + 5$, we see that g is the top half of a parabola.

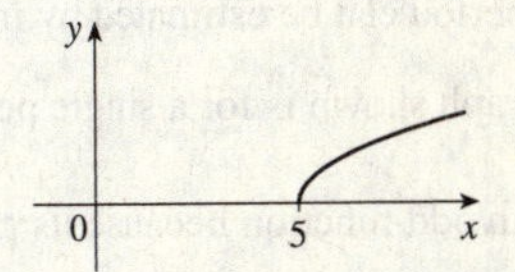

47. $G(x) = \dfrac{3x + |x|}{x}$. Since $|x| = \begin{cases} x & \text{if } x \geq 0 \\ -x & \text{if } x < 0 \end{cases}$, we have

$$G(x) = \begin{cases} \dfrac{3x + x}{x} & \text{if } x > 0 \\[2ex] \dfrac{3x - x}{x} & \text{if } x < 0 \end{cases} = \begin{cases} \dfrac{4x}{x} & \text{if } x > 0 \\[2ex] \dfrac{2x}{x} & \text{if } x < 0 \end{cases} = \begin{cases} 4 & \text{if } x > 0 \\ 2 & \text{if } x < 0 \end{cases}$$

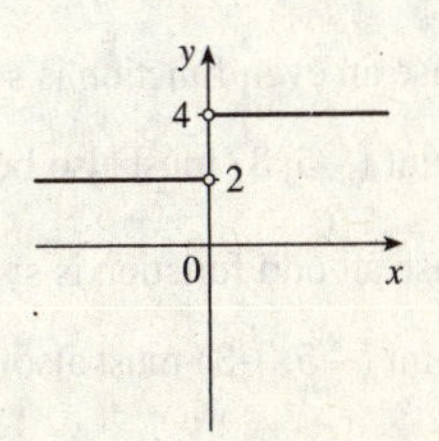

Note that G is not defined for $x = 0$. The domain is $(-\infty, 0) \cup (0, \infty)$.

© 2016 Cengage Learning. All Rights Reserved. May not be scanned, copied or duplicated, or posted to a publicly accessible website, in whole or in part.

49. $f(x) = \begin{cases} x+2 & \text{if } x < 0 \\ 1-x & \text{if } x \geq 0 \end{cases}$

The domain is $\mathbb{R}$.

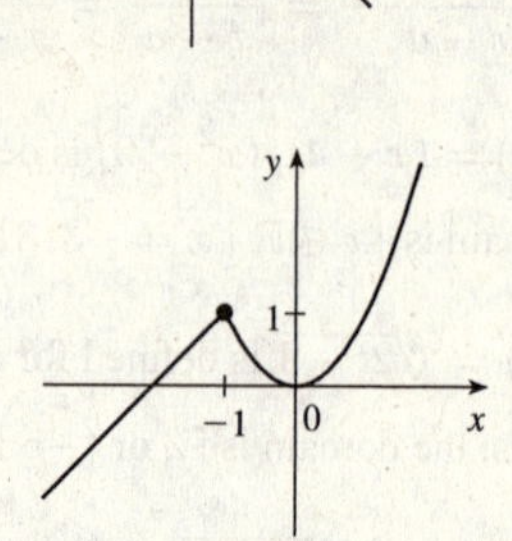

51. $f(x) = \begin{cases} x+2 & \text{if } x \leq -1 \\ x^2 & \text{if } x > -1 \end{cases}$

Note that for $x = -1$, both $x + 2$ and x^2 are equal to 1.

The domain is $\mathbb{R}$.

53. Let the length and width of the rectangle be L and W. Then the perimeter is $2L + 2W = 20$ and the area is $A = LW$. Solving the first equation for W in terms of L gives $W = \dfrac{20 - 2L}{2} = 10 - L$. Thus, $A(L) = L(10 - L) = 10L - L^2$. Since lengths are positive, the domain of A is $0 < L < 10$. If we further restrict L to be larger than W, then $5 < L < 10$ would be the domain.

55. Let the length of a side of the equilateral triangle be x. Then by the Pythagorean Theorem, the height y of the triangle satisfies $y^2 + \left(\frac{1}{2}x\right)^2 = x^2$, so that $y^2 = x^2 - \frac{1}{4}x^2 = \frac{3}{4}x^2$ and $y = \frac{\sqrt{3}}{2}x$. Using the formula for the area A of a triangle, $A = \frac{1}{2}(\text{base})(\text{height})$, we obtain $A(x) = \frac{1}{2}(x)\left(\frac{\sqrt{3}}{2}x\right) = \frac{\sqrt{3}}{4}x^2$, with domain $x > 0$.

57. Let each side of the base of the box have length x, and let the height of the box be h. Since the volume is 2, we know that $2 = hx^2$, so that $h = 2/x^2$, and the surface area is $S = x^2 + 4xh$. Thus, $S(x) = x^2 + 4x(2/x^2) = x^2 + (8/x)$, with domain $x > 0$.

59. We can summarize the total cost with a piecewise defined function.

$$T(x) = \begin{cases} 75x & \text{if } 0 < x \leq 2 \\ 150 + 50(x - 2) & \text{if } x > 2 \end{cases}$$

61. The period can be estimated by measuring the peak-to-peak distance on the graph. This is approximately 77 hours. Note that the graph shown is for a single person's temperature. The period for this species of malaria is, on average, 72 hours.

63. f is an odd function because its graph is symmetric about the origin. g is an even function because its graph is symmetric with respect to the y-axis.

65. (a) Because an even function is symmetric with respect to the y-axis, and the point $(5, 3)$ is on the graph of this even function, the point $(-5, 3)$ must also be on its graph.

(b) Because an odd function is symmetric with respect to the origin, and the point $(5, 3)$ is on the graph of this odd function, the point $(-5, -3)$ must also be on its graph.

© 2016 Cengage Learning. All Rights Reserved. May not be scanned, copied or duplicated, or posted to a publicly accessible website, in whole or in part.

67. $f(x) = \dfrac{x}{x^2+1}$.

$$f(-x) = \frac{-x}{(-x)^2+1} = \frac{-x}{x^2+1} = -\frac{x}{x^2+1} = -f(x).$$

So f is an odd function.

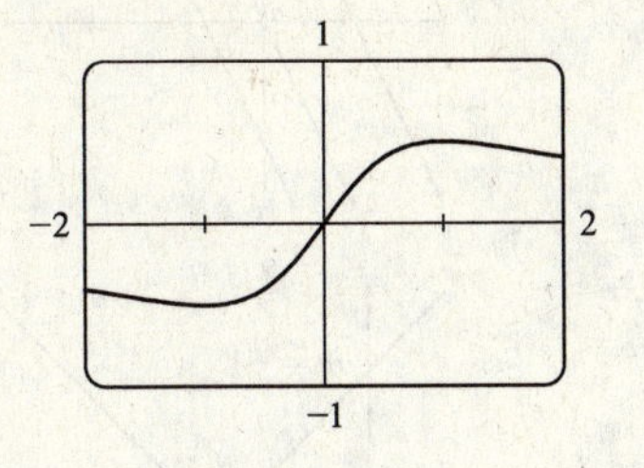

69. $f(x) = \dfrac{x}{x+1}$, so $f(-x) = \dfrac{-x}{-x+1} = \dfrac{x}{x-1}$.

Since this is neither $f(x)$ nor $-f(x)$, the function f is neither even nor odd.

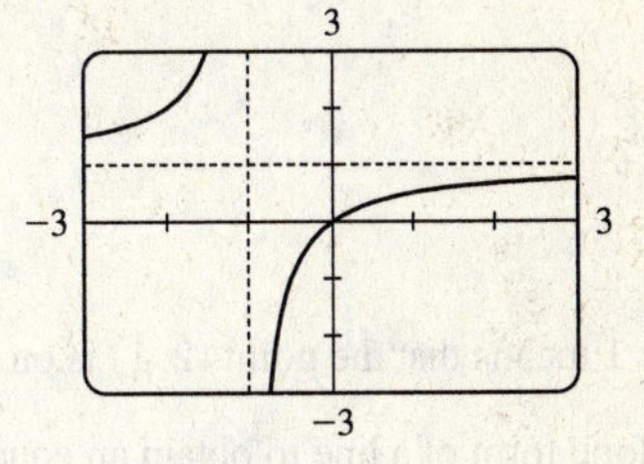

71. $f(x) = 1 + 3x^2 - x^4$.

$f(-x) = 1+3(-x)^2-(-x)^4 = 1+3x^2-x^4 = f(x)$.

So f is an even function.

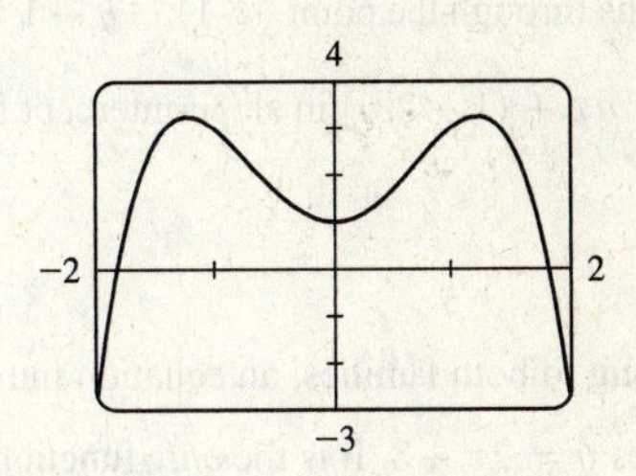

73. (i) If f and g are both even functions, then $f(-x) = f(x)$ and $g(-x) = g(x)$. Now

$(f+g)(-x) = f(-x) + g(-x) = f(x) + g(x) = (f+g)(x)$, so $f+g$ is an *even* function.

(ii) If f and g are both odd functions, then $f(-x) = -f(x)$ and $g(-x) = -g(x)$. Now

$(f+g)(-x) = f(-x) + g(-x) = -f(x) + [-g(x)] = -[f(x) + g(x)] = -(f+g)(x)$, so $f+g$ is an *odd* function.

(iii) If f is an even function and g is an odd function, then $(f+g)(-x) = f(-x) + g(-x) = f(x) + [-g(x)] = f(x) - g(x)$, which is not $(f+g)(x)$ nor $-(f+g)(x)$, so $f+g$ is *neither* even nor odd. (Exception: if f is the zero function, then $f+g$ will be *odd*. If g is the zero function, then $f+g$ will be *even*.)

1.2 Mathematical Models: A Catalog of Essential Functions

1. (a) $f(x) = \log_2 x$ is a logarithmic function.

(b) $g(x) = \sqrt[4]{x}$ is a root function with $n = 4$.

(c) $h(x) = \dfrac{2x^3}{1-x^2}$ is a rational function because it is a ratio of polynomials.

(d) $u(t) = 1 - 1.1t + 2.54t^2$ is a polynomial of degree 2 (also called a *quadratic function*).

(e) $v(t) = 5^t$ is an exponential function.

(f) $w(\theta) = \sin\theta\,\cos^2\theta$ is a trigonometric function.

3. We notice from the figure that g and h are even functions (symmetric with respect to the y-axis) and that f is an odd function (symmetric with respect to the origin). So (b) $[y = x^5]$ must be f. Since g is flatter than h near the origin, we must have (c) $[y = x^8]$ matched with g and (a) $[y = x^2]$ matched with h.

© 2016 Cengage Learning. All Rights Reserved. May not be scanned, copied or duplicated, or posted to a publicly accessible website, in whole or in part.

5. (a) An equation for the family of linear functions with slope 2 is $y = f(x) = 2x + b$, where b is the y-intercept.

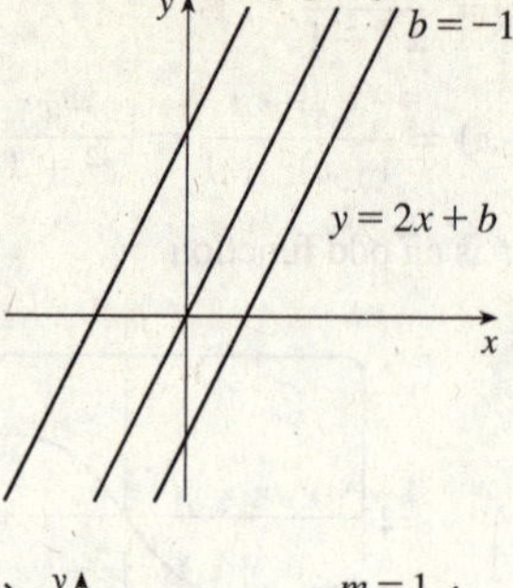

(b) $f(2) = 1$ means that the point $(2, 1)$ is on the graph of f. We can use the point-slope form of a line to obtain an equation for the family of linear functions through the point $(2, 1)$. $y - 1 = m(x - 2)$, which is equivalent to $y = mx + (1 - 2m)$ in slope-intercept form.

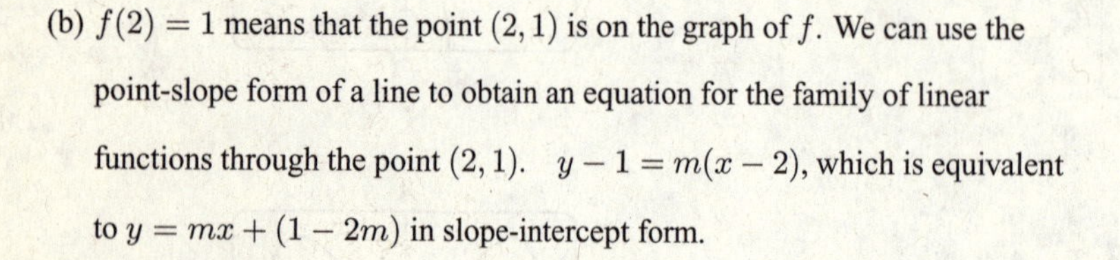

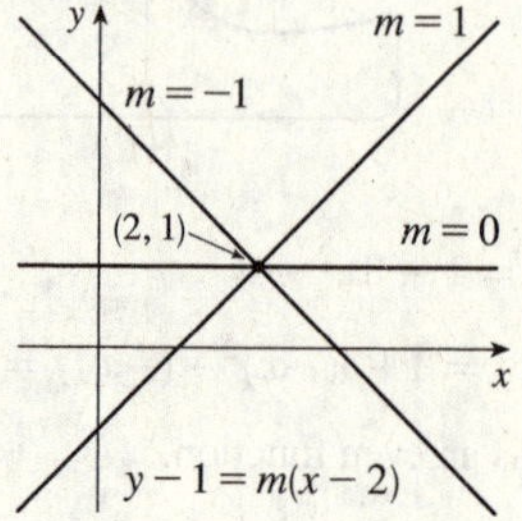

(c) To belong to both families, an equation must have slope $m = 2$, so the equation in part (b), $y = mx + (1 - 2m)$, becomes $y = 2x - 3$. It is the *only* function that belongs to both families.

7. All members of the family of linear functions $f(x) = c - x$ have graphs that are lines with slope -1. The y-intercept is c.

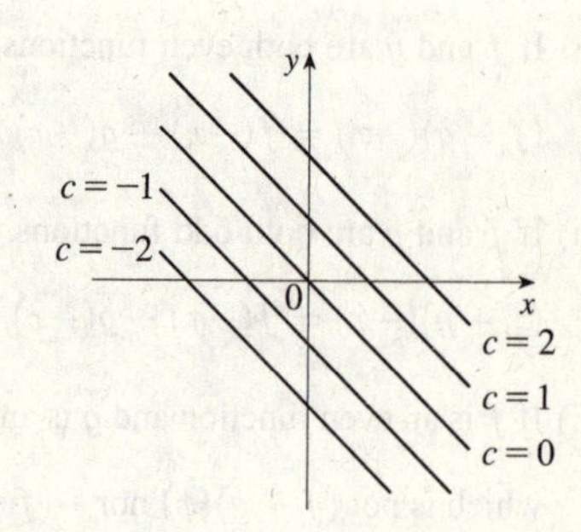

9. Since $f(-1) = f(0) = f(2) = 0$, f has zeros of -1, 0, and 2, so an equation for f is $f(x) = a[x - (-1)](x - 0)(x - 2)$, or $f(x) = ax(x + 1)(x - 2)$. Because $f(1) = 6$, we'll substitute 1 for x and 6 for $f(x)$.
$6 = a(1)(2)(-1) \;\Rightarrow\; -2a = 6 \;\Rightarrow\; a = -3$, so an equation for f is $f(x) = -3x(x + 1)(x - 2)$.

11. (a) $D = 200$, so $c = 0.0417D(a + 1) = 0.0417(200)(a + 1) = 8.34a + 8.34$. The slope is 8.34, which represents the change in mg of the dosage for a child for each change of 1 year in age.

(b) For a newborn, $a = 0$, so $c = 8.34$ mg.

13. (a)

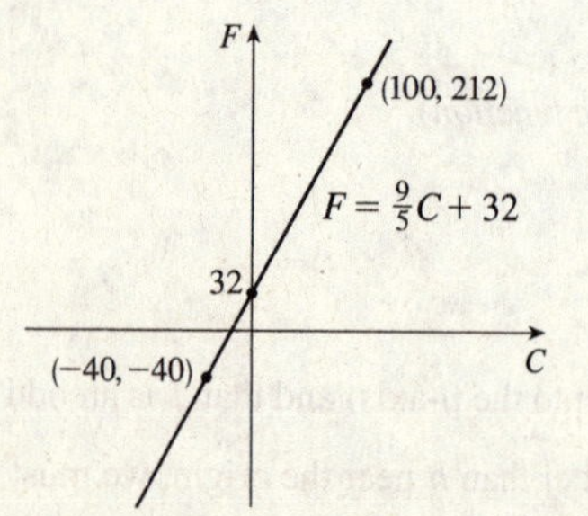

(b) The slope of $\frac{9}{5}$ means that F increases $\frac{9}{5}$ degrees for each increase of 1°C. (Equivalently, F increases by 9 when C increases by 5 and F decreases by 9 when C decreases by 5.) The F-intercept of 32 is the Fahrenheit temperature corresponding to a Celsius temperature of 0.

© 2016 Cengage Learning. All Rights Reserved. May not be scanned, copied or duplicated, or posted to a publicly accessible website, in whole or in part.

15. (a) Using N in place of x and T in place of y, we find the slope to be $\dfrac{T_2 - T_1}{N_2 - N_1} = \dfrac{80 - 70}{173 - 113} = \dfrac{10}{60} = \dfrac{1}{6}$. So a linear equation is $T - 80 = \frac{1}{6}(N - 173) \;\Leftrightarrow\; T - 80 = \frac{1}{6}N - \frac{173}{6} \;\Leftrightarrow\; T = \frac{1}{6}N + \frac{307}{6}$ $\left[\frac{307}{6} = 51.1\overline{6}\right]$.

(b) The slope of $\frac{1}{6}$ means that the temperature in Fahrenheit degrees increases one-sixth as rapidly as the number of cricket chirps per minute. Said differently, each increase of 6 cricket chirps per minute corresponds to an increase of 1°F.

(c) When $N = 150$, the temperature is given approximately by $T = \frac{1}{6}(150) + \frac{307}{6} = 76.1\overline{6}\,°\text{F} \approx 76\,°\text{F}$.

17. (a) The data appear to be periodic and a sine or cosine function would make the best model. A model of the form $f(x) = a\cos(bx) + c$ seems appropriate.

(b) The data appear to be decreasing in a linear fashion. A model of the form $f(x) = mx + b$ seems appropriate.

19. (a)

15

0 61,000

A linear model does seem appropriate.

(b) Using the points $(4000, 14.1)$ and $(60{,}000, 8.2)$, we obtain

$$y - 14.1 = \frac{8.2 - 14.1}{60{,}000 - 4000}(x - 4000) \text{ or, equivalently,}$$

$$y \approx -0.000105357x + 14.521429.$$

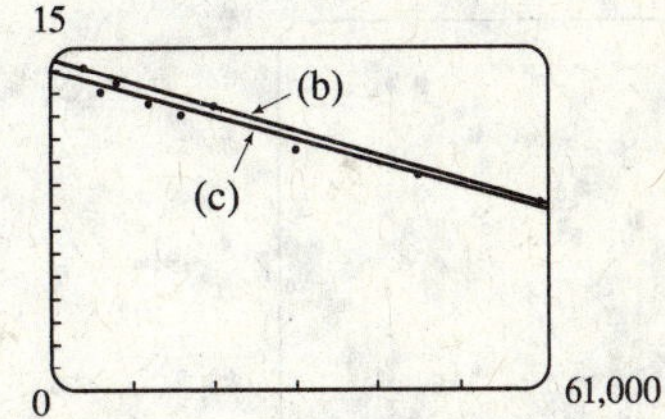

(c) Using a computing device, we obtain the least squares regression line $y = -0.0000997855x + 13.950764$.

The following commands and screens illustrate how to find the least squares regression line on a TI-84 Plus.

Enter the data into list one (L1) and list two (L2). Press [STAT] [1] to enter the editor.

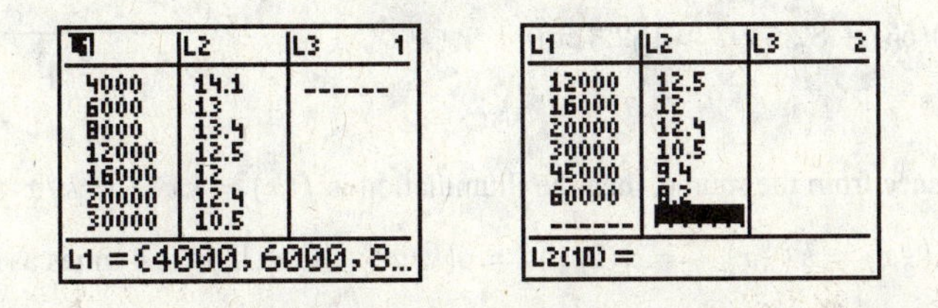

Find the regession line and store it in Y_1. Press [2nd] [QUIT] [STAT] [▸] [4] [VARS] [▸] [1] [1] [ENTER].

Note from the last figure that the regression line has been stored in Y_1 and that Plot1 has been turned on (Plot1 is highlighted). You can turn on Plot1 from the Y= menu by placing the cursor on Plot1 and pressing [ENTER] or by pressing [2nd] [STAT PLOT] [1] [ENTER].

[continued]

© 2016 Cengage Learning. All Rights Reserved. May not be scanned, copied or duplicated, or posted to a publicly accessible website, in whole or in part.

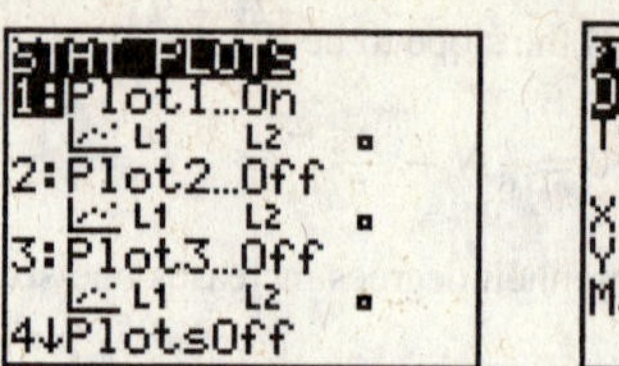

Now press ZOOM 9 to produce a graph of the data and the regression line. Note that choice 9 of the ZOOM menu automatically selects a window that displays all of the data.

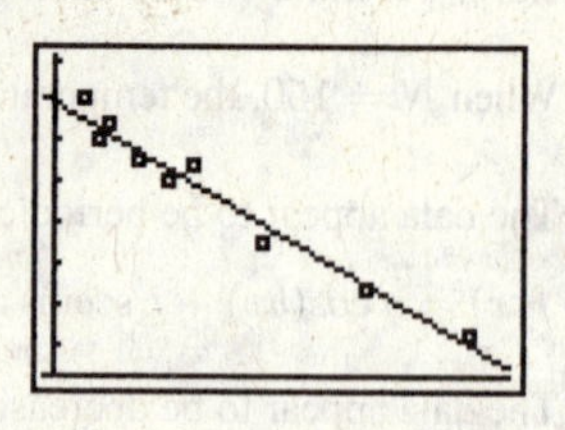

(d) When $x = 25{,}000$, $y \approx 11.456$; or about 11.5 per 100 population.

(e) When $x = 80{,}000$, $y \approx 5.968$; or about a 6% chance.

(f) When $x = 200{,}000$, y is negative, so the model does not apply.

21. (a)

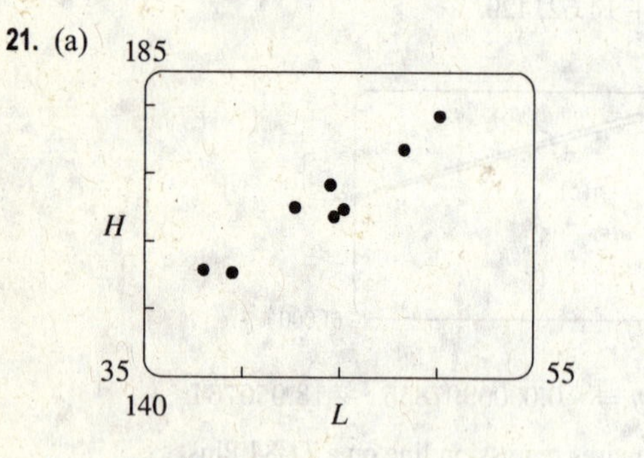

(b) Using a calculator to perform a linear regression gives $H = 1.8807L + 82.6497$ where H is the height in centimeters and L is the femur length in centimeters. This line, having slope 1.88 and H-intercept 82.65, is plotted below.

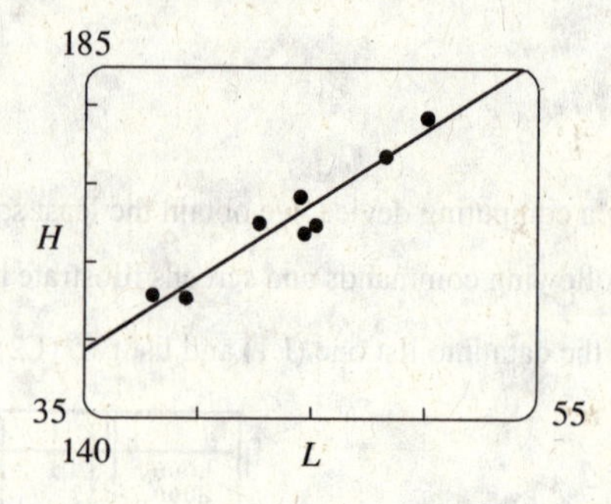

(c) The height of a person with $L = 53$ is $H(53) = (1.8807)(53) + 82.6497 \approx 182.3\,\text{cm}$.

23. If x is the original distance from the source, then the illumination is $f(x) = kx^{-2} = k/x^2$. Moving halfway to the lamp gives us an illumination of $f\left(\frac{1}{2}x\right) = k\left(\frac{1}{2}x\right)^{-2} = k(2/x)^2 = 4(k/x^2)$, so the light is 4 times as bright.

25. (a) Using a computing device, we obtain a power function $N = cA^b$, where $c \approx 3.1046$ and $b \approx 0.308$.

(b) If $A = 291$, then $N = cA^b \approx 17.8$, so you would expect to find 18 species of reptiles and amphibians on Dominica.

27. (a) Using a calculator to perform a 3rd-degree polynomial regression gives $L = 0.0155A^3 - 0.3725A^2 + 3.9461A + 1.2108$ where A is age and L is length. This polynomial is plotted along with a scatterplot of the data.

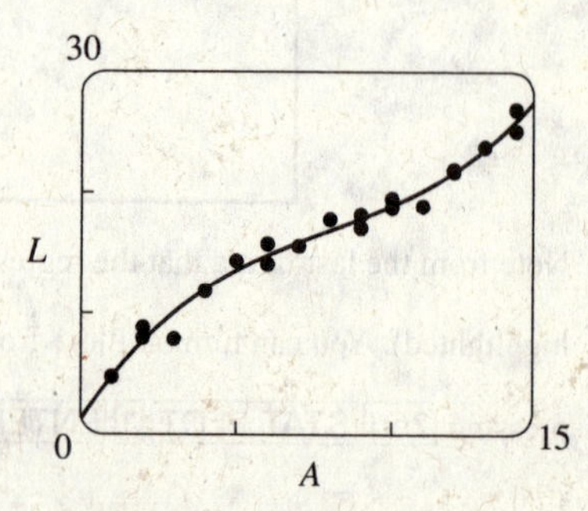

(b) A 5-year old rock bass has a length of $L(5) = (0.0155)(5)^3 - (0.3725)(5)^2 + (3.9461)(5) + 1.2108 \approx 13.6$ in

© 2016 Cengage Learning. All Rights Reserved. May not be scanned, copied or duplicated, or posted to a publicly accessible website, in whole or in part.

(c) Using computer algebra software to solve for A in the equation $20 = 0.0155A^3 - 0.3725A^2 + 3.9461A + 1.2108$ gives $A \approx 10.88$ years. Alternatively, the graph from part (a) can be used to estimate the age when $L = 20$ by drawing a horizontal line at $L = 20$ to the curve and observing the age at this point.

1.3 New Functions from Old Functions

1. (a) If the graph of f is shifted 3 units upward, its equation becomes $y = f(x) + 3$.

(b) If the graph of f is shifted 3 units downward, its equation becomes $y = f(x) - 3$.

(c) If the graph of f is shifted 3 units to the right, its equation becomes $y = f(x - 3)$.

(d) If the graph of f is shifted 3 units to the left, its equation becomes $y = f(x + 3)$.

(e) If the graph of f is reflected about the x-axis, its equation becomes $y = -f(x)$.

(f) If the graph of f is reflected about the y-axis, its equation becomes $y = f(-x)$.

(g) If the graph of f is stretched vertically by a factor of 3, its equation becomes $y = 3f(x)$.

(h) If the graph of f is shrunk vertically by a factor of 3, its equation becomes $y = \frac{1}{3}f(x)$.

3. (a) (graph 3) The graph of f is shifted 4 units to the right and has equation $y = f(x - 4)$.

(b) (graph 1) The graph of f is shifted 3 units upward and has equation $y = f(x) + 3$.

(c) (graph 4) The graph of f is shrunk vertically by a factor of 3 and has equation $y = \frac{1}{3}f(x)$.

(d) (graph 5) The graph of f is shifted 4 units to the left and reflected about the x-axis. Its equation is $y = -f(x + 4)$.

(e) (graph 2) The graph of f is shifted 6 units to the left and stretched vertically by a factor of 2. Its equation is $y = 2f(x + 6)$.

5. (a) To graph $y = f(2x)$ we shrink the graph of f horizontally by a factor of 2.

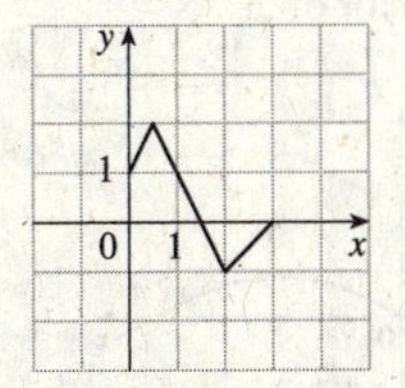

The point $(4, -1)$ on the graph of f corresponds to the point $\left(\frac{1}{2} \cdot 4, -1\right) = (2, -1)$.

(b) To graph $y = f\left(\frac{1}{2}x\right)$ we stretch the graph of f horizontally by a factor of 2.

The point $(4, -1)$ on the graph of f corresponds to the point $(2 \cdot 4, -1) = (8, -1)$.

(c) To graph $y = f(-x)$ we reflect the graph of f about the y-axis.

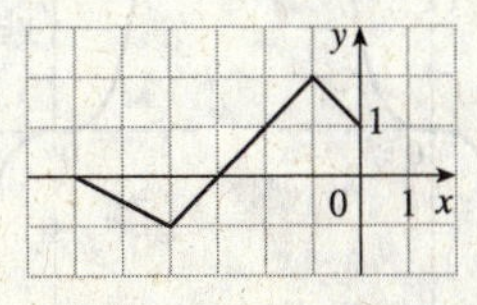

The point $(4, -1)$ on the graph of f corresponds to the point $(-1 \cdot 4, -1) = (-4, -1)$.

(d) To graph $y = -f(-x)$ we reflect the graph of f about the y-axis, then about the x-axis.

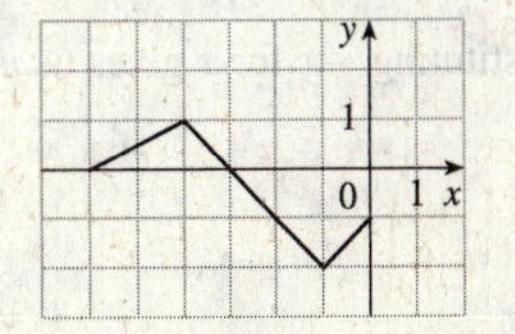

The point $(4, -1)$ on the graph of f corresponds to the point $(-1 \cdot 4, -1 \cdot -1) = (-4, 1)$.

© 2016 Cengage Learning. All Rights Reserved. May not be scanned, copied or duplicated, or posted to a publicly accessible website, in whole or in part.

7. $y = \dfrac{1}{x+2}$: Start with the graph of the reciprocal function $y = 1/x$ and shift 2 units to the left.

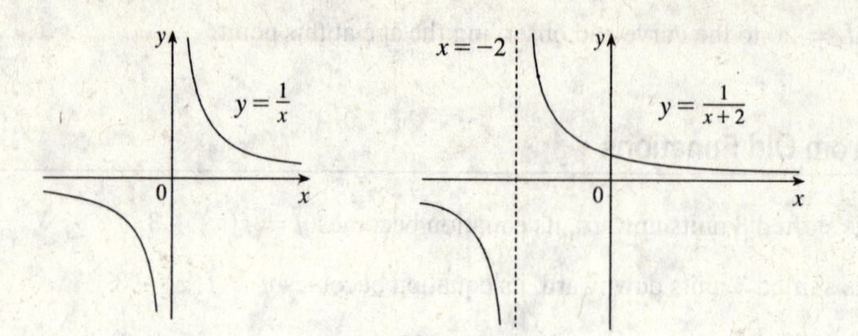

9. $y = -\sqrt[3]{x}$: Start with the graph of $y = \sqrt[3]{x}$ and reflect about the x-axis.

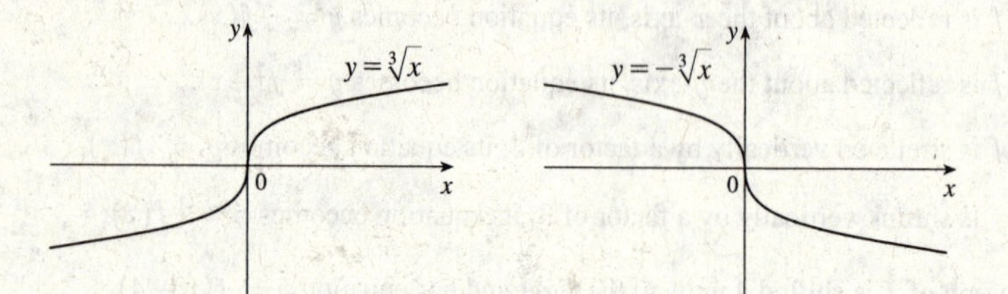

11. $y = \sqrt{x-2} - 1$: Start with the graph of $y = \sqrt{x}$, shift 2 units to the right, and then shift 1 unit downward.

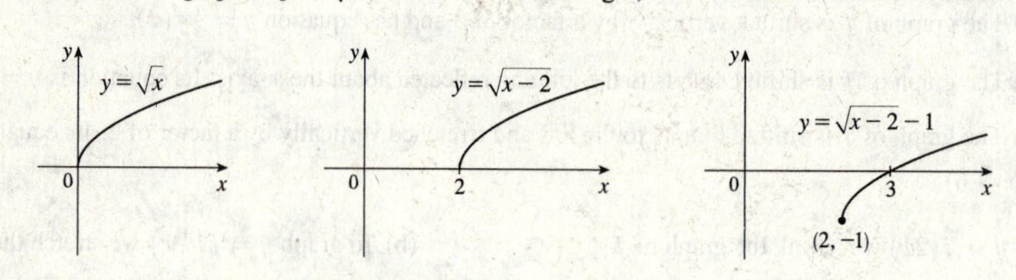

13. $y = \sin(x/2)$: Start with the graph of $y = \sin x$ and stretch horizontally by a factor of 2.

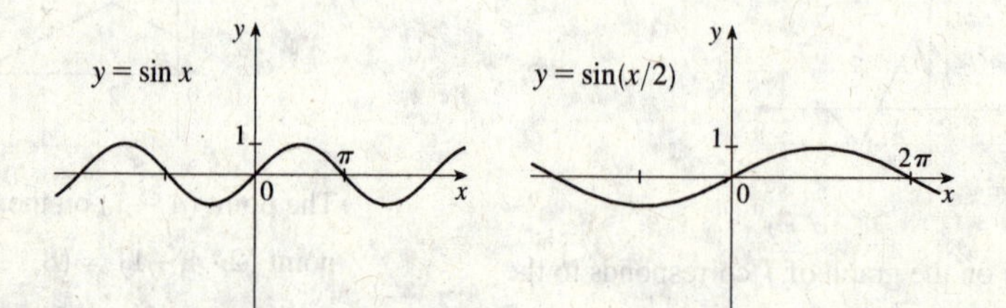

15. $y = -x^3$: Start with the graph of $y = x^3$ and reflect about the x-axis. Note: Reflecting about the y-axis gives the same result since substituting $-x$ for x gives us $y = (-x)^3 = -x^3$.

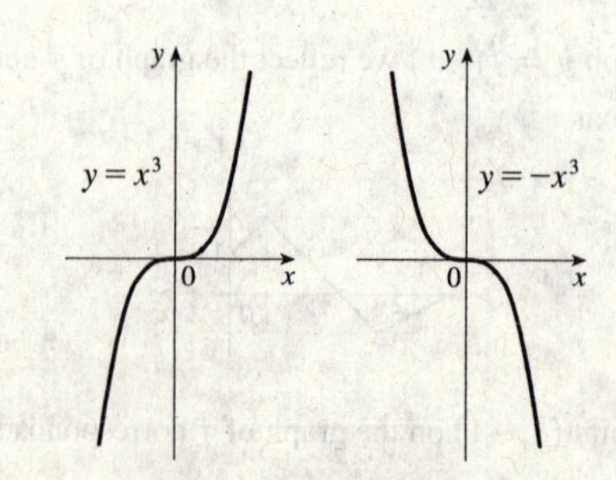

© 2016 Cengage Learning. All Rights Reserved. May not be scanned, copied or duplicated, or posted to a publicly accessible website, in whole or in part.

17. $y = \frac{1}{2}(1 - \cos x)$: Start with the graph of $y = \cos x$, reflect about the x-axis, shift 1 unit upward, and then shrink vertically by a factor of 2.

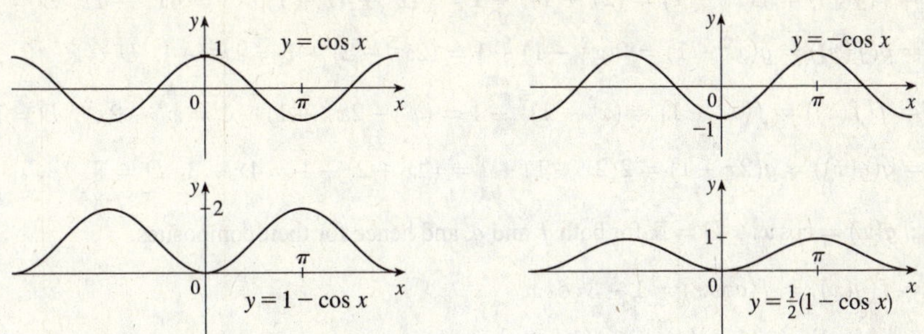

19. $y = 1 - 2x - x^2 = -(x^2 + 2x) + 1 = -(x^2 + 2x + 1) + 2 = -(x + 1)^2 + 2$: Start with the graph of $y = x^2$, reflect about the x-axis, shift 1 unit to the left, and then shift 2 units upward.

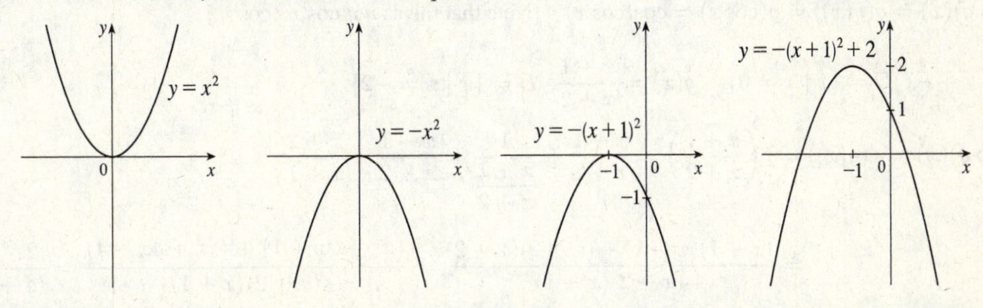

21. This is just like the solution to Example 4 except the amplitude of the curve (the 30°N curve in Figure 9 on June 21) is $14 - 12 = 2$. So the function is $L(t) = 12 + 2\sin\left[\frac{2\pi}{365}(t - 80)\right]$. March 31 is the 90th day of the year, so the model gives $L(90) \approx 12.34$ h. The daylight time (5:51 AM to 6:18 PM) is 12 hours and 27 minutes, or 12.45 h. The model value differs from the actual value by $\frac{12.45 - 12.34}{12.45} \approx 0.009$, less than 1%.

23. Let $D(t)$ be the water depth in meters at t hours after midnight. Apply the following transformations to the cosine function:

- Vertical stretch by factor 5 since the amplitude needs to be $\frac{12-2}{2} = 5$ m
- Horizontal stretch by factor $\frac{12}{2\pi} = \frac{6}{\pi}$ since the period needs to be 12 h
- Vertical shift 7 units upward since the function ranges between 2 and 12 which has a midpoint of $\frac{12+2}{2} = 7$ m
- Horizontal shift 6.75 units to right to position the maximum at $t = 6.75$ h (6:45AM)

Combining these transformations gives the water depth function $D(t) = 5\cos\left(\frac{\pi}{6}(t - 6.75)\right) + 7$.

25. Let $f(t)$ be the gene frequency after t years. The gene frequency dynamics can be modeled using a sine function with the following transformations:

- Vertical stretch by factor 30 since the amplitude needs to be $\frac{80-20}{2} = 30\%$
- Horizontal stretch by factor $\frac{3}{2\pi}$ since the period needs to be 3 years
- Vertical shift 50 units upward since the function ranges between 80 and 20 which has a midpoint of $\frac{80+20}{2} = 50$

Combining these transformations gives the gene frequency function $f(t) = 30\sin\left(\frac{2\pi}{3}t\right) + 50$.

27. $f(x) = x^3 + 2x^2$; $g(x) = 3x^2 - 1$. $D = \mathbb{R}$ for both f and g.

(a) $(f + g)(x) = (x^3 + 2x^2) + (3x^2 - 1) = x^3 + 5x^2 - 1$, $D = \mathbb{R}$.

(b) $(f - g)(x) = (x^3 + 2x^2) - (3x^2 - 1) = x^3 - x^2 + 1$, $D = \mathbb{R}$.

(c) $(fg)(x) = (x^3 + 2x^2)(3x^2 - 1) = 3x^5 + 6x^4 - x^3 - 2x^2$, $D = \mathbb{R}$.

(d) $\left(\frac{f}{g}\right)(x) = \frac{x^3 + 2x^2}{3x^2 - 1}$, $D = \left\{x \mid x \neq \pm\frac{1}{\sqrt{3}}\right\}$ since $3x^2 - 1 \neq 0$.

© 2016 Cengage Learning. All Rights Reserved. May not be scanned, copied or duplicated, or posted to a publicly accessible website, in whole or in part.

29. $f(x) = x^2 - 1$, $D = \mathbb{R}$; $g(x) = 2x + 1$, $D = \mathbb{R}$.

(a) $(f \circ g)(x) = f(g(x)) = f(2x+1) = (2x+1)^2 - 1 = (4x^2 + 4x + 1) - 1 = 4x^2 + 4x$, $D = \mathbb{R}$.

(b) $(g \circ f)(x) = g(f(x)) = g(x^2 - 1) = 2(x^2 - 1) + 1 = (2x^2 - 2) + 1 = 2x^2 - 1$, $D = \mathbb{R}$.

(c) $(f \circ f)(x) = f(f(x)) = f(x^2 - 1) = (x^2 - 1)^2 - 1 = (x^4 - 2x^2 + 1) - 1 = x^4 - 2x^2$, $D = \mathbb{R}$.

(d) $(g \circ g)(x) = g(g(x)) = g(2x+1) = 2(2x+1) + 1 = (4x+2) + 1 = 4x + 3$, $D = \mathbb{R}$.

31. $f(x) = 1 - 3x$; $g(x) = \cos x$. $D = \mathbb{R}$ for both f and g, and hence for their composites.

(a) $(f \circ g)(x) = f(g(x)) = f(\cos x) = 1 - 3\cos x$.

(b) $(g \circ f)(x) = g(f(x)) = g(1 - 3x) = \cos(1 - 3x)$.

(c) $(f \circ f)(x) = f(f(x)) = f(1 - 3x) = 1 - 3(1 - 3x) = 1 - 3 + 9x = 9x - 2$.

(d) $(g \circ g)(x) = g(g(x)) = g(\cos x) = \cos(\cos x)$ [Note that this is *not* $\cos x \cdot \cos x$.]

33. $f(x) = x + \dfrac{1}{x}$, $D = \{x \mid x \neq 0\}$; $g(x) = \dfrac{x+1}{x+2}$, $D = \{x \mid x \neq -2\}$

(a) $$(f \circ g)(x) = f(g(x)) = f\left(\frac{x+1}{x+2}\right) = \frac{x+1}{x+2} + \frac{1}{\dfrac{x+1}{x+2}} = \frac{x+1}{x+2} + \frac{x+2}{x+1}$$

$$= \frac{(x+1)(x+1) + (x+2)(x+2)}{(x+2)(x+1)} = \frac{(x^2 + 2x + 1) + (x^2 + 4x + 4)}{(x+2)(x+1)} = \frac{2x^2 + 6x + 5}{(x+2)(x+1)}$$

Since $g(x)$ is not defined for $x = -2$ and $f(g(x))$ is not defined for $x = -2$ and $x = -1$, the domain of $(f \circ g)(x)$ is $D = \{x \mid x \neq -2, -1\}$.

(b) $$(g \circ f)(x) = g(f(x)) = g\left(x + \frac{1}{x}\right) = \frac{\left(x + \dfrac{1}{x}\right) + 1}{\left(x + \dfrac{1}{x}\right) + 2} = \frac{\dfrac{x^2 + 1 + x}{x}}{\dfrac{x^2 + 1 + 2x}{x}} = \frac{x^2 + x + 1}{x^2 + 2x + 1} = \frac{x^2 + x + 1}{(x+1)^2}$$

Since $f(x)$ is not defined for $x = 0$ and $g(f(x))$ is not defined for $x = -1$, the domain of $(g \circ f)(x)$ is $D = \{x \mid x \neq -1, 0\}$.

(c) $$(f \circ f)(x) = f(f(x)) = f\left(x + \frac{1}{x}\right) = \left(x + \frac{1}{x}\right) + \frac{1}{x + \frac{1}{x}} = x + \frac{1}{x} + \frac{1}{\frac{x^2+1}{x}} = x + \frac{1}{x} + \frac{x}{x^2 + 1}$$

$$= \frac{x(x)(x^2+1) + 1\,(x^2+1) + x(x)}{x(x^2+1)} = \frac{x^4 + x^2 + x^2 + 1 + x^2}{x(x^2+1)}$$

$$= \frac{x^4 + 3x^2 + 1}{x(x^2+1)}, \quad D = \{x \mid x \neq 0\}$$

(d) $$(g \circ g)(x) = g(g(x)) = g\left(\frac{x+1}{x+2}\right) = \frac{\dfrac{x+1}{x+2} + 1}{\dfrac{x+1}{x+2} + 2} = \frac{\dfrac{x+1+1(x+2)}{x+2}}{\dfrac{x+1+2(x+2)}{x+2}} = \frac{x+1+x+2}{x+1+2x+4} = \frac{2x+3}{3x+5}$$

Since $g(x)$ is not defined for $x = -2$ and $g(g(x))$ is not defined for $x = -\frac{5}{3}$, the domain of $(g \circ g)(x)$ is $D = \left\{x \mid x \neq -2, -\frac{5}{3}\right\}$.

35. $(f \circ g \circ h)(x) = f(g(h(x))) = f(g(x^2)) = f(\sin(x^2)) = 3\sin(x^2) - 2$

© 2016 Cengage Learning. All Rights Reserved. May not be scanned, copied or duplicated, or posted to a publicly accessible website, in whole or in part.

37. $(f \circ g \circ h)(x) = f(g(h(x))) = f(g(x^3+2)) = f[(x^3+2)^2]$
$= f(x^6+4x^3+4) = \sqrt{(x^6+4x^3+4)-3} = \sqrt{x^6+4x^3+1}$

39. Let $g(x) = 2x + x^2$ and $f(x) = x^4$. Then $(f \circ g)(x) = f(g(x)) = f(2x+x^2) = (2x+x^2)^4 = F(x)$.

41. Let $g(x) = \sqrt[3]{x}$ and $f(x) = \dfrac{x}{1+x}$. Then $(f \circ g)(x) = f(g(x)) = f(\sqrt[3]{x}) = \dfrac{\sqrt[3]{x}}{1+\sqrt[3]{x}} = F(x)$.

43. Let $g(t) = t^2$ and $f(t) = \sec t \tan t$. Then $(f \circ g)(t) = f(g(t)) = f(t^2) = \sec(t^2)\tan(t^2) = v(t)$.

45. Let $h(x) = \sqrt{x}$, $g(x) = x - 1$, and $f(x) = \sqrt{x}$. Then
$(f \circ g \circ h)(x) = f(g(h(x))) = f(g(\sqrt{x})) = f(\sqrt{x}-1) = \sqrt{\sqrt{x}-1} = R(x)$.

47. Let $h(x) = \sqrt{x}$, $g(x) = \sec x$, and $f(x) = x^4$. Then
$(f \circ g \circ h)(x) = f(g(h(x))) = f(g(\sqrt{x})) = f(\sec\sqrt{x}) = (\sec\sqrt{x})^4 = \sec^4(\sqrt{x}) = H(x)$.

49. (a) $g(2) = 5$, because the point $(2,5)$ is on the graph of g. Thus, $f(g(2)) = f(5) = 4$, because the point $(5,4)$ is on the graph of f.

(b) $g(f(0)) = g(0) = 3$

(c) $(f \circ g)(0) = f(g(0)) = f(3) = 0$

(d) $(g \circ f)(6) = g(f(6)) = g(6)$. This value is not defined, because there is no point on the graph of g that has x-coordinate 6.

(e) $(g \circ g)(-2) = g(g(-2)) = g(1) = 4$

(f) $(f \circ f)(4) = f(f(4)) = f(2) = -2$

51. (a) Using the relationship *distance* = *rate* · *time* with the radius r as the distance, we have $r(t) = 60t$.

(b) $A = \pi r^2 \;\Rightarrow\; (A \circ r)(t) = A(r(t)) = \pi(60t)^2 = 3600\pi t^2$. This formula gives us the extent of the rippled area (in cm^2) at any time t.

53. (a) From the figure, we have a right triangle with legs 6 and d, and hypotenuse s.
By the Pythagorean Theorem, $d^2 + 6^2 = s^2 \;\Rightarrow\; s = f(d) = \sqrt{d^2+36}$.

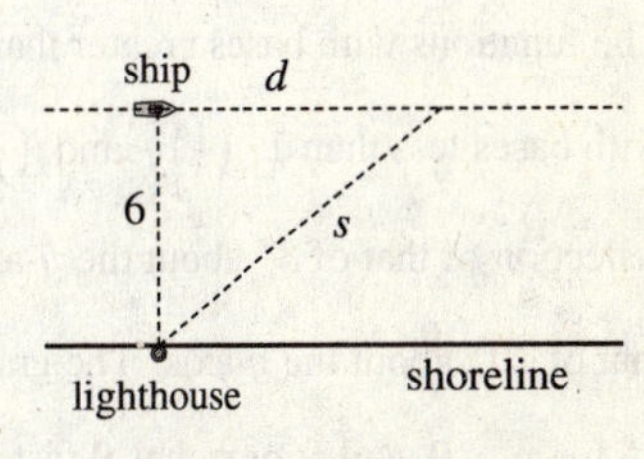

(b) Using $d = rt$, we get $d = (30 \text{ km/h})(t \text{ hours}) = 30t$ (in km). Thus, $d = g(t) = 30t$.

(c) $(f \circ g)(t) = f(g(t)) = f(30t) = \sqrt{(30t)^2+36} = \sqrt{900t^2+36}$. This function represents the distance between the lighthouse and the ship as a function of the time elapsed since noon.

55. (a) The diameter d of the tumor is increasing at a rate of g mm/year, so $d(t) = (g \text{ mm/year})(t \text{ year}) = gt$ (in mm).

(b) Using $S = 4\pi r^2 = \pi d^2$ for the surface area of a sphere, we get $(S \circ d)(t) = S(d(t)) = S(gt) = \pi(gt)^2 = \pi g^2 t^2$. Now, since P is proportional to the surface area, we have $P(S) = kS$ where k is a proportionality constant. Thus,
$(P \circ S \circ d)(t) = P(S(d(t))) = P(\pi g^2 t^2) = k\pi g^2 t^2$.
The result, $P = k\pi g^2 t^2$, gives the rate of enzyme production as a function of time.
Alternative Solution: If we assume the initial tumor size is nonzero so that $d(0) = d_0$, then $d(t) = d_0 + gt$. This gives $(P \circ S \circ d)(t) = k\pi(d_0 + gt)^2$.

© 2016 Cengage Learning. All Rights Reserved. May not be scanned, copied or duplicated, or posted to a publicly accessible website, in whole or in part.

57. If $f(x) = m_1x + b_1$ and $g(x) = m_2x + b_2$, then

$(f \circ g)(x) = f(g(x)) = f(m_2x + b_2) = m_1(m_2x + b_2) + b_1 = m_1m_2x + m_1b_2 + b_1$.

So $f \circ g$ is a linear function with slope m_1m_2.

59. $h(-x) = f(g(-x)) = f(-g(x))$. At this point, we can't simplify the expression, so we might try to find a counterexample to show that h is not an odd function. Let $g(x) = x$, an odd function, and $f(x) = x^2 + x$. Then $h(x) = x^2 + x$, which is neither even nor odd.

Now suppose f is an odd function. Then $f(-g(x)) = -f(g(x)) = -h(x)$. Hence, $h(-x) = -h(x)$, and so h is odd if both f and g are odd.

Now suppose f is an even function. Then $f(-g(x)) = f(g(x)) = h(x)$. Hence, $h(-x) = h(x)$, and so h is even if g is odd and f is even.

1.4 Exponential Functions

1. (a) $\dfrac{4^{-3}}{2^{-8}} = \dfrac{2^8}{4^3} = \dfrac{2^8}{(2^2)^3} = \dfrac{2^8}{2^6} = 2^{8-6} = 2^2 = 4$ (b) $\dfrac{1}{\sqrt[3]{x^4}} = \dfrac{1}{x^{4/3}} = x^{-4/3}$

3. (a) $b^8(2b)^4 = b^8 \cdot 2^4b^4 = 16b^{12}$ (b) $\dfrac{(6y^3)^4}{2y^5} = \dfrac{6^4(y^3)^4}{2y^5} = \dfrac{1296y^{12}}{2y^5} = 648y^7$

5. (a) $f(x) = b^x,\ b > 0$ (b) $\mathbb{R}$ (c) $(0, \infty)$ (d) See Figures 5(c), 5(b), and 5(a), respectively.

7. All of these graphs approach 0 as $x \to -\infty$, all of them pass through the point $(0, 1)$, and all of them are increasing and approach ∞ as $x \to \infty$. The larger the base, the faster the function increases for $x > 0$, and the faster it approaches 0 as $x \to -\infty$.

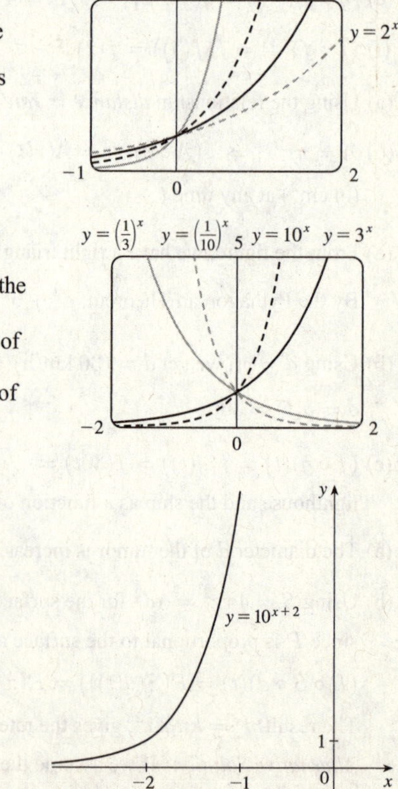

9. The functions with bases greater than 1 (3^x and 10^x) are increasing, while those with bases less than 1 $\left[\left(\frac{1}{3}\right)^x \text{ and } \left(\frac{1}{10}\right)^x\right]$ are decreasing. The graph of $\left(\frac{1}{3}\right)^x$ is the reflection of that of 3^x about the y-axis, and the graph of $\left(\frac{1}{10}\right)^x$ is the reflection of that of 10^x about the y-axis. The graph of 10^x increases more quickly than that of 3^x for $x > 0$, and approaches 0 faster as $x \to -\infty$.

11. We start with the graph of $y = 10^x$ (Figure 4) and shift it 2 units to the left to obtain the graph of $y = 10^{x+2}$.

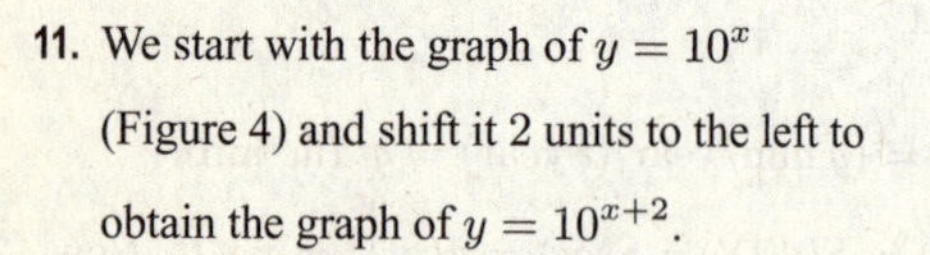

© 2016 Cengage Learning. All Rights Reserved. May not be scanned, copied or duplicated, or posted to a publicly accessible website, in whole or in part.

13. We start with the graph of $y = 2^x$ (Figure 4), reflect it about the y-axis, and then about the x-axis (or just rotate 180° to handle both reflections) to obtain the graph of $y = -2^{-x}$. In each graph, $y = 0$ is the horizontal asymptote.

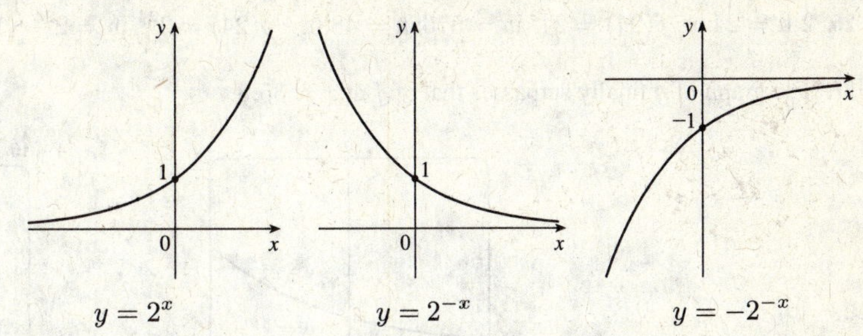

15. We start with the graph of $y = e^x$ (Figure 17) and reflect about the y-axis to get the graph of $y = e^{-x}$. Then we compress the graph vertically by a factor of 2 to obtain the graph of $y = \frac{1}{2}e^{-x}$ and then reflect about the x-axis to get the graph of $y = -\frac{1}{2}e^{-x}$. Finally, we shift the graph upward one unit to get the graph of $y = 1 - \frac{1}{2}e^{-x}$.

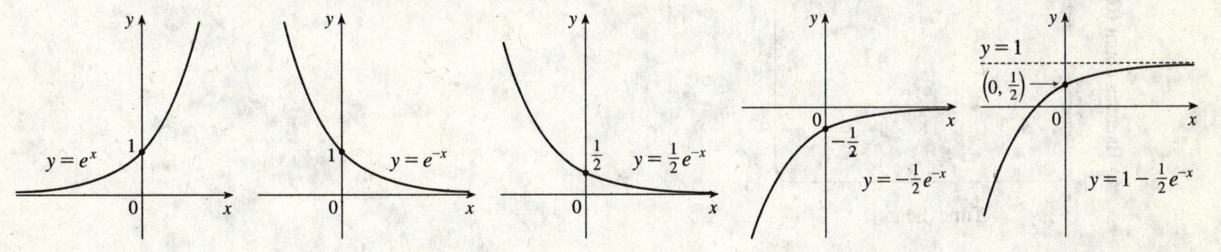

17. (a) To find the equation of the graph that results from shifting the graph of $y = e^x$ 2 units downward, we subtract 2 from the original function to get $y = e^x - 2$.

(b) To find the equation of the graph that results from shifting the graph of $y = e^x$ 2 units to the right, we replace x with $x - 2$ in the original function to get $y = e^{(x-2)}$.

(c) To find the equation of the graph that results from reflecting the graph of $y = e^x$ about the x-axis, we multiply the original function by -1 to get $y = -e^x$.

(d) To find the equation of the graph that results from reflecting the graph of $y = e^x$ about the y-axis, we replace x with $-x$ in the original function to get $y = e^{-x}$.

(e) To find the equation of the graph that results from reflecting the graph of $y = e^x$ about the x-axis and then about the y-axis, we first multiply the original function by -1 (to get $y = -e^x$) and then replace x with $-x$ in this equation to get $y = -e^{-x}$.

19. (a) The denominator is zero when $1 - e^{1-x^2} = 0 \Leftrightarrow e^{1-x^2} = 1 \Leftrightarrow 1 - x^2 = 0 \Leftrightarrow x = \pm 1$. Thus, the function $f(x) = \dfrac{1 - e^{x^2}}{1 - e^{1-x^2}}$ has domain $\{x \mid x \neq \pm 1\} = (-\infty, -1) \cup (-1, 1) \cup (1, \infty)$.

(b) The denominator is never equal to zero, so the function $f(x) = \dfrac{1 + x}{e^{\cos x}}$ has domain $\mathbb{R}$, or $(-\infty, \infty)$.

21. Use $y = Cb^x$ with the points $(1, 6)$ and $(3, 24)$. $6 = Cb^1$ $\left[C = \frac{6}{b}\right]$ and $24 = Cb^3 \Rightarrow 24 = \left(\frac{6}{b}\right)b^3 \Rightarrow$ $4 = b^2 \Rightarrow b = 2$ [since $b > 0$] and $C = \frac{6}{2} = 3$. The function is $f(x) = 3 \cdot 2^x$.

23. If $f(x) = 5^x$, then $\dfrac{f(x+h) - f(x)}{h} = \dfrac{5^{x+h} - 5^x}{h} = \dfrac{5^x 5^h - 5^x}{h} = \dfrac{5^x(5^h - 1)}{h} = 5^x\left(\dfrac{5^h - 1}{h}\right)$.

© 2016 Cengage Learning. All Rights Reserved. May not be scanned, copied or duplicated, or posted to a publicly accessible website, in whole or in part.

25. 2 ft $= 24$ in, $f(24) = 24^2$ in $= 576$ in $= 48$ ft. $g(24) = 2^{24}$ in $= 2^{24}/(12 \cdot 5280)$ mi ≈ 265 mi

27. The graph of g finally surpasses that of f at $x \approx 35.8$.

29. (a)

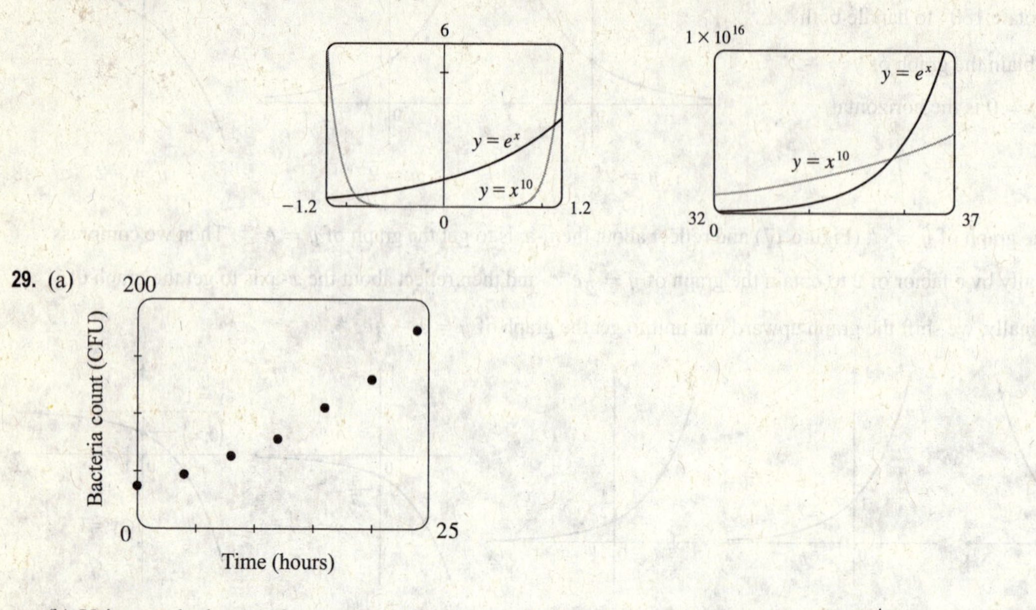

(b) Using a calculator to fit an exponential curve to the data gives $f(t) = (36.78) \cdot (1.07)^t$.

(c)

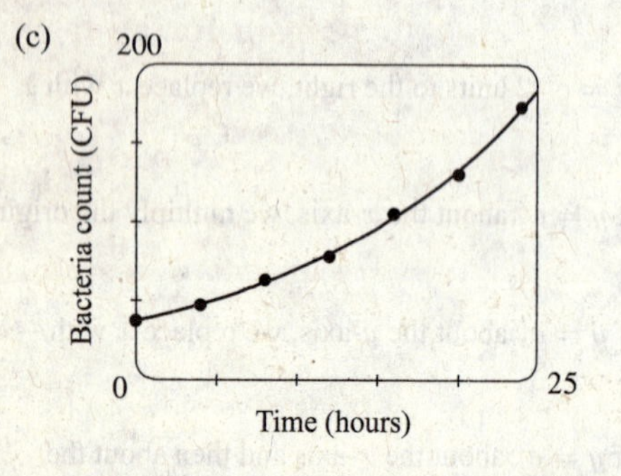

Using the TRACE feature of a calculator, we find that the bacteria count increases from 50 CFU to 100 CFU in about 10.8 hours. Therefore it takes approximately 10.8 hours for the bacteria count to double.

31. (a) Fifteen days represents 3 half-life periods (one half-life period is 5 days). $\cdot$ $200\left(\frac{1}{2}\right)^3 = 25$ mg

(b) In t hours, there will be $t/5$ half-life periods. The initial amount is 200 mg, so the amount remaining after t days is $y = 200\left(\frac{1}{2}\right)^{t/5}$, or equivalently, $y = 200 \cdot 2^{-t/5}$.

(c) $t = 3$ weeks $= 21$ days $\Rightarrow$ $y = 200 \cdot 2^{-21/5} \approx 10.9$ mg

(d) We graph $y_1 = 200 \cdot 2^{-t/5}$ and $y_2 = 1$. The two curves intersect at $t \approx 38.2$, so the mass will be reduced to 1 mg in about 38.2 days.

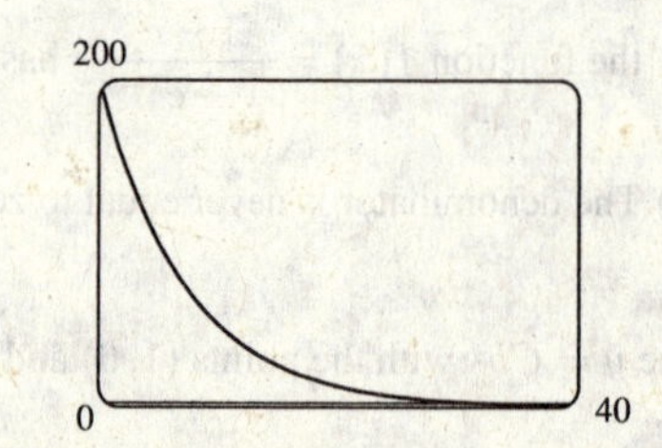

33. The half-life is approximately 3.5 days since the RNA load drops from 40 to 20 in that time.

35. Let $t = 0$ correspond to 1950 to get the model $P = ab^t$, where $a \approx 2614.086$ and $b \approx 1.01693$. To estimate the population in 1993, let $t = 43$ to obtain $P \approx 5381$ million. To predict the population in 2020, let $t = 70$ to obtain $P \approx 8466$ million.

© 2016 Cengage Learning. All Rights Reserved. May not be scanned, copied or duplicated, or posted to a publicly accessible website, in whole or in part.

37. 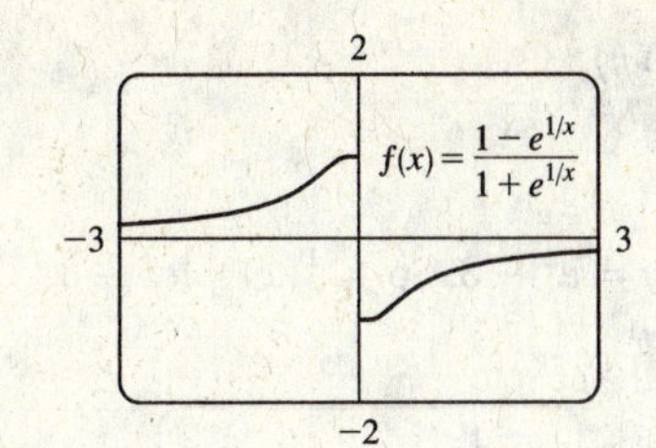

From the graph, it appears that f is an odd function (f is undefined for $x = 0$).

To prove this, we must show that $f(-x) = -f(x)$.

$$f(-x) = \frac{1 - e^{1/(-x)}}{1 + e^{1/(-x)}} = \frac{1 - e^{(-1/x)}}{1 + e^{(-1/x)}} = \frac{1 - \dfrac{1}{e^{1/x}}}{1 + \dfrac{1}{e^{1/x}}} \cdot \frac{e^{1/x}}{e^{1/x}} = \frac{e^{1/x} - 1}{e^{1/x} + 1}$$

$$= -\frac{1 - e^{1/x}}{1 + e^{1/x}} = -f(x)$$

so f is an odd function.

1.5 Logarithms; Semi-log and Log-log Plots

1. (a) See Definition 1.

(b) It must pass the Horizontal Line Test.

3. f is not one-to-one because $2 \neq 6$, but $f(2) = 2.0 = f(6)$.

5. We could draw a horizontal line that intersects the graph in more than one point. Thus, by the Horizontal Line Test, the function is not one-to-one.

7. No horizontal line intersects the graph more than once. Thus, by the Horizontal Line Test, the function is one-to-one.

9. The graph of $f(x) = x^2 - 2x$ is a parabola with axis of symmetry $x = -\dfrac{b}{2a} = -\dfrac{-2}{2(1)} = 1$. Pick any x-values equidistant from 1 to find two equal function values. For example, $f(0) = 0$ and $f(2) = 0$, so f is not one-to-one.

11. $g(x) = 1/x$. $x_1 \neq x_2 \;\Rightarrow\; 1/x_1 \neq 1/x_2 \;\Rightarrow\; g(x_1) \neq g(x_2)$, so g is one-to-one.

Geometric solution: The graph of g is the hyperbola shown in Figure 14 in Section 1.2. It passes the Horizontal Line Test, so g is one-to-one.

13. A football will attain every height h up to its maximum height twice: once on the way up, and again on the way down. Thus, even if t_1 does not equal t_2, $f(t_1)$ may equal $f(t_2)$, so f is not 1-1.

15. (a) Since f is 1-1, $f(6) = 17 \;\Leftrightarrow\; f^{-1}(17) = 6$.

(b) Since f is 1-1, $f^{-1}(3) = 2 \;\Leftrightarrow\; f(2) = 3$.

17. First, we must determine x such that $g(x) = 4$. By inspection, we see that if $x = 0$, then $g(x) = 4$. Since g is 1-1 (g is an increasing function), it has an inverse, and $g^{-1}(4) = 0$.

19. We solve $C = \frac{5}{9}(F - 32)$ for F: $\frac{9}{5}C = F - 32 \;\Rightarrow\; F = \frac{9}{5}C + 32$. This gives us a formula for the inverse function, that is, the Fahrenheit temperature F as a function of the Celsius temperature C. $F \geq -459.67 \;\Rightarrow\; \frac{9}{5}C + 32 \geq -459.67 \;\Rightarrow\; \frac{9}{5}C \geq -491.67 \;\Rightarrow\; C \geq -273.15$, the domain of the inverse function.

21. $y = f(x) = 1 + \sqrt{2 + 3x} \quad (y \geq 1) \;\Rightarrow\; y - 1 = \sqrt{2 + 3x} \;\Rightarrow\; (y - 1)^2 = 2 + 3x \;\Rightarrow\; (y - 1)^2 - 2 = 3x \;\Rightarrow\; x = \frac{1}{3}(y - 1)^2 - \frac{2}{3}$. Interchange x and y: $y = \frac{1}{3}(x - 1)^2 - \frac{2}{3}$. So $f^{-1}(x) = \frac{1}{3}(x - 1)^2 - \frac{2}{3}$. Note that the domain of f^{-1} is $x \geq 1$.

© 2016 Cengage Learning. All Rights Reserved. May not be scanned, copied or duplicated, or posted to a publicly accessible website, in whole or in part.

23. $y = f(x) = e^{2x-1} \Rightarrow \ln y = 2x - 1 \Rightarrow 1 + \ln y = 2x \Rightarrow x = \frac{1}{2}(1 + \ln y)$.

Interchange x and y: $y = \frac{1}{2}(1 + \ln x)$. So $f^{-1}(x) = \frac{1}{2}(1 + \ln x)$.

25. $y = f(x) = \ln(x + 3) \Rightarrow x + 3 = e^y \Rightarrow x = e^y - 3$. Interchange x and y: $y = e^x - 3$. So $f^{-1}(x) = e^x - 3$.

27. $y = f(x) = x^4 + 1 \Rightarrow y - 1 = x^4 \Rightarrow x = \sqrt[4]{y - 1}$ [not $\pm$ since $x \geq 0$]. Interchange x and y: $y = \sqrt[4]{x - 1}$. So $f^{-1}(x) = \sqrt[4]{x - 1}$. The graph of $y = \sqrt[4]{x - 1}$ is just the graph of $y = \sqrt[4]{x}$ shifted right one unit. From the graph, we see that f and f^{-1} are reflections about the line $y = x$.

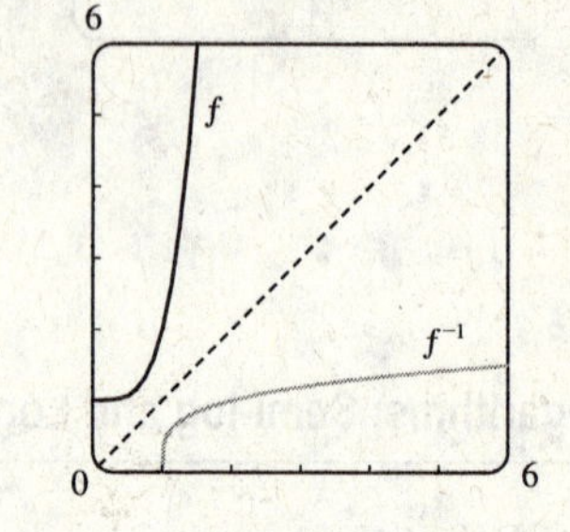

29. Reflect the graph of f about the line $y = x$. The points $(-1, -2)$, $(1, -1)$, $(2, 2)$, and $(3, 3)$ on f are reflected to $(-2, -1)$, $(-1, 1)$, $(2, 2)$, and $(3, 3)$ on f^{-1}.

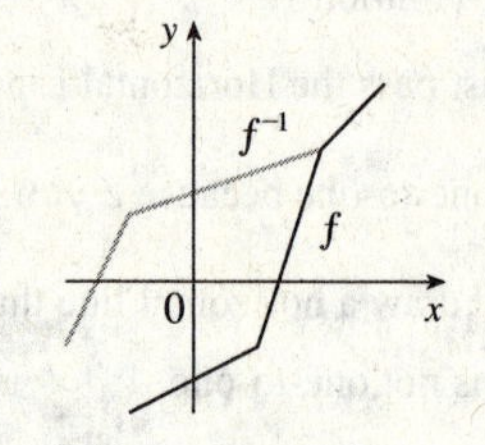

31. (a) $y = f(x) = \sqrt{1 - x^2}$ $(0 \leq x \leq 1$ and note that $y \geq 0) \Rightarrow y^2 = 1 - x^2 \Rightarrow x^2 = 1 - y^2 \Rightarrow x = \sqrt{1 - y^2}$.
So $f^{-1}(x) = \sqrt{1 - x^2}$, $0 \leq x \leq 1$. We see that f^{-1} and f are the same function.

(b) The graph of f is the portion of the circle $x^2 + y^2 = 1$ with $0 \leq x \leq 1$ and $0 \leq y \leq 1$ (quarter-circle in the first quadrant). The graph of f is symmetric with respect to the line $y = x$, so its reflection about $y = x$ is itself, that is, $f^{-1} = f$.

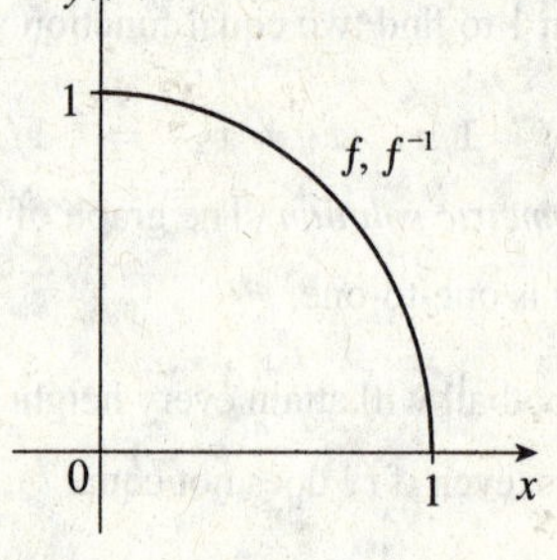

33. (a) It is defined as the inverse of the exponential function with base b, that is, $\log_b x = y \Leftrightarrow b^y = x$.

(b) $(0, \infty)$ (c) $\mathbb{R}$ (d) See Figure 11.

35. (a) $\log_5 125 = 3$ since $5^3 = 125$. (b) $\log_3 \frac{1}{27} = -3$ since $3^{-3} = \frac{1}{3^3} = \frac{1}{27}$.

37. (a)
$$\begin{aligned} \log_2 6 - \log_2 15 + \log_2 20 &= \log_2(\tfrac{6}{15}) + \log_2 20 && \text{[by Law 2]} \\ &= \log_2(\tfrac{6}{15} \cdot 20) && \text{[by Law 1]} \\ &= \log_2 8, \text{ and } \log_2 8 = 3 \text{ since } 2^3 = 8. \end{aligned}$$

(b)
$$\begin{aligned} \log_3 100 - \log_3 18 - \log_3 50 &= \log_3(\tfrac{100}{18}) - \log_3 50 = \log_3(\tfrac{100}{18 \cdot 50}) \\ &= \log_3(\tfrac{1}{9}), \text{ and } \log_3(\tfrac{1}{9}) = -2 \text{ since } 3^{-2} = \tfrac{1}{9}. \end{aligned}$$

© 2016 Cengage Learning. All Rights Reserved. May not be scanned, copied or duplicated, or posted to a publicly accessible website, in whole or in part.

39. $\ln 5 + 5\ln 3 = \ln 5 + \ln 3^5$ [by Law 3]

$= \ln(5 \cdot 3^5)$ [by Law 1]

$= \ln 1215$

41. $\frac{1}{3}\ln(x+2)^3 + \frac{1}{2}\left[\ln x - \ln(x^2+3x+2)^2\right] = \ln[(x+2)^3]^{1/3} + \frac{1}{2}\ln\dfrac{x}{(x^2+3x+2)^2}$ [by Laws 3, 2]

$= \ln(x+2) + \ln\dfrac{\sqrt{x}}{x^2+3x+2}$ [by Law 3]

$= \ln\dfrac{(x+2)\sqrt{x}}{(x+1)(x+2)}$ [by Law 1]

$= \ln\dfrac{\sqrt{x}}{x+1}$

Note that since $\ln x$ is defined for $x > 0$, we have $x+1$, $x+2$, and x^2+3x+2 all positive, and hence their logarithms are defined.

43. 3 ft = 36 in, so we need x such that $\log_2 x = 36 \Leftrightarrow x = 2^{36} = 68{,}719{,}476{,}736$. In miles, this is $68{,}719{,}476{,}736 \text{ in} \cdot \dfrac{1 \text{ ft}}{12 \text{ in}} \cdot \dfrac{1 \text{ mi}}{5280 \text{ ft}} \approx 1{,}084{,}587.7$ mi.

45. (a) Shift the graph of $y = \log_{10} x$ five units to the left to obtain the graph of $y = \log_{10}(x+5)$. Note the vertical asymptote of $x = -5$.

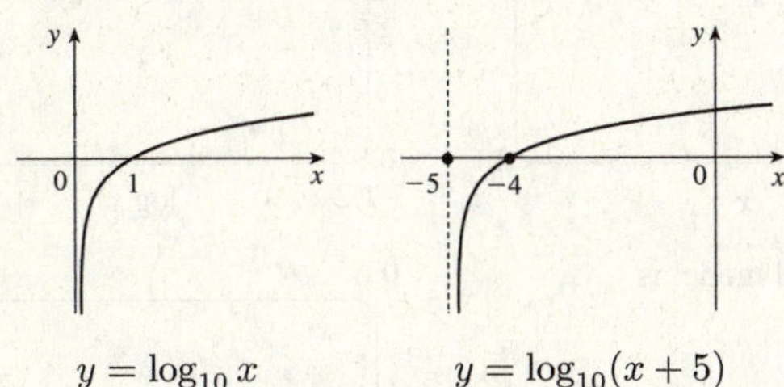

$y = \log_{10} x$ $y = \log_{10}(x+5)$

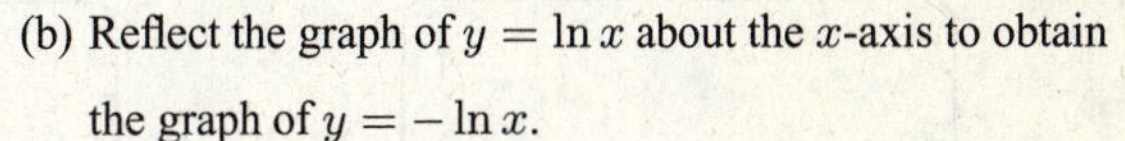

(b) Reflect the graph of $y = \ln x$ about the x-axis to obtain the graph of $y = -\ln x$.

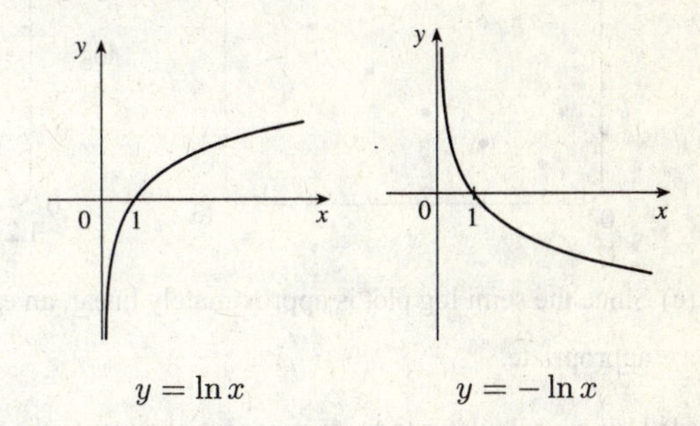

$y = \ln x$ $y = -\ln x$

47. (a) $e^{7-4x} = 6 \Leftrightarrow 7 - 4x = \ln 6 \Leftrightarrow 7 - \ln 6 = 4x \Leftrightarrow x = \frac{1}{4}(7 - \ln 6)$

(b) $\ln(3x - 10) = 2 \Leftrightarrow 3x - 10 = e^2 \Leftrightarrow 3x = e^2 + 10 \Leftrightarrow x = \frac{1}{3}(e^2 + 10)$

49. (a) $2^{x-5} = 3 \Leftrightarrow \log_2 3 = x - 5 \Leftrightarrow x = 5 + \log_2 3$.

Or: $2^{x-5} = 3 \Leftrightarrow \ln(2^{x-5}) = \ln 3 \Leftrightarrow (x-5)\ln 2 = \ln 3 \Leftrightarrow x - 5 = \dfrac{\ln 3}{\ln 2} \Leftrightarrow x = 5 + \dfrac{\ln 3}{\ln 2}$

(b) $\ln x + \ln(x-1) = \ln(x(x-1)) = 1 \Leftrightarrow x(x-1) = e^1 \Leftrightarrow x^2 - x - e = 0$. The quadratic formula (with $a = 1$, $b = -1$, and $c = -e$) gives $x = \frac{1}{2}\left(1 \pm \sqrt{1+4e}\right)$, but we reject the negative root since the natural logarithm is not defined for $x < 0$. So $x = \frac{1}{2}\left(1 + \sqrt{1+4e}\right)$.

51. (a) $\ln x < 0 \Rightarrow x < e^0 \Rightarrow x < 1$. Since the domain of $f(x) = \ln x$ is $x > 0$, the solution of the original inequality is $0 < x < 1$.

(b) $e^x > 5 \Rightarrow \ln e^x > \ln 5 \Rightarrow x > \ln 5$

© 2016 Cengage Learning. All Rights Reserved. May not be scanned, copied or duplicated, or posted to a publicly accessible website, in whole or in part.

53. (a) Solve for t in the equation: $c(t) = c_0 e^{-Kt/V} \Rightarrow 0.60 = 1.65e^{-340t/32941} \Leftrightarrow \ln\left(\frac{0.60}{1.65}\right) = \ln\left(e^{-340t/32941}\right) \Leftrightarrow$ $t = -\frac{32941}{340}\ln\left(\frac{0.60}{1.65}\right) \approx 98.0$ minutes

(b) Solve for T in the equation: $c(T) = c_0 e^{-KT/V} \Leftrightarrow \dfrac{c(T)}{c_0} = e^{-KT/V} \Leftrightarrow \ln\left(\dfrac{c(T)}{c_0}\right) = -KT/V \Leftrightarrow$

$$T = -\frac{V}{K}\ln\left(\frac{c(T)}{c_0}\right)$$

55. (a) We must have $e^x - 3 > 0 \Rightarrow e^x > 3 \Rightarrow x > \ln 3$. Thus, the domain of $f(x) = \ln(e^x - 3)$ is $(\ln 3, \infty)$.

(b) $y = \ln(e^x - 3) \Rightarrow e^y = e^x - 3 \Rightarrow e^x = e^y + 3 \Rightarrow x = \ln(e^y + 3)$, so $f^{-1}(x) = \ln(e^x + 3)$.

Now $e^x + 3 > 0 \Rightarrow e^x > -3$, which is true for any real x, so the domain of f^{-1} is $\mathbb{R}$.

57. (a) Find the inverse by solving for t: $n = 500 \cdot 4^t \Leftrightarrow \ln\left(\dfrac{n}{500}\right) = \ln(4^t) \Leftrightarrow \ln\left(\dfrac{n}{500}\right) = t\ln(4) \Leftrightarrow$

$t = \dfrac{\log(n/500)}{\log(4)}$. The inverse function gives the number of hours that have passed when the population size reaches n.

(b) Substituting $n = 10{,}000$ into the inverse function gives $t = \dfrac{\ln(10{,}000/500)}{\ln(4)} = \dfrac{\ln(20)}{\ln(4)} \approx 2.16$ hours.

59.

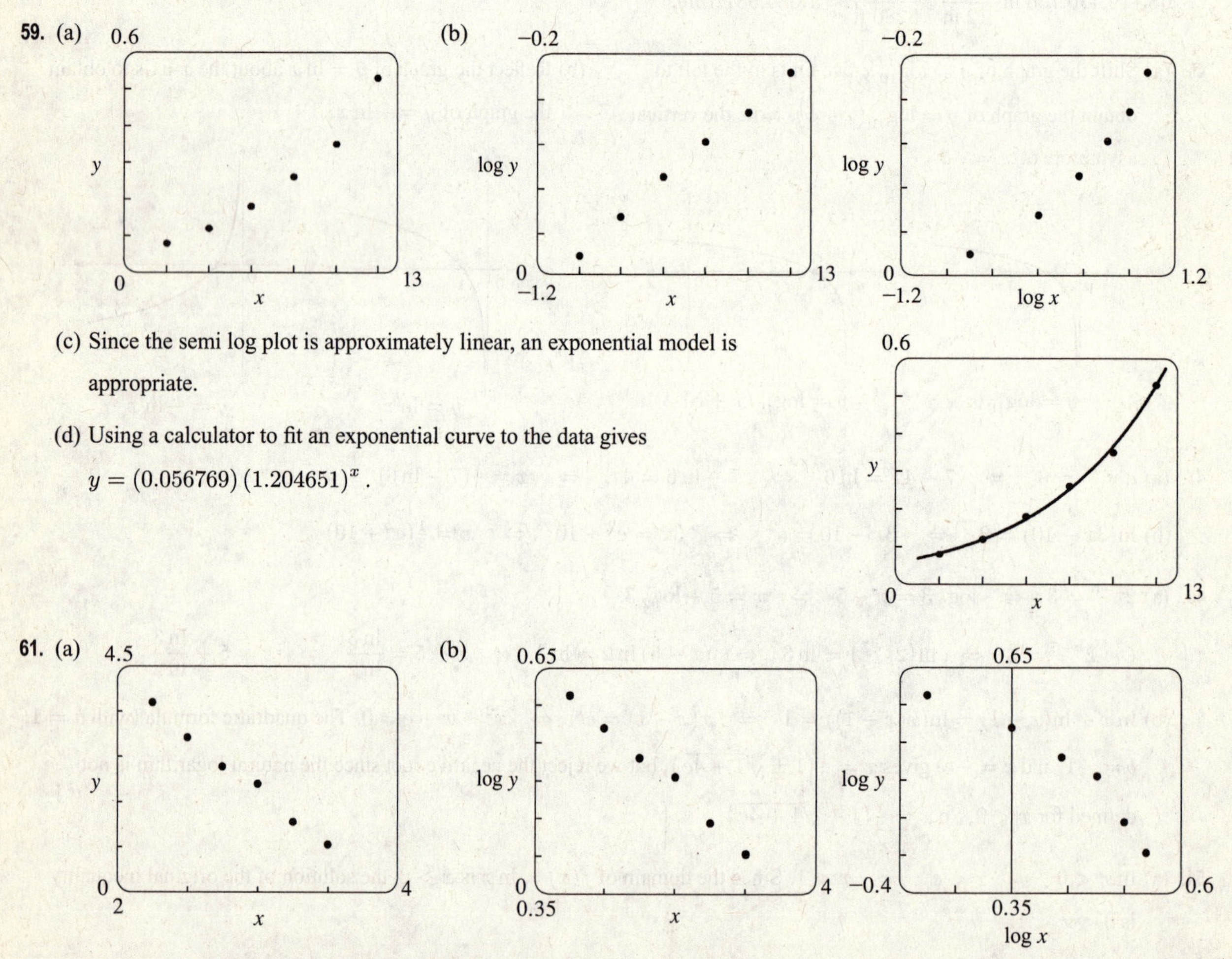

(c) Since the semi log plot is approximately linear, an exponential model is appropriate.

(d) Using a calculator to fit an exponential curve to the data gives $y = (0.056769)(1.204651)^x$.

61. (c) Since the scatter plot is approximately linear, a linear model is appropriate.

© 2016 Cengage Learning. All Rights Reserved. May not be scanned, copied or duplicated, or posted to a publicly accessible website, in whole or in part.

(d) Using a calculator to fit a line to the data gives $y = (-0.618857)\,x + 4.368000$.

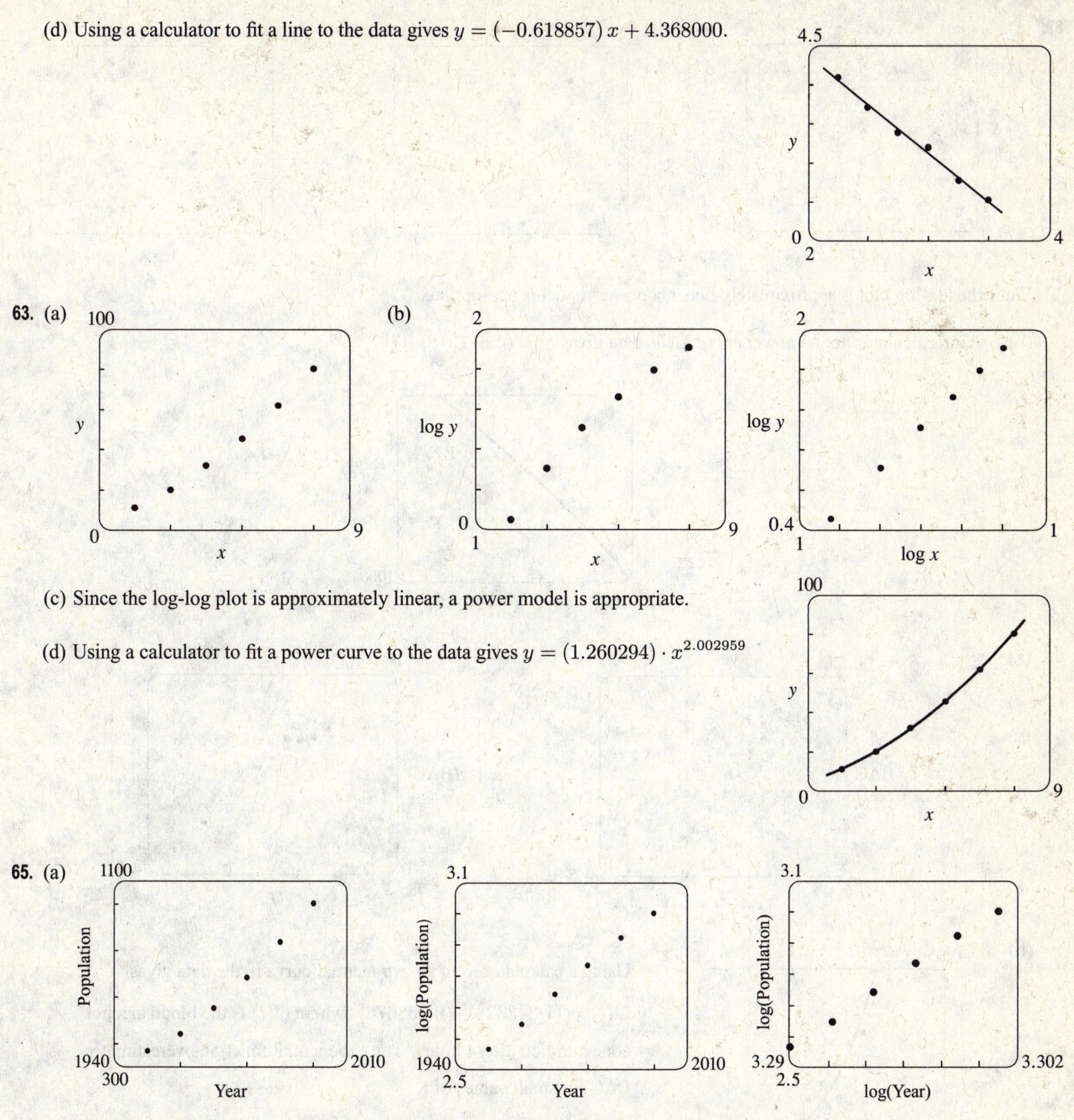

63. (a) (b)

(c) Since the log-log plot is approximately linear, a power model is appropriate.

(d) Using a calculator to fit a power curve to the data gives $y = (1.260294) \cdot x^{2.002959}$

65. (a)

Since the semi log plot is approximately linear, an exponential model is appropriate.

(b) Using a calculator to fit an exponential curve to the data gives $P = \left(2.276131 \cdot 10^{-15}\right) \cdot (1.020529)^{Y}$ where P is the population in millions and Y is the year. Alternatively, we could have defined Y to be the number of years since 1950.

(c) In 2010, the model predicts a population of $P = \left(2.276131 \cdot 10^{-15}\right) \cdot (1.020529)^{2010} \approx 1247$ million. The model overestimates the true population by $1247 - 1173 = 74$ million. Therefore, this exponential model does not generalize well to the future population growth in India.

© 2016 Cengage Learning. All Rights Reserved. May not be scanned, copied or duplicated, or posted to a publicly accessible website, in whole or in part.

67. (a)

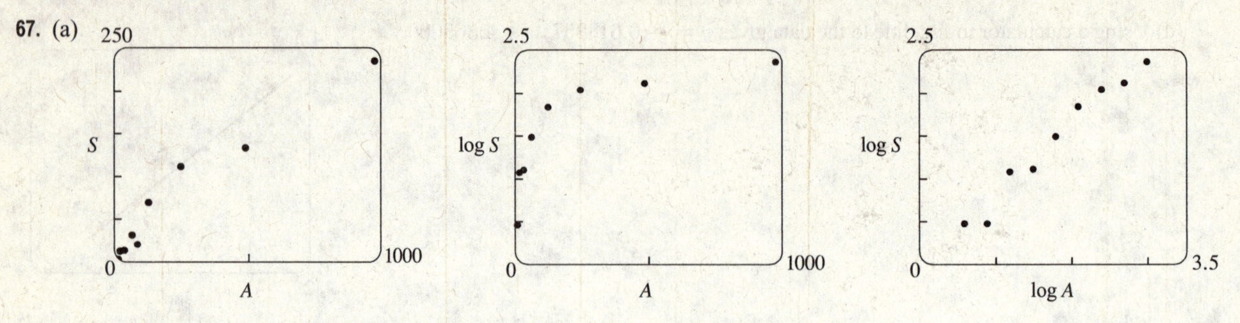

Since the log-log plot is approximately linear, a power model is appropriate.

(b) Using a calculator to fit a power curve to the data gives $S = (0.881518) \cdot A^{0.841701}$.

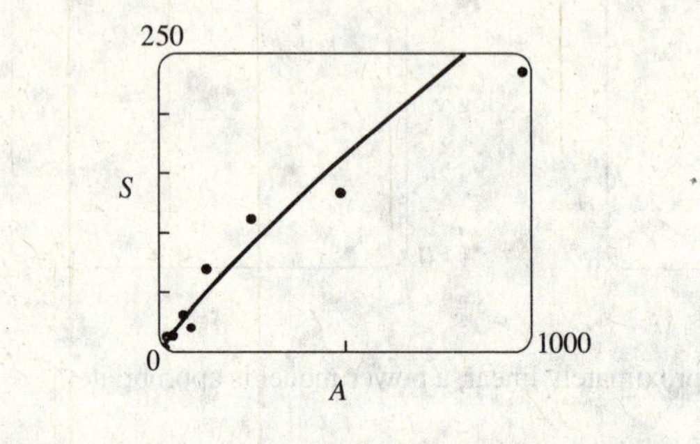

69. (a)

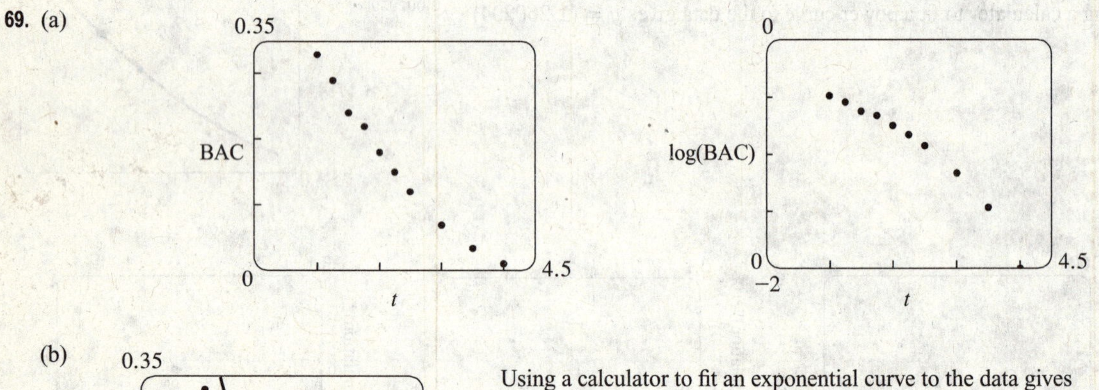

(b)

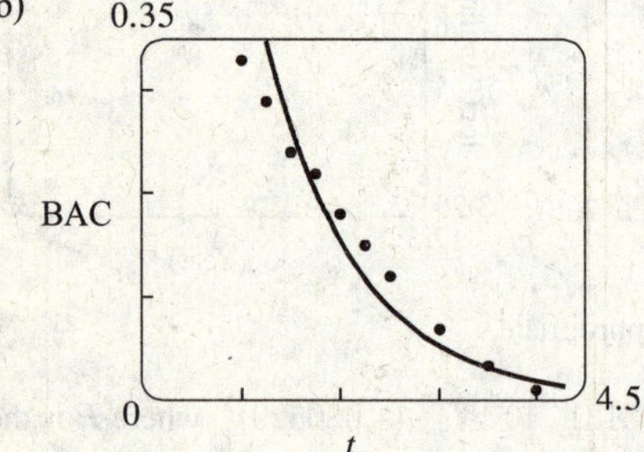

Using a calculator to fit an exponential curve to the data gives $C(t) = (1.343328) \cdot (0.338676)^t$ where $C(t)$ is the blood alcohol concentration after t hours. The exponential function overestimates BAC for small values of t.

(c) Solve for t in the equation: $C(t) \leq 0.08 \quad \Leftrightarrow \quad (1.343328) \cdot (0.338676)^t \leq 0.08 \quad \Leftrightarrow \quad \ln\left(0.338676^t\right) \leq \ln\left(\frac{0.08}{1.343328}\right)$

$\Leftrightarrow \quad t \ln(0.338676) \leq \ln\left(\frac{0.08}{1.343328}\right) \quad \Leftrightarrow \quad t \geq \dfrac{\ln(0.08/1.343328)}{\ln(0.338676)} \quad \left[\begin{array}{l}\text{inequality switched direction} \\ \text{because } \ln(0.338676) < 0\end{array}\right] \approx 2.61$ hr. Therefore,

the driver's blood alcohol concentration will be under the legal limit after approximately 2.6 hours.

© 2016 Cengage Learning. All Rights Reserved. May not be scanned, copied or duplicated, or posted to a publicly accessible website, in whole or in part.

1.6 Sequences and Difference Equations

1. $a_n = \dfrac{2n}{n^2+1}$, so the sequence is $\left\{\dfrac{2}{1+1}, \dfrac{4}{4+1}, \dfrac{6}{9+1}, \dfrac{8}{16+1}, \dfrac{10}{25+1}, \ldots\right\} = \left\{1, \dfrac{4}{5}, \dfrac{3}{5}, \dfrac{8}{17}, \dfrac{5}{13}, \ldots\right\}$.

3. $a_n = \dfrac{(-1)^{n-1}}{5^n}$, so the sequence is $\left\{\dfrac{1}{5^1}, \dfrac{-1}{5^2}, \dfrac{1}{5^3}, \dfrac{-1}{5^4}, \dfrac{1}{5^5}, \ldots\right\} = \left\{\dfrac{1}{5}, -\dfrac{1}{25}, \dfrac{1}{125}, -\dfrac{1}{625}, \dfrac{1}{3125}, \ldots\right\}$.

5.

n	$a_n = \dfrac{3n}{1+6n}$
1	0.4286
2	0.4615
3	0.4737
4	0.4800
5	0.4839
6	0.4865
7	0.4884
8	0.4898
9	0.4909
10	0.4918

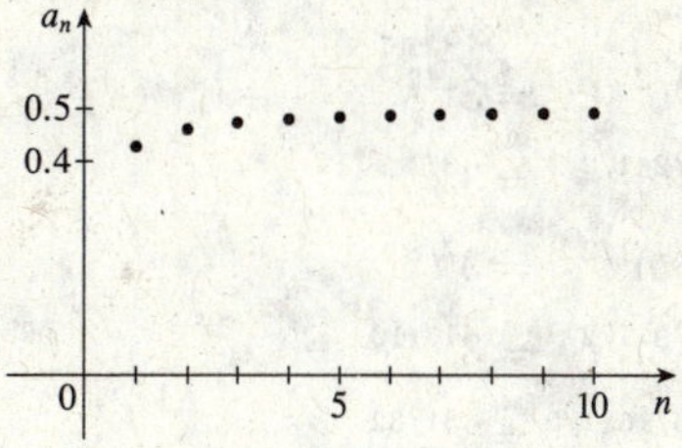

7.

n	$a_n = 1 + \left(-\frac{1}{2}\right)^n$
1	0.5000
2	1.2500
3	0.8750
4	1.0625
5	0.9688
6	1.0156
7	0.9922
8	1.0039
9	0.9980
10	1.0010

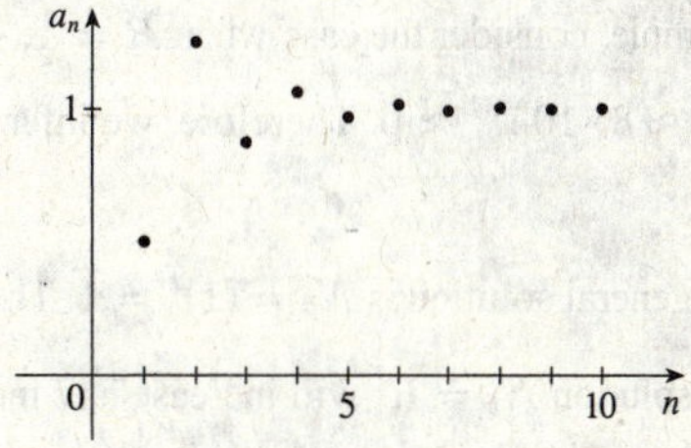

9. $\left\{1, \frac{1}{3}, \frac{1}{5}, \frac{1}{7}, \frac{1}{9}, \ldots\right\}$. The denominator of the *n*th term is the *n*th positive odd integer, so $a_n = \dfrac{1}{2n-1}$.

11. $\left\{-3, 2, -\frac{4}{3}, \frac{8}{9}, -\frac{16}{27}, \ldots\right\}$. The first term is -3 and each term is $-\frac{2}{3}$ times the preceding one, so $a_n = -3\left(-\frac{2}{3}\right)^{n-1}$.

13. $\left\{\frac{1}{2}, -\frac{4}{3}, \frac{9}{4}, -\frac{16}{5}, \frac{25}{6}, \ldots\right\}$. The numerator of the *n*th term is n^2 and its denominator is $n+1$. Including the alternating signs, we get $a_n = (-1)^{n+1}\dfrac{n^2}{n+1}$.

© 2016 Cengage Learning. All Rights Reserved. May not be scanned, copied or duplicated, or posted to a publicly accessible website, in whole or in part.

15. $a_1 = 1$

$a_2 = 5a_1 - 3 = 5(1) - 3 \quad = 2$

$a_3 = 5a_2 - 3 = 5(2) - 3 \quad = 7$

$a_4 = 5a_3 - 3 = 5(7) - 3 \quad = 32$

$a_5 = 5a_4 - 3 = 5(32) - 3 \quad = 157$

$a_6 = 5a_5 - 3 = 5(157) - 3 = 782$

17. $a_1 = 2$

$a_2 = a_1/(1 + a_1) = 2/(1 + 2) \quad = 2/3$

$a_3 = a_2/(1 + a_2) = (2/3)/(1 + 2/3) = 2/5$

$a_4 = a_3/(1 + a_3) = (2/5)/(1 + 2/5) = 2/7$

$a_5 = a_4/(1 + a_4) = (2/7)/(1 + 2/7) = 2/9$

$a_6 = a_5/(1 + a_5) = (2/9)/(1 + 2/9) = 2/11$

19. $a_1 = 1$

$a_2 = \sqrt{3a_1} = (3 \cdot 1)^{1/2} \quad = 3^{1/2}$

$a_3 = \sqrt{3a_2} = (3 \cdot 3^{1/2})^{1/2} \quad = 3^{3/4}$

$a_4 = \sqrt{3a_3} = (3 \cdot 3^{3/4})^{1/2} \quad = 3^{7/8}$

$a_5 = \sqrt{3a_4} = (3 \cdot 3^{7/8})^{1/2} \quad = 3^{15/16}$

$a_6 = \sqrt{3a_5} = (3 \cdot 3^{15/16})^{1/2} = 3^{31/32}$

21. $a_1 = 2$

$a_2 = 1$

$a_3 = a_2 - a_1 = 1 - 2 \quad = -1$

$a_4 = a_3 - a_2 = -1 - 1 \quad = -2$

$a_5 = a_4 - a_3 = -2 - (-1) = -1$

$a_6 = a_5 - a_4 = -1 - (-2) = 1$

23. Let a_n be the number of rabbit pairs in the nth month. Clearly $a_1 = 1 = a_2$. In the nth month, each pair that is 2 or more months old (that is, a_{n-2} pairs) will produce a new pair to add to the a_{n-1} pairs already present. Thus, $a_n = a_{n-1} + a_{n-2}$, so that $\{a_n\} = \{f_n\}$, the Fibonacci sequence.

25. The solution to the difference equation $N_{t+1} = RN_t$ as given in equation (2) is $N_t = N_0 R^t$. When $N_0 = 1$, the solution is $N_t = R^t$.

(a) The solution $N_t = R^t$ says that the tth term is found by multiplying R by itself t times. If $R < 1$, N_t will decrease as t increases. For example, consider the case when $R = \frac{1}{2}$, so that $N_1 = \frac{1}{2}, N_2 = \frac{1}{4}, N_3 = \frac{1}{8}$, and if t is very large, say 100, then $N_{100} = \frac{1}{2^{100}} \approx 8 \cdot 10^{-31} \approx 0$. Therefore, we infer that when $R < 1$, the value of N_t approaches zero as t becomes large.

(b) When $R = 1$, the general solution is $N_t = (1)^t = 1$. That is all terms in the sequence have a value of one.

(c) When $R > 1$, the solution $N_t = R^t$ will increase as t increases. For example, consider the case when $R = 2$, so that $N_1 = 2, N_2 = 4, N_3 = 8$, and if t is very large, say 100, then $N_{100} = 2^{100} \approx 10^{30}$. Therefore, we infer that when $R > 1$, the sequence grows indefinitely as t increases.

27-31 A calculator was used to compute the first 10 terms of each sequence and these (t, x_t) data points were then graphed.

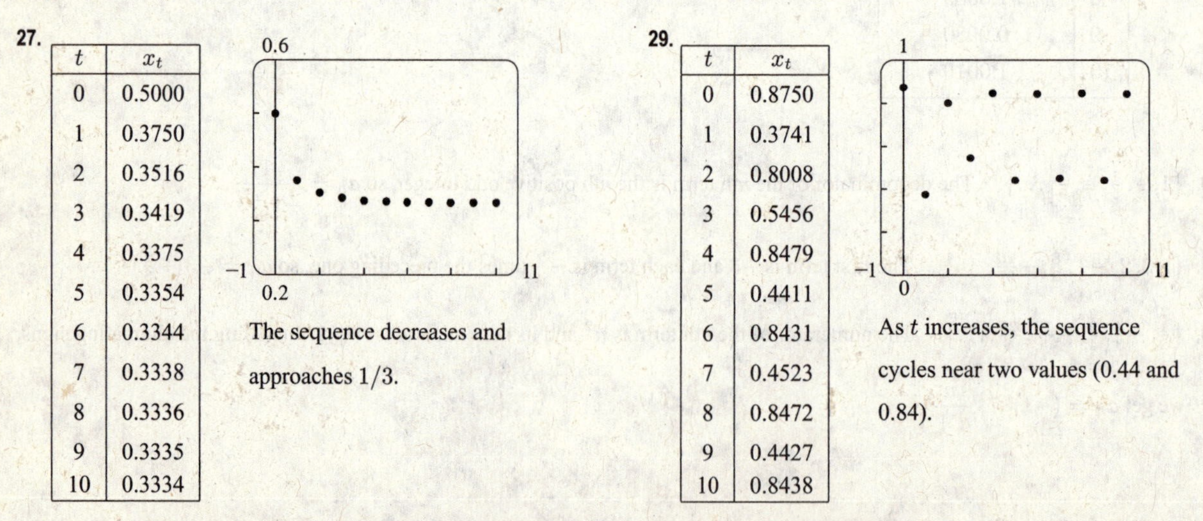

27.

t	x_t
0	0.5000
1	0.3750
2	0.3516
3	0.3419
4	0.3375
5	0.3354
6	0.3344
7	0.3338
8	0.3336
9	0.3335
10	0.3334

The sequence decreases and approaches $1/3$.

29.

t	x_t
0	0.8750
1	0.3741
2	0.8008
3	0.5456
4	0.8479
5	0.4411
6	0.8431
7	0.4523
8	0.8472
9	0.4427
10	0.8438

As t increases, the sequence cycles near two values (0.44 and 0.84).

© 2016 Cengage Learning. All Rights Reserved. May not be scanned, copied or duplicated, or posted to a publicly accessible website, in whole or in part.

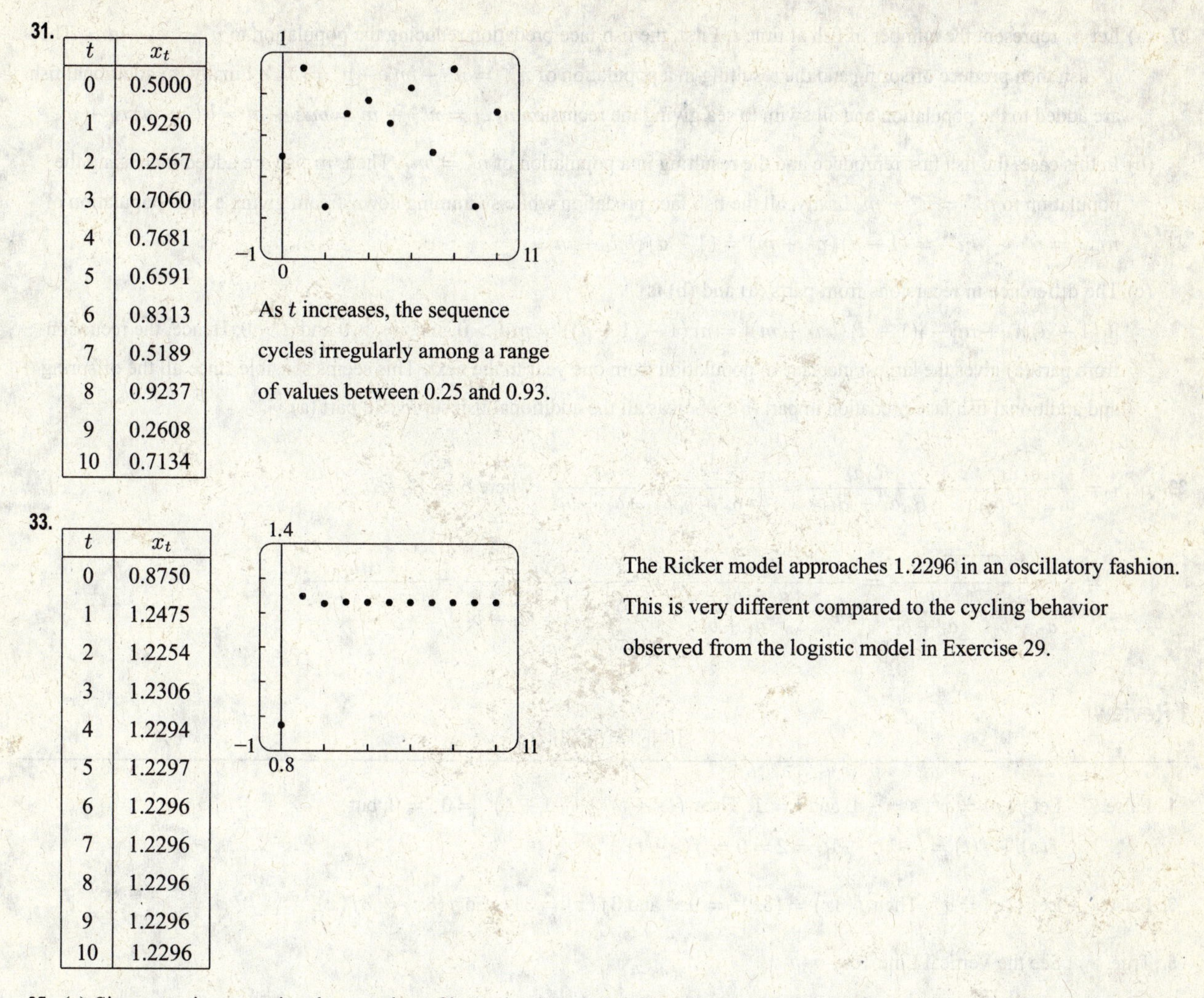

31.

t	x_t
0	0.5000
1	0.9250
2	0.2567
3	0.7060
4	0.7681
5	0.6591
6	0.8313
7	0.5189
8	0.9237
9	0.2608
10	0.7134

As t increases, the sequence cycles irregularly among a range of values between 0.25 and 0.93.

33.

t	x_t
0	0.8750
1	1.2475
2	1.2254
3	1.2306
4	1.2294
5	1.2297
6	1.2296
7	1.2296
8	1.2296
9	1.2296
10	1.2296

The Ricker model approaches 1.2296 in an oscillatory fashion. This is very different compared to the cycling behavior observed from the logistic model in Exercise 29.

35. (a) Since area is proportional to number of bacteria, the relationship between colony radius, r, and the population size can be found as follows: $A_{\text{circle}} = kN \quad \Leftrightarrow \quad \pi r^2 = kN \quad \Leftrightarrow \quad r = \sqrt{\frac{k}{\pi}N}$ where k is a proportionality constant.

Since I is proportional to the colony circumference, C, the input of new individuals is

$I = RC = R(2\pi r) = 2\pi R\sqrt{\frac{k}{\pi}N} = RK\sqrt{N}$ where $K = 2\pi\sqrt{\frac{k}{\pi}}$ is a constant. This gives the recursion equation

$N_{t+1} = N_t + RK\sqrt{N_t}$.

(b) A calculator was used to calculate and graph the first 10 terms of the sequence using $N_0 = 40$ for several different values of $R \cdot K$.

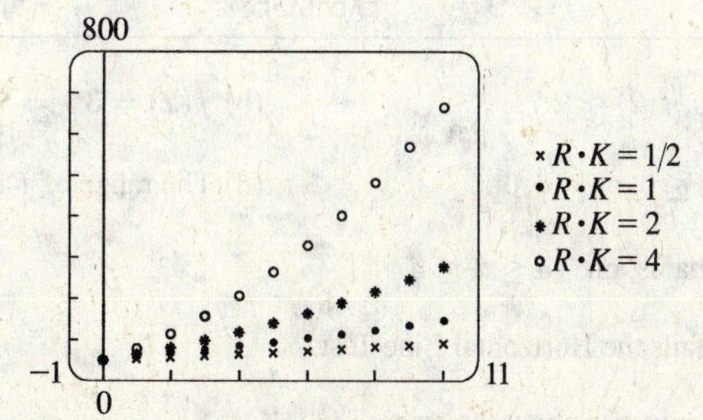

© 2016 Cengage Learning. All Rights Reserved. May not be scanned, copied or duplicated, or posted to a publicly accessible website, in whole or in part.

37. (a) Let n_t represent the number of fish at time t. First, the fish face predation reducing the population to $n^* = n_t - dn_t$. The n^* fish then produce offspring and die resulting in a population of $n^{**} = n^* + bn^* - n^* = bn^*$. Finally, m additional fish are added to the population and all swim to sea giving the recursion $n_{t+1} = n^{**} + m = bn^* + m = b(1-d)\,n_t + m$.

(b) In this case, the fish first reproduce and die resulting in a population of $n^* = bn_t$. Then, m fish are added increasing the population to $n^{**} = n^* + m$. Lastly, all the fish face predation while swimming downstream giving a final population of $n_{t+1} = n^{**} - dn^{**} = (1-d)(n^* + m) = (1-d)(bn_t + m)$.

(c) The difference in recursions from parts (a) and (b) is $[b(1-d)\,n_t + m] - [(1-d)(bn_t + m)] = m(1-(1-d)) = md > 0$ since $m > 0$ and $d > 0$. Hence, the recursion from part (a) gives the largest increase in population from one year to the next. This seems sensible since all the offspring and additional fish face predation in part (b), whereas all the additional fish survive in part (a).

39. $$p_{t+1} = \frac{a_{t+1}}{a_{t+1} + b_{t+1}} = \frac{R_a a_t}{R_a a_t + R_b b_t} = \frac{\frac{R_a}{R_b} a_t}{\frac{R_a}{R_b} a_t + b_t} = \frac{\alpha a_t}{\alpha a_t + b_t} \quad \text{where } \alpha = R_a/R_b$$

$$= \frac{\alpha \dfrac{a_t}{a_t + b_t}}{\alpha \dfrac{a_t}{a_t + b_t} + \dfrac{b_t}{a_t + b_t}} = \frac{\alpha p_t}{\alpha p_t + \dfrac{a_t + b_t - a_t}{a_t + b_t}} = \frac{\alpha p_t}{\alpha p_t + 1 - \dfrac{a_t}{a_t + b_t}} = \frac{\alpha p_t}{\alpha p_t + 1 - p_t}$$

1 Review

TRUE-FALSE QUIZ

1. False. Let $f(x) = x^2$, $s = -1$, and $t = 1$. Then $f(s+t) = (-1+1)^2 = 0^2 = 0$, but $f(s) + f(t) = (-1)^2 + 1^2 = 2 \neq 0 = f(s+t)$.

3. False. Let $f(x) = x^2$. Then $f(3x) = (3x)^2 = 9x^2$ and $3f(x) = 3x^2$. So $f(3x) \neq 3f(x)$.

5. True. See the Vertical Line Test.

7. False. Let $f(x) = x^3$. Then f is one-to-one and $f^{-1}(x) = \sqrt[3]{x}$. But $1/f(x) = 1/x^3$, which is not equal to $f^{-1}(x)$.

9. True. The function $\ln x$ is an increasing function on $(0, \infty)$.

11. False. Let $x = e^2$ and $a = e$. Then $\dfrac{\ln x}{\ln a} = \dfrac{\ln e^2}{\ln e} = \dfrac{2 \ln e}{\ln e} = 2$ and $\ln \dfrac{x}{a} = \ln \dfrac{e^2}{e} = \ln e = 1$, so in general the statement is false. What *is* true, however, is that $\ln \dfrac{x}{a} = \ln x - \ln a$.

EXERCISES

1. (a) When $x = 2$, $y \approx 2.7$. Thus, $f(2) \approx 2.7$.

(b) $f(x) = 3 \;\Rightarrow\; x \approx 2.3, 5.6$

(c) The domain of f is $-6 \leq x \leq 6$, or $[-6, 6]$.

(d) The range of f is $-4 \leq y \leq 4$, or $[-4, 4]$.

(e) f is increasing on $[-4, 4]$, that is, on $-4 \leq x \leq 4$.

(f) f is not one-to-one since it fails the Horizontal Line Test.

(g) f is odd since its graph is symmetric about the origin.

© 2016 Cengage Learning. All Rights Reserved. May not be scanned, copied or duplicated, or posted to a publicly accessible website, in whole or in part.

3. (a) $S(1000) \approx -36\,\text{m}$

(b) The sea level was lowest about $18,000$ years ago and highest about $121,000$ ago present.

(c) $\{S\,|-114 \le S \le 8\} = [-114, 8]$

(d) The drops in sea level around 150,000 and 18,000 years ago correspond to periods of glaciation during which large amounts of Earth's water was frozen in ice sheets.

5. $f(x) = x^2 - 2x + 3$, so $f(a+h) = (a+h)^2 - 2(a+h) + 3 = a^2 + 2ah + h^2 - 2a - 2h + 3$, and

$$\frac{f(a+h) - f(a)}{h} = \frac{(a^2 + 2ah + h^2 - 2a - 2h + 3) - (a^2 - 2a + 3)}{h} = \frac{h(2a + h - 2)}{h} = 2a + h - 2.$$

7. $f(x) = 2/(3x - 1)$. Domain: $3x - 1 \neq 0 \;\Rightarrow\; 3x \neq 1 \;\Rightarrow\; x \neq \frac{1}{3}$. $D = \left(-\infty, \frac{1}{3}\right) \cup \left(\frac{1}{3}, \infty\right)$

Range: all reals except 0 ($y = 0$ is the horizontal asymptote for f.) $R = (-\infty, 0) \cup (0, \infty)$

9. $h(x) = \ln(x + 6)$. Domain: $x + 6 > 0 \;\Rightarrow\; x > -6$. $D = (-6, \infty)$

Range: $x + 6 > 0$, so $\ln(x + 6)$ takes on all real numbers and, hence, the range is $\mathbb{R}$.

$R = (-\infty, \infty)$

11. (a) To obtain the graph of $y = f(x) + 8$, we shift the graph of $y = f(x)$ up 8 units.

(b) To obtain the graph of $y = f(x + 8)$, we shift the graph of $y = f(x)$ left 8 units.

(c) To obtain the graph of $y = 1 + 2f(x)$, we stretch the graph of $y = f(x)$ vertically by a factor of 2, and then shift the resulting graph 1 unit upward.

(d) To obtain the graph of $y = f(x - 2) - 2$, we shift the graph of $y = f(x)$ right 2 units (for the "-2" inside the parentheses), and then shift the resulting graph 2 units downward.

(e) To obtain the graph of $y = -f(x)$, we reflect the graph of $y = f(x)$ about the x-axis.

(f) To obtain the graph of $y = f^{-1}(x)$, we reflect the graph of $y = f(x)$ about the line $y = x$ (assuming f is one–to-one).

13. $y = -\sin 2x$: Start with the graph of $y = \sin x$, compress horizontally by a factor of 2, and reflect about the x-axis.

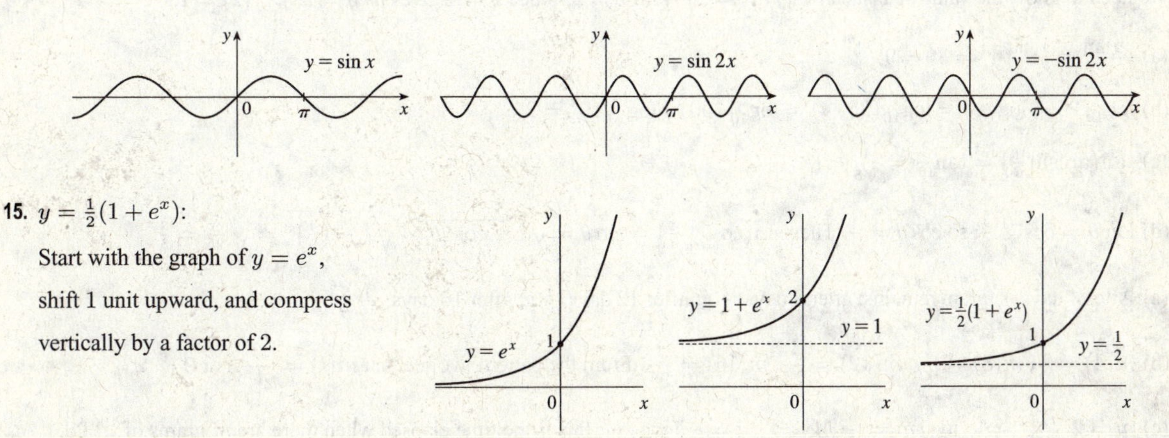

15. $y = \frac{1}{2}(1 + e^x)$:

Start with the graph of $y = e^x$, shift 1 unit upward, and compress vertically by a factor of 2.

© 2016 Cengage Learning. All Rights Reserved. May not be scanned, copied or duplicated, or posted to a publicly accessible website, in whole or in part.

17. $f(x) = \dfrac{1}{x+2}$:

Start with the graph of $f(x) = 1/x$ and shift 2 units to the left.

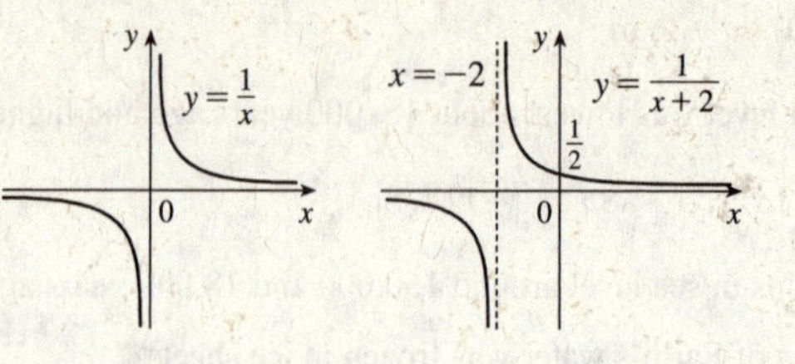

19. $f(x) = \begin{cases} -x & \text{if } x < 0 \\ e^x - 1 & \text{if } x \geq 0 \end{cases}$

On $(-\infty, 0)$, graph $y = -x$ (the line with slope -1 and y-intercept 0) with open endpoint $(0, 0)$.

On $[0, \infty)$, graph $y = e^x - 1$ (the graph of $y = e^x$ shifted 1 unit downward) with closed endpoint $(0, 0)$.

21. $f(x) = \ln x$, $\quad D = (0, \infty)$; $\quad g(x) = x^2 - 9$, $\quad D = \mathbb{R}$.

(a) $(f \circ g)(x) = f(g(x)) = f(x^2 - 9) = \ln(x^2 - 9)$.

Domain: $x^2 - 9 > 0 \quad \Rightarrow \quad x^2 > 9 \quad \Rightarrow \quad |x| > 3 \quad \Rightarrow \quad x \in (-\infty, -3) \cup (3, \infty)$

(b) $(g \circ f)(x) = g(f(x)) = g(\ln x) = (\ln x)^2 - 9$. Domain: $x > 0$, or $(0, \infty)$

(c) $(f \circ f)(x) = f(f(x)) = f(\ln x) = \ln(\ln x)$. Domain: $\ln x > 0 \quad \Rightarrow \quad x > e^0 = 1$, or $(1, \infty)$

(d) $(g \circ g)(x) = g(g(x)) = g(x^2 - 9) = (x^2 - 9)^2 - 9$. Domain: $x \in \mathbb{R}$, or $(-\infty, \infty)$

23.

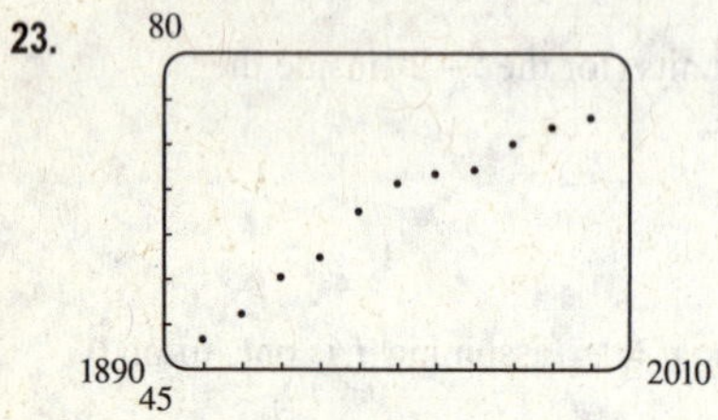

Many models appear to be plausible. Your choice depends on whether you think medical advances will keep increasing life expectancy, or if there is bound to be a natural leveling-off of life expectancy. A linear model, $y = 0.2493x - 423.4818$, gives us an estimate of 77.6 years for the year 2010.

25. We need to know the value of x such that $f(x) = 2x + \ln x = 2$. Since $x = 1$ gives us $y = 2$, $f^{-1}(2) = 1$.

27. (a) $e^{2 \ln 3} = (e^{\ln 3})^2 = 3^2 = 9$

(b) $\log_{10} 25 + \log_{10} 4 = \log_{10}(25 \cdot 4) = \log_{10} 100 = \log_{10} 10^2 = 2$

(c) $\tan\left(\arcsin \frac{1}{2}\right) = \tan \frac{\pi}{6} = \frac{1}{\sqrt{3}}$

(d) Let $\theta = \cos^{-1} \frac{4}{5}$, so $\cos \theta = \frac{4}{5}$. Then $\sin\left(\cos^{-1} \frac{4}{5}\right) = \sin \theta = \sqrt{1 - \cos^2 \theta} = \sqrt{1 - \left(\frac{4}{5}\right)^2} = \sqrt{\frac{9}{25}} = \frac{3}{5}$.

29. (a) After 4 days, $\frac{1}{2}$ gram remains; after 8 days, $\frac{1}{4}$ g; after 12 days, $\frac{1}{8}$ g; after 16 days, $\frac{1}{16}$ g.

(b) $m(4) = \dfrac{1}{2}$, $m(8) = \dfrac{1}{2^2}$, $m(12) = \dfrac{1}{2^3}$, $m(16) = \dfrac{1}{2^4}$. From the pattern, we see that $m(t) = \dfrac{1}{2^{t/4}}$, or $2^{-t/4}$.

(c) $m = 2^{-t/4} \quad \Rightarrow \quad \log_2 m = -t/4 \quad \Rightarrow \quad t = -4 \log_2 m$; this is the time elapsed when there are m grams of ^{100}Pd.

(d) $m = 0.01 \quad \Rightarrow \quad t = -4 \log_2 0.01 = -4\left(\dfrac{\ln 0.01}{\ln 2}\right) \approx 26.6$ days

© 2016 Cengage Learning. All Rights Reserved. May not be scanned, copied or duplicated, or posted to a publicly accessible website, in whole or in part.

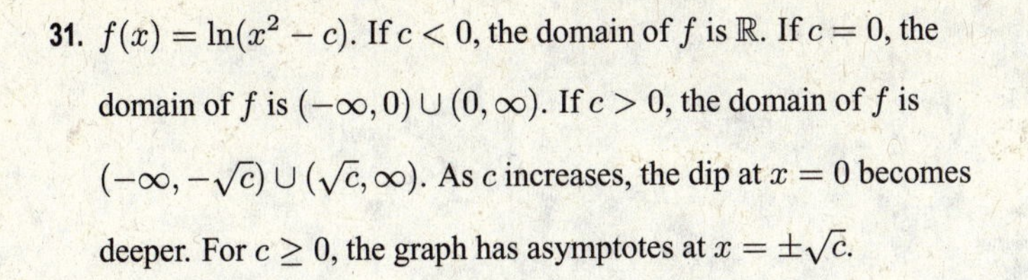

31. $f(x) = \ln(x^2 - c)$. If $c < 0$, the domain of f is $\mathbb{R}$. If $c = 0$, the domain of f is $(-\infty, 0) \cup (0, \infty)$. If $c > 0$, the domain of f is $(-\infty, -\sqrt{c}) \cup (\sqrt{c}, \infty)$. As c increases, the dip at $x = 0$ becomes deeper. For $c \geq 0$, the graph has asymptotes at $x = \pm\sqrt{c}$.

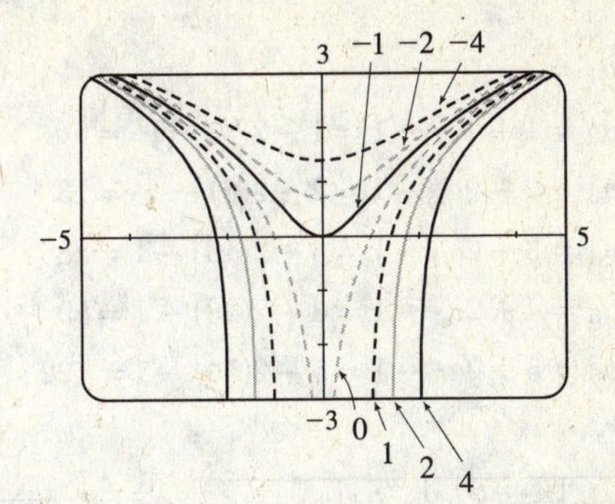

33. (a) (b)

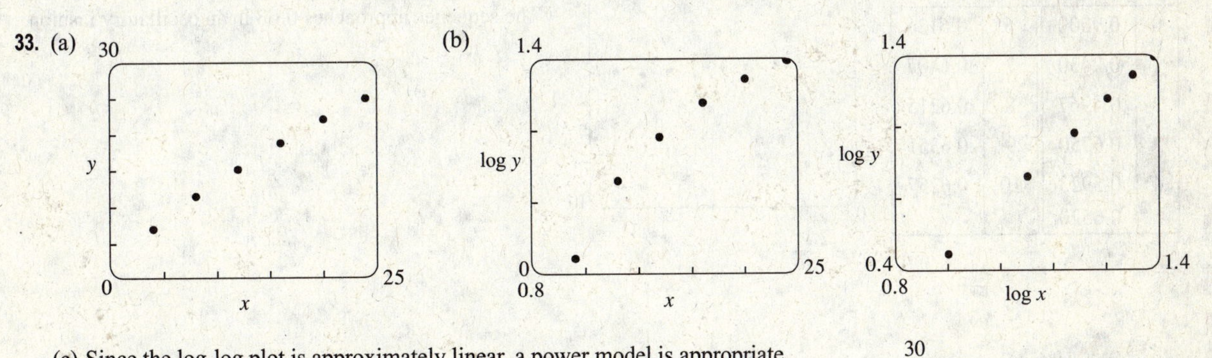

(c) Since the log-log plot is approximately linear, a power model is appropriate.

(d) Using computer software to fit a power curve to the data gives

$y = (2.608377) \cdot x^{0.712277}$.

35. (a)

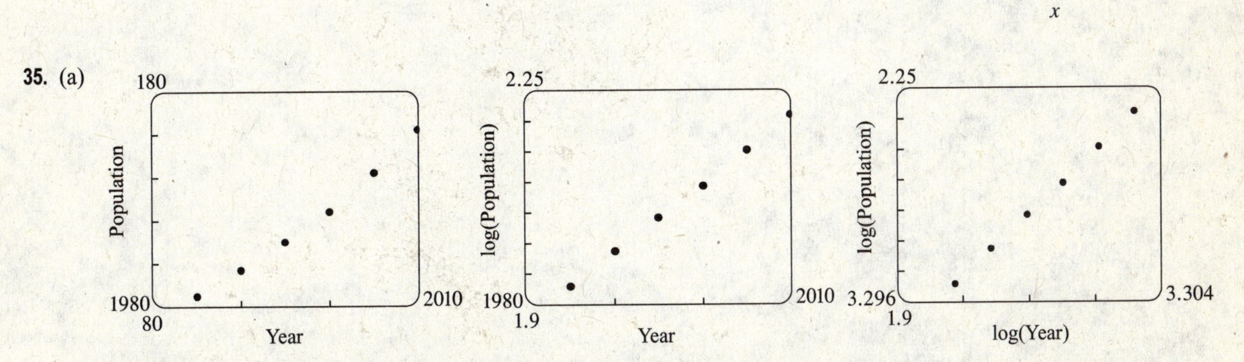

Both the semi-log and log-log plots are approximately linear, so an exponential or power model is appropriate.

(b) Using computer software to fit an exponential curve to the data gives $P = \left(6.6326 \cdot 10^{-21}\right) \cdot (1.025977)^Y$ where P is the population in millions and Y is the year. Alternatively, we could have defined Y to be the number of years since 1985.

(c) $P(2008) = \left(6.6326 \cdot 10^{-21}\right) \cdot (1.025977)^{2008} \approx 153$ million

$P(2020) = \left(6.6326 \cdot 10^{-21}\right) \cdot (1.025977)^{2020} \approx 209$ million

© 2016 Cengage Learning. All Rights Reserved. May not be scanned, copied or duplicated, or posted to a publicly accessible website, in whole or in part.

37. $a_1 = 3$

$a_2 = 1 + 2a_1 - 1 = 1 + 2(3) - 1 = 6$

$a_3 = 2 + 2a_2 - 1 = 2 + 2(6) - 1 = 13$

$a_4 = 3 + 2a_3 - 1 = 3 + 2(13) - 1 = 28$

$a_5 = 4 + 2a_4 - 1 = 4 + 2(28) - 1 = 59$

$a_6 = 5 + 2a_5 - 1 = 5 + 2(59) - 1 = 122$

39.

t	x_t	t	x_t
0	0.9000	6	0.6126
1	0.2430	7	0.6407
2	0.4967	8	0.6215
3	0.6750	9	0.6351
4	0.5923	10	0.6257
5	0.6520		

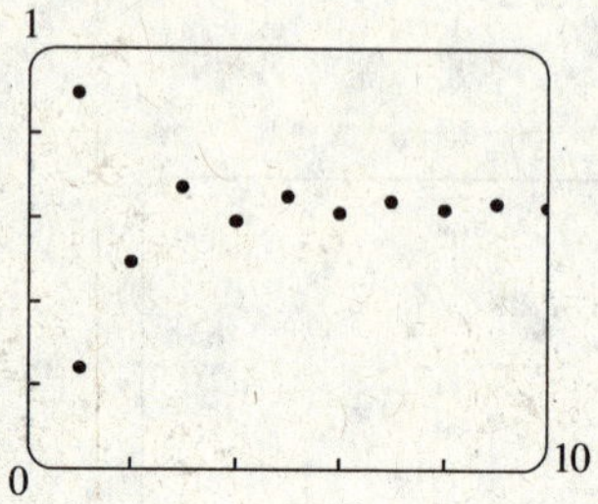

1

0

10

The sequence approaches 0.63 in an oscillatory fashion.

© 2016 Cengage Learning. All Rights Reserved. May not be scanned, copied or duplicated, or posted to a publicly accessible website, in whole or in part.

2 ☐ LIMITS

2.1 Limits of Sequences

1. (a) A sequence is an ordered list of numbers. It can also be defined as a function whose domain is the set of positive integers.

(b) The terms a_n approach 8 as n becomes large. In fact, we can make a_n as close to 8 as we like by taking n sufficiently large.

(c) The terms a_n become large as n becomes large. In fact, we can make a_n as large as we like by taking n sufficiently large.

3. The graph shows a decline in the world record for the men's 100-meter sprint as t increases. It is tempting to say that this sequence will approach zero, however, it is important to remember that the sequence represents data from a physical competition. Thus, the sequence likely has a nonzero limit as $t \to \infty$ since human physiology will ultimately limit how fast a human can sprint 100-meters. This means that there is a certain world record time which athletes can never surpass.

5.

n	a_n	n	a_n
1	0.2000	6	0.3000
2	0.2500	7	0.3043
3	0.2727	8	0.3077
4	0.2857	9	0.3103
5	0.2941	10	0.3125

The sequence appears to converge to a number between 0.30 and 0.35. Calculating the limit gives

$$\lim_{n\to\infty} a_n = \lim_{n\to\infty} \frac{n^2}{2n+3n^2} = \lim_{n\to\infty} \frac{\dfrac{n^2}{n^2}}{\dfrac{2n+3n^2}{n^2}} = \frac{\lim\limits_{n\to\infty} 1}{\lim\limits_{n\to\infty} \dfrac{2}{n} + \lim\limits_{n\to\infty} 3} = \frac{1}{0+3} = \frac{1}{3}.$$

This agrees with the value predicted from the data.

7.

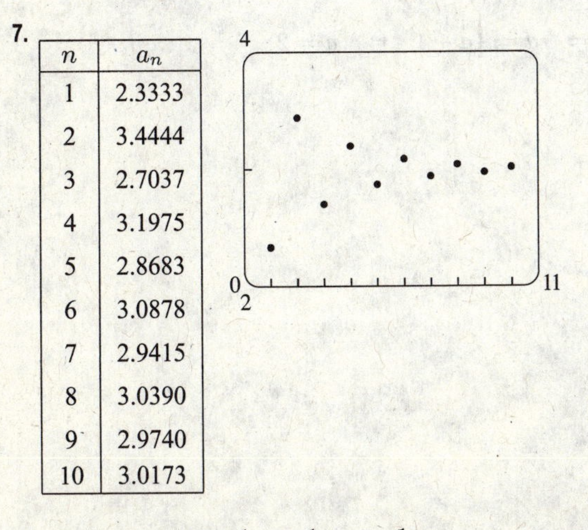

n	a_n
1	2.3333
2	3.4444
3	2.7037
4	3.1975
5	2.8683
6	3.0878
7	2.9415
8	3.0390
9	2.9740
10	3.0173

The sequence appears to converge to approximately 3. Calculating the limit gives $\lim\limits_{n\to\infty} a_n = \lim\limits_{n\to\infty} \left(3 + \left(-\frac{2}{3}\right)^n\right) = 3 + 0 = 3$. This agrees with the value predicted from the data.

9. $\lim\limits_{n\to\infty} a_n = \lim\limits_{n\to\infty} \dfrac{1}{3n^4} = \dfrac{1}{3} \lim\limits_{n\to\infty} \dfrac{1}{n^4} = 0$. Converges

© 2016 Cengage Learning. All Rights Reserved. May not be scanned, copied or duplicated, or posted to a publicly accessible website, in whole or in part.

11. $a_n = \dfrac{2n^2+n-1}{n^2} = 2 + \dfrac{1}{n} - \dfrac{1}{n^2}$ so $\lim\limits_{n\to\infty} a_n = \lim\limits_{n\to\infty} 2 + \lim\limits_{n\to\infty} \dfrac{1}{n} - \lim\limits_{n\to\infty} \dfrac{1}{n^2} = 2+0-0=2$ Converges

13. $\lim\limits_{n\to\infty} a_n = \lim\limits_{n\to\infty} \dfrac{3+5n}{2+7n} = \lim\limits_{n\to\infty} \dfrac{\frac{3+5n}{n}}{\frac{2+7n}{n}} = \lim\limits_{n\to\infty} \dfrac{\frac{3}{n}+5}{\frac{2}{n}+7} = \dfrac{\lim\limits_{n\to\infty} \frac{3}{n} + \lim\limits_{n\to\infty} 5}{\lim\limits_{n\to\infty} \frac{2}{n} + \lim\limits_{n\to\infty} 7} = \dfrac{0+5}{0+7} = \dfrac{5}{7}$ Converges

15. $a_n = 1-(0.2)^n$, so $\lim\limits_{n\to\infty} a_n = 1-0 = 1$ [by (3) with $r = 0.2$]. Converges

17. $a_n = \dfrac{n^2}{\sqrt{n^3+4n}} = \dfrac{n^2/\sqrt{n^3}}{\sqrt{n^3+4n}/\sqrt{n^3}} = \dfrac{\sqrt{n}}{\sqrt{1+4/n^2}}$, so $a_n \to \infty$ as $n \to \infty$ since $\lim\limits_{n\to\infty} \sqrt{n} = \infty$ and $\lim\limits_{n\to\infty} \sqrt{1+4/n^2} = 1$. Diverges

19. $a_n = \cos(n\pi/2) \Rightarrow a_1 = \cos(\pi/2) = 0,\quad a_2 = \cos(\pi) = -1,\quad a_3 = \cos(3\pi/2) = 0,\quad a_4 = \cos(2\pi) = 1,$ $a_5 = \cos(5\pi/2) = 0$. Observe that a_n cycles between the values 1, 0, and -1 as n increases. Hence the sequence does not converge.

21. $\lim\limits_{n\to\infty} a_n = \lim\limits_{n\to\infty} \dfrac{10^n}{1+9^n} = \lim\limits_{n\to\infty} \dfrac{\frac{10^n}{10^n}}{\frac{1+9^n}{10^n}} = \lim\limits_{n\to\infty} \dfrac{1}{\frac{1}{10^n} + \left(\frac{9}{10}\right)^n} = \dfrac{\lim\limits_{n\to\infty} 1}{\lim\limits_{n\to\infty} \left(\frac{1}{10}\right)^n + \lim\limits_{n\to\infty} \left(\frac{9}{10}\right)^n} = \infty$ because the denominator approaches 0 while the numerator remains constant. Diverges

23. $a_n = \ln(2n^2+1) - \ln(n^2+1) = \ln\left(\dfrac{2n^2+1}{n^2+1}\right) = \ln\left(\dfrac{2+1/n^2}{1+1/n^2}\right) \to \ln 2$ as $n \to \infty$. Converges

25. $a_n = \dfrac{e^n + e^{-n}}{e^{2n}-1} \cdot \dfrac{e^{-n}}{e^{-n}} = \dfrac{1+e^{-2n}}{e^n - e^{-n}} \to 0$ as $n \to \infty$ because $1+e^{-2n} \to 1$ and $e^n - e^{-n} \to \infty$. Converges

27.

n	a_n	n	a_n
1	1.0000	5	1.9375
2	1.5000	6	1.9688
3	1.7500	7	1.9844
4	1.8750	8	1.9922

The sequence appears to converge to 2. Assume the limit exists so that $\lim\limits_{n\to\infty} a_{n+1} = \lim\limits_{n\to\infty} a_n = a$, then $a_{n+1} = \frac{1}{2}a_n + 1 \Rightarrow$ $\lim\limits_{n\to\infty} a_{n+1} = \lim\limits_{n\to\infty} \left(\frac{1}{2}a_n + 1\right) \Rightarrow a = \frac{1}{2}a + 1 \Rightarrow a = 2$

Therefore, $\lim\limits_{n\to\infty} a_n = 2$.

29.

n	a_n	n	a_n
1	2.0000	5	17.0000
2	3.0000	6	33.0000
3	5.0000	7	65.0000
4	9.0000	8	129.0000

The sequence is divergent.

© 2016 Cengage Learning. All Rights Reserved. May not be scanned, copied or duplicated, or posted to a publicly accessible website, in whole or in part.

31.

n	a_n
1	1.0000
2	3.0000
3	1.5000
4	2.4000
5	1.7647
6	2.1702
7	1.8926
8	2.0742

The sequence appears to converge to 2. Assume the limit exists so that

$\lim\limits_{n\to\infty} a_{n+1} = \lim\limits_{n\to\infty} a_n = a$, then $a_{n+1} = \dfrac{6}{1+a_n} \Rightarrow \lim\limits_{n\to\infty} a_{n+1} = \lim\limits_{n\to\infty} \dfrac{6}{1+a_n} \Rightarrow$

$a = \dfrac{6}{1+a} \Rightarrow a^2 + a - 6 = 0 \Rightarrow (a-2)(a+3) = 0 \Rightarrow a = -3$ or $a = 2$

Therefore, if the limit exists it will be either -3 or 2, but since all terms of the sequence are positive, we see that $\lim\limits_{n\to\infty} a_n = 2$.

33.

n	a_n
1	1.0000
2	1.7321
3	1.9319
4	1.9829
5	1.9957
6	1.9989
7	1.9997
8	1.9999

The sequence appears to converge to 2. Assume the limit exists so that

$\lim\limits_{n\to\infty} a_{n+1} = \lim\limits_{n\to\infty} a_n = a$, then $a_{n+1} = \sqrt{2+a_n} \Rightarrow \lim\limits_{n\to\infty} a_{n+1} = \lim\limits_{n\to\infty} \sqrt{2+a_n} \Rightarrow$

$a = \sqrt{2+a} \Rightarrow a^2 - a - 2 = 0 \Rightarrow (a-2)(a+1) = 0 \Rightarrow a = -1$ or $a = 2$

Therefore, if the limit exists it will be either -1 or 2, but since all terms of the sequence are positive, we see that $\lim\limits_{n\to\infty} a_n = 2$.

35. (a) The quantity of the drug in the body after the first tablet is 100 mg. After the second tablet, there is 100 mg plus 20% of the first 100- mg tablet, that is, $[100 + 100(0.20)] = 120$ mg. After the third tablet, the quantity is $[100 + 120(0.20)] = 124$ mg.

(b) After the $n^{\text{th}} + 1$ tablet, there is 100 mg plus 20% of the nth tablet, so that $Q_{n+1} = 100 + (0.20)\,Q_n$

(c) From Formula (6), the solution to $Q_{n+1} = 100 + (0.20)\,Q_n$, $Q_0 = 0$ mg is

$$Q_n = (0.20)^n\,(0) + 100\left(\frac{1-0.20^n}{1-0.20}\right) = \frac{100}{0.80}\,(1-0.20^n) = 125\,(1-0.20^n)$$

(d) In the long run, we have $\lim\limits_{n\to\infty} Q_n = \lim\limits_{n\to\infty} 125\,(1-0.20^n) = 125\left(\lim\limits_{n\to\infty} 1 - \lim\limits_{n\to\infty} 0.20^n\right) = 125\,(1-0) = 125$ mg

37. (a) The quantity of the drug in the body after the first tablet is 150 mg. After the second tablet, there is 150 mg plus 5% of the first 150- mg tablet, that is, $[150 + 150(0.05)]$ mg. After the third tablet, the quantity is $[150 + 150(0.05) + 150(0.05)^2] = 157.875$ mg. After n tablets, the quantity (in mg) is

$150 + 150(0.05) + \cdots + 150(0.05)^{n-1}$. We can use Formula 5 to write this as $\dfrac{150(1-0.05^n)}{1-0.05} = \dfrac{3000}{19}(1-0.05^n)$.

(b) The number of milligrams remaining in the body in the long run is $\lim\limits_{n\to\infty}\left[\frac{3000}{19}(1-0.05^n)\right] = \frac{3000}{19}(1-0) \approx 157.895$, only 0.02 mg more than the amount after 3 tablets.

39. (a) Many people would guess that $x < 1$, but note that x consists of an infinite number of 9s.

(b) $x = 0.99999\ldots = \dfrac{9}{10} + \dfrac{9}{100} + \dfrac{9}{1000} + \dfrac{9}{10{,}000} + \cdots = \sum\limits_{n=1}^{\infty} \dfrac{9}{10^n}$, which is a geometric series with $a_1 = 0.9$ and $r = 0.1$. Its sum is $\dfrac{0.9}{1-0.1} = \dfrac{0.9}{0.9} = 1$, that is, $x = 1$.

© 2016 Cengage Learning. All Rights Reserved. May not be scanned, copied or duplicated, or posted to a publicly accessible website, in whole or in part.

(c) The number 1 has two decimal representations, $1.00000\ldots$ and $0.99999\ldots$.

(d) Except for 0, all rational numbers that have a terminating decimal representation can be written in more than one way. For example, 0.5 can be written as $0.49999\ldots$ as well as $0.50000\ldots$.

41. $0.\overline{8} = \dfrac{8}{10} + \dfrac{8}{10^2} + \cdots$ is a geometric series with $a = \dfrac{8}{10}$ and $r = \dfrac{1}{10}$. It converges to $\dfrac{a}{1-r} = \dfrac{8/10}{1-1/10} = \dfrac{8}{9}$.

43. $2.\overline{516} = 2 + \dfrac{516}{10^3} + \dfrac{516}{10^6} + \cdots$. Now $\dfrac{516}{10^3} + \dfrac{516}{10^6} + \cdots$ is a geometric series with $a = \dfrac{516}{10^3}$ and $r = \dfrac{1}{10^3}$. It converges to

$\dfrac{a}{1-r} = \dfrac{516/10^3}{1-1/10^3} = \dfrac{516/10^3}{999/10^3} = \dfrac{516}{999}$. Thus, $2.\overline{516} = 2 + \dfrac{516}{999} = \dfrac{2514}{999} = \dfrac{838}{333}$.

45. $1.5\overline{342} = 1.53 + \dfrac{42}{10^4} + \dfrac{42}{10^6} + \cdots$. Now $\dfrac{42}{10^4} + \dfrac{42}{10^6} + \cdots$ is a geometric series with $a = \dfrac{42}{10^4}$ and $r = \dfrac{1}{10^2}$.

It converges to $\dfrac{a}{1-r} = \dfrac{42/10^4}{1-1/10^2} = \dfrac{42/10^4}{99/10^2} = \dfrac{42}{9900}$.

Thus, $1.5\overline{342} = 1.53 + \dfrac{42}{9900} = \dfrac{153}{100} + \dfrac{42}{9900} = \dfrac{15{,}147}{9900} + \dfrac{42}{9900} = \dfrac{15{,}189}{9900}$ or $\dfrac{5063}{3300}$.

47.

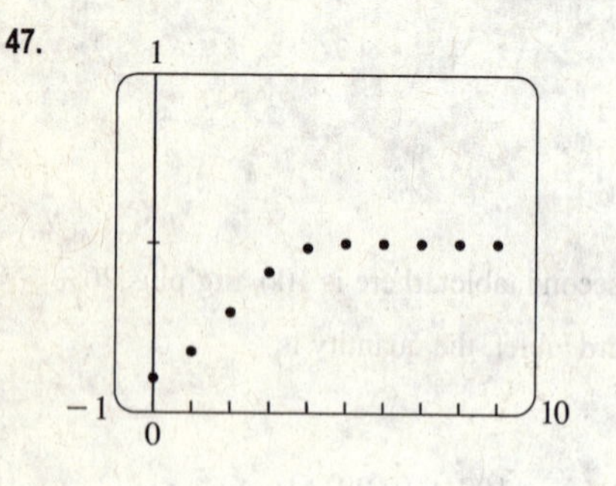

Computer software was used to plot the first 10 points of the recursion equation $x_{t+1} = 2x_t(1-x_t)$, $x_0 = 0.1$. The sequence appears to converge to a value of 0.5. Assume the limit exists so that $\lim\limits_{t\to\infty} x_{t+1} = \lim\limits_{t\to\infty} x_t = x$, then

$x_{t+1} = 2x_t(1-x_t) \;\Rightarrow\; \lim\limits_{t\to\infty} x_{t+1} = \lim\limits_{t\to\infty} 2x_t(1-x_t) \;\Rightarrow$
$x = 2x(1-x) \;\Rightarrow\; x(1-2x) = 0 \;\Rightarrow\; x = 0$ or $x = 1/2$. Therefore, if the limit exists it will be either 0 or $\frac{1}{2}$. Since the graph of the sequence appears to approach $\frac{1}{2}$, we see that $\lim\limits_{t\to\infty} x_t = \frac{1}{2}$.

49.

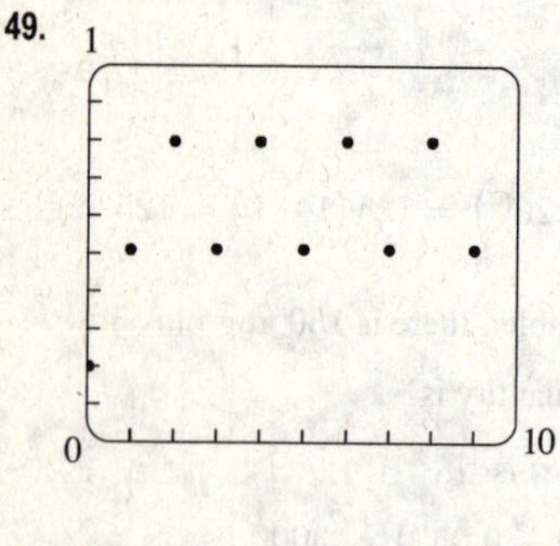

Computer software was used to plot the first 10 points of the recursion equation $x_{t+1} = 3.2x_t(1-x_t)$, $x_0 = 0.2$. The sequence does not appear to converge to a fixed value. Instead, the terms oscillate between values near 0.5 and 0.8.

51.

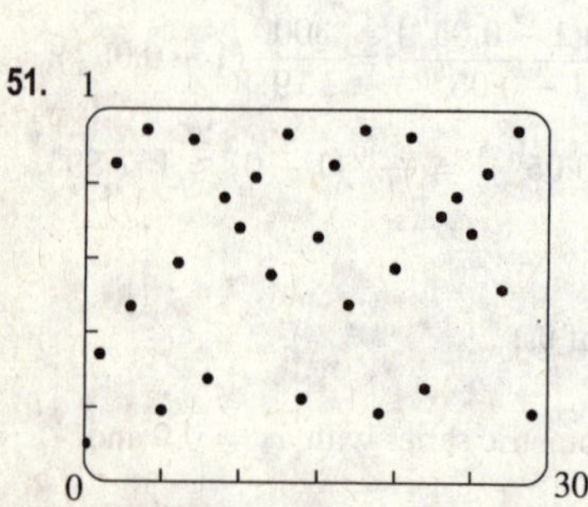

Computer software was used to plot the first 30 points of the recursion equation $x_{t+1} = 3.8x_t(1-x_t)$, $x_0 = 0.1$. The sequence does not appear to converge to a fixed value. The terms fluctuate substantially in value exhibiting chaotic behavior.

© 2016 Cengage Learning. All Rights Reserved. May not be scanned, copied or duplicated, or posted to a publicly accessible website, in whole or in part.

53. Computer software was used to plot the first 20 points of the recursion equation $x_{t+1} = \frac{1}{4}x_t(1 - x_t)$, with $x_0 = 0.2$ and $x_0 = 0.2001$. The plots indicate that the solutions are nearly identical, converging to zero as t increases.

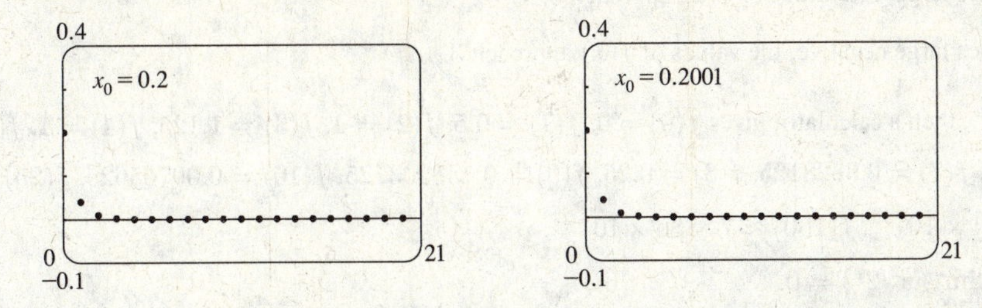

55.

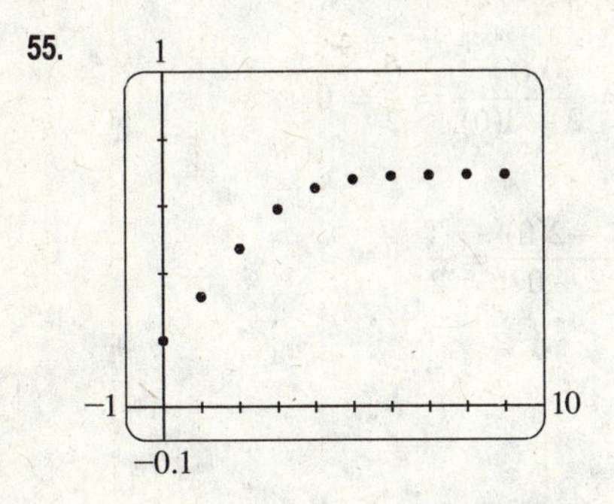

Computer software was used to plot the first 10 points of the recursion equation $x_{t+1} = 2x_t e^{-x_t}$, $x_0 = 0.2$. The sequence appears to converge to a value near 0.7. Assume the limit exists so that $\lim_{t\to\infty} x_{t+1} = \lim_{t\to\infty} x_t = x$, then

$x_{t+1} = 2x_t e^{-x_t} \quad \Rightarrow \quad \lim_{t\to\infty} x_{t+1} = \lim_{t\to\infty} 2x_t e^{-x_t} \quad \Rightarrow \quad x = 2xe^{-x} \quad \Rightarrow$

$x\left(1 - 2e^{-x}\right) = 0 \quad \Rightarrow \quad x = 0$ or $x = \ln 2 \approx 0.693$. Therefore, if the limit exists it will be either 0 or $\ln 2$. Since the graph of the sequence appears to approach $\ln 2$, we see that $\lim_{t\to\infty} x_t = \ln 2$.

57.

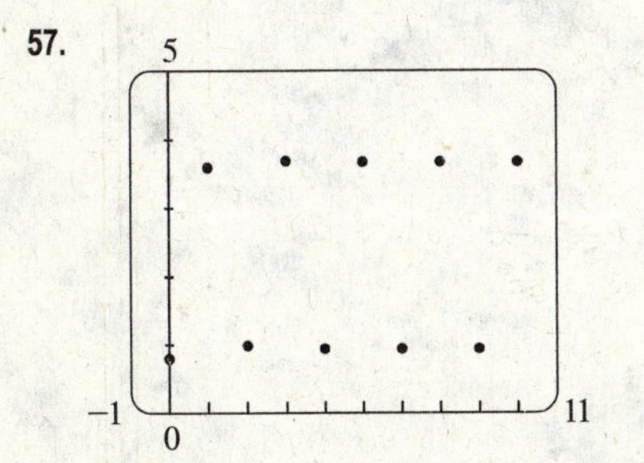

Computer software was used to plot the first 10 points of the recursion equation $x_{t+1} = 10x_t e^{-x_t}$, $x_0 = 0.8$. The sequence does not appear to converge to a fixed value of x_t. Instead, the terms oscillate between values near 0.9 and 3.7.

59. Let A_n represent the removed area of the Sierpinski carpet after the nth step of construction. In the first step, one square of area $\frac{1}{9}$ is removed so $A_1 = \frac{1}{9}$. In the second step, 8 squares each of area $\frac{1}{9}\left(\frac{1}{9}\right) = \frac{1}{9^2}$ are removed, so

$A_2 = A_1 + \frac{8}{9^2} = \frac{1}{9} + \frac{8}{9^2} = \frac{1}{9}\left(1 + \frac{8}{9}\right)$. In the third step, 8 squares are removed for each of the 8 squares removed in the previous step. So there are a total of $8 \cdot 8 = 8^2$ squares removed each having an area of $\frac{1}{9}\left(\frac{1}{9^2}\right) = \frac{1}{9^3}$. This gives

$A_3 = A_2 + \frac{8^2}{9^3} = \frac{1}{9}\left(1 + \frac{8}{9}\right) + \frac{8}{9^2} = \frac{1}{9}\left[1 + \frac{8}{9} + \left(\frac{8}{9}\right)^2\right]$. Observing the pattern in the first few terms of the sequence,

we deduce the general formula for the nth term to be $A_n = \frac{1}{9}\left[1 + \frac{8}{9} + \left(\frac{8}{9}\right)^2 + \ldots + \left(\frac{8}{9}\right)^{n-1}\right]$. The terms in parentheses represent the sum of a geometric sequence with $a = 1$ and $r = 8/9$. Using Equation (5), we can write

$A_n = \frac{1}{9}\left[\frac{1(1 - (8/9)^n)}{1 - 8/9}\right] = 1 - \left(\frac{8}{9}\right)^n$. As n increases, $\lim_{n\to\infty} A_n = \lim_{n\to\infty}\left[1 - \left(\frac{8}{9}\right)^n\right] = 1$. Hence the area of the removed squares is 1 implying that the Sierpinski carpet has zero area.

© 2016 Cengage Learning. All Rights Reserved. May not be scanned, copied or duplicated, or posted to a publicly accessible website, in whole or in part.

2.2 Limits of Functions at Infinity

1. (a) As x becomes large, the values of $f(x)$ approach 5.

(b) As x becomes large negative, the values of $f(x)$ approach 3.

3. If $f(x) = x^2/2^x$, then a calculator gives $f(0) = 0$, $f(1) = 0.5$, $f(2) = 1$, $f(3) = 1.125$, $f(4) = 1$, $f(5) = 0.78125$, $f(6) = 0.5625$, $f(7) = 0.3828125$, $f(8) = 0.25$, $f(9) = 0.158203125$, $f(10) = 0.09765625$, $f(20) \approx 0.00038147$, $f(50) \approx 2.2204 \times 10^{-12}$, $f(100) \approx 7.8886 \times 10^{-27}$.

It appears that $\lim\limits_{x\to\infty} (x^2/2^x) = 0$.

5. $$\lim_{x\to\infty} \frac{1}{2x+3} = \lim_{x\to\infty} \frac{1/x}{(2x+3)/x} = \frac{\lim\limits_{x\to\infty}(1/x)}{\lim\limits_{x\to\infty}(2+3/x)} = \frac{\lim\limits_{x\to\infty}(1/x)}{\lim\limits_{x\to\infty} 2 + 3\lim\limits_{x\to\infty}(1/x)} = \frac{0}{2+3(0)} = \frac{0}{2} = 0$$

7. $$\lim_{x\to\infty} \frac{3x-2}{2x+1} = \lim_{x\to\infty} \frac{(3x-2)/x}{(2x+1)/x} = \lim_{x\to\infty} \frac{3-2/x}{2+1/x} = \frac{\lim\limits_{x\to\infty} 3 - 2\lim\limits_{x\to\infty} 1/x}{\lim\limits_{x\to\infty} 2 + \lim\limits_{x\to\infty} 1/x} = \frac{3-2(0)}{2+0} = \frac{3}{2}$$

9. $$\lim_{x\to-\infty} \frac{1-x-x^2}{2x^2-7} = \lim_{x\to-\infty} \frac{(1-x-x^2)/x^2}{(2x^2-7)/x^2} = \frac{\lim\limits_{x\to-\infty}(1/x^2 - 1/x - 1)}{\lim\limits_{x\to-\infty}(2-7/x^2)}$$
$$= \frac{\lim\limits_{x\to-\infty}(1/x^2) - \lim\limits_{x\to-\infty}(1/x) - \lim\limits_{x\to-\infty} 1}{\lim\limits_{x\to-\infty} 2 - 7\lim\limits_{x\to-\infty}(1/x^2)} = \frac{0-0-1}{2-7(0)} = -\frac{1}{2}$$

11. $$\lim_{t\to-\infty} 0.6^t = \lim_{t\to-\infty} \left(\frac{3}{5}\right)^t = \lim_{t\to-\infty} \left(\frac{5}{3}\right)^{-t} = \infty \text{ since } 5/3 > 1 \text{ and } -t \to \infty \text{ as } t \to -\infty$$

13. $$\lim_{t\to\infty} \frac{\sqrt{t}+t^2}{2t-t^2} = \lim_{t\to\infty} \frac{(\sqrt{t}+t^2)/t^2}{(2t-t^2)/t^2} = \lim_{t\to\infty} \frac{1/t^{3/2}+1}{2/t-1} = \frac{0+1}{0-1} = -1$$

15. $$\lim_{x\to\infty} \frac{(2x^2+1)^2}{(x-1)^2(x^2+x)} = \lim_{x\to\infty} \frac{(2x^2+1)^2/x^4}{[(x-1)^2(x^2+x)]/x^4} = \lim_{x\to\infty} \frac{[(2x^2+1)/x^2]^2}{[(x^2-2x+1)/x^2][(x^2+x)/x^2]}$$
$$= \lim_{x\to\infty} \frac{(2+1/x^2)^2}{(1-2/x+1/x^2)(1+1/x)} = \frac{(2+0)^2}{(1-0+0)(1+0)} = 4$$

17. $$\lim_{x\to\infty} \left(\sqrt{9x^2+x} - 3x\right) = \lim_{x\to\infty} \frac{\left(\sqrt{9x^2+x}-3x\right)\left(\sqrt{9x^2+x}+3x\right)}{\sqrt{9x^2+x}+3x} = \lim_{x\to\infty} \frac{\left(\sqrt{9x^2+x}\right)^2 - (3x)^2}{\sqrt{9x^2+x}+3x}$$
$$= \lim_{x\to\infty} \frac{(9x^2+x)-9x^2}{\sqrt{9x^2+x}+3x} = \lim_{x\to\infty} \frac{x}{\sqrt{9x^2+x}+3x} \cdot \frac{1/x}{1/x}$$
$$= \lim_{x\to\infty} \frac{x/x}{\sqrt{9x^2/x^2+x/x^2}+3x/x} = \lim_{x\to\infty} \frac{1}{\sqrt{9+1/x}+3} = \frac{1}{\sqrt{9}+3} = \frac{1}{3+3} = \frac{1}{6}$$

19. $$\lim_{x\to\infty} \frac{6}{3+e^{-2x}} = \frac{6}{3+\lim\limits_{x\to\infty} e^{-2x}} = \frac{6}{3+0} = \frac{6}{3} = 2$$

21. $$\lim_{x\to\infty} \frac{x^4-3x^2+x}{x^3-x+2} = \lim_{x\to\infty} \frac{(x^4-3x^2+x)/x^3}{(x^3-x+2)/x^3} \quad \left[\begin{array}{c}\text{divide by the highest power}\\ \text{of } x \text{ in the denominator}\end{array}\right] \quad = \lim_{x\to\infty} \frac{x-3/x+1/x^2}{1-1/x^2+2/x^3} = \infty$$

since the numerator increases without bound and the denominator approaches 1 as $x \to \infty$.

© 2016 Cengage Learning. All Rights Reserved. May not be scanned, copied or duplicated, or posted to a publicly accessible website, in whole or in part.

23. $\lim\limits_{x\to-\infty}(x^4+x^5) = \lim\limits_{x\to-\infty} x^5(\frac{1}{x}+1)$ [factor out the largest power of x] $= -\infty$ because $x^5 \to -\infty$ and $1/x+1 \to 1$ as $x \to -\infty$.

Or: $\lim\limits_{x\to-\infty}(x^4+x^5) = \lim\limits_{x\to-\infty} x^4(1+x) = -\infty$.

25. As t increases, $1/t^2$ approaches zero, so $\lim\limits_{t\to\infty} e^{-1/t^2} = e^{-(0)} = 1$

27. $\lim\limits_{x\to\infty}\dfrac{1-e^x}{1+2e^x} = \lim\limits_{x\to\infty}\dfrac{(1-e^x)/e^x}{(1+2e^x)/e^x} = \lim\limits_{x\to\infty}\dfrac{1/e^x-1}{1/e^x+2} = \dfrac{0-1}{0+2} = -\dfrac{1}{2}$

29. $R(N) = SN/(c+N) \;\Rightarrow\; R(c) = Sc/(c+c) = S/2$. Hence, c is the nutrient concentration at which the growth rate is half of the maximum possible value. This is often referred to as the half-saturation constant.

31. $\lim\limits_{v\to\infty} N(v) = \lim\limits_{v\to\infty}\dfrac{8v}{1+2v+v^2} = \lim\limits_{v\to\infty}\dfrac{(8v)/v^2}{(1+2v+v^2)/v^2}$ [divide by the highest power of v in the denominator]

$= \lim\limits_{v\to\infty}\dfrac{8/v}{1/v^2+2/v+1} = \dfrac{0}{0+0+1} = 0$

Therefore, as the mortality rate increases, the number of new infections approaches zero.

33. $B(t) = \dfrac{8\times10^7}{1+3e^{-0.71t}} \;\Rightarrow\; \lim\limits_{t\to\infty} B(t) = \lim\limits_{t\to\infty}\dfrac{8\times10^7}{1+3e^{-0.71t}} = \dfrac{8\times10^7}{1+0} = 8\times10^7$. This means that in the long run the biomass of the Pacific halibut will tend to 8×10^7 kg.

35. $e^{-x} < 0.0001 \;\Rightarrow\; \ln(e^{-x}) < \ln(0.0001) \;\Rightarrow\; -x < \ln(0.0001) \;\Rightarrow\; x > -\ln(0.0001) \approx 9.21$, so x must be bigger than 9.21.

37. (a) $\lim\limits_{t\to\infty} v(t) = \lim\limits_{t\to\infty} v^*\left(1-e^{-gt/v^*}\right) = v^*(1-0) = v^*$

(b) Substituting the values $v^* = 7.5$ and $v(t) = 0.99 * 7.5$ into the velocity function gives $0.99 * 7.5 = (7.5)(1-e^{-gt/v^*})$

$\Rightarrow\; 0.99(7.5) = (7.5)\left(1-e^{-t(9.8)/(7.5)}\right) \;\Rightarrow\; e^{-t(9.8)/(7.5)} = 0.01 \;\Rightarrow\; t = -\dfrac{(7.5)\ln(0.01)}{9.8} = 3.52$ s

2.3 Limits of Functions at Finite Numbers

1. As x approaches 2, $f(x)$ approaches 5. [Or, the values of $f(x)$ can be made as close to 5 as we like by taking x sufficiently close to 2 (but $x \neq 2$).] Yes, the graph could have a hole at $(2,5)$ and be defined such that $f(2) = 3$.

3. (a) $\lim\limits_{x\to-3} f(x) = \infty$ means that the values of $f(x)$ can be made arbitrarily large (as large as we please) by taking x sufficiently close to -3 (but not equal to -3).

(b) $\lim\limits_{x\to4^+} f(x) = -\infty$ means that the values of $f(x)$ can be made arbitrarily large negative by taking x sufficiently close to 4 through values larger than 4.

5. (a) As x approaches 1, the values of $f(x)$ approach 2, so $\lim\limits_{x\to1} f(x) = 2$.

(b) As x approaches 3 from the left, the values of $f(x)$ approach 1, so $\lim\limits_{x\to3^-} f(x) = 1$.

(c) As x approaches 3 from the right, the values of $f(x)$ approach 4, so $\lim\limits_{x\to3^+} f(x) = 4$.

(d) $\lim\limits_{x\to3} f(x)$ does not exist since the left-hand limit does not equal the right-hand limit.

(e) When $x = 3$, $y = 3$, so $f(3) = 3$.

© 2016 Cengage Learning. All Rights Reserved. May not be scanned, copied or duplicated, or posted to a publicly accessible website, in whole or in part.

7. (a) $P(t)$ approaches 260 as x approaches 2 from the left, so $\lim_{t\to2^-} P(t) = 260$.

(b) $P(t)$ approaches 254 as x approaches 2 from the right, so $\lim_{t\to2^+} P(t) = 254$.

(c) $\lim_{t\to2} P(t)$ does not exist because $\lim_{t\to2^-} P(t) \neq \lim_{t\to2^+} P(t)$.

(d) $P(t)$ approaches 254 as x approaches 4 from the left, so $\lim_{t\to4^-} P(t) = 254$.

(e) $P(t)$ approaches 258 as x approaches 4 from the right, so $\lim_{t\to4^+} P(t) = 258$.

(f) $\lim_{t\to4} P(t)$ does not exist because $\lim_{t\to4^-} P(t) \neq \lim_{t\to4^+} P(t)$.

(g) $\lim_{x\to5} P(t) = 258$ because $\lim_{x\to5^-} P(t) = 258 = \lim_{x\to5^+} P(t)$

(h) On June 3 ($t = 2$), the population decreased by 6. This could have been a result of deaths, emigration, or a combination of the two. On June 5 ($t = 4$), the population increased by 4. This could have been a result of births, immigration, or a combination of the two.

9. (a) $\lim_{x\to0} g(x) = -\infty$ (b) $\lim_{x\to2^-} g(x) = -\infty$ (c) $\lim_{x\to2^+} g(x) = \infty$

(d) $\lim_{x\to\infty} g(x) = 2$ (e) $\lim_{x\to-\infty} g(x) = -1$ (f) Vertical: $x = 0$, $x = 2$; horizontal: $y = -1$, $y = 2$

11. $\lim_{x\to3^+} f(x) = 4$, $\lim_{x\to3^-} f(x) = 2$, $\lim_{x\to-2} f(x) = 2$, $f(3) = 3$, $f(-2) = 1$

13. $\lim_{x\to0} f(x) = -\infty$, $\lim_{x\to-\infty} f(x) = 5$, $\lim_{x\to\infty} f(x) = -5$

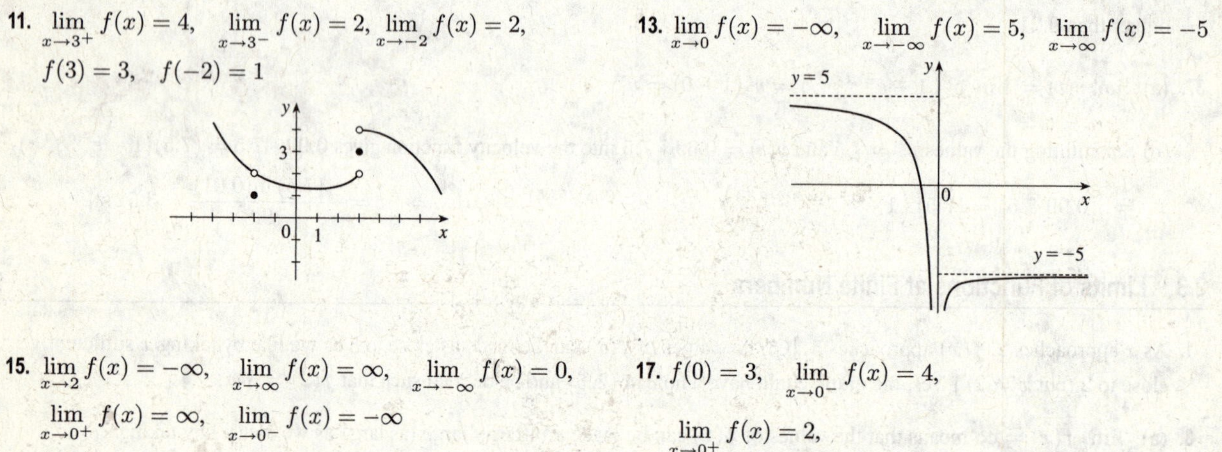

15. $\lim_{x\to2} f(x) = -\infty$, $\lim_{x\to\infty} f(x) = \infty$, $\lim_{x\to-\infty} f(x) = 0$, $\lim_{x\to0^+} f(x) = \infty$, $\lim_{x\to0^-} f(x) = -\infty$

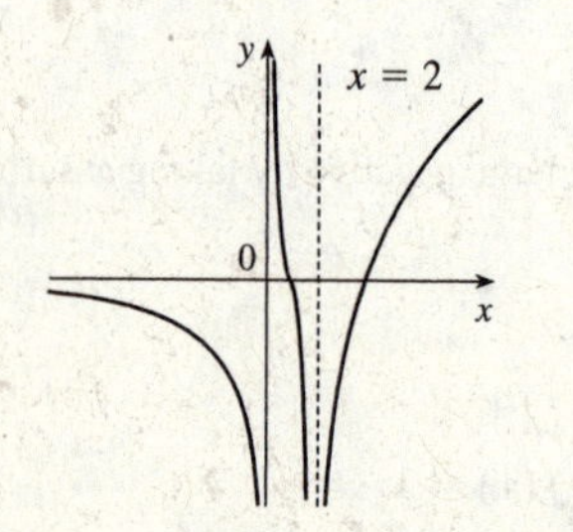

17. $f(0) = 3$, $\lim_{x\to0^-} f(x) = 4$,

$\lim_{x\to0^+} f(x) = 2$,

$\lim_{x\to-\infty} f(x) = -\infty$, $\lim_{x\to4^-} f(x) = -\infty$,

$\lim_{x\to4^+} f(x) = \infty$, $\lim_{x\to\infty} f(x) = 3$

© 2016 Cengage Learning. All Rights Reserved. May not be scanned, copied or duplicated, or posted to a publicly accessible website, in whole or in part.

19. For $f(x) = \dfrac{x^2 - 2x}{x^2 - x - 2}$:

x	$f(x)$
2.5	0.714286
2.1	0.677419
2.05	0.672131
2.01	0.667774
2.005	0.667221
2.001	0.666778

x	$f(x)$
1.9	0.655172
1.95	0.661017
1.99	0.665552
1.995	0.666110
1.999	0.666556

It appears that $\lim\limits_{x \to 2} \dfrac{x^2 - 2x}{x^2 - x - 2} = 0.\bar{6} = \frac{2}{3}$.

21. For $f(t) = \dfrac{e^{5t} - 1}{t}$:

t	$f(t)$
0.5	22.364988
0.1	6.487213
0.01	5.127110
0.001	5.012521
0.0001	5.001250

t	$f(t)$
−0.5	1.835830
−0.1	3.934693
−0.01	4.877058
−0.001	4.987521
−0.0001	4.998750

It appears that $\lim\limits_{t \to 0} \dfrac{e^{5t} - 1}{t} = 5$.

23. For $f(x) = \dfrac{\sqrt{x+4} - 2}{x}$:

x	$f(x)$
1	0.236068
0.5	0.242641
0.1	0.248457
0.05	0.249224
0.01	0.249844

x	$f(x)$
−1	0.267949
−0.5	0.258343
−0.1	0.251582
−0.05	0.250786
−0.01	0.250156

It appears that $\lim\limits_{x \to 0} \dfrac{\sqrt{x+4} - 2}{x} = 0.25 = \frac{1}{4}$.

25. For $f(x) = \dfrac{x^6 - 1}{x^{10} - 1}$:

x	$f(x)$
0.5	0.985337
0.9	0.719397
0.95	0.660186
0.99	0.612018
0.999	0.601200

x	$f(x)$
1.5	0.183369
1.1	0.484119
1.05	0.540783
1.01	0.588022
1.001	0.598800

It appears that $\lim\limits_{x \to 1} \dfrac{x^6 - 1}{x^{10} - 1} = 0.6 = \frac{3}{5}$.

27. (a) From the graphs, it seems that $\lim\limits_{x \to 0} \dfrac{\cos 2x - \cos x}{x^2} = -1.5$.

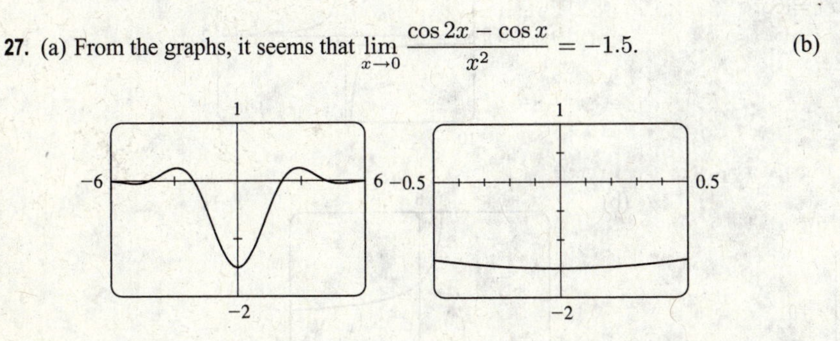

(b)

x	$f(x)$
±0.1	−1.493759
±0.01	−1.499938
±0.001	−1.499999
±0.0001	−1.500000

29. $\lim\limits_{x \to -3^+} \dfrac{x+2}{x+3} = -\infty$ since the numerator is negative and the denominator approaches 0 from the positive side as $x \to -3^+$.

31. $\lim\limits_{x \to 1} \dfrac{2-x}{(x-1)^2} = \infty$ since the numerator is positive and the denominator approaches 0 through positive values as $x \to 1$.

33. Let $t = x^2 - 9$. Then as $x \to 3^+$, $t \to 0^+$, and $\lim\limits_{x \to 3^+} \ln(x^2 - 9) = \lim\limits_{t \to 0^+} \ln t = -\infty$ by (8).

© 2016 Cengage Learning. All Rights Reserved. May not be scanned, copied or duplicated, or posted to a publicly accessible website, in whole or in part.

35. $\lim_{x\to 2\pi^-} x\csc x = \lim_{x\to 2\pi^-} \dfrac{x}{\sin x} = -\infty$ since the numerator is positive and the denominator approaches 0 through negative values as $x \to 2\pi^-$.

37. $\lim_{x\to 2^+} \dfrac{x^2-2x-8}{x^2-5x+6} = \lim_{x\to 2^+} \dfrac{(x-4)(x+2)}{(x-3)(x-2)} = \infty$ since the numerator is negative and the denominator approaches 0 through negative values as $x \to 2^+$.

39. (a) $f(x) = \dfrac{1}{x^3-1}$.

From these calculations, it seems that

$\lim_{x\to 1^-} f(x) = -\infty$ and $\lim_{x\to 1^+} f(x) = \infty$.

x	$f(x)$
0.5	−1.14
0.9	−3.69
0.99	−33.7
0.999	−333.7
0.9999	−3333.7
0.99999	−33,333.7

x	$f(x)$
1.5	0.42
1.1	3.02
1.01	33.0
1.001	333.0
1.0001	3333.0
1.00001	33,333.3

(b) If x is slightly smaller than 1, then $x^3 - 1$ will be a negative number close to 0, and the reciprocal of $x^3 - 1$, that is, $f(x)$, will be a negative number with large absolute value. So $\lim_{x\to 1^-} f(x) = -\infty$.

If x is slightly larger than 1, then $x^3 - 1$ will be a small positive number, and its reciprocal, $f(x)$, will be a large positive number. So $\lim_{x\to 1^+} f(x) = \infty$.

(c) It appears from the graph of f that

$\lim_{x\to 1^-} f(x) = -\infty$ and $\lim_{x\to 1^+} f(x) = \infty$.

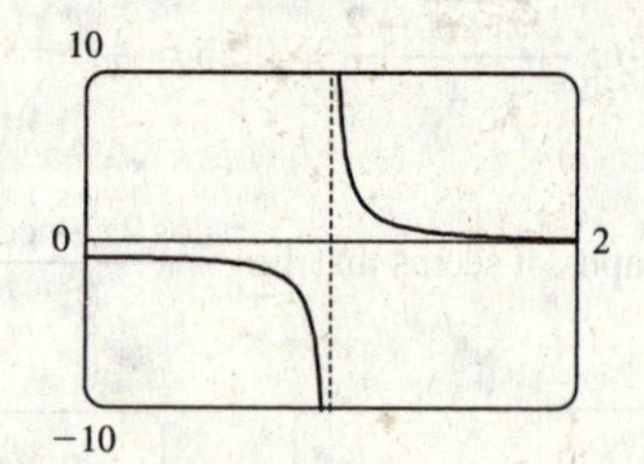

41. (a) Let $h(x) = (1+x)^{1/x}$.

x	$h(x)$
−0.001	2.71964
−0.0001	2.71842
−0.00001	2.71830
−0.000001	2.71828
0.000001	2.71828
0.00001	2.71827
0.0001	2.71815
0.001	2.71692

(b)

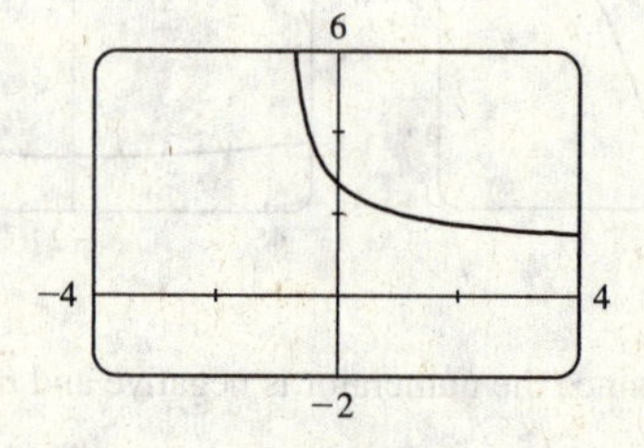

It appears that $\lim_{x\to 0} (1+x)^{1/x} \approx 2.71828$ which is approximately e.

In Section 3.7 we will see that the value of the limit is exactly e.

© 2016 Cengage Learning. All Rights Reserved. May not be scanned, copied or duplicated, or posted to a publicly accessible website, in whole or in part.

43. No matter how many times we zoom in toward the origin, the graphs of $f(x) = \sin(\pi/x)$ appear to consist of almost-vertical lines. This indicates more and more frequent oscillations as $x \to 0$.

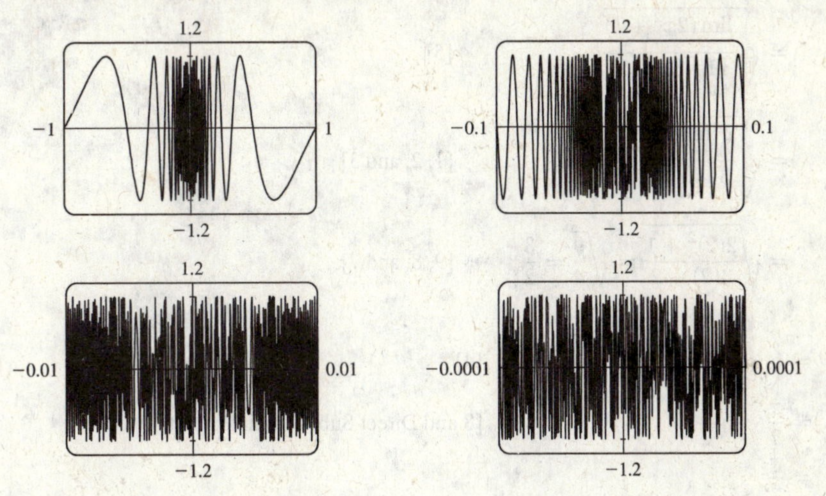

2.4 Limits: Algebraic Methods

1. (a) $\lim\limits_{x\to 2}[f(x)+5g(x)] = \lim\limits_{x\to 2} f(x) + \lim\limits_{x\to 2}[5g(x)]$ [Limit Law 1]

$= \lim\limits_{x\to 2} f(x) + 5\lim\limits_{x\to 2} g(x)$ [Limit Law 3]

$= 4 + 5(-2) = -6$

(b) $\lim\limits_{x\to 2}[g(x)]^3 = \left[\lim\limits_{x\to 2} g(x)\right]^3$ [Limit Law 6]

$= (-2)^3 = -8$

(c) $\lim\limits_{x\to 2}\sqrt{f(x)} = \sqrt{\lim\limits_{x\to 2} f(x)}$ [Limit Law 11]

$= \sqrt{4} = 2$

(d) $\lim\limits_{x\to 2}\dfrac{3f(x)}{g(x)} = \dfrac{\lim\limits_{x\to 2}[3f(x)]}{\lim\limits_{x\to 2} g(x)}$ [Limit Law 5]

$= \dfrac{3\lim\limits_{x\to 2} f(x)}{\lim\limits_{x\to 2} g(x)}$ [Limit Law 3]

$= \dfrac{3(4)}{-2} = -6$

(e) Because the limit of the denominator is 0, we can't use Limit Law 5. The given limit, $\lim\limits_{x\to 2}\dfrac{g(x)}{h(x)}$, does not exist because the denominator approaches 0 while the numerator approaches a nonzero number.

(f) $\lim\limits_{x\to 2}\dfrac{g(x)\,h(x)}{f(x)} = \dfrac{\lim\limits_{x\to 2}[g(x)\,h(x)]}{\lim\limits_{x\to 2} f(x)}$ [Limit Law 5]

$= \dfrac{\lim\limits_{x\to 2} g(x)\cdot\lim\limits_{x\to 2} h(x)}{\lim\limits_{x\to 2} f(x)}$ [Limit Law 4]

$= \dfrac{-2\cdot 0}{4} = 0$

3. $\lim\limits_{x\to -2}(3x^4+2x^2-x+1) = \lim\limits_{x\to -2} 3x^4 + \lim\limits_{x\to -2} 2x^2 - \lim\limits_{x\to -2} x + \lim\limits_{x\to -2} 1$ [Limit Laws 1 and 2]

$= 3\lim\limits_{x\to -2} x^4 + 2\lim\limits_{x\to -2} x^2 - \lim\limits_{x\to -2} x + \lim\limits_{x\to -2} 1$ [3]

$= 3(-2)^4 + 2(-2)^2 - (-2) + (1)$ [9, 8, and 7]

$= 48 + 8 + 2 + 1 = 59$

© 2016 Cengage Learning. All Rights Reserved. May not be scanned, copied or duplicated, or posted to a publicly accessible website, in whole or in part.

5. $$\lim_{x\to 2}\sqrt{\frac{2x^2+1}{3x-2}} = \sqrt{\lim_{x\to 2}\frac{2x^2+1}{3x-2}} \qquad \text{[Limit Law 11]}$$

$$= \sqrt{\frac{\lim_{x\to 2}(2x^2+1)}{\lim_{x\to 2}(3x-2)}} \qquad [5]$$

$$= \sqrt{\frac{2\lim_{x\to 2}x^2 + \lim_{x\to 2}1}{3\lim_{x\to 2}x - \lim_{x\to 2}2}} \qquad \text{[1, 2, and 3]}$$

$$= \sqrt{\frac{2(2)^2+1}{3(2)-2}} = \sqrt{\frac{9}{4}} = \frac{3}{2} \qquad \text{[9, 8, and 7]}$$

7. $$\lim_{\theta\to\pi/2}\theta\sin\theta = \left(\lim_{\theta\to\pi/2}\theta\right)\left(\lim_{\theta\to\pi/2}\sin\theta\right) \qquad [4]$$

$$= \frac{\pi}{2}\cdot\sin\frac{\pi}{2} \qquad \text{[8 and Direct Substitution Property]}$$

$$= \frac{\pi}{2}$$

9. $$\lim_{x\to 5}\frac{x^2-6x+5}{x-5} = \lim_{x\to 5}\frac{(x-5)(x-1)}{x-5} = \lim_{x\to 5}(x-1) = 5-1 = 4$$

11. $\lim_{x\to 5}\frac{x^2-5x+6}{x-5}$ does not exist since $x-5\to 0$, but $x^2-5x+6\to 6$ as $x\to 5$.

13. $$\lim_{t\to -3}\frac{t^2-9}{2t^2+7t+3} = \lim_{t\to -3}\frac{(t+3)(t-3)}{(2t+1)(t+3)} = \lim_{t\to -3}\frac{t-3}{2t+1} = \frac{-3-3}{2(-3)+1} = \frac{-6}{-5} = \frac{6}{5}$$

15. $$\lim_{h\to 0}\frac{(4+h)^2-16}{h} = \lim_{h\to 0}\frac{(16+8h+h^2)-16}{h} = \lim_{h\to 0}\frac{8h+h^2}{h} = \lim_{h\to 0}\frac{h(8+h)}{h} = \lim_{h\to 0}(8+h) = 8+0 = 8$$

17. By the formula for the sum of cubes, we have

$$\lim_{x\to -2}\frac{x+2}{x^3+8} = \lim_{x\to -2}\frac{x+2}{(x+2)(x^2-2x+4)} = \lim_{x\to -2}\frac{1}{x^2-2x+4} = \frac{1}{4+4+4} = \frac{1}{12}.$$

19. $$\lim_{x\to -4}\frac{\frac{1}{4}+\frac{1}{x}}{4+x} = \lim_{x\to -4}\frac{\frac{x+4}{4x}}{4+x} = \lim_{x\to -4}\frac{x+4}{4x(4+x)} = \lim_{x\to -4}\frac{1}{4x} = \frac{1}{4(-4)} = -\frac{1}{16}$$

21. $$\lim_{x\to 16}\frac{4-\sqrt{x}}{16x-x^2} = \lim_{x\to 16}\frac{(4-\sqrt{x})(4+\sqrt{x})}{(16x-x^2)(4+\sqrt{x})} = \lim_{x\to 16}\frac{16-x}{x(16-x)(4+\sqrt{x})}$$

$$= \lim_{x\to 16}\frac{1}{x(4+\sqrt{x})} = \frac{1}{16\left(4+\sqrt{16}\right)} = \frac{1}{16(8)} = \frac{1}{128}$$

23. $$\lim_{t\to 0}\left(\frac{1}{t\sqrt{1+t}}-\frac{1}{t}\right) = \lim_{t\to 0}\frac{1-\sqrt{1+t}}{t\sqrt{1+t}} = \lim_{t\to 0}\frac{\left(1-\sqrt{1+t}\right)\left(1+\sqrt{1+t}\right)}{t\sqrt{t+1}\left(1+\sqrt{1+t}\right)} = \lim_{t\to 0}\frac{-t}{t\sqrt{1+t}\left(1+\sqrt{1+t}\right)}$$

$$= \lim_{t\to 0}\frac{-1}{\sqrt{1+t}\left(1+\sqrt{1+t}\right)} = \frac{-1}{\sqrt{1+0}\left(1+\sqrt{1+0}\right)} = -\frac{1}{2}$$

© 2016 Cengage Learning. All Rights Reserved. May not be scanned, copied or duplicated, or posted to a publicly accessible website, in whole or in part.

25. (a)

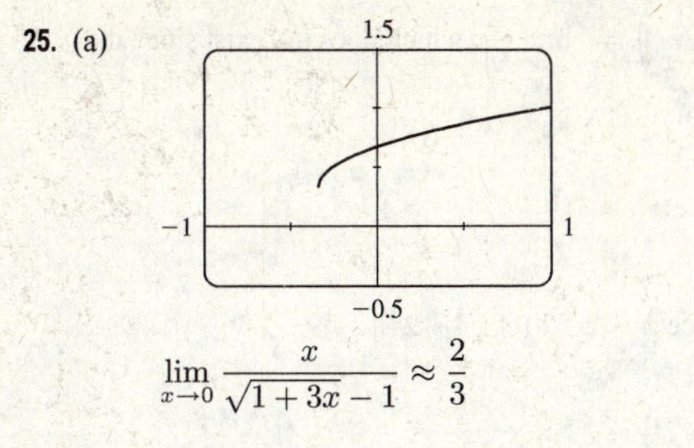

$\lim_{x\to 0} \frac{x}{\sqrt{1+3x}-1} \approx \frac{2}{3}$

(b)

x	$f(x)$
−0.001	0.6661663
−0.0001	0.6666167
−0.00001	0.6666617
−0.000001	0.6666662
0.000001	0.6666672
0.00001	0.6666717
0.0001	0.6667167
0.001	0.6671663

The limit appears to be $\frac{2}{3}$.

(c)
$$\begin{aligned}
\lim_{x\to 0}\left(\frac{x}{\sqrt{1+3x}-1}\cdot\frac{\sqrt{1+3x}+1}{\sqrt{1+3x}+1}\right) &= \lim_{x\to 0}\frac{x\left(\sqrt{1+3x}+1\right)}{(1+3x)-1} = \lim_{x\to 0}\frac{x\left(\sqrt{1+3x}+1\right)}{3x} \\
&= \frac{1}{3}\lim_{x\to 0}\left(\sqrt{1+3x}+1\right) && \text{[Limit Law 3]} \\
&= \frac{1}{3}\left[\sqrt{\lim_{x\to 0}(1+3x)}+\lim_{x\to 0}1\right] && \text{[1 and 11]} \\
&= \frac{1}{3}\left(\sqrt{\lim_{x\to 0}1+3\lim_{x\to 0}x}+1\right) && \text{[1, 3, and 7]} \\
&= \frac{1}{3}\left(\sqrt{1+3\cdot 0}+1\right) && \text{[7 and 8]} \\
&= \frac{1}{3}(1+1) = \frac{2}{3}
\end{aligned}$$

27. Let $f(x) = -x^2$, $g(x) = x^2\cos 20\pi x$ and $h(x) = x^2$. Then

$-1 \le \cos 20\pi x \le 1 \;\Rightarrow\; -x^2 \le x^2\cos 20\pi x \le x^2 \;\Rightarrow\; f(x) \le g(x) \le h(x).$

So since $\lim_{x\to 0} f(x) = \lim_{x\to 0} h(x) = 0$, by the Squeeze Theorem we have

$\lim_{x\to 0} g(x) = 0.$

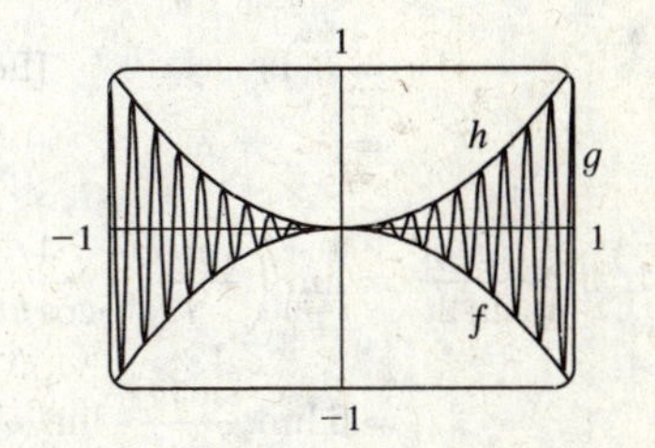

29. We have $\lim_{x\to 4}(4x-9) = 4(4)-9 = 7$ and $\lim_{x\to 4}(x^2-4x+7) = 4^2-4(4)+7 = 7$. Since $4x-9 \le f(x) \le x^2-4x+7$ for $x \ge 0$, $\lim_{x\to 4} f(x) = 7$ by the Squeeze Theorem.

31. $-1 \le \cos(2/x) \le 1 \;\Rightarrow\; -x^4 \le x^4\cos(2/x) \le x^4$. Since $\lim_{x\to 0}\left(-x^4\right) = 0$ and $\lim_{x\to 0} x^4 = 0$, we have $\lim_{x\to 0}\left[x^4\cos(2/x)\right] = 0$ by the Squeeze Theorem.

33. $|x-3| = \begin{cases} x-3 & \text{if } x-3 \ge 0 \\ -(x-3) & \text{if } x-3 < 0 \end{cases} = \begin{cases} x-3 & \text{if } x \ge 3 \\ 3-x & \text{if } x < 3 \end{cases}$

Thus, $\lim_{x\to 3^+}(2x+|x-3|) = \lim_{x\to 3^+}(2x+x-3) = \lim_{x\to 3^+}(3x-3) = 3(3)-3 = 6$ and

$\lim_{x\to 3^-}(2x+|x-3|) = \lim_{x\to 3^-}(2x+3-x) = \lim_{x\to 3^-}(x+3) = 3+3 = 6$. Since the left and right limits are equal,

$\lim_{x\to 3}(2x+|x-3|) = 6.$

© 2016 Cengage Learning. All Rights Reserved. May not be scanned, copied or duplicated, or posted to a publicly accessible website, in whole or in part.

35. Since $|x| = -x$ for $x < 0$, we have $\lim\limits_{x\to 0^-}\left(\frac{1}{x} - \frac{1}{|x|}\right) = \lim\limits_{x\to 0^-}\left(\frac{1}{x} - \frac{1}{-x}\right) = \lim\limits_{x\to 0^-}\frac{2}{x}$, which does not exist since the denominator approaches 0 and the numerator does not.

37. (a) (i) $$\lim_{x\to 2^+} g(x) = \lim_{x\to 2^+}\frac{x^2+x-6}{|x-2|} = \lim_{x\to 2^+}\frac{(x+3)(x-2)}{|x-2|}$$
$$= \lim_{x\to 2^+}\frac{(x+3)(x-2)}{x-2} \quad [\text{since } x-2>0 \text{ if } x\to 2^+]$$
$$= \lim_{x\to 2^+}(x+3) = 5$$

(ii) The solution is similar to the solution in part (i), but now $|x-2| = 2-x$ since $x-2<0$ if $x\to 2^-$.
Thus, $\lim\limits_{x\to 2^-} g(x) = \lim\limits_{x\to 2^-} -(x+3) = -5$.

(b) Since the right-hand and left-hand limits of g at $x = 2$ are not equal, $\lim\limits_{x\to 2} g(x)$ does not exist.

(c)

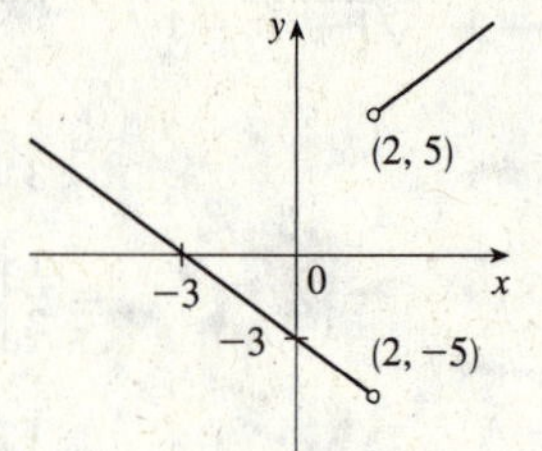

39. $$\lim_{x\to 0}\frac{\sin 3x}{x} = \lim_{x\to 0}\frac{3\sin 3x}{3x} \quad [\text{multiply numerator and denominator by 3}]$$
$$= 3\lim_{3x\to 0}\frac{\sin 3x}{3x} \quad [\text{as } x\to 0, 3x\to 0]$$
$$= 3\lim_{\theta\to 0}\frac{\sin\theta}{\theta} \quad [\text{let } \theta = 3x]$$
$$= 3(1) \quad [\text{Equation 6}]$$
$$= 3$$

41. $$\lim_{t\to 0}\frac{\tan 6t}{\sin 2t} = \lim_{t\to 0}\left(\frac{\sin 6t}{t}\cdot\frac{1}{\cos 6t}\cdot\frac{t}{\sin 2t}\right) = \lim_{t\to 0}\frac{6\sin 6t}{6t}\cdot\lim_{t\to 0}\frac{1}{\cos 6t}\cdot\lim_{t\to 0}\frac{2t}{2\sin 2t}$$
$$= 6\lim_{t\to 0}\frac{\sin 6t}{6t}\cdot\lim_{t\to 0}\frac{1}{\cos 6t}\cdot\frac{1}{2}\lim_{t\to 0}\frac{2t}{\sin 2t} = 6(1)\cdot\frac{1}{1}\cdot\frac{1}{2}(1) = 3$$

43. Divide numerator and denominator by θ. ($\sin\theta$ also works.)
$$\lim_{\theta\to 0}\frac{\sin\theta}{\theta+\tan\theta} = \lim_{\theta\to 0}\frac{\frac{\sin\theta}{\theta}}{1+\frac{\sin\theta}{\theta}\cdot\frac{1}{\cos\theta}} = \frac{\lim\limits_{\theta\to 0}\frac{\sin\theta}{\theta}}{1+\lim\limits_{\theta\to 0}\frac{\sin\theta}{\theta}\lim\limits_{\theta\to 0}\frac{1}{\cos\theta}} = \frac{1}{1+1\cdot 1} = \frac{1}{2}$$

45. (a) Since $p(x)$ is a polynomial, $p(x) = a_0 + a_1x + a_2x^2 + \cdots + a_nx^n$. Thus, by the Limit Laws,
$$\lim_{x\to a} p(x) = \lim_{x\to a}\left(a_0 + a_1x + a_2x^2 + \cdots + a_nx^n\right) = a_0 + a_1\lim_{x\to a} x + a_2\lim_{x\to a} x^2 + \cdots + a_n\lim_{x\to a} x^n$$
$$= a_0 + a_1a + a_2a^2 + \cdots + a_na^n = p(a)$$
Thus, for any polynomial p, $\lim\limits_{x\to a} p(x) = p(a)$.

(b) Let $r(x) = \frac{p(x)}{q(x)}$ where $p(x)$ and $q(x)$ are any polynomials, and suppose that $q(a) \neq 0$. Then
$$\lim_{x\to a} r(x) = \lim_{x\to a}\frac{p(x)}{q(x)} = \frac{\lim\limits_{x\to a} p(x)}{\lim\limits_{x\to a} q(x)} \quad [\text{Limit Law 5}] \quad = \frac{p(a)}{q(a)} \quad [\text{by part (a)}] \quad = r(a).$$

© 2016 Cengage Learning. All Rights Reserved. May not be scanned, copied or duplicated, or posted to a publicly accessible website, in whole or in part.

47. $\lim\limits_{h \to 0} \sin(a+h) = \lim\limits_{h \to 0} (\sin a \cos h + \cos a \sin h) = \lim\limits_{h \to 0} (\sin a \cos h) + \lim\limits_{h \to 0} (\cos a \sin h)$

$$= \left(\lim_{h \to 0} \sin a\right)\left(\lim_{h \to 0} \cos h\right) + \left(\lim_{h \to 0} \cos a\right)\left(\lim_{h \to 0} \sin h\right) = (\sin a)(1) + (\cos a)(0) = \sin a$$

49. $\lim\limits_{x \to 1} [f(x) - 8] = \lim\limits_{x \to 1} \left[\dfrac{f(x)-8}{x-1} \cdot (x-1)\right] = \lim\limits_{x \to 1} \dfrac{f(x)-8}{x-1} \cdot \lim\limits_{x \to 1} (x-1) = 10 \cdot 0 = 0.$

Thus, $\lim\limits_{x \to 1} f(x) = \lim\limits_{x \to 1} \{[f(x)-8]+8\} = \lim\limits_{x \to 1} [f(x)-8] + \lim\limits_{x \to 1} 8 = 0 + 8 = 8.$

Note: The value of $\lim\limits_{x \to 1} \dfrac{f(x)-8}{x-1}$ does not affect the answer since it's multiplied by 0. What's important is that $\lim\limits_{x \to 1} \dfrac{f(x)-8}{x-1}$ exists.

51. Since the denominator approaches 0 as $x \to -2$, the limit will exist only if the numerator also approaches 0 as $x \to -2$. In order for this to happen, we need $\lim\limits_{x \to -2} (3x^2 + ax + a + 3) = 0 \Leftrightarrow$ $3(-2)^2 + a(-2) + a + 3 = 0 \Leftrightarrow 12 - 2a + a + 3 = 0 \Leftrightarrow a = 15$. With $a = 15$, the limit becomes

$$\lim_{x \to -2} \frac{3x^2 + 15x + 18}{x^2 + x - 2} = \lim_{x \to -2} \frac{3(x+2)(x+3)}{(x-1)(x+2)} = \lim_{x \to -2} \frac{3(x+3)}{x-1} = \frac{3(-2+3)}{-2-1} = \frac{3}{-3} = -1.$$

2.5 Continuity

1. From Definition 1, $\lim\limits_{x \to 4} f(x) = f(4)$.

3. (a) f is discontinuous at -4 since $f(-4)$ is not defined and at -2, 2, and 4 since the limit does not exist (the left and right limits are not the same).

(b) f is continuous from the left at -2 since $\lim\limits_{x \to -2^-} f(x) = f(-2)$. f is continuous from the right at 2 and 4 since $\lim\limits_{x \to 2^+} f(x) = f(2)$ and $\lim\limits_{x \to 4^+} f(x) = f(4)$. It is continuous from neither side at -4 since $f(-4)$ is undefined.

5. The graph of $y = f(x)$ must have a discontinuity at $x = 2$ and must show that $\lim\limits_{x \to 2^+} f(x) = f(2)$.

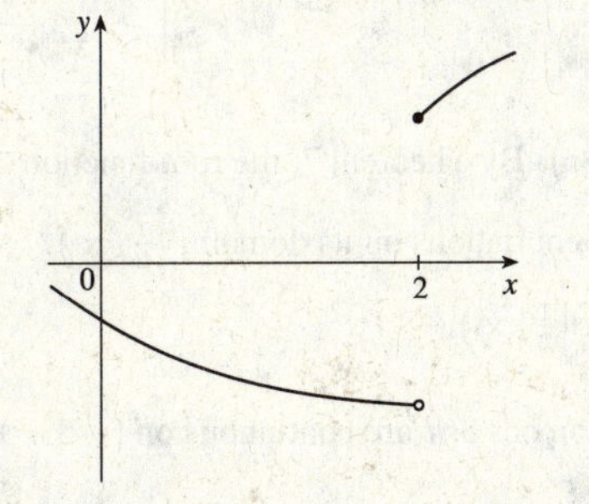

7. The graph of $y = f(x)$ must have a removable discontinuity (a hole) at $x = 3$ and a jump discontinuity at $x = 5$.

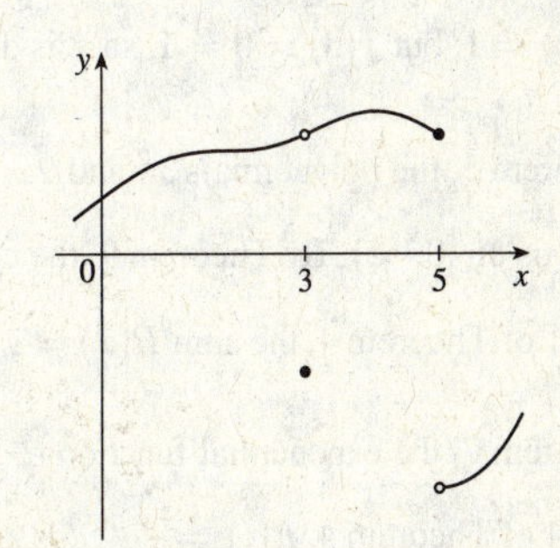

9. (a) C has discontinuities at 12, 24, and 36 hours since the limit does not exist at these points.

(b) C has jump discontinuities at the values of t listed in part (a) because the function jumps from one value to another at these points.

© 2016 Cengage Learning. All Rights Reserved. May not be scanned, copied or duplicated, or posted to a publicly accessible website, in whole or in part.

11. (a)

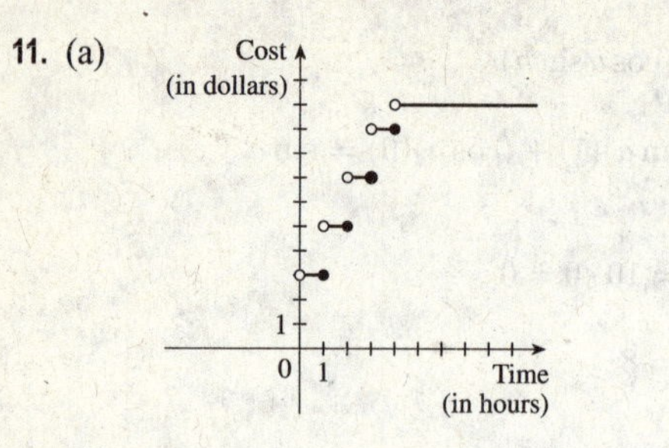

(b) There are discontinuities at times $t = 1, 2, 3$, and 4. A person parking in the lot would want to keep in mind that the charge will jump at the beginning of each hour.

13. Since f and g are continuous functions,

$$\begin{aligned}\lim_{x\to 3}[2f(x) - g(x)] &= 2\lim_{x\to 3} f(x) - \lim_{x\to 3} g(x) && \text{[by Limit Laws 2 and 3]}\\ &= 2f(3) - g(3) && \text{[by continuity of } f \text{ and } g \text{ at } x = 3]\\ &= 2\cdot 5 - g(3) = 10 - g(3)\end{aligned}$$

Since it is given that $\lim_{x\to 3}[2f(x) - g(x)] = 4$, we have $10 - g(3) = 4$, so $g(3) = 6$.

15. $\lim_{x\to -1} f(x) = \lim_{x\to -1}\left(x + 2x^3\right)^4 = \left(\lim_{x\to -1} x + 2\lim_{x\to -1} x^3\right)^4 = \left[-1 + 2(-1)^3\right]^4 = (-3)^4 = 81 = f(-1)$.

By the definition of continuity, f is continuous at $a = -1$.

17. $f(x) = \begin{cases} e^x & \text{if } x < 0 \\ x^2 & \text{if } x \geq 0 \end{cases}$

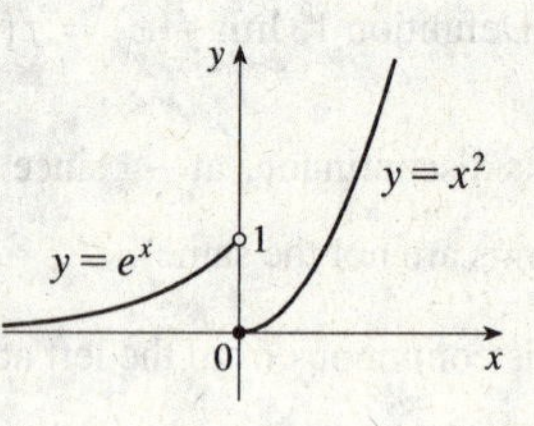

The left-hand limit of f at $a = 0$ is $\lim_{x\to 0^-} f(x) = \lim_{x\to 0^-} e^x = 1$. The right-hand limit of f at $a = 0$ is $\lim_{x\to 0^+} f(x) = \lim_{x\to 0^+} x^2 = 0$. Since these limits are not equal, $\lim_{x\to 0} f(x)$ does not exist and f is discontinuous at 0.

19. $f(x) = \begin{cases} \cos x & \text{if } x < 0 \\ 0 & \text{if } x = 0 \\ 1 - x^2 & \text{if } x > 0 \end{cases}$

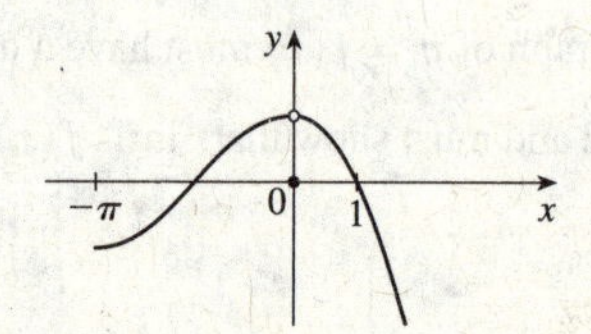

$\lim_{x\to 0} f(x) = 1$, but $f(0) = 0 \neq 1$, so f is discontinuous at 0.

21. By Theorem 5, the polynomials x^2 and $2x - 1$ are continuous on $(-\infty, \infty)$. By Theorem 7, the root function $\sqrt{x}$ is continuous on $[0, \infty)$. By Theorem 9, the composite function $\sqrt{2x - 1}$ is continuous on its domain, $\left[\frac{1}{2}, \infty\right)$. By part 1 of Theorem 4, the sum $R(x) = x^2 + \sqrt{2x - 1}$ is continuous on $\left[\frac{1}{2}, \infty\right)$.

23. By Theorem 7, the exponential function e^{-5t} and the trigonometric function $\cos 2\pi t$ are continuous on $(-\infty, \infty)$. By part 4 of Theorem 4, $L(t) = e^{-5t}\cos 2\pi t$ is continuous on $(-\infty, \infty)$.

25. By Theorem 5, the polynomial $t^4 - 1$ is continuous on $(-\infty, \infty)$. By Theorem 7, $\ln x$ is continuous on its domain, $(0, \infty)$. By Theorem 9, $\ln(t^4 - 1)$ is continuous on its domain, which is

$$\{t \mid t^4 - 1 > 0\} = \{t \mid t^4 > 1\} = \{t \mid |t| > 1\} = (-\infty, -1) \cup (1, \infty)$$

© 2016 Cengage Learning. All Rights Reserved. May not be scanned, copied or duplicated, or posted to a publicly accessible website, in whole or in part.

27. The function $y = \dfrac{1}{1+e^{1/x}}$ is discontinuous at $x = 0$ because the left- and right-hand limits at $x = 0$ are different.

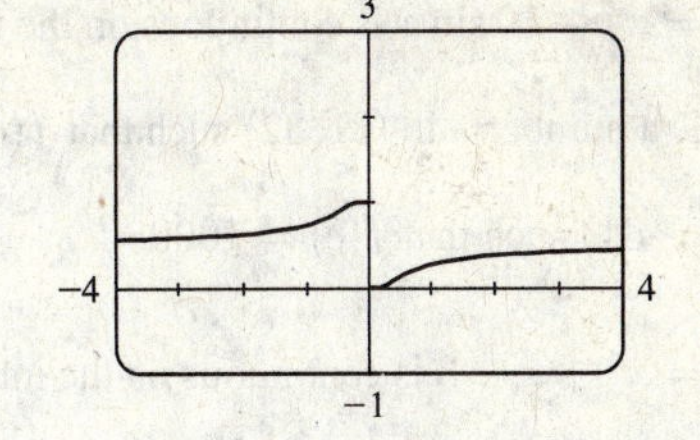

29. Because we are dealing with root functions, $5 + \sqrt{x}$ is continuous on $[0, \infty)$, $\sqrt{x+5}$ is continuous on $[-5, \infty)$, so the quotient $f(x) = \dfrac{5+\sqrt{x}}{\sqrt{5+x}}$ is continuous on $[0, \infty)$. Since f is continuous at $x = 4$, $\lim\limits_{x\to 4} f(x) = f(4) = \frac{7}{3}$.

31. Because $x^2 - x$ is continuous on $\mathbb{R}$, the composite function $f(x) = e^{x^2-x}$ is continuous on $\mathbb{R}$, so $\lim\limits_{x\to 1} f(x) = f(1) = e^{1-1} = e^0 = 1$.

33. $f(x) = \begin{cases} x^2 & \text{if } x < 1 \\ \sqrt{x} & \text{if } x \geq 1 \end{cases}$

By Theorem 5, since $f(x)$ equals the polynomial x^2 on $(-\infty, 1)$, f is continuous on $(-\infty, 1)$. By Theorem 7, since $f(x)$ equals the root function $\sqrt{x}$ on $(1, \infty)$, f is continuous on $(1, \infty)$. At $x = 1$, $\lim\limits_{x\to 1^-} f(x) = \lim\limits_{x\to 1^-} x^2 = 1$ and $\lim\limits_{x\to 1^+} f(x) = \lim\limits_{x\to 1^+} \sqrt{x} = 1$. Thus, $\lim\limits_{x\to 1} f(x)$ exists and equals 1. Also, $f(1) = \sqrt{1} = 1$. Thus, f is continuous at $x = 1$. We conclude that f is continuous on $(-\infty, \infty)$.

35. $f(x) = \begin{cases} x+2 & \text{if } x < 0 \\ e^x & \text{if } 0 \leq x \leq 1 \\ 2-x & \text{if } x > 1 \end{cases}$

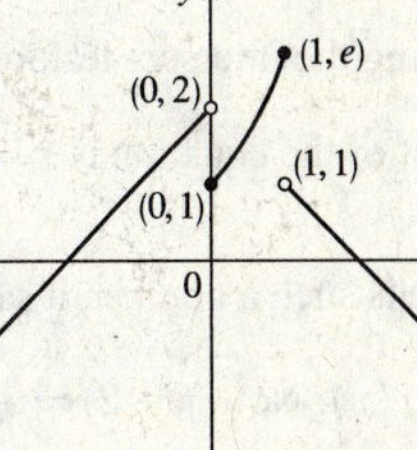

f is continuous on $(-\infty, 0)$ and $(1, \infty)$ since on each of these intervals it is a polynomial; it is continuous on $(0, 1)$ since it is an exponential.

Now $\lim\limits_{x\to 0^-} f(x) = \lim\limits_{x\to 0^-} (x+2) = 2$ and $\lim\limits_{x\to 0^+} f(x) = \lim\limits_{x\to 0^+} e^x = 1$, so f is discontinuous at 0. Since $f(0) = 1$, f is continuous from the right at 0. Also $\lim\limits_{x\to 1^-} f(x) = \lim\limits_{x\to 1^-} e^x = e$ and $\lim\limits_{x\to 1^+} f(x) = \lim\limits_{x\to 1^+} (2-x) = 1$, so f is discontinuous at 1. Since $f(1) = e$, f is continuous from the left at 1.

37. $f(x) = \begin{cases} cx^2 + 2x & \text{if } x < 2 \\ x^3 - cx & \text{if } x \geq 2 \end{cases}$

f is continuous on $(-\infty, 2)$ and $(2, \infty)$. Now $\lim\limits_{x\to 2^-} f(x) = \lim\limits_{x\to 2^-} (cx^2 + 2x) = 4c + 4$ and $\lim\limits_{x\to 2^+} f(x) = \lim\limits_{x\to 2^+} (x^3 - cx) = 8 - 2c$. So f is continuous $\Leftrightarrow$ $4c + 4 = 8 - 2c$ $\Leftrightarrow$ $6c = 4$ $\Leftrightarrow$ $c = \frac{2}{3}$. Thus, for f to be continuous on $(-\infty, \infty)$, $c = \frac{2}{3}$.

© 2016 Cengage Learning. All Rights Reserved. May not be scanned, copied or duplicated, or posted to a publicly accessible website, in whole or in part.

39. $f(x) = x^2 + 10\sin x$ is continuous on the interval $[31, 32]$, $f(31) \approx 957$, and $f(32) \approx 1030$. Since $957 < 1000 < 1030$, there is a number c in $(31, 32)$ such that $f(c) = 1000$ by the Intermediate Value Theorem. *Note:* There is also a number c in $(-32, -31)$ such that $f(c) = 1000$.

41. $f(x) = x^4 + x - 3$ is continuous on the interval $[1, 2]$, $f(1) = -1$, and $f(2) = 15$. Since $-1 < 0 < 15$, there is a number c in $(1, 2)$ such that $f(c) = 0$ by the Intermediate Value Theorem. Thus, there is a root of the equation $x^4 + x - 3 = 0$ in the interval $(1, 2)$.

43. The equation $e^x = 3 - 2x$ is equivalent to the equation $e^x + 2x - 3 = 0$. $f(x) = e^x + 2x - 3$ is continuous on the interval $[0, 1]$, $f(0) = -2$, and $f(1) = e - 1 \approx 1.72$. Since $-2 < 0 < e - 1$, there is a number c in $(0, 1)$ such that $f(c) = 0$ by the Intermediate Value Theorem. Thus, there is a root of the equation $e^x + 2x - 3 = 0$, or $e^x = 3 - 2x$, in the interval $(0, 1)$.

45. (a) $f(x) = \cos x - x^3$ is continuous on the interval $[0, 1]$, $f(0) = 1 > 0$, and $f(1) = \cos 1 - 1 \approx -0.46 < 0$. Since $1 > 0 > -0.46$, there is a number c in $(0, 1)$ such that $f(c) = 0$ by the Intermediate Value Theorem. Thus, there is a root of the equation $\cos x - x^3 = 0$, or $\cos x = x^3$, in the interval $(0, 1)$.

(b) $f(0.86) \approx 0.016 > 0$ and $f(0.87) \approx -0.014 < 0$, so there is a root between 0.86 and 0.87, that is, in the interval $(0.86, 0.87)$.

47. (a) Let $f(x) = 100e^{-x/100} - 0.01x^2$. Then $f(0) = 100 > 0$ and $f(100) = 100e^{-1} - 100 \approx -63.2 < 0$. So by the Intermediate Value Theorem, there is a number c in $(0, 100)$ such that $f(c) = 0$. This implies that $100e^{-c/100} = 0.01c^2$.

(b) Using the intersect feature of the graphing device, we find that the root of the equation is $x = 70.347$, correct to three decimal places.

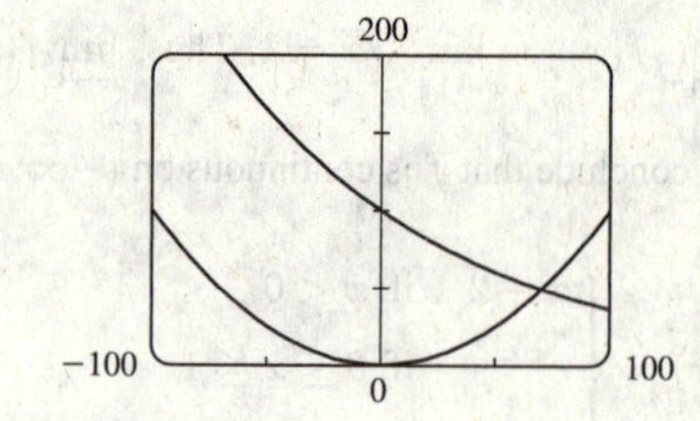

49. If there is such a number, it satisfies the equation $x^3 + 1 = x \Leftrightarrow x^3 - x + 1 = 0$. Let the left-hand side of this equation be called $f(x)$. Now $f(-2) = -5 < 0$, and $f(-1) = 1 > 0$. Note also that $f(x)$ is a polynomial, and thus continuous. So by the Intermediate Value Theorem, there is a number c between -2 and -1 such that $f(c) = 0$, so that $c = c^3 + 1$.

51. Define $u(t)$ to be the monk's distance from the monastery, as a function of time t (in hours), on the first day, and define $d(t)$ to be his distance from the monastery, as a function of time, on the second day. Let D be the distance from the monastery to the top of the mountain. From the given information we know that $u(0) = 0$, $u(12) = D$, $d(0) = D$ and $d(12) = 0$. Now consider the function $u - d$, which is clearly continuous. We calculate that $(u - d)(0) = -D$ and $(u - d)(12) = D$. So by the Intermediate Value Theorem, there must be some time t_0 between 0 and 12 such that $(u - d)(t_0) = 0 \Leftrightarrow u(t_0) = d(t_0)$. So at time t_0 after 7:00 AM, the monk will be at the same place on both days.

© 2016 Cengage Learning. All Rights Reserved. May not be scanned, copied or duplicated, or posted to a publicly accessible website, in whole or in part.

2 Review

TRUE-FALSE QUIZ

1. True. If $\lim_{n\to\infty} a_n = L$, then as $n \to \infty$, $2n+1 \to \infty$, so $a_{2n+1} \to L$.

3. False. Limit Law 2 applies only if the individual limits exist (these don't).

5. True. Limit Law 5 applies.

7. False. Consider $\lim_{x\to 5} \dfrac{x(x-5)}{x-5}$ or $\lim_{x\to 5} \dfrac{\sin(x-5)}{x-5}$. The first limit exists and is equal to 5. By Equation 6 in Section 2.4, we know that the latter limit exists (and it is equal to 1).

9. True. A polynomial is continuous everywhere, so $\lim_{x\to b} p(x)$ exists and is equal to $p(b)$.

11. True. For example, the function $f(x) = \dfrac{\sqrt{4x^2+1}}{x-5}$ has two different horizontal asymptotes since $\lim_{x\to\infty} f(x) = 2$ and $\lim_{x\to-\infty} f(x) = -2$. The horizontal asymptotes are $y = 2$ and $y = -2$.

13. False. Consider $f(x) = \begin{cases} 1/(x-1) & \text{if } x \neq 1 \\ 2 & \text{if } x = 1 \end{cases}$

15. True. Use Theorem 2.5.7 with $a = 2$, $b = 5$, and $g(x) = 4x^2 - 11$. Note that $f(4) = 3$ is not needed.

EXERCISES

1. $\left\{\dfrac{2+n^3}{1+2n^3}\right\}$ converges since $\lim_{n\to\infty} \dfrac{2+n^3}{1+2n^3} = \lim_{n\to\infty} \dfrac{2/n^3+1}{1/n^3+2} = \dfrac{1}{2}$.

3. $\lim_{n\to\infty} a_n = \lim_{n\to\infty} \dfrac{n^3}{1+n^2} = \lim_{n\to\infty} \dfrac{n}{1/n^2+1} = \infty$, so the sequence diverges.

5. $a_{n+1} = \frac{1}{3}a_n + 3$, $a_1 = 1$, $a_2 \approx 3.3333$, $a_3 \approx 4.1111$, $a_4 \approx 4.3704$, $a_5 \approx 4.4568$, $a_6 \approx 4.4856$, $a_7 \approx 4.4952$, $a_8 \approx 4.4984$

n	a_n	n	a_n
1	1.0000	5	4.4568
2	3.3333	6	4.4856
3	4.1111	7	4.4952
4	4.3704	8	4.4984

The sequence appears to converge to 4.5. Assume the limit exists so that $\lim_{n\to\infty} a_{n+1} = \lim_{n\to\infty} a_n = a$, then $a_{n+1} = \frac{1}{3}a_n + 3 \Rightarrow \lim_{n\to\infty} a_{n+1} = \lim_{n\to\infty} \left(\frac{1}{3}a_n + 3\right) \Rightarrow a = \frac{1}{3}a + 3 \Rightarrow a = \frac{9}{2} = 4.5$. This agrees with the value estimated from the data table.

7. $1.2345345345\ldots = 1.2 + 0.0\overline{345} = \dfrac{12}{10} + \dfrac{345/10{,}000}{1-1/1000} = \dfrac{12}{10} + \dfrac{345}{9990} = \dfrac{4111}{3330}$

9. (a) (i) $\lim_{x\to 2^+} f(x) = 3$ (ii) $\lim_{x\to -3^+} f(x) = 0$

(iii) $\lim_{x\to -3} f(x)$ does not exist since the left and right limits are not equal. (The left limit is -2.)

(iv) $\lim_{x\to 4} f(x) = 2$

(v) $\lim_{x\to 0} f(x) = \infty$ (vi) $\lim_{x\to 2^-} f(x) = -\infty$

(vii) $\lim_{x\to\infty} f(x) = 4$ (viii) $\lim_{x\to-\infty} f(x) = -1$

© 2016 Cengage Learning. All Rights Reserved. May not be scanned, copied or duplicated, or posted to a publicly accessible website, in whole or in part.

(b) The equations of the horizontal asymptotes are $y = -1$ and $y = 4$.

(c) The equations of the vertical asymptotes are $x = 0$ and $x = 2$.

(d) f is discontinuous at $x = -3, 0, 2$, and 4. The discontinuities are jump, infinite, infinite, and removable, respectively.

11. $\displaystyle \lim_{x\to\infty} \frac{1-x}{2+5x} = \lim_{x\to\infty} \frac{1/x-1}{2/x+5} = \lim_{x\to\infty} \frac{0-1}{0+5} = -\frac{1}{5}$

13. Since the exponential function is continuous, $\displaystyle \lim_{x\to 1} e^{x^3-x} = e^{1-1} = e^0 = 1$.

15. $\displaystyle \lim_{x\to -3} \frac{x^2-9}{x^2+2x-3} = \lim_{x\to -3} \frac{(x+3)(x-3)}{(x+3)(x-1)} = \lim_{x\to -3} \frac{x-3}{x-1} = \frac{-3-3}{-3-1} = \frac{-6}{-4} = \frac{3}{2}$

17. $\displaystyle \lim_{h\to 0} \frac{(h-1)^3+1}{h} = \lim_{h\to 0} \frac{(h^3-3h^2+3h-1)+1}{h} = \lim_{h\to 0} \frac{h^3-3h^2+3h}{h} = \lim_{h\to 0} (h^2-3h+3) = 3$

Another solution: Factor the numerator as a sum of two cubes and then simplify.

$$\begin{aligned}\lim_{h\to 0} \frac{(h-1)^3+1}{h} &= \lim_{h\to 0} \frac{(h-1)^3+1^3}{h} = \lim_{h\to 0} \frac{[(h-1)+1]\left[(h-1)^2-1(h-1)+1^2\right]}{h} \\ &= \lim_{h\to 0} \left[(h-1)^2-h+2\right] = 1-0+2 = 3\end{aligned}$$

19. $\displaystyle \lim_{r\to 9} \frac{\sqrt{r}}{(r-9)^4} = \infty$ since $(r-9)^4 \to 0^+$ as $r \to 9$ and $\dfrac{\sqrt{r}}{(r-9)^4} > 0$ for $r \neq 9$.

21. $\displaystyle \lim_{u\to 1} \frac{u^4-1}{u^3+5u^2-6u} = \lim_{u\to 1} \frac{(u^2+1)(u^2-1)}{u(u^2+5u-6)} = \lim_{u\to 1} \frac{(u^2+1)(u+1)(u-1)}{u(u+6)(u-1)} = \lim_{u\to 1} \frac{(u^2+1)(u+1)}{u(u+6)} = \frac{2(2)}{1(7)} = \frac{4}{7}$

23. Let $t = \sin x$. Then as $x \to \pi^-$, $\sin x \to 0^+$, so $t \to 0^+$. Thus, $\displaystyle \lim_{x\to \pi^-} \ln(\sin x) = \lim_{t\to 0^+} \ln t = -\infty$.

25. Since x is positive, $\sqrt{x^2} = |x| = x$. Thus,

$$\lim_{x\to\infty} \frac{\sqrt{x^2-9}}{2x-6} = \lim_{x\to\infty} \frac{\sqrt{x^2-9}/\sqrt{x^2}}{(2x-6)/x} = \lim_{x\to\infty} \frac{\sqrt{1-9/x^2}}{2-6/x} = \frac{\sqrt{1-0}}{2-0} = \frac{1}{2}$$

27.
$$\begin{aligned}\lim_{x\to\infty} \left(\sqrt{x^2+4x+1}-x\right) &= \lim_{x\to\infty} \left[\frac{\sqrt{x^2+4x+1}-x}{1} \cdot \frac{\sqrt{x^2+4x+1}+x}{\sqrt{x^2+4x+1}+x}\right] = \lim_{x\to\infty} \frac{(x^2+4x+1)-x^2}{\sqrt{x^2+4x+1}+x} \\ &= \lim_{x\to\infty} \frac{(4x+1)/x}{\left(\sqrt{x^2+4x+1}+x\right)/x} \qquad \left[\text{divide by } x = \sqrt{x^2} \text{ for } x > 0\right] \\ &= \lim_{x\to\infty} \frac{4+1/x}{\sqrt{1+4/x+1/x^2}+1} = \frac{4+0}{\sqrt{1+0+0}+1} = \frac{4}{2} = 2\end{aligned}$$

29. $\displaystyle \lim_{[\mathrm{S}]\to\infty} v = \lim_{[\mathrm{S}]\to\infty} \frac{0.50[\mathrm{S}]}{3.0\times 10^{-4}+[\mathrm{S}]} = \lim_{[\mathrm{S}]\to\infty} \frac{0.50}{3.0\times 10^{-4}/[\mathrm{S}]+1} = \frac{0.50}{0+1} = 0.50$. As the concentration grows larger the enzymatic reaction rate will approach 0.50.

© 2016 Cengage Learning. All Rights Reserved. May not be scanned, copied or duplicated, or posted to a publicly accessible website, in whole or in part.

31. (a) $f(x) = \sqrt{-x}$ if $x < 0$, $f(x) = 3 - x$ if $0 \le x < 3$, $f(x) = (x-3)^2$ if $x > 3$.

(i) $\lim\limits_{x\to 0^+} f(x) = \lim\limits_{x\to 0^+} (3 - x) = 3$

(ii) $\lim\limits_{x\to 0^-} f(x) = \lim\limits_{x\to 0^-} \sqrt{-x} = 0$

(iii) Because of (i) and (ii), $\lim\limits_{x\to 0} f(x)$ does not exist.

(iv) $\lim\limits_{x\to 3^-} f(x) = \lim\limits_{x\to 3^-} (3 - x) = 0$

(v) $\lim\limits_{x\to 3^+} f(x) = \lim\limits_{x\to 3^+} (x - 3)^2 = 0$

(vi) Because of (iv) and (v), $\lim\limits_{x\to 3} f(x) = 0$.

(b) f is discontinuous at 0 since $\lim\limits_{x\to 0} f(x)$ does not exist.

f is discontinuous at 3 since $f(3)$ does not exist.

(c)

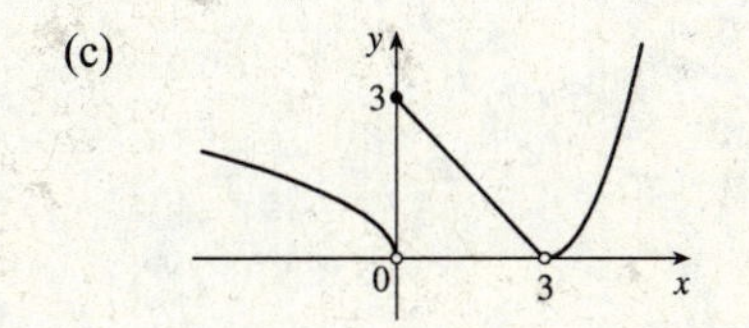

33. $f(x) = 2x^3 + x^2 + 2$ is a polynomial, so it is continuous on $[-2, -1]$ and $f(-2) = -10 < 0 < 1 = f(-1)$. So by the Intermediate Value Theorem there is a number c in $(-2, -1)$ such that $f(c) = 0$, that is, the equation $2x^3 + x^2 + 2 = 0$ has a root in $(-2, -1)$.

© 2016 Cengage Learning. All Rights Reserved. May not be scanned, copied or duplicated, or posted to a publicly accessible website, in whole or in part.

3 □ DERIVATIVES

3.1 Derivatives and Rates of Change

1. (a) This is just the slope of the line through two points: $m_{PQ} = \dfrac{\Delta y}{\Delta x} = \dfrac{f(x) - f(3)}{x - 3}$.

(b) This is the limit of the slope of the secant line PQ as Q approaches P: $m = \lim\limits_{x \to 3} \dfrac{f(x) - f(3)}{x - 3}$.

3. (a) (i) Using Definition 2 with $f(x) = 4x - x^2$ and $P(1, 3)$,

$$m = \lim_{x \to a} \frac{f(x) - f(a)}{x - a} = \lim_{x \to 1} \frac{(4x - x^2) - 3}{x - 1} = \lim_{x \to 1} \frac{-(x^2 - 4x + 3)}{x - 1} = \lim_{x \to 1} \frac{-(x - 1)(x - 3)}{x - 1}$$

$$= \lim_{x \to 1} (3 - x) = 3 - 1 = 2$$

(ii) Using Equation 3 with $f(x) = 4x - x^2$ and $P(1, 3)$,

$$m = \lim_{h \to 0} \frac{f(a + h) - f(a)}{h} = \lim_{h \to 0} \frac{f(1 + h) - f(1)}{h} = \lim_{h \to 0} \frac{\left[4(1 + h) - (1 + h)^2\right] - 3}{h}$$

$$= \lim_{h \to 0} \frac{4 + 4h - 1 - 2h - h^2 - 3}{h} = \lim_{h \to 0} \frac{-h^2 + 2h}{h} = \lim_{h \to 0} \frac{h(-h + 2)}{h} = \lim_{h \to 0} (-h + 2) = 2$$

(b) An equation of the tangent line is $y - f(a) = f'(a)(x - a) \;\Rightarrow\; y - f(1) = f'(1)(x - 1) \;\Rightarrow\; y - 3 = 2(x - 1)$, or $y = 2x + 1$.

(c)

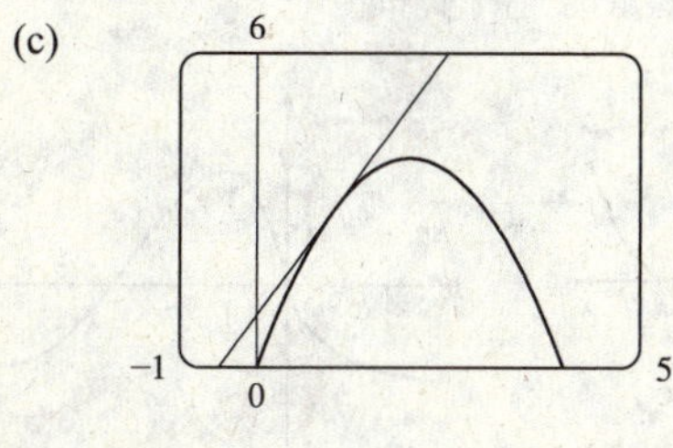

The graph of $y = 2x + 1$ is tangent to the graph of $y = 4x - x^2$ at the point $(1, 3)$. Now zoom in toward the point $(1, 3)$ until the parabola and the tangent line are indistiguishable.

5. Using (2) with $f(x) = 4x - 3x^2$ and $P(2, -4)$ [we could also use (3)],

$$m = \lim_{x \to a} \frac{f(x) - f(a)}{x - a} = \lim_{x \to 2} \frac{(4x - 3x^2) - (-4)}{x - 2} = \lim_{x \to 2} \frac{-3x^2 + 4x + 4}{x - 2}$$

$$= \lim_{x \to 2} \frac{(-3x - 2)(x - 2)}{x - 2} = \lim_{x \to 2} (-3x - 2) = -3(2) - 2 = -8$$

Tangent line: $y - (-4) = -8(x - 2) \;\Leftrightarrow\; y + 4 = -8x + 16 \;\Leftrightarrow\; y = -8x + 12$.

7. Using (2), $m = \lim\limits_{x \to 1} \dfrac{\sqrt{x} - \sqrt{1}}{x - 1} = \lim\limits_{x \to 1} \dfrac{(\sqrt{x} - 1)(\sqrt{x} + 1)}{(x - 1)(\sqrt{x} + 1)} = \lim\limits_{x \to 1} \dfrac{x - 1}{(x - 1)(\sqrt{x} + 1)} = \lim\limits_{x \to 1} \dfrac{1}{\sqrt{x} + 1} = \dfrac{1}{2}$.

Tangent line: $y - 1 = \frac{1}{2}(x - 1) \;\Leftrightarrow\; y = \frac{1}{2}x + \frac{1}{2}$

© 2016 Cengage Learning. All Rights Reserved. May not be scanned, copied or duplicated, or posted to a publicly accessible website, in whole or in part.

9. (a) Using (3) with $y = f(x) = 3 + 4x^2 - 2x^3$,

$$\begin{aligned}
m &= \lim_{h\to 0}\frac{f(a+h)-f(a)}{h} = \lim_{h\to 0}\frac{3+4(a+h)^2-2(a+h)^3-(3+4a^2-2a^3)}{h} \\
&= \lim_{h\to 0}\frac{3+4(a^2+2ah+h^2)-2(a^3+3a^2h+3ah^2+h^3)-3-4a^2+2a^3}{h} \\
&= \lim_{h\to 0}\frac{3+4a^2+8ah+4h^2-2a^3-6a^2h-6ah^2-2h^3-3-4a^2+2a^3}{h} \\
&= \lim_{h\to 0}\frac{8ah+4h^2-6a^2h-6ah^2-2h^3}{h} = \lim_{h\to 0}\frac{h(8a+4h-6a^2-6ah-2h^2)}{h} \\
&= \lim_{h\to 0}(8a+4h-6a^2-6ah-2h^2) = 8a-6a^2
\end{aligned}$$

(b) At $(1, 5)$: $m = 8(1) - 6(1)^2 = 2$, so an equation of the tangent line is $y - 5 = 2(x - 1) \quad\Leftrightarrow\quad y = 2x + 3$.

At $(2, 3)$: $m = 8(2) - 6(2)^2 = -8$, so an equation of the tangent line is $y - 3 = -8(x - 2) \quad\Leftrightarrow\quad y = -8x + 19$.

(c)

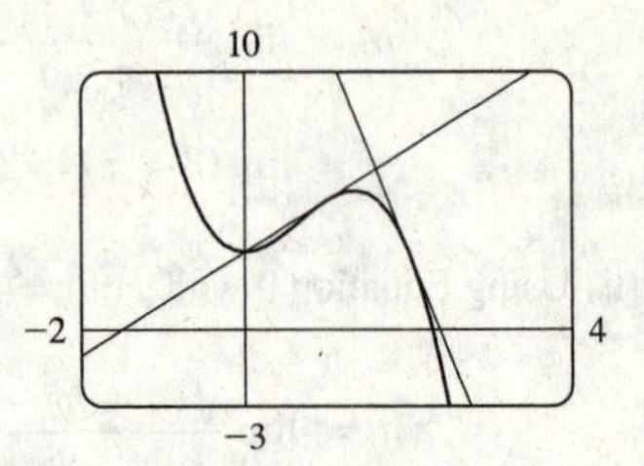

11. $g'(0)$ is the only negative value. The slope at $x = 4$ is smaller than the slope at $x = 2$ and both are smaller than the slope at $x = -2$. Thus, $g'(0) < 0 < g'(4) < g'(2) < g'(-2)$.

13. For the tangent line $y = 4x - 5$: when $x = 2$, $y = 4(2) - 5 = 3$ and its slope is 4 (the coefficient of x). At the point of tangency, these values are shared with the curve $y = f(x)$; that is, $f(2) = 3$ and $f'(2) = 4$.

15. We begin by drawing a curve through the origin with a slope of 3 to satisfy $f(0) = 0$ and $f'(0) = 3$. Since $f'(1) = 0$, we will round off our figure so that there is a horizontal tangent directly over $x = 1$. Last, we make sure that the curve has a slope of -1 as we pass over $x = 2$. Two of the many possibilities are shown.

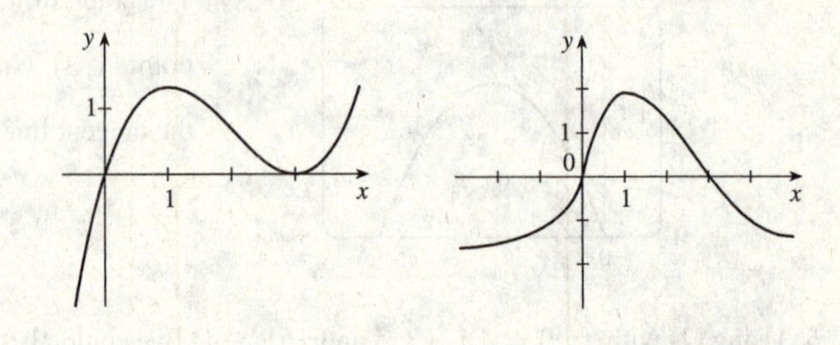

17. Using (4) with $f(x) = 3x^2 - x^3$ and $a = 1$,

$$\begin{aligned}
f'(1) &= \lim_{h\to 0}\frac{f(1+h)-f(1)}{h} = \lim_{h\to 0}\frac{[3(1+h)^2-(1+h)^3]-2}{h} \\
&= \lim_{h\to 0}\frac{(3+6h+3h^2)-(1+3h+3h^2+h^3)-2}{h} = \lim_{h\to 0}\frac{3h-h^3}{h} = \lim_{h\to 0}\frac{h(3-h^2)}{h} \\
&= \lim_{h\to 0}(3-h^2) = 3-0 = 3
\end{aligned}$$

Tangent line: $y - 2 = 3(x - 1) \quad\Leftrightarrow\quad y - 2 = 3x - 3 \quad\Leftrightarrow\quad y = 3x - 1$

© 2016 Cengage Learning. All Rights Reserved. May not be scanned, copied or duplicated, or posted to a publicly accessible website, in whole or in part.

19. (a) Using (4) with $F(x) = 5x/(1+x^2)$ and the point $(2, 2)$, we have

(b)

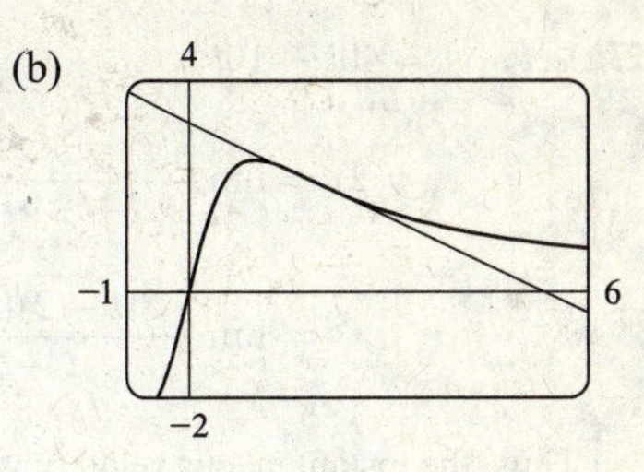

$$F'(2) = \lim_{h\to 0}\frac{F(2+h)-F(2)}{h} = \lim_{h\to 0}\frac{\dfrac{5(2+h)}{1+(2+h)^2}-2}{h}$$

$$= \lim_{h\to 0}\frac{\dfrac{5h+10}{h^2+4h+5}-2}{h} = \lim_{h\to 0}\frac{\dfrac{5h+10-2(h^2+4h+5)}{h^2+4h+5}}{h}$$

$$= \lim_{h\to 0}\frac{-2h^2-3h}{h(h^2+4h+5)} = \lim_{h\to 0}\frac{h(-2h-3)}{h(h^2+4h+5)} = \lim_{h\to 0}\frac{-2h-3}{h^2+4h+5} = \frac{-3}{5}$$

So an equation of the tangent line at $(2, 2)$ is $y - 2 = -\frac{3}{5}(x-2)$ or $y = -\frac{3}{5}x + \frac{16}{5}$.

21. Use (4) with $f(x) = 3x^2 - 4x + 1$.

$$f'(a) = \lim_{h\to 0}\frac{f(a+h)-f(a)}{h} = \lim_{h\to 0}\frac{[3(a+h)^2-4(a+h)+1]-(3a^2-4a+1)]}{h}$$

$$= \lim_{h\to 0}\frac{3a^2+6ah+3h^2-4a-4h+1-3a^2+4a-1}{h} = \lim_{h\to 0}\frac{6ah+3h^2-4h}{h}$$

$$= \lim_{h\to 0}\frac{h(6a+3h-4)}{h} = \lim_{h\to 0}(6a+3h-4) = 6a-4$$

23. Use (4) with $f(t) = (2t+1)/(t+3)$.

$$f'(a) = \lim_{h\to 0}\frac{f(a+h)-f(a)}{h} = \lim_{h\to 0}\frac{\dfrac{2(a+h)+1}{(a+h)+3}-\dfrac{2a+1}{a+3}}{h} = \lim_{h\to 0}\frac{(2a+2h+1)(a+3)-(2a+1)(a+h+3)}{h(a+h+3)(a+3)}$$

$$= \lim_{h\to 0}\frac{(2a^2+6a+2ah+6h+a+3)-(2a^2+2ah+6a+a+h+3)}{h(a+h+3)(a+3)}$$

$$= \lim_{h\to 0}\frac{5h}{h(a+h+3)(a+3)} = \lim_{h\to 0}\frac{5}{(a+h+3)(a+3)} = \frac{5}{(a+3)^2}$$

25. Use (4) with $f(x) = \sqrt{1-2x}$.

$$f'(a) = \lim_{h\to 0}\frac{f(a+h)-f(a)}{h} = \lim_{h\to 0}\frac{\sqrt{1-2(a+h)}-\sqrt{1-2a}}{h}$$

$$= \lim_{h\to 0}\frac{\sqrt{1-2(a+h)}-\sqrt{1-2a}}{h}\cdot\frac{\sqrt{1-2(a+h)}+\sqrt{1-2a}}{\sqrt{1-2(a+h)}+\sqrt{1-2a}} = \lim_{h\to 0}\frac{\left(\sqrt{1-2(a+h)}\right)^2-\left(\sqrt{1-2a}\right)^2}{h\left(\sqrt{1-2(a+h)}+\sqrt{1-2a}\right)}$$

$$= \lim_{h\to 0}\frac{(1-2a-2h)-(1-2a)}{h\left(\sqrt{1-2(a+h)}+\sqrt{1-2a}\right)} = \lim_{h\to 0}\frac{-2h}{h\left(\sqrt{1-2(a+h)}+\sqrt{1-2a}\right)}$$

$$= \lim_{h\to 0}\frac{-2}{\sqrt{1-2(a+h)}+\sqrt{1-2a}} = \frac{-2}{\sqrt{1-2a}+\sqrt{1-2a}} = \frac{-2}{2\sqrt{1-2a}} = \frac{-1}{\sqrt{1-2a}}$$

© 2016 Cengage Learning. All Rights Reserved. May not be scanned, copied or duplicated, or posted to a publicly accessible website, in whole or in part.

27. Let $s(t) = 40t - 16t^2$.

$$v(2) = \lim_{t \to 2} \frac{s(t) - s(2)}{t - 2} = \lim_{t \to 2} \frac{(40t - 16t^2) - 16}{t - 2} = \lim_{t \to 2} \frac{-16t^2 + 40t - 16}{t - 2} = \lim_{t \to 2} \frac{-8(2t^2 - 5t + 2)}{t - 2}$$

$$= \lim_{t \to 2} \frac{-8(t - 2)(2t - 1)}{t - 2} = -8 \lim_{t \to 2}(2t - 1) = -8(3) = -24$$

Thus, the instantaneous velocity when $t = 2$ is -24 ft/s.

29. $$v(a) = \lim_{h \to 0} \frac{s(a + h) - s(a)}{h} = \lim_{h \to 0} \frac{\dfrac{1}{(a + h)^2} - \dfrac{1}{a^2}}{h} = \lim_{h \to 0} \frac{\dfrac{a^2 - (a + h)^2}{a^2(a + h)^2}}{h} = \lim_{h \to 0} \frac{a^2 - (a^2 + 2ah + h^2)}{ha^2(a + h)^2}$$

$$= \lim_{h \to 0} \frac{-(2ah + h^2)}{ha^2(a + h)^2} = \lim_{h \to 0} \frac{-h(2a + h)}{ha^2(a + h)^2} = \lim_{h \to 0} \frac{-(2a + h)}{a^2(a + h)^2} = \frac{-2a}{a^2 \cdot a^2} = \frac{-2}{a^3} \text{ m/s}$$

So $v(1) = \frac{-2}{1^3} = -2$ m/s, $v(2) = \frac{-2}{2^3} = -\frac{1}{4}$ m/s, and $v(3) = \frac{-2}{3^3} = -\frac{2}{27}$ m/s.

31. (a) (i) Average rate of change over $[1990, 2000] = \dfrac{P(2000) - P(1990)}{2000 - 1990} = \dfrac{282.2 - 249.6}{10} = 3.26$ million/year

(ii) Average rate of change over $[1995, 2000] = \dfrac{P(2000) - P(1995)}{2000 - 1995} = \dfrac{282.2 - 266.3}{5} = 3.18$ million/year

(iii) Average rate of change over $[2000, 2005] = \dfrac{P(2005) - P(2000)}{2005 - 2000} = \dfrac{295.8 - 282.2}{5} = 2.72$ million/year

(iv) Average rate of change over $[2000, 2010] = \dfrac{P(2010) - P(2000)}{2010 - 2000} = \dfrac{308.3 - 282.2}{10} = 2.61$ million/year

(b) $P'(2000)$ lies in between the average rates of change over $[1995, 2000]$ and $[2000, 2005]$, that is, in between 2.72 and 3.18. Using the average of these values as an estimate gives $P'(2000) \approx (2.72 + 3.18)/2 = 2.95$ million/year. Hence, the US population was growing at a rate of approximately 3 million people per year in 2000.

33. (a) (i) Average rate of change over $[1.0, 2.0] = \dfrac{C(2.0) - C(1.0)}{2.0 - 1.0} = \dfrac{0.18 - 0.33}{1} = -0.15$ (mg/mL)/hour

(ii) Average rate of change over $[1.5, 2.0] = \dfrac{C(2.0) - C(1.5)}{2.0 - 1.5} = \dfrac{0.18 - 0.24}{0.5} = -0.12$ (mg/mL)/hour

(iii) Average rate of change over $[2.0, 2.5] = \dfrac{C(2.5) - C(2.0)}{2.5 - 2.0} = \dfrac{0.12 - 0.18}{0.5} = -0.12$ (mg/mL)/hour

(iv) Average rate of change over $[2.0, 3.0] = \dfrac{C(3.0) - C(2.0)}{3.0 - 2.0} = \dfrac{0.07 - 0.18}{1} = -0.11$ (mg/mL)/hour

The units are measured in mg/mL per hour.

(b) $C'(2.0)$ lies in between the average rates of change over $[1.5, 2.0]$ and $[2.0, 2.5]$, that is, in between -0.12 and -0.12. Therefore $C'(2.0) \approx -0.12$ (mg/mL)/hour meaning that the blood alcohol concentration was decreasing at a rate of approximately 0.12 mg/mL per hour after 2.0 hours.

35. $T'(12)$ is the rate of change of temperature 12 hours after midnight on May 7, 2012. $T'(12)$ can be estimated using the average rates of of change over $[10, 12]$ and $[12, 14]$. These are $\dfrac{62 - 57}{12 - 10} = 2.5$ and $\dfrac{68 - 62}{14 - 12} = 3$, respectively. Computing the average of these two values gives the estimate $T'(12) \approx (2.5 + 3)/2 = 2.75\,^\circ\text{F/hour}$.

© 2016 Cengage Learning. All Rights Reserved. May not be scanned, copied or duplicated, or posted to a publicly accessible website, in whole or in part.

37. The sketch shows the graph for a room temperature of 72° and a refrigerator temperature of 38°. The initial rate of change is greater in magnitude than the rate of change after an hour.

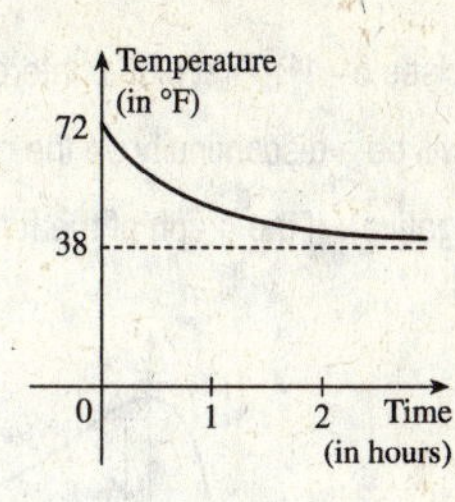

39. (a) $f'(x)$ is the rate of change of the production cost with respect to the number of ounces of gold produced. Its units are dollars per ounce.

(b) After 800 ounces of gold have been produced, the rate at which the production cost is increasing is \$17/ounce. So the cost of producing the 800th (or 801st) ounce is about \$17.

(c) In the short term, the values of $f'(x)$ will decrease because more efficient use is made of start-up costs as x increases. But eventually $f'(x)$ might increase due to large-scale operations.

41. (a) $S'(T)$ is the rate at which the oxygen solubility changes with respect to the water temperature. Its units are (mg/L)/°C.

(b) For $T = 16°$C, it appears that the tangent line to the curve goes through the points $(0, 14)$ and $(32, 6)$. So $S'(16) \approx \dfrac{6 - 14}{32 - 0} = -\dfrac{8}{32} = -0.25$ (mg/L)/°C. This means that as the temperature increases past 16°C, the oxygen solubility is decreasing at a rate of 0.25 (mg/L)/°C.

3.2 The Derivative as a Function

1. It appears that f is an odd function, so f' will be an even function—that is, $f'(-a) = f'(a)$.

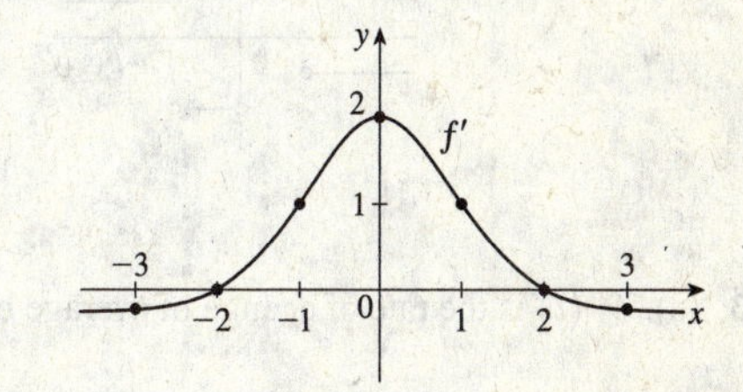

(a) $f'(-3) \approx -0.2$

(b) $f'(-2) \approx 0$ (c) $f'(-1) \approx 1$ (d) $f'(0) \approx 2$

(e) $f'(1) \approx 1$ (f) $f'(2) \approx 0$ (g) $f'(3) \approx -0.2$

3. (a)$'$ = II, since from left to right, the slopes of the tangents to graph (a) start out negative, become 0, then positive, then 0, then negative again. The actual function values in graph II follow the same pattern.

(b)$'$ = IV, since from left to right, the slopes of the tangents to graph (b) start out at a fixed positive quantity, then suddenly become negative, then positive again. The discontinuities in graph IV indicate sudden changes in the slopes of the tangents.

(c)$'$ = I, since the slopes of the tangents to graph (c) are negative for $x < 0$ and positive for $x > 0$, as are the function values of graph I.

(d)$'$ = III, since from left to right, the slopes of the tangents to graph (d) are positive, then 0, then negative, then 0, then positive, then 0, then negative again, and the function values in graph III follow the same pattern.

© 2016 Cengage Learning. All Rights Reserved. May not be scanned, copied or duplicated, or posted to a publicly accessible website, in whole or in part.

Hints for Exercises 5–11: First plot x-intercepts on the graph of f' for any horizontal tangents on the graph of f. Look for any corners on the graph of f—there will be a discontinuity on the graph of f'. On any interval where f has a tangent with positive (or negative) slope, the graph of f' will be positive (or negative). If the graph of the function is linear, the graph of f' will be a horizontal line.

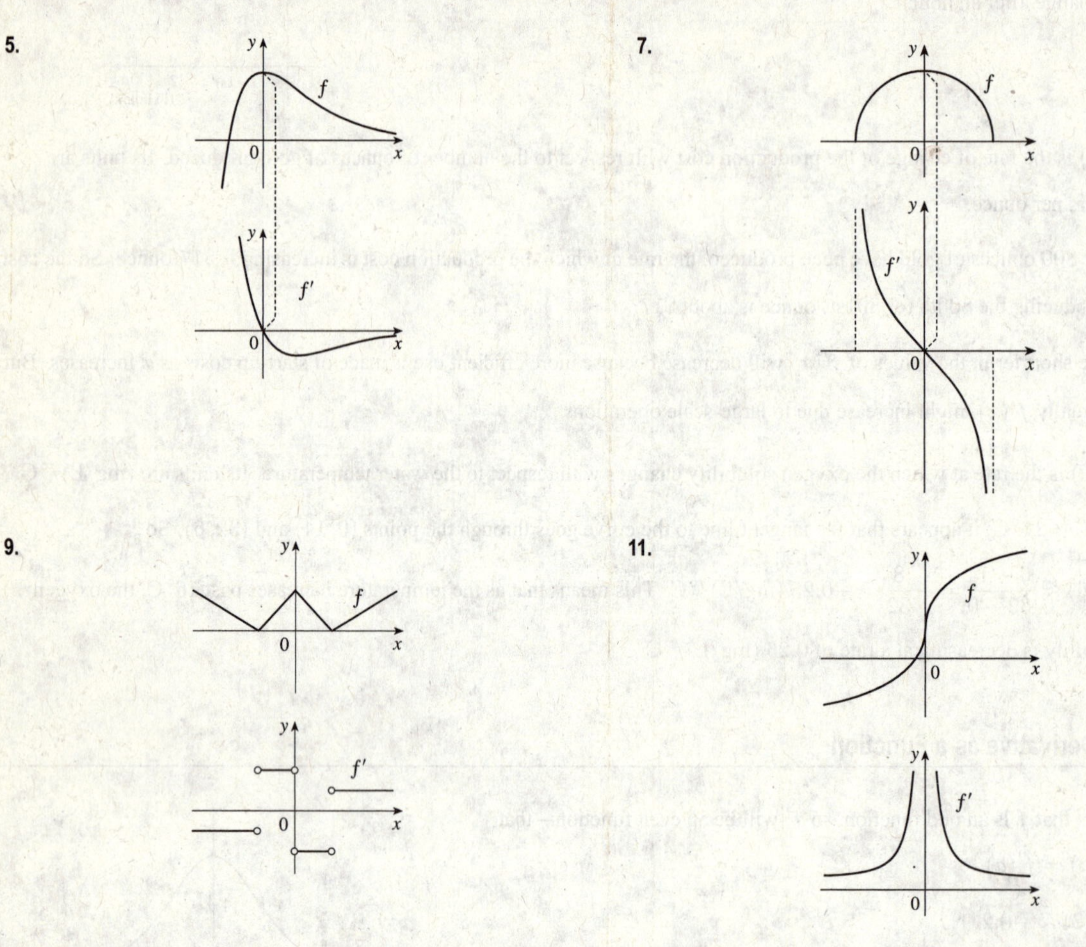

13. (a) $W'(t)$ is the rate of change of average body weight with respect to time for tadpoles raised in a density of 80 tadpoles/L.

(b) The slopes of the tangent lines on the graph of $y = W(t)$ start out relatively small and keep increasing, reaching a maximum at about $t = 4$. Then the slope decreases until it reaches zero at about $t = 6.5$; The function $y = W(t)$ obtains its maximum value at this point. Finally, the curve decreases making the slope negative for $t > 6.5$.

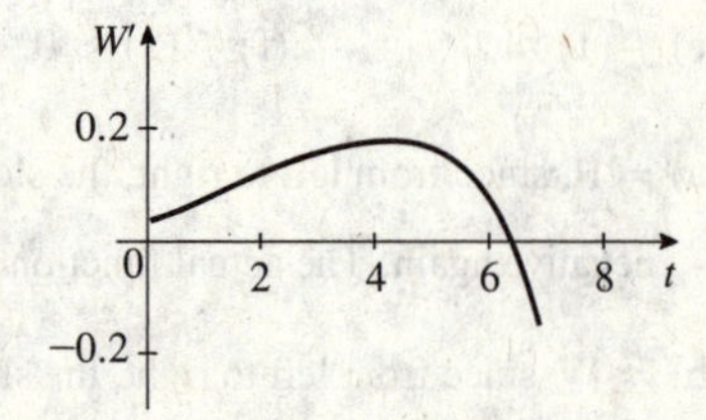

15. It appears that there are horizontal tangents on the graph of M for $t = 1963$ and $t = 1971$. Thus, there are zeros for those values of t on the graph of M'. The derivative is negative for the years 1963 to 1971.

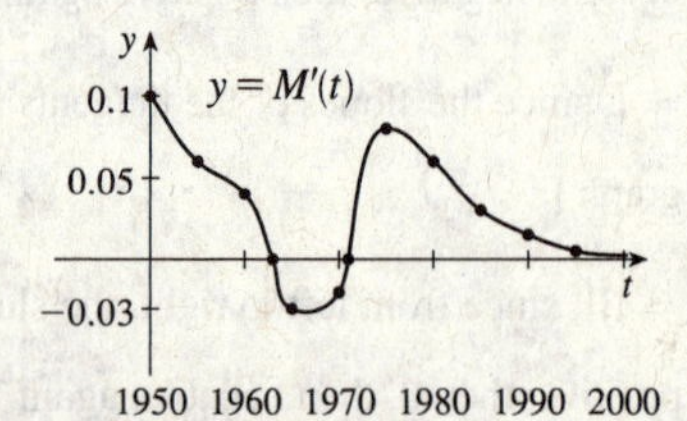

© 2016 Cengage Learning. All Rights Reserved. May not be scanned, copied or duplicated, or posted to a publicly accessible website, in whole or in part.

17.

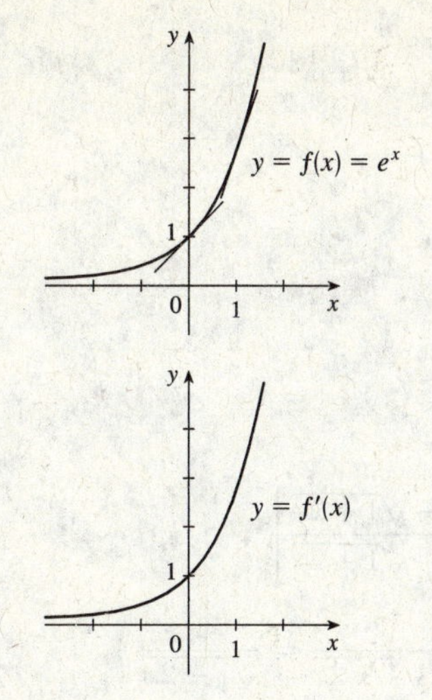

The slope at 0 appears to be 1 and the slope at 1 appears to be 2.7. As x decreases, the slope gets closer to 0. Since the graphs are so similar, we might guess that $f'(x) = e^x$.

19. (a) By zooming in, we estimate that $f'(0) = 0$, $f'\left(\frac{1}{2}\right) = 1$, $f'(1) = 2$, and $f'(2) = 4$.

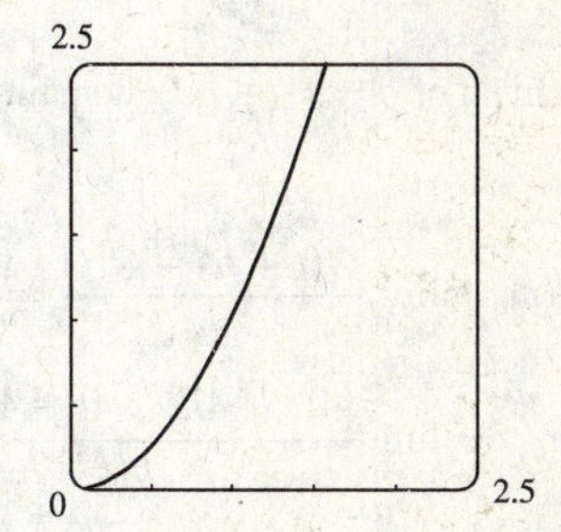

(b) By symmetry, $f'(-x) = -f'(x)$. So $f'\left(-\frac{1}{2}\right) = -1$, $f'(-1) = -2$, and $f'(-2) = -4$.

(c) It appears that $f'(x)$ is twice the value of x, so we guess that $f'(x) = 2x$.

(d) $$f'(x) = \lim_{h\to 0} \frac{f(x+h) - f(x)}{h} = \lim_{h\to 0} \frac{(x+h)^2 - x^2}{h}$$
$$= \lim_{h\to 0} \frac{(x^2 + 2hx + h^2) - x^2}{h} = \lim_{h\to 0} \frac{2hx + h^2}{h} = \lim_{h\to 0} \frac{h(2x+h)}{h} = \lim_{h\to 0}(2x+h) = 2x$$

21. $$f'(x) = \lim_{h\to 0} \frac{f(x+h) - f(x)}{h} = \lim_{h\to 0} \frac{\left[\frac{1}{2}(x+h) - \frac{1}{3}\right] - \left(\frac{1}{2}x - \frac{1}{3}\right)}{h} = \lim_{h\to 0} \frac{\frac{1}{2}x + \frac{1}{2}h - \frac{1}{3} - \frac{1}{2}x + \frac{1}{3}}{h}$$
$$= \lim_{h\to 0} \frac{\frac{1}{2}h}{h} = \lim_{h\to 0} \tfrac{1}{2} = \tfrac{1}{2}$$

Domain of f = domain of $f' = \mathbb{R}$.

23. $$f'(t) = \lim_{h\to 0} \frac{f(t+h) - f(t)}{h} = \lim_{h\to 0} \frac{\left[5(t+h) - 9(t+h)^2\right] - (5t - 9t^2)}{h}$$
$$= \lim_{h\to 0} \frac{5t + 5h - 9(t^2 + 2th + h^2) - 5t + 9t^2}{h} = \lim_{h\to 0} \frac{5t + 5h - 9t^2 - 18th - 9h^2 - 5t + 9t^2}{h}$$
$$= \lim_{h\to 0} \frac{5h - 18th - 9h^2}{h} = \lim_{h\to 0} \frac{h(5 - 18t - 9h)}{h} = \lim_{h\to 0}(5 - 18t - 9h) = 5 - 18t$$

Domain of f = domain of $f' = \mathbb{R}$.

© 2016 Cengage Learning. All Rights Reserved. May not be scanned, copied or duplicated, or posted to a publicly accessible website, in whole or in part.

25. $f'(x) = \lim\limits_{h\to 0} \dfrac{f(x+h)-f(x)}{h} = \lim\limits_{h\to 0} \dfrac{[(x+h)^2 - 2(x+h)^3] - (x^2 - 2x^3)}{h}$

$= \lim\limits_{h\to 0} \dfrac{x^2 + 2xh + h^2 - 2x^3 - 6x^2h - 6xh^2 - 2h^3 - x^2 + 2x^3}{h}$

$= \lim\limits_{h\to 0} \dfrac{2xh + h^2 - 6x^2h - 6xh^2 - 2h^3}{h} = \lim\limits_{h\to 0} \dfrac{h(2x + h - 6x^2 - 6xh - 2h^2)}{h}$

$= \lim\limits_{h\to 0} (2x + h - 6x^2 - 6xh - 2h^2) = 2x - 6x^2$

Domain of f = domain of $f' = \mathbb{R}$.

27. $g'(x) = \lim\limits_{h\to 0} \dfrac{g(x+h)-g(x)}{h} = \lim\limits_{h\to 0} \dfrac{\sqrt{1+2(x+h)} - \sqrt{1+2x}}{h} \left[\dfrac{\sqrt{1+2(x+h)} + \sqrt{1+2x}}{\sqrt{1+2(x+h)} + \sqrt{1+2x}}\right]$

$= \lim\limits_{h\to 0} \dfrac{(1+2x+2h) - (1+2x)}{h\left[\sqrt{1+2(x+h)} + \sqrt{1+2x}\right]} = \lim\limits_{h\to 0} \dfrac{2}{\sqrt{1+2x+2h} + \sqrt{1+2x}} = \dfrac{2}{2\sqrt{1+2x}} = \dfrac{1}{\sqrt{1+2x}}$

Domain of $g = \left[-\frac{1}{2}, \infty\right)$, domain of $g' = \left(-\frac{1}{2}, \infty\right)$.

29. $G'(t) = \lim\limits_{h\to 0} \dfrac{G(t+h)-G(t)}{h} = \lim\limits_{h\to 0} \dfrac{\dfrac{4(t+h)}{(t+h)+1} - \dfrac{4t}{t+1}}{h} = \lim\limits_{h\to 0} \dfrac{\dfrac{4(t+h)(t+1) - 4t(t+h+1)}{(t+h+1)(t+1)}}{h}$

$= \lim\limits_{h\to 0} \dfrac{(4t^2 + 4ht + 4t + 4h) - (4t^2 + 4ht + 4t)}{h(t+h+1)(t+1)} = \lim\limits_{h\to 0} \dfrac{4h}{h(t+h+1)(t+1)}$

$= \lim\limits_{h\to 0} \dfrac{4}{(t+h+1)(t+1)} = \dfrac{4}{(t+1)^2}$

Domain of G = domain of $G' = (-\infty, -1) \cup (-1, \infty)$.

31. $f'(x) = \lim\limits_{h\to 0} \dfrac{f(x+h)-f(x)}{h} = \lim\limits_{h\to 0} \dfrac{(x+h)^4 - x^4}{h} = \lim\limits_{h\to 0} \dfrac{(x^4 + 4x^3h + 6x^2h^2 + 4xh^3 + h^4) - x^4}{h}$

$= \lim\limits_{h\to 0} \dfrac{4x^3h + 6x^2h^2 + 4xh^3 + h^4}{h} = \lim\limits_{h\to 0} (4x^3 + 6x^2h + 4xh^2 + h^3) = 4x^3$

Domain of f = domain of $f' = \mathbb{R}$.

33. (a) $f'(x) = \lim\limits_{h\to 0} \dfrac{f(x+h)-f(x)}{h} = \lim\limits_{h\to 0} \dfrac{[(x+h)^4 + 2(x+h)] - (x^4 + 2x)}{h}$

$= \lim\limits_{h\to 0} \dfrac{x^4 + 4x^3h + 6x^2h^2 + 4xh^3 + h^4 + 2x + 2h - x^4 - 2x}{h}$

$= \lim\limits_{h\to 0} \dfrac{4x^3h + 6x^2h^2 + 4xh^3 + h^4 + 2h}{h} = \lim\limits_{h\to 0} \dfrac{h(4x^3 + 6x^2h + 4xh^2 + h^3 + 2)}{h}$

$= \lim\limits_{h\to 0} (4x^3 + 6x^2h + 4xh^2 + h^3 + 2) = 4x^3 + 2$

(b) Notice that $f'(x) = 0$ when f has a horizontal tangent, $f'(x)$ is positive when the tangents have positive slope, and $f'(x)$ is negative when the tangents have negative slope.

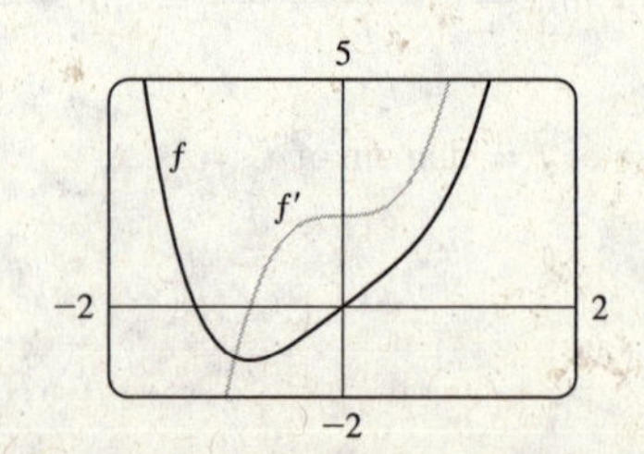

© 2016 Cengage Learning. All Rights Reserved. May not be scanned, copied or duplicated, or posted to a publicly accessible website, in whole or in part.

35. (a) $N'(t)$ is the rate of change of the number of malarial parasites with respect to time. Its units are (# parasites/μL)/day.

(b) The average rate of change over $[1, 2]$ is $\dfrac{2357 - 228}{2 - 1} = 2129$. Repeating this calculation for each interval gives the values found in the table (left). To estimate $N'(t)$, we average the rates of change for intervals adjacent to t. For example, $N'(2) \approx (2,129 + 10,393)/2 = 6,261$. A table of approximate $N'(t)$ values is shown far-right.

Interval	Rate of change
$[1, 2]$	$2,129$
$[2, 3]$	$10,393$
$[3, 4]$	$13,911$
$[4, 5]$	$345,670$
$[5, 6]$	$1,845,110$
$[6, 7]$	$4,530,959$

$\Rightarrow$

t	$N'(t)$
2	$6,261$
3	$12,152$
4	$179,791$
5	$1,095,390$
6	$3,188,035$

37. f is not differentiable at $x = -4$, because the graph has a corner there, and at $x = 0$, because there is a discontinuity there.

39. f is not differentiable at $x = -1$, because the graph has a vertical tangent there, and at $x = 4$, because the graph has a corner there.

41. As we zoom in toward $(-1, 0)$, the curve appears more and more like a straight line, so $f(x) = x + \sqrt{|x|}$ is differentiable at $x = -1$. But no matter how much we zoom in toward the origin, the curve doesn't straighten out—we can't eliminate the sharp point (a cusp). So f is not differentiable at $x = 0$.

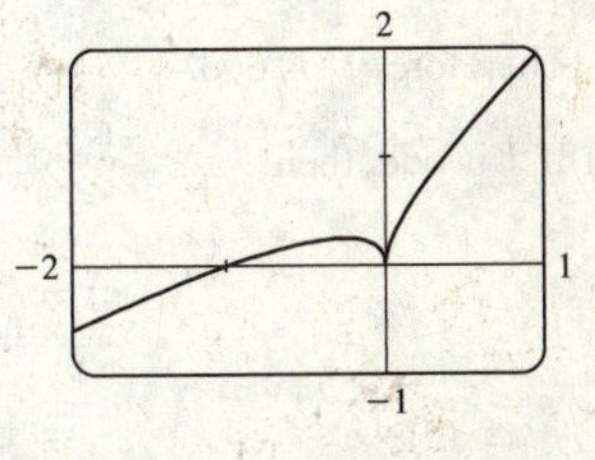

43. $a = f, b = f', c = f''$. We can see this because where a has a horizontal tangent, $b = 0$, and where b has a horizontal tangent, $c = 0$. We can immediately see that c can be neither f nor f', since at the points where c has a horizontal tangent, neither a nor b is equal to 0.

45. $$f'(x) = \lim_{h\to 0} \frac{f(x+h) - f(x)}{h} = \lim_{h\to 0} \frac{[3(x+h)^2 + 2(x+h) + 1] - (3x^2 + 2x + 1)}{h}$$
$$= \lim_{h\to 0} \frac{(3x^2 + 6xh + 3h^2 + 2x + 2h + 1) - (3x^2 + 2x + 1)}{h} = \lim_{h\to 0} \frac{6xh + 3h^2 + 2h}{h}$$
$$= \lim_{h\to 0} \frac{h(6x + 3h + 2)}{h} = \lim_{h\to 0} (6x + 3h + 2) = 6x + 2$$

$$f''(x) = \lim_{h\to 0} \frac{f'(x+h) - f'(x)}{h} = \lim_{h\to 0} \frac{[6(x+h) + 2] - (6x + 2)}{h} = \lim_{h\to 0} \frac{(6x + 6h + 2) - (6x + 2)}{h}$$
$$= \lim_{h\to 0} \frac{6h}{h} = \lim_{h\to 0} 6 = 6$$

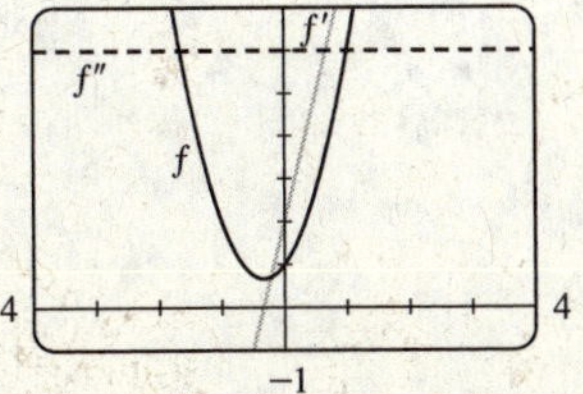

We see from the graph that our answers are reasonable because the graph of f' is that of a linear function and the graph of f'' is that of a constant function.

© 2016 Cengage Learning. All Rights Reserved. May not be scanned, copied or duplicated, or posted to a publicly accessible website, in whole or in part.

47. (a) Since $f'(x) < 0$ on $(1,4)$, f is decreasing on this interval. Since $f'(x) > 0$ on $(0,1)$ and $(4,5)$, f is increasing on these intervals.

(b) Since $f(0) = 0$, start at the origin. Draw an increasing function on $(0,1)$ with a local maximum at $x = 1$. Now draw a decreasing function on $(1,4)$ and the steepest slope should occur at $x = 2.5$ since that's where the smallest value of f' occurs. Last, draw an increasing function on $(4,5)$ making sure you have a local minimum at $x = 4$.

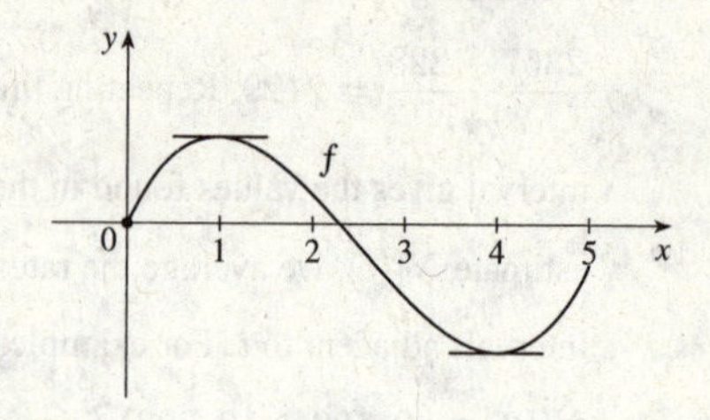

49. (a) If f is even, then

$$\begin{aligned} f'(-x) &= \lim_{h\to 0}\frac{f(-x+h)-f(-x)}{h} = \lim_{h\to 0}\frac{f[-(x-h)]-f(-x)}{h} \\ &= \lim_{h\to 0}\frac{f(x-h)-f(x)}{h} = -\lim_{h\to 0}\frac{f(x-h)-f(x)}{-h} \qquad [\text{let } \Delta x = -h] \\ &= -\lim_{\Delta x\to 0}\frac{f(x+\Delta x)-f(x)}{\Delta x} = -f'(x) \end{aligned}$$

Therefore, f' is odd.

(b) If f is odd, then

$$\begin{aligned} f'(-x) &= \lim_{h\to 0}\frac{f(-x+h)-f(-x)}{h} = \lim_{h\to 0}\frac{f[-(x-h)]-f(-x)}{h} \\ &= \lim_{h\to 0}\frac{-f(x-h)+f(x)}{h} = \lim_{h\to 0}\frac{f(x-h)-f(x)}{-h} \qquad [\text{let } \Delta x = -h] \\ &= \lim_{\Delta x\to 0}\frac{f(x+\Delta x)-f(x)}{\Delta x} = f'(x) \end{aligned}$$

Therefore, f' is even.

3.3 Basic Differentiation Formulas

1. (a) e is the number such that $\lim\limits_{h\to 0}\dfrac{e^h-1}{h} = 1$.

(b)

x	$\dfrac{2.7^x-1}{x}$
-0.001	0.9928
-0.0001	0.9932
0.001	0.9937
0.0001	0.9933

x	$\dfrac{2.8^x-1}{x}$
-0.001	1.0291
-0.0001	1.0296
0.001	1.0301
0.0001	1.0297

From the tables (to two decimal places),

$\lim\limits_{h\to 0}\dfrac{2.7^h-1}{h} = 0.99$ and $\lim\limits_{h\to 0}\dfrac{2.8^h-1}{h} = 1.03$.

Since $0.99 < 1 < 1.03$, $2.7 < e < 2.8$.

3. $f(x) = 186.5$ is a constant function, so its derivative is 0, that is, $f'(x) = 0$.

5. $f(x) = 5x - 1 \;\Rightarrow\; f'(x) = 5 - 0 = 5$

7. $f(x) = x^3 - 4x + 6 \;\Rightarrow\; f'(x) = 3x^2 - 4(1) + 0 = 3x^2 - 4$

9. $f(x) = x - 3\sin x \;\Rightarrow\; f'(x) = 1 - 3\cos x$

© 2016 Cengage Learning. All Rights Reserved. May not be scanned, copied or duplicated, or posted to a publicly accessible website, in whole or in part.

11. $f(t) = \frac{1}{4}(t^4 + 8) \Rightarrow f'(t) = \frac{1}{4}(t^4 + 8)' = \frac{1}{4}(4t^{4-1} + 0) = t^3$

13. $A(s) = -\dfrac{12}{s^5} = -12s^{-5} \Rightarrow A'(s) = -12(-5s^{-6}) = 60s^{-6}$ or $60/s^6$

15. $g(t) = 2t^{-3/4} \Rightarrow g'(t) = 2(-\frac{3}{4}t^{-7/4}) = -\frac{3}{2}t^{-7/4}$

17. $y = 3e^x + \dfrac{4}{\sqrt[3]{x}} = 3e^x + 4x^{-1/3} \Rightarrow y' = 3(e^x) + 4(-\frac{1}{3})x^{-4/3} = 3e^x - \frac{4}{3}x^{-4/3}$

19. $F(x) = (\frac{1}{2}x)^5 = (\frac{1}{2})^5 x^5 = \frac{1}{32}x^5 \Rightarrow F'(x) = \frac{1}{32}(5x^4) = \frac{5}{32}x^4$

21. $y = \dfrac{x^2 + 4x + 3}{\sqrt{x}} = x^{3/2} + 4x^{1/2} + 3x^{-1/2} \Rightarrow$

$y' = \frac{3}{2}x^{1/2} + 4(\frac{1}{2})x^{-1/2} + 3(-\frac{1}{2})x^{-3/2} = \frac{3}{2}\sqrt{x} + \dfrac{2}{\sqrt{x}} - \dfrac{3}{2x\sqrt{x}}$ $\left[\text{note that } x^{3/2} = x^{2/2} \cdot x^{1/2} = x\sqrt{x}\right]$

The last expression can be written as $\dfrac{3x^2}{2x\sqrt{x}} + \dfrac{4x}{2x\sqrt{x}} - \dfrac{3}{2x\sqrt{x}} = \dfrac{3x^2 + 4x - 3}{2x\sqrt{x}}$.

23. $y = 4\pi^2 \Rightarrow y' = 0$ since $4\pi^2$ is a constant.

25. $g(y) = \dfrac{A}{y^{10}} + B\cos y = Ay^{-10} + B\cos y \Rightarrow g'(y) = A(-10)y^{-11} + B(-\sin y) = -\dfrac{10A}{y^{11}} - B\sin y$

27. $f(x) = k(a - x)(b - x) \Rightarrow f'(x) = k(b - x)(-1) + k(a - x)(-1) = k(2x - a - b)$

29. $u = \sqrt[5]{t} + 4\sqrt{t^5} = t^{1/5} + 4t^{5/2} \Rightarrow u' = \frac{1}{5}t^{-4/5} + 4\left(\frac{5}{2}t^{3/2}\right) = \frac{1}{5}t^{-4/5} + 10t^{3/2}$ or $1/\left(5\sqrt[5]{t^4}\right) + 10\sqrt{t^3}$

31. $G(y) = \dfrac{A}{y^{10}} + Be^y = Ay^{-10} + Be^y \Rightarrow G'(y) = -10Ay^{-11} + Be^y = -\dfrac{10A}{y^{11}} + Be^y$

33. $y = \sqrt[4]{x} = x^{1/4} \Rightarrow y' = \frac{1}{4}x^{-3/4} = \dfrac{1}{4\sqrt[4]{x^3}}$. At $(1, 1)$, $y' = \frac{1}{4}$ and an equation of the tangent line is

$y - 1 = \frac{1}{4}(x - 1)$ or $y = \frac{1}{4}x + \frac{3}{4}$.

35. $y = 6\cos x \Rightarrow y' = -6\sin x$. At $(\pi/3, 3)$, $y' = -6\sin(\pi/3) = -6(\sqrt{3}/2) = -3\sqrt{3}$ and an equation of the tangent line is $y - 3 = -3\sqrt{3}(x - \pi/3)$ or $y = -3\sqrt{3}x + 3 + \pi\sqrt{3}$. The slope of the normal line is $1/(3\sqrt{3})$ (the negative reciprocal of $-3\sqrt{3}$) and an equation of the normal line is $y - 3 = \dfrac{1}{3\sqrt{3}}\left(x - \dfrac{\pi}{3}\right)$ or $y = \dfrac{1}{3\sqrt{3}}x + 3 - \dfrac{\pi}{9\sqrt{3}}$.

37. $y = x^4 + 2e^x \Rightarrow y' = 4x^3 + 2e^x$. At $(0, 2)$, $y' = 2$ and an equation of the tangent line is $y - 2 = 2(x - 0)$ or $y = 2x + 2$. The slope of the normal line is $-\frac{1}{2}$ (the negative reciprocal of 2) and an equation of the normal line is $y - 2 = -\frac{1}{2}(x - 0)$ or $y = -\frac{1}{2}x + 2$.

39. $y = f(x) = x + \sqrt{x} \Rightarrow f'(x) = 1 + \frac{1}{2}x^{-1/2}$.

So the slope of the tangent line at $(1, 2)$ is $f'(1) = 1 + \frac{1}{2}(1) = \frac{3}{2}$

and its equation is $y - 2 = \frac{3}{2}(x - 1)$ or $y = \frac{3}{2}x + \frac{1}{2}$.

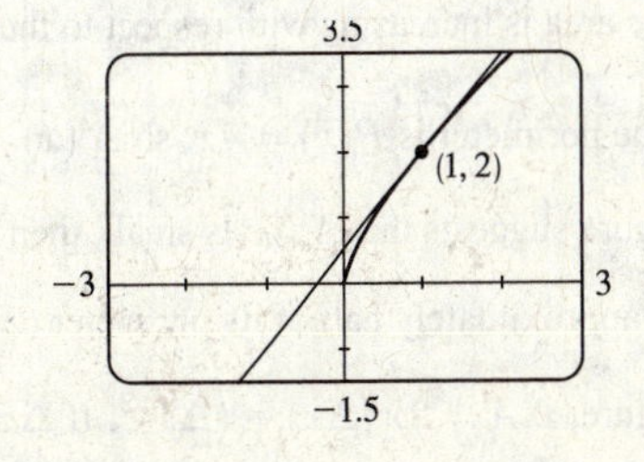

© 2016 Cengage Learning. All Rights Reserved. May not be scanned, copied or duplicated, or posted to a publicly accessible website, in whole or in part.

41. $f(x) = 3x^{15} - 5x^3 + 3 \Rightarrow f'(x) = 45x^{14} - 15x^2$.

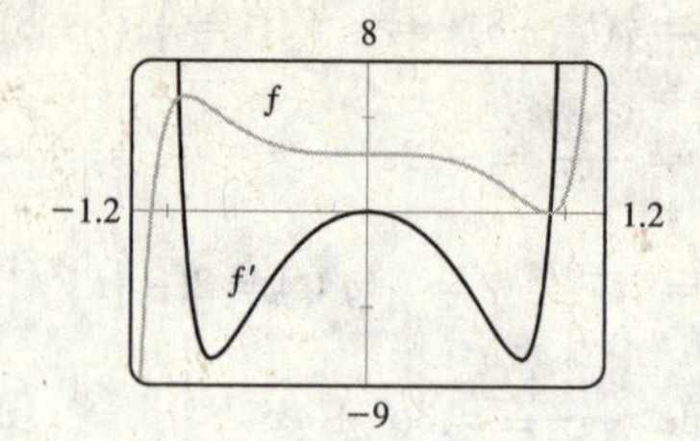

Notice that $f'(x) = 0$ when f has a horizontal tangent, f' is positive when f is increasing, and f' is negative when f is decreasing.

43. $f(x) = x^4 - 3x^3 + 16x \Rightarrow f'(x) = 4x^3 - 9x^2 + 16 \Rightarrow f''(x) = 12x^2 - 18x$

45. $g(t) = 2\cos t - 3\sin t \Rightarrow g'(t) = -2\sin t - 3\cos t \Rightarrow g''(t) = -2\cos t + 3\sin t$

47. $L(A) = 0.0155A^3 - 0.372A^2 + 3.95A + 1.21 \Rightarrow$

$$\frac{dL}{dA} = 0.0155\,(3A^2) - 0.372\,(2A^1) + 3.95\,(A^0) + 0 = 0.0465A^2 - 0.744A + 3.95 \Rightarrow$$

$\left.\frac{dL}{dA}\right|_{A=12} = 0.0465\,(12)^2 - 0.744\,(12) + 3.95 = 1.718$. Therefore, a 12-year old rock fish grows at a rate of 1.718 inches/year.

49. (a) Using $v = \frac{P}{4\eta l}(R^2 - r^2)$ with $R = 0.01$, $l = 3$, $P = 3000$, and $\eta = 0.027$, we have v as a function of r:

$v(r) = \frac{3000}{4(0.027)3}(0.01^2 - r^2)$. $v(0) = 0.\overline{925}$ cm/s, $v(0.005) = 0.69\overline{4}$ cm/s, $v(0.01) = 0$.

(b) $v(r) = \frac{P}{4\eta l}(R^2 - r^2) \Rightarrow v'(r) = \frac{P}{4\eta l}(-2r) = -\frac{Pr}{2\eta l}$. When $l = 3$, $P = 3000$, and $\eta = 0.027$, we have

$v'(r) = -\frac{3000r}{2(0.027)3}$. $v'(0) = 0$, $v'(0.005) = -92.\overline{592}$ (cm/s)/cm, and $v'(0.01) = -185.\overline{185}$ (cm/s)/cm.

(c) The velocity is greatest where $r = 0$ (at the center) and the velocity is changing most where $r = R = 0.01$ cm (at the edge).

51. (a) $s(t) = t^3 - 4.5t^2 - 7t \Rightarrow v(t) = s'(t) = 3t^2 - 9t - 7 \Rightarrow a(t) = v'(t) = 6t - 9$

(b) $v(t) = 3t^2 - 9t - 7 = 5 \Leftrightarrow 3t^2 - 9t - 12 = 0 \Leftrightarrow 3(t-4)(t+1) = 0 \Leftrightarrow t = 4$ or -1. Since $t \geq 0$, the particle reaches a velocity of 5 m/s at $t = 4$ s.

(c) $a(t) = 6t - 9 = 0 \Leftrightarrow t = 1.5$. The acceleration changes from negative to positive, so the velocity changes from decreasing to increasing. Thus, at $t = 1.5$ s, the velocity has its minimum value.

53. (a) $A(x) = x^2 \Rightarrow A'(x) = 2x$. $A'(15) = 30\ \text{mm}^2/\text{mm}$ is the rate at which the area is increasing with respect to the side length as x reaches 15 mm.

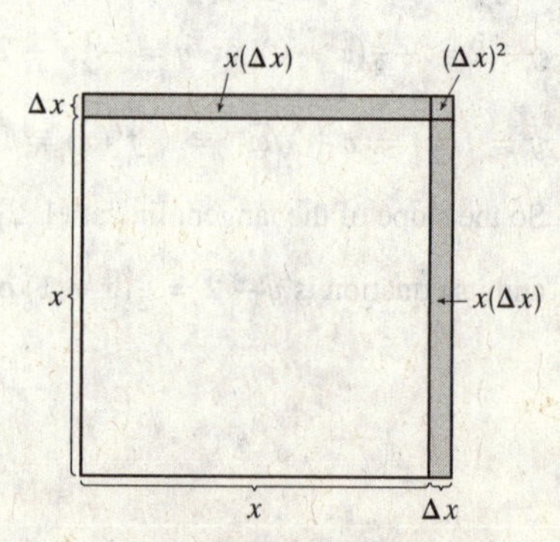

(b) The perimeter is $P(x) = 4x$, so $A'(x) = 2x = \frac{1}{2}(4x) = \frac{1}{2}P(x)$. The figure suggests that if Δx is small, then the change in the area of the square is approximately half of its perimeter (2 of the 4 sides) times Δx. From the figure, $\Delta A = 2x\,(\Delta x) + (\Delta x)^2$. If Δx is small, then $\Delta A \approx 2x\,(\Delta x)$ and so $\Delta A/\Delta x \approx 2x$.

© 2016 Cengage Learning. All Rights Reserved. May not be scanned, copied or duplicated, or posted to a publicly accessible website, in whole or in part.

55. (a) Using $A(r) = \pi r^2$, we find that the average rate of change is:

(i) $\dfrac{A(3) - A(2)}{3 - 2} = \dfrac{9\pi - 4\pi}{1} = 5\pi$ (ii) $\dfrac{A(2.5) - A(2)}{2.5 - 2} = \dfrac{6.25\pi - 4\pi}{0.5} = 4.5\pi$

(iii) $\dfrac{A(2.1) - A(2)}{2.1 - 2} = \dfrac{4.41\pi - 4\pi}{0.1} = 4.1\pi$

(b) $A(r) = \pi r^2 \Rightarrow A'(r) = 2\pi r$, so $A'(2) = 4\pi$.

(c) The circumference is $C(r) = 2\pi r = A'(r)$. The figure suggests that if Δr is small, then the change in the area of the circle (a ring around the outside) is approximately equal to its circumference times Δr. Straightening out this ring gives us a shape that is approximately rectangular with length $2\pi r$ and width Δr, so $\Delta A \approx 2\pi r(\Delta r)$. Algebraically, $\Delta A = A(r + \Delta r) - A(r) = \pi(r + \Delta r)^2 - \pi r^2 = 2\pi r(\Delta r) + \pi(\Delta r)^2$. So we see that if Δr is small, then $\Delta A \approx 2\pi r(\Delta r)$ and therefore, $\Delta A/\Delta r \approx 2\pi r$.

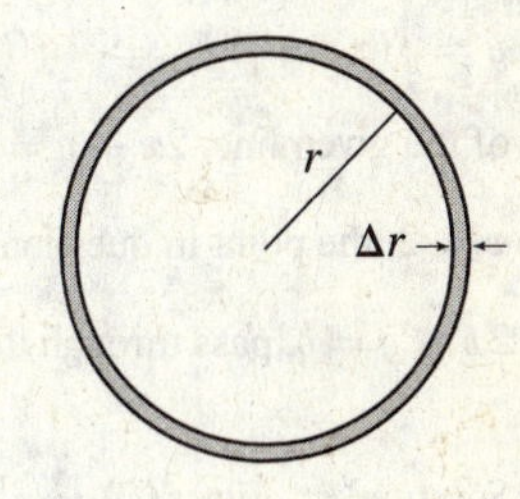

57. $\dfrac{d}{dx}(\sin x) = \cos x \Rightarrow \dfrac{d^2}{dx^2}(\sin x) = -\sin x \Rightarrow \dfrac{d^3}{dx^3}(\sin x) = -\cos x \Rightarrow \dfrac{d^4}{dx^4}(\sin x) = \sin x$.

The derivatives of $\sin x$ occur in a cycle of four. Since $99 = 4(24) + 3$, we have $\dfrac{d^{99}}{dx^{99}}(\sin x) = \dfrac{d^3}{dx^3}(\sin x) = -\cos x$.

59. $f(x) = x + 2\sin x$ has a horizontal tangent when $f'(x) = 0 \Leftrightarrow 1 + 2\cos x = 0 \Leftrightarrow \cos x = -\frac{1}{2} \Leftrightarrow$ $x = \frac{2\pi}{3} + 2\pi n$ or $\frac{4\pi}{3} + 2\pi n$, where n is an integer. Note that $\frac{4\pi}{3}$ and $\frac{2\pi}{3}$ are $\pm\frac{\pi}{3}$ units from π. This allows us to write the solutions in the more compact equivalent form $(2n + 1)\pi \pm \frac{\pi}{3}$, n an integer.

61. $y = 6x^3 + 5x - 3 \Rightarrow m = y' = 18x^2 + 5$, but $x^2 \geq 0$ for all x, so $m \geq 5$ for all x.

63. The slope of the line $12x - y = 1$ (or $y = 12x - 1$) is 12, so the slope of both lines tangent to the curve is 12. $y = 1 + x^3 \Rightarrow y' = 3x^2$. Thus, $3x^2 = 12 \Rightarrow x^2 = 4 \Rightarrow x = \pm 2$, which are the x-coordinates at which the tangent lines have slope 12. The points on the curve are $(2, 9)$ and $(-2, -7)$, so the tangent line equations are $y - 9 = 12(x - 2)$ or $y = 12x - 15$ and $y + 7 = 12(x + 2)$ or $y = 12x + 17$.

65. The slope of $y = x^2 - 5x + 4$ is given by $m = y' = 2x - 5$. The slope of $x - 3y = 5 \Leftrightarrow y = \frac{1}{3}x - \frac{5}{3}$ is $\frac{1}{3}$, so the desired normal line must have slope $\frac{1}{3}$, and hence, the tangent line to the parabola must have slope -3. This occurs if $2x - 5 = -3 \Rightarrow 2x = 2 \Rightarrow x = 1$. When $x = 1$, $y = 1^2 - 5(1) + 4 = 0$, and an equation of the normal line is $y - 0 = \frac{1}{3}(x - 1)$ or $y = \frac{1}{3}x - \frac{1}{3}$.

67.

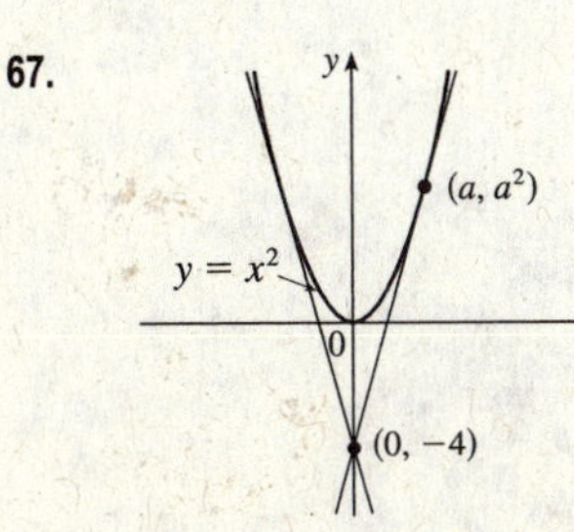

Let (a, a^2) be a point on the parabola at which the tangent line passes through the point $(0, -4)$. The tangent line has slope $2a$ and equation $y - (-4) = 2a(x - 0) \Leftrightarrow y = 2ax - 4$. Since (a, a^2) also lies on the line, $a^2 = 2a(a) - 4$, or $a^2 = 4$. So $a = \pm 2$ and the points are $(2, 4)$ and $(-2, 4)$.

© 2016 Cengage Learning. All Rights Reserved. May not be scanned, copied or duplicated, or posted to a publicly accessible website, in whole or in part.

69. $f'(x) = \lim\limits_{h \to 0} \dfrac{f(x+h) - f(x)}{h} = \lim\limits_{h \to 0} \dfrac{\dfrac{1}{x+h} - \dfrac{1}{x}}{h} = \lim\limits_{h \to 0} \dfrac{x - (x+h)}{hx(x+h)} = \lim\limits_{h \to 0} \dfrac{-h}{hx(x+h)} = \lim\limits_{h \to 0} \dfrac{-1}{x(x+h)} = -\dfrac{1}{x^2}$

71. Substituting $x = 1$ and $y = 1$ into $y = ax^2 + bx$ gives us $a + b = 1$ **(1)**. The slope of the tangent line $y = 3x - 2$ is 3 and the slope of the tangent to the parabola at (x, y) is $y' = 2ax + b$. At $x = 1$, $y' = 3 \Rightarrow 3 = 2a + b$ **(2)**. Subtracting **(1)** from **(2)** gives us $2 = a$ and it follows that $b = -1$. The parabola has equation $y = 2x^2 - x$.

73. $y = f(x) = ax^2 \Rightarrow f'(x) = 2ax$. So the slope of the tangent to the parabola at $x = 2$ is $m = 2a(2) = 4a$. The slope of the given line, $2x + y = b \Leftrightarrow y = -2x + b$, is seen to be -2, so we must have $4a = -2 \Leftrightarrow a = -\frac{1}{2}$. So when $x = 2$, the point in question has y-coordinate $-\frac{1}{2} \cdot 2^2 = -2$. Now we simply require that the given line, whose equation is $2x + y = b$, pass through the point $(2, -2)$: $2(2) + (-2) = b \Leftrightarrow b = 2$. So we must have $a = -\frac{1}{2}$ and $b = 2$.

75. *Solution 1:* Let $f(x) = x^{1000}$. Then, by the definition of a derivative, $f'(1) = \lim\limits_{x \to 1} \dfrac{f(x) - f(1)}{x - 1} = \lim\limits_{x \to 1} \dfrac{x^{1000} - 1}{x - 1}$.

But this is just the limit we want to find, and we know (from the Power Rule) that $f'(x) = 1000x^{999}$, so

$f'(1) = 1000(1)^{999} = 1000$. So $\lim\limits_{x \to 1} \dfrac{x^{1000} - 1}{x - 1} = 1000$.

Solution 2: Note that $(x^{1000} - 1) = (x - 1)(x^{999} + x^{998} + x^{997} + \cdots + x^2 + x + 1)$. So

$$\begin{aligned} \lim_{x \to 1} \frac{x^{1000} - 1}{x - 1} &= \lim_{x \to 1} \frac{(x-1)(x^{999} + x^{998} + x^{997} + \cdots + x^2 + x + 1)}{x - 1} = \lim_{x \to 1} (x^{999} + x^{998} + x^{997} + \cdots + x^2 + x + 1) \\ &= \underbrace{1 + 1 + 1 + \cdots + 1 + 1 + 1}_{1000 \text{ ones}} = 1000, \text{ as above.} \end{aligned}$$

3.4 The Product and Quotient Rules

1. Product Rule: $f(x) = (1 + 2x^2)(x - x^2) \Rightarrow$

$$f'(x) = (1 + 2x^2)(1 - 2x) + (x - x^2)(4x) = 1 - 2x + 2x^2 - 4x^3 + 4x^2 - 4x^3 = 1 - 2x + 6x^2 - 8x^3.$$

Multiplying first: $f(x) = (1 + 2x^2)(x - x^2) = x - x^2 + 2x^3 - 2x^4 \Rightarrow f'(x) = 1 - 2x + 6x^2 - 8x^3$ (equivalent).

3. By the Product Rule, $f(x) = (x^3 + 2x)e^x \Rightarrow$

$$\begin{aligned} f'(x) &= (x^3 + 2x)(e^x)' + e^x(x^3 + 2x)' = (x^3 + 2x)e^x + e^x(3x^2 + 2) \\ &= e^x[(x^3 + 2x) + (3x^2 + 2)] = e^x(x^3 + 3x^2 + 2x + 2) \end{aligned}$$

5. $g(t) = t^3 \cos t \Rightarrow g'(t) = t^3(-\sin t) + (\cos t) \cdot 3t^2 = 3t^2 \cos t - t^3 \sin t$ or $t^2(3\cos t - t \sin t)$

7. $F(y) = \left(\dfrac{1}{y^2} - \dfrac{3}{y^4}\right)(y + 5y^3) = (y^{-2} - 3y^{-4})(y + 5y^3) \overset{\text{PR}}{\Rightarrow}$

$$\begin{aligned} F'(y) &= (y^{-2} - 3y^{-4})(1 + 15y^2) + (y + 5y^3)(-2y^{-3} + 12y^{-5}) \\ &= (y^{-2} + 15 - 3y^{-4} - 45y^{-2}) + (-2y^{-2} + 12y^{-4} - 10 + 60y^{-2}) \\ &= 5 + 14y^{-2} + 9y^{-4} \text{ or } 5 + 14/y^2 + 9/y^4 \end{aligned}$$

© 2016 Cengage Learning. All Rights Reserved. May not be scanned, copied or duplicated, or posted to a publicly accessible website, in whole or in part.

9. $f(x) = \sin x + \frac{1}{2}\cot x \;\Rightarrow\; f'(x) = \cos x - \frac{1}{2}\csc^2 x$

11. $h(\theta) = \theta\csc\theta - \cot\theta \;\Rightarrow\; h'(\theta) = \theta(-\csc\theta\,\cot\theta) + (\csc\theta)\cdot 1 - \left(-\csc^2\theta\right) = \csc\theta - \theta\csc\theta\,\cot\theta + \csc^2\theta$

13. By the Quotient Rule, $y = \dfrac{e^x}{x^2} \;\Rightarrow\; y' = \dfrac{x^2\dfrac{d}{dx}(e^x) - e^x\dfrac{d}{dx}(x^2)}{(x^2)^2} = \dfrac{x^2(e^x) - e^x(2x)}{x^4} = \dfrac{xe^x(x-2)}{x^4} = \dfrac{e^x(x-2)}{x^3}$.

15. $g(x) = \dfrac{3x-1}{2x+1} \;\overset{\text{QR}}{\Rightarrow}\; g'(x) = \dfrac{(2x+1)(3)-(3x-1)(2)}{(2x+1)^2} = \dfrac{6x+3-6x+2}{(2x+1)^2} = \dfrac{5}{(2x+1)^2}$

17. $y = \dfrac{x^3}{1-x^2} \;\overset{\text{QR}}{\Rightarrow}\; y' = \dfrac{(1-x^2)(3x^2)-x^3(-2x)}{(1-x^2)^2} = \dfrac{x^2(3-3x^2+2x^2)}{(1-x^2)^2} = \dfrac{x^2(3-x^2)}{(1-x^2)^2}$

19. $y = \dfrac{t^2+2}{t^4-3t^2+1} \;\overset{\text{QR}}{\Rightarrow}$

$$y' = \frac{(t^4-3t^2+1)(2t)-(t^2+2)(4t^3-6t)}{(t^4-3t^2+1)^2} = \frac{2t[(t^4-3t^2+1)-(t^2+2)(2t^2-3)]}{(t^4-3t^2+1)^2}$$

$$= \frac{2t(t^4-3t^2+1-2t^4-4t^2+3t^2+6)}{(t^4-3t^2+1)^2} = \frac{2t(-t^4-4t^2+7)}{(t^4-3t^2+1)^2}$$

21. $y = (r^2-2r)e^r \;\overset{\text{PR}}{\Rightarrow}\; y' = (r^2-2r)(e^r) + e^r(2r-2) = e^r(r^2-2r+2r-2) = e^r(r^2-2)$

23. $f(\theta) = \dfrac{\sec\theta}{1+\sec\theta} \;\Rightarrow$

$$f'(\theta) = \frac{(1+\sec\theta)(\sec\theta\,\tan\theta)-(\sec\theta)(\sec\theta\,\tan\theta)}{(1+\sec\theta)^2} = \frac{(\sec\theta\,\tan\theta)\,[(1+\sec\theta)-\sec\theta]}{(1+\sec\theta)^2} = \frac{\sec\theta\,\tan\theta}{(1+\sec\theta)^2}$$

25. $y = \dfrac{\sin x}{x^2} \;\Rightarrow\; y' = \dfrac{x^2\cos x - (\sin x)(2x)}{(x^2)^2} = \dfrac{x(x\cos x - 2\sin x)}{x^4} = \dfrac{x\cos x - 2\sin x}{x^3}$

27. $y = \dfrac{v^3 - 2v\sqrt{v}}{v} = v^2 - 2\sqrt{v} = v^2 - 2v^{1/2} \;\Rightarrow\; y' = 2v - 2\left(\frac{1}{2}\right)v^{-1/2} = 2v - v^{-1/2}$.

We can change the form of the answer as follows: $2v - v^{-1/2} = 2v - \dfrac{1}{\sqrt{v}} = \dfrac{2v\sqrt{v}-1}{\sqrt{v}} = \dfrac{2v^{3/2}-1}{\sqrt{v}}$

29. $f(t) = \dfrac{2t}{2+\sqrt{t}} \;\overset{\text{QR}}{\Rightarrow}\; f'(t) = \dfrac{(2+t^{1/2})(2)-2t\left(\frac{1}{2}t^{-1/2}\right)}{(2+\sqrt{t})^2} = \dfrac{4+2t^{1/2}-t^{1/2}}{(2+\sqrt{t})^2} = \dfrac{4+t^{1/2}}{(2+\sqrt{t})^2}$ or $\dfrac{4+\sqrt{t}}{(2+\sqrt{t})^2}$

31. $f(x) = \dfrac{A}{B+Ce^x} \;\overset{\text{QR}}{\Rightarrow}\; f'(x) = \dfrac{(B+Ce^x)\cdot 0 - A(Ce^x)}{(B+Ce^x)^2} = -\dfrac{ACe^x}{(B+Ce^x)^2}$

33. $f(x) = \dfrac{x}{x+c/x} \;\Rightarrow\; f'(x) = \dfrac{(x+c/x)(1)-x(1-c/x^2)}{\left(x+\dfrac{c}{x}\right)^2} = \dfrac{x+c/x-x+c/x}{\left(\dfrac{x^2+c}{x}\right)^2} = \dfrac{2c/x}{\dfrac{(x^2+c)^2}{x^2}}\cdot\dfrac{x^2}{x^2} = \dfrac{2cx}{(x^2+c)^2}$

35. $y = \dfrac{2x}{x+1} \;\Rightarrow\; y' = \dfrac{(x+1)(2)-(2x)(1)}{(x+1)^2} = \dfrac{2}{(x+1)^2}$.

At $(1,1)$, $y' = \frac{1}{2}$, and an equation of the tangent line is $y - 1 = \frac{1}{2}(x-1)$, or $y = \frac{1}{2}x + \frac{1}{2}$.

© 2016 Cengage Learning. All Rights Reserved. May not be scanned, copied or duplicated, or posted to a publicly accessible website, in whole or in part.

37. $y = 2xe^x \;\Rightarrow\; y' = 2(x \cdot e^x + e^x \cdot 1) = 2e^x(x+1)$.

At $(0,0)$, $y' = 2e^0(0+1) = 2 \cdot 1 \cdot 1 = 2$, and an equation of the tangent line is $y - 0 = 2(x-0)$, or $y = 2x$. The slope of the normal line is $-\frac{1}{2}$, so an equation of the normal line is $y - 0 = -\frac{1}{2}(x-0)$, or $y = -\frac{1}{2}x$.

39. $f(x) = x^4e^x \;\Rightarrow\; f'(x) = x^4e^x + e^x \cdot 4x^3 = (x^4 + 4x^3)e^x \;\left[\text{or } x^3e^x(x+4)\right] \;\Rightarrow$

$f''(x) = (x^4 + 4x^3)e^x + e^x(4x^3 + 12x^2) = (x^4 + 4x^3 + 4x^3 + 12x^2)e^x$

$= (x^4 + 8x^3 + 12x^2)e^x \;\left[\text{or } x^2e^x(x+2)(x+6)\right]$

41. $H(\theta) = \theta\sin\theta \;\Rightarrow\; H'(\theta) = \theta\,(\cos\theta) + (\sin\theta)\cdot 1 = \theta\cos\theta + \sin\theta \;\Rightarrow$

$H''(\theta) = \theta\,(-\sin\theta) + (\cos\theta)\cdot 1 + \cos\theta = -\theta\sin\theta + 2\cos\theta$

43. $f(x) = xe^x \;\Rightarrow\; f'(x) = e^x(1) + x\,(e^x) = e^x\,(1+x) \;\Rightarrow\; f''(x) = e^x\,(1) + (1+x)\,e^x = e^x\,(2+x) \;\Rightarrow$

$f'''(x) = e^x\,(1) + (2+x)\,e^x = e^x\,(3+x) \;\Rightarrow\; f^{(4)}(x) = e^x\,(1) + (3+x)\,e^x = e^x\,(4+x)$.

The pattern suggests that $f^{(n)}(x) = e^x\,(n+x)$. (We could use mathematical induction to prove this formula.)

45. $\dfrac{d}{dx}(\csc x) = \dfrac{d}{dx}\left(\dfrac{1}{\sin x}\right) = \dfrac{(\sin x)(0) - 1(\cos x)}{\sin^2 x} = \dfrac{-\cos x}{\sin^2 x} = -\dfrac{1}{\sin x}\cdot\dfrac{\cos x}{\sin x} = -\csc x\,\cot x$

47. $\dfrac{d}{dx}(\cot x) = \dfrac{d}{dx}\left(\dfrac{\cos x}{\sin x}\right) = \dfrac{(\sin x)(-\sin x) - (\cos x)(\cos x)}{\sin^2 x} = -\dfrac{\sin^2 x + \cos^2 x}{\sin^2 x} = -\dfrac{1}{\sin^2 x} = -\csc^2 x$

49. We are given that $f(5) = 1$, $f'(5) = 6$, $g(5) = -3$, and $g'(5) = 2$.

(a) $(fg)'(5) = f(5)g'(5) + g(5)f'(5) = (1)(2) + (-3)(6) = 2 - 18 = -16$

(b) $\left(\dfrac{f}{g}\right)'(5) = \dfrac{g(5)f'(5) - f(5)g'(5)}{[g(5)]^2} = \dfrac{(-3)(6) - (1)(2)}{(-3)^2} = -\dfrac{20}{9}$

(c) $\left(\dfrac{g}{f}\right)'(5) = \dfrac{f(5)g'(5) - g(5)f'(5)}{[f(5)]^2} = \dfrac{(1)(2) - (-3)(6)}{(1)^2} = 20$

51. (a) From the graphs of f and g, we obtain the following values: $f(1) = 2$ since the point $(1,2)$ is on the graph of f; $g(1) = 1$ since the point $(1,1)$ is on the graph of g; $f'(1) = 2$ since the slope of the line segment between $(0,0)$ and $(2,4)$ is $\dfrac{4-0}{2-0} = 2$; $g'(1) = -1$ since the slope of the line segment between $(-2,4)$ and $(2,0)$ is $\dfrac{0-4}{2-(-2)} = -1$.

Now $u(x) = f(x)g(x)$, so $u'(1) = f(1)g'(1) + g(1)\,f'(1) = 2\cdot(-1) + 1\cdot 2 = 0$.

(b) $v(x) = f(x)/g(x)$, so $v'(5) = \dfrac{g(5)f'(5) - f(5)g'(5)}{[g(5)]^2} = \dfrac{2\left(-\frac{1}{3}\right) - 3\cdot\frac{2}{3}}{2^2} = \dfrac{-\frac{8}{3}}{4} = -\dfrac{2}{3}$

53. (a) $y = xg(x) \;\Rightarrow\; y' = xg'(x) + g(x)\cdot 1 = xg'(x) + g(x)$

(b) $y = \dfrac{x}{g(x)} \;\Rightarrow\; y' = \dfrac{g(x)\cdot 1 - xg'(x)}{[g(x)]^2} = \dfrac{g(x) - xg'(x)}{[g(x)]^2}$

(c) $y = \dfrac{g(x)}{x} \;\Rightarrow\; y' = \dfrac{xg'(x) - g(x)\cdot 1}{(x)^2} = \dfrac{xg'(x) - g(x)}{x^2}$

© 2016 Cengage Learning. All Rights Reserved. May not be scanned, copied or duplicated, or posted to a publicly accessible website, in whole or in part.

55. $v = \dfrac{0.14[\mathrm{S}]}{0.015 + [\mathrm{S}]} \quad \Rightarrow \quad \dfrac{dv}{d[\mathrm{S}]} = \dfrac{(0.015 + [\mathrm{S}])\,(0.14) - (0.14[\mathrm{S}])\,(1)}{(0.015 + [\mathrm{S}])^2} = \dfrac{0.0021}{(0.015 + [\mathrm{S}])^2}$

$dv/d[\mathrm{S}]$ is the rate of change of the enzymatic reaction rate with respect to the concentration of the substrate.

57. $PV = nRT \quad \Rightarrow \quad T = \dfrac{PV}{nR} = \dfrac{PV}{(10)(0.0821)} = \dfrac{1}{0.821}(PV)$. Using the Product Rule, we have

$$\frac{dT}{dt} = \frac{1}{0.821}\left[P(t)V'(t) + V(t)P'(t)\right] = \frac{1}{0.821}\left[(8)(-0.15) + (10)(0.10)\right] \approx -0.2436 \text{ K/min.}$$

59. If $y = f(x) = \dfrac{x}{x+1}$, then $f'(x) = \dfrac{(x+1)(1) - x(1)}{(x+1)^2} = \dfrac{1}{(x+1)^2}$. When $x = a$, the equation of the tangent line is

$y - \dfrac{a}{a+1} = \dfrac{1}{(a+1)^2}(x - a)$. This line passes through $(1, 2)$ when $2 - \dfrac{a}{a+1} = \dfrac{1}{(a+1)^2}(1 - a) \quad \Leftrightarrow$

$2(a+1)^2 - a(a+1) = 1 - a \quad \Leftrightarrow \quad 2a^2 + 4a + 2 - a^2 - a - 1 + a = 0 \quad \Leftrightarrow \quad a^2 + 4a + 1 = 0.$

The quadratic formula gives the roots of this equation as $a = \dfrac{-4 \pm \sqrt{4^2 - 4(1)(1)}}{2(1)} = \dfrac{-4 \pm \sqrt{12}}{2} = -2 \pm \sqrt{3}$,

so there are two such tangent lines. Since

$$f(-2 \pm \sqrt{3}) = \frac{-2 \pm \sqrt{3}}{-2 \pm \sqrt{3} + 1} = \frac{-2 \pm \sqrt{3}}{-1 \pm \sqrt{3}} \cdot \frac{-1 \mp \sqrt{3}}{-1 \mp \sqrt{3}}$$
$$= \frac{2 \pm 2\sqrt{3} \mp \sqrt{3} - 3}{1 - 3} = \frac{-1 \pm \sqrt{3}}{-2} = \frac{1 \mp \sqrt{3}}{2},$$

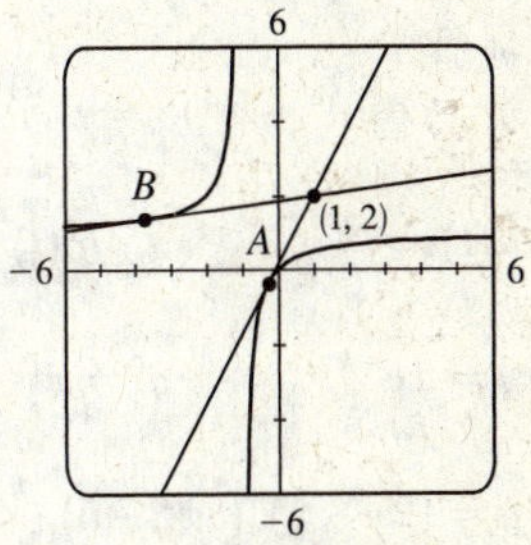

the lines touch the curve at $A\left(-2 + \sqrt{3}, \frac{1 - \sqrt{3}}{2}\right) \approx (-0.27, -0.37)$

and $B\left(-2 - \sqrt{3}, \frac{1 + \sqrt{3}}{2}\right) \approx (-3.73, 1.37)$.

6
B
A
(1, 2)
−6
6
−6

61. (a) $(fgh)' = [(fg)h]' = (fg)'h + (fg)h' = (f'g + fg')h + (fg)h' = f'gh + fg'h + fgh'$

(b) Putting $f = g = h$ in part (a), we have $\dfrac{d}{dx}[f(x)]^3 = (fff)' = f'ff + ff'f + fff' = 3fff' = 3[f(x)]^2 f'(x)$.

(c) $\dfrac{d}{dx}\left(e^{3x}\right) = \dfrac{d}{dx}\left(e^x\right)^3 = 3(e^x)^2 e^x = 3e^{2x}e^x = 3e^{3x}$

63. (a) $\dfrac{d}{dx}\left(\dfrac{1}{g(x)}\right) = \dfrac{g(x) \cdot \frac{d}{dx}(1) - 1 \cdot \frac{d}{dx}[g(x)]}{[g(x)]^2}$ [Quotient Rule] $= \dfrac{g(x) \cdot 0 - 1 \cdot g'(x)}{[g(x)]^2} = \dfrac{0 - g'(x)}{[g(x)]^2} = -\dfrac{g'(x)}{[g(x)]^2}$

(b) $y = \dfrac{1}{x^4 + x^2 + 1} \quad \Rightarrow \quad y' = -\dfrac{2x(2x^2 + 1)}{(x^4 + x^2 + 1)^2}$

(c) $\dfrac{d}{dx}\left(x^{-n}\right) = \dfrac{d}{dx}\left(\dfrac{1}{x^n}\right) = -\dfrac{(x^n)'}{(x^n)^2}$ [by the Reciprocal Rule] $= -\dfrac{nx^{n-1}}{x^{2n}} = -nx^{n-1-2n} = -nx^{-n-1}$

© 2016 Cengage Learning. All Rights Reserved. May not be scanned, copied or duplicated, or posted to a publicly accessible website, in whole or in part.

3.5 The Chain Rule

1. Let $u = g(x) = 1 + 4x$ and $y = f(u) = \sqrt[3]{u}$. Then $\dfrac{dy}{dx} = \dfrac{dy}{du}\dfrac{du}{dx} = \left(\tfrac{1}{3}u^{-2/3}\right)(4) = \dfrac{4}{3\sqrt[3]{(1+4x)^2}}$.

3. Let $u = g(x) = \pi x$ and $y = f(u) = \tan u$. Then $\dfrac{dy}{dx} = \dfrac{dy}{du}\dfrac{du}{dx} = (\sec^2 u)(\pi) = \pi \sec^2 \pi x$.

5. Let $u = g(x) = \sqrt{x}$ and $y = f(u) = e^u$. Then $\dfrac{dy}{dx} = \dfrac{dy}{du}\dfrac{du}{dx} = (e^u)\left(\tfrac{1}{2}x^{-1/2}\right) = e^{\sqrt{x}} \cdot \dfrac{1}{2\sqrt{x}} = \dfrac{e^{\sqrt{x}}}{2\sqrt{x}}$.

7. $F(x) = (x^4 + 3x^2 - 2)^5 \quad\Rightarrow\quad F'(x) = 5(x^4 + 3x^2 - 2)^4 \cdot \dfrac{d}{dx}(x^4 + 3x^2 - 2) = 5(x^4 + 3x^2 - 2)^4(4x^3 + 6x)$
$\left[\text{or } 10x(x^4 + 3x^2 - 2)^4(2x^2 + 3)\right]$

9. $F(x) = \sqrt{1 - 2x} = (1 - 2x)^{1/2} \quad\Rightarrow\quad F'(x) = \tfrac{1}{2}(1 - 2x)^{-1/2}(-2) = -\dfrac{1}{\sqrt{1 - 2x}}$

11. $f(z) = \dfrac{1}{z^2 + 1} = (z^2 + 1)^{-1} \quad\Rightarrow\quad f'(z) = -1(z^2 + 1)^{-2}(2z) = -\dfrac{2z}{(z^2 + 1)^2}$

13. $y = \cos(a^3 + x^3) \quad\Rightarrow\quad y' = -\sin(a^3 + x^3) \cdot 3x^2 \quad [a^3 \text{ is just a constant}] \quad = -3x^2 \sin(a^3 + x^3)$

15. $h(t) = t^3 - 3^t \quad\Rightarrow\quad h'(t) = 3t^2 - 3^t \ln 3 \quad$ [by Formula 5]

17. $y = xe^{-kx} \quad\Rightarrow\quad y' = x\left[e^{-kx}(-k)\right] + e^{-kx} \cdot 1 = e^{-kx}(-kx + 1) \quad \left[\text{or } (1 - kx)e^{-kx}\right]$

19. $y = (2x - 5)^4(8x^2 - 5)^{-3} \quad\Rightarrow$

$$\begin{aligned} y' &= 4(2x - 5)^3(2)(8x^2 - 5)^{-3} + (2x - 5)^4(-3)(8x^2 - 5)^{-4}(16x) \\ &= 8(2x - 5)^3(8x^2 - 5)^{-3} - 48x(2x - 5)^4(8x^2 - 5)^{-4} \end{aligned}$$

[This simplifies to $8(2x - 5)^3(8x^2 - 5)^{-4}(-4x^2 + 30x - 5)$.]

21. $y = e^{x\cos x} \quad\Rightarrow\quad y' = e^{x\cos x} \cdot \dfrac{d}{dx}(x\cos x) = e^{x\cos x}\,[x(-\sin x) + (\cos x)\cdot 1] = e^{x\cos x}(\cos x - x\sin x)$

23. $y = \left(\dfrac{x^2 + 1}{x^2 - 1}\right)^3 \quad\Rightarrow$

$$\begin{aligned} y' &= 3\left(\frac{x^2 + 1}{x^2 - 1}\right)^2 \cdot \frac{d}{dx}\left(\frac{x^2 + 1}{x^2 - 1}\right) = 3\left(\frac{x^2 + 1}{x^2 - 1}\right)^2 \cdot \frac{(x^2 - 1)(2x) - (x^2 + 1)(2x)}{(x^2 - 1)^2} \\ &= 3\left(\frac{x^2 + 1}{x^2 - 1}\right)^2 \cdot \frac{2x[x^2 - 1 - (x^2 + 1)]}{(x^2 - 1)^2} = 3\left(\frac{x^2 + 1}{x^2 - 1}\right)^2 \cdot \frac{2x(-2)}{(x^2 - 1)^2} = \frac{-12x(x^2 + 1)^2}{(x^2 - 1)^4} \end{aligned}$$

25. $y = \sec^2 x + \tan^2 x = (\sec x)^2 + (\tan x)^2 \quad\Rightarrow$

$$y' = 2(\sec x)(\sec x \tan x) + 2(\tan x)(\sec^2 x) = 2\sec^2 x \tan x + 2\sec^2 x \tan x = 4\sec^2 x \tan x$$

© 2016 Cengage Learning. All Rights Reserved. May not be scanned, copied or duplicated, or posted to a publicly accessible website, in whole or in part.

27. $y = \dfrac{r}{\sqrt{r^2+1}} \Rightarrow$

$$y' = \frac{\sqrt{r^2+1}\,(1) - r\cdot\frac{1}{2}(r^2+1)^{-1/2}(2r)}{\left(\sqrt{r^2+1}\right)^2} = \frac{\sqrt{r^2+1} - \dfrac{r^2}{\sqrt{r^2+1}}}{\left(\sqrt{r^2+1}\right)^2} = \frac{\dfrac{\sqrt{r^2+1}\,\sqrt{r^2+1} - r^2}{\sqrt{r^2+1}}}{\left(\sqrt{r^2+1}\right)^2}$$

$$= \frac{(r^2+1) - r^2}{\left(\sqrt{r^2+1}\right)^3} = \frac{1}{(r^2+1)^{3/2}} \text{ or } (r^2+1)^{-3/2}$$

Another solution: Write y as a product and make use of the Product Rule. $y = r(r^2+1)^{-1/2} \Rightarrow$

$$y' = r\cdot -\tfrac{1}{2}(r^2+1)^{-3/2}(2r) + (r^2+1)^{-1/2}\cdot 1 = (r^2+1)^{-3/2}[-r^2 + (r^2+1)^1] = (r^2+1)^{-3/2}(1) = (r^2+1)^{-3/2}.$$

The step that students usually have trouble with is factoring out $(r^2+1)^{-3/2}$. But this is no different than factoring out x^2 from $x^2 + x^5$; that is, we are just factoring out a factor with the *smallest* exponent that appears on it. In this case, $-\frac{3}{2}$ is smaller than $-\frac{1}{2}$.

29. $y = \sin(\tan 2x) \Rightarrow y' = \cos(\tan 2x)\cdot \dfrac{d}{dx}(\tan 2x) = \cos(\tan 2x)\cdot \sec^2(2x)\cdot \dfrac{d}{dx}(2x) = 2\cos(\tan 2x)\sec^2(2x)$

31. Using Formula 5 and the Chain Rule, $y = 2^{\sin \pi x} \Rightarrow$

$$y' = 2^{\sin \pi x}(\ln 2)\cdot \frac{d}{dx}(\sin \pi x) = 2^{\sin \pi x}(\ln 2)\cdot \cos \pi x \cdot \pi = 2^{\sin \pi x}(\pi \ln 2)\cos \pi x$$

33. $y = \cot^2(\sin\theta) = [\cot(\sin\theta)]^2 \Rightarrow$

$$y' = 2[\cot(\sin\theta)]\cdot \frac{d}{d\theta}[\cot(\sin\theta)] = 2\cot(\sin\theta)\cdot[-\csc^2(\sin\theta)\cdot\cos\theta] = -2\cos\theta\,\cot(\sin\theta)\,\csc^2(\sin\theta)$$

35. $y = \cos\sqrt{\sin(\tan\pi x)} = \cos(\sin(\tan\pi x))^{1/2} \Rightarrow$

$$y' = -\sin(\sin(\tan\pi x))^{1/2}\cdot\frac{d}{dx}(\sin(\tan\pi x))^{1/2} = -\sin(\sin(\tan\pi x))^{1/2}\cdot\tfrac{1}{2}(\sin(\tan\pi x))^{-1/2}\cdot\frac{d}{dx}(\sin(\tan\pi x))$$

$$= \frac{-\sin\sqrt{\sin(\tan\pi x)}}{2\sqrt{\sin(\tan\pi x)}}\cdot\cos(\tan\pi x)\cdot\frac{d}{dx}\tan\pi x = \frac{-\sin\sqrt{\sin(\tan\pi x)}}{2\sqrt{\sin(\tan\pi x)}}\cdot\cos(\tan\pi x)\cdot\sec^2(\pi x)\cdot\pi$$

$$= \frac{-\pi\cos(\tan\pi x)\sec^2(\pi x)\sin\sqrt{\sin(\tan\pi x)}}{2\sqrt{\sin(\tan\pi x)}}$$

37. $y = \cos(x^2) \Rightarrow y' = -\sin(x^2)\cdot 2x = -2x\sin(x^2) \Rightarrow$

$$y'' = -2x\cos(x^2)\cdot 2x + \sin(x^2)\cdot(-2) = -4x^2\cos(x^2) - 2\sin(x^2)$$

39. $y = e^{\alpha x}\sin\beta x \Rightarrow y' = e^{\alpha x}\cdot\beta\cos\beta x + \sin\beta x\cdot\alpha e^{\alpha x} = e^{\alpha x}(\beta\cos\beta x + \alpha\sin\beta x) \Rightarrow$

$$y'' = e^{\alpha x}(-\beta^2\sin\beta x + \alpha\beta\cos\beta x) + (\beta\cos\beta x + \alpha\sin\beta x)\cdot\alpha e^{\alpha x}$$

$$= e^{\alpha x}(-\beta^2\sin\beta x + \alpha\beta\cos\beta x + \alpha\beta\cos\beta x + \alpha^2\sin\beta x) = e^{\alpha x}(\alpha^2\sin\beta x - \beta^2\sin\beta x + 2\alpha\beta\cos\beta x)$$

$$= e^{\alpha x}\left[(\alpha^2-\beta^2)\sin\beta x + 2\alpha\beta\cos\beta x\right]$$

© 2016 Cengage Learning. All Rights Reserved. May not be scanned, copied or duplicated, or posted to a publicly accessible website, in whole or in part.

41. $y = (1 + 2x)^{10} \Rightarrow y' = 10(1 + 2x)^9 \cdot 2 = 20(1 + 2x)^9$.

At $(0, 1)$, $y' = 20(1 + 0)^9 = 20$, and an equation of the tangent line is $y - 1 = 20(x - 0)$, or $y = 20x + 1$.

43. $y = \sin(\sin x) \Rightarrow y' = \cos(\sin x) \cdot \cos x$. At $(\pi, 0)$, $y' = \cos(\sin \pi) \cdot \cos \pi = \cos(0) \cdot (-1) = 1(-1) = -1$, and an equation of the tangent line is $y - 0 = -1(x - \pi)$, or $y = -x + \pi$.

45. $F(x) = f(g(x)) \Rightarrow F'(x) = f'(g(x)) \cdot g'(x)$, so $F'(5) = f'(g(5)) \cdot g'(5) = f'(-2) \cdot 6 = 4 \cdot 6 = 24$

47. (a) $h(x) = f(g(x)) \Rightarrow h'(x) = f'(g(x)) \cdot g'(x)$, so $h'(1) = f'(g(1)) \cdot g'(1) = f'(2) \cdot 6 = 5 \cdot 6 = 30$.

(b) $H(x) = g(f(x)) \Rightarrow H'(x) = g'(f(x)) \cdot f'(x)$, so $H'(1) = g'(f(1)) \cdot f'(1) = g'(3) \cdot 4 = 9 \cdot 4 = 36$.

49. (a) $F(x) = f(e^x) \Rightarrow F'(x) = f'(e^x) \dfrac{d}{dx}(e^x) = f'(e^x)e^x$

(b) $G(x) = e^{f(x)} \Rightarrow G'(x) = e^{f(x)} \dfrac{d}{dx} f(x) = e^{f(x)} f'(x)$

51. $r(x) = f(g(h(x))) \Rightarrow r'(x) = f'(g(h(x))) \cdot g'(h(x)) \cdot h'(x)$, so

$r'(1) = f'(g(h(1))) \cdot g'(h(1)) \cdot h'(1) = f'(g(2)) \cdot g'(2) \cdot 4 = f'(3) \cdot 5 \cdot 4 = 6 \cdot 5 \cdot 4 = 120$

53. The use of $D, D^2, \ldots, D^n$ is just a derivative notation (see text page 157). In general, $Df(2x) = 2f'(2x)$, $D^2 f(2x) = 4f''(2x), \ldots, D^n f(2x) = 2^n f^{(n)}(2x)$. Since $f(x) = \cos x$ and $50 = 4(12) + 2$, we have $f^{(50)}(x) = f^{(2)}(x) = -\cos x$, so $D^{50} \cos 2x = -2^{50} \cos 2x$.

55. $s(t) = 10 + \frac{1}{4}\sin(10\pi t) \Rightarrow v(t) = s'(t) = 0 + \frac{1}{4}[\cos(10\pi t)](10\pi) = \frac{5}{2}\pi\cos(10\pi t) \Rightarrow$

$a(t) = v'(t) = \frac{5}{2}\pi[-\sin(10\pi t)](10\pi) = -25\pi^2 \sin(10\pi t)$

57. $m(t) = \frac{1}{2}e^{-t}(\sin t - \cos t) + \frac{1}{2} \Rightarrow m'(t) = \frac{1}{2}e^{-t}(\cos t + \sin t) + \left(-\frac{1}{2}e^{-t}\right)(\sin t - \cos t) = e^{-t}\cos t$

Therefore, the rate of change of mRNA concentration as a function of time is $m'(t) = e^{-t}\cos t$.

59. (a) $C(t) = 0.0225te^{-0.0467t} \Rightarrow$

$C'(t) = 0.0225\left(e^{-0.0467t}(1) + t\left(-0.0467e^{-0.0467t}\right)\right) = 0.0225e^{-0.0467t}(1 - 0.0467t)$

After 10 minutes, $C'(10) = 0.0225e^{-0.0467(10)}(1 - 0.0467 \cdot 10) \approx 0.00752$ (mg/mL)/min. Hence, the BAC is increasing at a rate of about 0.00752 (mg/mL)/min.

(b) After another 30 minutes, $C'(40) = 0.0225e^{-0.0467(40)}(1 - 0.0467 \cdot 40) \approx -0.003016$ (mg/mL)/min. Hence, the BAC is decreasing at a rate of about 0.003016 (mg/mL)/min.

61. (a) $\lim\limits_{t\to\infty} p(t) = \lim\limits_{t\to\infty} \dfrac{1}{1 + ae^{-kt}} = \dfrac{1}{1 + a \cdot 0} = 1$, since $k > 0 \Rightarrow -kt \to -\infty \Rightarrow e^{-kt} \to 0$.

(b) $p(t) = (1 + ae^{-kt})^{-1} \Rightarrow \dfrac{dp}{dt} = -(1 + ae^{-kt})^{-2}(-kae^{-kt}) = \dfrac{kae^{-kt}}{(1 + ae^{-kt})^2}$

(c) From the graph of $p(t) = (1 + 10e^{-0.5t})^{-1}$, it seems that $p(t) = 0.8$ (indicating that 80% of the population has heard the rumor) when $t \approx 7.4$ hours.

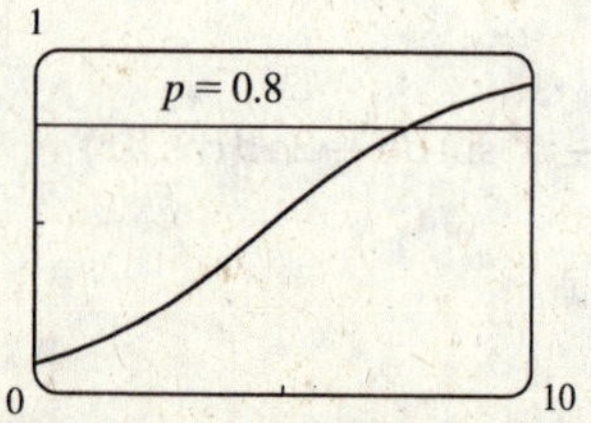

© 2016 Cengage Learning. All Rights Reserved. May not be scanned, copied or duplicated, or posted to a publicly accessible website, in whole or in part.

63. (a) $\dfrac{d}{dx}(xy + 2x + 3x^2) = \dfrac{d}{dx}(4) \;\Rightarrow\; (x \cdot y' + y \cdot 1) + 2 + 6x = 0 \;\Rightarrow\; xy' = -y - 2 - 6x \;\Rightarrow$

$y' = \dfrac{-y - 2 - 6x}{x}$ or $y' = -6 - \dfrac{y + 2}{x}$.

(b) $xy + 2x + 3x^2 = 4 \;\Rightarrow\; xy = 4 - 2x - 3x^2 \;\Rightarrow\; y = \dfrac{4 - 2x - 3x^2}{x} = \dfrac{4}{x} - 2 - 3x$, so $y' = -\dfrac{4}{x^2} - 3$.

(c) From part (a), $y' = \dfrac{-y - 2 - 6x}{x} = \dfrac{-(4/x - 2 - 3x) - 2 - 6x}{x} = \dfrac{-4/x - 3x}{x} = -\dfrac{4}{x^2} - 3$.

65. $\dfrac{d}{dx}(x^3 + y^3) = \dfrac{d}{dx}(1) \;\Rightarrow\; 3x^2 + 3y^2 \cdot y' = 0 \;\Rightarrow\; 3y^2\, y' = -3x^2 \;\Rightarrow\; y' = -\dfrac{x^2}{y^2}$

67. $\dfrac{d}{dx}(x^2 + xy - y^2) = \dfrac{d}{dx}(4) \;\Rightarrow\; 2x + x \cdot y' + y \cdot 1 - 2y\, y' = 0 \;\Rightarrow$

$xy' - 2y\, y' = -2x - y \;\Rightarrow\; (x - 2y)\, y' = -2x - y \;\Rightarrow\; y' = \dfrac{-2x - y}{x - 2y} = \dfrac{2x + y}{2y - x}$

69. $\dfrac{d}{dx}\left[x^4(x + y)\right] = \dfrac{d}{dx}\left[y^2(3x - y)\right] \;\Rightarrow\; x^4(1 + y') + (x + y) \cdot 4x^3 = y^2(3 - y') + (3x - y) \cdot 2y\, y' \;\Rightarrow$

$x^4 + x^4\, y' + 4x^4 + 4x^3 y = 3y^2 - y^2\, y' + 6xy\, y' - 2y^2\, y' \;\Rightarrow\; x^4\, y' + 3y^2\, y' - 6xy\, y' = 3y^2 - 5x^4 - 4x^3 y \;\Rightarrow$

$(x^4 + 3y^2 - 6xy)\, y' = 3y^2 - 5x^4 - 4x^3 y \;\Rightarrow\; y' = \dfrac{3y^2 - 5x^4 - 4x^3 y}{x^4 + 3y^2 - 6xy}$

71. $\dfrac{d}{dx}(4 \cos x \sin y) = \dfrac{d}{dx}(1) \;\Rightarrow\; 4\left[\cos x \cdot \cos y \cdot y' + \sin y \cdot (-\sin x)\right] = 0 \;\Rightarrow$

$y'(4 \cos x \cos y) = 4 \sin x \sin y \;\Rightarrow\; y' = \dfrac{4 \sin x \sin y}{4 \cos x \cos y} = \tan x \tan y$

73. $\dfrac{d}{dx}(e^{x/y}) = \dfrac{d}{dx}(x - y) \;\Rightarrow\; e^{x/y} \cdot \dfrac{d}{dx}\left(\dfrac{x}{y}\right) = 1 - y' \;\Rightarrow$

$e^{x/y} \cdot \dfrac{y \cdot 1 - x \cdot y'}{y^2} = 1 - y' \;\Rightarrow\; e^{x/y} \cdot \dfrac{1}{y} - \dfrac{xe^{x/y}}{y^2} \cdot y' = 1 - y' \;\Rightarrow\; y' - \dfrac{xe^{x/y}}{y^2} \cdot y' = 1 - \dfrac{e^{x/y}}{y} \;\Rightarrow$

$$y'\left(1 - \frac{xe^{x/y}}{y^2}\right) = \frac{y - e^{x/y}}{y} \;\Rightarrow\; y' = \frac{\dfrac{y - e^{x/y}}{y}}{\dfrac{y^2 - xe^{x/y}}{y^2}} = \frac{y(y - e^{x/y})}{y^2 - xe^{x/y}}$$

75. $\dfrac{d}{dx}(e^y \cos x) = \dfrac{d}{dx}\left[1 + \sin(xy)\right] \;\Rightarrow\; e^y(-\sin x) + \cos x \cdot e^y \cdot y' = \cos(xy) \cdot (xy' + y \cdot 1) \;\Rightarrow$

$-e^y \sin x + e^y \cos x \cdot y' = x \cos(xy) \cdot y' + y \cos(xy) \;\Rightarrow\; e^y \cos x \cdot y' - x \cos(xy) \cdot y' = e^y \sin x + y \cos(xy) \;\Rightarrow$

$\left[e^y \cos x - x \cos(xy)\right] y' = e^y \sin x + y \cos(xy) \;\Rightarrow\; y' = \dfrac{e^y \sin x + y \cos(xy)}{e^y \cos x - x \cos(xy)}$

77. $x^2 + xy + y^2 = 3 \;\Rightarrow\; 2x + x\,y' + y \cdot 1 + 2yy' = 0 \;\Rightarrow\; x\,y' + 2y\,y' = -2x - y \;\Rightarrow\; y'(x + 2y) = -2x - y \;\Rightarrow$

$y' = \dfrac{-2x - y}{x + 2y}$. When $x = 1$ and $y = 1$, we have $y' = \dfrac{-2 - 1}{1 + 2 \cdot 1} = \dfrac{-3}{3} = -1$, so an equation of the tangent line is

$y - 1 = -1(x - 1)$ or $y = -x + 2$.

© 2016 Cengage Learning. All Rights Reserved. May not be scanned, copied or duplicated, or posted to a publicly accessible website, in whole or in part.

79. $x^2 + y^2 = (2x^2 + 2y^2 - x)^2 \Rightarrow 2x + 2y\,y' = 2(2x^2 + 2y^2 - x)(4x + 4y\,y' - 1)$. When $x = 0$ and $y = \frac{1}{2}$, we have $0 + y' = 2(\frac{1}{2})(2y' - 1) \Rightarrow y' = 2y' - 1 \Rightarrow y' = 1$, so an equation of the tangent line is $y - \frac{1}{2} = 1(x - 0)$ or $y = x + \frac{1}{2}$.

81. $\rho e^{-qA} = 1 - A \Rightarrow \dfrac{d}{d\rho}\left(\rho e^{-qA}\right) = \dfrac{d}{d\rho}(1 - A) \Rightarrow e^{-qA}(1) + \rho e^{-qA}\left(-q\dfrac{dA}{d\rho}\right) = -\dfrac{dA}{d\rho} \Rightarrow$

$$\frac{dA}{d\rho} = \frac{-e^{qA}}{1 - \rho q e^{-qA}} = \frac{1}{\rho q - e^{qA}}$$

83. $V = x^3 \Rightarrow \dfrac{dV}{dt} = \dfrac{dV}{dx}\dfrac{dx}{dt} = 3x^2\dfrac{dx}{dt}$

85. Let s denote the side of a square. The square's area A is given by $A = s^2$. Differentiating with respect to t gives us $\dfrac{dA}{dt} = 2s\dfrac{ds}{dt}$. When $A = 16$, $s = 4$. Substitution 4 for s and 6 for $\dfrac{ds}{dt}$ gives us $\dfrac{dA}{dt} = 2(4)(6) = 48\text{ cm}^2/\text{s}$.

87. Differentiating both sides of $PV = C$ with respect to t and using the Product Rule gives us $P\dfrac{dV}{dt} + V\dfrac{dP}{dt} = 0 \Rightarrow$ $\dfrac{dV}{dt} = -\dfrac{V}{P}\dfrac{dP}{dt}$. When $V = 600$, $P = 150$ and $\dfrac{dP}{dt} = 20$, so we have $\dfrac{dV}{dt} = -\dfrac{600}{150}(20) = -80$. Thus, the volume is decreasing at a rate of $80\text{ cm}^3/\text{min}$.

89. (a) $m = \pi r^2 L\left[\rho - (\rho - 1)k^2\right] \Rightarrow \dfrac{dm}{dt} = \pi r^2 L\left(\dfrac{d\rho}{dt} - k^2\dfrac{d\rho}{dt}\right) = \pi r^2 L\left(1 - k^2\right)\dfrac{d\rho}{dt}$

(b) $m = \pi r^2 L\left[\rho - (\rho - 1)k^2\right] \Rightarrow \dfrac{dm}{dt} = \pi r^2 L\left[0 - 2(\rho - 1)k\dfrac{dk}{dt}\right] = -2\pi r^2 L(\rho - 1)k\dfrac{dk}{dt}$

91. $r = \frac{1}{2}\left(1 + \sqrt{1 + 8s}\right) \Rightarrow \dfrac{dr}{da} = \dfrac{dr}{ds}\dfrac{ds}{da} = \frac{1}{2}\left(\frac{1}{2}(1 + 8s)^{-1/2} \cdot 8\right)\dfrac{ds}{da} = \dfrac{2}{\sqrt{1 + 8s}}\dfrac{ds}{da}$

93. $P = 4\eta l v/R^2 = 4\eta l v R^{-2} \Rightarrow \dfrac{dP}{dx} = \dfrac{dP}{dR}\dfrac{dR}{dx} = 4\eta l v\left(-2R^{-3}\right)\dfrac{dR}{dx} = -\dfrac{8\eta l v}{R^3}R'(x)$

95. Since $\theta^\circ = \left(\frac{\pi}{180}\right)\theta$ rad, we have $\dfrac{d}{d\theta}(\sin\theta^\circ) = \dfrac{d}{d\theta}\left(\sin\frac{\pi}{180}\theta\right) = \frac{\pi}{180}\cos\frac{\pi}{180}\theta = \frac{\pi}{180}\cos\theta^\circ$.

3.6 Exponential Growth and Decay

1. The relative growth rate is $\dfrac{1}{P}\dfrac{dP}{dt} = 0.7944$, so $\dfrac{dP}{dt} = 0.7944P$ and, by Theorem 2, $P(t) = P(0)e^{0.7944t} = 2e^{0.7944t}$.

Thus, $P(6) = 2e^{0.7944(6)} \approx 234.99$ or about 235 members.

3. (a) By Theorem 2, $P(t) = P(0)e^{kt} = 100e^{kt}$. Now $P(1) = 100e^{k(1)} = 420 \Rightarrow e^k = \frac{420}{100} \Rightarrow k = \ln 4.2$.

So $P(t) = 100e^{(\ln 4.2)t} = 100(4.2)^t$.

(b) $P(3) = 100(4.2)^3 = 7408.8 \approx 7409$ bacteria

(c) $dP/dt = kP \Rightarrow P'(3) = k \cdot P(3) = (\ln 4.2)\left(100(4.2)^3\right)$ [from part (a)] $\approx 10{,}632$ bacteria/h

(d) $P(t) = 100(4.2)^t = 10{,}000 \Rightarrow (4.2)^t = 100 \Rightarrow t = (\ln 100)/(\ln 4.2) \approx 3.2$ hours

© 2016 Cengage Learning. All Rights Reserved. May not be scanned, copied or duplicated, or posted to a publicly accessible website, in whole or in part.

5. (a) Let the population (in millions) in the year t be $P(t)$. Since the initial time is the year 1750, we substitute $t - 1750$ for t in Theorem 2, so the exponential model gives $P(t) = P(1750)e^{k(t-1750)}$. Then $P(1800) = 980 = 790e^{k(1800-1750)} \Rightarrow$ $\frac{980}{790} = e^{k(50)} \Rightarrow \ln \frac{980}{790} = 50k \Rightarrow k = \frac{1}{50} \ln \frac{980}{790} \approx 0.0043104$. So with this model, we have $P(1900) = 790e^{k(1900-1750)} \approx 1508$ million, and $P(1950) = 790e^{k(1950-1750)} \approx 1871$ million. Both of these estimates are much too low.

(b) In this case, the exponential model gives $P(t) = P(1850)e^{k(t-1850)} \Rightarrow P(1900) = 1650 = 1260e^{k(1900-1850)} \Rightarrow$ $\ln \frac{1650}{1260} = k(50) \Rightarrow k = \frac{1}{50} \ln \frac{1650}{1260} \approx 0.005393$. So with this model, we estimate $P(1950) = 1260e^{k(1950-1850)} \approx 2161$ million. This is still too low, but closer than the estimate of $P(1950)$ in part (a).

(c) The exponential model gives $P(t) = P(1900)e^{k(t-1900)} \Rightarrow P(1950) = 2560 = 1650e^{k(1950-1900)} \Rightarrow$ $\ln \frac{2560}{1650} = k(50) \Rightarrow k = \frac{1}{50} \ln \frac{2560}{1650} \approx 0.008785$. With this model, we estimate $P(2000) = 1650e^{k(2000-1900)} \approx 3972$ million. This is much too low. The discrepancy is explained by the fact that the world birth rate (average yearly number of births per person) is about the same as always, whereas the mortality rate (especially the infant mortality rate) is much lower, owing mostly to advances in medical science and to the wars in the first part of the twentieth century. The exponential model assumes, among other things, that the birth and mortality rates will remain constant.

7. (a) If $y(t)$ is the mass (in mg) remaining after t years, then $y(t) = y(0)e^{kt} = 100e^{kt}$.

$y(30) = 100e^{30k} = \frac{1}{2}(100) \Rightarrow e^{30k} = \frac{1}{2} \Rightarrow k = -(\ln 2)/30 \Rightarrow y(t) = 100e^{-(\ln 2)t/30} = 100 \cdot 2^{-t/30}$

(b) $y(100) = 100 \cdot 2^{-100/30} \approx 9.92$ mg

(c) $y(t) = 100 \cdot 2^{-t/30} \Rightarrow y'(t) = 100 \cdot 2^{-t/30} \cdot (-\frac{1}{30} \ln(2)) \Rightarrow y'(100) = -\dfrac{10 \ln 2}{3} \cdot 2^{-100/30} \approx -0.229$ mg/year

(d) $100e^{-(\ln 2)t/30} = 1 \Rightarrow -(\ln 2)t/30 = \ln \frac{1}{100} \Rightarrow t = -30 \frac{\ln 0.01}{\ln 2} \approx 199.3$ years

9. (a) If $y(t)$ is the mass after t days and $y(0) = A$, then $y(t) = Ae^{kt}$.

$y(1) = Ae^{k} = 0.945A \Rightarrow e^{k} = 0.945 \Rightarrow k = \ln 0.945$.

Then $Ae^{(\ln 0.945)t} = \frac{1}{2}A \Leftrightarrow \ln e^{(\ln 0.945)t} = \ln \frac{1}{2} \Leftrightarrow (\ln 0.945)t = \ln \frac{1}{2} \Leftrightarrow t = -\frac{\ln 2}{\ln 0.945} \approx 12.25$ years.

(b) $Ae^{(\ln 0.945)t} = 0.20A \Leftrightarrow (\ln 0.945)t = \ln \frac{1}{5} \Leftrightarrow t = -\frac{\ln 5}{\ln 0.945} \approx 28.45$ years

11. Let $y(t)$ be the amount of ^{14}C at time t.Thus, $y(t) = y(0)e^{-kt}$ and k is determined by using the half-life:

$y(5730) = \frac{1}{2}y(0) \Rightarrow y(0)e^{-k(5730)} = \frac{1}{2}y(0) \Rightarrow e^{-5730k} = \frac{1}{2} \Rightarrow -5730k = \ln \frac{1}{2} \Rightarrow k = -\dfrac{\ln \frac{1}{2}}{5730} = \dfrac{\ln 2}{5730}$.

The fraction of ^{14}C remaining in a 68 million year old dinosaur is $\dfrac{y(68 \cdot 10^6)}{y(0)} = e^{-(68 \cdot 10^6) \ln 2/5730} \approx 0$. With a 0.1% threshold, detecting ^{14}C requires that $\dfrac{y(t)}{y(0)} \geq 0.001 \Rightarrow e^{-(\ln 2/5730)t} \geq 0.001 \Rightarrow -\dfrac{\ln 2}{5730} t \geq \ln 0.001 \Rightarrow$

$t \leq -\dfrac{5730 \cdot \ln 0.001}{\ln 2} \approx 57{,}104$ years.

© 2016 Cengage Learning. All Rights Reserved. May not be scanned, copied or duplicated, or posted to a publicly accessible website, in whole or in part.

13. (a) Using Newton's Law of Cooling, $\dfrac{dT}{dt} = k(T - T_s)$, we have $\dfrac{dT}{dt} = k(T - 75)$. Now let $y = T - 75$, so

$y(0) = T(0) - 75 = 185 - 75 = 110$, so y is a solution of the initial-value problem $dy/dt = ky$ with $y(0) = 110$ and by Theorem 2 we have $y(t) = y(0)e^{kt} = 110e^{kt}$.

$y(30) = 110e^{30k} = 150 - 75 \;\Rightarrow\; e^{30k} = \frac{75}{110} = \frac{15}{22} \;\Rightarrow\; k = \frac{1}{30}\ln\frac{15}{22}$, so $y(t) = 110e^{\frac{1}{30}t\ln\left(\frac{15}{22}\right)}$ and

$y(45) = 110e^{\frac{45}{30}\ln\left(\frac{15}{22}\right)} \approx 62°\text{F}$. Thus, $T(45) \approx 62 + 75 = 137°\text{F}$.

(b) $T(t) = 100 \;\Rightarrow\; y(t) = 25$. $y(t) = 110e^{\frac{1}{30}t\ln\left(\frac{15}{22}\right)} = 25 \;\Rightarrow\; e^{\frac{1}{30}t\ln\left(\frac{15}{22}\right)} = \frac{25}{110} \;\Rightarrow\; \frac{1}{30}t\ln\frac{15}{22} = \ln\frac{25}{110} \;\Rightarrow$

$$t = \frac{30\ln\frac{25}{110}}{\ln\frac{15}{22}} \approx 116 \text{ min.}$$

15. $\dfrac{dT}{dt} = k(T - 20)$. Letting $y = T - 20$, we get $\dfrac{dy}{dt} = ky$, so $y(t) = y(0)e^{kt}$. $y(0) = T(0) - 20 = 5 - 20 = -15$, so

$y(25) = y(0)e^{25k} = -15e^{25k}$, and $y(25) = T(25) - 20 = 10 - 20 = -10$, so $-15e^{25k} = -10 \;\Rightarrow\; e^{25k} = \frac{2}{3}$. Thus,

$25k = \ln\left(\frac{2}{3}\right)$ and $k = \frac{1}{25}\ln\left(\frac{2}{3}\right)$, so $y(t) = y(0)e^{kt} = -15e^{(1/25)\ln(2/3)t}$. More simply, $e^{25k} = \frac{2}{3} \;\Rightarrow\; e^k = \left(\frac{2}{3}\right)^{1/25} \;\Rightarrow$

$e^{kt} = \left(\frac{2}{3}\right)^{t/25} \;\Rightarrow\; y(t) = -15 \cdot \left(\frac{2}{3}\right)^{t/25}$.

(a) $T(50) = 20 + y(50) = 20 - 15 \cdot \left(\frac{2}{3}\right)^{50/25} = 20 - 15 \cdot \left(\frac{2}{3}\right)^2 = 20 - \frac{20}{3} = 13.\bar{3}\,°\text{C}$

(b) $15 = T(t) = 20 + y(t) = 20 - 15 \cdot \left(\frac{2}{3}\right)^{t/25} \;\Rightarrow\; 15 \cdot \left(\frac{2}{3}\right)^{t/25} = 5 \;\Rightarrow\; \left(\frac{2}{3}\right)^{t/25} = \frac{1}{3} \;\Rightarrow$

$(t/25)\ln\left(\frac{2}{3}\right) = \ln\left(\frac{1}{3}\right) \;\Rightarrow\; t = 25\ln\left(\frac{1}{3}\right)/\ln\left(\frac{2}{3}\right) \approx 67.74$ min.

17. (a) Let $P(h)$ be the pressure at altitude h. Then $dP/dh = kP \;\Rightarrow\; P(h) = P(0)e^{kh} = 101.3e^{kh}$.

$P(1000) = 101.3e^{1000k} = 87.14 \;\Rightarrow\; 1000k = \ln\left(\frac{87.14}{101.3}\right) \;\Rightarrow\; k = \frac{1}{1000}\ln\left(\frac{87.14}{101.3}\right) \;\Rightarrow$

$P(h) = 101.3\,e^{\frac{1}{1000}h\ln\left(\frac{87.14}{101.3}\right)}$, so $P(3000) = 101.3e^{3\ln\left(\frac{87.14}{101.3}\right)} \approx 64.5$ kPa.

(b) $P(6187) = 101.3\,e^{\frac{6187}{1000}\ln\left(\frac{87.14}{101.3}\right)} \approx 39.9$ kPa

3.7 Derivatives of the Logarithmic and Inverse Tangent Functions

1. The differentiation formula for logarithmic functions, $\dfrac{d}{dx}(\log_b x) = \dfrac{1}{x\ln b}$, is simplest when $b = e$ because $\ln e = 1$.

3. $f(x) = \sin(\ln x) \;\Rightarrow\; f'(x) = \cos(\ln x) \cdot \dfrac{d}{dx}\ln x = \cos(\ln x) \cdot \dfrac{1}{x} = \dfrac{\cos(\ln x)}{x}$

5. $f(x) = \log_2(1 - 3x) \;\Rightarrow\; f'(x) = \dfrac{1}{(1 - 3x)\ln 2}\dfrac{d}{dx}(1 - 3x) = \dfrac{-3}{(1 - 3x)\ln 2}$ or $\dfrac{3}{(3x - 1)\ln 2}$

7. $f(x) = \sqrt[5]{\ln x} = (\ln x)^{1/5} \;\Rightarrow\; f'(x) = \frac{1}{5}(\ln x)^{-4/5}\dfrac{d}{dx}(\ln x) = \dfrac{1}{5(\ln x)^{4/5}} \cdot \dfrac{1}{x} = \dfrac{1}{5x\sqrt[5]{(\ln x)^4}}$

9. $f(x) = \sin x\,\ln(5x) \;\Rightarrow\; f'(x) = \sin x \cdot \dfrac{1}{5x} \cdot \dfrac{d}{dx}(5x) + \ln(5x) \cdot \cos x = \dfrac{\sin x \cdot 5}{5x} + \cos x\ln(5x) = \dfrac{\sin x}{x} + \cos x\,\ln(5x)$

11. $F(t) = \ln\dfrac{(2t + 1)^3}{(3t - 1)^4} = \ln(2t + 1)^3 - \ln(3t - 1)^4 = 3\ln(2t + 1) - 4\ln(3t - 1) \;\Rightarrow$

$F'(t) = 3 \cdot \dfrac{1}{2t + 1} \cdot 2 - 4 \cdot \dfrac{1}{3t - 1} \cdot 3 = \dfrac{6}{2t + 1} - \dfrac{12}{3t - 1}$, or combined, $\dfrac{-6(t + 3)}{(2t + 1)(3t - 1)}$.

© 2016 Cengage Learning. All Rights Reserved. May not be scanned, copied or duplicated, or posted to a publicly accessible website, in whole or in part.

13. $g(x) = \ln\left(x\sqrt{x^2-1}\right) = \ln x + \ln(x^2-1)^{1/2} = \ln x + \frac{1}{2}\ln(x^2-1) \Rightarrow$

$$g'(x) = \frac{1}{x} + \frac{1}{2}\cdot\frac{1}{x^2-1}\cdot 2x = \frac{1}{x} + \frac{x}{x^2-1} = \frac{x^2-1+x\cdot x}{x(x^2-1)} = \frac{2x^2-1}{x(x^2-1)}$$

15. $y = \ln\left|2-x-5x^2\right| \Rightarrow y' = \dfrac{1}{2-x-5x^2}\cdot(-1-10x) = \dfrac{-10x-1}{2-x-5x^2}$ or $\dfrac{10x+1}{5x^2+x-2}$

17. $y = \ln(e^{-x}+xe^{-x}) = \ln(e^{-x}(1+x)) = \ln(e^{-x}) + \ln(1+x) = -x + \ln(1+x) \Rightarrow$

$$y' = -1 + \frac{1}{1+x} = \frac{-1-x+1}{1+x} = -\frac{x}{1+x}$$

19. $y = 2x\log_{10}\sqrt{x} = 2x\log_{10}x^{1/2} = 2x\cdot\frac{1}{2}\log_{10}x = x\log_{10}x \Rightarrow y' = x\cdot\dfrac{1}{x\ln 10} + \log_{10}x\cdot 1 = \dfrac{1}{\ln 10} + \log_{10}x$

Note: $\dfrac{1}{\ln 10} = \dfrac{\ln e}{\ln 10} = \log_{10}e$, so the answer could be written as $\dfrac{1}{\ln 10} + \log_{10}x = \log_{10}e + \log_{10}x = \log_{10}ex$.

21. $y = x^2\ln(2x) \Rightarrow y' = x^2\cdot\dfrac{1}{2x}\cdot 2 + \ln(2x)\cdot(2x) = x + 2x\ln(2x) \Rightarrow$

$$y'' = 1 + 2x\cdot\frac{1}{2x}\cdot 2 + \ln(2x)\cdot 2 = 1 + 2 + 2\ln(2x) = 3 + 2\ln(2x)$$

23. $f(x) = \dfrac{x}{1-\ln(x-1)} \Rightarrow$

$$f'(x) = \frac{[1-\ln(x-1)]\cdot 1 - x\cdot\dfrac{-1}{x-1}}{[1-\ln(x-1)]^2} = \frac{\dfrac{(x-1)[1-\ln(x-1)]+x}{x-1}}{[1-\ln(x-1)]^2} = \frac{x-1-(x-1)\ln(x-1)+x}{(x-1)[1-\ln(x-1)]^2}$$

$$= \frac{2x-1-(x-1)\ln(x-1)}{(x-1)[1-\ln(x-1)]^2}$$

$\text{Dom}(f) = \{x \mid x-1>0 \text{ and } 1-\ln(x-1)\neq 0\} = \{x \mid x>1 \text{ and } \ln(x-1)\neq 1\}$

$= \{x \mid x>1 \text{ and } x-1\neq e^1\} = \{x \mid x>1 \text{ and } x\neq 1+e\} = (1,1+e)\cup(1+e,\infty)$

25. $y = \ln(x^2-3x+1) \Rightarrow y' = \dfrac{1}{x^2-3x+1}\cdot(2x-3) \Rightarrow y'(3) = \frac{1}{1}\cdot 3 = 3$, so an equation of a tangent line at $(3,0)$ is $y-0 = 3(x-3)$, or $y = 3x-9$.

27. $f(x) = \dfrac{\ln x}{x^2} \Rightarrow f'(x) = \dfrac{x^2(1/x)-(\ln x)(2x)}{(x^2)^2} = \dfrac{x-2x\ln x}{x^4} = \dfrac{x(1-2\ln x)}{x^4} = \dfrac{1-2\ln x}{x^3}$,

so $f'(1) = \dfrac{1-2\ln 1}{1^3} = \dfrac{1-2\cdot 0}{1} = 1$.

29. $t = \ln\left(\dfrac{3c+\sqrt{9c^2-8c}}{2}\right) = \ln\left(3c+\sqrt{9c^2-8c}\right) - \ln 2 \Rightarrow$

$$\frac{dt}{dc} = \frac{1}{3c+\sqrt{9c^2-8c}}\cdot\left(3+\tfrac{1}{2}\left(9c^2-8c\right)^{-1/2}(18c-8)\right) = \frac{3+(9c-4)/\sqrt{9c^2-8c}}{3c+\sqrt{9c^2-8c}}.$$

This gives the rate of change of dialysis duration as the initial urea concentration increases.

31. $a = \dfrac{5370}{\ln 2}\ln\left(\dfrac{N_0}{N}\right) = \dfrac{5370}{\ln 2}(\ln N_0 - \ln N) \Rightarrow \dfrac{da}{dN} = 0 - \dfrac{5370}{\ln 2}\cdot\dfrac{1}{N}(1) = -\dfrac{5370}{(\ln 2)N}$

This gives the rate of change of the estimated age with respect to an increase in the measured amount of ^{14}C.

© 2016 Cengage Learning. All Rights Reserved. May not be scanned, copied or duplicated, or posted to a publicly accessible website, in whole or in part.

33. $y = (2x+1)^5(x^4-3)^6 \quad\Rightarrow\quad \ln y = \ln\big((2x+1)^5(x^4-3)^6\big) \quad\Rightarrow\quad \ln y = 5\ln(2x+1) + 6\ln(x^4-3) \quad\Rightarrow$

$$\frac{1}{y}y' = 5\cdot\frac{1}{2x+1}\cdot 2 + 6\cdot\frac{1}{x^4-3}\cdot 4x^3 \quad\Rightarrow$$

$$y' = y\left(\frac{10}{2x+1} + \frac{24x^3}{x^4-3}\right) = (2x+1)^5(x^4-3)^6\left(\frac{10}{2x+1} + \frac{24x^3}{x^4-3}\right).$$

[The answer could be simplified to $y' = 2(2x+1)^4(x^4-3)^5(29x^4+12x^3-15)$, but this is unnecessary.]

35. $y = \dfrac{\sin^2 x\,\tan^4 x}{(x^2+1)^2} \quad\Rightarrow\quad \ln y = \ln(\sin^2 x\,\tan^4 x) - \ln(x^2+1)^2 \quad\Rightarrow$

$$\ln y = \ln(\sin x)^2 + \ln(\tan x)^4 - \ln(x^2+1)^2 \quad\Rightarrow\quad \ln y = 2\,\ln|\sin x| + 4\,\ln|\tan x| - 2\,\ln(x^2+1) \quad\Rightarrow$$

$$\frac{1}{y}y' = 2\cdot\frac{1}{\sin x}\cdot\cos x + 4\cdot\frac{1}{\tan x}\cdot\sec^2 x - 2\cdot\frac{1}{x^2+1}\cdot 2x \quad\Rightarrow\quad y' = \frac{\sin^2 x\,\tan^4 x}{(x^2+1)^2}\left(2\cot x + \frac{4\sec^2 x}{\tan x} - \frac{4x}{x^2+1}\right)$$

37. $y = x^x \quad\Rightarrow\quad \ln y = \ln x^x \quad\Rightarrow\quad \ln y = x\ln x \quad\Rightarrow\quad y'/y = x(1/x) + (\ln x)\cdot 1 \quad\Rightarrow\quad y' = y(1+\ln x) \quad\Rightarrow$

$$y' = x^x(1+\ln x)$$

39. $y = (\cos x)^x \quad\Rightarrow\quad \ln y = \ln(\cos x)^x \quad\Rightarrow\quad \ln y = x\ln\cos x \quad\Rightarrow\quad \dfrac{1}{y}\,y' = x\cdot\dfrac{1}{\cos x}\cdot(-\sin x) + \ln\cos x\cdot 1 \quad\Rightarrow$

$$y' = y\left(\ln\cos x - \frac{x\sin x}{\cos x}\right) \quad\Rightarrow\quad y' = (\cos x)^x(\ln\cos x - x\tan x)$$

41. $y = (\tan x)^{1/x} \quad\Rightarrow\quad \ln y = \ln(\tan x)^{1/x} \quad\Rightarrow\quad \ln y = \dfrac{1}{x}\ln\tan x \quad\Rightarrow$

$$\frac{1}{y}y' = \frac{1}{x}\cdot\frac{1}{\tan x}\cdot\sec^2 x + \ln\tan x\cdot\left(-\frac{1}{x^2}\right) \quad\Rightarrow\quad y' = y\left(\frac{\sec^2 x}{x\tan x} - \frac{\ln\tan x}{x^2}\right) \quad\Rightarrow$$

$$y' = (\tan x)^{1/x}\left(\frac{\sec^2 x}{x\tan x} - \frac{\ln\tan x}{x^2}\right) \quad\text{or}\quad y' = (\tan x)^{1/x}\cdot\frac{1}{x}\left(\csc x\sec x - \frac{\ln\tan x}{x}\right)$$

43. $y = (\tan^{-1} x)^2 \quad\Rightarrow\quad y' = 2(\tan^{-1} x)^1\cdot\dfrac{d}{dx}(\tan^{-1} x) = 2\tan^{-1} x\cdot\dfrac{1}{1+x^2} = \dfrac{2\tan^{-1} x}{1+x^2}$

45. $y = \arctan(\cos\theta) \quad\Rightarrow\quad y' = \dfrac{1}{1+(\cos\theta)^2}\,(-\sin\theta) = -\dfrac{\sin\theta}{1+\cos^2\theta}$

47. $y = \tan^{-1}\left(x - \sqrt{x^2+1}\right) \quad\Rightarrow$

$$y' = \frac{1}{1+\left(x-\sqrt{x^2+1}\right)^2}\left(1 - \frac{x}{\sqrt{x^2+1}}\right) = \frac{1}{1+x^2-2x\sqrt{x^2+1}+x^2+1}\left(\frac{\sqrt{x^2+1}-x}{\sqrt{x^2+1}}\right)$$

$$= \frac{\sqrt{x^2+1}-x}{2\left(1+x^2-x\sqrt{x^2+1}\right)\sqrt{x^2+1}} = \frac{\sqrt{x^2+1}-x}{2\left[\sqrt{x^2+1}\,(1+x^2) - x(x^2+1)\right]} = \frac{\sqrt{x^2+1}-x}{2\left[(1+x^2)\left(\sqrt{x^2+1}-x\right)\right]}$$

$$= \frac{1}{2(1+x^2)}$$

49. Let $t = e^x$. As $x\to\infty$, $t\to\infty$. $\lim\limits_{x\to\infty}\arctan(e^x) = \lim\limits_{t\to\infty}\arctan t = \frac{\pi}{2}$ by (8).

© 2016 Cengage Learning. All Rights Reserved. May not be scanned, copied or duplicated, or posted to a publicly accessible website, in whole or in part.

51. $y = \ln(x^2 + y^2) \Rightarrow y' = \dfrac{1}{x^2 + y^2}\dfrac{d}{dx}(x^2 + y^2) \Rightarrow y' = \dfrac{2x + 2yy'}{x^2 + y^2} \Rightarrow x^2y' + y^2y' = 2x + 2yy' \Rightarrow$

$x^2y' + y^2y' - 2yy' = 2x \Rightarrow (x^2 + y^2 - 2y)y' = 2x \Rightarrow y' = \dfrac{2x}{x^2 + y^2 - 2y}$

53. $f(x) = \ln(x - 1) \Rightarrow f'(x) = \dfrac{1}{(x - 1)} = (x - 1)^{-1} \Rightarrow f''(x) = -(x - 1)^{-2} \Rightarrow f'''(x) = 2(x - 1)^{-3} \Rightarrow$

$f^{(4)}(x) = -2 \cdot 3(x - 1)^{-4} \Rightarrow \cdots \Rightarrow f^{(n)}(x) = (-1)^{n-1} \cdot 2 \cdot 3 \cdot 4 \cdot \cdots \cdot (n - 1)(x - 1)^{-n} = (-1)^{n-1}\dfrac{(n - 1)!}{(x - 1)^n}$

55. If $f(x) = \ln(1 + x)$, then $f'(x) = \dfrac{1}{1 + x}$, so $f'(0) = 1$.

Thus, $\displaystyle\lim_{x \to 0}\frac{\ln(1 + x)}{x} = \lim_{x \to 0}\frac{f(x)}{x} = \lim_{x \to 0}\frac{f(x) - f(0)}{x - 0} = f'(0) = 1.$

3.8 Linear Approximations and Taylor Polynomials

1. $f(x) = x^4 + 3x^2 \Rightarrow f'(x) = 4x^3 + 6x$, so $f(-1) = 4$ and $f'(-1) = -10$.

Thus, $L(x) = f(-1) + f'(-1)(x - (-1)) = 4 + (-10)(x + 1) = -10x - 6$.

3. $f(x) = \cos x \Rightarrow f'(x) = -\sin x$, so $f\left(\frac{\pi}{2}\right) = 0$ and $f'\left(\frac{\pi}{2}\right) = -1$.

Thus, $L(x) = f\left(\frac{\pi}{2}\right) + f'\left(\frac{\pi}{2}\right)\left(x - \frac{\pi}{2}\right) = 0 - 1\left(x - \frac{\pi}{2}\right) = -x + \frac{\pi}{2}$.

5. $f(x) = \sqrt{1 - x} \Rightarrow f'(x) = \dfrac{-1}{2\sqrt{1 - x}}$, so $f(0) = 1$ and $f'(0) = -\frac{1}{2}$.

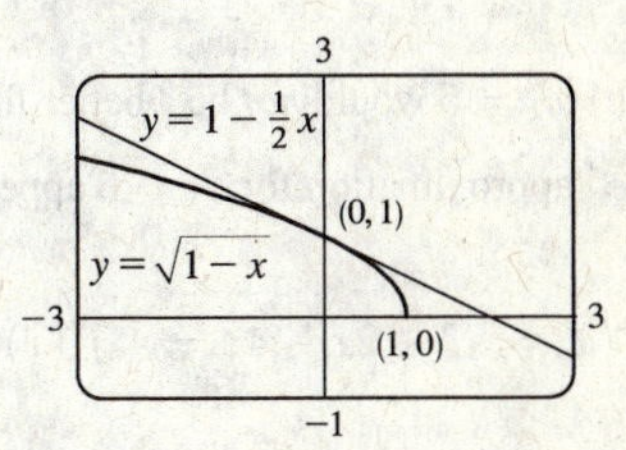

Therefore,

$\sqrt{1 - x} = f(x) \approx f(0) + f'(0)(x - 0) = 1 + \left(-\frac{1}{2}\right)(x - 0) = 1 - \frac{1}{2}x.$

So $\sqrt{0.9} = \sqrt{1 - 0.1} \approx 1 - \frac{1}{2}(0.1) = 0.95$

and $\sqrt{0.99} = \sqrt{1 - 0.01} \approx 1 - \frac{1}{2}(0.01) = 0.995$.

7. $f(x) = \sqrt[3]{1 - x} = (1 - x)^{1/3} \Rightarrow f'(x) = -\frac{1}{3}(1 - x)^{-2/3}$, so $f(0) = 1$ and $f'(0) = -\frac{1}{3}$. Thus, $f(x) \approx f(0) + f'(0)(x - 0) = 1 - \frac{1}{3}x$. We need $\sqrt[3]{1 - x} - 0.1 < 1 - \frac{1}{3}x < \sqrt[3]{1 - x} + 0.1$, which is true when $-1.204 < x < 0.706$.

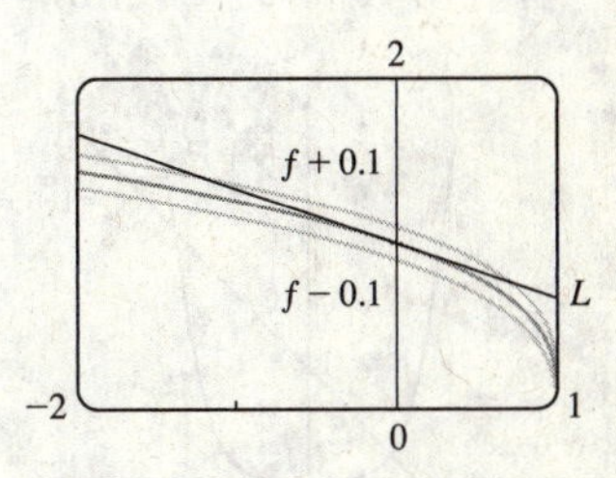

9. $f(x) = \dfrac{1}{(1 + 2x)^4} = (1 + 2x)^{-4} \Rightarrow$

$f'(x) = -4(1 + 2x)^{-5}(2) = \dfrac{-8}{(1 + 2x)^5}$, so $f(0) = 1$ and $f'(0) = -8$.

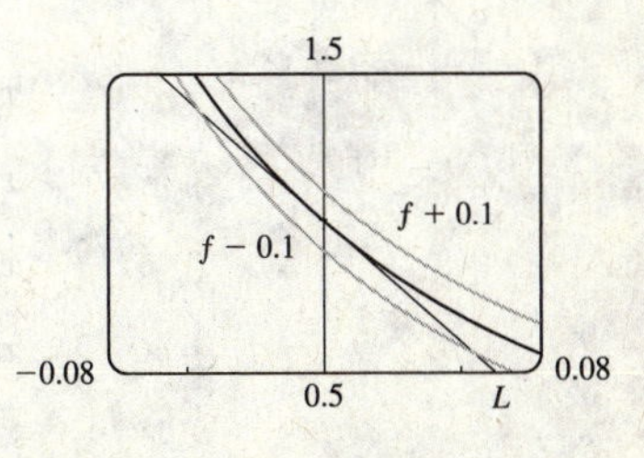

Thus, $f(x) \approx f(0) + f'(0)(x - 0) = 1 + (-8)(x - 0) = 1 - 8x$.

We need $\dfrac{1}{(1 + 2x)^4} - 0.1 < 1 - 8x < \dfrac{1}{(1 + 2x)^4} + 0.1$, which is true when $-0.045 < x < 0.055$.

© 2016 Cengage Learning. All Rights Reserved. May not be scanned, copied or duplicated, or posted to a publicly accessible website, in whole or in part.

11. To estimate $(2.001)^5$, we'll find the linearization of $f(x) = x^5$ at $a = 2$. Since $f'(x) = 5x^4$, $f(2) = 32$, and $f'(2) = 80$, we have $L(x) = 32 + 80(x - 2) = 80x - 128$. Thus, $x^5 \approx 80x - 128$ when x is near 2 , so $(2.001)^5 \approx 80(2.001) - 128 = 160.08 - 128 = 32.08$.

13. $y = f(x) = \ln x \Rightarrow f'(x) = 1/x$, so $f(1) = 0$ and $f'(1) = 1$. The linear approximation of f at 1 is $f(1) + f'(1)(x - 1) = 0 + 1(x - 1) = x - 1$. Now $f(1.05) = \ln 1.05 \approx 1.05 - 1 = 0.05$, so the approximation is reasonable.

15. $f(s) = \dfrac{p\,(1+s)}{1+sp} \Rightarrow f'(s) = \dfrac{(1+sp)\,(p) - p\,(1+s)\,(p)}{(1+sp)^2} = \dfrac{p\,(1-p)}{(1+sp)^2}$. Since s is small, a linearization at $a = 0$ can be used. So $f\,(0) = p$ and $f'\,(0) = p\,(1-p)$. Thus, $L(s) = f(0) + f'(0)(s - 0) = p + p(1-p)(s-0) = p + p(1-p)s$.

17. $F = kR^4 \Rightarrow dF = 4kR^3\,dR \Rightarrow \dfrac{dF}{F} = \dfrac{4kR^3\,dR}{kR^4} = 4\left(\dfrac{dR}{R}\right)$. Thus, the relative change in F is about 4 times the relative change in R. So a 5% increase in the radius corresponds to a 20% increase in blood flow.

19. (a)

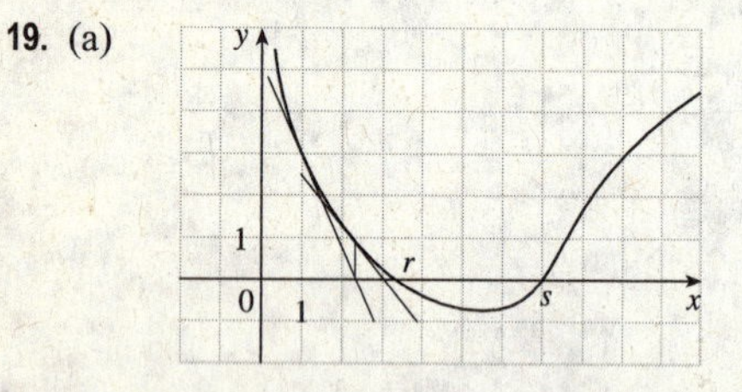

The tangent line at $x = 1$ intersects the x-axis at $x \approx 2.3$, so $x_2 \approx 2.3$. The tangent line at $x = 2.3$ intersects the x-axis at $x \approx 3$, so $x_3 \approx 3.0$.

(b) $x_1 = 5$ would *not* be a better first approximation than $x_1 = 1$ since the tangent line is nearly horizontal. In fact, the second approximation for $x_1 = 5$ appears to be to the left of $x = 1$.

21. $f(x) = x^3 + 2x - 4 \Rightarrow f'(x) = 3x^2 + 2$, so $x_{n+1} = x_n - \dfrac{x_n^3 + 2x_n - 4}{3x_n^2 + 2}$. Now $x_1 = 1 \Rightarrow$

$$x_2 = 1 - \frac{1 + 2 - 4}{3 \cdot 1^2 + 2} = 1 - \frac{-1}{5} = 1.2 \Rightarrow x_3 = 1.2 - \frac{(1.2)^3 + 2(1.2) - 4}{3(1.2)^2 + 2} \approx 1.1797.$$

23.

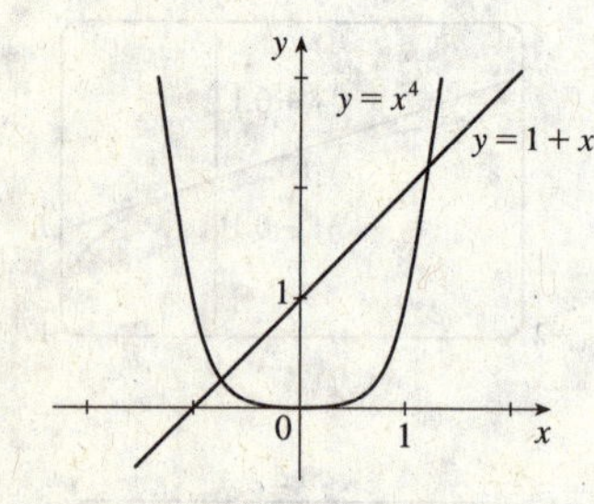

From the graph, we see that there appear to be points of intersection near $x = -0.7$ and $x = 1.2$. Solving $x^4 = 1 + x$ is the same as solving $f(x) = x^4 - x - 1 = 0$. $f(x) = x^4 - x - 1 \Rightarrow f'(x) = 4x^3 - 1$, so $x_{n+1} = x_n - \dfrac{x_n^4 - x_n - 1}{4x_n^3 - 1}$.

$x_1 = -0.7$	$x_1 = 1.2$
$x_2 \approx -0.725253$	$x_2 \approx 1.221380$
$x_3 \approx -0.724493$	$x_3 \approx 1.220745$
$x_4 \approx -0.724492 \approx x_5$	$x_4 \approx 1.220744 \approx x_5$

To six decimal places, the roots of the equation are -0.724492 and 1.220744.

© 2016 Cengage Learning. All Rights Reserved. May not be scanned, copied or duplicated, or posted to a publicly accessible website, in whole or in part.

25.

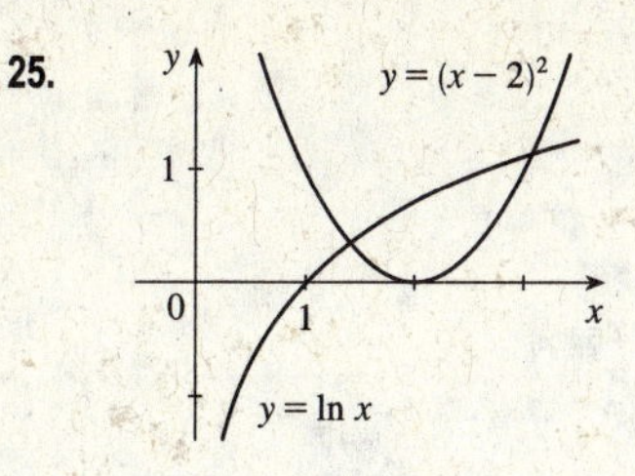

From the graph, we see that there appear to be points of intersection near $x = 1.5$ and $x = 3$. Solving $(x-2)^2 = \ln x$ is the same as solving $f(x) = (x-2)^2 - \ln x = 0$. $f(x) = (x-2)^2 - \ln x \;\Rightarrow$

$f'(x) = 2(x-2) - 1/x$, so $x_{n+1} = x_n - \dfrac{(x_n-2)^2 - \ln x_n}{2(x_n-2) - 1/x_n}$.

$x_1 = 1.5$	$x_1 = 3$
$x_2 \approx 1.406721$	$x_2 \approx 3.059167$
$x_3 \approx 1.412370$	$x_3 \approx 3.057106$
$x_4 \approx 1.412391 \approx x_5$	$x_4 \approx 3.057104 \approx x_5$

To six decimal places, the roots of the equation are 1.412391 and 3.057104.

27.

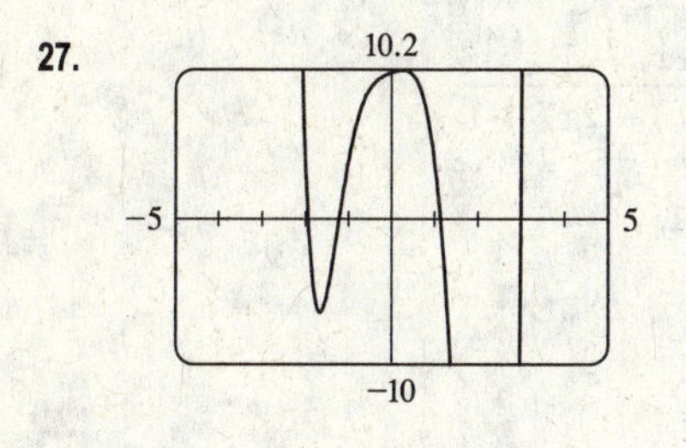

$f(x) = x^6 - x^5 - 6x^4 - x^2 + x + 10 \;\Rightarrow$

$f'(x) = 6x^5 - 5x^4 - 24x^3 - 2x + 1 \;\Rightarrow$

$$x_{n+1} = x_n - \frac{x_n^6 - x_n^5 - 6x_n^4 - x_n^2 + x_n + 10}{6x_n^5 - 5x_n^4 - 24x_n^3 - 2x_n + 1}.$$

From the graph of f, there appear to be roots near -1.9, -1.2, 1.1, and 3.

$x_1 = -1.9$	$x_1 = -1.2$	$x_1 = 1.1$	$x_1 = 3$
$x_2 \approx -1.94278290$	$x_2 \approx -1.22006245$	$x_2 \approx 1.14111662$	$x_2 \approx 2.99$
$x_3 \approx -1.93828380$	$x_3 \approx -1.21997997 \approx x_4$	$x_3 \approx 1.13929741$	$x_3 \approx 2.98984106$
$x_4 \approx -1.93822884$		$x_4 \approx 1.13929375 \approx x_5$	$x_4 \approx 2.98984102 \approx x_5$
$x_5 \approx -1.93822883 \approx x_6$			

To eight decimal places, the roots of the equation are -1.93822883, -1.21997997, 1.13929375, and 2.98984102.

29.

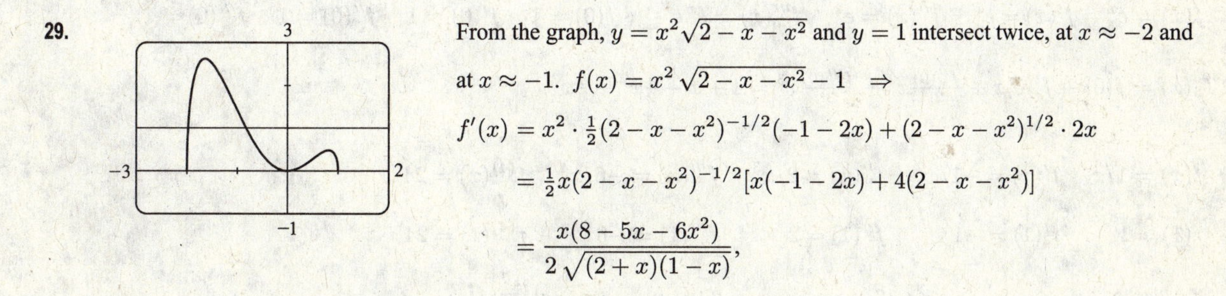

From the graph, $y = x^2\sqrt{2 - x - x^2}$ and $y = 1$ intersect twice, at $x \approx -2$ and at $x \approx -1$. $f(x) = x^2\sqrt{2 - x - x^2} - 1 \;\Rightarrow$

$$\begin{aligned} f'(x) &= x^2 \cdot \tfrac{1}{2}(2 - x - x^2)^{-1/2}(-1 - 2x) + (2 - x - x^2)^{1/2} \cdot 2x \\ &= \tfrac{1}{2}x(2 - x - x^2)^{-1/2}[x(-1-2x) + 4(2 - x - x^2)] \\ &= \frac{x(8 - 5x - 6x^2)}{2\sqrt{(2+x)(1-x)}}, \end{aligned}$$

so $x_{n+1} = x_n - \dfrac{x_n^2\sqrt{2 - x_n - x_n^2} - 1}{\dfrac{x_n(8 - 5x_n - 6x_n^2)}{2\sqrt{(2+x_n)(1-x_n)}}}$. Trying $x_1 = -2$ won't work because $f'(-2)$ is undefined, so we'll try $x_1 = -1.95$.

[continued]

© 2016 Cengage Learning. All Rights Reserved. May not be scanned, copied or duplicated, or posted to a publicly accessible website, in whole or in part.

$$\begin{aligned} x_1 &= -1.95 \\ x_2 &\approx -1.98580357 \\ x_3 &\approx -1.97899778 \\ x_4 &\approx -1.97807848 \\ x_5 &\approx -1.97806682 \\ x_6 &\approx -1.97806681 \approx x_7 \end{aligned} \qquad \begin{aligned} x_1 &= -0.8 \\ x_2 &\approx -0.82674444 \\ x_3 &\approx -0.82646236 \\ x_4 &\approx -0.82646233 \approx x_5 \end{aligned}$$

To eight decimal places, the roots of the equation are -1.97806681 and -0.82646233.

31.

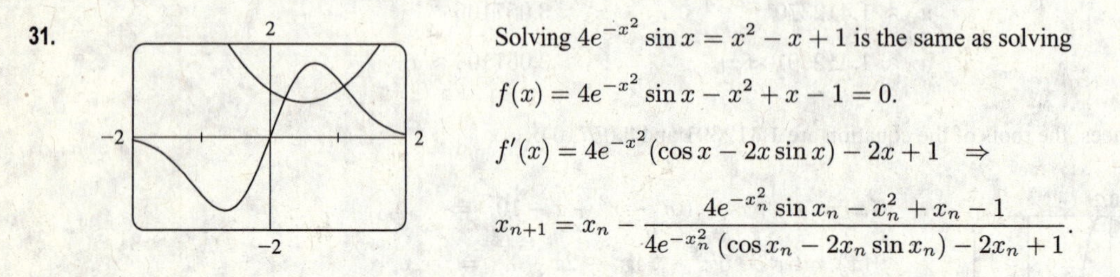

Solving $4e^{-x^2}\sin x = x^2 - x + 1$ is the same as solving

$$f(x) = 4e^{-x^2}\sin x - x^2 + x - 1 = 0.$$

$$f'(x) = 4e^{-x^2}(\cos x - 2x\sin x) - 2x + 1 \quad \Rightarrow$$

$$x_{n+1} = x_n - \frac{4e^{-x_n^2}\sin x_n - x_n^2 + x_n - 1}{4e^{-x_n^2}(\cos x_n - 2x_n\sin x_n) - 2x_n + 1}.$$

From the figure, we see that the graphs intersect at approximately $x = 0.2$ and $x = 1.1$.

$$\begin{aligned} x_1 &= 0.2 \\ x_2 &\approx 0.21883273 \\ x_3 &\approx 0.21916357 \\ x_4 &\approx 0.21916368 \approx x_5 \end{aligned} \qquad \begin{aligned} x_1 &= 1.1 \\ x_2 &\approx 1.08432830 \\ x_3 &\approx 1.08422462 \approx x_4 \end{aligned}$$

To eight decimal places, the roots of the equation are 0.21916368 and 1.08422462.

33. $f(x) = x^3 - 3x + 6 \;\Rightarrow\; f'(x) = 3x^2 - 3$. If $x_1 = 1$, then $f'(x_1) = 0$ and the tangent line used for approximating x_2 is horizontal. Attempting to find x_2 results in trying to divide by zero.

35. $f(x) = e^x \quad f'(x) = e^x \quad f''(x) = e^x \quad f'''(x) = e^x; \qquad f(0) = 1 \quad f'(0) = 1 \quad f''(0) = 1 \quad f'''(0) = 1$

$T_3(x) = f(0) + f'(0)x + \frac{f''(0)}{2!}x^2 + \frac{f'''(0)}{3!}x^3 = 1 + x + \frac{1}{2}x^2 + \frac{1}{6}x^3$

37. $f(x) = 1/x \quad f'(x) = -1/x^2 \quad f''(x) = 2/x^3 \quad f'''(x) = -6/x^4 \quad f^{(4)}(x) = 24/x^5$

$f(1) = 1 \quad f'(1) = -1 \quad f''(1) = 2 \quad f'''(1) = -6 \quad f^{(4)}(1) = 24$

$$\begin{aligned} T_3(x) &= f(1) + f'(1)(x-1) + \tfrac{f''(1)}{2!}(x-1)^2 + \tfrac{f'''(1)}{3!}(x-1)^3 + \tfrac{f^{(4)}(1)}{4!}(x-1)^4 \\ &= 1 - (x-1) + (x-1)^2 - (x-1)^3 + (x-1)^4 \end{aligned}$$

© 2016 Cengage Learning. All Rights Reserved. May not be scanned, copied or duplicated, or posted to a publicly accessible website, in whole or in part.

39. From Example 1 in Section 3.8, we have $f(1) = 2$, $f'(1) = \frac{1}{4}$, and

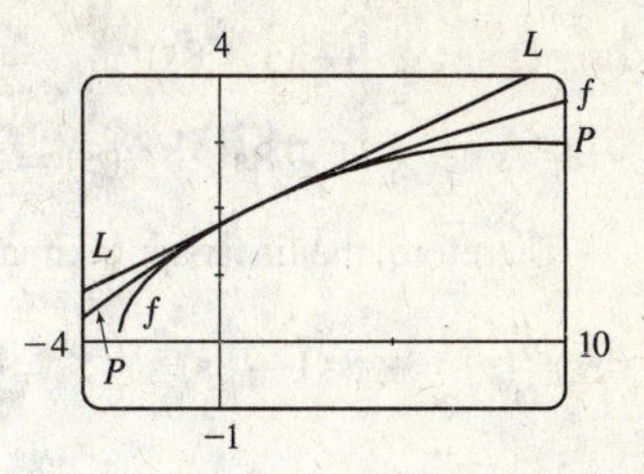

$f'(x) = \frac{1}{2}(x+3)^{-1/2}$. So $f''(x) = -\frac{1}{4}(x+3)^{-3/2} \Rightarrow f''(1) = -\frac{1}{32}$.

The quadratic approximation $P(x)$ is

$\sqrt{x+3} \approx f(1) + f'(1)(x-1) + \frac{1}{2}f''(1)(x-1)^2 = 2 + \frac{1}{4}(x-1) - \frac{1}{64}(x-1)^2$.

The figure shows the function $f(x) = \sqrt{x+3}$ together with its linear approximation $L(x) = \frac{1}{4}x + \frac{7}{4}$ and its quadratic approximation $P(x)$. You can see that $P(x)$ is a better approximation than $L(x)$ and this is borne out by the numerical values in the following chart.

	from $L(x)$	actual value	from $P(x)$
$\sqrt{3.98}$	1.9950	1.99499373...	1.99499375
$\sqrt{4.05}$	2.0125	2.01246118...	2.01246094
$\sqrt{4.2}$	2.0500	2.04939015...	2.04937500

41. $T_n(x) = f(a) + f'(a)(x-a) + \dfrac{f''(a)}{2!}(x-a)^2 + \cdots + \dfrac{f^{(n)}(a)}{n!}(x-a)^n$. To compute the coefficients in this equation we need to calculate the derivatives of f at 0:

$$\begin{aligned} f(x) &= \sin x & f(0) &= \sin 0 = 0 \\ f'(x) &= \cos x & f'(0) &= \cos 0 = 1 \\ f''(x) &= -\sin x & f''(0) &= 0 \\ f'''(x) &= -\cos x & f'''(0) &= -1 \\ f^{(4)}(x) &= \sin x & f^{(4)}(0) &= 0 \\ f^{(5)}(x) &= \cos x & f^{(5)}(0) &= 1 \end{aligned}$$

From the original expression for $T_n(x)$, with $a = 0$, we have

$$\begin{aligned} T_1(x) &= 0 + x = x \\ T_2(x) &= 0 + x + 0 \cdot x^2 = x \\ T_3(x) &= 0 + x + 0 \cdot x^2 - \tfrac{1}{3!}x^3 = x - \tfrac{1}{6}x^3 \\ T_4(x) &= 0 + x + 0 \cdot x^2 - \tfrac{1}{3!}x^3 + 0 \cdot x^4 = x - \tfrac{1}{6}x^3 \\ T_5(x) &= 0 + x + 0 \cdot x^2 - \tfrac{1}{3!}x^3 + 0 \cdot x^4 + \tfrac{1}{5!}x^5 = x - \tfrac{1}{6}x^3 + \tfrac{1}{120}x^5 \end{aligned}$$

We graph T_1, T_2, T_3, T_4, T_5, and f:

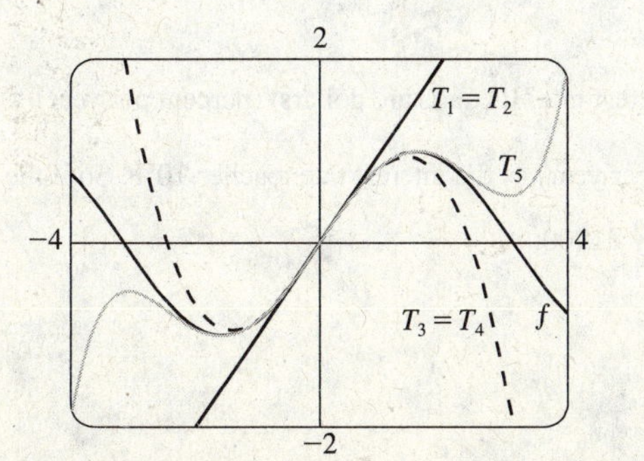

Notice that $T_1(x)$ is a good approximation to $\sin x$ near 0, $T_3(x)$ is a good approximation on a larger interval, and $T_5(x)$ is a better approximation. Each successive Taylor polynomial is a good approximation on a larger interval than the previous one.

© 2016 Cengage Learning. All Rights Reserved. May not be scanned, copied or duplicated, or posted to a publicly accessible website, in whole or in part.

43. (a) $r(s) = \frac{1}{2} + \frac{1}{2}(1+8s)^{1/2} \Rightarrow r(0) = 1$

$r'(s) = \frac{1}{4}(1+8s)^{-1/2}(8) = 2(1+8s)^{-1/2} \Rightarrow r'(0) = 2$

Therefore, the linear approximation near $s = 0$ is $T_1(s) = r(0) + r'(0)(s-0) = 1 + 2s$.

(b) $r''(s) = -8(1+8s)^{-3/2} \Rightarrow r''(0) = -8$. Hence, the second order Taylor polynomial centered at

$$s = 0 \text{ is } T_2(s) = r(0) + r'(0)(s-0) + \frac{r''(0)}{2!}(s-0)^2 = 1 + 2s - 4s^2$$

3 Review

TRUE-FALSE QUIZ

1. False. See the note after Theorem 4 in Section 3.2.

3. True. This is the Sum Rule.

5. True. This is the Chain Rule.

7. False. $\frac{d}{dx} f(\sqrt{x}) = f'(\sqrt{x}) \cdot \frac{1}{2}x^{-1/2} = \frac{f'(\sqrt{x})}{2\sqrt{x}}$, which is not $\frac{f'(x)}{2\sqrt{x}}$.

9. False. $\frac{d}{dx}(10^x) = 10^x \ln 10$, which is not equal to $x10^{x-1}$.

11. True. $\frac{d}{dx}(\tan^2 x) = 2\tan x \sec^2 x$, and $\frac{d}{dx}(\sec^2 x) = 2\sec x(\sec x \tan x) = 2\tan x \sec^2 x$.

Or: $\frac{d}{dx}(\sec^2 x) = \frac{d}{dx}(1 + \tan^2 x) = \frac{d}{dx}(\tan^2 x)$.

13. True. $g(x) = x^5 \Rightarrow g'(x) = 5x^4 \Rightarrow g'(2) = 5(2)^4 = 80$, and by the definition of the derivative,

$$\lim_{x \to 2} \frac{g(x) - g(2)}{x - 2} = g'(2) = 80.$$

EXERCISES

1. Estimating the slopes of the tangent lines at $x = 2$, 3, and 5, we obtain approximate values 0.4, 2, and 0.1. Since the graph is concave downward at $x = 5$, $f''(5)$ is negative. Arranging the numbers in increasing order, we have: $f''(5) < 0 < f'(5) < f'(2) < 1 < f'(3)$.

3. (a) $f'(r)$ is the rate at which the total cost changes with respect to the interest rate. Its units are dollars/(percent per year).

(b) The total cost of paying off the loan is increasing by \$1200/(percent per year) as the interest rate reaches 10%. So if the interest rate goes up from 10% to 11%, the cost goes up approximately \$1200.

(c) As r increases, C increases. So $f'(r)$ will always be positive.

© 2016 Cengage Learning. All Rights Reserved. May not be scanned, copied or duplicated, or posted to a publicly accessible website, in whole or in part.

5.

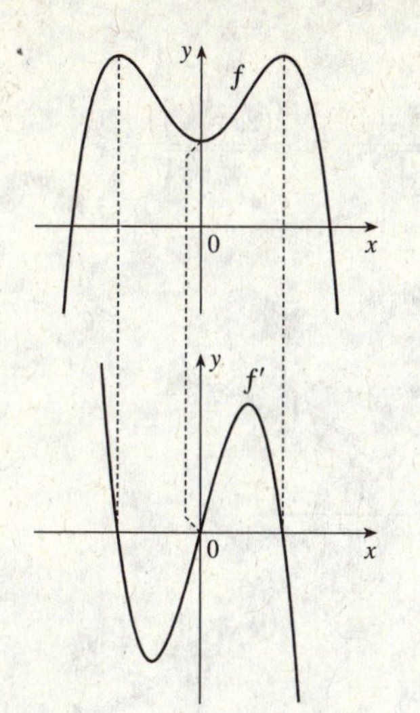

7.

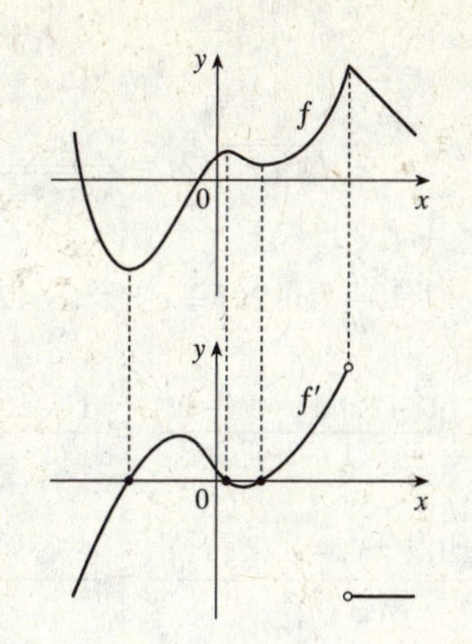

9. (a) The derivative $H'(t)$ is the rate at which the patient's heart rate changes with respect to time.

(b) The slope of the tangent line on the graph of $y = H(t)$ starts out with a value of approximately zero and decreases, reaching a minimum at $t \approx 1$. The slope then increases, reaching zero at $t \approx 3$, and continues to increase until it is again zero at $t \approx 4$. The slope decreases for a very short interval and then increases, returning to zero at $t \approx 4.2$. The slope then transitions through intervals of increase, decrease, and increase sequentially.

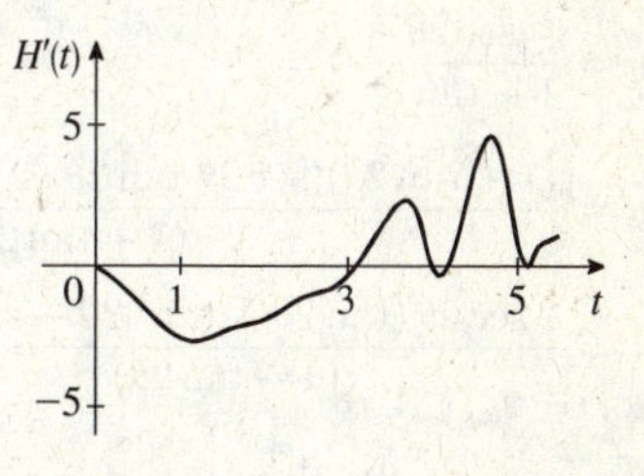

11. (a) $f'(x) = \lim\limits_{h\to 0} \dfrac{f(x+h) - f(x)}{h} = \lim\limits_{h\to 0} \dfrac{\sqrt{3-5(x+h)} - \sqrt{3-5x}}{h} \dfrac{\sqrt{3-5(x+h)} + \sqrt{3-5x}}{\sqrt{3-5(x+h)} + \sqrt{3-5x}}$

$= \lim\limits_{h\to 0} \dfrac{[3-5(x+h)] - (3-5x)}{h\left(\sqrt{3-5(x+h)} + \sqrt{3-5x}\right)} = \lim\limits_{h\to 0} \dfrac{-5}{\sqrt{3-5(x+h)} + \sqrt{3-5x}} = \dfrac{-5}{2\sqrt{3-5x}}$

(b) Domain of f: (the radicand must be nonnegative) $3 - 5x \geq 0 \;\Rightarrow$ $5x \leq 3 \;\Rightarrow\; x \in \left(-\infty, \frac{3}{5}\right]$

Domain of f': exclude $\frac{3}{5}$ because it makes the denominator zero; $x \in \left(-\infty, \frac{3}{5}\right)$

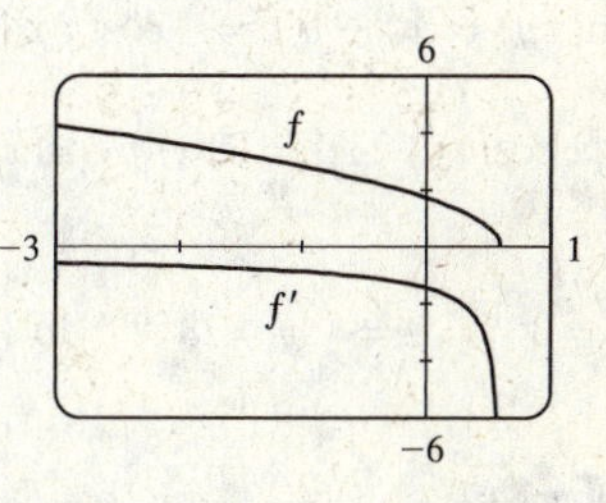

(c) Our answer to part (a) is reasonable because $f'(x)$ is always negative and f is always decreasing.

13. f is not differentiable: at $x = -4$ because f is not continuous, at $x = -1$ because f has a corner, at $x = 2$ because f is not continuous, and at $x = 5$ because f has a vertical tangent.

15. $y = (x^4 - 3x^2 + 5)^3 \;\Rightarrow$

$y' = 3(x^4 - 3x^2 + 5)^2 \dfrac{d}{dx}(x^4 - 3x^2 + 5) = 3(x^4 - 3x^2 + 5)^2(4x^3 - 6x) = 6x(x^4 - 3x^2 + 5)^2(2x^2 - 3)$

17. $y = \sqrt{x} + \dfrac{1}{\sqrt[3]{x^4}} = x^{1/2} + x^{-4/3} \;\Rightarrow\; y' = \frac{1}{2}x^{-1/2} - \frac{4}{3}x^{-7/3} = \dfrac{1}{2\sqrt{x}} - \dfrac{4}{3\sqrt[3]{x^7}}$

© 2016 Cengage Learning. All Rights Reserved. May not be scanned, copied or duplicated, or posted to a publicly accessible website, in whole or in part.

19. $y = 2x\sqrt{x^2+1} \;\Rightarrow$

$$y' = 2x \cdot \tfrac{1}{2}(x^2+1)^{-1/2}(2x) + \sqrt{x^2+1}\,(2) = \frac{2x^2}{\sqrt{x^2+1}} + 2\sqrt{x^2+1} = \frac{2x^2+2(x^2+1)}{\sqrt{x^2+1}} = \frac{2(2x^2+1)}{\sqrt{x^2+1}}$$

21. $y = e^{\sin 2\theta} \;\Rightarrow\; y' = e^{\sin 2\theta}\dfrac{d}{d\theta}(\sin 2\theta) = e^{\sin 2\theta}(\cos 2\theta)(2) = 2\cos 2\theta\, e^{\sin 2\theta}$

23. $y = \dfrac{t}{1-t^2} \;\Rightarrow\; y' = \dfrac{(1-t^2)(1) - t(-2t)}{(1-t^2)^2} = \dfrac{1-t^2+2t^2}{(1-t^2)^2} = \dfrac{t^2+1}{(1-t^2)^2}$

25. $y = \dfrac{e^{1/x}}{x^2} \;\Rightarrow\; y' = \dfrac{x^2(e^{1/x})' - e^{1/x}\,(x^2)'}{(x^2)^2} = \dfrac{x^2(e^{1/x})(-1/x^2) - e^{1/x}(2x)}{x^4} = \dfrac{-e^{1/x}(1+2x)}{x^4}$

27. $\dfrac{d}{dx}(xy^4 + x^2y) = \dfrac{d}{dx}(x+3y) \;\Rightarrow\; x \cdot 4y^3y' + y^4 \cdot 1 + x^2 \cdot y' + y \cdot 2x = 1 + 3y' \;\Rightarrow$

$$y'(4xy^3 + x^2 - 3) = 1 - y^4 - 2xy \;\Rightarrow\; y' = \frac{1-y^4-2xy}{4xy^3+x^2-3}$$

29. $y = \dfrac{\sec 2\theta}{1+\tan 2\theta} \;\Rightarrow$

$$y' = \frac{(1+\tan 2\theta)(\sec 2\theta\,\tan 2\theta \cdot 2) - (\sec 2\theta)(\sec^2 2\theta \cdot 2)}{(1+\tan 2\theta)^2} = \frac{2\sec 2\theta\,[(1+\tan 2\theta)\tan 2\theta - \sec^2 2\theta]}{(1+\tan 2\theta)^2}$$

$$= \frac{2\sec 2\theta\,(\tan 2\theta + \tan^2 2\theta - \sec^2 2\theta)}{(1+\tan 2\theta)^2} = \frac{2\sec 2\theta\,(\tan 2\theta - 1)}{(1+\tan 2\theta)^2} \qquad [1+\tan^2 x = \sec^2 x]$$

31. $y = e^{cx}(c\sin x - \cos x) \;\Rightarrow$

$$y' = e^{cx}(c\cos x + \sin x) + ce^{cx}(c\sin x - \cos x) = e^{cx}(c^2\sin x - c\cos x + c\cos x + \sin x)$$

$$= e^{cx}(c^2\sin x + \sin x) = e^{cx}\sin x\,(c^2+1)$$

33. $y = \log_5(1+2x) \;\Rightarrow\; y' = \dfrac{1}{(1+2x)\ln 5}\dfrac{d}{dx}(1+2x) = \dfrac{2}{(1+2x)\ln 5}$

35. $\sin(xy) = x^2 - y \;\Rightarrow\; \cos(xy)(xy' + y \cdot 1) = 2x - y' \;\Rightarrow\; x\cos(xy)y' + y' = 2x - y\cos(xy) \;\Rightarrow$

$$y'[x\cos(xy)+1] = 2x - y\cos(xy) \;\Rightarrow\; y' = \frac{2x - y\cos(xy)}{x\cos(xy)+1}$$

37. $y = 3^{x\ln x} \;\Rightarrow\; y' = 3^{x\ln x}(\ln 3)\dfrac{d}{dx}(x\ln x) = 3^{x\ln x}(\ln 3)\left(x \cdot \dfrac{1}{x} + \ln x \cdot 1\right) = 3^{x\ln x}(\ln 3)(1+\ln x)$

39. $y = \ln\sin x - \tfrac{1}{2}\sin^2 x \;\Rightarrow\; y' = \dfrac{1}{\sin x} \cdot \cos x - \tfrac{1}{2} \cdot 2\sin x \cdot \cos x = \cot x - \sin x\,\cos x$

41. $y = x\tan^{-1}(4x) \;\Rightarrow\; y' = x \cdot \dfrac{1}{1+(4x)^2} \cdot 4 + \tan^{-1}(4x) \cdot 1 = \dfrac{4x}{1+16x^2} + \tan^{-1}(4x)$

43. $y = \ln|\sec 5x + \tan 5x| \;\Rightarrow$

$$y' = \frac{1}{\sec 5x + \tan 5x}(\sec 5x\,\tan 5x \cdot 5 + \sec^2 5x \cdot 5) = \frac{5\sec 5x\,(\tan 5x + \sec 5x)}{\sec 5x + \tan 5x} = 5\sec 5x$$

45. $y = \tan^2(\sin\theta) = [\tan(\sin\theta)]^2 \;\Rightarrow\; y' = 2[\tan(\sin\theta)] \cdot \sec^2(\sin\theta) \cdot \cos\theta$

© 2016 Cengage Learning. All Rights Reserved. May not be scanned, copied or duplicated, or posted to a publicly accessible website, in whole or in part.

47. $y = \sin(\tan\sqrt{1+x^3}) \Rightarrow y' = \cos(\tan\sqrt{1+x^3})(\sec^2\sqrt{1+x^3})\left[3x^2/(2\sqrt{1+x^3})\right]$

49. $y = \cos\left(e^{\sqrt{\tan 3x}}\right) \Rightarrow$

$$y' = -\sin\left(e^{\sqrt{\tan 3x}}\right)\cdot\left(e^{\sqrt{\tan 3x}}\right)' = -\sin\left(e^{\sqrt{\tan 3x}}\right)e^{\sqrt{\tan 3x}}\cdot\tfrac{1}{2}(\tan 3x)^{-1/2}\cdot\sec^2(3x)\cdot 3$$

$$= \frac{-3\sin\left(e^{\sqrt{\tan 3x}}\right)e^{\sqrt{\tan 3x}}\sec^2(3x)}{2\sqrt{\tan 3x}}$$

51. $f(t) = \sqrt{4t+1} \Rightarrow f'(t) = \frac{1}{2}(4t+1)^{-1/2}\cdot 4 = 2(4t+1)^{-1/2} \Rightarrow$

$f''(t) = 2(-\frac{1}{2})(4t+1)^{-3/2}\cdot 4 = -4/(4t+1)^{3/2}$, so $f''(2) = -4/9^{3/2} = -\frac{4}{27}$.

53. $f(x) = 2^x \Rightarrow f'(x) = 2^x\ln 2 \Rightarrow f''(x) = (2^x\ln 2)\ln 2 = 2^x(\ln 2)^2 \Rightarrow$

$f'''(x) = (2^x\ln 2)(\ln 2)^2 = 2^x(\ln 2)^3 \Rightarrow \cdots \Rightarrow f^{(n)}(x) = (2^x\ln 2)(\ln 2)^{n-1} = 2^x(\ln 2)^n$

55. $y = 4\sin^2 x \Rightarrow y' = 4\cdot 2\sin x\cos x$. At $(\frac{\pi}{6}, 1)$, $y' = 8\cdot\frac{1}{2}\cdot\frac{\sqrt{3}}{2} = 2\sqrt{3}$, so an equation of the tangent line is $y - 1 = 2\sqrt{3}\,(x - \frac{\pi}{6})$, or $y = 2\sqrt{3}\,x + 1 - \pi\sqrt{3}/3$.

57. $y = (2+x)e^{-x} \Rightarrow y' = (2+x)(-e^{-x}) + e^{-x}\cdot 1 = e^{-x}[-(2+x)+1] = e^{-x}(-x-1)$.

At $(0,2)$, $y' = 1(-1) = -1$, so an equation of the tangent line is $y - 2 = -1(x-0)$, or $y = -x+2$.

The slope of the normal line is 1, so an equation of the normal line is $y - 2 = 1(x-0)$, or $y = x+2$.

59. (a) $f(x) = x\sqrt{5-x} \Rightarrow$

$$f'(x) = x\left[\frac{1}{2}(5-x)^{-1/2}(-1)\right] + \sqrt{5-x} = \frac{-x}{2\sqrt{5-x}} + \sqrt{5-x}\cdot\frac{2\sqrt{5-x}}{2\sqrt{5-x}} = \frac{-x}{2\sqrt{5-x}} + \frac{2(5-x)}{2\sqrt{5-x}}$$

$$= \frac{-x+10-2x}{2\sqrt{5-x}} = \frac{10-3x}{2\sqrt{5-x}}$$

(b) At $(1,2)$: $f'(1) = \frac{7}{4}$.

So an equation of the tangent line is $y - 2 = \frac{7}{4}(x-1)$ or $y = \frac{7}{4}x + \frac{1}{4}$.

At $(4,4)$: $f'(4) = -\frac{2}{2} = -1$.

So an equation of the tangent line is $y - 4 = -1(x-4)$ or $y = -x+8$.

(c)

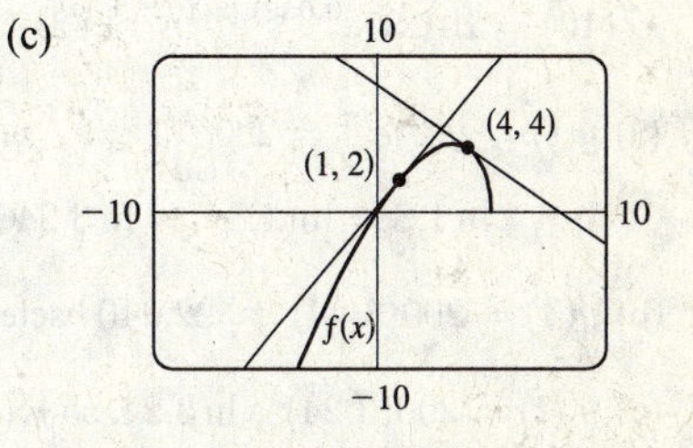

(d)

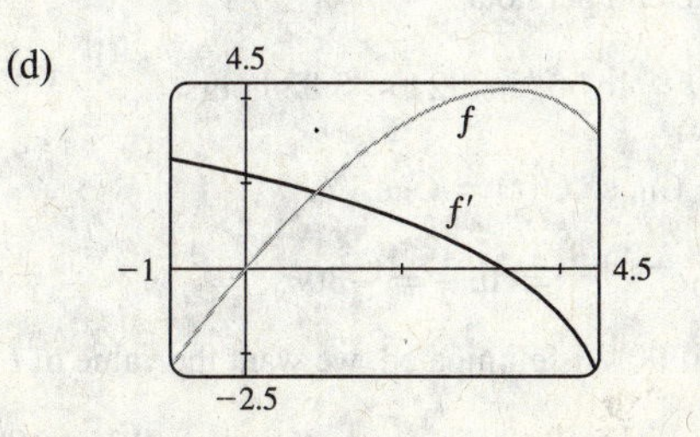

The graphs look reasonable, since f' is positive where f has tangents with positive slope, and f' is negative where f has tangents with negative slope.

61. (a) $h(x) = f(x)\,g(x) \Rightarrow h'(x) = f(x)\,g'(x) + g(x)\,f'(x) \Rightarrow$

$h'(2) = f(2)\,g'(2) + g(2)\,f'(2) = (3)(4) + (5)(-2) = 12 - 10 = 2$

(b) $F(x) = f(g(x)) \Rightarrow F'(x) = f'(g(x))\,g'(x) \Rightarrow F'(2) = f'(g(2))\,g'(2) = f'(5)(4) = 11\cdot 4 = 44$

© 2016 Cengage Learning. All Rights Reserved. May not be scanned, copied or duplicated, or posted to a publicly accessible website, in whole or in part.

63. $f(x) = x^2 g(x) \;\Rightarrow\; f'(x) = x^2 g'(x) + g(x)(2x) = x[xg'(x) + 2g(x)]$

65. $f(x) = [g(x)]^2 \;\Rightarrow\; f'(x) = 2[g(x)] \cdot g'(x) = 2g(x)\,g'(x)$

67. $f(x) = g(e^x) \;\Rightarrow\; f'(x) = g'(e^x)\,e^x$

69. $f(x) = \ln|g(x)| \;\Rightarrow\; f'(x) = \dfrac{1}{g(x)}g'(x) = \dfrac{g'(x)}{g(x)}$

71. $y = [\ln(x+4)]^2 \;\Rightarrow\; y' = 2[\ln(x+4)]^1 \cdot \dfrac{1}{x+4} \cdot 1 = 2\,\dfrac{\ln(x+4)}{x+4}$ and $y' = 0 \;\Leftrightarrow\; \ln(x+4) = 0 \;\Leftrightarrow$ $x + 4 = e^0 \;\Rightarrow\; x + 4 = 1 \;\Leftrightarrow\; x = -3$, so the tangent is horizontal at the point $(-3, 0)$.

73. $x^2 + 2y^2 = 1 \;\Rightarrow\; 2x + 4yy' = 0 \;\Rightarrow\; y' = -x/(2y) = 1 \;\Leftrightarrow\; x = -2y$. Since the points lie on the ellipse, we have $(-2y)^2 + 2y^2 = 1 \;\Rightarrow\; 6y^2 = 1 \;\Rightarrow\; y = \pm\frac{1}{\sqrt{6}}$. The points are $\left(-\frac{2}{\sqrt{6}}, \frac{1}{\sqrt{6}}\right)$ and $\left(\frac{2}{\sqrt{6}}, -\frac{1}{\sqrt{6}}\right)$.

75. $s(t) = Ae^{-ct}\cos(\omega t + \delta) \;\Rightarrow$

$v(t) = s'(t) = A\{e^{-ct}[-\omega\sin(\omega t + \delta)] + \cos(\omega t + \delta)(-ce^{-ct})\} = -Ae^{-ct}[\omega\sin(\omega t + \delta) + c\cos(\omega t + \delta)] \;\Rightarrow$

$$\begin{aligned} a(t) = v'(t) &= -A\{e^{-ct}[\omega^2\cos(\omega t + \delta) - c\omega\sin(\omega t + \delta)] + [\omega\sin(\omega t + \delta) + c\cos(\omega t + \delta)](-ce^{-ct})\} \\ &= -Ae^{-ct}[\omega^2\cos(\omega t + \delta) - c\omega\sin(\omega t + \delta) - c\omega\sin(\omega t + \delta) - c^2\cos(\omega t + \delta)] \\ &= -Ae^{-ct}[(\omega^2 - c^2)\cos(\omega t + \delta) - 2c\omega\sin(\omega t + \delta)] = Ae^{-ct}[(c^2 - \omega^2)\cos(\omega t + \delta) + 2c\omega\sin(\omega t + \delta)] \end{aligned}$$

77. (a) $V = \frac{1}{3}\pi r^2 h \;\Rightarrow\; dV/dh = \frac{1}{3}\pi r^2$ [r constant]

(b) $V = \frac{1}{3}\pi r^2 h \;\Rightarrow\; dV/dr = \frac{2}{3}\pi r h$ [h constant]

79. The rate of increase of expenditures t years after 1970 is given by $E'(t) = 101.35e^{0.088128t}(0.088128) = (8.9317728)\,e^{0.088128t}$. Hence, the rate of increase of expenditures in 1980 was $E'(10) = 101.35e^{0.088128(10)} \approx \$21.56$ billion/year, and in 2000, $E'(30) = 101.35e^{0.088128(30)} \approx \$125.64$ billion/year.

81. (a) $y(t) = y(0)e^{kt} = 200e^{kt} \;\Rightarrow\; y(0.5) = 200e^{0.5k} = 360 \;\Rightarrow\; e^{0.5k} = 1.8 \;\Rightarrow\; 0.5k = \ln 1.8 \;\Rightarrow$ $k = 2\ln 1.8 = \ln(1.8)^2 = \ln 3.24 \;\Rightarrow\; y(t) = 200e^{(\ln 3.24)t} = 200(3.24)^t$

(b) $y(4) = 200(3.24)^4 \approx 22{,}040$ bacteria

(c) $y'(t) = 200(3.24)^t \cdot \ln 3.24$, so $y'(4) = 200(3.24)^4 \cdot \ln 3.24 \approx 25{,}910$ bacteria per hour

(d) $200(3.24)^t = 10{,}000 \;\Rightarrow\; (3.24)^t = 50 \;\Rightarrow\; t\ln 3.24 = \ln 50 \;\Rightarrow\; t = \ln 50/\ln 3.24 \approx 3.33$ hours

83. (a) $C'(t) = -kC(t) \;\Rightarrow\; C(t) = C(0)e^{-kt}$ by Theorem 3.6.2. But $C(0) = C_0$, so $C(t) = C_0e^{-kt}$.

(b) $C(30) = \frac{1}{2}C_0$ since the concentration is reduced by half. Thus, $\frac{1}{2}C_0 = C_0e^{-30k} \;\Rightarrow \ln\frac{1}{2} = -30k \;\Rightarrow$ $k = -\frac{1}{30}\ln\frac{1}{2} = \frac{1}{30}\ln 2$. Since 10% of the original concentration remains if 90% is eliminated, we want the value of t such that $C(t) = \frac{1}{10}C_0$. Therefore, $\frac{1}{10}C_0 = C_0e^{-t(\ln 2)/30} \;\Rightarrow\; \ln 0.1 = -t(\ln 2)/30 \;\Rightarrow\; t = -\frac{30}{\ln 2}\ln 0.1 \approx 100$ h.

85. If $x =$ edge length, then $V = x^3 \;\Rightarrow\; dV/dt = 3x^2\,dx/dt = 10 \;\Rightarrow\; dx/dt = 10/(3x^2)$ and $S = 6x^2 \;\Rightarrow$ $dS/dt = (12x)\,dx/dt = 12x[10/(3x^2)] = 40/x$. When $x = 30$, $dS/dt = \frac{40}{30} = \frac{4}{3}$ cm^2/min.

© 2016 Cengage Learning. All Rights Reserved. May not be scanned, copied or duplicated, or posted to a publicly accessible website, in whole or in part.

87. $f(x) = x^5 - x^4 + 3x^2 - 3x - 2 \Rightarrow f'(x) = 5x^4 - 4x^3 + 6x - 3$, so $x_{n+1} = x_n - \dfrac{x_n^5 - x_n^4 + 3x_n^2 - 3x_n - 2}{5x_n^4 - 4x_n^3 + 6x_n - 3}$.

Now $x_1 = 1 \Rightarrow x_2 = 1.5 \Rightarrow x_3 \approx 1.343860 \Rightarrow x_4 \approx 1.300320 \Rightarrow x_5 \approx 1.297396 \Rightarrow$ $x_6 \approx 1.297383 \approx x_7$, so the root in $[1, 2]$ is 1.297383, to six decimal places.

89. (a) $f(x) = \sqrt[3]{1+3x} = (1+3x)^{1/3} \Rightarrow f'(x) = (1+3x)^{-2/3}$, so the linearization of f at $a = 0$ is $L(x) = f(0) + f'(0)(x-0) = 1^{1/3} + 1^{-2/3}x = 1 + x$. Thus, $\sqrt[3]{1+3x} \approx 1 + x \Rightarrow$ $\sqrt[3]{1.03} = \sqrt[3]{1+3(0.01)} \approx 1 + (0.01) = 1.01$.

(b) The linear approximation is $\sqrt[3]{1+3x} \approx 1 + x$, so for the required accuracy we want $\sqrt[3]{1+3x} - 0.1 < 1 + x < \sqrt[3]{1+3x} + 0.1$. From the graph, it appears that this is true when $-0.235 < x < 0.401$.

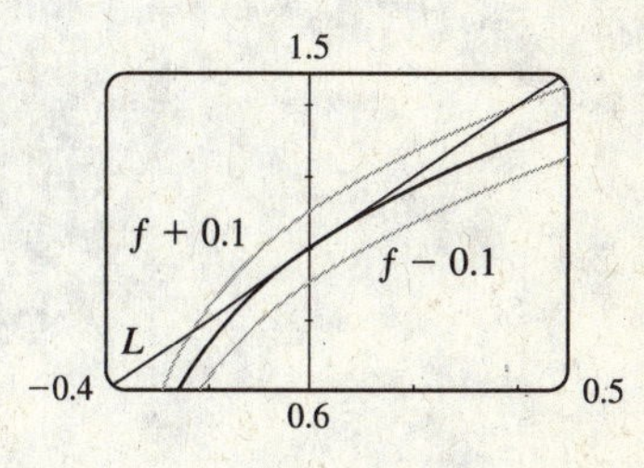

91. (a) $t(c) = \ln\left(\dfrac{3c + \sqrt{9c^2 - 8c}}{2}\right) = \ln\left(3c + \sqrt{9c^2 - 8c}\right) - \ln 2 \Rightarrow t(1) = \ln\left(\dfrac{3+1}{2}\right) = \ln 2$

$$t'(c) = \frac{1}{3c + \sqrt{9c^2 - 8c}} \cdot \left(3 + \tfrac{1}{2}\left(9c^2 - 8c\right)^{-1/2}(18c - 8)\right) = \frac{3 + (9c-4)\left(9c^2 - 8c\right)^{-1/2}}{3c + \sqrt{9c^2 - 8c}} \Rightarrow t'(1) = \frac{3+5}{3+1} = 2$$

Therefore, the linear approximation near $c = 1$ is $T_1(c) = t(1) + t'(1)(c-1) = \ln 2 + 2(c-1)$.

(b) $$t''(c) = \frac{\left(3c + \sqrt{9c^2 - 8c}\right)\left(\dfrac{9}{\sqrt{9c^2 - 8c}} - \dfrac{(9c-4)^2}{(9c^2 - 8c)^{3/2}}\right) - \left(3 + \dfrac{(9c-4)}{\sqrt{9c^2 - 8c}}\right)\left(3 + \dfrac{(9c-4)}{\sqrt{9c^2 - 8c}}\right)}{\left(3c + \sqrt{9c^2 - 8c}\right)^2} \Rightarrow$$

$$t''(1) = \frac{(3+1)(9-25) - (3+5)^2}{(3+1)^2} = -8.$$

Hence, the second order Taylor polynomial is $T_2(c) = t(1) + t'(1)(c-1) + \dfrac{t''(1)}{2!}(c-1)^2 = \ln 2 + 2(c-1) - 4(c-1)^2$.

93. $$\lim_{\theta \to \pi/3} \frac{\cos\theta - 0.5}{\theta - \pi/3} = \left[\frac{d}{d\theta}\cos\theta\right]_{\theta = \pi/3} = -\sin\frac{\pi}{3} = -\frac{\sqrt{3}}{2}$$

© 2016 Cengage Learning. All Rights Reserved. May not be scanned, copied or duplicated, or posted to a publicly accessible website, in whole or in part.

4 □ APPLICATIONS OF DIFFERENTIATION

4.1 Maximum and Minimum Values

1. A function f has an **absolute minimum** at $x = c$ if $f(c)$ is the smallest function value on the entire domain of f, whereas f has a **local minimum** at c if $f(c)$ is the smallest function value when x is near c.

3. Absolute maximum at s, absolute minimum at r, local maximum at c, local minima at b and r, neither a maximum nor a minimum at a and d.

5. Local maximum values: $f(0.18) \approx 0.36\,\text{mV}$, $f(0.30) \approx 1.2\,\text{mV}$, $f(0.57) \approx 0.31\,\text{mV}$

Absolute maximum value: $f(0.30) \approx 1.2\,\text{mV}$

Local minimum values: $f(0.29) \approx -0.01\,\text{mV}$, $f(0.32) \approx -0.76\,\text{mV}$

Absolute minimum value: $f(0.32) \approx -0.76\,\text{mV}$

7. Observe that each increment on the t-axis represents an increase of 7 days.

Local maximum values: $D(\text{Oct }23, 1918) \approx 91$, $D(\text{Jan }22, 1919) \approx 15$

Absolute maximum value: $D(\text{Oct }23, 1918) \approx 91$

Local minimum values: $D(\text{Nov }27, 1918) \approx 4$, $D(\text{Feb }13, 1919) \approx 7$

Absolute minimum value: $D(\text{Sept }11, 1918) \approx 0$

9. Absolute minimum at 2, absolute maximum at 3, local minimum at 4

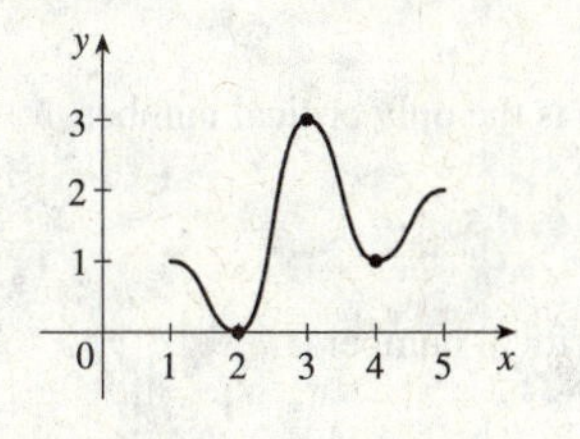

11. Absolute maximum at 5, absolute minimum at 2, local maximum at 3, local minima at 2 and 4

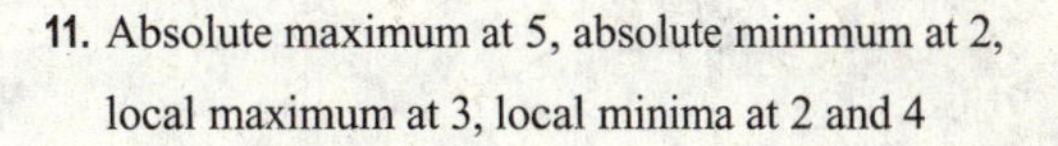

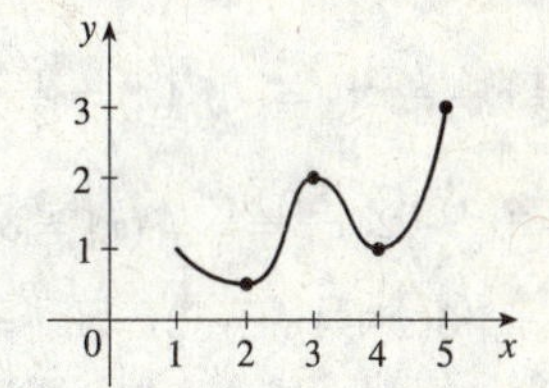

13. (a) (b) (c)

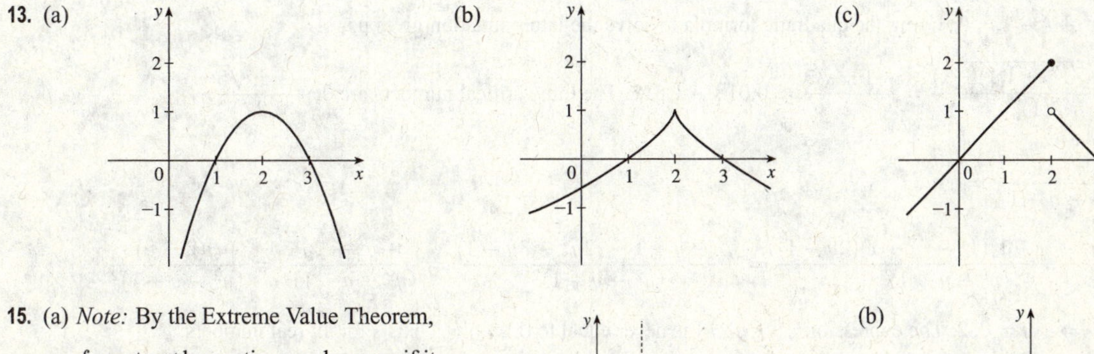

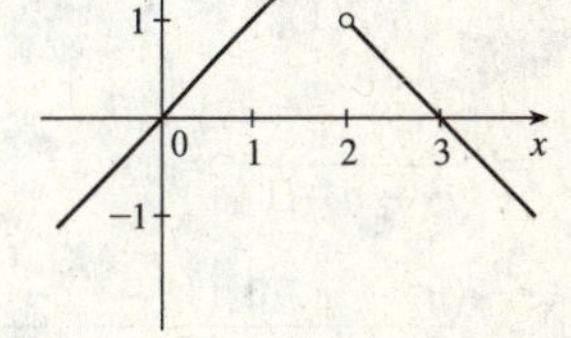

15. (a) *Note:* By the Extreme Value Theorem, f must *not* be continuous; because if it were, it would attain an absolute minimum.

(b)

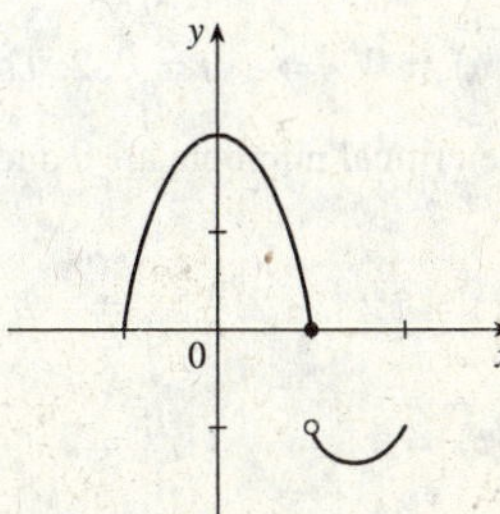

© 2016 Cengage Learning. All Rights Reserved. May not be scanned, copied or duplicated, or posted to a publicly accessible website, in whole or in part.

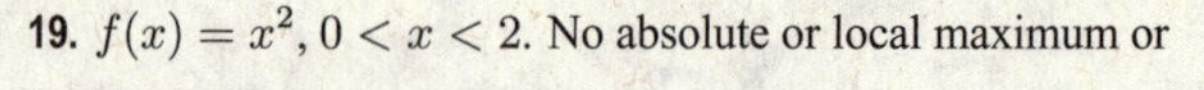

17. $f(x) = \frac{1}{2}(3x - 1)$, $x \leq 3$. Absolute maximum $f(3) = 4$; no local maximum. No absolute or local minimum.

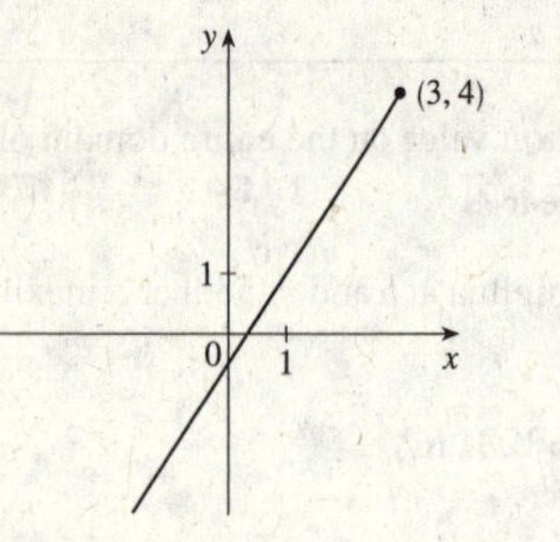

19. $f(x) = x^2$, $0 < x < 2$. No absolute or local maximum or minimum value.

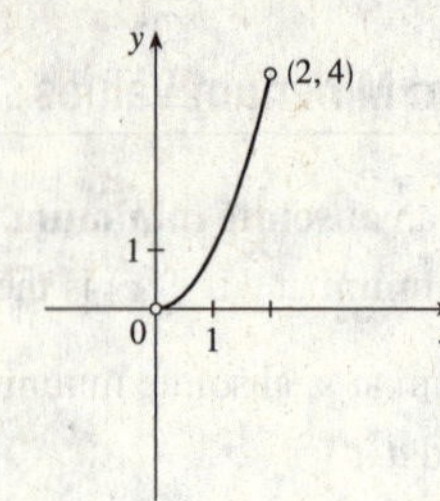

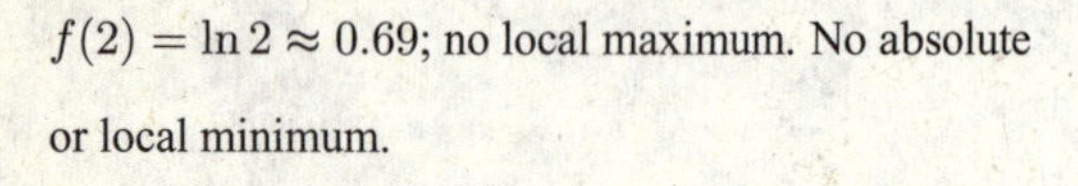

21. $f(x) = \ln x$, $0 < x \leq 2$. Absolute maximum $f(2) = \ln 2 \approx 0.69$; no local maximum. No absolute or local minimum.

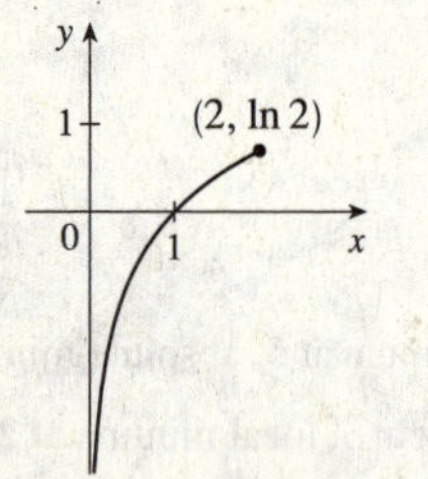

23. $f(x) = 1 - \sqrt{x}$. Absolute maximum $f(0) = 1$; no local maximum. No absolute or local minimum.

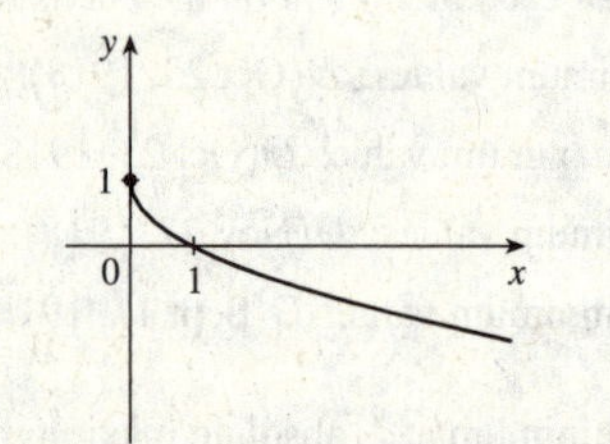

25. $f(x) = 4 + \frac{1}{3}x - \frac{1}{2}x^2 \quad \Rightarrow \quad f'(x) = \frac{1}{3} - x$. $\quad f'(x) = 0 \quad \Rightarrow \quad x = \frac{1}{3}$. This is the only critical number.

27. $f(x) = x^3 + 3x^2 - 24x \quad \Rightarrow \quad f'(x) = 3x^2 + 6x - 24 = 3(x^2 + 2x - 8)$.

$f'(x) = 0 \quad \Rightarrow \quad 3(x + 4)(x - 2) = 0 \quad \Rightarrow \quad x = -4, 2$. These are the only critical numbers.

29. $s(t) = 3t^4 + 4t^3 - 6t^2 \quad \Rightarrow \quad s'(t) = 12t^3 + 12t^2 - 12t$. $\quad s'(t) = 0 \quad \Rightarrow \quad 12t(t^2 + t - 1) \quad \Rightarrow$

$t = 0$ or $t^2 + t - 1 = 0$. Using the quadratic formula to solve the latter equation gives us

$t = \dfrac{-1 \pm \sqrt{1^2 - 4(1)(-1)}}{2(1)} = \dfrac{-1 \pm \sqrt{5}}{2} \approx 0.618, -1.618$. The three critical numbers are 0, $\dfrac{-1 \pm \sqrt{5}}{2}$.

31. $g(y) = \dfrac{y - 1}{y^2 - y + 1} \quad \Rightarrow$

$$g'(y) = \frac{(y^2 - y + 1)(1) - (y - 1)(2y - 1)}{(y^2 - y + 1)^2} = \frac{y^2 - y + 1 - (2y^2 - 3y + 1)}{(y^2 - y + 1)^2} = \frac{-y^2 + 2y}{(y^2 - y + 1)^2} = \frac{y(2 - y)}{(y^2 - y + 1)^2}.$$

$g'(y) = 0 \quad \Rightarrow \quad y = 0, 2$. The expression $y^2 - y + 1$ is never equal to 0, so $g'(y)$ exists for all real numbers. The critical numbers are 0 and 2.

33. $h(t) = t^{3/4} - 2t^{1/4} \quad \Rightarrow \quad h'(t) = \frac{3}{4}t^{-1/4} - \frac{2}{4}t^{-3/4} = \frac{1}{4}t^{-3/4}(3t^{1/2} - 2) = \dfrac{3\sqrt{t} - 2}{4\sqrt[4]{t^3}}$.

$h'(t) = 0 \quad \Rightarrow \quad 3\sqrt{t} = 2 \quad \Rightarrow \quad \sqrt{t} = \frac{2}{3} \quad \Rightarrow \quad t = \frac{4}{9}$. $h'(t)$ does not exist at $t = 0$, so the critical numbers are 0 and $\frac{4}{9}$.

© 2016 Cengage Learning. All Rights Reserved. May not be scanned, copied or duplicated, or posted to a publicly accessible website, in whole or in part.

35. $F(x) = x^{4/5}(x-4)^2 \;\Rightarrow$

$$F'(x) = x^{4/5} \cdot 2(x-4) + (x-4)^2 \cdot \tfrac{4}{5}x^{-1/5} = \tfrac{1}{5}x^{-1/5}(x-4)[5 \cdot x \cdot 2 + (x-4) \cdot 4]$$

$$= \frac{(x-4)(14x-16)}{5x^{1/5}} = \frac{2(x-4)(7x-8)}{5x^{1/5}}$$

$F'(x) = 0 \;\Rightarrow\; x = 4, \frac{8}{7}$. $F'(0)$ does not exist. Thus, the three critical numbers are 0, $\frac{8}{7}$, and 4.

37. $f(\theta) = 2\cos\theta + \sin^2\theta \;\Rightarrow\; f'(\theta) = -2\sin\theta + 2\sin\theta\,\cos\theta$. $f'(\theta) = 0 \;\Rightarrow\; 2\sin\theta\,(\cos\theta - 1) = 0 \;\Rightarrow\; \sin\theta = 0$ or $\cos\theta = 1 \;\Rightarrow\; \theta = n\pi$ [n an integer] or $\theta = 2n\pi$. The solutions $\theta = n\pi$ include the solutions $\theta = 2n\pi$, so the critical numbers are $\theta = n\pi$.

39. $f(x) = x^2 e^{-3x} \;\Rightarrow\; f'(x) = x^2(-3e^{-3x}) + e^{-3x}(2x) = xe^{-3x}(-3x + 2)$. $f'(x) = 0 \;\Rightarrow\; x = 0, \frac{2}{3}$ [e^{-3x} is never equal to 0]. $f'(x)$ always exists, so the critical numbers are 0 and $\frac{2}{3}$.

41. $f(x) = 12 + 4x - x^2$, $[0, 5]$. $f'(x) = 4 - 2x = 0 \;\Leftrightarrow\; x = 2$. $f(0) = 12$, $f(2) = 16$, and $f(5) = 7$. So $f(2) = 16$ is the absolute maximum value and $f(5) = 7$ is the absolute minimum value.

43. $f(x) = 2x^3 - 3x^2 - 12x + 1$, $[-2, 3]$. $f'(x) = 6x^2 - 6x - 12 = 6(x^2 - x - 2) = 6(x-2)(x+1) = 0 \;\Leftrightarrow$ $x = 2, -1$. $f(-2) = -3$, $f(-1) = 8$, $f(2) = -19$, and $f(3) = -8$. So $f(-1) = 8$ is the absolute maximum value and $f(2) = -19$ is the absolute minimum value.

45. $f(x) = x^4 - 2x^2 + 3$, $[-2, 3]$. $f'(x) = 4x^3 - 4x = 4x(x^2 - 1) = 4x(x+1)(x-1) = 0 \;\Leftrightarrow\; x = -1, 0, 1$. $f(-2) = 11$, $f(-1) = 2$, $f(0) = 3$, $f(1) = 2$, $f(3) = 66$. So $f(3) = 66$ is the absolute maximum value and $f(\pm 1) = 2$ is the absolute minimum value.

47. $f(t) = t\sqrt{4 - t^2}$, $[-1, 2]$.

$$f'(t) = t \cdot \tfrac{1}{2}(4 - t^2)^{-1/2}(-2t) + (4 - t^2)^{1/2} \cdot 1 = \frac{-t^2}{\sqrt{4 - t^2}} + \sqrt{4 - t^2} = \frac{-t^2 + (4 - t^2)}{\sqrt{4 - t^2}} = \frac{4 - 2t^2}{\sqrt{4 - t^2}}.$$

$f'(t) = 0 \;\Rightarrow\; 4 - 2t^2 = 0 \;\Rightarrow\; t^2 = 2 \;\Rightarrow\; t = \pm\sqrt{2}$, but $t = -\sqrt{2}$ is not in the given interval, $[-1, 2]$. $f'(t)$ does not exist if $4 - t^2 = 0 \;\Rightarrow\; t = \pm 2$, but -2 is not in the given interval. $f(-1) = -\sqrt{3}$, $f(\sqrt{2}) = 2$, and $f(2) = 0$. So $f(\sqrt{2}) = 2$ is the absolute maximum value and $f(-1) = -\sqrt{3}$ is the absolute minimum value.

49. $f(x) = xe^{-x^2/8}$, $[-1, 4]$. $f'(x) = x \cdot e^{-x^2/8} \cdot (-\frac{x}{4}) + e^{-x^2/8} \cdot 1 = e^{-x^2/8}(-\frac{x^2}{4} + 1)$. Since $e^{-x^2/8}$ is never 0, $f'(x) = 0 \;\Rightarrow\; -x^2/4 + 1 = 0 \;\Rightarrow\; 1 = x^2/4 \;\Rightarrow\; x^2 = 4 \;\Rightarrow\; x = \pm 2$, but -2 is not in the given interval, $[-1, 4]$. $f(-1) = -e^{-1/8} \approx -0.88$, $f(2) = 2e^{-1/2} \approx 1.21$, and $f(4) = 4e^{-2} \approx 0.54$. So $f(2) = 2e^{-1/2} = 2/\sqrt{e}$ is the absolute maximum value and $f(-1) = -e^{-1/8} = -1/\sqrt[8]{e}$ is the absolute minimum value.

51. $f(x) = \ln(x^2 + x + 1)$, $[-1, 1]$. $f'(x) = \dfrac{1}{x^2 + x + 1} \cdot (2x + 1) = 0 \;\Leftrightarrow\; x = -\frac{1}{2}$. Since $x^2 + x + 1 > 0$ for all x, the domain of f and f' is $\mathbb{R}$. $f(-1) = \ln 1 = 0$, $f(-\frac{1}{2}) = \ln\frac{3}{4} \approx -0.29$, and $f(1) = \ln 3 \approx 1.10$. So $f(1) = \ln 3 \approx 1.10$ is the absolute maximum value and $f(-\frac{1}{2}) = \ln\frac{3}{4} \approx -0.29$ is the absolute minimum value.

© 2016 Cengage Learning. All Rights Reserved. May not be scanned, copied or duplicated, or posted to a publicly accessible website, in whole or in part.

53. $f(t) = 2\cos t + \sin 2t,\ [0, \pi/2]$.

$f'(t) = -2\sin t + \cos 2t \cdot 2 = -2\sin t + 2(1 - 2\sin^2 t) = -2(2\sin^2 t + \sin t - 1) = -2(2\sin t - 1)(\sin t + 1)$.

$f'(t) = 0 \;\Rightarrow\; \sin t = \frac{1}{2}$ or $\sin t = -1 \;\Rightarrow\; t = \frac{\pi}{6}$. $f(0) = 2$, $f(\frac{\pi}{6}) = \sqrt{3} + \frac{1}{2}\sqrt{3} = \frac{3}{2}\sqrt{3} \approx 2.60$, and $f(\frac{\pi}{2}) = 0$.

So $f(\frac{\pi}{6}) = \frac{3}{2}\sqrt{3}$ is the absolute maximum value and $f(\frac{\pi}{2}) = 0$ is the absolute minimum value.

55. $f(x) = x^a(1 - x)^b,\ 0 \le x \le 1, a > 0, b > 0$.

$$\begin{aligned} f'(x) &= x^a \cdot b(1-x)^{b-1}(-1) + (1-x)^b \cdot ax^{a-1} = x^{a-1}(1-x)^{b-1}[x \cdot b(-1) + (1-x) \cdot a] \\ &= x^{a-1}(1-x)^{b-1}(a - ax - bx) \end{aligned}$$

At the endpoints, we have $f(0) = f(1) = 0$ [the minimum value of f]. In the interval $(0, 1)$, $f'(x) = 0 \;\Leftrightarrow\; x = \dfrac{a}{a+b}$.

$$f\left(\frac{a}{a+b}\right) = \left(\frac{a}{a+b}\right)^a \left(1 - \frac{a}{a+b}\right)^b = \frac{a^a}{(a+b)^a}\left(\frac{a+b-a}{a+b}\right)^b = \frac{a^a}{(a+b)^a} \cdot \frac{b^b}{(a+b)^b} = \frac{a^a b^b}{(a+b)^{a+b}}.$$

So $f\left(\dfrac{a}{a+b}\right) = \dfrac{a^a b^b}{(a+b)^{a+b}}$ is the absolute maximum value.

57. $P(v) = \dfrac{10 + v + v^2}{1 + v}, 0 \le v \le 9$. $P'(v) = \dfrac{(1+v)(1+2v) - (10 + v + v^2)(1)}{(1+v)^2} = \dfrac{v^2 + 2v - 9}{(1+v)^2}$. $P'(v) = 0 \;\Leftrightarrow$

$\dfrac{v^2 + 2v - 9}{(1+v)^2} = 0 \;\Leftrightarrow\; v^2 + 2v - 9 = 0 \;\Leftrightarrow\; v = \dfrac{-2 \pm \sqrt{2^2 - 4(-9)}}{2} = -1 \pm \sqrt{10} \approx -4.16,\ 2.16$, but -4.16 is not

in the given interval, $[0, 9]$. $P(0) = 10$, $P(2.16) \approx 5.32$, and $P(9) = 10$. So the smallest population size is $P(2.16) \approx 5.32$ and the largest is $P(0) = P(9) = 10$.

59. (a) $v(r) = k(r_0 - r)r^2 = kr_0r^2 - kr^3 \;\Rightarrow\; v'(r) = 2kr_0r - 3kr^2$.

$v'(r) = 0 \;\Rightarrow\; kr(2r_0 - 3r) = 0 \;\Rightarrow\; r = 0$ or $\frac{2}{3}r_0$ (but 0 is not in the interval). Evaluating v at $\frac{1}{2}r_0$, $\frac{2}{3}r_0$, and r_0, we get $v(\frac{1}{2}r_0) = \frac{1}{8}kr_0^3$, $v(\frac{2}{3}r_0) = \frac{4}{27}kr_0^3$, and $v(r_0) = 0$. Since $\frac{4}{27} > \frac{1}{8}$, v attains its maximum value at $r = \frac{2}{3}r_0$. This supports the statement in the text.

(b) From part (a), the maximum value of v is $\frac{4}{27}kr_0^3$.

(c)

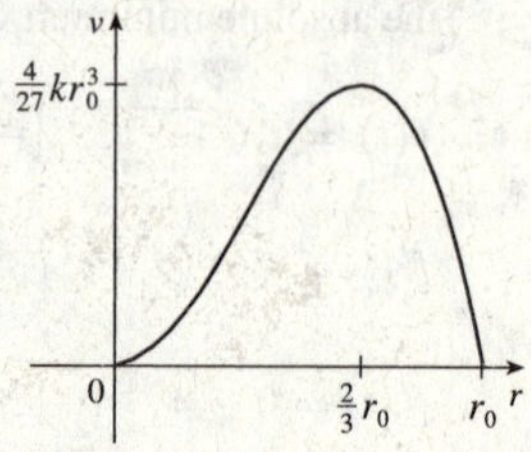

61. The density is defined as $\rho = \dfrac{\text{mass}}{\text{volume}} = \dfrac{1000}{V(T)}$ (in g/cm^3). But a critical point of ρ will also be a critical point of V [since $\dfrac{d\rho}{dT} = -1000V^{-2}\dfrac{dV}{dT}$ and V is never 0], and V is easier to differentiate than ρ.

$V(T) = 999.87 - 0.06426T + 0.0085043T^2 - 0.0000679T^3 \;\Rightarrow\; V'(T) = -0.06426 + 0.0170086T - 0.0002037T^2$.

Setting this equal to 0 and using the quadratic formula to find T, we get

$T = \dfrac{-0.0170086 \pm \sqrt{0.0170086^2 - 4 \cdot 0.0002037 \cdot 0.06426}}{2(-0.0002037)} \approx 3.9665°\text{C}$ or $79.5318°\text{C}$. Since we are only interested

in the region $0°\text{C} \le T \le 30°\text{C}$, we check the density ρ at the endpoints and at $3.9665°\text{C}$: $\rho(0) \approx \dfrac{1000}{999.87} \approx 1.00013$;

$\rho(30) \approx \dfrac{1000}{1003.7628} \approx 0.99625$; $\rho(3.9665) \approx \dfrac{1000}{999.7447} \approx 1.000255$. So water has its maximum density at about $3.9665°\text{C}$.

© 2016 Cengage Learning. All Rights Reserved. May not be scanned, copied or duplicated, or posted to a publicly accessible website, in whole or in part.

4.2 How Derivatives Affect the Shape of a Graph

1. $\dfrac{f(8)-f(0)}{8-0} = \dfrac{6-4}{8} = \dfrac{1}{4}$. The values of c which satisfy $f'(c) = \frac{1}{4}$ seem to be about $c = 0.8, 3.2, 4.4$, and 6.1.

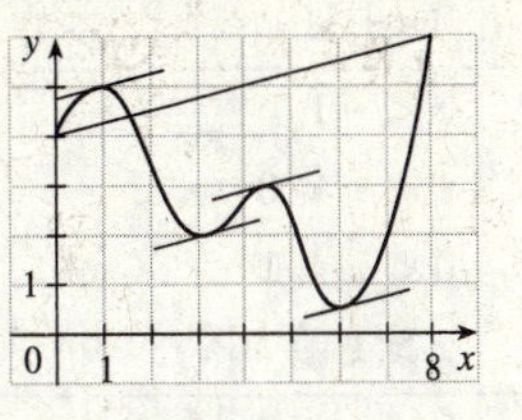

3. If $3 \le f'(x) \le 5$ for all x, then by the Mean Value Theorem, $f(8) - f(2) = f'(c) \cdot (8-2)$ for some c in $[2, 8]$. (f is differentiable for all x, so, in particular, f is differentiable on $(2, 8)$ and continuous on $[2, 8]$. Thus, the hypotheses of the Mean Value Theorem are satisfied.) Since $f(8) - f(2) = 6f'(c)$ and $3 \le f'(c) \le 5$, it follows that $6 \cdot 3 \le 6f'(c) \le 6 \cdot 5 \;\Rightarrow\; 18 \le f(8) - f(2) \le 30$.

5. (a) f is increasing on $(1, 3)$ and $(4, 6)$.

(b) f is decreasing on $(0, 1)$ and $(3, 4)$.

(c) f is concave upward on $(0, 2)$.

(d) f is concave downward on $(2, 4)$ and $(4, 6)$.

(e) The point of inflection is $(2, 3)$.

7. (a) See the First Derivative Test.

(b) See the Second Derivative Test and the note that precedes Example 9.

9. (a) There is an IP at $x = 3$ because the graph of f changes from CD to CU there. There is an IP at $x = 5$ because the graph of f changes from CU to CD there.

(b) There is an IP at $x = 2$ and at $x = 6$ because $f'(x)$ has a maximum value there, and so $f''(x)$ changes from positive to negative there. There is an IP at $x = 4$ because $f'(x)$ has a minimum value there and so $f''(x)$ changes from negative to positive there.

(c) There is an inflection point at $x = 1$ because $f''(x)$ changes from negative to positive there, and so the graph of f changes from concave downward to concave upward. There is an inflection point at $x = 7$ because $f''(x)$ changes from positive to negative there, and so the graph of f changes from concave upward to concave downward.

11. (a) $f(x) = 2x^3 + 3x^2 - 36x \;\Rightarrow\; f'(x) = 6x^2 + 6x - 36 = 6(x^2 + x - 6) = 6(x+3)(x-2)$.

We don't need to include the "6" in the chart to determine the sign of $f'(x)$.

Interval	$x+3$	$x-2$	$f'(x)$	f
$x < -3$	$-$	$-$	$+$	increasing on $(-\infty, -3)$
$-3 < x < 2$	$+$	$-$	$-$	decreasing on $(-3, 2)$
$x > 2$	$+$	$+$	$+$	increasing on $(2, \infty)$

(b) f changes from increasing to decreasing at $x = -3$ and from decreasing to increasing at $x = 2$. Thus, $f(-3) = 81$ is a local maximum value and $f(2) = -44$ is a local minimum value.

(c) $f'(x) = 6x^2 + 6x - 36 \;\Rightarrow\; f''(x) = 12x + 6$. $f''(x) = 0$ at $x = -\frac{1}{2}$, $f''(x) > 0 \;\Leftrightarrow\; x > -\frac{1}{2}$, and $f''(x) < 0 \;\Leftrightarrow\; x < -\frac{1}{2}$. Thus, f is concave upward on $\left(-\frac{1}{2}, \infty\right)$ and concave downward on $\left(-\infty, -\frac{1}{2}\right)$. There is an inflection point at $\left(-\frac{1}{2}, f\left(-\frac{1}{2}\right)\right) = \left(-\frac{1}{2}, \frac{37}{2}\right)$.

© 2016 Cengage Learning. All Rights Reserved. May not be scanned, copied or duplicated, or posted to a publicly accessible website, in whole or in part.

13. (a) $f(x) = x^4 - 2x^2 + 3 \Rightarrow f'(x) = 4x^3 - 4x = 4x(x^2 - 1) = 4x(x+1)(x-1)$.

Interval	$x+1$	x	$x-1$	$f'(x)$	f
$x < -1$	$-$	$-$	$-$	$-$	decreasing on $(-\infty, -1)$
$-1 < x < 0$	$+$	$-$	$-$	$+$	increasing on $(-1, 0)$
$0 < x < 1$	$+$	$+$	$-$	$-$	decreasing on $(0, 1)$
$x > 1$	$+$	$+$	$+$	$+$	increasing on $(1, \infty)$

(b) f changes from increasing to decreasing at $x = 0$ and from decreasing to increasing at $x = -1$ and $x = 1$. Thus, $f(0) = 3$ is a local maximum value and $f(\pm 1) = 2$ are local minimum values.

(c) $f''(x) = 12x^2 - 4 = 12\left(x^2 - \frac{1}{3}\right) = 12\left(x + 1/\sqrt{3}\right)\left(x - 1/\sqrt{3}\right)$. $f''(x) > 0 \Leftrightarrow x < -1/\sqrt{3}$ or $x > 1/\sqrt{3}$ and $f''(x) < 0 \Leftrightarrow -1/\sqrt{3} < x < 1/\sqrt{3}$. Thus, f is concave upward on $\left(-\infty, -\sqrt{3}/3\right)$ and $\left(\sqrt{3}/3, \infty\right)$ and concave downward on $\left(-\sqrt{3}/3, \sqrt{3}/3\right)$. There are inflection points at $\left(\pm\sqrt{3}/3, \frac{22}{9}\right)$.

15. (a) $f(x) = \sin x + \cos x$, $0 \le x \le 2\pi$. $f'(x) = \cos x - \sin x = 0 \Rightarrow \cos x = \sin x \Rightarrow 1 = \dfrac{\sin x}{\cos x} \Rightarrow$ $\tan x = 1 \Rightarrow x = \frac{\pi}{4}$ or $\frac{5\pi}{4}$. Thus, $f'(x) > 0 \Leftrightarrow \cos x - \sin x > 0 \Leftrightarrow \cos x > \sin x \Leftrightarrow 0 < x < \frac{\pi}{4}$ or $\frac{5\pi}{4} < x < 2\pi$ and $f'(x) < 0 \Leftrightarrow \cos x < \sin x \Leftrightarrow \frac{\pi}{4} < x < \frac{5\pi}{4}$. So f is increasing on $\left(0, \frac{\pi}{4}\right)$ and $\left(\frac{5\pi}{4}, 2\pi\right)$ and f is decreasing on $\left(\frac{\pi}{4}, \frac{5\pi}{4}\right)$.

(b) f changes from increasing to decreasing at $x = \frac{\pi}{4}$ and from decreasing to increasing at $x = \frac{5\pi}{4}$. Thus, $f\left(\frac{\pi}{4}\right) = \sqrt{2}$ is a local maximum value and $f\left(\frac{5\pi}{4}\right) = -\sqrt{2}$ is a local minimum value.

(c) $f''(x) = -\sin x - \cos x = 0 \Rightarrow -\sin x = \cos x \Rightarrow \tan x = -1 \Rightarrow x = \frac{3\pi}{4}$ or $\frac{7\pi}{4}$. Divide the interval $(0, 2\pi)$ into subintervals with these numbers as endpoints and complete a second derivative chart.

Interval	$f''(x) = -\sin x - \cos x$	Concavity
$\left(0, \frac{3\pi}{4}\right)$	$f''\left(\frac{\pi}{2}\right) = -1 < 0$	downward
$\left(\frac{3\pi}{4}, \frac{7\pi}{4}\right)$	$f''(\pi) = 1 > 0$	upward
$\left(\frac{7\pi}{4}, 2\pi\right)$	$f''\left(\frac{11\pi}{6}\right) = \frac{1}{2} - \frac{1}{2}\sqrt{3} < 0$	downward

There are inflection points at $\left(\frac{3\pi}{4}, 0\right)$ and $\left(\frac{7\pi}{4}, 0\right)$.

17. (a) $f(x) = e^{2x} + e^{-x} \Rightarrow f'(x) = 2e^{2x} - e^{-x}$. $f'(x) > 0 \Leftrightarrow 2e^{2x} > e^{-x} \Leftrightarrow e^{3x} > \frac{1}{2} \Leftrightarrow 3x > \ln\frac{1}{2} \Leftrightarrow$ $x > \frac{1}{3}(\ln 1 - \ln 2) \Leftrightarrow x > -\frac{1}{3}\ln 2$ $[\approx -0.23]$ and $f'(x) < 0$ if $x < -\frac{1}{3}\ln 2$. So f is increasing on $\left(-\frac{1}{3}\ln 2, \infty\right)$ and f is decreasing on $\left(-\infty, -\frac{1}{3}\ln 2\right)$.

(b) f changes from decreasing to increasing at $x = -\frac{1}{3}\ln 2$. Thus,

$$f\left(-\tfrac{1}{3}\ln 2\right) = f\left(\ln\sqrt[3]{1/2}\right) = e^{2\ln\sqrt[3]{1/2}} + e^{-\ln\sqrt[3]{1/2}} = e^{\ln\sqrt[3]{1/4}} + e^{\ln\sqrt[3]{2}} = \sqrt[3]{1/4} + \sqrt[3]{2} = 2^{-2/3} + 2^{1/3} \quad [\approx 1.89]$$

is a local minimum value.

(c) $f''(x) = 4e^{2x} + e^{-x} > 0$ [the sum of two positive terms]. Thus, f is concave upward on $(-\infty, \infty)$ and there is no point of inflection.

© 2016 Cengage Learning. All Rights Reserved. May not be scanned, copied or duplicated, or posted to a publicly accessible website, in whole or in part.

19. (a) $y = f(x) = \dfrac{\ln x}{\sqrt{x}}$. (Note that f is only defined for $x > 0$.)

$$f'(x) = \frac{\sqrt{x}\,(1/x) - \ln x\left(\frac{1}{2}x^{-1/2}\right)}{x} = \frac{\dfrac{1}{\sqrt{x}} - \dfrac{\ln x}{2\sqrt{x}}}{x} \cdot \frac{2\sqrt{x}}{2\sqrt{x}} = \frac{2 - \ln x}{2x^{3/2}} > 0 \;\Leftrightarrow\; 2 - \ln x > 0 \;\Leftrightarrow$$

$\ln x < 2 \;\Leftrightarrow\; x < e^2$. Therefore f is increasing on $(0, e^2)$ and decreasing on (e^2, ∞).

(b) f changes from increasing to decreasing at $x = e^2$, so $f(e^2) = \dfrac{\ln e^2}{\sqrt{e^2}} = \dfrac{2}{e}$ is a local maximum value.

(c) $$f''(x) = \frac{2x^{3/2}(-1/x) - (2 - \ln x)(3x^{1/2})}{(2x^{3/2})^2} = \frac{-2x^{1/2} + 3x^{1/2}(\ln x - 2)}{4x^3} = \frac{x^{1/2}(-2 + 3\ln x - 6)}{4x^3} = \frac{3\ln x - 8}{4x^{5/2}}$$

$f''(x) = 0 \;\Leftrightarrow\; \ln x = \frac{8}{3} \;\Leftrightarrow\; x = e^{8/3}$. $f''(x) > 0 \;\Leftrightarrow\; x > e^{8/3}$, so f is concave upward on $(e^{8/3}, \infty)$ and concave downward on $(0, e^{8/3})$. There is an inflection point at $\left(e^{8/3}, \frac{8}{3}e^{-4/3}\right) \approx (14.39, 0.70)$.

21. $f(x) = x + \sqrt{1-x} \;\Rightarrow\; f'(x) = 1 + \frac{1}{2}(1-x)^{-1/2}(-1) = 1 - \dfrac{1}{2\sqrt{1-x}}$. Note that f is defined for $1 - x \geq 0$; that is, for $x \leq 1$. $f'(x) = 0 \;\Rightarrow\; 2\sqrt{1-x} = 1 \;\Rightarrow\; \sqrt{1-x} = \frac{1}{2} \;\Rightarrow\; 1 - x = \frac{1}{4} \;\Rightarrow\; x = \frac{3}{4}$. f' does not exist at $x = 1$, but we can't have a local maximum or minimum at an endpoint.

First Derivative Test: $f'(x) > 0 \;\Rightarrow\; x < \frac{3}{4}$ and $f'(x) < 0 \;\Rightarrow\; \frac{3}{4} < x < 1$. Since f' changes from positive to negative at $x = \frac{3}{4}$, $f\left(\frac{3}{4}\right) = \frac{5}{4}$ is a local maximum value.

Second Derivative Test: $f''(x) = -\frac{1}{2}\left(-\frac{1}{2}\right)(1-x)^{-3/2}(-1) = -\dfrac{1}{4\left(\sqrt{1-x}\right)^3}$.

$f''\left(\frac{3}{4}\right) = -2 < 0 \;\Rightarrow\; f\left(\frac{3}{4}\right) = \frac{5}{4}$ is a local maximum value.

Preference: The First Derivative Test may be slightly easier to apply in this case.

23. (a) By the Second Derivative Test, if $f'(2) = 0$ and $f''(2) = -5 < 0$, f has a local maximum at $x = 2$.

(b) If $f'(6) = 0$, we know that f has a horizontal tangent at $x = 6$. Knowing that $f''(6) = 0$ does not provide any additional information since the Second Derivative Test fails. For example, the first and second derivatives of $y = (x-6)^4$, $y = -(x-6)^4$, and $y = (x-6)^3$ all equal zero for $x = 6$, but the first has a local minimum at $x = 6$, the second has a local maximum at $x = 6$, and the third has an inflection point at $x = 6$.

25. (a) $f(x) = 2x^3 - 3x^2 - 12x \;\Rightarrow\; f'(x) = 6x^2 - 6x - 12 = 6(x^2 - x - 2) = 6(x-2)(x+1)$.

$f'(x) > 0 \;\Leftrightarrow\; x < -1$ or $x > 2$ and $f'(x) < 0 \;\Leftrightarrow\; -1 < x < 2$. So f is increasing on $(-\infty, -1)$ and $(2, \infty)$, and f is decreasing on $(-1, 2)$.

(b) Since f changes from increasing to decreasing at $x = -1$, $f(-1) = 7$ is a local maximum value. Since f changes from decreasing to increasing at $x = 2$, $f(2) = -20$ is a local minimum value.

(c) $f''(x) = 6(2x - 1) \;\Rightarrow\; f''(x) > 0$ on $\left(\frac{1}{2}, \infty\right)$ and $f''(x) < 0$ on $\left(-\infty, \frac{1}{2}\right)$. So f is concave upward on $\left(\frac{1}{2}, \infty\right)$ and concave downward on $\left(-\infty, \frac{1}{2}\right)$. There is a change in concavity at $x = \frac{1}{2}$, and we have an inflection point at $\left(\frac{1}{2}, -\frac{13}{2}\right)$.

(d)

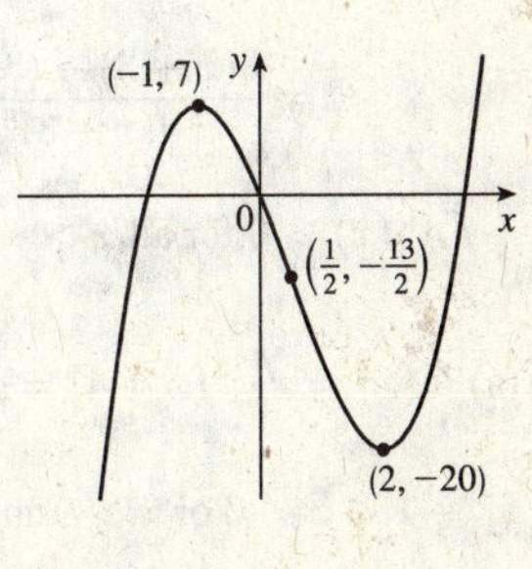

© 2016 Cengage Learning. All Rights Reserved. May not be scanned, copied or duplicated, or posted to a publicly accessible website, in whole or in part.

27. (a) $f(x) = 2 + 2x^2 - x^4 \Rightarrow f'(x) = 4x - 4x^3 = 4x(1 - x^2) = 4x(1 + x)(1 - x)$. $f'(x) > 0 \Leftrightarrow x < -1$ or $0 < x < 1$ and $f'(x) < 0 \Leftrightarrow -1 < x < 0$ or $x > 1$. So f is increasing on $(-\infty, -1)$ and $(0, 1)$ and f is decreasing on $(-1, 0)$ and $(1, \infty)$.

(b) f changes from increasing to decreasing at $x = -1$ and $x = 1$, so $f(-1) = 3$ and $f(1) = 3$ are local maximum values. f changes from decreasing to increasing at $x = 0$, so $f(0) = 2$ is a local minimum value.

(c) $f''(x) = 4 - 12x^2 = 4(1 - 3x^2)$. $f''(x) = 0 \Leftrightarrow 1 - 3x^2 = 0 \Leftrightarrow x^2 = \frac{1}{3} \Leftrightarrow x = \pm 1/\sqrt{3}$. $f''(x) > 0$ on $(-1/\sqrt{3}, 1/\sqrt{3})$ and $f''(x) < 0$ on $(-\infty, -1/\sqrt{3})$ and $(1/\sqrt{3}, \infty)$. So f is concave upward on $(-1/\sqrt{3}, 1/\sqrt{3})$ and f is concave downward on $(-\infty, -1/\sqrt{3})$ and $(1/\sqrt{3}, \infty)$. $f(\pm 1/\sqrt{3}) = 2 + \frac{2}{3} - \frac{1}{9} = \frac{23}{9}$. There are points of inflection at $(\pm 1/\sqrt{3}, \frac{23}{9})$.

(d)

29. (a) $h(x) = (x + 1)^5 - 5x - 2 \Rightarrow h'(x) = 5(x + 1)^4 - 5$. $h'(x) = 0 \Leftrightarrow 5(x + 1)^4 = 5 \Leftrightarrow (x + 1)^4 = 1 \Rightarrow (x + 1)^2 = 1 \Rightarrow x + 1 = 1$ or $x + 1 = -1 \Rightarrow x = 0$ or $x = -2$. $h'(x) > 0 \Leftrightarrow x < -2$ or $x > 0$ and $h'(x) < 0 \Leftrightarrow -2 < x < 0$. So h is increasing on $(-\infty, -2)$ and $(0, \infty)$ and h is decreasing on $(-2, 0)$.

(b) $h(-2) = 7$ is a local maximum value and $h(0) = -1$ is a local minimum value.

(c) $h''(x) = 20(x + 1)^3 = 0 \Leftrightarrow x = -1$. $h''(x) > 0 \Leftrightarrow x > -1$ and $h''(x) < 0 \Leftrightarrow x < -1$, so h is CU on $(-1, \infty)$ and h is CD on $(-\infty, -1)$. There is a point of inflection at $(-1, h(-1)) = (-1, 3)$.

(d)

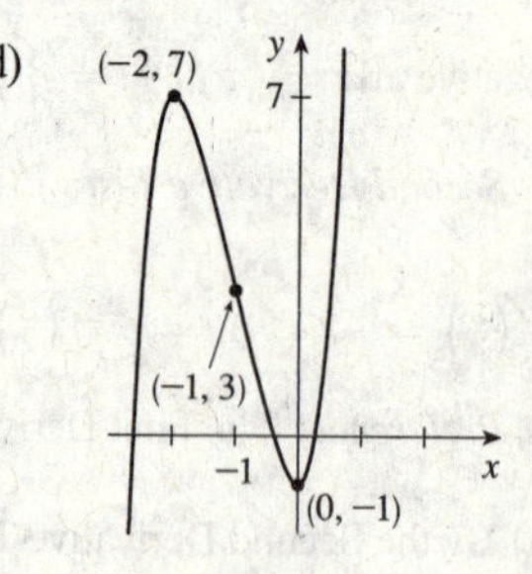

31. (a) $A(x) = x\sqrt{x + 3} \Rightarrow A'(x) = x \cdot \frac{1}{2}(x+3)^{-1/2} + \sqrt{x + 3} \cdot 1 = \dfrac{x}{2\sqrt{x + 3}} + \sqrt{x + 3} = \dfrac{x + 2(x + 3)}{2\sqrt{x + 3}} = \dfrac{3x + 6}{2\sqrt{x + 3}}$.

The domain of A is $[-3, \infty)$. $A'(x) > 0$ for $x > -2$ and $A'(x) < 0$ for $-3 < x < -2$, so A is increasing on $(-2, \infty)$ and decreasing on $(-3, -2)$.

(b) $A(-2) = -2$ is a local minimum value.

(c)
$$A''(x) = \frac{2\sqrt{x + 3} \cdot 3 - (3x + 6) \cdot \dfrac{1}{\sqrt{x + 3}}}{\left(2\sqrt{x + 3}\right)^2}$$
$$= \frac{6(x + 3) - (3x + 6)}{4(x + 3)^{3/2}} = \frac{3x + 12}{4(x + 3)^{3/2}} = \frac{3(x + 4)}{4(x + 3)^{3/2}}$$

$A''(x) > 0$ for all $x > -3$, so A is concave upward on $(-3, \infty)$. There is no inflection point.

(d)

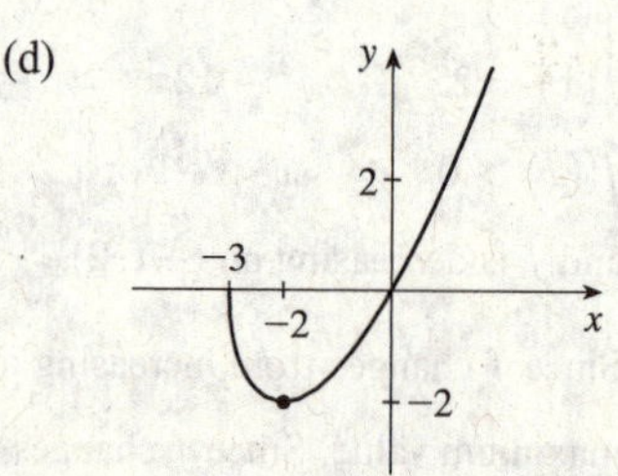

33. (a) $C(x) = x^{1/3}(x + 4) = x^{4/3} + 4x^{1/3} \Rightarrow C'(x) = \frac{4}{3}x^{1/3} + \frac{4}{3}x^{-2/3} = \frac{4}{3}x^{-2/3}(x + 1) = \dfrac{4(x + 1)}{3\sqrt[3]{x^2}}$. $C'(x) > 0$ if $-1 < x < 0$ or $x > 0$ and $C'(x) < 0$ for $x < -1$, so C is increasing on $(-1, \infty)$ and C is decreasing on $(-\infty, -1)$.

© 2016 Cengage Learning. All Rights Reserved. May not be scanned, copied or duplicated, or posted to a publicly accessible website, in whole or in part.

(b) $C(-1) = -3$ is a local minimum value.

(c) $C''(x) = \frac{4}{9}x^{-2/3} - \frac{8}{9}x^{-5/3} = \frac{4}{9}x^{-5/3}(x-2) = \dfrac{4(x-2)}{9\sqrt[3]{x^5}}$.

$C''(x) < 0$ for $0 < x < 2$ and $C''(x) > 0$ for $x < 0$ and $x > 2$, so C is concave downward on $(0, 2)$ and concave upward on $(-\infty, 0)$ and $(2, \infty)$.

There are inflection points at $(0, 0)$ and $\left(2, 6\sqrt[3]{2}\right) \approx (2, 7.56)$.

(d)

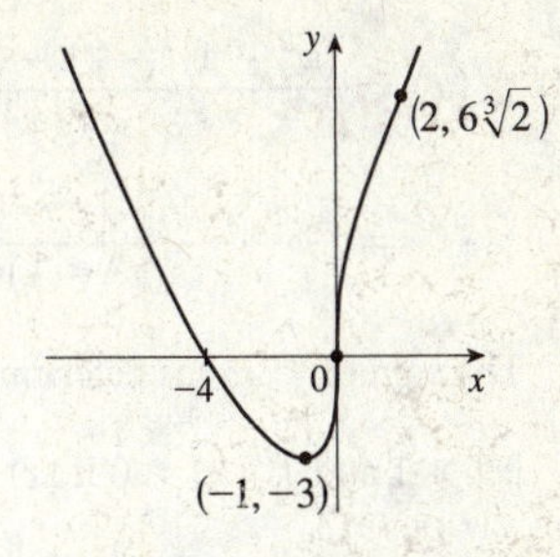

35. (a) $f(\theta) = 2\cos\theta + \cos^2\theta,\ 0 \le \theta \le 2\pi \quad\Rightarrow\quad f'(\theta) = -2\sin\theta + 2\cos\theta\,(-\sin\theta) = -2\sin\theta\,(1+\cos\theta)$.

$f'(\theta) = 0 \quad\Leftrightarrow\quad \theta = 0, \pi$, and 2π. $f'(\theta) > 0 \quad\Leftrightarrow\quad \pi < \theta < 2\pi$ and $f'(\theta) < 0 \quad\Leftrightarrow\quad 0 < \theta < \pi$. So f is increasing on $(\pi, 2\pi)$ and f is decreasing on $(0, \pi)$.

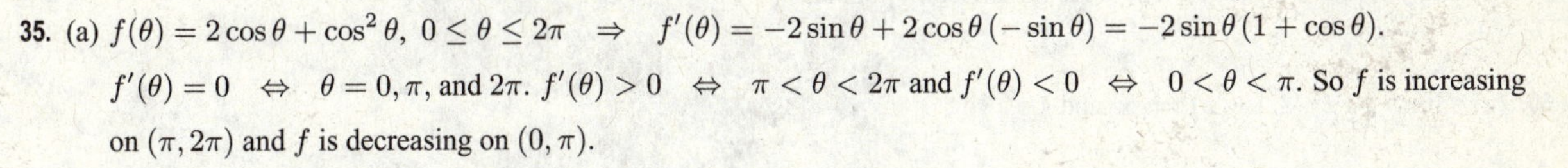

(b) $f(\pi) = -1$ is a local minimum value.

(c) $f'(\theta) = -2\sin\theta\,(1+\cos\theta) \quad\Rightarrow$

$$\begin{aligned} f''(\theta) &= -2\sin\theta\,(-\sin\theta) + (1+\cos\theta)(-2\cos\theta) = 2\sin^2\theta - 2\cos\theta - 2\cos^2\theta \\ &= 2(1-\cos^2\theta) - 2\cos\theta - 2\cos^2\theta = -4\cos^2\theta - 2\cos\theta + 2 \\ &= -2(2\cos^2\theta + \cos\theta - 1) = -2(2\cos\theta - 1)(\cos\theta + 1) \end{aligned}$$

Since $-2(\cos\theta + 1) < 0$ [for $\theta \ne \pi$], $f''(\theta) > 0 \quad\Rightarrow\quad 2\cos\theta - 1 < 0 \quad\Rightarrow\quad \cos\theta < \frac{1}{2} \quad\Rightarrow\quad \frac{\pi}{3} < \theta < \frac{5\pi}{3}$ and $f''(\theta) < 0 \quad\Rightarrow\quad \cos\theta > \frac{1}{2} \quad\Rightarrow\quad 0 < \theta < \frac{\pi}{3}$ or $\frac{5\pi}{3} < \theta < 2\pi$. So f is CU on $\left(\frac{\pi}{3}, \frac{5\pi}{3}\right)$ and f is CD on $\left(0, \frac{\pi}{3}\right)$ and $\left(\frac{5\pi}{3}, 2\pi\right)$. There are points of inflection at $\left(\frac{\pi}{3}, f\left(\frac{\pi}{3}\right)\right) = \left(\frac{\pi}{3}, \frac{5}{4}\right)$ and $\left(\frac{5\pi}{3}, f\left(\frac{5\pi}{3}\right)\right) = \left(\frac{5\pi}{3}, \frac{5}{4}\right)$.

(d)

37. $f(x) = \dfrac{x^2}{x^2-1} = \dfrac{x^2}{(x+1)(x-1)}$ has domain $(-\infty, -1) \cup (-1, 1) \cup (1, \infty)$.

(a) $\displaystyle\lim_{x\to\pm\infty} f(x) = \lim_{x\to\pm\infty} \frac{x^2/x^2}{(x^2-1)/x^2} = \lim_{x\to\pm\infty} \frac{1}{1-1/x^2} = \frac{1}{1-0} = 1$, so $y = 1$ is a HA.

$\displaystyle\lim_{x\to -1^-} \frac{x^2}{x^2-1} = \infty$ since $x^2 \to 1$ and $(x^2-1) \to 0^+$ as $x \to -1^-$, so $x = -1$ is a VA.

$\displaystyle\lim_{x\to 1^+} \frac{x^2}{x^2-1} = \infty$ since $x^2 \to 1$ and $(x^2-1) \to 0^+$ as $x \to 1^+$, so $x = 1$ is a VA.

(b) $f(x) = \dfrac{x^2}{x^2-1} \quad\Rightarrow\quad f'(x) = \dfrac{(x^2-1)(2x) - x^2(2x)}{(x^2-1)^2} = \dfrac{2x[(x^2-1) - x^2]}{(x^2-1)^2} = \dfrac{-2x}{(x^2-1)^2}$. Since $(x^2-1)^2$ is positive for all x in the domain of f, the sign of the derivative is determined by the sign of $-2x$. Thus, $f'(x) > 0$ if $x < 0$ $(x \ne -1)$ and $f'(x) < 0$ if $x > 0$ $(x \ne 1)$. So f is increasing on $(-\infty, -1)$ and $(-1, 0)$, and f is decreasing on $(0, 1)$ and $(1, \infty)$.

(c) $f'(x) = 0 \quad\Rightarrow\quad x = 0$ and $f(0) = 0$ is a local maximum value.

© 2016 Cengage Learning. All Rights Reserved. May not be scanned, copied or duplicated, or posted to a publicly accessible website, in whole or in part.

(d) $f''(x) = \dfrac{(x^2-1)^2(-2) - (-2x)\cdot 2(x^2-1)(2x)}{[(x^2-1)^2]^2}$

$= \dfrac{2(x^2-1)[-(x^2-1)+4x^2]}{(x^2-1)^4} = \dfrac{2(3x^2+1)}{(x^2-1)^3}$.

The sign of $f''(x)$ is determined by the denominator; that is, $f''(x) > 0$ if $|x| > 1$ and $f''(x) < 0$ if $|x| < 1$. Thus, f is CU on $(-\infty, -1)$ and $(1, \infty)$, and f is CD on $(-1, 1)$. There are no inflection points.

(e)

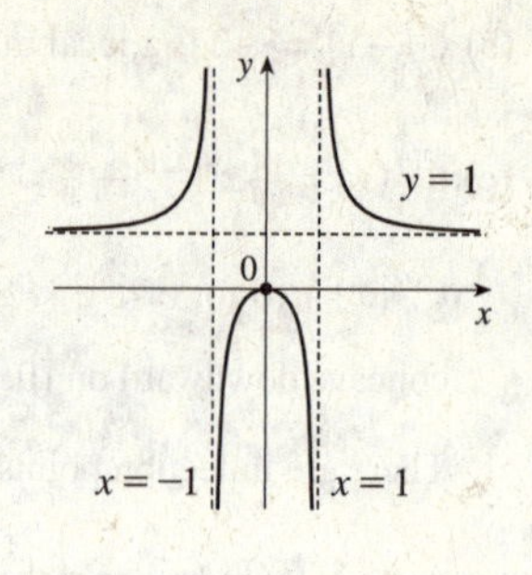

39. (a) $\lim\limits_{x\to-\infty}\left(\sqrt{x^2+1}-x\right) = \infty$ and

$\lim\limits_{x\to\infty}\left(\sqrt{x^2+1}-x\right) = \lim\limits_{x\to\infty}\left(\sqrt{x^2+1}-x\right)\dfrac{\sqrt{x^2+1}+x}{\sqrt{x^2+1}+x} = \lim\limits_{x\to\infty}\dfrac{1}{\sqrt{x^2+1}+x} = 0$, so $y = 0$ is a HA.

(b) $f(x) = \sqrt{x^2+1} - x \;\Rightarrow\; f'(x) = \dfrac{x}{\sqrt{x^2+1}} - 1$. Since $\dfrac{x}{\sqrt{x^2+1}} < 1$ for all x, $f'(x) < 0$, so f is decreasing on $\mathbb{R}$.

(c) No minimum or maximum

(d) $f''(x) = \dfrac{(x^2+1)^{1/2}(1) - x\cdot\frac{1}{2}(x^2+1)^{-1/2}(2x)}{\left(\sqrt{x^2+1}\right)^2}$

$= \dfrac{(x^2+1)^{1/2} - \dfrac{x^2}{(x^2+1)^{1/2}}}{x^2+1} = \dfrac{(x^2+1)-x^2}{(x^2+1)^{3/2}} = \dfrac{1}{(x^2+1)^{3/2}} > 0$,

so f is CU on $\mathbb{R}$. No IP

(e)

41. $f(x) = \ln(1-\ln x)$ is defined when $x > 0$ (so that $\ln x$ is defined) and $1 - \ln x > 0$ [so that $\ln(1-\ln x)$ is defined]. The second condition is equivalent to $1 > \ln x \;\Leftrightarrow\; x < e$, so f has domain $(0, e)$.

(a) As $x \to 0^+$, $\ln x \to -\infty$, so $1 - \ln x \to \infty$ and $f(x) \to \infty$. As $x \to e^-$, $\ln x \to 1^-$, so $1 - \ln x \to 0^+$ and $f(x) \to -\infty$. Thus, $x = 0$ and $x = e$ are vertical asymptotes. There is no horizontal asymptote.

(b) $f'(x) = \dfrac{1}{1-\ln x}\left(-\dfrac{1}{x}\right) = -\dfrac{1}{x(1-\ln x)} < 0$ on $(0, e)$. Thus, f is decreasing on its domain, $(0, e)$.

(c) $f'(x) \neq 0$ on $(0, e)$, so f has no local maximum or minimum value.

(d) $f''(x) = -\dfrac{-[x(1-\ln x)]'}{[x(1-\ln x)]^2} = \dfrac{x(-1/x) + (1-\ln x)}{x^2(1-\ln x)^2}$

$= -\dfrac{\ln x}{x^2(1-\ln x)^2}$

so $f''(x) > 0 \;\Leftrightarrow\; \ln x < 0 \;\Leftrightarrow\; 0 < x < 1$. Thus, f is CU on $(0, 1)$ and CD on $(1, e)$. There is an inflection point at $(1, 0)$.

(e)

© 2016 Cengage Learning. All Rights Reserved. May not be scanned, copied or duplicated, or posted to a publicly accessible website, in whole or in part.

43. (a) $\lim\limits_{x\to\pm\infty} e^{-1/(x+1)} = 1$ since $-1/(x+1) \to 0$, so $y = 1$ is a HA. $\lim\limits_{x\to-1^+} e^{-1/(x+1)} = 0$ since $-1/(x+1) \to -\infty$,

$\lim\limits_{x\to-1^-} e^{-1/(x+1)} = \infty$ since $-1/(x+1) \to \infty$, so $x = -1$ is a VA.

(b) $f(x) = e^{-1/(x+1)} \Rightarrow f'(x) = e^{-1/(x+1)}\left[-(-1)\dfrac{1}{(x+1)^2}\right]$ [Reciprocal Rule] $= e^{-1/(x+1)}/(x+1)^2 \Rightarrow$

$f'(x) > 0$ for all x except -1, so f is increasing on $(-\infty, -1)$ and $(-1, \infty)$.

(c) There is no local maximum or minimum.

(d) $f''(x) = \dfrac{(x+1)^2 e^{-1/(x+1)}\left[1/(x+1)^2\right] - e^{-1/(x+1)}\left[2(x+1)\right]}{\left[(x+1)^2\right]^2}$

$= \dfrac{e^{-1/(x+1)}\left[1-(2x+2)\right]}{(x+1)^4} = -\dfrac{e^{-1/(x+1)}(2x+1)}{(x+1)^4} \Rightarrow$

$f''(x) > 0 \Leftrightarrow 2x + 1 < 0 \Leftrightarrow x < -\frac{1}{2}$, so f is CU on $(-\infty, -1)$ and $\left(-1, -\frac{1}{2}\right)$, and CD on $\left(-\frac{1}{2}, \infty\right)$. f has an IP at $\left(-\frac{1}{2}, e^{-2}\right)$.

(e)

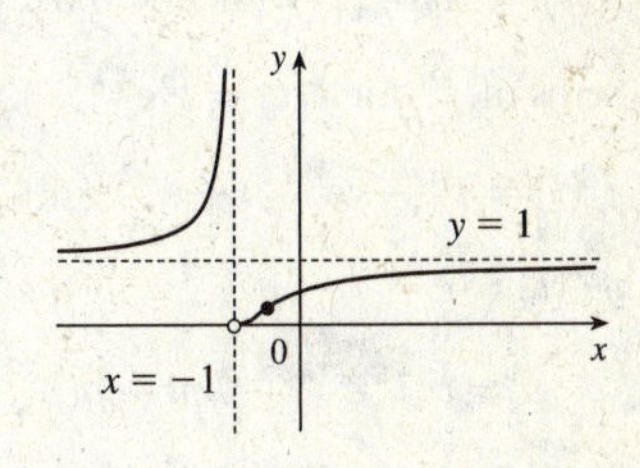

45. The nonnegative factors $(x+1)^2$ and $(x-6)^4$ do not affect the sign of $f'(x) = (x+1)^2(x-3)^5(x-6)^4$.

So $f'(x) > 0 \Rightarrow (x-3)^5 > 0 \Rightarrow x - 3 > 0 \Rightarrow x > 3$. Thus, f is increasing on the interval $(3, \infty)$.

47. (a) I'm very unhappy. It's uncomfortably hot and $f'(3) = 2$ indicates that the temperature is increasing, and $f''(3) = 4$ indicates that the rate of increase is increasing. (The temperature is rapidly getting warmer.)

(b) I'm still unhappy, but not as unhappy as in part (a). It's uncomfortably hot and $f'(3) = 2$ indicates that the temperature is increasing, but $f''(3) = -4$ indicates that the rate of increase is decreasing. (The temperature is slowly getting warmer.)

(c) I'm somewhat happy. It's uncomfortably hot and $f'(3) = -2$ indicates that the temperature is decreasing, but $f''(3) = 4$ indicates that the rate of change is increasing. (The rate of change is negative but it's becoming less negative. The temperature is slowly getting cooler.)

(d) I'm very happy. It's uncomfortably hot and $f'(3) = -2$ indicates that the temperature is decreasing, and $f''(3) = -4$ indicates that the rate of change is decreasing, that is, becoming more negative. (The temperature is rapidly getting cooler.)

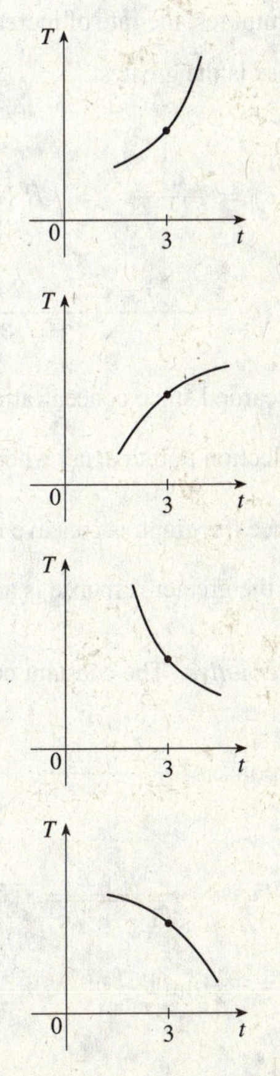

© 2016 Cengage Learning. All Rights Reserved. May not be scanned, copied or duplicated, or posted to a publicly accessible website, in whole or in part.

49. At first the depth increases slowly because the base of the mug is wide. But as the mug narrows, the coffee rises more quickly. Thus, the depth d increases at an increasing rate and its graph is concave upward. The rate of increase of d has a maximum where the mug is narrowest; that is, when the mug is half full. It is there that the inflection point (IP) occurs. Then the rate of increase of d starts to decrease as the mug widens and the graph becomes concave down.

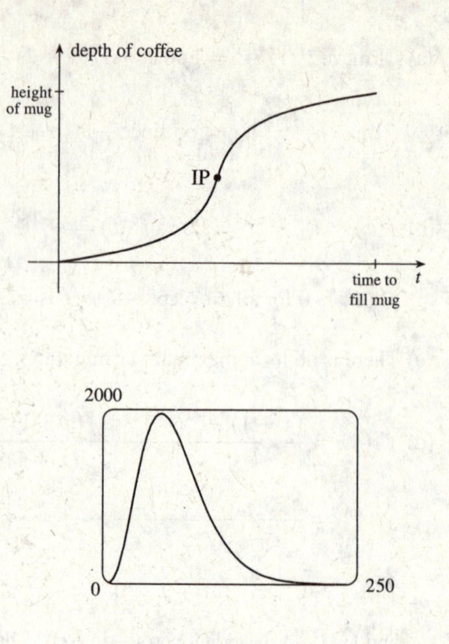

51. $S(t) = At^p e^{-kt}$ with $A = 0.01$, $p = 4$, and $k = 0.07$. We will find the zeros of f'' for $f(t) = t^p e^{-kt}$.

$$f'(t) = t^p(-ke^{-kt}) + e^{-kt}(pt^{p-1}) = e^{-kt}(-kt^p + pt^{p-1})$$

$$\begin{aligned} f''(t) &= e^{-kt}(-kpt^{p-1} + p(p-1)t^{p-2}) + (-kt^p + pt^{p-1})(-ke^{-kt}) \\ &= t^{p-2}e^{-kt}[-kpt + p(p-1) + k^2t^2 - kpt] \\ &= t^{p-2}e^{-kt}(k^2t^2 - 2kpt + p^2 - p) \end{aligned}$$

Using the given values of p and k gives us $f''(t) = t^2e^{-0.07t}(0.0049t^2 - 0.56t + 12)$. So $S''(t) = 0.01f''(t)$ and its zeros are $t = 0$ and the solutions of $0.0049t^2 - 0.56t + 12 = 0$, which are $t_1 = \frac{200}{7} \approx 28.57$ and $t_2 = \frac{600}{7} \approx 85.71$. At t_1 minutes, the rate of increase of the level of medication in the bloodstream is at its greatest and at t_2 minutes, the rate of decrease is the greatest.

53. (a) $R(c) = \dfrac{c^2}{3 + c^2} \Rightarrow R'(c) = \dfrac{2c\,(3 + c^2) - 2c^3}{(3 + c^2)^2} = \dfrac{6c}{(3 + c^2)^2} \Rightarrow$

$R''(c) = \dfrac{6\,(3 + c^2)^2 - 24c^2\,(3 + c^2)}{(3 + c^2)^4} = \dfrac{18\,(1 - c^2)}{(3 + c^2)^3} \Rightarrow R''(c) = 0$ when $c = 1$ and the negative solution is discarded since concentrations must be positive. Since $R''(c) > 0$ when $0 < c < 1$ and $R''(c) < 0$ when $c > 1$, the inflection point occurs when $c = 1$.

(b) Since the graph is concave down when $c > 1$, the average of $R(1.5)$ and $R(2.5)$ will be less than $R(2)$ [see Example 6]. So the greater response is achieved with the constant concentration treatment.

Alternative: The constant concentration treatment has a response of $\dfrac{2^2}{3 + 2^2} \approx 0.57$, whereas the other treatment has a response of $\dfrac{1}{2}\left[\dfrac{1.5^2}{3 + 1.5^2} + \dfrac{2.5^2}{3 + 2.5^2}\right] \approx 0.552$. Therefore, the constant concentration treatment yields a better response.

55. $m = f(v) = \dfrac{m_0}{\sqrt{1 - v^2/c^2}}$. The m-intercept is $f(0) = m_0$. There are no v-intercepts. $\lim\limits_{v \to c^-} f(v) = \infty$, so $v = c$ is a VA.

$f'(v) = -\frac{1}{2}m_0(1 - v^2/c^2)^{-3/2}(-2v/c^2) = \dfrac{m_0 v}{c^2(1 - v^2/c^2)^{3/2}} = \dfrac{m_0 v}{\dfrac{c^2(c^2 - v^2)^{3/2}}{c^3}} = \dfrac{m_0 cv}{(c^2 - v^2)^{3/2}} > 0$, so f is

© 2016 Cengage Learning. All Rights Reserved. May not be scanned, copied or duplicated, or posted to a publicly accessible website, in whole or in part.

increasing on $(0, c)$. There are no local extreme values.

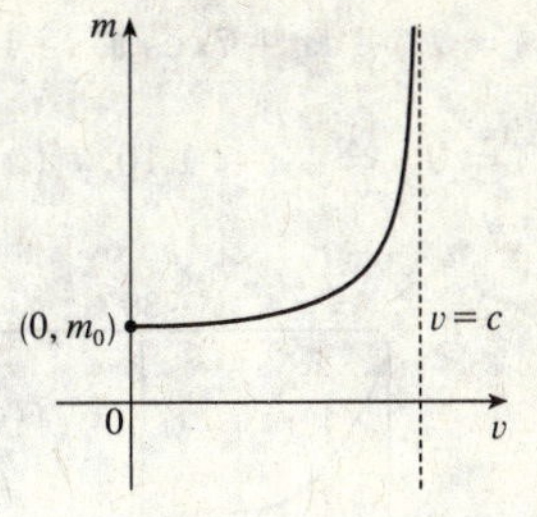

$$f''(v) = \frac{(c^2 - v^2)^{3/2}(m_0c) - m_0cv \cdot \frac{3}{2}(c^2 - v^2)^{1/2}(-2v)}{[(c^2 - v^2)^{3/2}]^2}$$

$$= \frac{m_0c(c^2 - v^2)^{1/2}[(c^2 - v^2) + 3v^2]}{(c^2 - v^2)^3} = \frac{m_0c(c^2 + 2v^2)}{(c^2 - v^2)^{5/2}} > 0,$$

so f is CU on $(0, c)$. There are no inflection points.

57. $f(x) = ax^3 + bx^2 + cx + d \Rightarrow f'(x) = 3ax^2 + 2bx + c$.

We are given that $f(1) = 0$ and $f(-2) = 3$, so $f(1) = a + b + c + d = 0$ and $f(-2) = -8a + 4b - 2c + d = 3$. Also $f'(1) = 3a + 2b + c = 0$ and $f'(-2) = 12a - 4b + c = 0$ by Fermat's Theorem. Solving these four equations, we get $a = \frac{2}{9}, b = \frac{1}{3}, c = -\frac{4}{3}, d = \frac{7}{9}$, so the function is $f(x) = \frac{1}{9}(2x^3 + 3x^2 - 12x + 7)$.

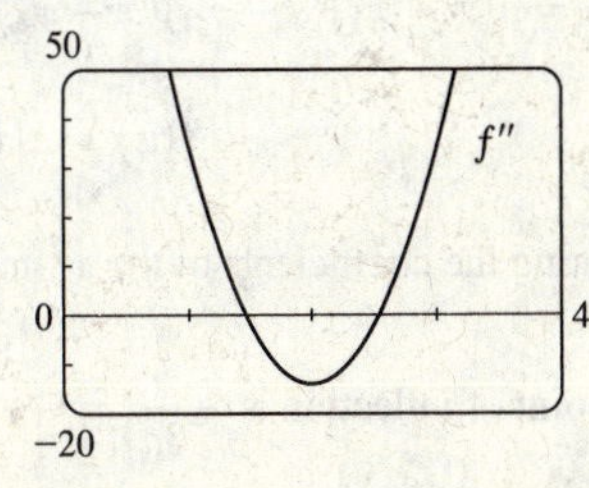

59. $f(x) = 4x^4 - 32x^3 + 89x^2 - 95x + 29 \Rightarrow f'(x) = 16x^3 - 96x^2 + 178x - 95 \Rightarrow f''(x) = 48x^2 - 192x + 178$.

$f(x) = 0 \Leftrightarrow x \approx 0.5,\ 1.60$; $f'(x) = 0 \Leftrightarrow x \approx 0.92,\ 2.5,\ 2.58$ and $f''(x) = 0 \Leftrightarrow x \approx 1.46,\ 2.54$.

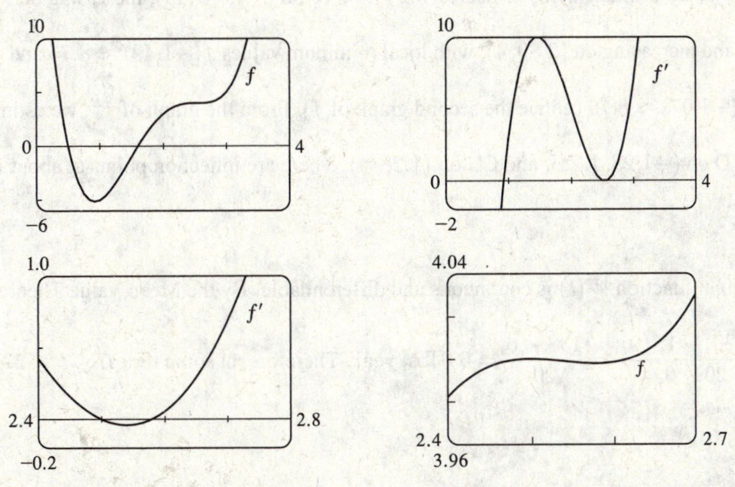

From the graphs of f', we estimate that $f' < 0$ and that f is decreasing on $(-\infty, 0.92)$ and $(2.5, 2.58)$, and that $f' > 0$ and f is increasing on $(0.92, 2.5)$ and $(2.58, \infty)$ with local minimum values $f(0.92) \approx -5.12$ and $f(2.58) \approx 3.998$ and local maximum value $f(2.5) = 4$. The graphs of f' make it clear that f has a maximum and a minimum near $x = 2.5$, shown more clearly in the fourth graph.

From the graph of f'', we estimate that $f'' > 0$ and that f is CU on $(-\infty, 1.46)$ and $(2.54, \infty)$, and that $f'' < 0$ and f is CD on $(1.46, 2.54)$. There are inflection points at about $(1.46, -1.40)$ and $(2.54, 3.999)$.

© 2016 Cengage Learning. All Rights Reserved. May not be scanned, copied or duplicated, or posted to a publicly accessible website, in whole or in part.

61. $f(x) = x^2 - 4x + 7\cos x$, $-4 \le x \le 4$. $f'(x) = 2x - 4 - 7\sin x \Rightarrow f''(x) = 2 - 7\cos x$.

$f(x) = 0 \Leftrightarrow x \approx 1.10$; $f'(x) = 0 \Leftrightarrow x \approx -1.49, -1.07$, or 2.89; $f''(x) = 0 \Leftrightarrow x = \pm\cos^{-1}\left(\frac{2}{7}\right) \approx \pm 1.28$.

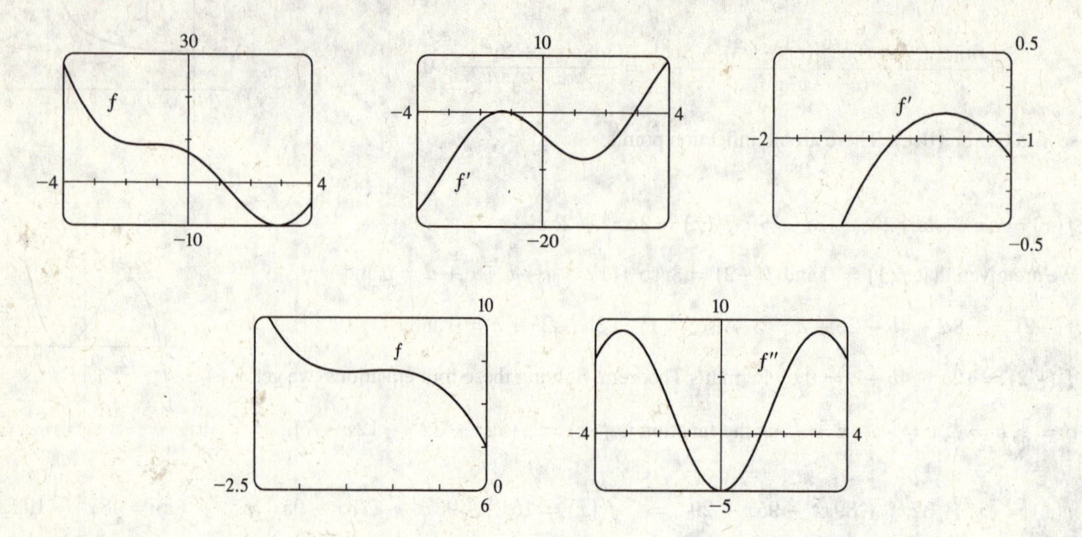

From the graphs of f', we estimate that f is decreasing ($f' < 0$) on $(-4, -1.49)$, increasing on $(-1.49, -1.07)$, decreasing on $(-1.07, 2.89)$, and increasing on $(2.89, 4)$, with local minimum values $f(-1.49) \approx 8.75$ and $f(2.89) \approx -9.99$ and local maximum value $f(-1.07) \approx 8.79$ (notice the second graph of f). From the graph of f'', we estimate that f is CU ($f'' > 0$) on $(-4, -1.28)$, CD on $(-1.28, 1.28)$, and CU on $(1.28, 4)$. There are inflection points at about $(-1.28, 8.77)$ and $(1.28, -1.48)$.

63. An individual's weight function $W(t)$ is continuous and differentiable. By the Mean Value Theorem, there is a number c such that $W'(c) = \dfrac{W(20) - W(0)}{20 - 0} = \dfrac{138 - 6}{20} = 6.6\ \text{lbs/year}$. Therefore, at some time $0 \le t \le 20$, she was growing at $6.6\ \text{lbs/year}$.

65. Let the cubic function be $f(x) = ax^3 + bx^2 + cx + d \Rightarrow f'(x) = 3ax^2 + 2bx + c \Rightarrow f''(x) = 6ax + 2b$.

So f is CU when $6ax + 2b > 0 \Leftrightarrow x > -b/(3a)$, CD when $x < -b/(3a)$, and so the only point of inflection occurs when $x = -b/(3a)$. If the graph has three x-intercepts x_1, x_2 and x_3, then the expression for $f(x)$ must factor as $f(x) = a(x - x_1)(x - x_2)(x - x_3)$. Multiplying these factors together gives us

$$f(x) = a[x^3 - (x_1 + x_2 + x_3)x^2 + (x_1x_2 + x_1x_3 + x_2x_3)x - x_1x_2x_3]$$

Equating the coefficients of the x^2-terms for the two forms of f gives us $b = -a(x_1 + x_2 + x_3)$. Hence, the x-coordinate of the point of inflection is $-\dfrac{b}{3a} = -\dfrac{-a(x_1 + x_2 + x_3)}{3a} = \dfrac{x_1 + x_2 + x_3}{3}$.

© 2016 Cengage Learning. All Rights Reserved. May not be scanned, copied or duplicated, or posted to a publicly accessible website, in whole or in part.

4.3 L'Hospital's Rule: Comparing Rates of Growth

1. This limit has the form $\frac{0}{0}$. We can simply factor and simplify to evaluate the limit.

$$\lim_{x\to 1}\frac{x^2-1}{x^2-x}=\lim_{x\to 1}\frac{(x+1)(x-1)}{x(x-1)}=\lim_{x\to 1}\frac{x+1}{x}=\frac{1+1}{1}=2$$

3. This limit has the form $\frac{0}{0}$. $\displaystyle\lim_{x\to(\pi/2)^+}\frac{\cos x}{1-\sin x}\overset{\text{H}}{=}\lim_{x\to(\pi/2)^+}\frac{-\sin x}{-\cos x}=\lim_{x\to(\pi/2)^+}\tan x=-\infty.$

5. This limit has the form $\frac{0}{0}$. $\displaystyle\lim_{t\to 0}\frac{e^t-1}{t^3}\overset{\text{H}}{=}\lim_{t\to 0}\frac{e^t}{3t^2}=\infty$ since $e^t\to 1$ and $3t^2\to 0^+$ as $t\to 0$.

7. This limit has the form $\frac{\infty}{\infty}$. $\displaystyle\lim_{x\to\infty}\frac{\ln x}{\sqrt{x}}\overset{\text{H}}{=}\lim_{x\to\infty}\frac{1/x}{\frac{1}{2}x^{-1/2}}=\lim_{x\to\infty}\frac{2}{\sqrt{x}}=0$

9. $\displaystyle\lim_{x\to 0^+}[(\ln x)/x]=-\infty$ since $\ln x\to-\infty$ as $x\to 0^+$ and dividing by small values of x just increases the magnitude of the quotient $(\ln x)/x$. L'Hospital's Rule does not apply.

11. This limit has the form $\frac{0}{0}$.

$$\lim_{x\to 0}\frac{\sqrt{1+2x}-\sqrt{1-4x}}{x}\overset{\text{H}}{=}\lim_{x\to 0}\frac{\frac{1}{2}(1+2x)^{-1/2}\cdot 2-\frac{1}{2}(1-4x)^{-1/2}(-4)}{1}$$

$$=\lim_{x\to 0}\left(\frac{1}{\sqrt{1+2x}}+\frac{2}{\sqrt{1-4x}}\right)=\frac{1}{\sqrt{1}}+\frac{2}{\sqrt{1}}=3$$

13. This limit has the form $\frac{0}{0}$. $\displaystyle\lim_{t\to 0}\frac{5^t-3^t}{t}\overset{\text{H}}{=}\lim_{t\to 0}\frac{5^t\ln 5-3^t\ln 3}{1}=\ln 5-\ln 3=\ln\tfrac{5}{3}$

15. This limit has the form $\frac{0}{0}$. $\displaystyle\lim_{x\to 0}\frac{e^x-1-x}{x^2}\overset{\text{H}}{=}\lim_{x\to 0}\frac{e^x-1}{2x}\overset{\text{H}}{=}\lim_{x\to 0}\frac{e^x}{2}=\frac{1}{2}$

17. This limit has the form $\frac{0}{0}$. $\displaystyle\lim_{x\to 1}\frac{1-x+\ln x}{1+\cos\pi x}\overset{\text{H}}{=}\lim_{x\to 1}\frac{-1+1/x}{-\pi\sin\pi x}\overset{\text{H}}{=}\lim_{x\to 1}\frac{-1/x^2}{-\pi^2\cos\pi x}=\frac{-1}{-\pi^2(-1)}=-\frac{1}{\pi^2}$

19. This limit has the form $\frac{0}{0}$. $\displaystyle\lim_{x\to 1}\frac{x^a-ax+a-1}{(x-1)^2}\overset{\text{H}}{=}\lim_{x\to 1}\frac{ax^{a-1}-a}{2(x-1)}\overset{\text{H}}{=}\lim_{x\to 1}\frac{a(a-1)x^{a-2}}{2}=\frac{a(a-1)}{2}$

21. This limit has the form $\frac{0}{0}$. $\displaystyle\lim_{x\to 0}\frac{\cos x-1+\frac{1}{2}x^2}{x^4}\overset{\text{H}}{=}\lim_{x\to 0}\frac{-\sin x+x}{4x^3}\overset{\text{H}}{=}\lim_{x\to 0}\frac{-\cos x+1}{12x^2}\overset{\text{H}}{=}\lim_{x\to 0}\frac{\sin x}{24x}\overset{\text{H}}{=}\lim_{x\to 0}\frac{\cos x}{24}=\frac{1}{24}$

23. This limit has the form $\infty\cdot 0$. We'll change it to the form $\frac{0}{0}$.

$$\lim_{x\to\infty}x\sin(\pi/x)=\lim_{x\to\infty}\frac{\sin(\pi/x)}{1/x}\overset{\text{H}}{=}\lim_{x\to\infty}\frac{\cos(\pi/x)(-\pi/x^2)}{-1/x^2}=\pi\lim_{x\to\infty}\cos(\pi/x)=\pi(1)=\pi$$

25. This limit has the form $\infty\cdot 0$. We'll change it to the form $\frac{0}{0}$.

$$\lim_{x\to 0}\cot 2x\sin 6x=\lim_{x\to 0}\frac{\sin 6x}{\tan 2x}\overset{\text{H}}{=}\lim_{x\to 0}\frac{6\cos 6x}{2\sec^2 2x}=\frac{6(1)}{2(1)^2}=3$$

© 2016 Cengage Learning. All Rights Reserved. May not be scanned, copied or duplicated, or posted to a publicly accessible website, in whole or in part.

27. This limit has the form $\infty \cdot 0$. $\lim\limits_{x\to\infty} x^3e^{-x^2} = \lim\limits_{x\to\infty} \dfrac{x^3}{e^{x^2}} \overset{\text{H}}{=} \lim\limits_{x\to\infty} \dfrac{3x^2}{2xe^{x^2}} = \lim\limits_{x\to\infty} \dfrac{3x}{2e^{x^2}} \overset{\text{H}}{=} \lim\limits_{x\to\infty} \dfrac{3}{4xe^{x^2}} = 0$

29. This limit has the form $\infty - \infty$.

$$\lim_{x\to1}\left(\frac{x}{x-1} - \frac{1}{\ln x}\right) = \lim_{x\to1} \frac{x\ln x - (x-1)}{(x-1)\ln x} \overset{\text{H}}{=} \lim_{x\to1} \frac{x(1/x) + \ln x - 1}{(x-1)(1/x) + \ln x} = \lim_{x\to1} \frac{\ln x}{1 - (1/x) + \ln x}$$

$$\overset{\text{H}}{=} \lim_{x\to1} \frac{1/x}{1/x^2 + 1/x} \cdot \frac{x^2}{x^2} = \lim_{x\to1} \frac{x}{1+x} = \frac{1}{1+1} = \frac{1}{2}$$

31. We will multiply and divide by the conjugate of the expression to change the form of the expression.

$$\lim_{x\to\infty}\left(\sqrt{x^2+x} - x\right) = \lim_{x\to\infty}\left(\frac{\sqrt{x^2+x}-x}{1} \cdot \frac{\sqrt{x^2+x}+x}{\sqrt{x^2+x}+x}\right) = \lim_{x\to\infty} \frac{(x^2+x) - x^2}{\sqrt{x^2+x}+x}$$

$$= \lim_{x\to\infty} \frac{x}{\sqrt{x^2+x}+x} = \lim_{x\to\infty} \frac{1}{\sqrt{1+1/x}+1} = \frac{1}{\sqrt{1}+1} = \frac{1}{2}$$

As an alternate solution, write $\sqrt{x^2+x} - x$ as $\sqrt{x^2+x} - \sqrt{x^2}$, factor out $\sqrt{x^2}$, rewrite as $\left(\sqrt{1+1/x} - 1\right)/(1/x)$, and apply l'Hospital's Rule.

33. The limit has the form $\infty - \infty$ and we will change the form to a product by factoring out x.

$$\lim_{x\to\infty}(x - \ln x) = \lim_{x\to\infty} x\left(1 - \frac{\ln x}{x}\right) = \infty \text{ since } \lim_{x\to\infty} \frac{\ln x}{x} \overset{\text{H}}{=} \lim_{x\to\infty} \frac{1/x}{1} = 0.$$

35. $\lim\limits_{x\to\infty} xe^{-x} = \lim\limits_{x\to\infty}(x/e^x) \overset{\text{H}}{=} \lim\limits_{x\to\infty}(1/e^x) = 0$, so $y = 0$ is a HA. $\lim\limits_{x\to-\infty} xe^{-x} = -\infty$. $f(x) = xe^{-x} \Rightarrow$

$f'(x) = x(-e^{-x}) + e^{-x} \cdot 1 = e^{-x}(1-x) > 0 \Leftrightarrow 1 - x > 0 \Leftrightarrow x < 1$,

so f is increasing on $(-\infty, 1)$ and decreasing on $(1, \infty)$. By the FDT, $f(1) = 1/e$ is a local maximum.

$f''(x) = e^{-x}(-1) + (1-x)(-e^{-x}) = e^{-x}(-1-1+x) = e^{-x}(x-2) > 0 \Leftrightarrow$

$x - 2 > 0 \Leftrightarrow x > 2$, so f is CU on $(2, \infty)$ and CD on $(-\infty, 2)$. IP is $(2, 2/e^2)$.

37.

$\lim\limits_{x\to\infty} \dfrac{\ln x}{x} \overset{\text{H}}{=} \lim\limits_{x\to\infty} \dfrac{1/x}{1} = 0$, so $y = 0$ is a HA. Also $\lim\limits_{x\to0^+} \dfrac{\ln x}{x} = -\infty$

since $\ln x \to -\infty$ and $x \to 0^+$, so $x = 0$ is a VA.

$f(x) = \dfrac{\ln x}{x} \Rightarrow f'(x) = \dfrac{x(1/x) - (\ln x)(1)}{x^2} = \dfrac{1 - \ln x}{x^2} = 0$

when $\ln x = 1 \Leftrightarrow x = e$.

$f'(x) > 0 \Leftrightarrow 1 - \ln x > 0 \Leftrightarrow \ln x < 1 \Leftrightarrow 0 < x < e$. $f'(x) < 0 \Leftrightarrow x > e$.

So f is increasing on $(0, e)$ and decreasing on (e, ∞). By the FDT, $f(e) = 1/e$ is a local maximum.

$f''(x) = \dfrac{x^2(-1/x) - (1 - \ln x)(2x)}{(x^2)^2} = \dfrac{x(-1 - 2 + 2\ln x)}{x^4} = \dfrac{2\ln x - 3}{x^3}$, so $f''(x) > 0 \Leftrightarrow 2\ln x - 3 > 0 \Leftrightarrow$

$\ln x > \frac{3}{2} \Leftrightarrow x > e^{3/2}$. $f''(x) < 0 \Leftrightarrow 0 < x < e^{3/2}$. So f is CU on $\left(e^{3/2}, \infty\right)$ and CD on $\left(0, e^{3/2}\right)$. There is an inflection point at $\left(e^{3/2}, \frac{3}{2}e^{-3/2}\right)$.

© 2016 Cengage Learning. All Rights Reserved. May not be scanned, copied or duplicated, or posted to a publicly accessible website, in whole or in part.

39. Observe that for each function $\lim_{x\to\infty} y = \infty$, so computing the ratio of any two functions results in a $\frac{\infty}{\infty}$ limit as $x \to \infty$.

Comparing pairs of functions: $\displaystyle\lim_{x\to\infty} \frac{e^{3x}}{e^{2x}} = \lim_{x\to\infty} e^x = \infty$ so e^{3x} grows faster than e^{2x}.

$$\lim_{x\to\infty} \frac{e^{2x}}{x^5} \overset{\text{H}}{=} \lim_{x\to\infty} \frac{2e^{2x}}{5x^4} \overset{\text{H}}{=} \lim_{x\to\infty} \frac{2^2 e^{2x}}{(5\cdot 4)\,x^3} \overset{\text{H}}{=} \cdots \overset{\text{H}}{=} \lim_{x\to\infty} \frac{2^5 e^{2x}}{5!} = \infty \quad \text{so } e^{2x} \text{ grows faster than } x^5.$$

$$\lim_{x\to\infty} \frac{x^5}{\ln(x^{10})} \overset{\text{H}}{=} \lim_{x\to\infty} \frac{5x^4}{(1/x^{10})\,(10x^9)} = \lim_{x\to\infty} \tfrac{1}{2}x^5 = \infty \quad \text{so } x^5 \text{ grows faster than } \ln(x^{10}).$$

Therefore, the ranking from fastest to slowest is: $y = e^{3x}$, $y = e^{2x}$, $y = x^5$, $y = \ln(x^{10})$

41. Observe that for each function $\lim_{x\to\infty} y = \infty$, so computing the ratio of any two functions results in a $\frac{\infty}{\infty}$ limit as $x \to \infty$.

Comparing pairs of functions: $\displaystyle\lim_{x\to\infty} \frac{\sqrt{x}}{\sqrt[3]{x}} = \lim_{x\to\infty} \frac{x^{1/2}}{x^{1/3}} = \lim_{x\to\infty} x^{1/6} = \infty$ so $\sqrt{x}$ grows faster than $\sqrt[3]{x}$.

$$\lim_{x\to\infty} \frac{\sqrt[3]{x}}{(\ln x)^3} \overset{\text{H}}{=} \lim_{x\to\infty} \frac{\frac{1}{3}x^{-2/3}}{3(\ln x)^2 x^{-1}} = \lim_{x\to\infty} \frac{x^{1/3}}{9(\ln x)^2} \overset{\text{H}}{=} \lim_{x\to\infty} \frac{x^{1/3}}{54(\ln x)} \overset{\text{H}}{=} \lim_{x\to\infty} \frac{x^{1/3}}{162} = \infty \text{ so } \sqrt[3]{x} \text{ grows faster than } (\ln x)^3.$$

$$\lim_{x\to\infty} \frac{(\ln x)^3}{(\ln x)^2} = \lim_{x\to\infty} (\ln x) = \infty \quad \text{so } (\ln x)^3 \text{ grows faster than } (\ln x)^2.$$

Therefore, the ranking from fastest to slowest is: $y = \sqrt{x}$, $y = \sqrt[3]{x}$, $y = (\ln x)^3$, $y = (\ln x)^2$

43. The dominant terms in the numerator and denominator are $e^{0.1x}$ and x^3 respectively. Since exponential functions grow faster than power functions, we expect the limit to approach infinity. Note that this is a $\frac{\infty}{\infty}$ limit, so we can apply l'Hospital's Rule:

$$\lim_{x\to\infty} \frac{e^{-2x} + x + e^{0.1x}}{x^3 - x^2} \overset{\text{H}}{=} \lim_{x\to\infty} \frac{-2e^{-2x} + 1 + 0.1e^{0.1x}}{3x^2 - 2x} \overset{\text{H}}{=} \lim_{x\to\infty} \frac{4e^{-2x} + (0.1)^2\, e^{0.1x}}{6x - 2}$$

$$\overset{\text{H}}{=} \lim_{x\to\infty} \frac{-8e^{-2x} + (0.1)^3\, e^{0.1x}}{6} = \infty$$

45. Observe that for each function $\lim_{x\to\infty} y = 0$, so computing the ratio of any two functions results in a $\frac{0}{0}$ limit as $x \to \infty$.

Comparing pairs of functions:

$$\lim_{x\to\infty} \frac{e^{-x}}{1/x^2} = \lim_{x\to\infty} \frac{x^2}{e^x} \overset{\text{H}}{=} \lim_{x\to\infty} \frac{2x}{e^x} \overset{\text{H}}{=} \lim_{x\to\infty} \frac{2}{e^x} = 0 \quad \text{so } e^{-x} \text{ approaches zero faster than } 1/x^2.$$

$$\lim_{x\to\infty} \frac{1/x^2}{1/x} = \lim_{x\to\infty} \frac{x}{x^2} = \lim_{x\to\infty} \frac{1}{x} = 0 \quad \text{so } 1/x^2 \text{ approaches zero faster than } 1/x.$$

$$\lim_{x\to\infty} \frac{1/x}{x^{-1/2}} = \lim_{x\to\infty} \frac{x^{1/2}}{x} = \lim_{x\to\infty} \frac{1}{x^{1/2}} = 0 \quad \text{so } 1/x \text{ approaches zero faster than } x^{-1/2}.$$

Therefore, the ranking from fastest to slowest is: $y = e^{-x}$, $y = \dfrac{1}{x^2}$, $y = \dfrac{1}{x}$, $y = x^{-1/2}$

47. $\displaystyle\lim_{x\to\infty} \frac{x}{\sqrt{x^2+1}} \overset{\text{H}}{=} \lim_{x\to\infty} \frac{1}{\frac{1}{2}(x^2+1)^{-1/2}(2x)} = \lim_{x\to\infty} \frac{\sqrt{x^2+1}}{x}$. Repeated applications of l'Hospital's Rule result in the original limit or the limit of the reciprocal of the function. Another method is to try dividing the numerator and denominator by x: $\displaystyle\lim_{x\to\infty} \frac{x}{\sqrt{x^2+1}} = \lim_{x\to\infty} \frac{x/x}{\sqrt{x^2/x^2 + 1/x^2}} = \lim_{x\to\infty} \frac{1}{\sqrt{1+1/x^2}} = \frac{1}{1} = 1$

© 2016 Cengage Learning. All Rights Reserved. May not be scanned, copied or duplicated, or posted to a publicly accessible website, in whole or in part.

49. (a) $dN/dt = rN \;\Rightarrow\; \dfrac{dN/dt}{N} = r$ so the per capita growth rate is r.

(b) $dN/dt = rN(1 - N/K) \;\Rightarrow\; \dfrac{dN/dt}{N} = r(1 - N/K) \;\Rightarrow\; \lim_{N\to 0} \dfrac{dN/dt}{N} = \lim_{N\to 0} r(1 - N/K) = r$ so the per capita growth rate is r when N is small.

(c) $dN/dt = f(N) \;\Rightarrow\; \dfrac{dN/dt}{N} = \dfrac{f(N)}{N} \;\Rightarrow\; \lim_{N\to 0} \dfrac{dN/dt}{N} = \lim_{N\to 0} \dfrac{f(N)}{N}$ is a $\frac{0}{0}$ limit since $f(0) = 0$.

So $\lim_{N\to 0} \dfrac{dN/dt}{N} \overset{\text{H}}{=} \lim_{N\to 0} \dfrac{f'(N)}{1} = f'(0)$ Therefore, the per capita growth rate is $f'(0)$ when N is small.

51. (a) With $\rho = 0.05$ and $r = 3$, the percentage of luminance is

$$P = \frac{100\left(1 - 10^{-\rho r^2}\right)}{\rho r^2 \ln 10} = \frac{100\left(1 - 10^{-0.05(3)^2}\right)}{0.05\,(3)^2 \ln 10} = \frac{100\left(1 - 10^{-0.45}\right)}{0.45 \ln 10} \approx 62.3\%$$

(b) With $\rho = 0.05$ and $r = 2$, the percentage of luminance is

$$P = \frac{100\left(1 - 10^{-\rho r^2}\right)}{\rho r^2 \ln 10} = \frac{100\left(1 - 10^{-0.05(2)^2}\right)}{0.05\,(2)^2 \ln 10} = \frac{100\left(1 - 10^{-0.2}\right)}{0.2 \ln 10} \approx 80.1\%$$

This percentage of luminance is greater than that found in part (a) with $r = 3$. This makes sense since Stiles and Crawford found that light entering closer to the center of the pupil measures brighter than light entering farther away from the pupil's center.

(c) $\frac{0}{0}$ limit:

$$\lim_{r\to 0} P = \lim_{r\to 0} \frac{100\left(1 - 10^{-\rho r^2}\right)}{\rho r^2 \ln 10} \overset{\text{H}}{=} \lim_{r\to 0} \frac{-100\left(10^{-\rho r^2}\right)(-2\rho r)\ln 10}{2\rho r \ln 10} = 100 \lim_{r\to 0} 10^{-\rho r^2} = 100\left(10^0\right) = 100\%$$

We expect this result since all the light entering at the center of the pupil will be sensed at the retina (see Figure caption).

53. Taking the limit as $\alpha \to \beta$ results in a $\frac{0}{0}$ limit. Thus, we use L'Hospital's Rule as follows.

$$\begin{aligned} \lim_{\alpha\to\beta} \frac{e^{-\alpha t} - e^{-\beta t}}{\beta - \alpha} &\overset{\text{H}}{=} \lim_{\alpha\to\beta} \frac{\frac{d}{d\alpha}\left(e^{-\alpha t} - e^{-\beta t}\right)}{\frac{d}{d\alpha}(\beta - \alpha)} \\ &= \lim_{\alpha\to\beta} \frac{-te^{-\alpha t}}{-1} \\ &= \lim_{\alpha\to\beta} te^{-\alpha t} \\ &= te^{-\beta t} \end{aligned}$$

So the drug concentration is $te^{-\beta t}$ (or equivalently $te^{-\alpha t}$) when $\alpha \approx \beta$.

55. $\displaystyle \lim_{x\to\infty} \frac{e^x}{x^n} \overset{\text{H}}{=} \lim_{x\to\infty} \frac{e^x}{nx^{n-1}} \overset{\text{H}}{=} \lim_{x\to\infty} \frac{e^x}{n(n-1)x^{n-2}} \overset{\text{H}}{=} \cdots \overset{\text{H}}{=} \lim_{x\to\infty} \frac{e^x}{n!} = \infty$

57. We see that both numerator and denominator approach 0, so we can use l'Hospital's Rule:

$$\begin{aligned} \lim_{x\to a} \frac{\sqrt{2a^3x - x^4} - a\sqrt[3]{aax}}{a - \sqrt[4]{ax^3}} &\overset{\text{H}}{=} \lim_{x\to a} \frac{\frac{1}{2}(2a^3x - x^4)^{-1/2}(2a^3 - 4x^3) - a\left(\frac{1}{3}\right)(aax)^{-2/3}a^2}{-\frac{1}{4}(ax^3)^{-3/4}(3ax^2)} \\ &= \frac{\frac{1}{2}(2a^3a - a^4)^{-1/2}(2a^3 - 4a^3) - \frac{1}{3}a^3(a^2a)^{-2/3}}{-\frac{1}{4}(aa^3)^{-3/4}(3aa^2)} \\ &= \frac{(a^4)^{-1/2}(-a^3) - \frac{1}{3}a^3(a^3)^{-2/3}}{-\frac{3}{4}a^3(a^4)^{-3/4}} = \frac{-a - \frac{1}{3}a}{-\frac{3}{4}} = \frac{4}{3}\left(\frac{4}{3}a\right) = \frac{16}{9}a \end{aligned}$$

© 2016 Cengage Learning. All Rights Reserved. May not be scanned, copied or duplicated, or posted to a publicly accessible website, in whole or in part.

4.4 Optimization Problems

1. (a)

First Number	Second Number	Product
1	22	22
2	21	42
3	20	60
4	19	76
5	18	90
6	17	102
7	16	112
8	15	120
9	14	126
10	13	130
11	12	132

We needn't consider pairs where the first number is larger than the second, since we can just interchange the numbers in such cases. The answer appears to be 11 and 12, but we have considered only integers in the table.

(b) Call the two numbers x and y. Then $x + y = 23$, so $y = 23 - x$. Call the product P. Then $P = xy = x(23 - x) = 23x - x^2$, so we wish to maximize the function $P(x) = 23x - x^2$. Since $P'(x) = 23 - 2x$, we see that $P'(x) = 0 \quad \Leftrightarrow \quad x = \frac{23}{2} = 11.5$. Thus, the maximum value of P is $P(11.5) = (11.5)^2 = 132.25$ and it occurs when $x = y = 11.5$.

Or: Note that $P''(x) = -2 < 0$ for all x, so P is everywhere concave downward and the local maximum at $x = 11.5$ must be an absolute maximum.

3. The two numbers are x and $\dfrac{100}{x}$, where $x > 0$. Minimize $f(x) = x + \dfrac{100}{x}$. $f'(x) = 1 - \dfrac{100}{x^2} = \dfrac{x^2 - 100}{x^2}$. The critical number is $x = 10$. Since $f'(x) < 0$ for $0 < x < 10$ and $f'(x) > 0$ for $x > 10$, there is an absolute minimum at $x = 10$. The numbers are 10 and 10.

5. If the rectangle has dimensions x and y, then its perimeter is $2x + 2y = 100$ m, so $y = 50 - x$. Thus, the area is $A = xy = x(50 - x)$. We wish to maximize the function $A(x) = x(50 - x) = 50x - x^2$, where $0 < x < 50$. Since $A'(x) = 50 - 2x = -2(x - 25)$, $A'(x) > 0$ for $0 < x < 25$ and $A'(x) < 0$ for $25 < x < 50$. Thus, A has an absolute maximum at $x = 25$, and $A(25) = 25^2 = 625\text{ m}^2$. The dimensions of the rectangle that maximize its area are $x = y = 25$ m. (The rectangle is a square.)

7. We need to maximize Y for $N \geq 0$. $Y(N) = \dfrac{kN}{1 + N^2} \quad \Rightarrow$

$Y'(N) = \dfrac{(1 + N^2)k - kN(2N)}{(1 + N^2)^2} = \dfrac{k(1 - N^2)}{(1 + N^2)^2} = \dfrac{k(1 + N)(1 - N)}{(1 + N^2)^2}$. $Y'(N) > 0$ for $0 < N < 1$ and $Y'(N) < 0$ for $N > 1$. Thus, Y has an absolute maximum of $Y(1) = \frac{1}{2}k$ at $N = 1$.

9. (a)

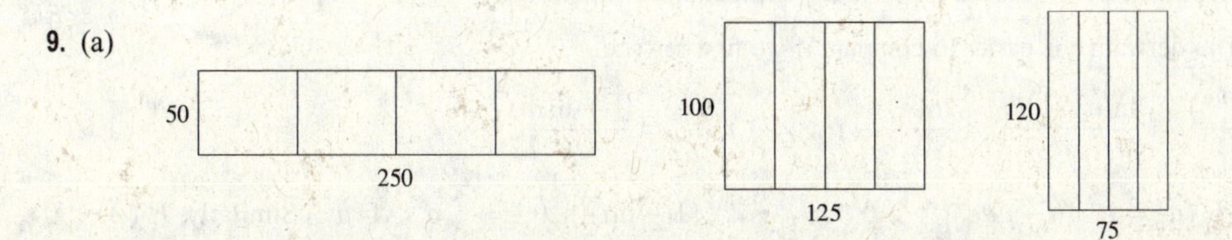

The areas of the three figures are 12,500, 12,500, and 9000 ft^2. There appears to be a maximum area of at least 12,500 ft^2.

© 2016 Cengage Learning. All Rights Reserved. May not be scanned, copied or duplicated, or posted to a publicly accessible website, in whole or in part.

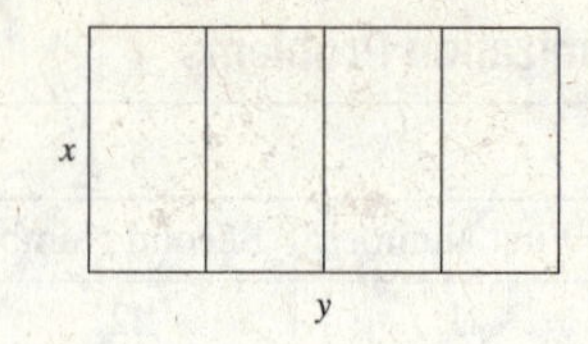

(b) Let x denote the length of each of two sides and three dividers.

Let y denote the length of the other two sides.

(c) Area $A = \text{length} \times \text{width} = y \cdot x$

(d) Length of fencing $= 750 \quad \Rightarrow \quad 5x + 2y = 750$

(e) $5x + 2y = 750 \quad \Rightarrow \quad y = 375 - \frac{5}{2}x \quad \Rightarrow \quad A(x) = \left(375 - \frac{5}{2}x\right)x = 375x - \frac{5}{2}x^2$

(f) $A'(x) = 375 - 5x = 0 \quad \Rightarrow \quad x = 75$. Since $A''(x) = -5 < 0$ there is an absolute maximum when $x = 75$. Then $y = \frac{375}{2} = 187.5$. The largest area is $75\left(\frac{375}{2}\right) = 14{,}062.5\ \text{ft}^2$. These values of x and y are between the values in the first and second figures in part (a). Our original estimate was low.

11. Let b be the length of the base of the box and h the height. The surface area is $1200 = b^2 + 4hb \quad \Rightarrow \quad h = (1200 - b^2)/(4b)$. The volume is $V = b^2h = b^2(1200 - b^2)/4b = 300b - b^3/4 \quad \Rightarrow \quad V'(b) = 300 - \frac{3}{4}b^2$.
$V'(b) = 0 \quad \Rightarrow \quad 300 = \frac{3}{4}b^2 \quad \Rightarrow \quad b^2 = 400 \quad \Rightarrow \quad b = \sqrt{400} = 20$. Since $V'(b) > 0$ for $0 < b < 20$ and $V'(b) < 0$ for $b > 20$, there is an absolute maximum when $b = 20$ by the First Derivative Test for Absolute Extreme Values (see page 253). If $b = 20$, then $h = (1200 - 20^2)/(4 \cdot 20) = 10$, so the largest possible volume is $b^2h = (20)^2(10) = 4000\ \text{cm}^3$.

13. (a) Let the rectangle have sides x and y and area A, so $A = xy$ or $y = A/x$. The problem is to minimize the perimeter $= 2x + 2y = 2x + 2A/x = P(x)$. Now $P'(x) = 2 - 2A/x^2 = 2(x^2 - A)/x^2$. So the critical number is $x = \sqrt{A}$. Since $P'(x) < 0$ for $0 < x < \sqrt{A}$ and $P'(x) > 0$ for $x > \sqrt{A}$, there is an absolute minimum at $x = \sqrt{A}$. The sides of the rectangle are $\sqrt{A}$ and $A/\sqrt{A} = \sqrt{A}$, so the rectangle is a square.

(b) Let p be the perimeter and x and y the lengths of the sides, so $p = 2x + 2y \quad \Rightarrow \quad 2y = p - 2x \quad \Rightarrow \quad y = \frac{1}{2}p - x$. The area is $A(x) = x\left(\frac{1}{2}p - x\right) = \frac{1}{2}px - x^2$. Now $A'(x) = 0 \quad \Rightarrow \quad \frac{1}{2}p - 2x = 0 \quad \Rightarrow \quad 2x = \frac{1}{2}p \quad \Rightarrow \quad x = \frac{1}{4}p$. Since $A''(x) = -2 < 0$, there is an absolute maximum for A when $x = \frac{1}{4}p$ by the Second Derivative Test. The sides of the rectangle are $\frac{1}{4}p$ and $\frac{1}{2}p - \frac{1}{4}p = \frac{1}{4}p$, so the rectangle is a square.

15. The distance d from the origin $(0, 0)$ to a point $(x, 2x + 3)$ on the line is given by $d = \sqrt{(x - 0)^2 + (2x + 3 - 0)^2}$ and the square of the distance is $S = d^2 = x^2 + (2x + 3)^2$. $S' = 2x + 2(2x + 3)2 = 10x + 12$ and $S' = 0 \quad \Leftrightarrow \quad x = -\frac{6}{5}$. Now $S'' = 10 > 0$, so we know that S has a minimum at $x = -\frac{6}{5}$. Thus, the y-value is $2\left(-\frac{6}{5}\right) + 3 = \frac{3}{5}$ and the point is $\left(-\frac{6}{5}, \frac{3}{5}\right)$.

17. (a) $r(a) = \dfrac{\ln(ae^{-\mu a})}{a} \quad \Rightarrow \quad r'(a) = \dfrac{a(ae^{-\mu a})^{-1}(e^{-\mu a} - \mu a e^{-\mu a}) - \ln(ae^{-\mu a})}{a^2} = \dfrac{1 - \mu a - \ln(e^{\ln a - \mu a})}{a^2} = \dfrac{1 - \ln a}{a^2}$

And $r'(a) > 0 \;\Rightarrow\; 1 - \ln a > 0 \;\Rightarrow\; 1 > \ln a \;\Rightarrow \Rightarrow\; a < e$ and $r'(a) < 0$ when $a > e$, so the first derivative test indicates there is a maximum at $a = e$. Hence $a = e$ is the optimal age of maturity.

Alternative Solution: The derivative is easier to compute if we first rewrite

$$r(a) = \frac{\ln(a) + \ln(e^{-\mu a})}{a} = \frac{\ln(a) - \mu a}{a} = \frac{\ln(a)}{a} - \mu \quad \Rightarrow \quad r'(a) = \frac{1 - \ln a}{a^2}.$$

(b) $R(a) = ae^{-\mu a} \quad \Rightarrow \quad R'(a) = e^{-\mu a} - \mu a e^{-\mu a} > 0 \quad \Rightarrow \quad e^{-\mu a}(1 - \mu a) > 0 \quad \Rightarrow \quad a < 1/\mu$. Similarly, $R'(a) < 0$ when $a > 1/\mu$ so there is a maximum at $a = 1/\mu$. Hence $a = 1/\mu$ is the optimal age of maturity.

© 2016 Cengage Learning. All Rights Reserved. May not be scanned, copied or duplicated, or posted to a publicly accessible website, in whole or in part.

19. (a) If $c(x) = \dfrac{C(x)}{x}$, then, by the Quotient Rule, we have $c'(x) = \dfrac{xC'(x) - C(x)}{x^2}$. Now $c'(x) = 0$ when $xC'(x) - C(x) = 0$ and this gives $C'(x) = \dfrac{C(x)}{x} = c(x)$. Therefore, the marginal cost equals the average cost.

(b) (i) $C(x) = 16{,}000 + 200x + 4x^{3/2}$, $C(1000) = 16{,}000 + 200{,}000 + 40{,}000\sqrt{10} \approx 216{,}000 + 126{,}491$, so $C(1000) \approx \$342{,}491$. $c(x) = C(x)/x = \dfrac{16{,}000}{x} + 200 + 4x^{1/2}$, $c(1000) \approx \$342.49/\text{unit}$. $C'(x) = 200 + 6x^{1/2}$, $C'(1000) = 200 + 60\sqrt{10} \approx \$389.74/\text{unit}$.

(ii) We must have $C'(x) = c(x) \Leftrightarrow 200 + 6x^{1/2} = \dfrac{16{,}000}{x} + 200 + 4x^{1/2} \Leftrightarrow 2x^{3/2} = 16{,}000 \Leftrightarrow$ $x = (8{,}000)^{2/3} = 400$ units. To check that this is a minimum, we calculate $c'(x) = \dfrac{-16{,}000}{x^2} + \dfrac{2}{\sqrt{x}} = \dfrac{2}{x^2}\left(x^{3/2} - 8000\right)$. This is negative for $x < (8000)^{2/3} = 400$, zero at $x = 400$, and positive for $x > 400$, so c is decreasing on $(0, 400)$ and increasing on $(400, \infty)$. Thus, c has an absolute minimum at $x = 400$. [*Note:* $c''(x)$ is *not* positive for all $x > 0$.]

(iii) The minimum average cost is $c(400) = 40 + 200 + 80 = \$320/\text{unit}$.

21. (a) In Example 5, we saw that if the number of fish harvested is $H = hN$ where h is a measure of "fishing effort", then the stabilized population size is $N = K\left(1 - \dfrac{h}{r}\right)$. The profit P is the difference between revenue and expenses so

$$P = pH - CH = h(p - C)N = K(p - C)\left(h - \frac{h^2}{r}\right) \Rightarrow \frac{dP}{dh} = K(p - C)\left(1 - \frac{2h}{r}\right) = 0 \text{ when } h = \frac{r}{2}.$$

Also, dP/dh is positive when $h < \frac{1}{2}r$ and negative when $h > \frac{1}{2}r$. So the largest profit is achieved when $h = \frac{1}{2}r$ (which also maximizes the harvest size).

(b) When the unit cost is $C = \alpha/h$, the profit function is

$$P = (p - C)H = \left(p - \frac{\alpha}{h}\right)H = \left(p - \frac{\alpha}{h}\right)hN = K(ph - \alpha)\left(1 - \frac{h}{r}\right) \Rightarrow$$

$$\frac{dP}{dh} = K\left[p\left(1 - \frac{h}{r}\right) + (ph - \alpha)\left(-\frac{1}{r}\right)\right] = K\left[p + \frac{\alpha}{r} - \frac{2p}{r}h\right] = 0 \Rightarrow h = \frac{rp + \alpha}{2p} = \frac{r}{2} + \frac{\alpha}{2p}$$

The second derivative is $\dfrac{d^2P}{dh^2} = -\dfrac{2Kp}{r} < 0$ (since all constants are positive). Therefore, the second derivative test indicates that the fishing effort $h = \dfrac{r}{2} + \dfrac{\alpha}{2p}$ maximizes the profit.

Using this fishing effort, the stabilized population size is $N = K\left(1 - \dfrac{h}{r}\right) = K\left[1 - \dfrac{1}{r}\left(\dfrac{r}{2} + \dfrac{\alpha}{2p}\right)\right] = \dfrac{K}{2}\left(1 - \dfrac{\alpha}{rp}\right)$.

(c) The fixed cost model in part (a) gave a fishing effort $h_f = \dfrac{r}{2}$ and the variable cost model in part (b) gave $h_v = \dfrac{r}{2} + \dfrac{\alpha}{2p}$.

Therefore the harvest effort increases by $\dfrac{\alpha}{2p}$ when the unit cost is inversely proportional to the fishing effort. Assuming $\alpha < rp$, the stabilized population size in the fixed-cost model $\left(N = \dfrac{K}{2}\right)$ is larger than the population size in the variable-cost model $\left(N = \dfrac{K}{2}\left(1 - \dfrac{\alpha}{rp}\right)\right)$. Because extra harvesting results in lower unit costs, it is more profitable to collect a larger harvest in part (b).

© 2016 Cengage Learning. All Rights Reserved. May not be scanned, copied or duplicated, or posted to a publicly accessible website, in whole or in part.

23. (a) The nectar function starts at zero and increases with the amount of time spent on the flower. The function is concave down implying that there are diminishing returns for longer time periods.

(b) In Example 2, we saw that the average amount of nectar harvested per second over one cycle is $f(t) = \dfrac{N(t)}{t+T}$ since it takes T seconds of travel followed by t seconds of foraging. Computing the derivative to find the condition for optimality:

$$f'(t) = \frac{(t+T)\,N'(t) - N(t)}{(t+T)^2} \;\Rightarrow\; f'(t) = 0 \;\Leftrightarrow\; (t+T)\,N'(t) - N(t) = 0 \;\Leftrightarrow\; N'(t) = \frac{N(t)}{t+T}.$$

This is a required condition for optimality, but does not guarantee $Q(t)$ will be a maximum. We verify $Q(t)$ is a maximum in part (c).

(c) We compute the derivative of $f'(t)$ to apply the second derivative test:

$$\begin{aligned}
f''(t) &= \frac{(t+T)^2\,[(t+T)\,N''(t) + N'(t) - N'(t)] - 2\,(t+T)\,[(t+T)\,N'(t) - N(t)]}{(t+T)^4} \\
&= \frac{(t+T)^3\,N''(t) - 2\,(t+T)\,[(t+T)\,N'(t) - N(t)]}{(t+T)^4} \\
&= \frac{(t+T)^2\,N''(t) - 2\,[(t+T)\,N'(t) - N(t)]}{(t+T)^3} \\
&= \frac{(t+T)^2\,N''(t) - 2\,(0)}{(t+T)^3} \qquad \text{since we know } (t+T)\,N'(t) = N(t) \text{ from part (b)} \\
&= \frac{N''(t)}{t+T}
\end{aligned}$$

Recall $N''(t) < 0$ so $f''(t) = \dfrac{N''(t)}{t+T} < 0$ when the condition from part (b) is satisfied. Thus, from the second derivative test, $f(t)$ will be a maximum.

25.

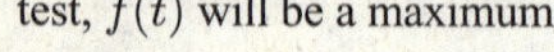

The total time is

$$\begin{aligned}
T(x) &= (\text{time from } A \text{ to } C) + (\text{time from } C \text{ to } B) \\
&= \frac{\sqrt{a^2+x^2}}{v_1} + \frac{\sqrt{b^2+(d-x)^2}}{v_2}, \quad 0 < x < d
\end{aligned}$$

$$T'(x) = \frac{x}{v_1\sqrt{a^2+x^2}} - \frac{d-x}{v_2\sqrt{b^2+(d-x)^2}} = \frac{\sin\theta_1}{v_1} - \frac{\sin\theta_2}{v_2}$$

The minimum occurs when $T'(x) = 0 \;\Rightarrow\; \dfrac{\sin\theta_1}{v_1} = \dfrac{\sin\theta_2}{v_2}$.

[*Note:* $T''(x) > 0$]

27. $S = 6sh - \frac{3}{2}s^2\cot\theta + 3s^2\frac{\sqrt{3}}{2}\csc\theta$

(a) $\dfrac{dS}{d\theta} = \frac{3}{2}s^2\csc^2\theta - 3s^2\frac{\sqrt{3}}{2}\csc\theta\cot\theta$ or $\frac{3}{2}s^2\csc\theta\left(\csc\theta - \sqrt{3}\cot\theta\right)$.

(b) $\dfrac{dS}{d\theta} = 0$ when $\csc\theta - \sqrt{3}\cot\theta = 0 \;\Rightarrow\; \dfrac{1}{\sin\theta} - \sqrt{3}\dfrac{\cos\theta}{\sin\theta} = 0 \;\Rightarrow\; \cos\theta = \frac{1}{\sqrt{3}}$. The First Derivative Test shows that the minimum surface area occurs when $\theta = \cos^{-1}\left(\frac{1}{\sqrt{3}}\right) \approx 55°$.

© 2016 Cengage Learning. All Rights Reserved. May not be scanned, copied or duplicated, or posted to a publicly accessible website, in whole or in part.

(c)

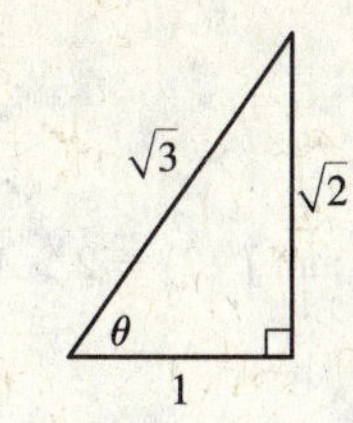

If $\cos\theta = \frac{1}{\sqrt{3}}$, then $\cot\theta = \frac{1}{\sqrt{2}}$ and $\csc\theta = \frac{\sqrt{3}}{\sqrt{2}}$, so the surface area is

$$S = 6sh - \tfrac{3}{2}s^2\tfrac{1}{\sqrt{2}} + 3s^2\tfrac{\sqrt{3}}{2}\tfrac{\sqrt{3}}{\sqrt{2}} = 6sh - \tfrac{3}{2\sqrt{2}}s^2 + \tfrac{9}{2\sqrt{2}}s^2$$

$$= 6sh + \tfrac{6}{2\sqrt{2}}s^2 = 6s\left(h + \tfrac{1}{2\sqrt{2}}s\right)$$

29. (a)

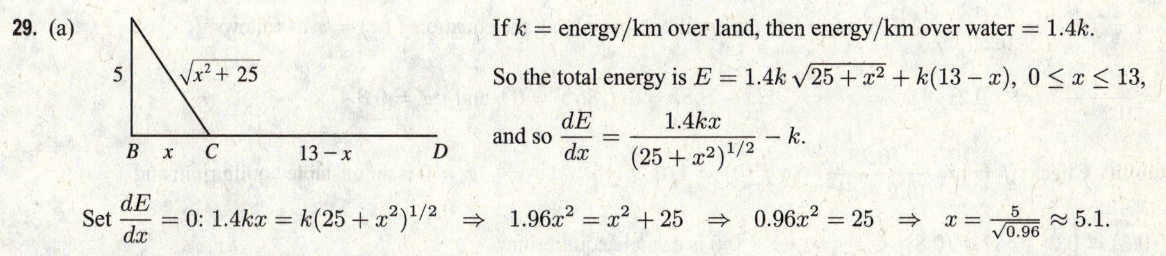

If k = energy/km over land, then energy/km over water $= 1.4k$.

So the total energy is $E = 1.4k\sqrt{25+x^2} + k(13-x),\ 0 \le x \le 13$,

and so $\dfrac{dE}{dx} = \dfrac{1.4kx}{(25+x^2)^{1/2}} - k$.

Set $\dfrac{dE}{dx} = 0$: $1.4kx = k(25+x^2)^{1/2} \Rightarrow 1.96x^2 = x^2 + 25 \Rightarrow 0.96x^2 = 25 \Rightarrow x = \frac{5}{\sqrt{0.96}} \approx 5.1$.

Testing against the value of E at the endpoints: $E(0) = 1.4k(5) + 13k = 20k$, $E(5.1) \approx 17.9k$, $E(13) \approx 19.5k$.

Thus, to minimize energy, the bird should fly to a point about 5.1 km from B.

(b) If W/L is large, the bird would fly to a point C that is closer to B than to D to minimize the energy used flying over water. If W/L is small, the bird would fly to a point C that is closer to D than to B to minimize the distance of the flight.

$E = W\sqrt{25+x^2} + L(13-x) \Rightarrow \dfrac{dE}{dx} = \dfrac{Wx}{\sqrt{25+x^2}} - L = 0$ when $\dfrac{W}{L} = \dfrac{\sqrt{25+x^2}}{x}$. By the same sort of argument as in part (a), this ratio will give the minimal expenditure of energy if the bird heads for the point x km from B.

(c) For flight direct to D, $x = 13$, so from part (b), $W/L = \frac{\sqrt{25+13^2}}{13} \approx 1.07$. There is no value of W/L for which the bird should fly directly to B. But note that $\lim\limits_{x\to 0^+}(W/L) = \infty$, so if the point at which E is a minimum is close to B, then W/L is large.

(d) Assuming that the birds instinctively choose the path that minimizes the energy expenditure, we can use the equation for $dE/dx = 0$ from part (a) with $1.4k = c$, $x = 4$, and $k = 1$: $c(4) = 1 \cdot (25+4^2)^{1/2} \Rightarrow c = \sqrt{41}/4 \approx 1.6$.

4.5 Recursions: Equilibria and Stability

1.

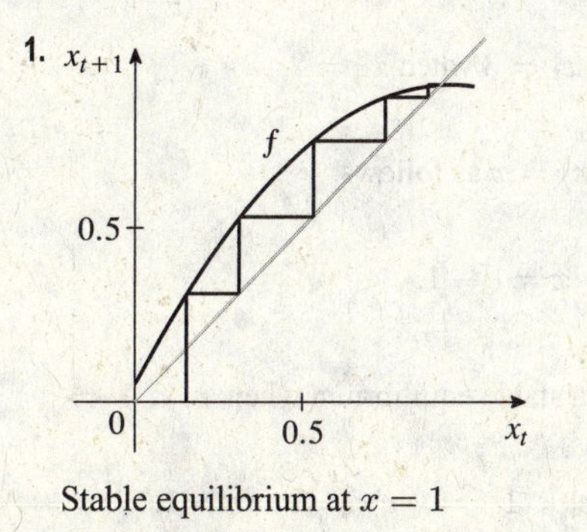

Stable equilibrium at $x = 1$

3.

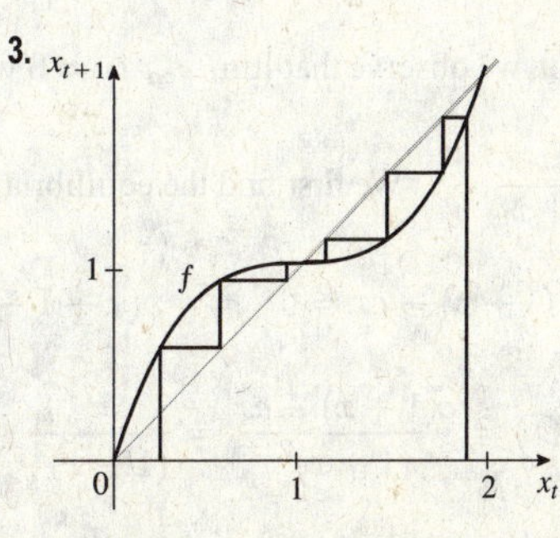

Unstable equilibria at $x = 0$ and $x = 2$, stable equilibrium at $x = 1$

© 2016 Cengage Learning. All Rights Reserved. May not be scanned, copied or duplicated, or posted to a publicly accessible website, in whole or in part.

5. $x_{t+1} = f(x_t) = \frac{1}{2}x_t^2$. We first find the equilibria by solving the equation $f(x) = x$ as follows:

$x = \frac{1}{2}x^2 \quad \Leftrightarrow \quad x^2 - 2x = 0 \quad \Leftrightarrow \quad x(x-2) = 0.$ So $x = 0$ and $x = 2$.

Stability Check: $f'(x) = x$. So $f'(0) = 0 \quad \Rightarrow \quad |f'(0)| < 1 \quad \Rightarrow \quad x = 0$ is a stable equilibrium and $f'(2) = 2 > 1 \quad \Rightarrow$ $x = 2$ is an unstable equilibrium.

7. $x_{t+1} = f(x_t) = \dfrac{x_t}{0.2 + x_t}$. We first find the equilibria by solving the equation $f(x) = x$ as follows:

$x = \dfrac{x}{0.2 + x} \quad \Leftrightarrow \quad 0.2x + x^2 = x \quad \Leftrightarrow \quad x(x - 0.8) = 0.$ So $x = 0$ and $x = 0.8$.

Stability Check: $f'(x) = \dfrac{0.2}{(0.2 + x)^2}$. So $f'(0) = 1/0.2 = 5 > 1 \quad \Rightarrow \quad x = 0$ is an unstable equilibrium and

$f'(0.8) = 0.2 \quad \Rightarrow \quad |f'(0.8)| < 1 \quad \Rightarrow \quad x = 0.8$ is a stable equilibrium.

9. $x_{t+1} = f(x_t) = 10x_t e^{-2x_t}$. We first find the equilibria by solving the equation $f(x) = x$ as follows:

$x = 10xe^{-2x} \quad \Leftrightarrow \quad x\left(1 - 10e^{-2x}\right).$ So $x = 0$ and $1 = 10e^{-2x} \quad \Leftrightarrow \quad x = \frac{1}{2}\ln(10)$.

Stability Check: $f'(x) = 10\left(e^{-2x} - 2xe^{-2x}\right) = 10e^{-2x}(1 - 2x)$. So $f'(0) = 10 > 1 \quad \Rightarrow \quad x = 0$ is an unstable equilibrium and $f'(\frac{1}{2}\ln(10)) = 1 - \ln(10) \approx -1.3 \quad \Rightarrow \quad |f'(\frac{1}{2}\ln(10))| > 1 \quad \Rightarrow \quad x = \frac{1}{2}\ln(10)$ is an unstable equilibrium.

11. $x_{t+1} = f(x_t) = \dfrac{4x_t^2}{x_t^2 + 3}$. Solving for x in the equation $f(x) = x$ gives

$x = \dfrac{4x^2}{x^2 + 3} \quad \Leftrightarrow \quad x^3 - 4x^2 + 3x = 0 \quad \Leftrightarrow \quad x(x-1)(x-3) = 0.$

So $\hat{x} = 0$ or $\hat{x} = 1$ or $\hat{x} = 3$ are the equilibria

Stability Check: $f'(x) = \dfrac{8x\left(x^2 + 3\right) - 8x^3}{(x^2 + 3)^2} = \dfrac{24x}{(x^2 + 3)^2}$

So $f'(0) = 0 \quad \Rightarrow \quad |f'(0)| < 1 \quad \Rightarrow \quad x = 0$ is a stable equilibrium

and $f'(1) = \dfrac{24}{16} = 1.5 \quad \Rightarrow \quad |f'(1)| > 1 \quad \Rightarrow \quad x = 1$ is an unstable equilibrium

and $f'(3) = \dfrac{72}{144} = 0.5 \quad \Rightarrow \quad |f'(3)| < 1 \quad \Rightarrow \quad x = 3$ is a stable equilibrium

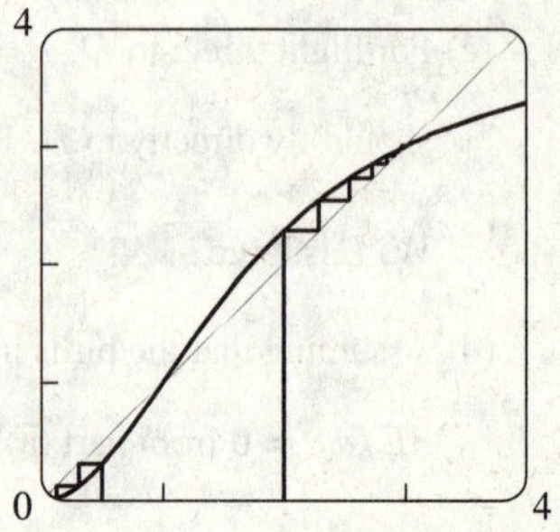

From the cobweb plot, we observe that $\lim_{t\to\infty} x_t = 0$ when $x_0 = 0.5$ and $\lim_{t\to\infty} x_t = 3$ when $x_0 = 2$.

13. $x_{t+1} = f(x_t) = \dfrac{cx_t}{1 + x_t}$. We first find the equilibria by solving the equation $f(x) = x$ as follows:

$x = \dfrac{cx}{1 + x} \quad \Leftrightarrow \quad x(1 + x) - cx = 0 \quad \Leftrightarrow \quad x(x + 1 - c) = 0.$ So $x = 0$ and $x = c - 1$.

Stability Check: $f'(x) = \dfrac{c(1 + x) - cx}{(1 + x)^2} = \dfrac{c}{(1 + x)^2}$. So $f'(0) = c$ and $x = 0$ is a stable equilibrium when $|c| < 1 \Rightarrow$ $-1 < c < 1$. Also $f'(c - 1) = 1/c$ so $x = c - 1$ is a stable equilibrium when $|1/c| < 1 \quad \Rightarrow \quad c < -1$ or $c > 1$.

© 2016 Cengage Learning. All Rights Reserved. May not be scanned, copied or duplicated, or posted to a publicly accessible website, in whole or in part.

15. (a) There is $0.1Q_n$ mg of the drug left over just before taking the nth $+ 1$ tablet. Hence $Q_{n+1} = f(Q_n) = 0.1Q_n + 200$.

(b) We find the equilibria by solving the equation $f(Q) = Q$ as follows:

$$Q = 0.1Q + 200 \quad \Rightarrow \quad 0.9Q = 200 \quad \Rightarrow \quad Q = \frac{2000}{9} \approx 222.22$$

(c)

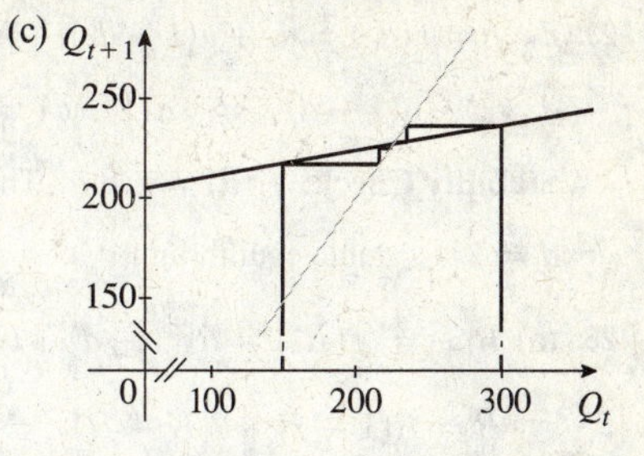

17. Computer software was used to construct a cobweb plot and a plot of the first 10 terms of the sequence. From Example 4, $\hat{x} = 0$ will be a stable equilibrium since $|c| < 1$ and the other equilibrium is $\hat{x} = 1 - \frac{1}{0.8} = -\frac{1}{4}$. But this is not relevant since $x \geq 0$, so the only equilibrium is $\hat{x} = 0$. Both graphs confirm this is the case.

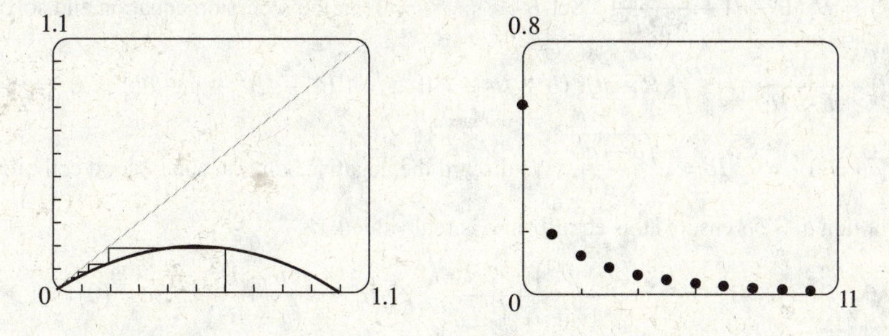

19. Computer software was used to construct a cobweb plot and a plot of the first 10 terms of the sequence. From Example 4, $\hat{x} = 0$ will be an unstable equilibrium since $c > 1$ and $\hat{x} = 1 - \frac{1}{2.7} = \frac{17}{27} \approx 0.63$ is a stable (oscillating) equilibrium since $1 < c < 3$. Both graphs confirm this is the case.

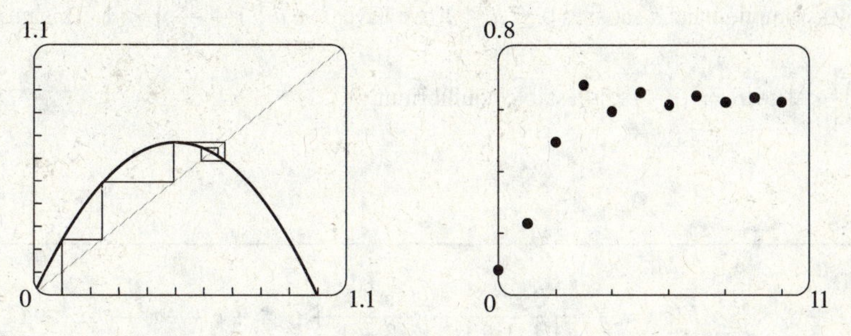

21. $N_{t+1} = f(N_t) = N_t + rN_t\left(1 - \frac{N_t}{K}\right) - hN_t$. We first find the equilibria by solving the equation $f(N) = N$ as follows:

$$N = N + rN\left(1 - \frac{N}{K}\right) - hN \quad \Leftrightarrow \quad N\left[r\left(1 - \frac{N}{K}\right) - h\right] = 0. \quad \text{So } N = 0 \quad \text{and} \quad r\left(1 - \frac{N}{K}\right) - h = 0 \Leftrightarrow$$

$$N = K\left(1 - \frac{h}{r}\right).$$

Stability Check: $f'(N) = 1 + r\left(1 - 2\frac{N}{K}\right) - h$. So $f'(0) = 1 + r - h$ and $N = 0$ is a stable equilibrium when $|1 + r - h| < 1 \Rightarrow -1 < 1 + r - h < 1 \Rightarrow h - 2 < r < h$. But since h and r are positive and $0 \leq h \leq 1$, $N = 0$ is stable if $r < h$. Also $f'\left(K\left(1 - \frac{h}{r}\right)\right) = 1 + r - 2r\left(1 - \frac{h}{r}\right) - h = 1 + h - r$ so $N = K\left(1 - \frac{h}{r}\right)$ is a stable equilibrium when $|1 + h - r| < 1 \quad \Rightarrow \quad -1 < 1 + h - r < 1 \quad \Rightarrow \quad h < r < h + 2$.

© 2016 Cengage Learning. All Rights Reserved. May not be scanned, copied or duplicated, or posted to a publicly accessible website, in whole or in part.

23. $d_{t+1} = f(d_t) = d_t + a(1-d_t)$. We first find the equilibria by solving the equation $f(d) = d$ as follows:

$d = d + a(1-d) \iff a(1-d) = 0$ So $d = 1$ is the only equilibrium.

Stability Check: $f'(d) = 1 - a$. The discovery rate satisfies $0 < a < 1$ so $0 < 1 - a < 1 \Rightarrow |f'(d)| < 1$. Therefore, $d = 1$ is a stable equilibrium.

25. (a) $R_{t+1} = f(R_t) = R_t(1-d) + \theta(K - R_t)$. We first find the equilibria by solving the equation $f(R) = R$ as follows:

$$R = R(1-d) + \theta(K-R) \iff R(\theta + d) = \theta K \iff R = \frac{\theta K}{\theta + d}$$

Stability Check: $f'(R) = 1 - d - \theta$. So $|f'(R)| < 1$ when $-1 < 1 - d - \theta < 1 \Rightarrow -2 < -(d+\theta) < 0 \Rightarrow$ $0 < d + \theta < 2$. So $R = \dfrac{\theta K}{\theta + d}$ is a stable equilibrium when $0 < d + \theta < 2$.

(b) $R_{t+1} = f(R_t) = R_t(1-d) + \dfrac{aR_t}{b + R_t^2}$ Set $R_{t+1} = R_t = R$ in the recursion equation and solve for the equilibria:

$$R = R(1-d) + \frac{aR}{b+R^2} \iff aR - dR\left(b + R^2\right) = 0 \iff R\left[-dR^2 + a - db\right] = 0 \Rightarrow R = 0 \text{ and}$$

$-dR^2 + a - db = 0 \Rightarrow R = \sqrt{\dfrac{a - db}{d}}$. We discard the negative solution since blood cell production is positive. Also note the assumption $a > bd$ ensures this equilibrium is real valued.

Stability Check: $f'(R) = 1 - d + a\dfrac{(b+R^2) - 2R^2}{(b+R^2)^2} = 1 - d + a\dfrac{b - R^2}{(b+R^2)^2}$. So $f'(0) = 1 - d + \dfrac{a}{b}$ and since $\dfrac{a}{b} > d$ by assumption, we have $f'(0) > 1$. Thus $R = 0$ is an unstable equilibrium. Also,

$$f'\left(\sqrt{\tfrac{a-db}{d}}\right) = 1 - d + a\frac{b - \frac{a-db}{d}}{\left(b + \frac{a-db}{d}\right)^2} = 1 - d + a\frac{2b - a/d}{(a/d)^2} = 1 - 2d + \frac{2bd^2}{a} = 1 - 2d\left(1 - \frac{bd}{a}\right). \text{ Since}$$

$0 < \dfrac{bd}{a} < 1$ by assumption and d satisfies $0 \le d \le 1$, we have $0 < d\left(1 - \dfrac{bd}{a}\right) < 1$. This guarantees that $\left|f'\left(\sqrt{\tfrac{a-db}{d}}\right)\right| < 1$ so $R = \sqrt{\tfrac{a-db}{d}}$ is a stable equilibrium.

4.6 Antiderivatives

1. $f(x) = \frac{1}{2} + \frac{3}{4}x^2 - \frac{4}{5}x^3 \Rightarrow F(x) = \frac{1}{2}x + \dfrac{3}{4}\dfrac{x^{2+1}}{2+1} - \dfrac{4}{5}\dfrac{x^{3+1}}{3+1} + C = \frac{1}{2}x + \frac{1}{4}x^3 - \frac{1}{5}x^4 + C$

Check: $F'(x) = \frac{1}{2} + \frac{1}{4}(3x^2) - \frac{1}{5}(4x^3) + 0 = \frac{1}{2} + \frac{3}{4}x^2 - \frac{4}{5}x^3 = f(x)$

3. $f(x) = (x+1)(2x-1) = 2x^2 + x - 1 \Rightarrow F(x) = 2\left(\frac{1}{3}x^3\right) + \frac{1}{2}x^2 - x + C = \frac{2}{3}x^3 + \frac{1}{2}x^2 - x + C$

5. $f(x) = 5x^{1/4} - 7x^{3/4} \Rightarrow F(x) = 5\dfrac{x^{1/4+1}}{\frac{1}{4}+1} - 7\dfrac{x^{3/4+1}}{\frac{3}{4}+1} + C = 5\dfrac{x^{5/4}}{5/4} - 7\dfrac{x^{7/4}}{7/4} + C = 4x^{5/4} - 4x^{7/4} + C$

7. $f(x) = 6\sqrt{x} - \sqrt[6]{x} = 6x^{1/2} - x^{1/6} \Rightarrow$

$$F(x) = 6\frac{x^{1/2+1}}{\frac{1}{2}+1} - \frac{x^{1/6+1}}{\frac{1}{6}+1} + C = 6\frac{x^{3/2}}{3/2} - \frac{x^{7/6}}{7/6} + C = 4x^{3/2} - \tfrac{6}{7}x^{7/6} + C$$

9. $f(x) = \sqrt{2}$ is a constant function, so $F(x) = \sqrt{2}\,x + C$.

© 2016 Cengage Learning. All Rights Reserved. May not be scanned, copied or duplicated, or posted to a publicly accessible website, in whole or in part.

11. $c(t) = \dfrac{3}{t^2} = 3t^{-2} \quad\Rightarrow\quad C(t) = 3\left(\dfrac{t^{-1}}{-1}\right) + K = -\dfrac{3}{t} + K$ where K is a constant.

13. $g(\theta) = \cos\theta - 5\sin\theta \quad\Rightarrow\quad G(\theta) = \sin\theta - 5(-\cos\theta) + C = \sin\theta + 5\cos\theta + C$

15. $v(s) = 4s + 3e^s \quad\Rightarrow\quad V(s) = 2s^2 + 3e^s + C$

17. $f(u) = \dfrac{u^4 + 3\sqrt{u}}{u^2} = \dfrac{u^4}{u^2} + \dfrac{3u^{1/2}}{u^2} = u^2 + 3u^{-3/2} \quad\Rightarrow$

$$F(u) = \frac{u^3}{3} + 3\,\frac{u^{-3/2+1}}{-3/2+1} + C = \frac{1}{3}u^3 + 3\frac{u^{-1/2}}{-1/2} + C = \frac{1}{3}u^3 - \frac{6}{\sqrt{u}} + C$$

19. $f(t) = \dfrac{t^4 - t^2 + 1}{t^2} = t^2 - 1 + t^{-2} \quad\Rightarrow\quad F(t) = \dfrac{t^3}{3} - t + \dfrac{t^{-1}}{-1} + C = \frac{1}{3}t^3 - t - \dfrac{1}{t} + C$

21. $\dfrac{dy}{dt} = t^2 + 1 \quad\Rightarrow\quad y(t) = \frac{1}{3}t^3 + t + C$ and $y(0) = 6 \quad\Rightarrow\quad \frac{1}{3}(0)^3 + 0 + C = 6 \quad\Rightarrow\quad C = 6.$

So the solution is $y(t) = \frac{1}{3}t^3 + t + 6,\ t \geq 0$

23. $\dfrac{dP}{dt} = 2e^{3t} \quad\Rightarrow\quad P(t) = \frac{2}{3}e^{3t} + C$ and $P(0) = 1 \quad\Rightarrow\quad \frac{2}{3}e^{3(0)} + C = 1 \quad\Rightarrow\quad C = \frac{1}{3}$

So the solution is $P(t) = \frac{2}{3}e^{3t} + \frac{1}{3},\ t \geq 0$

25. $\dfrac{dr}{d\theta} = \cos\theta + \sec\theta\tan\theta \quad\Rightarrow\quad r(\theta) = \sin\theta + \sec\theta + C$ and $r(\pi/3) = 4 \quad\Rightarrow\quad \sin\left(\frac{\pi}{3}\right) + \sec\left(\frac{\pi}{3}\right) + C = 4 \quad\Rightarrow$

$C = 2 - \frac{\sqrt{3}}{2}$. So the solution is $r(\theta) = \sin\theta + \sec\theta + 2 - \frac{\sqrt{3}}{2},\ 0 < \theta < \pi/2$

27. $\dfrac{du}{dt} = \sqrt{t} + \dfrac{2}{\sqrt{t}} = t^{1/2} + 2t^{-1/2} \quad\Rightarrow\quad u(t) = \dfrac{t^{3/2}}{3/2} + 2\dfrac{t^{1/2}}{1/2} + C = \frac{2}{3}t^{3/2} + 4t^{1/2} + C$ and $u(1) = 5 \quad\Rightarrow$

$\frac{2}{3} + 4 + C = 5 \quad\Rightarrow\quad C = \frac{1}{3}$. So the solution is $u(t) = \frac{2}{3}t^{3/2} + 4t^{1/2} + \frac{1}{3},\ t > 0.$

29. $f''(x) = 6x + 12x^2 \quad\Rightarrow\quad f'(x) = 6 \cdot \dfrac{x^2}{2} + 12 \cdot \dfrac{x^3}{3} + C = 3x^2 + 4x^3 + C \quad\Rightarrow$

$f(x) = 3 \cdot \dfrac{x^3}{3} + 4 \cdot \dfrac{x^4}{4} + Cx + D = x^3 + x^4 + Cx + D$ [C and D are just arbitrary constants]

31. $f''(x) = \frac{2}{3}x^{2/3} \quad\Rightarrow\quad f'(x) = \dfrac{2}{3}\left(\dfrac{x^{5/3}}{5/3}\right) + C = \frac{2}{5}x^{5/3} + C \quad\Rightarrow\quad f(x) = \dfrac{2}{5}\left(\dfrac{x^{8/3}}{8/3}\right) + Cx + D = \frac{3}{20}x^{8/3} + Cx + D$

33. $f'(x) = 1 - 6x \quad\Rightarrow\quad f(x) = x - 3x^2 + C$. $f(0) = C$ and $f(0) = 8 \quad\Rightarrow\quad C = 8$, so $f(x) = x - 3x^2 + 8$.

35. $f'(x) = \sqrt{x}(6 + 5x) = 6x^{1/2} + 5x^{3/2} \quad\Rightarrow\quad f(x) = 4x^{3/2} + 2x^{5/2} + C.$

$f(1) = 6 + C$ and $f(1) = 10 \quad\Rightarrow\quad C = 4$, so $f(x) = 4x^{3/2} + 2x^{5/2} + 4$.

© 2016 Cengage Learning. All Rights Reserved. May not be scanned, copied or duplicated, or posted to a publicly accessible website, in whole or in part.

37. $f''(\theta) = \sin\theta + \cos\theta \;\Rightarrow\; f'(\theta) = -\cos\theta + \sin\theta + C$. $f'(0) = -1 + C$ and $f'(0) = 4 \;\Rightarrow\; C = 5$, so $f'(\theta) = -\cos\theta + \sin\theta + 5$ and hence, $f(\theta) = -\sin\theta - \cos\theta + 5\theta + D$. $f(0) = -1 + D$ and $f(0) = 3 \;\Rightarrow\; D = 4$, so $f(\theta) = -\sin\theta - \cos\theta + 5\theta + 4$.

39. $f''(x) = 2 - 12x \;\Rightarrow\; f'(x) = 2x - 6x^2 + C \;\Rightarrow\; f(x) = x^2 - 2x^3 + Cx + D$.
$f(0) = D$ and $f(0) = 9 \;\Rightarrow\; D = 9$. $f(2) = 4 - 16 + 2C + 9 = 2C - 3$ and $f(2) = 15 \;\Rightarrow\; 2C = 18 \;\Rightarrow$ $C = 9$, so $f(x) = x^2 - 2x^3 + 9x + 9$.

41. Let $N(t)$ represent the number of bacteria after t hours. Then $\dfrac{dN}{dt} = 3.4657e^{0.1386t} \;\Rightarrow$

$N(t) = \dfrac{3.4657}{0.1386}e^{0.1386t} + C \approx 25.005e^{0.1386t} + C$ and $N(0) = 25 \;\Rightarrow\; C \approx -0.005$.

So the population size is given by $N(t) \approx 25.005e^{0.1386t} - 0.005$ and after four hours $N(4) \approx 43.526 \approx 44$ bacteria.

43. $v(t) = s'(t) = \sin t - \cos t \;\Rightarrow\; s(t) = -\cos t - \sin t + C$. $s(0) = -1 + C$ and $s(0) = 0 \;\Rightarrow\; C = 1$, so $s(t) = -\cos t - \sin t + 1$.

45. (a) We first observe that since the stone is dropped 450 m above the ground, $v(0) = 0$ and $s(0) = 450$.
$v'(t) = a(t) = -9.8 \;\Rightarrow\; v(t) = -9.8t + C$. Now $v(0) = 0 \;\Rightarrow\; C = 0$, so $v(t) = -9.8t \;\Rightarrow$ $s(t) = -4.9t^2 + D$. Last, $s(0) = 450 \;\Rightarrow\; D = 450 \;\Rightarrow\; s(t) = 450 - 4.9t^2$.

(b) The stone reaches the ground when $s(t) = 0$. $450 - 4.9t^2 = 0 \;\Rightarrow\; t^2 = 450/4.9 \;\Rightarrow\; t_1 = \sqrt{450/4.9} \approx 9.58$ s.

(c) The velocity with which the stone strikes the ground is $v(t_1) = -9.8\sqrt{450/4.9} \approx -93.9$ m/s.

47.

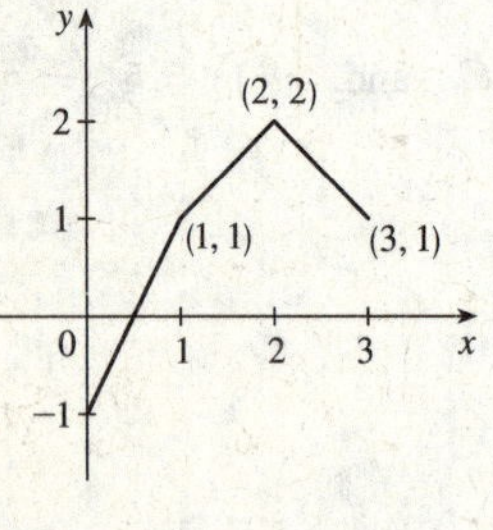

$$f'(x) = \begin{cases} 2 & \text{if } 0 \le x < 1 \\ 1 & \text{if } 1 < x < 2 \\ -1 & \text{if } 2 < x \le 3 \end{cases} \;\Rightarrow\; f(x) = \begin{cases} 2x + C & \text{if } 0 \le x < 1 \\ x + D & \text{if } 1 < x < 2 \\ -x + E & \text{if } 2 < x \le 3 \end{cases}$$

$f(0) = -1 \;\Rightarrow\; 2(0) + C = -1 \;\Rightarrow\; C = -1$. Starting at the point $(0, -1)$ and moving to the right on a line with slope 2 gets us to the point $(1, 1)$.

The slope for $1 < x < 2$ is 1, so we get to the point $(2, 2)$. Here we have used the fact that f is continuous. We can include the point $x = 1$ on either the first or the second part of f. The line connecting $(1, 1)$ to $(2, 2)$ is $y = x$, so $D = 0$. The slope for $2 < x \le 3$ is -1, so we get to $(3, 1)$. $f(3) = 1 \;\Rightarrow\; -3 + E = 1 \;\Rightarrow\; E = 4$. Thus

$$f(x) = \begin{cases} 2x - 1 & \text{if } 0 \le x \le 1 \\ x & \text{if } 1 < x < 2 \\ -x + 4 & \text{if } 2 \le x \le 3 \end{cases}$$

Note that $f'(x)$ does not exist at $x = 1$ or at $x = 2$.

© 2016 Cengage Learning. All Rights Reserved. May not be scanned, copied or duplicated, or posted to a publicly accessible website, in whole or in part.

4 Review

TRUE-FALSE QUIZ

1. False. For example, take $f(x) = x^3$, then $f'(x) = 3x^2$ and $f'(0) = 0$, but $f(0) = 0$ is not a maximum or minimum; $(0, 0)$ is an inflection point.

3. False. For example, $f(x) = x$ is continuous on $(0, 1)$ but attains neither a maximum nor a minimum value on $(0, 1)$. Don't confuse this with f being continuous on the *closed* interval $[a, b]$, which would make the statement true.

5. True. This is an example of part (b) of the I/D Test.

7. False. $f'(x) = g'(x) \Rightarrow f(x) = g(x) + C$. For example, if $f(x) = x + 2$ and $g(x) = x + 1$, then $f'(x) = g'(x) = 1$, but $f(x) \neq g(x)$.

9. True. The graph of one such function is sketched.

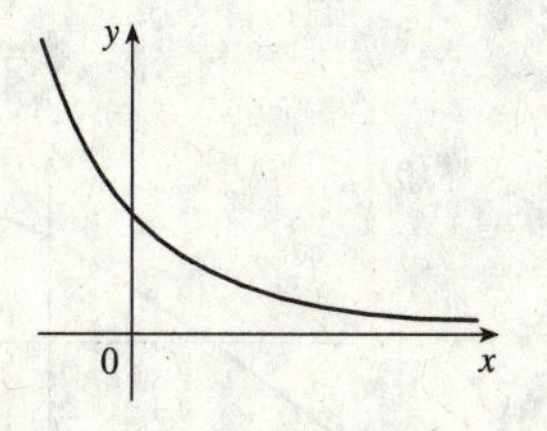

11. True. Let $x_1 < x_2$ where $x_1, x_2 \in I$. Then $f(x_1) < f(x_2)$ and $g(x_1) < g(x_2)$ [since f and g are increasing on I], so $(f + g)(x_1) = f(x_1) + g(x_1) < f(x_2) + g(x_2) = (f + g)(x_2)$.

13. False. Take $f(x) = x$ and $g(x) = x - 1$. Then both f and g are increasing on $(0, 1)$. But $f(x)\,g(x) = x(x - 1)$ is not increasing on $(0, 1)$.

15. True. Let $x_1, x_2 \in I$ and $x_1 < x_2$. Then $f(x_1) < f(x_2)$ [f is increasing] $\Rightarrow \dfrac{1}{f(x_1)} > \dfrac{1}{f(x_2)}$ [f is positive] $\Rightarrow$ $g(x_1) > g(x_2) \Rightarrow g(x) = 1/f(x)$ is decreasing on I.

17. True. If f is periodic, then there is a number p such that $f(x + p) = f(p)$ for all x. Differentiating gives $f'(x) = f'(x + p) \cdot (x + p)' = f'(x + p) \cdot 1 = f'(x + p)$, so f' is periodic.

EXERCISES

1. $f(x) = x^3 - 6x^2 + 9x + 1$, $[2, 4]$. $f'(x) = 3x^2 - 12x + 9 = 3(x^2 - 4x + 3) = 3(x - 1)(x - 3)$. $f'(x) = 0 \Rightarrow$ $x = 1$ or $x = 3$, but 1 is not in the interval. $f'(x) > 0$ for $3 < x < 4$ and $f'(x) < 0$ for $2 < x < 3$, so $f(3) = 1$ is a local minimum value. Checking the endpoints, we find $f(2) = 3$ and $f(4) = 5$. Thus, $f(3) = 1$ is the absolute minimum value and $f(4) = 5$ is the absolute maximum value.

3. $f(x) = \dfrac{3x - 4}{x^2 + 1}$, $[-2, 2]$. $f'(x) = \dfrac{(x^2 + 1)(3) - (3x - 4)(2x)}{(x^2 + 1)^2} = \dfrac{-(3x^2 - 8x - 3)}{(x^2 + 1)^2} = \dfrac{-(3x + 1)(x - 3)}{(x^2 + 1)^2}$.
$f'(x) = 0 \Rightarrow x = -\frac{1}{3}$ or $x = 3$, but 3 is not in the interval. $f'(x) > 0$ for $-\frac{1}{3} < x < 2$ and $f'(x) < 0$ for $-2 < x < -\frac{1}{3}$, so $f\left(-\frac{1}{3}\right) = \frac{-5}{10/9} = -\frac{9}{2}$ is a local minimum value. Checking the endpoints, we find $f(-2) = -2$ and $f(2) = \frac{2}{5}$. Thus, $f\left(-\frac{1}{3}\right) = -\frac{9}{2}$ is the absolute minimum value and $f(2) = \frac{2}{5}$ is the absolute maximum value.

© 2016 Cengage Learning. All Rights Reserved. May not be scanned, copied or duplicated, or posted to a publicly accessible website, in whole or in part.

5. $f(x) = x + \sin 2x$, $[0, \pi]$. $f'(x) = 1 + 2\cos 2x = 0 \Leftrightarrow \cos 2x = -\frac{1}{2} \Leftrightarrow 2x = \frac{2\pi}{3}$ or $\frac{4\pi}{3} \Leftrightarrow x = \frac{\pi}{3}$ or $\frac{2\pi}{3}$.

$f''(x) = -4\sin 2x$, so $f''\left(\frac{\pi}{3}\right) = -4\sin\frac{2\pi}{3} = -2\sqrt{3} < 0$ and $f''\left(\frac{2\pi}{3}\right) = -4\sin\frac{4\pi}{3} = 2\sqrt{3} > 0$, so

$f\left(\frac{\pi}{3}\right) = \frac{\pi}{3} + \frac{\sqrt{3}}{2} \approx 1.91$ is a local maximum value and $f\left(\frac{2\pi}{3}\right) = \frac{2\pi}{3} - \frac{\sqrt{3}}{2} \approx 1.23$ is a local minimum value. Also $f(0) = 0$ and $f(\pi) = \pi$, so $f(0) = 0$ is the absolute minimum value and $f(\pi) = \pi$ is the absolute maximum value.

7. (a) $f(x) = 2 - 2x - x^3$ is a polynomial, so there is no asymptote.

(b) $f'(x) = -2 - 3x^2 = -1(3x^2 + 2) < 0$, so f is decreasing on $\mathbb{R}$.

(c) No local extrema

(d) $f''(x) = -6x < 0$ on $(0, \infty)$ and $f''(x) > 0$ on $(-\infty, 0)$, so f is CD on $(0, \infty)$ and CU on $(-\infty, 0)$. IP at $(0, 2)$

(e)

9. (a) $f(x) = x + \sqrt{1 - x}$ has no asymptote.

(b) $f'(x) = 1 - 1/\left(2\sqrt{1 - x}\right) = 0 \Leftrightarrow 2\sqrt{1 - x} = 1 \Leftrightarrow 1 - x = \frac{1}{4} \Leftrightarrow x = \frac{3}{4}$ and $f'(x) > 0 \Leftrightarrow x < \frac{3}{4}$, so f is increasing on $\left(-\infty, \frac{3}{4}\right)$ and decreasing on $\left(\frac{3}{4}, 1\right)$.

(c) $f\left(\frac{3}{4}\right) = \frac{3}{4} + \sqrt{1 - \frac{3}{4}} = \frac{3}{4} + \sqrt{\frac{1}{4}} = \frac{3}{4} + \frac{1}{2} = \frac{5}{4}$ is a local maximum.

(d) $f''(x) = -\dfrac{1}{4(1 - x)^{3/2}} < 0$ on the domain of f, so f is CD on $(-\infty, 1)$. No IP

(e)

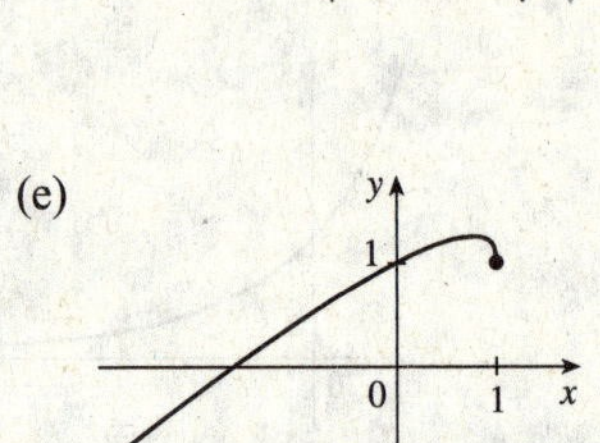

11. (a) $y = f(x) = \sin^2 x - 2\cos x$ has no asymptote.

(b) $y' = 2\sin x\,\cos x + 2\sin x = 2\sin x\,(\cos x + 1)$. $y' = 0 \Leftrightarrow \sin x = 0$ or $\cos x = -1 \Leftrightarrow x = n\pi$ or $x = (2n + 1)\pi$. $y' > 0$ when $\sin x > 0$, since $\cos x + 1 \geq 0$ for all x. Therefore, $y' > 0$ (and so f is increasing) on $(2n\pi, (2n + 1)\pi)$; $y' < 0$ (and so f is decreasing) on $((2n - 1)\pi, 2n\pi)$ or equivalently, $((2n + 1)\pi, (2n + 2)\pi)$.

(c) Local maxima are $f((2n + 1)\pi) = 2$; local minima are $f(2n\pi) = -2$.

(d) $y' = \sin 2x + 2\sin x \Rightarrow$

$$\begin{aligned} y'' &= 2\cos 2x + 2\cos x = 2(2\cos^2 x - 1) + 2\cos x \\ &= 4\cos^2 x + 2\cos x - 2 = 2(2\cos^2 x + \cos x - 1) \\ &= 2(2\cos x - 1)(\cos x + 1) \end{aligned}$$

$y'' = 0 \Leftrightarrow \cos x = \frac{1}{2}$ or $-1 \Leftrightarrow x = 2n\pi \pm \frac{\pi}{3}$ or $x = (2n + 1)\pi$.

$y'' > 0$ (and so f is CU) on $\left(2n\pi - \frac{\pi}{3}, 2n\pi + \frac{\pi}{3}\right)$; $y'' \leq 0$ (and so f is CD) on $\left(2n\pi + \frac{\pi}{3}, 2n\pi + \frac{5\pi}{3}\right)$. IPs at $\left(2n\pi \pm \frac{\pi}{3}, -\frac{1}{4}\right)$

(e)

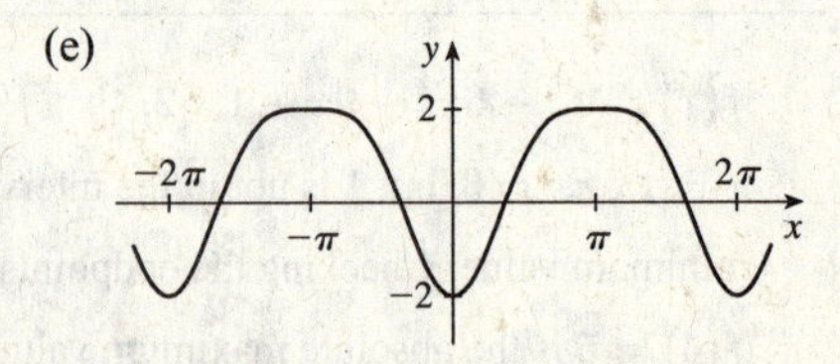

© 2016 Cengage Learning. All Rights Reserved. May not be scanned, copied or duplicated, or posted to a publicly accessible website, in whole or in part.

13. (a) $\lim_{x\to\pm\infty}\left(e^x+e^{-3x}\right)=\infty$, no asymptote.

(e)

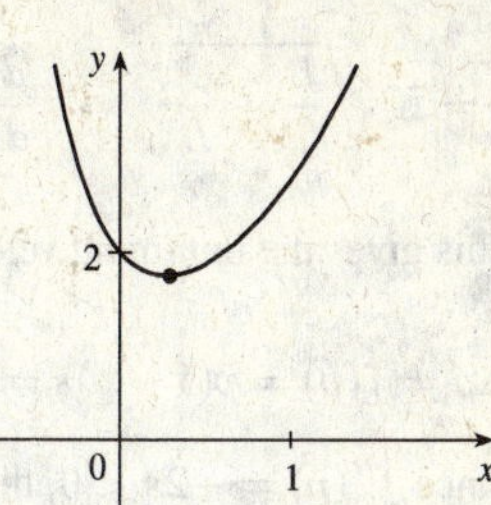

(b) $y=f(x)=e^x+e^{-3x} \;\Rightarrow\; f'(x)=e^x-3e^{-3x}=e^{-3x}(e^{4x}-3)>0 \;\Leftrightarrow$
$e^{4x}>3 \;\Leftrightarrow\; 4x>\ln 3 \;\Leftrightarrow\; x>\frac{1}{4}\ln 3$, so f is increasing on $\left(\frac{1}{4}\ln 3,\infty\right)$ and decreasing on $\left(-\infty,\frac{1}{4}\ln 3\right)$.

(c) $f\left(\frac{1}{4}\ln 3\right)=3^{1/4}+3^{-3/4}\approx 1.75$ is a local and absolute minimum.

(d) $f''(x)=e^x+9e^{-3x}>0$, so f is CU on $(-\infty,\infty)$. No IP

15. (a) $C(t)=2.5\left(e^{-0.3t}-e^{-0.7t}\right) \;\Rightarrow\; C'(t)=2.5\left(-0.3e^{-0.3t}+0.7e^{-0.7t}\right)=0 \;\Rightarrow\; e^{0.4t}=\dfrac{0.7}{0.3} \;\Rightarrow$
$t=\dfrac{\ln(7/3)}{0.4}\approx 2.118$ h. Also, $C'(t)>0$ when $t<\dfrac{\ln(7/3)}{0.4}$ and $C'(t)<0$ when $t>\dfrac{\ln(7/3)}{0.4}$. So the maximum concentration is $C(2.118)\approx 0.7567\ \mu$g/mL

(b) $C''(t)=2.5\left(0.09e^{-0.3t}-0.49e^{-0.7t}\right)=0 \;\Rightarrow\; e^{0.4t}=\dfrac{0.49}{0.09} \;\Rightarrow\; t=\dfrac{\ln(49/9)}{0.4}\approx 4.236$ h. Also, $C''(t)>0$ (CU) when $t>4.236$ and $C''(t)<0$ (CD) when $t<4.236$, so $t\approx 4.236$ is a point of inflection. Hence, the rate of change of concentration begins to increase after 4.236 hours.

17. Let $P(t)$ be the population after t weeks. We will assume that $P(t)$ can be approximated by a differentiable function (though in reality population size changes by discrete increments and is not differentiable). Using $a=0$, $b=5$, $P(0)=300$, and $P'(c)=120$ in the Mean Value Theorem gives

$$P(b)-P(a)=P'(c)\,(b-a) \;\Rightarrow\; P(5)-300\le 120\,(5-0) \;\Rightarrow\; P(5)\le 300+120(5)=900$$

Since the population grows at a rate of *at most* $P'=120$, the population after 5 weeks is at most 900.

19. This limit has the form $\frac{0}{0}$. $\displaystyle\lim_{x\to 0}\frac{\tan\pi x}{\ln(1+x)}\overset{\text{H}}{=}\lim_{x\to 0}\frac{\pi\sec^2\pi x}{1/(1+x)}=\frac{\pi\cdot 1^2}{1/1}=\pi$

21. This limit has the form $\frac{0}{0}$. $\displaystyle\lim_{x\to 0}\frac{e^{4x}-1-4x}{x^2}\overset{\text{H}}{=}\lim_{x\to 0}\frac{4e^{4x}-4}{2x}\overset{\text{H}}{=}\lim_{x\to 0}\frac{16e^{4x}}{2}=\lim_{x\to 0}8e^{4x}=8\cdot 1=8$

23. This limit has the form $\infty\cdot 0$. $\displaystyle\lim_{x\to\infty}x^3e^{-x}=\lim_{x\to\infty}\frac{x^3}{e^x}\overset{\text{H}}{=}\lim_{x\to\infty}\frac{3x^2}{e^x}\overset{\text{H}}{=}\lim_{x\to\infty}\frac{6x}{e^x}\overset{\text{H}}{=}\lim_{x\to\infty}\frac{6}{e^x}=0$

25. This limit has the form $\infty-\infty$.

$$\lim_{x\to 1^+}\left(\frac{x}{x-1}-\frac{1}{\ln x}\right)=\lim_{x\to 1^+}\left(\frac{x\ln x-x+1}{(x-1)\ln x}\right)\overset{\text{H}}{=}\lim_{x\to 1^+}\frac{x\cdot(1/x)+\ln x-1}{(x-1)\cdot(1/x)+\ln x}=\lim_{x\to 1^+}\frac{\ln x}{1-1/x+\ln x}$$
$$\overset{\text{H}}{=}\lim_{x\to 1^+}\frac{1/x}{1/x^2+1/x}=\frac{1}{1+1}=\frac{1}{2}$$

27. Call the two integers x and y. Then $x+4y=1000$, so $x=1000-4y$. Their product is $P=xy=(1000-4y)y$, so our problem is to maximize the function $P(y)=1000y-4y^2$, where $0<y<250$ and y is an integer. $P'(y)=1000-8y$, so $P'(y)=0 \;\Leftrightarrow\; y=125$. $P''(y)=-8<0$, so $P(125)=62{,}500$ is an absolute maximum. Since the optimal y turned out to be an integer, we have found the desired pair of numbers, namely $x=1000-4(125)=500$ and $y=125$.

© 2016 Cengage Learning. All Rights Reserved. May not be scanned, copied or duplicated, or posted to a publicly accessible website, in whole or in part.

29. $v = K\sqrt{\dfrac{L}{C} + \dfrac{C}{L}} \;\Rightarrow\; \dfrac{dv}{dL} = \dfrac{K}{2\sqrt{(L/C) + (C/L)}}\left(\dfrac{1}{C} - \dfrac{C}{L^2}\right) = 0 \;\Leftrightarrow\; \dfrac{1}{C} = \dfrac{C}{L^2} \;\Leftrightarrow\; L^2 = C^2 \;\Leftrightarrow\; L = C.$

This gives the minimum velocity since $v' < 0$ for $0 < L < C$ and $v' > 0$ for $L > C$.

31. $\Delta p = f(p) = p(1-p)s = sp - sp^2 \;\Rightarrow\; f'(p) = s - 2sp = 0 \;\Rightarrow\; p = 1/2.$

Since $f''(p) = -2s < 0$, the rate of evolution Δp is largest when $p = 1/2$.

33. $x_{t+1} = f(x_t) = \dfrac{4x_t}{1 + 5x_t}$. We first find the equilibria by solving the equation $f(x) = x$ as follows:

$x = \dfrac{4x}{1+5x} \;\Leftrightarrow\; 5x^2 - 3x = 0 \;\Leftrightarrow\; x(5x-3) = 0$. So $x = 0$ and $x = \dfrac{3}{5}$ are the equilibria.

Stability Check: $f'(x) = \dfrac{4(1+5x) - 20x}{(1+5x)^2} = \dfrac{4}{(1+5x)^2}$. So $f'(0) = 4 > 1 \;\Rightarrow\; x = 0$ is an unstable equilibrium and

$f'(3/5) = 1/4 \;\Rightarrow\; |f'(3/5)| < 1 \;\Rightarrow\; x = 3/5$ is a stable equilibrium.

35. $x_{t+1} = f(x_t) = \dfrac{6x_t^2}{x_t^2 + 8}$. Solving for x in the equation $f(x) = x$ gives

$x = \dfrac{6x^2}{x^2+8} \;\Leftrightarrow\; x^3 - 6x^2 + 8x = 0 \;\Leftrightarrow\; x(x-2)(x-4) = 0.$

So $x = 0$ or $x = 2$ or $x = 4$ are the equilibria

Stability Check: $f'(x) = \dfrac{12x(x^2+8) - 12x^3}{(x^2+8)^2} = \dfrac{96x}{(x^2+8)^2}$

So $f'(0) = 0 \;\Rightarrow\; |f'(0)| < 1 \;\Rightarrow\; x = 0$ is a stable equilibrium

and $f'(2) = \dfrac{192}{144} \approx 1.33 \;\Rightarrow\; |f'(2)| > 1 \;\Rightarrow\; x = 2$ is an unstable equilibrium

and $f'(4) = \dfrac{384}{576} \approx 0.67 \;\Rightarrow\; |f'(4)| < 1 \;\Rightarrow\; x = 4$ is a stable equilibrium.

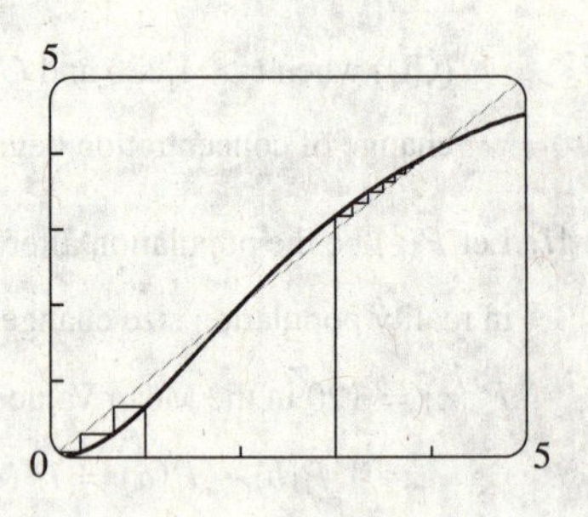

From the cobweb plot, we observe that $\lim_{t\to\infty} x_t = 0$ when $x_0 = 1$ and $\lim_{t\to\infty} x_t = 4$ when $x_0 = 3$.

37. $f(x) = \sin x + \sec x \tan x \;\Rightarrow\; F(x) = -\cos x + \sec x + C$

39. $q(t) = 2 + (t+1)(t^2 - 1) = t^3 + t^2 - t + 1 \;\Rightarrow\; Q(t) = \frac{1}{4}t^4 + \frac{1}{3}t^3 - \frac{1}{2}t^2 + t + C$

41. $\dfrac{dy}{dt} = 1 - e^{\pi t} \;\Rightarrow\; y(t) = t - \frac{1}{\pi}e^{\pi t} + C$ and $y(0) = 0 \;\Rightarrow\; C = 1/\pi$. So the solution is $y(t) = t - \frac{1}{\pi}e^{\pi t} + \frac{1}{\pi}$.

43. $f''(x) = 1 - 6x + 48x^2 \;\Rightarrow\; f'(x) = x - 3x^2 + 16x^3 + C$. $f'(0) = C$ and $f'(0) = 2 \;\Rightarrow\; C = 2$, so

$f'(x) = x - 3x^2 + 16x^3 + 2$ and hence, $f(x) = \frac{1}{2}x^2 - x^3 + 4x^4 + 2x + D$.

$f(0) = D$ and $f(0) = 1 \;\Rightarrow\; D = 1$, so $f(x) = \frac{1}{2}x^2 - x^3 + 4x^4 + 2x + 1$.

45. $a(t) = v'(t) = \sin t + 3\cos t \;\Rightarrow\; v(t) = -\cos t + 3\sin t + C$.

$v(0) = -1 + 0 + C$ and $v(0) = 2 \;\Rightarrow\; C = 3$, so $v(t) = -\cos t + 3\sin t + 3$ and $s(t) = -\sin t - 3\cos t + 3t + D$.

$s(0) = -3 + D$ and $s(0) = 0 \;\Rightarrow\; D = 3$, and $s(t) = -\sin t - 3\cos t + 3t + 3$.

© 2016 Cengage Learning. All Rights Reserved. May not be scanned, copied or duplicated, or posted to a publicly accessible website, in whole or in part.

47. Let $y = f(x) = e^{-x^2}$. The area of the rectangle under the curve from $-x$ to x is $A(x) = 2xe^{-x^2}$ where $x \geq 0$. We maximize $A(x)$: $A'(x) = 2e^{-x^2} - 4x^2e^{-x^2} = 2e^{-x^2}(1 - 2x^2) = 0 \Rightarrow x = \frac{1}{\sqrt{2}}$. This gives a maximum since $A'(x) > 0$ for $0 \leq x < \frac{1}{\sqrt{2}}$ and $A'(x) < 0$ for $x > \frac{1}{\sqrt{2}}$. We next determine the points of inflection of $f(x)$. Notice that $f'(x) = -2xe^{-x^2} = -A(x)$. So $f''(x) = -A'(x)$ and hence, $f''(x) < 0$ for $-\frac{1}{\sqrt{2}} < x < \frac{1}{\sqrt{2}}$ and $f''(x) > 0$ for $x < -\frac{1}{\sqrt{2}}$ and $x > \frac{1}{\sqrt{2}}$. So $f(x)$ changes concavity at $x = \pm\frac{1}{\sqrt{2}}$, and the two vertices of the rectangle of largest area are at the inflection points.

© 2016 Cengage Learning. All Rights Reserved. May not be scanned, copied or duplicated, or posted to a publicly accessible website, in whole or in part.

5 ☐ INTEGRALS

5.1 Areas, Distances and Pathogenesis

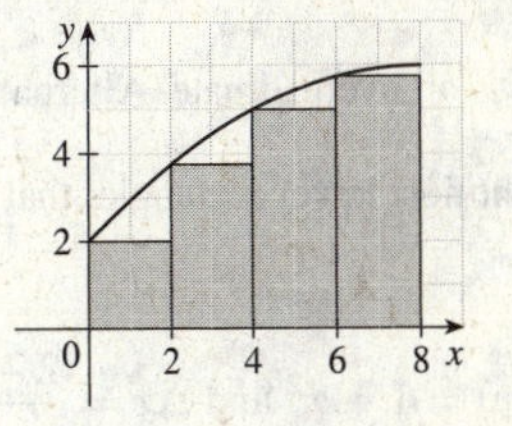

1. (a) Since f is *increasing*, we can obtain a *lower* estimate by using *left* endpoints. We are instructed to use four rectangles, so $n = 4$.

$$L_4 = \sum_{i=1}^{4} f(x_{i-1})\,\Delta x \quad \left[\Delta x = \frac{b-a}{n} = \frac{8-0}{4} = 2\right]$$

$$= f(x_0)\cdot 2 + f(x_1)\cdot 2 + f(x_2)\cdot 2 + f(x_3)\cdot 2$$

$$= 2[f(0) + f(2) + f(4) + f(6)]$$

$$= 2(2 + 3.75 + 5 + 5.75) = 2(16.5) = 33$$

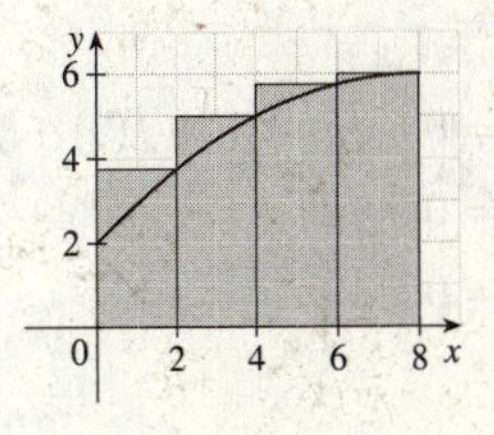

Since f is *increasing*, we can obtain an *upper* estimate by using *right* endpoints.

$$R_4 = \sum_{i=1}^{4} f(x_i)\,\Delta x$$

$$= f(x_1)\cdot 2 + f(x_2)\cdot 2 + f(x_3)\cdot 2 + f(x_4)\cdot 2$$

$$= 2[f(2) + f(4) + f(6) + f(8)]$$

$$= 2(3.75 + 5 + 5.75 + 6) = 2(20.5) = 41$$

Comparing R_4 to L_4, we see that we have added the area of the rightmost upper rectangle, $f(8)\cdot 2$, to the sum and subtracted the area of the leftmost lower rectangle, $f(0)\cdot 2$, from the sum.

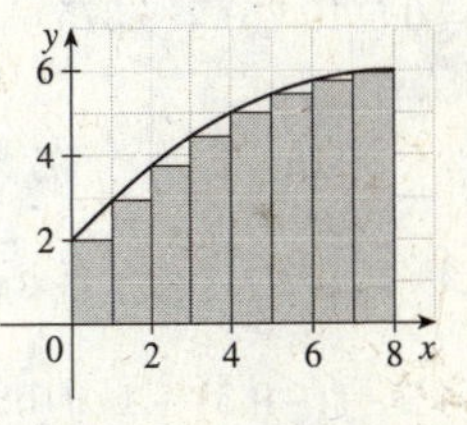

(b) $$L_8 = \sum_{i=1}^{8} f(x_{i-1})\Delta x \quad \left[\Delta x = \frac{8-0}{8} = 1\right]$$

$$= 1[f(x_0) + f(x_1) + \cdots + f(x_7)]$$

$$= f(0) + f(1) + \cdots + f(7)$$

$$\approx 2 + 3.0 + 3.75 + 4.4 + 5 + 5.4 + 5.75 + 5.9$$

$$= 35.2$$

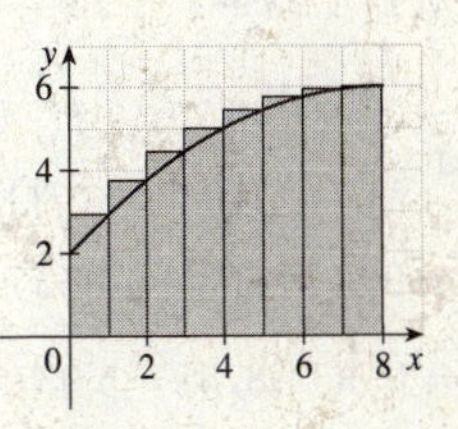

$$R_8 = \sum_{i=1}^{8} f(x_i)\Delta x = f(1) + f(2) + \cdots + f(8)$$

$$= L_8 + 1\cdot f(8) - 1\cdot f(0) \quad \left[\begin{array}{c}\text{add rightmost upper rectangle,}\\ \text{subtract leftmost lower rectangle}\end{array}\right]$$

$$= 35.2 + 6 - 2 = 39.2$$

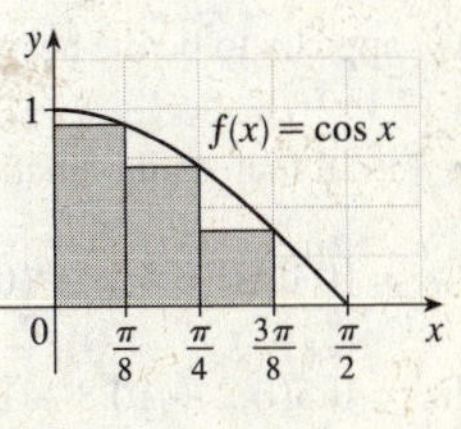

3. (a) $$R_4 = \sum_{i=1}^{4} f(x_i)\,\Delta x \quad \left[\Delta x = \frac{\pi/2 - 0}{4} = \frac{\pi}{8}\right] \quad = \left[\sum_{i=1}^{4} f(x_i)\right]\Delta x$$

$$= [f(x_1) + f(x_2) + f(x_3) + f(x_4)]\,\Delta x$$

$$= \left[\cos\tfrac{\pi}{8} + \cos\tfrac{2\pi}{8} + \cos\tfrac{3\pi}{8} + \cos\tfrac{4\pi}{8}\right]\tfrac{\pi}{8}$$

$$\approx (0.9239 + 0.7071 + 0.3827 + 0)\tfrac{\pi}{8} \approx 0.7908$$

Since f is *decreasing* on $[0, \pi/2]$, an *underestimate* is obtained by using the *right* endpoint approximation, R_4.

© 2016 Cengage Learning. All Rights Reserved. May not be scanned, copied or duplicated, or posted to a publicly accessible website, in whole or in part.

(b) $L_4 = \sum_{i=1}^{4} f(x_{i-1})\,\Delta x = \left[\sum_{i=1}^{4} f(x_{i-1})\right]\Delta x$

$= [f(x_0) + f(x_1) + f(x_2) + f(x_3)]\,\Delta x$

$= \left[\cos 0 + \cos\frac{\pi}{8} + \cos\frac{2\pi}{8} + \cos\frac{3\pi}{8}\right]\frac{\pi}{8}$

$\approx (1 + 0.9239 + 0.7071 + 0.3827)\frac{\pi}{8} \approx 1.1835$

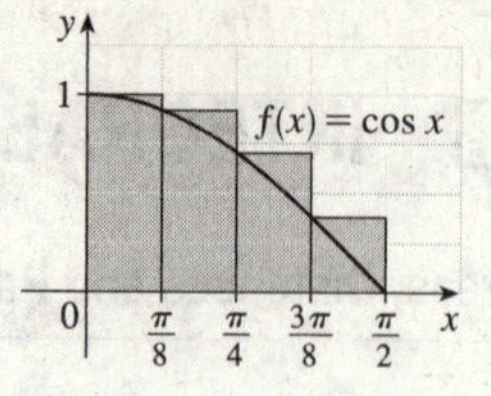

L_4 is an overestimate. Alternatively, we could just add the area of the leftmost upper rectangle and subtract the area of the rightmost lower rectangle; that is, $L_4 = R_4 + f(0)\cdot\frac{\pi}{8} - f\left(\frac{\pi}{2}\right)\cdot\frac{\pi}{8}$.

5. (a) $f(x) = 1 + x^2$ and $\Delta x = \dfrac{2-(-1)}{3} = 1 \;\Rightarrow$

$R_3 = 1\cdot f(0) + 1\cdot f(1) + 1\cdot f(2) = 1\cdot 1 + 1\cdot 2 + 1\cdot 5 = 8.$

$\Delta x = \dfrac{2-(-1)}{6} = 0.5 \;\Rightarrow$

$R_6 = 0.5[f(-0.5) + f(0) + f(0.5) + f(1) + f(1.5) + f(2)]$

$= 0.5(1.25 + 1 + 1.25 + 2 + 3.25 + 5)$

$= 0.5(13.75) = 6.875$

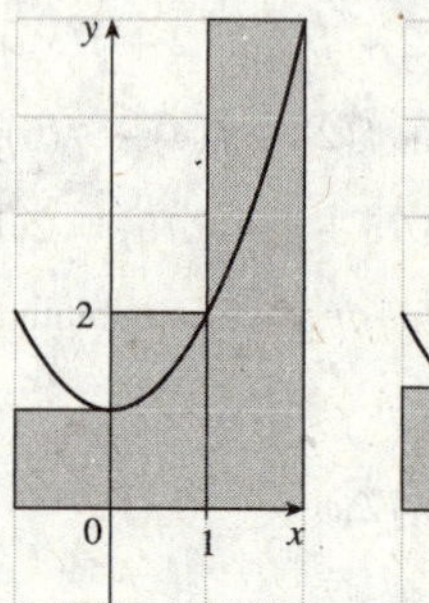

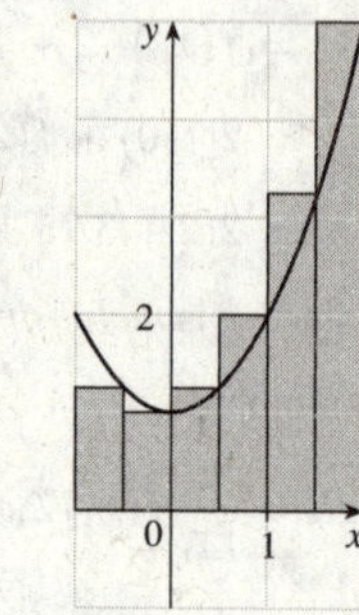

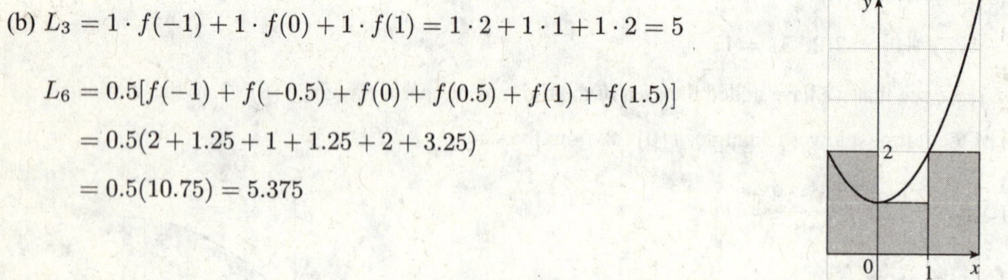

(b) $L_3 = 1\cdot f(-1) + 1\cdot f(0) + 1\cdot f(1) = 1\cdot 2 + 1\cdot 1 + 1\cdot 2 = 5$

$L_6 = 0.5[f(-1) + f(-0.5) + f(0) + f(0.5) + f(1) + f(1.5)]$

$= 0.5(2 + 1.25 + 1 + 1.25 + 2 + 3.25)$

$= 0.5(10.75) = 5.375$

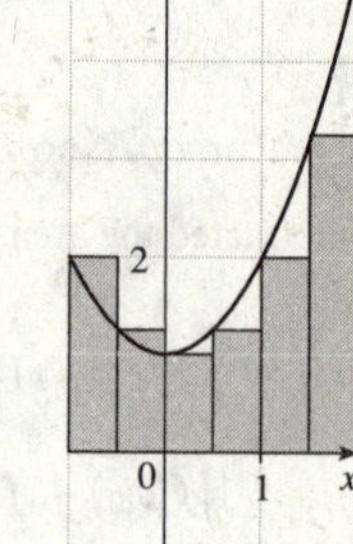

(c) $M_3 = 1\cdot f(-0.5) + 1\cdot f(0.5) + 1\cdot f(1.5)$

$= 1\cdot 1.25 + 1\cdot 1.25 + 1\cdot 3.25 = 5.75$

$M_6 = 0.5[f(-0.75) + f(-0.25) + f(0.25) + f(0.75) + f(1.25) + f(1.75)]$

$= 0.5(1.5625 + 1.0625 + 1.0625 + 1.5625 + 2.5625 + 4.0625)$

$= 0.5(11.875) = 5.9375$

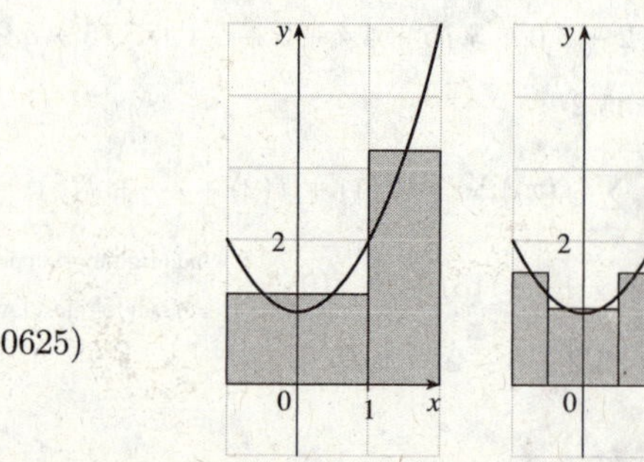

(d) M_6 appears to be the best estimate.

7. Since v is an increasing function, L_6 will give us a lower estimate and R_6 will give us an upper estimate.

$L_6 = (0\text{ ft/s})(0.5\text{ s}) + (6.2)(0.5) + (10.8)(0.5) + (14.9)(0.5) + (18.1)(0.5) + (19.4)(0.5) = 0.5(69.4) = 34.7\text{ ft}$

$R_6 = 0.5(6.2 + 10.8 + 14.9 + 18.1 + 19.4 + 20.2) = 0.5(89.6) = 44.8\text{ ft}$

© 2016 Cengage Learning. All Rights Reserved. May not be scanned, copied or duplicated, or posted to a publicly accessible website, in whole or in part.

9. $f(t) = -t(t-21)(t+1)$ and $\Delta t = \dfrac{12-0}{6} = 2 \;\Rightarrow$

$$\begin{aligned} M_6 &= 2\cdot f(1) + 2\cdot f(3) + 2\cdot f(5) + 2\cdot f(7) + 2\cdot f(9) + 2\cdot f(11) \\ &= 2\cdot 40 + 2\cdot 216 + 2\cdot 480 + 2\cdot 784 + 2\cdot 1080 + 2\cdot 1320 \\ &= 7,840 \text{ (infected cells/mL)} \cdot \text{days} \end{aligned}$$

11. Using a two-week interval $\Delta t = 14$ to estimate the area using left and right endpoints gives:

$$L_6 = (0.0079)(14) + (0.0638)(14) + (0.1944)(14) + (0.4435)(14) + (0.5620)(14) + (0.4630)(14) = 24.2844 \approx 24 \text{ people}$$

$$R_6 = (0.0638)(14) + (0.1944)(14) + (0.4435)(14) + (0.5620)(14) + (0.4630)(14) + (0.2897)(14) = 28.2296 \approx 28 \text{ people}$$

13. For a decreasing function, using left endpoints gives us an overestimate and using right endpoints results in an underestimate. We will use M_6 to get an estimate. $\Delta t = 1$, so

$$M_6 = 1[v(0.5) + v(1.5) + v(2.5) + v(3.5) + v(4.5) + v(5.5)] \approx 55 + 40 + 28 + 18 + 10 + 4 = 155 \text{ ft}$$

For a very rough check on the above calculation, we can draw a line from $(0, 70)$ to $(6, 0)$ and calculate the area of the triangle: $\frac{1}{2}(70)(6) = 210$. This is clearly an overestimate, so our midpoint estimate of 155 is reasonable.

15. $f(x) = \dfrac{2x}{x^2+1}$, $1 \le x \le 3$. $\Delta x = (3-1)/n = 2/n$ and $x_i = 1 + i\Delta x = 1 + 2i/n$.

$$A = \lim_{n\to\infty} R_n = \lim_{n\to\infty} \sum_{i=1}^{n} f(x_i)\Delta x = \lim_{n\to\infty} \sum_{i=1}^{n} \frac{2(1+2i/n)}{(1+2i/n)^2+1} \cdot \frac{2}{n}.$$

17. $f(x) = x\cos x$, $0 \le x \le \frac{\pi}{2}$. $\Delta x = (\frac{\pi}{2} - 0)/n = \frac{\pi}{2}/n$ and $x_i = 0 + i\,\Delta x = \frac{\pi}{2}i/n$.

$$A = \lim_{n\to\infty} R_n = \lim_{n\to\infty} \sum_{i=1}^{n} f(x_i)\,\Delta x = \lim_{n\to\infty} \sum_{i=1}^{n} \frac{i\pi}{2n} \cos\left(\frac{i\pi}{2n}\right) \cdot \frac{\pi}{2n}.$$

19. (a) Since f is an increasing function, L_n is an underestimate of A [lower sum] and R_n is an overestimate of A [upper sum]. Thus, A, L_n, and R_n are related by the inequality $L_n < A < R_n$.

(b)
$$\begin{aligned} R_n &= f(x_1)\Delta x + f(x_2)\Delta x + \cdots + f(x_n)\Delta x \\ L_n &= f(x_0)\Delta x + f(x_1)\Delta x + \cdots + f(x_{n-1})\Delta x \\ R_n - L_n &= f(x_n)\Delta x - f(x_0)\Delta x = \Delta x[f(x_n) - f(x_0)] = \frac{b-a}{n}[f(b) - f(a)] \end{aligned}$$

(c) $A > L_n$, so $R_n - A < R_n - L_n$; that is, $R_n - A < \dfrac{b-a}{n}[f(b) - f(a)]$.

21. (a) $y = f(x) = x^5$. $\Delta x = \dfrac{2-0}{n} = \dfrac{2}{n}$ and $x_i = 0 + i\,\Delta x = \dfrac{2i}{n}$.

$$A = \lim_{n\to\infty} R_n = \lim_{n\to\infty} \sum_{i=1}^{n} f(x_i)\,\Delta x = \lim_{n\to\infty} \sum_{i=1}^{n} \left(\frac{2i}{n}\right)^5 \cdot \frac{2}{n} = \lim_{n\to\infty} \sum_{i=1}^{n} \frac{32i^5}{n^5} \cdot \frac{2}{n} = \lim_{n\to\infty} \frac{64}{n^6} \sum_{i=1}^{n} i^5.$$

(b) $\displaystyle\sum_{i=1}^{n} i^5 \overset{\text{CAS}}{=} \frac{n^2(n+1)^2(2n^2+2n-1)}{12}$

(c)
$$\begin{aligned} \lim_{n\to\infty} \frac{64}{n^6} \cdot \frac{n^2(n+1)^2(2n^2+2n-1)}{12} &= \frac{64}{12} \lim_{n\to\infty} \frac{(n^2+2n+1)(2n^2+2n-1)}{n^2 \cdot n^2} \\ &= \frac{16}{3} \lim_{n\to\infty} \left(1 + \frac{2}{n} + \frac{1}{n^2}\right)\left(2 + \frac{2}{n} - \frac{1}{n^2}\right) = \tfrac{16}{3} \cdot 1 \cdot 2 = \tfrac{32}{3} \end{aligned}$$

© 2016 Cengage Learning. All Rights Reserved. May not be scanned, copied or duplicated, or posted to a publicly accessible website, in whole or in part.

23. $y = f(x) = \cos x$. $\Delta x = \dfrac{b-0}{n} = \dfrac{b}{n}$ and $x_i = 0 + i\,\Delta x = \dfrac{bi}{n}$.

$$A = \lim_{n\to\infty} R_n = \lim_{n\to\infty} \sum_{i=1}^{n} f(x_i)\,\Delta x = \lim_{n\to\infty} \sum_{i=1}^{n} \cos\left(\frac{bi}{n}\right)\cdot\frac{b}{n} \overset{\text{CAS}}{=} \lim_{n\to\infty}\left[\frac{b\sin\left(b\left(\dfrac{1}{2n}+1\right)\right)}{2n\sin\left(\dfrac{b}{2n}\right)} - \frac{b}{2n}\right] \overset{\text{CAS}}{=} \sin b$$

If $b = \frac{\pi}{2}$, then $A = \sin\frac{\pi}{2} = 1$.

5.2 The Definite Integral

1. $f(x) = 3 - \frac{1}{2}x$, $2 \le x \le 14$. $\Delta x = \dfrac{b-a}{n} = \dfrac{14-2}{6} = 2$.

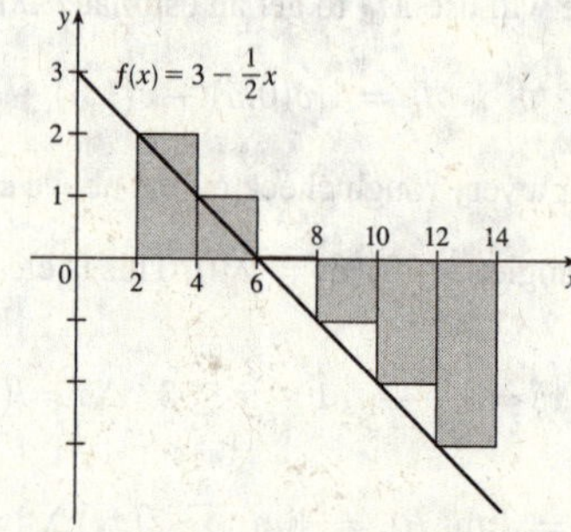

Since we are using left endpoints, $x_i^* = x_{i-1}$.

$$\begin{aligned} L_6 &= \sum_{i=1}^{6} f(x_{i-1})\,\Delta x \\ &= (\Delta x)\,[f(x_0) + f(x_1) + f(x_2) + f(x_3) + f(x_4) + f(x_5)] \\ &= 2[f(2) + f(4) + f(6) + f(8) + f(10) + f(12)] \\ &= 2[2 + 1 + 0 + (-1) + (-2) + (-3)] = 2(-3) = -6 \end{aligned}$$

The Riemann sum represents the sum of the areas of the two rectangles above the x-axis minus the sum of the areas of the three rectangles below the x-axis; that is, the *net area* of the rectangles with respect to the x-axis.

3. $f(x) = e^x - 2$, $0 \le x \le 2$. $\Delta x = \dfrac{b-a}{n} = \dfrac{2-0}{4} = \dfrac{1}{2}$.

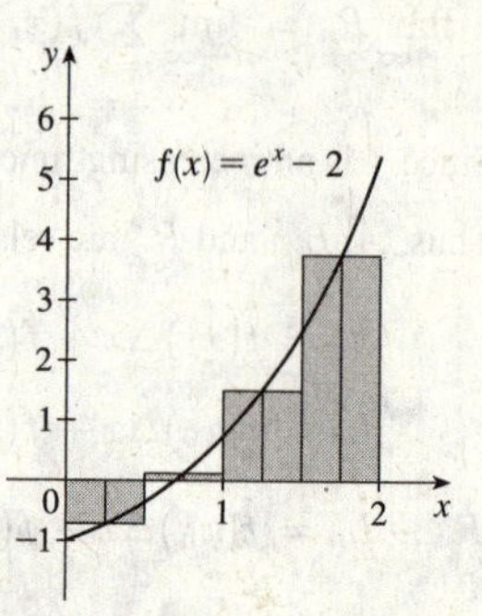

Since we are using midpoints, $x_i^* = \overline{x}_i = \frac{1}{2}(x_{i-1} + x_i)$.

$$\begin{aligned} M_4 &= \sum_{i=1}^{4} f(\overline{x}_i)\,\Delta x = (\Delta x)\,[f(\overline{x}_1) + f(\overline{x}_2) + f(\overline{x}_3) + f(\overline{x}_4)] \\ &= \tfrac{1}{2}\left[f\left(\tfrac{1}{4}\right) + f\left(\tfrac{3}{4}\right) + f\left(\tfrac{5}{4}\right) + f\left(\tfrac{7}{4}\right)\right] \\ &= \tfrac{1}{2}\left[(e^{1/4} - 2) + (e^{3/4} - 2) + (e^{5/4} - 2) + (e^{7/4} - 2)\right] \\ &\approx 2.322986 \end{aligned}$$

The Riemann sum represents the sum of the areas of the three rectangles above the x-axis minus the area of the rectangle below the x-axis; that is, the *net area* of the rectangles with respect to the x-axis.

5. $\Delta x = (b-a)/n = (8-0)/4 = 8/4 = 2$.

(a) Using the right endpoints to approximate $\int_0^8 f(x)\,dx$, we have

$$\sum_{i=1}^{4} f(x_i)\,\Delta x = 2[f(2) + f(4) + f(6) + f(8)] \approx 2[1 + 2 + (-2) + 1] = 4.$$

(b) Using the left endpoints to approximate $\int_0^8 f(x)\,dx$, we have

$$\sum_{i=1}^{4} f(x_{i-1})\,\Delta x = 2[f(0) + f(2) + f(4) + f(6)] \approx 2[2 + 1 + 2 + (-2)] = 6.$$

© 2016 Cengage Learning. All Rights Reserved. May not be scanned, copied or duplicated, or posted to a publicly accessible website, in whole or in part.

(c) Using the midpoint of each subinterval to approximate $\int_0^8 f(x)\,dx$, we have

$$\sum_{i=1}^{4} f(\overline{x}_i)\,\Delta x = 2[f(1) + f(3) + f(5) + f(7)] \approx 2[3 + 2 + 1 + (-1)] = 10.$$

7. Since f is increasing, $L_5 \le \int_{10}^{30} f(x)\,dx \le R_5$.

$$\begin{aligned}\text{Lower estimate } &= L_5 = \sum_{i=1}^{5} f(x_{i-1})\Delta x = 4[f(10) + f(14) + f(18) + f(22) + f(26)] \\ &= 4[-12 + (-6) + (-2) + 1 + 3] = 4(-16) = -64\end{aligned}$$

$$\begin{aligned}\text{Upper estimate } &= R_5 = \sum_{i=1}^{5} f(x)\Delta x = 4[f(14) + f(18) + f(22) + f(26) + f(30)] \\ &= 4[-6 + (-2) + 1 + 3 + 8] = 4(4) = 16\end{aligned}$$

9. $\Delta x = (10 - 2)/4 = 2$, so the endpoints are 2, 4, 6, 8, and 10, and the midpoints are 3, 5, 7, and 9. The Midpoint Rule gives $\int_2^{10} \sqrt{x^3 + 1}\,dx \approx \sum_{i=1}^{4} f(\overline{x}_i)\,\Delta x = 2\left(\sqrt{3^3 + 1} + \sqrt{5^3 + 1} + \sqrt{7^3 + 1} + \sqrt{9^3 + 1}\right) \approx 124.1644$.

11. $\Delta x = (1 - 0)/5 = 0.2$, so the endpoints are 0, 0.2, 0.4, 0.6, 0.8, and 1, and the midpoints are 0.1, 0.3, 0.5, 0.7, and 0.9. The Midpoint Rule gives

$$\int_0^1 \sin(x^2)\,dx \approx \sum_{i=1}^{5} f(\overline{x}_i)\,\Delta x = 0.2\left[\sin(0.1)^2 + \sin(0.3)^2 + \sin(0.5)^2 + \sin(0.7)^2 + \sin(0.9)^2\right] \approx 0.3084.$$

13. Using the Midpoint Rule with $n = 5$ and $\Delta t = \frac{100-0}{5} = 20$ gives

$$\begin{aligned}\int_0^{100} C(t)\,dt &\approx \sum_{i=1}^{5} C(\bar{t}_i)\,\Delta t = 20[C(10) + C(30) + C(50) + C(70) + C(90)] \\ &= 20\,(1.3 + 2.2 + 2.5 + 2.3 + 1.6) \\ &= 198\ (\mu\text{g/mL}) \cdot \text{min}\end{aligned}$$

15. On $[2, 6]$, $\lim\limits_{n\to\infty} \sum_{i=1}^{n} x_i \ln(1 + x_i^2)\,\Delta x = \int_2^6 x \ln(1 + x^2)\,dx$.

17. On $[1, 8]$, $\lim\limits_{n\to\infty} \sum_{i=1}^{n} \sqrt{2x_i^* + (x_i^*)^2}\,\Delta x = \int_1^8 \sqrt{2x + x^2}\,dx$.

19. Note that $\Delta x = \dfrac{5 - (-1)}{n} = \dfrac{6}{n}$ and $x_i = -1 + i\,\Delta x = -1 + \dfrac{6i}{n}$.

$$\begin{aligned}\int_{-1}^{5} (1 + 3x)\,dx &= \lim_{n\to\infty} \sum_{i=1}^{n} f(x_i)\,\Delta x = \lim_{n\to\infty} \sum_{i=1}^{n} \left[1 + 3\left(-1 + \frac{6i}{n}\right)\right]\frac{6}{n} = \lim_{n\to\infty} \frac{6}{n} \sum_{i=1}^{n} \left[-2 + \frac{18i}{n}\right] \\ &= \lim_{n\to\infty} \frac{6}{n}\left[\sum_{i=1}^{n}(-2) + \sum_{i=1}^{n} \frac{18i}{n}\right] = \lim_{n\to\infty} \frac{6}{n}\left[-2n + \frac{18}{n}\sum_{i=1}^{n} i\right] \\ &= \lim_{n\to\infty} \frac{6}{n}\left[-2n + \frac{18}{n} \cdot \frac{n(n+1)}{2}\right] = \lim_{n\to\infty}\left[-12 + \frac{108}{n^2} \cdot \frac{n(n+1)}{2}\right] \\ &= \lim_{n\to\infty}\left[-12 + 54\frac{n+1}{n}\right] = \lim_{n\to\infty}\left[-12 + 54\left(1 + \frac{1}{n}\right)\right] = -12 + 54 \cdot 1 = 42\end{aligned}$$

© 2016 Cengage Learning. All Rights Reserved. May not be scanned, copied or duplicated, or posted to a publicly accessible website, in whole or in part.

21. Note that $\Delta x = \dfrac{2-0}{n} = \dfrac{2}{n}$ and $x_i = 0 + i\,\Delta x = \dfrac{2i}{n}$.

$$\int_0^2 (2-x^2)\,dx = \lim_{n\to\infty} \sum_{i=1}^n f(x_i)\,\Delta x = \lim_{n\to\infty} \sum_{i=1}^n \left(2 - \frac{4i^2}{n^2}\right)\left(\frac{2}{n}\right) = \lim_{n\to\infty} \frac{2}{n}\left[\sum_{i=1}^n 2 - \frac{4}{n^2}\sum_{i=1}^n i^2\right]$$

$$= \lim_{n\to\infty} \frac{2}{n}\left(2n - \frac{4}{n^2}\sum_{i=1}^n i^2\right) = \lim_{n\to\infty}\left[4 - \frac{8}{n^3}\cdot\frac{n(n+1)(2n+1)}{6}\right]$$

$$= \lim_{n\to\infty}\left(4 - \frac{4}{3}\cdot\frac{n+1}{n}\cdot\frac{2n+1}{n}\right) = \lim_{n\to\infty}\left[4 - \frac{4}{3}\left(1+\frac{1}{n}\right)\left(2+\frac{1}{n}\right)\right] = 4 - \tfrac{4}{3}\cdot 1\cdot 2 = \tfrac{4}{3}$$

23. Note that $\Delta x = \dfrac{2-1}{n} = \dfrac{1}{n}$ and $x_i = 1 + i\,\Delta x = 1 + i(1/n) = 1 + i/n$.

$$\int_1^2 x^3\,dx = \lim_{n\to\infty} \sum_{i=1}^n f(x_i)\,\Delta x = \lim_{n\to\infty} \sum_{i=1}^n \left(1+\frac{i}{n}\right)^3\left(\frac{1}{n}\right) = \lim_{n\to\infty} \frac{1}{n}\sum_{i=1}^n \left(\frac{n+i}{n}\right)^3$$

$$= \lim_{n\to\infty} \frac{1}{n^4}\sum_{i=1}^n \left(n^3 + 3n^2 i + 3ni^2 + i^3\right) = \lim_{n\to\infty} \frac{1}{n^4}\left[\sum_{i=1}^n n^3 + \sum_{i=1}^n 3n^2 i + \sum_{i=1}^n 3ni^2 + \sum_{i=1}^n i^3\right]$$

$$= \lim_{n\to\infty} \frac{1}{n^4}\left[n\cdot n^3 + 3n^2\sum_{i=1}^n i + 3n\sum_{i=1}^n i^2 + \sum_{i=1}^n i^3\right]$$

$$= \lim_{n\to\infty}\left[1 + \frac{3}{n^2}\cdot\frac{n(n+1)}{2} + \frac{3}{n^3}\cdot\frac{n(n+1)(2n+1)}{6} + \frac{1}{n^4}\cdot\frac{n^2(n+1)^2}{4}\right]$$

$$= \lim_{n\to\infty}\left[1 + \frac{3}{2}\cdot\frac{n+1}{n} + \frac{1}{2}\cdot\frac{n+1}{n}\cdot\frac{2n+1}{n} + \frac{1}{4}\cdot\frac{(n+1)^2}{n^2}\right]$$

$$= \lim_{n\to\infty}\left[1 + \frac{3}{2}\left(1+\frac{1}{n}\right) + \frac{1}{2}\left(1+\frac{1}{n}\right)\left(2+\frac{1}{n}\right) + \frac{1}{4}\left(1+\frac{1}{n}\right)^2\right] = 1 + \frac{3}{2} + \tfrac{1}{2}\cdot 2 + \tfrac{1}{4} = 3.75$$

25. $f(x) = \dfrac{x}{1+x^5}$, $a = 2$, $b = 6$, and $\Delta x = \dfrac{6-2}{n} = \dfrac{4}{n}$. Using Theorem 4, we get $x_i^* = x_i = 2 + i\,\Delta x = 2 + \dfrac{4i}{n}$,

so $$\int_2^6 \frac{x}{1+x^5}\,dx = \lim_{n\to\infty} R_n = \lim_{n\to\infty} \sum_{i=1}^n \frac{2+\dfrac{4i}{n}}{1+\left(2+\dfrac{4i}{n}\right)^5}\cdot\frac{4}{n}.$$

27. (a) Think of $\int_0^2 f(x)\,dx$ as the area of a trapezoid with bases 1 and 3 and height 2. The area of a trapezoid is $A = \frac{1}{2}(b+B)h$, so $\int_0^2 f(x)\,dx = \frac{1}{2}(1+3)2 = 4$.

(b) $\int_0^5 f(x)\,dx = \int_0^2 f(x)\,dx + \int_2^3 f(x)\,dx + \int_3^5 f(x)\,dx$

trapezoid rectangle triangle

$= \frac{1}{2}(1+3)2 + 3\cdot 1 + \frac{1}{2}\cdot 2\cdot 3 = 4 + 3 + 3 = 10$

(c) $\int_5^7 f(x)\,dx$ is the negative of the area of the triangle with base 2 and height 3. $\int_5^7 f(x)\,dx = -\frac{1}{2}\cdot 2\cdot 3 = -3$.

(d) $\int_7^9 f(x)\,dx$ is the negative of the area of a trapezoid with bases 3 and 2 and height 2, so it equals $-\frac{1}{2}(B+b)h = -\frac{1}{2}(3+2)2 = -5$. Thus,

$\int_0^9 f(x)\,dx = \int_0^5 f(x)\,dx + \int_5^7 f(x)\,dx + \int_7^9 f(x)\,dx = 10 + (-3) + (-5) = 2$.

© 2016 Cengage Learning. All Rights Reserved. May not be scanned, copied or duplicated, or posted to a publicly accessible website, in whole or in part.

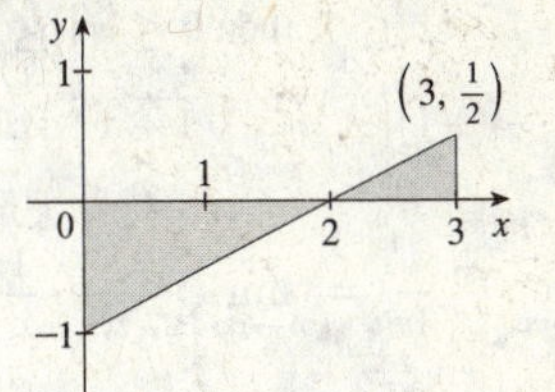

29. $\int_0^3 \left(\frac{1}{2}x - 1\right) dx$ can be interpreted as the area of the triangle above the x-axis minus the area of the triangle below the x-axis; that is,

$\frac{1}{2}(1)\left(\frac{1}{2}\right) - \frac{1}{2}(2)(1) = \frac{1}{4} - 1 = -\frac{3}{4}$.

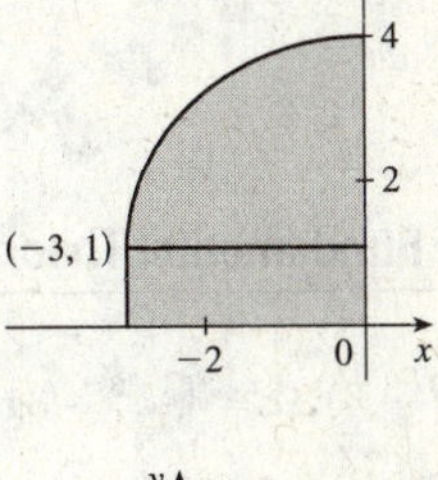

31. $\int_{-3}^{0} \left(1 + \sqrt{9 - x^2}\right) dx$ can be interpreted as the area under the graph of $f(x) = 1 + \sqrt{9 - x^2}$ between $x = -3$ and $x = 0$. This is equal to one-quarter the area of the circle with radius 3, plus the area of the rectangle, so

$\int_{-3}^{0} \left(1 + \sqrt{9 - x^2}\right) dx = \frac{1}{4}\pi \cdot 3^2 + 1 \cdot 3 = 3 + \frac{9}{4}\pi$.

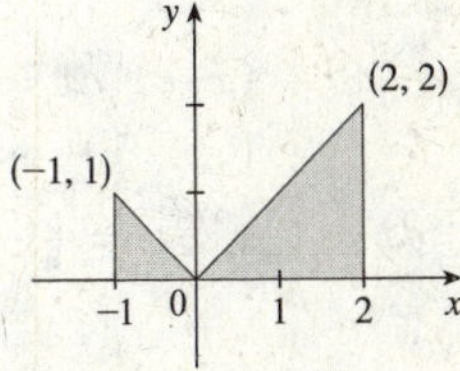

33. $\int_{-1}^{2} |x| \, dx$ can be interpreted as the sum of the areas of the two shaded triangles; that is, $\frac{1}{2}(1)(1) + \frac{1}{2}(2)(2) = \frac{1}{2} + \frac{4}{2} = \frac{5}{2}$.

35. $\int_{\pi}^{\pi} \sin^2 x \cos^4 x \, dx = 0$ since the limits of intergration are equal.

37. $\int_{-2}^{2} f(x)\,dx + \int_{2}^{5} f(x)\,dx - \int_{-2}^{-1} f(x)\,dx = \int_{-2}^{5} f(x)\,dx + \int_{-1}^{-2} f(x)\,dx$ [by Property 5 and reversing limits]

$= \int_{-1}^{5} f(x)\,dx$ [Property 5]

39. $\int_0^9 [2f(x) + 3g(x)]\,dx = 2\int_0^9 f(x)\,dx + 3\int_0^9 g(x)\,dx = 2(37) + 3(16) = 122$

41. $\int_0^3 f(x)\,dx$ is clearly less than -1 and has the smallest value. The slope of the tangent line of f at $x = 1$, $f'(1)$, has a value between -1 and 0, so it has the next smallest value. The largest value is $\int_3^8 f(x)\,dx$, followed by $\int_4^8 f(x)\,dx$, which has a value about 1 unit less than $\int_3^8 f(x)\,dx$. Still positive, but with a smaller value than $\int_4^8 f(x)\,dx$, is $\int_0^8 f(x)\,dx$. Ordering these quantities from smallest to largest gives us

$$\int_0^3 f(x)\,dx < f'(1) < \int_0^8 f(x)\,dx < \int_4^8 f(x)\,dx < \int_3^8 f(x)\,dx \text{ or } B < E < A < D < C$$

43. $I = \int_{-4}^{2} [f(x) + 2x + 5]\,dx = \int_{-4}^{2} f(x)\,dx + 2\int_{-4}^{2} x\,dx + \int_{-4}^{2} 5\,dx = I_1 + 2I_2 + I_3$

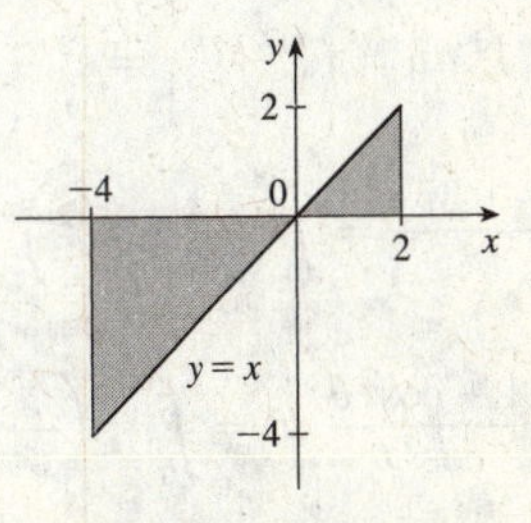

$I_1 = -3$ [area below x-axis] $+ 3 - 3 = -3$

$I_2 = -\frac{1}{2}(4)(4)$ [area of triangle, see figure] $+ \frac{1}{2}(2)(2)$

$= -8 + 2 = -6$

$I_3 = 5[2 - (-4)] = 5(6) = 30$

Thus, $I = -3 + 2(-6) + 30 = 15$.

© 2016 Cengage Learning. All Rights Reserved. May not be scanned, copied or duplicated, or posted to a publicly accessible website, in whole or in part.

45. If $-1 \le x \le 1$, then $0 \le x^2 \le 1$ and $1 \le 1 + x^2 \le 2$, so $1 \le \sqrt{1+x^2} \le \sqrt{2}$ and

$1[1-(-1)] \le \int_{-1}^{1} \sqrt{1+x^2}\, dx \le \sqrt{2}\,[1-(-1)]$ [Property 8]; that is, $2 \le \int_{-1}^{1} \sqrt{1+x^2}\, dx \le 2\sqrt{2}$.

47. $\lim\limits_{n\to\infty} \sum\limits_{i=1}^{n} \dfrac{i^4}{n^5} = \lim\limits_{n\to\infty} \sum\limits_{i=1}^{n} \dfrac{i^4}{n^4} \cdot \dfrac{1}{n} = \lim\limits_{n\to\infty} \sum\limits_{i=1}^{n} \left(\dfrac{i}{n}\right)^4 \dfrac{1}{n}$. At this point, we need to recognize the limit as being of the form

$\lim\limits_{n\to\infty} \sum\limits_{i=1}^{n} f(x_i)\, \Delta x$, where $\Delta x = (1-0)/n = 1/n$, $x_i = 0 + i\,\Delta x = i/n$, and $f(x) = x^4$. Thus, the definite integral is $\int_0^1 x^4\, dx$.

5.3 The Fundamental Theorem of Calculus

1. $\int_{-2}^{3} (x^2 - 3)\, dx = \left[\frac{1}{3}x^3 - 3x\right]_{-2}^{3} = (9-9) - \left(-\frac{8}{3} + 6\right) = \frac{8}{3} - \frac{18}{3} = -\frac{10}{3}$

3. $\int_0^2 \left(x^4 - \frac{3}{4}x^2 + \frac{2}{3}x - 1\right) dx = \left[\frac{1}{5}x^5 - \frac{1}{4}x^3 + \frac{1}{3}x^2 - x\right]_0^2 = \left(\frac{32}{5} - 2 + \frac{4}{3} - 2\right) - 0 = \frac{96 - 30 + 20 - 30}{15} = \frac{56}{15}$

5. $\int_0^1 x^{4/5}\, dx = \left[\frac{5}{9}x^{9/5}\right]_0^1 = \frac{5}{9} - 0 = \frac{5}{9}$

7. $\int_{-1}^{0} (2x - e^x)\, dx = \left[x^2 - e^x\right]_{-1}^{0} = (0-1) - (1 - e^{-1}) = -2 + 1/e$

9. $\int_1^2 (1+2y)^2\, dy = \int_1^2 (1 + 4y + 4y^2)\, dy = \left[y + 2y^2 + \frac{4}{3}y^3\right]_1^2 = \left(2 + 8 + \frac{32}{3}\right) - \left(1 + 2 + \frac{4}{3}\right) = \frac{62}{3} - \frac{13}{3} = \frac{49}{3}$

11.
$$\int_1^9 \frac{x-1}{\sqrt{x}}\, dx = \int_1^9 \left(\frac{x}{\sqrt{x}} - \frac{1}{\sqrt{x}}\right) dx = \int_1^9 (x^{1/2} - x^{-1/2})\, dx = \left[\tfrac{2}{3}x^{3/2} - 2x^{1/2}\right]_1^9$$
$$= \left(\tfrac{2}{3}\cdot 27 - 2\cdot 3\right) - \left(\tfrac{2}{3} - 2\right) = 12 - \left(-\tfrac{4}{3}\right) = \tfrac{40}{3}$$

13. $\int_0^1 x\left(\sqrt[3]{x} + \sqrt[4]{x}\right) dx = \int_0^1 (x^{4/3} + x^{5/4})\, dx = \left[\frac{3}{7}x^{7/3} + \frac{4}{9}x^{9/4}\right]_0^1 = \left(\frac{3}{7} + \frac{4}{9}\right) - 0 = \frac{55}{63}$

15. $\int_0^{\pi/4} \sec^2 t\, dt = \left[\tan t\right]_0^{\pi/4} = \tan\frac{\pi}{4} - \tan 0 = 1 - 0 = 1$

17. $\displaystyle\int_1^9 \frac{1}{2x}\, dx = \frac{1}{2}\int_1^9 \frac{1}{x}\, dx = \tfrac{1}{2}\Big[\ln|x|\Big]_1^9 = \tfrac{1}{2}(\ln 9 - \ln 1) = \tfrac{1}{2}\ln 9 - 0 = \ln 9^{1/2} = \ln 3$

19. $\displaystyle\int_0^1 (x^e + e^x)\, dx = \left[\frac{x^{e+1}}{e+1} + e^x\right]_0^1 = \left(\frac{1}{e+1} + e\right) - (0+1) = \frac{1}{e+1} + e - 1$

21. $\int_{-1}^{1} e^{u+1}\, du = \left[e^{u+1}\right]_{-1}^{1} = e^2 - e^0 = e^2 - 1$ [or start with $e^{u+1} = e^u e^1$]

23. $\displaystyle\int_1^2 \frac{v^3 + 3v^6}{v^4} = \int_1^2 \left(\frac{1}{v} + 3v^2\right) dv = \left[\ln|v| + v^3\right]_1^2 = (\ln 2 + 8) - (\ln 1 + 1) = \ln 2 + 7$

25.
$$\int_0^{\pi/4} \frac{1 + \cos^2\theta}{\cos^2\theta}\, d\theta = \int_0^{\pi/4} \left(\frac{1}{\cos^2\theta} + \frac{\cos^2\theta}{\cos^2\theta}\right) d\theta = \int_0^{\pi/4} (\sec^2\theta + 1)\, d\theta$$
$$= \left[\tan\theta + \theta\right]_0^{\pi/4} = \left(\tan\tfrac{\pi}{4} + \tfrac{\pi}{4}\right) - (0 + 0) = 1 + \tfrac{\pi}{4}$$

© 2016 Cengage Learning. All Rights Reserved. May not be scanned, copied or duplicated, or posted to a publicly accessible website, in whole or in part.

27. $\displaystyle\int_0^{1/\sqrt{3}} \frac{t^2-1}{t^4-1}\,dt = \int_0^{1/\sqrt{3}} \frac{t^2-1}{(t^2+1)(t^2-1)}\,dt = \int_0^{1/\sqrt{3}} \frac{1}{t^2+1}\,dt = \left[\arctan t\right]_0^{1/\sqrt{3}} = \arctan(1/\sqrt{3}) - \arctan 0$

$= \frac{\pi}{6} - 0 = \frac{\pi}{6}$

29. $f(x) = 1/x^2$ is not continuous on the interval $[-1, 3]$, so the Evaluation Theorem does not apply. In fact, f has an infinite discontinuity at $x = 0$, so $\int_{-1}^{3} (1/x^2)\,dx$ does not exist.

31. It appears that the area under the graph is about $\frac{2}{3}$ of the area of the viewing rectangle, or about $\frac{2}{3}\pi \approx 2.1$. The actual area is

$\int_0^{\pi} \sin x\,dx = [-\cos x]_0^{\pi} = (-\cos\pi) - (-\cos 0) = -(-1) + 1 = 2.$

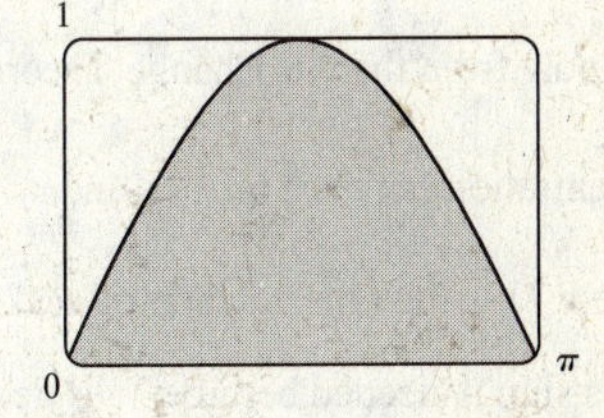

33. $\int_{-1}^{2} x^3\,dx = \left[\frac{1}{4}x^4\right]_{-1}^{2} = 4 - \frac{1}{4} = \frac{15}{4} = 3.75$

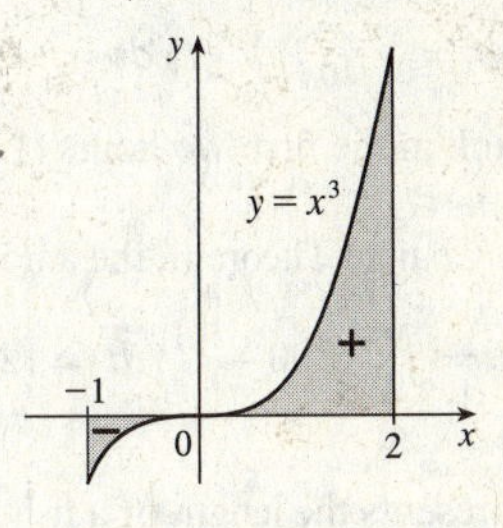

35. $\displaystyle\frac{d}{dx}\left[\sin x - \tfrac{1}{3}\sin^3 x + C\right] = \frac{d}{dx}\left[\sin x - \tfrac{1}{3}(\sin x)^3 + C\right] = \cos x - \tfrac{1}{3}\cdot 3(\sin x)^2(\cos x) + 0$

$= \cos x(1 - \sin^2 x) = \cos x(\cos^2 x) = \cos^3 x$

37. $\int\left(\cos x + \frac{1}{2}x\right)dx = \sin x + \frac{1}{4}x^2 + C$. The members of the family in the figure correspond to $C = -5$, 0, 5, and 10.

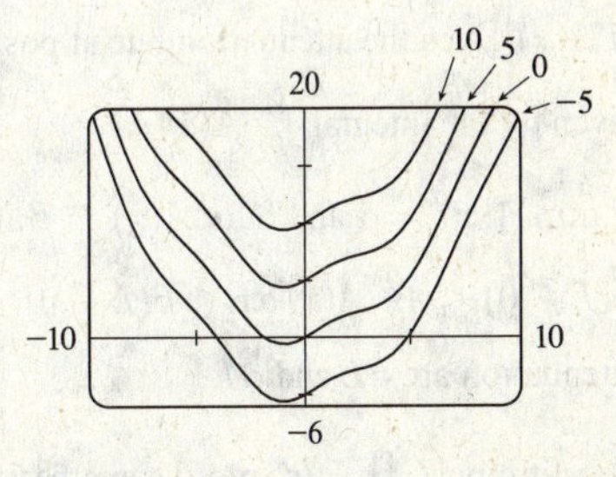

39. $\displaystyle\int (1-t)(2+t^2)\,dt = \int(2 - 2t + t^2 - t^3)\,dt = 2t - 2\frac{t^2}{2} + \frac{t^3}{3} - \frac{t^4}{4} + C = 2t - t^2 + \tfrac{1}{3}t^3 - \tfrac{1}{4}t^4 + C$

41. $\int(1 + \tan^2\alpha)\,d\alpha = \int\sec^2\alpha\,d\alpha = \tan\alpha + C$

43. $\displaystyle\int \frac{\sin x}{1-\sin^2 x}\,dx = \int\frac{\sin x}{\cos^2 x}\,dx = \int \frac{1}{\cos x}\cdot\frac{\sin x}{\cos x}\,dx = \int \sec x\,\tan x\,dx = \sec x + C$

45. $f(t) = -t(t-21)(t+1) = -t^3 + 20t^2 + 21t \quad\Rightarrow$

$$\int_0^{12} f(t)\,dt = \int_0^{12}\left(-t^3 + 20t^2 + 21t\right)dt = \left[-\tfrac{1}{4}t^4 + \tfrac{20}{3}t^3 + \tfrac{21}{2}t^2\right]_0^{12}$$

$$= \left[-\tfrac{1}{4}(12)^4 + \tfrac{20}{3}(12)^3 + \tfrac{21}{2}(12)^2\right] - [0 + 0 + 0] = 7{,}848$$

Hence, the threshold amount of infection is 7,848 (number of cells per mL)·days.

© 2016 Cengage Learning. All Rights Reserved. May not be scanned, copied or duplicated, or posted to a publicly accessible website, in whole or in part.

47. If $w'(t)$ is the rate of change of weight in pounds per year, then $w(t)$ represents the weight in pounds of the child at age t. We know from the Net Change Theorem that $\int_5^{10} w'(t)\,dt = w(10) - w(5)$, so the integral represents the increase in the child's weight (in pounds) between the ages of 5 and 10.

49. If $f(x)$ gives the density of sea urchins at point x along a coastline, then $\int_a^b f(x)\,dx$ represents the total number of sea urchins between points a and b.

51. We know from the Net Change Theorem that $\displaystyle\int_{t_1}^{t_2} \frac{d[\mathrm{C}]}{dt}\,dt = [\mathrm{C}]\,(t_2) - [\mathrm{C}]\,(t_1)$, so the integral represents the change in concentration between time t_1 and t_2.

53. Since $r(t)$ is the rate at which oil leaks, we can write $r(t) = -V'(t)$, where $V(t)$ is the volume of oil at time t. [Note that the minus sign is needed because V is decreasing, so $V'(t)$ is negative, but $r(t)$ is positive.] Thus, by the Net Change Theorem, $\int_0^{120} r(t)\,dt = -\int_0^{120} V'(t)\,dt = -[V(120) - V(0)] = V(0) - V(120)$, which is the number of gallons of oil that leaked from the tank in the first two hours (120 minutes).

55. By the Net Change Theorem, the amount of water that flows from the tank during the first 10 minutes is $\int_0^{10} r(t)\,dt = \int_0^{10} (200 - 4t)\,dt = \left[200t - 2t^2\right]_0^{10} = (2000 - 200) - 0 = 1800$ liters.

57. If $L(a)$ represents the length of a fish length at age a, then the growth rate is $\dfrac{dL}{da} = 29e^{-a}$. So the increase in length after the first five years is $\int_0^5 \left(29e^{-a}\right)\,da = \left[-29e^{-a}\right]_0^5 = -29\left[e^{-5} - e^0\right] = 28.8046$, and since the initial length is 1 cm, the fish length after five years is approximately $1 + 28.8 = 29.8$ cm.

59. (a) If $A(x)$ gives the attenuation rate at position x in the tissue, the total attenuation along the path from $x = 0$ to $x = L$ is given by the integral $\int_0^L A(x)\,dx$.

(b) If $\min A(x) = \alpha$ and $\max A(x) = \beta$, then $\alpha \le A(x) \le \beta$ and by Property 5.2.8 we have $\alpha(L - 0) \le \int_0^L A(x)\,dx \le \beta(L - 0) \quad \Rightarrow \quad \alpha L \le$ Total Attenuation $\le \beta L$. So the upper and lower bounds on total attenuation are βL and αL.

61. (a) Substituting $I(t) = kt$ into the expression for $P(I)$ gives $P(t) = P_{\max}\left(1 - e^{-akt}\right)$

(b) Since $P(t)$ is the rate of primary production, the total amount of primary production over the first five units of time is:

$$\int_0^5 P(t)\,dt = P_{\max}\int_0^5 (1 - e^{-akt})\,dt = P_{\max}\left[t + \frac{1}{ak}e^{-akt}\right]_0^5$$

$$= P_{\max}\left[\left(5 + \frac{1}{ak}e^{-ak(5)}\right) - \left(0 + \frac{1}{ak}e^{-ak(0)}\right)\right] = P_{\max}\left[5 + \frac{1}{ak}e^{-5ka} - \frac{1}{ak}\right]$$

(c) Since $P(t)$ is the rate of primary production, the total amount of primary production over the first t units of time is:

$$\int_0^t P(s)\,ds = P_{\max}\int_0^t (1 - e^{-aks})\,ds = P_{\max}\left[s + \frac{1}{ak}e^{-aks}\right]_0^t$$

$$= P_{\max}\left[\left(t + \frac{1}{ak}e^{-akt}\right) - \left(0 + \frac{1}{ak}e^{-ak(0)}\right)\right] = P_{\max}\left[t + \frac{1}{ak}e^{-akt} - \frac{1}{ak}\right]$$

(d) By FTC1, the rate of change of total primary production at time t is $\dfrac{d}{dt}\displaystyle\int_0^t P(s)\,ds = P(t) = P_{\max}\left(1 - e^{-akt}\right)$.

© 2016 Cengage Learning. All Rights Reserved. May not be scanned, copied or duplicated, or posted to a publicly accessible website, in whole or in part.

63. (a) Let $f(x) = \sqrt{x} \;\Rightarrow\; f'(x) = 1/(2\sqrt{x}) > 0$ for $x > 0 \;\Rightarrow\; f$ is increasing on $(0, \infty)$. If $x \geq 0$, then $x^3 \geq 0$, so $1 + x^3 \geq 1$ and since f is increasing, this means that $f(1 + x^3) \geq f(1) \;\Rightarrow\; \sqrt{1 + x^3} \geq 1$ for $x \geq 0$. Next let $g(t) = t^2 - t \;\Rightarrow\; g'(t) = 2t - 1 \;\Rightarrow\; g'(t) > 0$ when $t \geq 1$. Thus, g is increasing on $(1, \infty)$. And since $g(1) = 0$, $g(t) \geq 0$ when $t \geq 1$. Now let $t = \sqrt{1 + x^3}$, where $x \geq 0$. $\sqrt{1 + x^3} \geq 1$ (from above) $\;\Rightarrow\; t \geq 1 \;\Rightarrow\; g(t) \geq 0 \;\Rightarrow\; (1 + x^3) - \sqrt{1 + x^3} \geq 0$ for $x \geq 0$. Therefore, $1 \leq \sqrt{1 + x^3} \leq 1 + x^3$ for $x \geq 0$.

(b) From part (a) and Property 7: $\int_0^1 1\,dx \leq \int_0^1 \sqrt{1 + x^3}\,dx \leq \int_0^1 (1 + x^3)\,dx \;\Leftrightarrow$

$\left[x\right]_0^1 \leq \int_0^1 \sqrt{1 + x^3}\,dx \leq \left[x + \frac{1}{4}x^4\right]_0^1 \;\Leftrightarrow\; 1 \leq \int_0^1 \sqrt{1 + x^3}\,dx \leq 1 + \frac{1}{4} = 1.25.$

65. (a) $g(x) = \int_0^x f(t)\,dt.$

$g(0) = \int_0^0 f(t)\,dt = 0$

$g(1) = \int_0^1 f(t)\,dt = 1 \cdot 2 = 2$ [rectangle],

$g(2) = \int_0^2 f(t)\,dt = \int_0^1 f(t)\,dt + \int_1^2 f(t)\,dt = g(1) + \int_1^2 f(t)\,dt$

$= 2 + 1 \cdot 2 + \frac{1}{2} \cdot 1 \cdot 2 = 5$ [rectangle plus triangle],

$g(3) = \int_0^3 f(t)\,dt = g(2) + \int_2^3 f(t)\,dt = 5 + \frac{1}{2} \cdot 1 \cdot 4 = 7,$

$g(6) = g(3) + \int_3^6 f(t)\,dt$ [the integral is negative since f lies under the x-axis]

$= 7 + \left[-\left(\frac{1}{2} \cdot 2 \cdot 2 + 1 \cdot 2\right)\right] = 7 - 4 = 3$

(d)

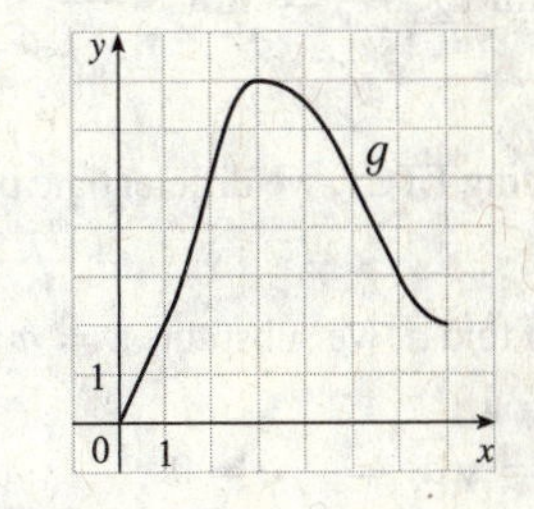

(b) g is increasing on $(0, 3)$ because as x increases from 0 to 3, we keep adding more area.

(c) g has a maximum value when we start subtracting area; that is, at $x = 3$.

67. $y = 1 + t^2$, $g(x)$

(a) By FTC1, $g(x) = \int_0^x (1 + t^2)\,dt \;\Rightarrow\; g'(x) = f(x) = 1 + x^2.$

(b) By FTC2, $g(x) = \int_0^x (1 + t^2)\,dt = \left[t + \frac{1}{3}t^3\right]_0^x = \left(x + \frac{1}{3}x^3\right) - 0 \;\Rightarrow\; g'(x) = 1 + x^2.$

69. $f(t) = \dfrac{1}{t^3 + 1}$ and $g(x) = \displaystyle\int_1^x \frac{1}{t^3 + 1}\,dt$, so by FTC1, $g'(x) = f(x) = \dfrac{1}{x^3 + 1}$. Note that the lower limit, 1, could be any real number greater than -1 and not affect this answer.

71. $f(t) = t^2 \sin t$ and $g(y) = \int_2^y t^2 \sin t\,dt$, so by FTC1, $g'(y) = f(y) = y^2 \sin y$.

73. $F(x) = \displaystyle\int_x^0 \sqrt{1 + \sec t}\,dt = -\int_0^x \sqrt{1 + \sec t}\,dt \;\Rightarrow\; F'(x) = -\frac{d}{dx}\int_0^x \sqrt{1 + \sec t}\,dt = -\sqrt{1 + \sec x}$

75. Let $u = \dfrac{1}{x}$. Then $\dfrac{du}{dx} = -\dfrac{1}{x^2}$. Also, $\dfrac{dh}{dx} = \dfrac{dh}{du}\dfrac{du}{dx}$, so

$$h'(x) = \frac{d}{dx}\int_2^{1/x} \arctan t\,dt = \frac{d}{du}\int_2^u \arctan t\,dt \cdot \frac{du}{dx} = \arctan u \,\frac{du}{dx} = -\frac{\arctan(1/x)}{x^2}.$$

© 2016 Cengage Learning. All Rights Reserved. May not be scanned, copied or duplicated, or posted to a publicly accessible website, in whole or in part.

77. Let $u = \tan x$. Then $\dfrac{du}{dx} = \sec^2 x$. Also, $\dfrac{dy}{dx} = \dfrac{dy}{du}\dfrac{du}{dx}$, so

$$y' = \frac{d}{dx}\int_0^{\tan x} \sqrt{t+\sqrt{t}}\,dt = \frac{d}{du}\int_0^{u} \sqrt{t+\sqrt{t}}\,dt \cdot \frac{du}{dx} = \sqrt{u+\sqrt{u}}\,\frac{du}{dx} = \sqrt{\tan x + \sqrt{\tan x}}\,\sec^2 x.$$

79. By FTC2, $\int_1^4 f'(x)\,dx = f(4) - f(1)$, so $17 = f(4) - 12 \quad\Rightarrow\quad f(4) = 17 + 12 = 29.$

81. The second derivative is the derivative of the first derivative, so we'll apply the Net Change Theorem with $F = h'$.

$\int_1^2 h''(u)\,du = \int_1^2 (h')'(u)\,du = h'(2) - h'(1) = 5 - 2 = 3$. The other information is unnecessary.

83. $\displaystyle \lim_{n\to\infty} \sum_{i=1}^{n} \frac{i^3}{n^4} = \lim_{n\to\infty} \frac{1-0}{n} \sum_{i=1}^{n} \left(\frac{i}{n}\right)^3 = \int_0^1 x^3\,dx = \left[\frac{x^4}{4}\right]_0^1 = \frac{1}{4}$

85. Using FTC1, we differentiate both sides of $6 + \displaystyle\int_a^x \frac{f(t)}{t^2}\,dt = 2\sqrt{x}$ to get $\dfrac{f(x)}{x^2} = 2\dfrac{1}{2\sqrt{x}} \quad\Rightarrow\quad f(x) = x^{3/2}$.

To find a, we substitute $x = a$ in the original equation to obtain $6 + \displaystyle\int_a^a \frac{f(t)}{t^2}\,dt = 2\sqrt{a} \quad\Rightarrow\quad 6 + 0 = 2\sqrt{a} \quad\Rightarrow$

$3 = \sqrt{a} \quad\Rightarrow\quad a = 9.$

5.4 The Substitution Rule

1. Let $u = -x$. Then $du = -\,dx$, so $dx = -\,du$. Thus, $\int e^{-x}dx = \int e^u(-du) = -e^u + C = -e^{-x} + C$. Don't forget that it is often very easy to check an indefinite integration by differentiating your answer. In this case,

$\dfrac{d}{dx}(-e^{-x} + C) = -[e^{-x}(-1)] = e^{-x}$, the desired result.

3. Let $u = x^3 + 1$. Then $du = 3x^2\,dx$ and $x^2\,dx = \frac{1}{3}\,du$, so

$$\int x^2\sqrt{x^3+1}\,dx = \int \sqrt{u}\left(\tfrac{1}{3}\,du\right) = \frac{1}{3}\frac{u^{3/2}}{3/2} + C = \frac{1}{3}\cdot\frac{2}{3}u^{3/2} + C = \tfrac{2}{9}(x^3+1)^{3/2} + C.$$

5. Let $u = \cos\theta$. Then $du = -\sin\theta\,d\theta$ and $\sin\theta\,d\theta = -du$, so

$$\int \cos^3\theta\,\sin\theta\,d\theta = \int u^3\,(-du) = -\frac{u^4}{4} + C = -\tfrac{1}{4}\cos^4\theta + C.$$

7. Let $u = x^2$. Then $du = 2x\,dx$ and $x\,dx = \frac{1}{2}\,du$, so $\int x\sin(x^2)\,dx = \int \sin u\left(\frac{1}{2}\,du\right) = -\frac{1}{2}\cos u + C = -\frac{1}{2}\cos(x^2) + C$.

9. Let $u = 3x - 2$. Then $du = 3\,dx$ and $dx = \frac{1}{3}\,du$, so $\int (3x-2)^{20}\,dx = \int u^{20}\left(\frac{1}{3}\,du\right) = \frac{1}{3}\cdot\frac{1}{21}u^{21} + C = \frac{1}{63}(3x-2)^{21} + C$.

11. Let $u = \pi t$. Then $du = \pi\,dt$ and $dt = \frac{1}{\pi}\,du$, so $\int \sin\pi t\,dt = \int \sin u\left(\frac{1}{\pi}\,du\right) = \frac{1}{\pi}(-\cos u) + C = -\frac{1}{\pi}\cos\pi t + C$.

13. Let $u = \ln x$. Then $du = \dfrac{dx}{x}$, so $\displaystyle\int \frac{(\ln x)^2}{x}\,dx = \int u^2\,du = \tfrac{1}{3}u^3 + C = \tfrac{1}{3}(\ln x)^3 + C$.

15. Let $u = 5 - 3x$. Then $du = -3\,dx$ and $dx = -\frac{1}{3}\,du$, so

$$\int \frac{dx}{5-3x} = \int \frac{1}{u}\left(-\tfrac{1}{3}\,du\right) = -\tfrac{1}{3}\ln|u| + C = -\tfrac{1}{3}\ln|5-3x| + C.$$

© 2016 Cengage Learning. All Rights Reserved. May not be scanned, copied or duplicated, or posted to a publicly accessible website, in whole or in part.

17. Let $u = 3ax + bx^3$. Then $du = (3a + 3bx^2)\,dx = 3(a + bx^2)\,dx$, so

$$\int \frac{a + bx^2}{\sqrt{3ax + bx^3}}\,dx = \int \frac{\frac{1}{3}\,du}{u^{1/2}} = \frac{1}{3}\int u^{-1/2}\,du = \tfrac{1}{3} \cdot 2u^{1/2} + C = \tfrac{2}{3}\sqrt{3ax + bx^3} + C.$$

19. Let $u = 1 + e^x$. Then $du = e^x\,dx$, so $\int e^x\sqrt{1 + e^x}\,dx = \int \sqrt{u}\,du = \frac{2}{3}u^{3/2} + C = \frac{2}{3}(1 + e^x)^{3/2} + C$.

Or: Let $u = \sqrt{1 + e^x}$. Then $u^2 = 1 + e^x$ and $2u\,du = e^x\,dx$, so

$\int e^x\sqrt{1 + e^x}\,dx = \int u \cdot 2u\,du = \frac{2}{3}u^3 + C = \frac{2}{3}(1 + e^x)^{3/2} + C$.

21. Let $u = \sin x$. Then $du = \cos x\,dx$, so $\displaystyle\int \frac{\cos x}{\sin^2 x}\,dx = \int \frac{1}{u^2}\,du = \int u^{-2}\,du = \frac{u^{-1}}{-1} + C = -\frac{1}{u} + C = -\frac{1}{\sin x} + C$

[or $-\csc x + C$].

23. Let $u = x^3 + 3x$. Then $du = (3x^2 + 3)\,dx$ and $\frac{1}{3}\,du = (x^2 + 1)\,dx$, so

$\int (x^2 + 1)(x^3 + 3x)^4\,dx = \int u^4\left(\frac{1}{3}\,du\right) = \frac{1}{3} \cdot \frac{1}{5}u^5 + C = \frac{1}{15}(x^3 + 3x)^5 + C$.

25. Let $u = \cot x$. Then $du = -\csc^2 x\,dx$ and $\csc^2 x\,dx = -du$, so

$$\int \sqrt{\cot x}\,\csc^2 x\,dx = \int \sqrt{u}\,(-du) = -\frac{u^{3/2}}{3/2} + C = -\tfrac{2}{3}(\cot x)^{3/2} + C.$$

27. Let $u = e^{2r}$. Then $du = 2e^{2r}\,dr$ and $e^{2r}\,dr = \frac{1}{2}\,du$, so

$$\int e^{2r}\sin(e^{2r})\,dr = \int \sin u\left(\tfrac{1}{2}\,du\right) = -\tfrac{1}{2}\cos u + C = -\tfrac{1}{2}\cos\left(e^{2r}\right) + C.$$

29. Let $u = \sec x$. Then $du = \sec x\,\tan x\,dx$, so

$\int \sec^3 x\,\tan x\,dx = \int \sec^2 x\,(\sec x\,\tan x)\,dx = \int u^2\,du = \frac{1}{3}u^3 + C = \frac{1}{3}\sec^3 x + C$.

31. Let $u = 2x + 5$. Then $du = 2\,dx$ and $x = \frac{1}{2}(u - 5)$, so

$$\begin{aligned}\int x(2x + 5)^8\,dx &= \int \tfrac{1}{2}(u - 5)u^8\left(\tfrac{1}{2}\,du\right) = \tfrac{1}{4}\int (u^9 - 5u^8)\,du \\ &= \tfrac{1}{4}\left(\tfrac{1}{10}u^{10} - \tfrac{5}{9}u^9\right) + C = \tfrac{1}{40}(2x + 5)^{10} - \tfrac{5}{36}(2x + 5)^9 + C\end{aligned}$$

33. $\displaystyle\int \frac{\sin 2x}{1 + \cos^2 x}\,dx = 2\int \frac{\sin x\cos x}{1 + \cos^2 x}\,dx = 2I$. Let $u = \cos x$. Then $du = -\sin x\,dx$, so

$\displaystyle 2I = -2\int \frac{u\,du}{1 + u^2} = -2 \cdot \tfrac{1}{2}\ln(1 + u^2) + C = -\ln(1 + u^2) + C = -\ln(1 + \cos^2 x) + C.$

Or: Let $u = 1 + \cos^2 x$.

35. Let $u = 1 + x^2$. Then $du = 2x\,dx$, so

$$\begin{aligned}\int \frac{1 + x}{1 + x^2}\,dx &= \int \frac{1}{1 + x^2}\,dx + \int \frac{x}{1 + x^2}\,dx = \tan^{-1} x + \int \frac{\frac{1}{2}\,du}{u} = \tan^{-1} x + \tfrac{1}{2}\ln|u| + C \\ &= \tan^{-1} x + \tfrac{1}{2}\ln\left|1 + x^2\right| + C = \tan^{-1} x + \tfrac{1}{2}\ln\left(1 + x^2\right) + C \quad [\text{since } 1 + x^2 > 0].\end{aligned}$$

37. Let $u = \frac{\pi}{2}t$, so $du = \frac{\pi}{2}\,dt$. When $t = 0$, $u = 0$; when $t = 1$, $u = \frac{\pi}{2}$. Thus,

$$\int_0^1 \cos(\pi t/2)\,dt = \int_0^{\pi/2} \cos u\left(\tfrac{2}{\pi}\,du\right) = \tfrac{2}{\pi}\left[\sin u\right]_0^{\pi/2} = \tfrac{2}{\pi}\left(\sin\tfrac{\pi}{2} - \sin 0\right) = \tfrac{2}{\pi}(1 - 0) = \tfrac{2}{\pi}$$

© 2016 Cengage Learning. All Rights Reserved. May not be scanned, copied or duplicated, or posted to a publicly accessible website, in whole or in part.

39. Let $u = 1 + 7x$, so $du = 7\,dx$. When $x = 0$, $u = 1$; when $x = 1$, $u = 8$. Thus,

$$\int_0^1 \sqrt[3]{1+7x}\,dx = \int_1^8 u^{1/3}\left(\tfrac{1}{7}\,du\right) = \tfrac{1}{7}\left[\tfrac{3}{4}u^{4/3}\right]_1^8 = \tfrac{3}{28}\left(8^{4/3} - 1^{4/3}\right) = \tfrac{3}{28}(16-1) = \tfrac{45}{28}$$

41. Let $u = 1 + 2x^3$, so $du = 6x^2\,dx$. When $x = 0$, $u = 1$; when $x = 1$, $u = 3$. Thus,

$\int_0^1 x^2\left(1+2x^3\right)^5\,dx = \int_1^3 u^5\left(\tfrac{1}{6}\,du\right) = \tfrac{1}{6}\left[\tfrac{1}{6}u^6\right]_1^3 = \tfrac{1}{36}\left(3^6 - 1^6\right) = \tfrac{1}{36}(729-1) = \tfrac{728}{36} = \tfrac{182}{9}$.

43. Let $u = \sqrt{x}$, so $du = \dfrac{1}{2\sqrt{x}}\,dx$. When $x = 1$, $u = 1$; when $x = 4$, $u = 2$.

Thus, $\displaystyle\int_1^4 \frac{e^{\sqrt{x}}}{\sqrt{x}}\,dx = \int_1^2 e^u(2\,du) = 2\left[e^u\right]_1^2 = 2(e^2 - e)$.

45. $\int_{-\pi/4}^{\pi/4}(x^3 + x^4\tan x)\,dx = 0$ by Theorem 6(b), since $f(x) = x^3 + x^4\tan x$ is an odd function.

47. Let $u = x - 1$, so $u + 1 = x$ and $du = dx$. When $x = 1$, $u = 0$; when $x = 2$, $u = 1$. Thus,

$$\int_1^2 x\sqrt{x-1}\,dx = \int_0^1 (u+1)\sqrt{u}\,du = \int_0^1 \left(u^{3/2} + u^{1/2}\right)du = \left[\tfrac{2}{5}u^{5/2} + \tfrac{2}{3}u^{3/2}\right]_0^1 = \tfrac{2}{5} + \tfrac{2}{3} = \tfrac{16}{15}.$$

49. Let $u = e^z + z$, so $du = (e^z + 1)\,dz$. When $z = 0$, $u = 1$; when $z = 1$, $u = e + 1$. Thus,

$$\int_0^1 \frac{e^z+1}{e^z+z}\,dz = \int_1^{e+1} \frac{1}{u}\,du = \Big[\ln|u|\Big]_1^{e+1} = \ln|e+1| - \ln|1| = \ln(e+1).$$

51. Let $u = 1 + \sqrt{x}$, so $du = \dfrac{1}{2\sqrt{x}}\,dx \;\Rightarrow\; 2\sqrt{x}\,du = dx \;\Rightarrow\; 2(u-1)\,du = dx$. When $x = 0$, $u = 1$; when $x = 1$, $u = 2$. Thus,

$$\int_0^1 \frac{dx}{(1+\sqrt{x})^4} = \int_1^2 \frac{1}{u^4}\cdot[2(u-1)\,du] = 2\int_1^2\left(\frac{1}{u^3} - \frac{1}{u^4}\right)du = 2\left[-\frac{1}{2u^2} + \frac{1}{3u^3}\right]_1^2$$

$$= 2\left[\left(-\tfrac{1}{8} + \tfrac{1}{24}\right) - \left(-\tfrac{1}{2} + \tfrac{1}{3}\right)\right] = 2\left(\tfrac{1}{12}\right) = \tfrac{1}{6}$$

53. First write the integral as a sum of two integrals:

$I = \int_{-2}^2 (x+3)\sqrt{4-x^2}\,dx = I_1 + I_2 = \int_{-2}^2 x\sqrt{4-x^2}\,dx + \int_{-2}^2 3\sqrt{4-x^2}\,dx$. $I_1 = 0$ by Theorem 6(b), since $f(x) = x\sqrt{4-x^2}$ is an odd function and we are integrating from $x = -2$ to $x = 2$. We interpret I_2 as three times the area of a semicircle with radius 2, so $I = 0 + 3\cdot\tfrac{1}{2}\left(\pi\cdot 2^2\right) = 6\pi$.

55. **First Figure** Let $u = \sqrt{x}$, so $x = u^2$ and $dx = 2u\,du$. When $x = 0$, $u = 0$; when $x = 1$, $u = 1$. Thus,

$A_1 = \int_0^1 e^{\sqrt{x}}\,dx = \int_0^1 e^u(2u\,du) = 2\int_0^1 ue^u\,du$.

Second Figure $A_2 = \int_0^1 2xe^x\,dx = 2\int_0^1 ue^u\,du$.

Third Figure Let $u = \sin x$, so $du = \cos x\,dx$. When $x = 0$, $u = 0$; when $x = \frac{\pi}{2}$, $u = 1$. Thus,

$A_3 = \int_0^{\pi/2} e^{\sin x}\sin 2x\,dx = \int_0^{\pi/2} e^{\sin x}(2\sin x\,\cos x)\,dx = \int_0^1 e^u(2u\,du) = 2\int_0^1 ue^u\,du$.

Since $A_1 = A_2 = A_3$, all three areas are equal.

© 2016 Cengage Learning. All Rights Reserved. May not be scanned, copied or duplicated, or posted to a publicly accessible website, in whole or in part.

57. The rate is measured in liters per minute. Integrating from $t = 0$ minutes to $t = 60$ minutes will give us the total amount of oil that leaks out (in liters) during the first hour.

$\int_0^{60} r(t)\,dt = \int_0^{60} 100e^{-0.01t}\,dt \qquad [u = -0.01t, du = -0.01dt]$

$= 100\int_0^{-0.6} e^u(-100\,du) = -10{,}000\left[e^u\right]_0^{-0.6} = -10{,}000(e^{-0.6} - 1) \approx 4511.9 \approx 4512$ liters

59. The volume of inhaled air in the lungs at time t is

$$V(t) = \int_0^t f(u)\,du = \int_0^t \frac{1}{2}\sin\left(\frac{2\pi}{5}u\right)du = \int_0^{2\pi t/5} \frac{1}{2}\sin v\left(\frac{5}{2\pi}\,dv\right) \qquad \left[\text{substitute } v = \tfrac{2\pi}{5}u,\ dv = \tfrac{2\pi}{5}\,du\right]$$

$$= \frac{5}{4\pi}\left[-\cos v\right]_0^{2\pi t/5} = \frac{5}{4\pi}\left[-\cos\left(\frac{2\pi}{5}t\right) + 1\right] = \frac{5}{4\pi}\left[1 - \cos\left(\frac{2\pi}{5}t\right)\right] \text{ liters}$$

61. Since $g(t)$ is the tumor growth rate, the increase in tumor volume in the first year (12 months) is given by:

$$\int_0^{12} g(t)\,dt = \int_0^{12} 2^{1-e^{-t}}e^{-t}\ln 2\,dt. \quad \text{Let } u = 1 - e^{-t}. \text{ Then } du = e^{-t}dt, \text{ so}$$

$$\int_0^{12} g(t)\,dt = \int_0^{1-e^{-12}} 2^u \ln 2\,du = \int_0^{1-e^{-12}} \frac{d}{du}\left(2^u\right)du = \left[2^u\right]_0^{1-e^{-12}} = 2^{1-e^{-12}} - 1 \approx 1.0\text{ mm}^3$$

63. If maturity is reached after 20 days, the number of degree days required is given by the integral:

$$\int_0^{20} T(t)\,dt = \int_0^{20} 15\left(1 + \sin\frac{2\pi t}{60}\right)dt = 15\left(\int_0^{20} dt + \int_0^{20} \sin\frac{2\pi t}{60}\,dt\right)$$

$$= 15\left(t\big]_0^{20} + \int_0^{2\pi/3} \sin u\left(\tfrac{60}{2\pi}\,du\right)\right) \qquad \left[\text{substitute } u = \tfrac{2\pi}{60}t,\ du = \tfrac{2\pi}{60}\,dt\right]$$

$$= 15\left(20 - 0 + \tfrac{60}{2\pi}\int_0^{2\pi/3} \sin u\,du\right) = 15\left(20 + \tfrac{60}{2\pi}\left[-\cos u\right]_0^{2\pi/3}\right) = 15\left(20 + \tfrac{60}{2\pi}\left[-\cos\tfrac{2\pi}{3} + \cos 0\right]\right)$$

$$= 300 + \frac{675}{\pi} \approx 515 \text{ degree days}$$

Since $T(t)$ is measured in degrees and t is measured in days, the units of $\int_0^{20} T(t)\,dt$ are degree days.

65. Let $u = 2x$. Then $du = 2\,dx$, so $\int_0^2 f(2x)\,dx = \int_0^4 f(u)\left(\tfrac{1}{2}\,du\right) = \tfrac{1}{2}\int_0^4 f(u)\,du = \tfrac{1}{2}(10) = 5$.

67. Let $u = 1 - x$. Then $x = 1 - u$ and $dx = -du$, so

$\int_0^1 x^a(1-x)^b\,dx = \int_1^0 (1-u)^a u^b(-du) = \int_0^1 u^b(1-u)^a\,du = \int_0^1 x^b(1-x)^a\,dx$.

5.5 Integration by Parts

1. Let $u = \ln x$, $dv = x^2\,dx \Rightarrow du = \frac{1}{x}\,dx$, $v = \frac{1}{3}x^3$. Then by Formula 2,

$\int x^2 \ln x\,dx = (\ln x)\left(\tfrac{1}{3}x^3\right) - \int\left(\tfrac{1}{3}x^3\right)\left(\tfrac{1}{x}\right)dx = \tfrac{1}{3}x^3\ln x - \tfrac{1}{3}\int x^2\,dx = \tfrac{1}{3}x^3\ln x - \tfrac{1}{3}\left(\tfrac{1}{3}x^3\right) + C$

$= \tfrac{1}{3}x^3\ln x - \tfrac{1}{9}x^3 + C \quad \left[\text{or } \tfrac{1}{3}x^3\left(\ln x - \tfrac{1}{3}\right) + C\right]$

3. Let $u = x$, $dv = \cos 5x\,dx \Rightarrow du = dx$, $v = \frac{1}{5}\sin 5x$. Then by Formula 2,

$\int x\cos 5x\,dx = \tfrac{1}{5}x\sin 5x - \int\tfrac{1}{5}\sin 5x\,dx = \tfrac{1}{5}x\sin 5x + \tfrac{1}{25}\cos 5x + C$.

© 2016 Cengage Learning. All Rights Reserved. May not be scanned, copied or duplicated, or posted to a publicly accessible website, in whole or in part.

5. Let $u = r$, $dv = e^{r/2}\,dr \;\Rightarrow\; du = dr$, $v = 2e^{r/2}$.

Then $\int re^{r/2}\,dr = 2re^{r/2} - \int 2e^{r/2}\,dr = 2re^{r/2} - 4e^{r/2} + C = 2(r-2)e^{r/2} + C$.

7. Let $u = x^2$, $dv = \sin \pi x\,dx \;\Rightarrow\; du = 2x\,dx$ and $v = -\frac{1}{\pi}\cos \pi x$. Then

$I = \int x^2 \sin \pi x\,dx = -\frac{1}{\pi}x^2 \cos \pi x + \frac{2}{\pi}\int x \cos \pi x\,dx$ $(\star)$. Next let $U = x$, $dV = \cos \pi x\,dx \;\Rightarrow\; dU = dx$,

$V = \frac{1}{\pi}\sin \pi x$, so $\int x \cos \pi x\,dx = \frac{1}{\pi}x \sin \pi x - \frac{1}{\pi}\int \sin \pi x\,dx = \frac{1}{\pi}x \sin \pi x + \frac{1}{\pi^2}\cos \pi x + C_1$.

Substituting for $\int x \cos \pi x\,dx$ in $(\star)$, we get

$I = -\frac{1}{\pi}x^2 \cos \pi x + \frac{2}{\pi}\left(\frac{1}{\pi}x \sin \pi x + \frac{1}{\pi^2}\cos \pi x + C_1\right) = -\frac{1}{\pi}x^2 \cos \pi x + \frac{2}{\pi^2}x \sin \pi x + \frac{2}{\pi^3}\cos \pi x + C$, where $C = \frac{2}{\pi}C_1$.

9. Let $u = \ln \sqrt[3]{x}$, $dv = dx \;\Rightarrow\; du = \dfrac{1}{\sqrt[3]{x}}\left(\dfrac{1}{3}x^{-2/3}\right)dx = \dfrac{1}{3x}\,dx$, $v = x$. Then

$$\int \ln \sqrt[3]{x}\,dx = x \ln \sqrt[3]{x} - \int x \cdot \frac{1}{3x}\,dx = x \ln \sqrt[3]{x} - \frac{1}{3}x + C.$$

Second solution: Rewrite $\int \ln \sqrt[3]{x}\,dx = \frac{1}{3}\int \ln x\,dx$, and apply Example 2.

Third solution: Substitute $y = \sqrt[3]{x}$, to obtain $\int \ln \sqrt[3]{x}\,dx = 3\int y^2 \ln y\,dy$, and apply Exercise 1.

11. First let $u = \sin 3\theta$, $dv = e^{2\theta}\,d\theta \;\Rightarrow\; du = 3\cos 3\theta\,d\theta$, $v = \frac{1}{2}e^{2\theta}$. Then

$I = \int e^{2\theta}\sin 3\theta\,d\theta = \frac{1}{2}e^{2\theta}\sin 3\theta - \frac{3}{2}\int e^{2\theta}\cos 3\theta\,d\theta$. Next let $U = \cos 3\theta$, $dV = e^{2\theta}\,d\theta \;\Rightarrow\; dU = -3\sin 3\theta\,d\theta$,

$V = \frac{1}{2}e^{2\theta}$ to get $\int e^{2\theta}\cos 3\theta\,d\theta = \frac{1}{2}e^{2\theta}\cos 3\theta + \frac{3}{2}\int e^{2\theta}\sin 3\theta\,d\theta$. Substituting in the previous formula gives

$I = \frac{1}{2}e^{2\theta}\sin 3\theta - \frac{3}{4}e^{2\theta}\cos 3\theta - \frac{9}{4}\int e^{2\theta}\sin 3\theta\,d\theta = \frac{1}{2}e^{2\theta}\sin 3\theta - \frac{3}{4}e^{2\theta}\cos 3\theta - \frac{9}{4}I \;\Rightarrow$

$\frac{13}{4}I = \frac{1}{2}e^{2\theta}\sin 3\theta - \frac{3}{4}e^{2\theta}\cos 3\theta + C_1$. Hence, $I = \frac{1}{13}e^{2\theta}(2\sin 3\theta - 3\cos 3\theta) + C$, where $C = \frac{4}{13}C_1$.

13. Let $u = t$, $dv = \sin 3t\,dt \;\Rightarrow\; du = dt$, $v = -\frac{1}{3}\cos 3t$. Then

$\int_0^\pi t \sin 3t\,dt = \left[-\frac{1}{3}t \cos 3t\right]_0^\pi + \frac{1}{3}\int_0^\pi \cos 3t\,dt = \left(\frac{1}{3}\pi - 0\right) + \frac{1}{9}\left[\sin 3t\right]_0^\pi = \frac{\pi}{3}$.

15. Let $u = \ln x$, $dv = x^{-2}\,dx \;\Rightarrow\; du = \dfrac{1}{x}\,dx$, $v = -x^{-1}$. By (6),

$$\int_1^2 \frac{\ln x}{x^2}\,dx = \left[-\frac{\ln x}{x}\right]_1^2 + \int_1^2 x^{-2}\,dx = -\tfrac{1}{2}\ln 2 + \ln 1 + \left[-\frac{1}{x}\right]_1^2 = -\tfrac{1}{2}\ln 2 + 0 - \tfrac{1}{2} + 1 = \tfrac{1}{2} - \tfrac{1}{2}\ln 2.$$

17. Let $u = y$, $dv = \dfrac{dy}{e^{2y}} = e^{-2y}dy \;\Rightarrow\; du = dy$, $v = -\frac{1}{2}e^{-2y}$. Then

$$\int_0^1 \frac{y}{e^{2y}}\,dy = \left[-\tfrac{1}{2}ye^{-2y}\right]_0^1 + \tfrac{1}{2}\int_0^1 e^{-2y}dy = \left(-\tfrac{1}{2}e^{-2} + 0\right) - \tfrac{1}{4}\left[e^{-2y}\right]_0^1 = -\tfrac{1}{2}e^{-2} - \tfrac{1}{4}e^{-2} + \tfrac{1}{4} = \tfrac{1}{4} - \tfrac{3}{4}e^{-2}.$$

19. Let $u = (\ln x)^2$, $dv = dx \;\Rightarrow\; du = \dfrac{2}{x}\ln x\,dx$, $v = x$. By (6), $I = \int_1^2 (\ln x)^2\,dx = \left[x(\ln x)^2\right]_1^2 - 2\int_1^2 \ln x\,dx$.

To evaluate the last integral, let $U = \ln x$, $dV = dx \;\Rightarrow\; dU = \dfrac{1}{x}\,dx$, $V = x$. Thus,

$$\begin{aligned} I &= \left[x(\ln x)^2\right]_1^2 - 2\left(\left[x \ln x\right]_1^2 - \textstyle\int_1^2 dx\right) = \left[x(\ln x)^2 - 2x \ln x + 2x\right]_1^2 \\ &= \left(2(\ln 2)^2 - 4\ln 2 + 4\right) - (0 - 0 + 2) = 2(\ln 2)^2 - 4\ln 2 + 2 \end{aligned}$$

© 2016 Cengage Learning. All Rights Reserved. May not be scanned, copied or duplicated, or posted to a publicly accessible website, in whole or in part.

21. Let $y = \sqrt{x}$, so that $dy = \frac{1}{2}x^{-1/2}\,dx = \dfrac{1}{2\sqrt{x}}\,dx = \dfrac{1}{2y}\,dx$. Thus, $\int \cos\sqrt{x}\,dx = \int \cos y\,(2y\,dy) = 2\int y\cos y\,dy$. Now use parts with $u = y$, $dv = \cos y\,dy$, $du = dy$, $v = \sin y$ to get $\int y\cos y\,dy = y\sin y - \int \sin y\,dy = y\sin y + \cos y + C_1$, so $\int \cos\sqrt{x}\,dx = 2y\sin y + 2\cos y + C = 2\sqrt{x}\sin\sqrt{x} + 2\cos\sqrt{x} + C$.

23. Let $x = \theta^2$, so that $dx = 2\theta\,d\theta$. Thus, $\displaystyle\int_{\sqrt{\pi/2}}^{\sqrt{\pi}} \theta^3\cos(\theta^2)\,d\theta = \int_{\sqrt{\pi/2}}^{\sqrt{\pi}} \theta^2\cos(\theta^2)\cdot\tfrac{1}{2}(2\theta\,d\theta) = \tfrac{1}{2}\int_{\pi/2}^{\pi} x\cos x\,dx$. Now use parts with $u = x$, $dv = \cos x\,dx$, $du = dx$, $v = \sin x$ to get

$$\begin{aligned}\tfrac{1}{2}\int_{\pi/2}^{\pi} x\cos x\,dx &= \tfrac{1}{2}\left(\left[x\sin x\right]_{\pi/2}^{\pi} - \int_{\pi/2}^{\pi}\sin x\,dx\right) = \tfrac{1}{2}\left[x\sin x + \cos x\right]_{\pi/2}^{\pi}\\ &= \tfrac{1}{2}(\pi\sin\pi + \cos\pi) - \tfrac{1}{2}\left(\tfrac{\pi}{2}\sin\tfrac{\pi}{2} + \cos\tfrac{\pi}{2}\right) = \tfrac{1}{2}(\pi\cdot 0 - 1) - \tfrac{1}{2}\left(\tfrac{\pi}{2}\cdot 1 + 0\right) = -\tfrac{1}{2} - \tfrac{\pi}{4}\end{aligned}$$

25. Let $y = 1 + x$, so that $dy = dx$. Thus, $\int x\ln(1+x)\,dx = \int (y-1)\ln y\,dy$. Now use parts with $u = \ln y$, $dv = (y-1)\,dy$, $du = \frac{1}{y}\,dy$, $v = \frac{1}{2}y^2 - y$ to get

$$\begin{aligned}\textstyle\int (y-1)\ln y\,dy &= \left(\tfrac{1}{2}y^2 - y\right)\ln y - \textstyle\int\left(\tfrac{1}{2}y - 1\right)dy = \tfrac{1}{2}y(y-2)\ln y - \tfrac{1}{4}y^2 + y + C\\ &= \tfrac{1}{2}(1+x)(x-1)\ln(1+x) - \tfrac{1}{4}(1+x)^2 + 1 + x + C,\end{aligned}$$

which can be written as $\frac{1}{2}(x^2-1)\ln(1+x) - \frac{1}{4}x^2 + \frac{1}{2}x + \frac{3}{4} + C$.

27. (a) $\displaystyle\int \sin^n x\,dx = \int \left(\sin^{n-1}x\right)(\sin x)\,dx$. Let $u = \sin^{n-1}x$, $dv = \sin x\,dx \;\Rightarrow\; du = (n-1)\sin^{n-2}x\,\cos x\,dx$, $v = -\cos x$. So integration by parts gives

$$\begin{aligned}\int \sin^n x\,dx &= -\cos x\,\sin^{n-1}x + (n-1)\int \sin^{n-2}x\,\cos^2 x\,dx\\ &= -\cos x\,\sin^{n-1}x + (n-1)\int \sin^{n-2}x\,\left(1 - \sin^2 x\right)dx \qquad \left[\text{since } \cos^2 x = 1 - \sin^2 x\right]\\ &= -\cos x\,\sin^{n-1}x + (n-1)\int \sin^{n-2}x\,dx - (n-1)\int \sin^n x\,dx\end{aligned}$$

Rearranging to solve for $\int \sin^n x\,dx$ in the algebraic equation above (see Example 4), we have

$$n\int \sin^n x\,dx = -\cos x\,\sin^{n-1}x + (n-1)\int \sin^{n-2}x\,dx$$

Or $\displaystyle\int \sin^n x\,dx = -\frac{1}{n}\cos x\,\sin^{n-1}x + \frac{n-1}{n}\int \sin^{n-2}x\,dx$

(b) Take $n = 2$ in the reduction formula from part (a) to get $\displaystyle\int \sin^2 x\,dx = -\frac{1}{2}\cos x\sin x + \frac{1}{2}\int 1\,dx = \frac{x}{2} - \frac{\sin 2x}{4} + C$ where in the last step we have used the double-angle formula (15a) in Appendix C.

(c) $\int \sin^4 x\,dx = -\frac{1}{4}\cos x\sin^3 x + \frac{3}{4}\int \sin^2 x\,dx = -\frac{1}{4}\cos x\sin^3 x + \frac{3}{8}x - \frac{3}{16}\sin 2x + C$.

29. Let $u = (\ln x)^n$, $dv = dx \;\Rightarrow\; du = n(\ln x)^{n-1}(dx/x)$, $v = x$. By Formula 2,
$\int (\ln x)^n\,dx = x(\ln x)^n - \int nx(\ln x)^{n-1}(dx/x) = x(\ln x)^n - n\int (\ln x)^{n-1}\,dx$.

© 2016 Cengage Learning. All Rights Reserved. May not be scanned, copied or duplicated, or posted to a publicly accessible website, in whole or in part.

31. By repeated applications of the reduction formula in Exercise 29,

$$\begin{aligned}\int(\ln x)^3\,dx &= x\,(\ln x)^3 - 3\int(\ln x)^2\,dx = x(\ln x)^3 - 3\big[x(\ln x)^2 - 2\int(\ln x)^1\,dx\big]\\ &= x\,(\ln x)^3 - 3x(\ln x)^2 + 6\big[x(\ln x)^1 - 1\int(\ln x)^0\,dx\big]\\ &= x\,(\ln x)^3 - 3x(\ln x)^2 + 6x\ln x - 6\int 1\,dx = x\,(\ln x)^3 - 3x(\ln x)^2 + 6x\ln x - 6x + C\end{aligned}$$

33. Let $u = 11.4t$, $dv = e^{-t}\,dt \Rightarrow du = 11.4\,dt$, $v = -e^{-t}$. So integration by parts gives

$$\int_0^4 C(t)\,dt = \int_0^4 11.4te^{-t}\,dt = -11.4te^{-t}\big]_0^4 + 11.4\int_0^4 e^{-t}\,dt = -45.6e^{-4} + 11.4\left[-e^{-t}\right]_0^4$$

$$= -45.6e^{-4} - 11.4\left[e^{-4} - 1\right] \approx 10.3560\ (\mu\text{g/mL})\cdot\text{h}$$

35. $\displaystyle\int_0^1 p(t)\,dt = \int_0^1 \left[\tfrac{1}{2} - \tfrac{1}{2}e^{-t}(\sin t + \cos t)\right]\,dt = \left[\tfrac{1}{2}t\right]_0^1 - \tfrac{1}{2}\int_0^1 e^{-t}(\sin t + \cos t)\,dt$

Let $u = \sin t + \cos t$, $dv = e^{-t}\,dt \Rightarrow du = (\cos t - \sin t)\,dt$, $v = -e^{-t}$. So integration by parts gives

$\displaystyle\int e^{-t}(\sin t + \cos t)\,dt = -e^{-t}(\sin t + \cos t) + \int e^{-t}(\cos t - \sin t)\,dt$. Now, let $U = \cos t - \sin t$,

$dV = e^{-t}\,dt \Rightarrow dU = -(\sin t + \cos t)\,dt$, $V = -e^{-t}$. So integrating by parts again gives

$\displaystyle\int e^{-t}(\sin t + \cos t)\,dt = -e^{-t}(\sin t + \cos t) - e^{-t}(\cos t - \sin t) - \int e^{-t}(\sin t + \cos t)\,dt \Rightarrow$

$\displaystyle 2\int e^{-t}(\sin t + \cos t)\,dt = -2e^{-t}\cos t \Rightarrow \int e^{-t}(\sin t + \cos t)\,dt = -e^{-t}\cos t + C$. Substituting this back into the

original expression, we have $\displaystyle\int_0^1 p(t)\,dt = \tfrac{1}{2} - \tfrac{1}{2}\left[-e^{-t}\cos t\right]_0^1 = \tfrac{1}{2} + \tfrac{1}{2}e^{-1}\cos 1 - \tfrac{1}{2} = \tfrac{1}{2}e^{-1}\cos 1 \approx 0.09938$

37. For $I = \int_1^4 xf''(x)\,dx$, let $u = x$, $dv = f''(x)\,dx \Rightarrow du = dx$, $v = f'(x)$. Then

$I = \left[xf'(x)\right]_1^4 - \int_1^4 f'(x)\,dx = 4f'(4) - 1\cdot f'(1) - [f(4) - f(1)] = 4\cdot 3 - 1\cdot 5 - (7 - 2) = 12 - 5 - 5 = 2.$

We used the fact that f'' is continuous to guarantee that I exists.

5.6 Partial Fractions

1. (a) $\dfrac{1}{x^2 - 1} = \dfrac{1}{(x+1)(x-1)} = \dfrac{A}{x+1} + \dfrac{B}{x-1}$ (b) $\dfrac{2}{x^2 + x} = \dfrac{2}{x(x+1)} = \dfrac{A}{x} + \dfrac{B}{x+1}$

3. $\displaystyle\int\frac{x}{x-6}\,dx = \int\frac{(x-6)+6}{x-6}\,dx = \int\left(1 + \frac{6}{x-6}\right)dx = x + 6\ln|x-6| + C$

5. $\dfrac{x-9}{(x+5)(x-2)} = \dfrac{A}{x+5} + \dfrac{B}{x-2}$. Multiply both sides by $(x+5)(x-2)$ to get $x - 9 = A(x-2) + B(x+5)\,(*)$, or equivalently, $x - 9 = (A+B)x - 2A + 5B$. Equating coefficients of x on each side of the equation gives us $1 = A + B$ **(1)** and equating constants gives us $-9 = -2A + 5B$ **(2)**. Adding two times **(1)** to **(2)** gives us $-7 = 7B \Leftrightarrow B = -1$ and hence, $A = 2$. [Alternatively, to find the coefficients A and B, we may use substitution as follows: substitute 2 for x in $(*)$ to get $-7 = 7B \Leftrightarrow B = -1$, then substitute -5 for x in $(*)$ to get $-14 = -7A \Leftrightarrow A = 2$.] Thus,

$$\int\frac{x-9}{(x+5)(x-2)}\,dx = \int\left(\frac{2}{x+5} + \frac{-1}{x-2}\right)dx = 2\ln|x+5| - \ln|x-2| + C.$$

© 2016 Cengage Learning. All Rights Reserved. May not be scanned, copied or duplicated, or posted to a publicly accessible website, in whole or in part.

7. $\dfrac{1}{x^2-1}=\dfrac{1}{(x+1)(x-1)}=\dfrac{A}{x+1}+\dfrac{B}{x-1}$. Multiply both sides by $(x+1)(x-1)$ to get $1=A(x-1)+B(x+1)$.

Substituting 1 for x gives $1=2B \Leftrightarrow B=\frac{1}{2}$. Substituting -1 for x gives $1=-2A \Leftrightarrow A=-\frac{1}{2}$. Thus,

$$\int_2^3 \frac{1}{x^2-1}\,dx=\int_2^3\left(\frac{-1/2}{x+1}+\frac{1/2}{x-1}\right)dx=\left[-\tfrac{1}{2}\ln|x+1|+\tfrac{1}{2}\ln|x-1|\right]_2^3$$
$$=\left(-\tfrac{1}{2}\ln 4+\tfrac{1}{2}\ln 2\right)-\left(-\tfrac{1}{2}\ln 3+\tfrac{1}{2}\ln 1\right)=\tfrac{1}{2}(\ln 2+\ln 3-\ln 4)\quad\left[\text{or } \tfrac{1}{2}\ln\tfrac{3}{2}\right]$$

9. $\displaystyle\int\frac{ax}{x^2-bx}\,dx=\int\frac{ax}{x(x-b)}\,dx=\int\frac{a}{x-b}\,dx=a\ln|x-b|+C$

11. $\dfrac{2}{2x^2+3x+1}=\dfrac{2}{(2x+1)(x+1)}=\dfrac{A}{2x+1}+\dfrac{B}{x+1}$. Multiply both sides by $(2x+1)(x+1)$ to get

$2=A(x+1)+B(2x+1)$. The coefficients of x must be equal and the constant terms are also equal, so $A+2B=0$ and $A+B=2$. Subtracting the second equation from the first gives $B=-2$, and hence, $A=4$. Thus,

$$\int_0^1\frac{2}{2x^2+3x+1}\,dx=\int_0^1\left(\frac{4}{2x+1}-\frac{2}{x+1}\right)dx=\left[\frac{4}{2}\ln|2x+1|-2\ln|x+1|\right]_0^1=(2\ln 3-2\ln 2)-0=2\ln\frac{3}{2}.$$

Another method: Substituting -1 for x in the equation $2=A(x+1)+B(2x+1)$ gives $2=-B \Leftrightarrow B=-2$.

Substituting $-\frac{1}{2}$ for x gives $2=\frac{1}{2}A \Leftrightarrow A=4$.

13. $\dfrac{4y^2-7y-12}{y(y+2)(y-3)}=\dfrac{A}{y}+\dfrac{B}{y+2}+\dfrac{C}{y-3} \Rightarrow 4y^2-7y-12=A(y+2)(y-3)+By(y-3)+Cy(y+2)$. Setting

$y=0$ gives $-12=-6A$, so $A=2$. Setting $y=-2$ gives $18=10B$, so $B=\frac{9}{5}$. Setting $y=3$ gives $3=15C$,

so $C=\frac{1}{5}$. Now

$$\int_1^2\frac{4y^2-7y-12}{y(y+2)(y-3)}\,dy=\int_1^2\left(\frac{2}{y}+\frac{9/5}{y+2}+\frac{1/5}{y-3}\right)dy=\left[2\ln|y|+\tfrac{9}{5}\ln|y+2|+\tfrac{1}{5}\ln|y-3|\right]_1^2$$
$$=2\ln 2+\tfrac{9}{5}\ln 4+\tfrac{1}{5}\ln 1-2\ln 1-\tfrac{9}{5}\ln 3-\tfrac{1}{5}\ln 2$$
$$=2\ln 2+\tfrac{18}{5}\ln 2-\tfrac{1}{5}\ln 2-\tfrac{9}{5}\ln 3=\tfrac{27}{5}\ln 2-\tfrac{9}{5}\ln 3=\tfrac{9}{5}(3\ln 2-\ln 3)=\tfrac{9}{5}\ln\tfrac{8}{3}$$

15. Let $u=\sqrt{x}$, so $u^2=x$ and $dx=2u\,du$. Thus,

$$\int_9^{16}\frac{\sqrt{x}}{x-4}\,dx=\int_3^4\frac{u}{u^2-4}2u\,du=2\int_3^4\frac{u^2}{u^2-4}\,du=2\int_3^4\left(1+\frac{4}{u^2-4}\right)du\qquad\text{[by long division]}$$
$$=2+8\int_3^4\frac{du}{(u+2)(u-2)}\quad(*)$$

Multiply $\dfrac{1}{(u+2)(u-2)}=\dfrac{A}{u+2}+\dfrac{B}{u-2}$ by $(u+2)(u-2)$ to get $1=A(u-2)+B(u+2)$. Equating coefficients we

get $A+B=0$ and $-2A+2B=1$. Solving gives us $B=\frac{1}{4}$ and $A=-\frac{1}{4}$, so $\dfrac{1}{(u+2)(u-2)}=\dfrac{-1/4}{u+2}+\dfrac{1/4}{u-2}$ and $(*)$ is

$$2+8\int_3^4\left(\frac{-1/4}{u+2}+\frac{1/4}{u-2}\right)du=2+8\left[-\tfrac{1}{4}\ln|u+2|+\tfrac{1}{4}\ln|u-2|\right]_3^4$$
$$=2+\Big[2\ln|u-2|-2\ln|u+2|\Big]_3^4=2+2\left[\ln\left|\frac{u-2}{u+2}\right|\right]_3^4$$
$$=2+2\left(\ln\tfrac{2}{6}-\ln\tfrac{1}{5}\right)=2+2\ln\tfrac{2/6}{1/5}$$
$$=2+2\ln\tfrac{5}{3}\ \text{ or }\ 2+\ln\left(\tfrac{5}{3}\right)^2=2+\ln\tfrac{25}{9}$$

© 2016 Cengage Learning. All Rights Reserved. May not be scanned, copied or duplicated, or posted to a publicly accessible website, in whole or in part.

17. Let $u = e^x$. Then $x = \ln u$, $dx = \dfrac{du}{u}$ $\Rightarrow$

$$\int \frac{e^{2x}\,dx}{e^{2x}+3e^x+2} = \int \frac{u^2\,(du/u)}{u^2+3u+2} = \int \frac{u\,du}{(u+1)(u+2)} = \int \left[\tfrac{-1}{u+1} + \tfrac{2}{u+2}\right] du$$

$$= 2\ln|u+2| - \ln|u+1| + C = \ln\left[\frac{(e^x+2)^2}{e^x+1}\right] + C$$

19. $\dfrac{5x^2+3x-2}{x^3+2x^2} = \dfrac{5x^2+3x-2}{x^2\,(x+2)} = \dfrac{A}{x} + \dfrac{B}{x^2} + \dfrac{C}{x+2}$. Multiply by $x^2(x+2)$ to get

$5x^2+3x-2 = Ax(x+2) + B(x+2) + Cx^2$. Set $x = -2$ to get $C = 3$, and take

$x = 0$ to get $B = -1$. Equating the coefficients of x^2 gives $5 = A + C \;\Rightarrow\; A = 2$. So

$$\int \frac{5x^2+3x-2}{x^3+2x^2}\,dx = \int \left(\frac{2}{x} - \frac{1}{x^2} + \frac{3}{x+2}\right) dx = 2\ln|x| + \frac{1}{x} + 3\ln|x+2| + C.$$

21. $f(x) = \dfrac{2x^2+x+1}{x(x^2+1)} = \dfrac{A}{x} + \dfrac{Bx+C}{x^2+1} \;\Rightarrow\; 2x^2+x+1 = A(x^2+1) + (Bx+C)\,x = (A+B)\,x^2 + Cx + A.$

Setting $x = 0$ gives $A = 1$. Equating coefficients of x^2 gives $A + B = 2$, so $B = 1$, and equating coefficients of x gives $C = 1$. Hence

$$\int f(x)\,dx = \int \frac{2x^2+x+1}{x(x^2+1)}\,dx = \int \left(\frac{1}{x} + \frac{x+1}{x^2+1}\right) dx = \ln|x| + \int \left(\frac{1}{x^2+1} + \frac{x}{x^2+1}\right) dx$$

$$= \ln|x| + \tan^{-1} x + \tfrac{1}{2}\int \frac{1}{u}\,du \qquad [\text{Substitution: } u = x^2+1,\ du = 2x\,dx]$$

$$= \ln|x| + \tan^{-1} x + \tfrac{1}{2}\ln|u| + C = \ln|x| + \tan^{-1} x + \tfrac{1}{2}\ln\left|x^2+1\right| + C$$

23. There are only finitely many values of x where $Q(x) = 0$ (assuming that Q is not the zero polynomial). At all other values of x, $F(x)/Q(x) = G(x)/Q(x)$, so $F(x) = G(x)$. In other words, the values of F and G agree at all except perhaps finitely many values of x. By continuity of F and G, the polynomials F and G must agree at those values of x too.

More explicitly: If a is a value of x such that $Q(a) = 0$, then $Q(x) \neq 0$ for all x sufficiently close to a. Thus,

$$\begin{aligned} F(a) &= \lim_{x\to a} F(x) && [\text{by continuity of } F] \\ &= \lim_{x\to a} G(x) && [\text{whenever } Q(x) \neq 0] \\ &= G(a) && [\text{by continuity of } G] \end{aligned}$$

5.7 Integration Using Tables and Computer Algebra Systems

1. Let $u = \pi x$, so that $du = \pi\,dx$. Then

$$\int \tan^3(\pi x)\,dx = \int \tan^3 u\left(\tfrac{1}{\pi}\,du\right) = \tfrac{1}{\pi}\int \tan^3 u\,du \overset{69}{=} \tfrac{1}{\pi}\left[\tfrac{1}{2}\tan^2 u + \ln|\cos u|\right] + C$$

$$= \tfrac{1}{2\pi}\tan^2(\pi x) + \tfrac{1}{\pi}\ln|\cos(\pi x)| + C$$

3. Let $u = 2x$ and $a = 3$. Then $du = 2\,dx$ and

$$\int \frac{dx}{x^2\sqrt{4x^2+9}} = \int \frac{\frac{1}{2}\,du}{\dfrac{u^2}{4}\sqrt{u^2+a^2}} = 2\int \frac{du}{u^2\sqrt{a^2+u^2}} \overset{28}{=} -2\,\frac{\sqrt{a^2+u^2}}{a^2u} + C$$

$$= -2\,\frac{\sqrt{4x^2+9}}{9\cdot 2x} + C = -\frac{\sqrt{4x^2+9}}{9x} + C$$

© 2016 Cengage Learning. All Rights Reserved. May not be scanned, copied or duplicated, or posted to a publicly accessible website, in whole or in part.

5. Let $u = e^x$, so that $du = e^x\,dx$ and $e^{2x} = u^2$. Then

$$\int e^{2x}\arctan(e^x)\,dx = \int u^2 \arctan u \left(\frac{du}{u}\right) = \int u \arctan u\,du$$

$$\overset{92}{=} \frac{u^2+1}{2}\arctan u - \frac{u}{2} + C = \frac{1}{2}(e^{2x}+1)\arctan(e^x) - \frac{1}{2}e^x + C$$

7. $\int x^3 \sin x\,dx \overset{84}{=} -x^3\cos x + 3\int x^2 \cos x\,dx$, $\int x^2\cos x\,dx \overset{85}{=} x^2 \sin x - 2\int x\sin x\,dx$, and

$\int x \sin x\,dx \overset{82}{=} \sin x - x\cos x + C$. Substituting, we get

$\int x^3\sin x\,dx = -x^3\cos x + 3\left[x^2\sin x - 2(\sin x - x\cos x)\right] + C = -x^3\cos x + 3x^2\sin x - 6\sin x + 6x\cos x + C.$

So $\int_0^\pi x^3\sin x\,dx = \left[-x^3\cos x + 3x^2\sin x - 6\sin x + 6x\cos x\right]_0^\pi = \left(-\pi^3\cdot -1 + 6\pi\cdot -1\right) - (0) = \pi^3 - 6\pi.$

9. $\displaystyle\int \frac{\tan^3(1/z)}{z^2}\,dz \quad \begin{bmatrix} u = 1/z, \\ du = -dz/z^2 \end{bmatrix} \quad = -\int \tan^3 u\,du \overset{69}{=} -\tfrac{1}{2}\tan^2 u - \ln|\cos u| + C = -\tfrac{1}{2}\tan^2\!\left(\tfrac{1}{z}\right) - \ln\left|\cos\!\left(\tfrac{1}{z}\right)\right| + C$

11. Let $u = \sin x$. Then $du = \cos x\,dx$, so

$$\int \sin^2 x\,\cos x\,\ln(\sin x)\,dx = \int u^2 \ln u\,du \overset{101}{=} \frac{u^{2+1}}{(2+1)^2}\left[(2+1)\ln u - 1\right] + C = \tfrac{1}{9}u^3(3\ln u - 1) + C$$

$$= \tfrac{1}{9}\sin^3 x\,[3\ln(\sin x) - 1] + C$$

13. Let $u = e^x$ and $a = \sqrt{3}$. Then $du = e^x\,dx$ and

$$\int \frac{e^x}{3 - e^{2x}}\,dx = \int \frac{du}{a^2 - u^2} \overset{19}{=} \frac{1}{2a}\ln\left|\frac{u+a}{u-a}\right| + C = \frac{1}{2\sqrt{3}}\ln\left|\frac{e^x + \sqrt{3}}{e^x - \sqrt{3}}\right| + C.$$

15. $\displaystyle\int \frac{x^4\,dx}{\sqrt{x^{10}-2}} = \int \frac{x^4\,dx}{\sqrt{(x^5)^2 - 2}} = \frac{1}{5}\int \frac{du}{\sqrt{u^2-2}} \quad \begin{bmatrix} u = x^5, \\ du = 5x^4\,dx \end{bmatrix}$

$$\overset{43}{=} \tfrac{1}{5}\ln\left|u + \sqrt{u^2-2}\,\right| + C = \tfrac{1}{5}\ln\left|x^5 + \sqrt{x^{10}-2}\,\right| + C$$

17. Let $u = \ln x$ and $a = 2$. Then $du = dx/x$ and

$$\int \frac{\sqrt{4 + (\ln x)^2}}{x}\,dx = \int \sqrt{a^2 + u^2}\,du \overset{21}{=} \frac{u}{2}\sqrt{a^2+u^2} + \frac{a^2}{2}\ln\left(u + \sqrt{a^2+u^2}\right) + C$$

$$= \tfrac{1}{2}(\ln x)\sqrt{4 + (\ln x)^2} + 2\ln\left[\ln x + \sqrt{4 + (\ln x)^2}\right] + C$$

19. (a) $\displaystyle\frac{d}{du}\left[\frac{1}{b^3}\left(a + bu - \frac{a^2}{a+bu} - 2a\ln|a+bu|\right) + C\right] = \frac{1}{b^3}\left[b + \frac{ba^2}{(a+bu)^2} - \frac{2ab}{(a+bu)}\right]$

$$= \frac{1}{b^3}\left[\frac{b(a+bu)^2 + ba^2 - (a+bu)2ab}{(a+bu)^2}\right] = \frac{1}{b^3}\left[\frac{b^3u^2}{(a+bu)^2}\right] = \frac{u^2}{(a+bu)^2}$$

(b) Let $t = a + bu \;\Rightarrow\; dt = b\,du$. Note that $u = \dfrac{t-a}{b}$ and $du = \dfrac{1}{b}\,dt$.

$$\int \frac{u^2\,du}{(a+bu)^2} = \frac{1}{b^3}\int \frac{(t-a)^2}{t^2}\,dt = \frac{1}{b^3}\int \frac{t^2 - 2at + a^2}{t^2}\,dt = \frac{1}{b^3}\int\left(1 - \frac{2a}{t} + \frac{a^2}{t^2}\right)dt$$

$$= \frac{1}{b^3}\left(t - 2a\ln|t| - \frac{a^2}{t}\right) + C = \frac{1}{b^3}\left(a + bu - \frac{a^2}{a+bu} - 2a\ln|a+bu|\right) + C$$

© 2016 Cengage Learning. All Rights Reserved. May not be scanned, copied or duplicated, or posted to a publicly accessible website, in whole or in part.

21. Maple and Mathematica both give $\int \sec^4 x\,dx = \frac{2}{3}\tan x + \frac{1}{3}\tan x \sec^2 x$, while Derive gives the second term as $\dfrac{\sin x}{3\cos^3 x} = \dfrac{1}{3}\dfrac{\sin x}{\cos x}\dfrac{1}{\cos^2 x} = \dfrac{1}{3}\tan x \sec^2 x$. Using Formula 77, we get

$\int \sec^4 x\,dx = \frac{1}{3}\tan x \sec^2 x + \frac{2}{3}\int \sec^2 x\,dx = \frac{1}{3}\tan x \sec^2 x + \frac{2}{3}\tan x + C$.

23. Maple gives $\int x\sqrt{1+2x}\,dx = \frac{1}{10}(1+2x)^{5/2} - \frac{1}{6}(1+2x)^{3/2}$, Mathematica gives $\sqrt{1+2x}\left(\frac{2}{5}x^2 + \frac{1}{15}x - \frac{1}{15}\right)$, and Derive gives $\frac{1}{15}(1+2x)^{3/2}(3x-1)$. The first two expressions can be simplified to Derive's result. If we use Formula 54, we get

$$\int x\sqrt{1+2x}\,dx = \frac{2}{15(2)^2}(3\cdot 2x - 2\cdot 1)(1+2x)^{3/2} + C = \tfrac{1}{30}(6x-2)(1+2x)^{3/2} + C = \tfrac{1}{15}(3x-1)(1+2x)^{3/2}.$$

25. Maple gives $\int \tan^5 x\,dx = \frac{1}{4}\tan^4 x - \frac{1}{2}\tan^2 x + \frac{1}{2}\ln(1+\tan^2 x)$, Mathematica gives $\int \tan^5 x\,dx = \frac{1}{4}[-1 - 2\cos(2x)]\sec^4 x - \ln(\cos x)$, and Derive gives $\int \tan^5 x\,dx = \frac{1}{4}\tan^4 x - \frac{1}{2}\tan^2 x - \ln(\cos x)$. These expressions are equivalent, and none includes absolute value bars or a constant of integration. Note that Mathematica's and Derive's expressions suggest that the integral is undefined where $\cos x < 0$, which is not the case. Using Formula 75, $\int \tan^5 x\,dx = \frac{1}{5-1}\tan^{5-1} x - \int \tan^{5-2} x\,dx = \frac{1}{4}\tan^4 x - \int \tan^3 x\,dx$. Using Formula 69, $\int \tan^3 x\,dx = \frac{1}{2}\tan^2 x + \ln|\cos x| + C$, so $\int \tan^5 x\,dx = \frac{1}{4}\tan^4 x - \frac{1}{2}\tan^2 x - \ln|\cos x| + C$.

27. Derive, Maple, and Mathematica all give $\displaystyle\int \frac{1}{\sqrt{1+\sqrt[3]{x}}}\,dx = \frac{2}{5}\sqrt{\sqrt[3]{x}+1}\left(3\sqrt[3]{x^2} - 4\sqrt[3]{x} + 8\right)$. [Maple adds a constant of $-\frac{16}{5}$.] We'll change the form of the integral by letting $u = \sqrt[3]{x}$, so that $u^3 = x$ and $3u^2\,du = dx$. Then

$$\begin{aligned}\int \frac{1}{\sqrt{1+\sqrt[3]{x}}}\,dx &= \int \frac{3u^2\,du}{\sqrt{1+u}} \overset{56}{=} 3\left[\frac{2}{15(1)^3}\left(8(1)^2 + 3(1)^2u^2 - 4(1)(1)u\right)\sqrt{1+u}\right] + C \\ &= \tfrac{2}{5}(8 + 3u^2 - 4u)\sqrt{1+u} + C = \tfrac{2}{5}\left(8 + 3\sqrt[3]{x^2} - 4\sqrt[3]{x}\right)\sqrt{1+\sqrt[3]{x}} + C\end{aligned}$$

5.8 Improper Integrals

1. The area under the graph of $y = 1/x^3 = x^{-3}$ between $x = 1$ and $x = t$ is $A(t) = \int_1^t x^{-3}\,dx = \left[-\frac{1}{2}x^{-2}\right]_1^t = -\frac{1}{2}t^{-2} - \left(-\frac{1}{2}\right) = \frac{1}{2} - 1/(2t^2)$. So the area for $1 \le x \le 10$ is $A(10) = 0.5 - 0.005 = 0.495$, the area for $1 \le x \le 100$ is $A(100) = 0.5 - 0.00005 = 0.49995$, and the area for $1 \le x \le 1000$ is $A(1000) = 0.5 - 0.0000005 = 0.4999995$. The total area under the curve for $x \ge 1$ is $\lim\limits_{t\to\infty} A(t) = \lim\limits_{t\to\infty}\left[\frac{1}{2} - 1/(2t^2)\right] = \frac{1}{2}$.

3. $$\begin{aligned}\int_3^\infty \frac{1}{(x-2)^{3/2}}\,dx &= \lim_{t\to\infty}\int_3^t (x-2)^{-3/2}\,dx = \lim_{t\to\infty}\left[-2(x-2)^{-1/2}\right]_3^t \qquad [u = x-2,\ du = dx] \\ &= \lim_{t\to\infty}\left(\frac{-2}{\sqrt{t-2}} + \frac{2}{\sqrt{1}}\right) = 0 + 2 = 2. \qquad \text{Convergent}\end{aligned}$$

5. $$\begin{aligned}\int_{-\infty}^{-1} \frac{1}{\sqrt{2-w}}\,dw &= \lim_{t\to-\infty}\int_t^{-1} \frac{1}{\sqrt{2-w}}\,dw = \lim_{t\to-\infty}\left[-2\sqrt{2-w}\right]_t^{-1} \qquad [u = 2-w,\ du = -dw] \\ &= \lim_{t\to-\infty}\left[-2\sqrt{3} + 2\sqrt{2-t}\right] = \infty. \qquad \text{Divergent}\end{aligned}$$

© 2016 Cengage Learning. All Rights Reserved. May not be scanned, copied or duplicated, or posted to a publicly accessible website, in whole or in part.

7. $\int_4^\infty e^{-y/2}\,dy = \lim\limits_{t\to\infty}\int_4^t e^{-y/2}\,dy = \lim\limits_{t\to\infty}\left[-2e^{-y/2}\right]_4^t = \lim\limits_{t\to\infty}(-2e^{-t/2}+2e^{-2}) = 0 + 2e^{-2} = 2e^{-2}$.

Convergent

9. $\int_{2\pi}^\infty \sin\theta\,d\theta = \lim\limits_{t\to\infty}\int_{2\pi}^t \sin\theta\,d\theta = \lim\limits_{t\to\infty}\left[-\cos\theta\right]_{2\pi}^t = \lim\limits_{t\to\infty}(-\cos t + 1)$. This limit does not exist, so the integral is divergent. Divergent

11. $\int_{-\infty}^\infty xe^{-x^2}\,dx = \int_{-\infty}^0 xe^{-x^2}\,dx + \int_0^\infty xe^{-x^2}\,dx$.

$\int_{-\infty}^0 xe^{-x^2}\,dx = \lim\limits_{t\to-\infty}\left(-\frac{1}{2}\right)\left[e^{-x^2}\right]_t^0 = \lim\limits_{t\to-\infty}\left(-\frac{1}{2}\right)\left(1-e^{-t^2}\right) = -\frac{1}{2}\cdot 1 = -\frac{1}{2}$, and

$\int_0^\infty xe^{-x^2}\,dx = \lim\limits_{t\to\infty}\left(-\frac{1}{2}\right)\left[e^{-x^2}\right]_0^t = \lim\limits_{t\to\infty}\left(-\frac{1}{2}\right)\left(e^{-t^2}-1\right) = -\frac{1}{2}\cdot(-1) = \frac{1}{2}$.

Therefore, $\int_{-\infty}^\infty xe^{-x^2}\,dx = -\frac{1}{2}+\frac{1}{2} = 0$. Convergent

13. $$\int_1^\infty \frac{x+1}{x^2+2x}\,dx = \lim_{t\to\infty}\int_1^t \frac{\frac{1}{2}(2x+2)}{x^2+2x}\,dx = \frac{1}{2}\lim_{t\to\infty}\left[\ln(x^2+2x)\right]_1^t = \frac{1}{2}\lim_{t\to\infty}\left[\ln(t^2+2t)-\ln 3\right] = \infty.$$

Divergent

15. $$\int_0^\infty se^{-5s}\,ds = \lim_{t\to\infty}\int_0^t se^{-5s}\,ds = \lim_{t\to\infty}\left[-\tfrac{1}{5}se^{-5s} - \tfrac{1}{25}e^{-5s}\right] \quad \left[\begin{array}{l}\text{by integration by}\\ \text{parts with } u = s\end{array}\right]$$

$$= \lim_{t\to\infty}\left(-\tfrac{1}{5}te^{-5t} - \tfrac{1}{25}e^{-5t} + \tfrac{1}{25}\right) = 0 - 0 + \tfrac{1}{25} \quad \text{[by l'Hospital's Rule]}$$

$= \frac{1}{25}$. Convergent

17. $$\int_1^\infty \frac{\ln x}{x}\,dx = \lim_{t\to\infty}\left[\frac{(\ln x)^2}{2}\right]_1^t \quad \left[\begin{array}{l}\text{by substitution with}\\ u=\ln x,\ du = dx/x\end{array}\right] = \lim_{t\to\infty}\frac{(\ln t)^2}{2} = \infty.$$ Divergent

19. $$\int_{-\infty}^\infty \frac{x^2}{9+x^6}\,dx = \int_{-\infty}^0 \frac{x^2}{9+x^6}\,dx + \int_0^\infty \frac{x^2}{9+x^6}\,dx = 2\int_0^\infty \frac{x^2}{9+x^6}\,dx \quad \text{[since the integrand is even]}.$$

Now $$\int \frac{x^2\,dx}{9+x^6}\ \left[\begin{array}{l}u = x^3\\ du = 3x^2\,dx\end{array}\right] = \int \frac{\frac{1}{3}\,du}{9+u^2}\ \left[\begin{array}{l}u = 3v\\ du = 3\,dv\end{array}\right] = \int \frac{\frac{1}{3}(3\,dv)}{9+9v^2} = \frac{1}{9}\int \frac{dv}{1+v^2}$$

$$= \frac{1}{9}\tan^{-1}v + C = \frac{1}{9}\tan^{-1}\left(\frac{u}{3}\right) + C = \frac{1}{9}\tan^{-1}\left(\frac{x^3}{3}\right) + C,$$

so $$2\int_0^\infty \frac{x^2}{9+x^6}\,dx = 2\lim_{t\to\infty}\int_0^t \frac{x^2}{9+x^6}\,dx = 2\lim_{t\to\infty}\left[\frac{1}{9}\tan^{-1}\left(\frac{x^3}{3}\right)\right]_0^t = 2\lim_{t\to\infty}\frac{1}{9}\tan^{-1}\left(\frac{t^3}{3}\right) = \frac{2}{9}\cdot\frac{\pi}{2} = \frac{\pi}{9}.$$

Convergent

21. $$\int_e^\infty \frac{1}{x(\ln x)^3}\,dx = \lim_{t\to\infty}\int_e^t \frac{1}{x(\ln x)^3}\,dx = \lim_{t\to\infty}\int_1^{\ln t} u^{-3}\,du \quad \left[\begin{array}{l}u = \ln x,\\ du = dx/x\end{array}\right] = \lim_{t\to\infty}\left[-\frac{1}{2u^2}\right]_1^{\ln t}$$

$$= \lim_{t\to\infty}\left[-\frac{1}{2(\ln t)^2} + \frac{1}{2}\right] = 0 + \frac{1}{2} = \frac{1}{2}.$$ Convergent

© 2016 Cengage Learning. All Rights Reserved. May not be scanned, copied or duplicated, or posted to a publicly accessible website, in whole or in part.

23. 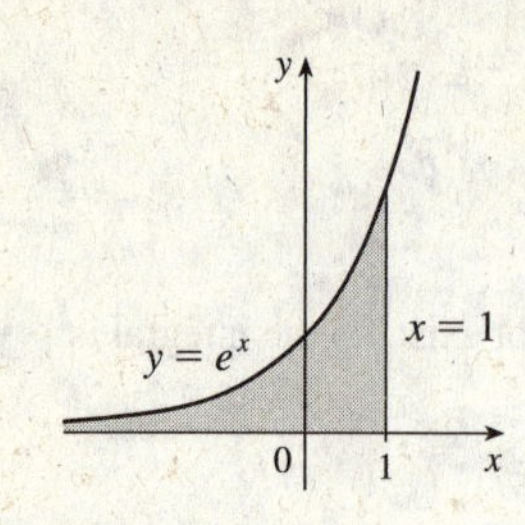

Area $= \int_{-\infty}^{1} e^x\,dx = \lim\limits_{t\to-\infty} \left[e^x\right]_t^1 = e - \lim\limits_{t\to-\infty} e^t = e$

25. (a) $C(t) = 23te^{-2t} \quad\Rightarrow\quad C'(t) = 23\left(e^{-2t} - 2te^{-2t}\right) > 0 \quad\Rightarrow\quad 23e^{-2t}\left(1-2t\right) > 0 \quad\Rightarrow\quad t < 1/2$. So $C'(t) > 0$ when $t < 1/2$ and $C'(t) < 0$ when $t > 1/2$. Thus, the maximum drug concentration occurs when $t = 1/2$ and the max value is $C(1/2) = 23(\frac{1}{2})e^{-1} \approx 4.23\,\mu\text{g/mL}$

(b) Integrating by parts with $u = t$, $du = dt$, $dv = e^{-2t}dt$, $v = -\frac{1}{2}e^{-2t}$, we have

$$\int_0^\infty C(t)\,dt = 23\lim_{b\to\infty}\int_0^b te^{-2t}\,dt = 23\lim_{b\to\infty}\left(-\tfrac{1}{2}te^{-2t}\Big]_0^b + \tfrac{1}{2}\int_0^b e^{-2t}\,dt\right) = 23\lim_{b\to\infty}\left(-\tfrac{1}{2}\frac{b}{e^{2b}} + \tfrac{1}{2}\left[-\tfrac{1}{2}e^{-2t}\right]_0^b\right)$$

$$= \tfrac{23}{2}\left[-\lim_{b\to\infty}\frac{b}{e^{2b}} + \lim_{b\to\infty}\left(-\tfrac{1}{2}e^{-2b} + \tfrac{1}{2}\right)\right] \overset{\text{H}}{=} \tfrac{23}{2}\left[-\lim_{b\to\infty}\frac{1}{2e^{2b}} + \left(0 + \tfrac{1}{2}\right)\right] = \tfrac{23}{2}\left[0 + \tfrac{1}{2}\right]$$

$$= \frac{23}{4} = 5.75\,(\mu\text{g/mL}) \times \text{hours}$$

This is the long-term "availability" of a single drug dose.

27. $\int_0^\infty P(I(x))\,dx = \int_0^\infty aI(x)\,dx = a\lim\limits_{b\to\infty}\int_0^b e^{-kx}\,dx = a\lim\limits_{b\to\infty}\left[\dfrac{1}{-k}e^{-kx}\right]_0^b = \dfrac{a}{k}\lim\limits_{b\to\infty}\left[1 - e^{-kb}\right] = \dfrac{a}{k}$

This is the rate of photosynthesis per unit area in a water column of infinite depth. In practice, this is a good approximation for an entire water column of unit area in the deep ocean.

29. We would expect a small percentage of bulbs to burn out in the first few hundred hours, most of the bulbs to burn out after close to 700 hours, and a few overachievers to burn on and on.

(a) 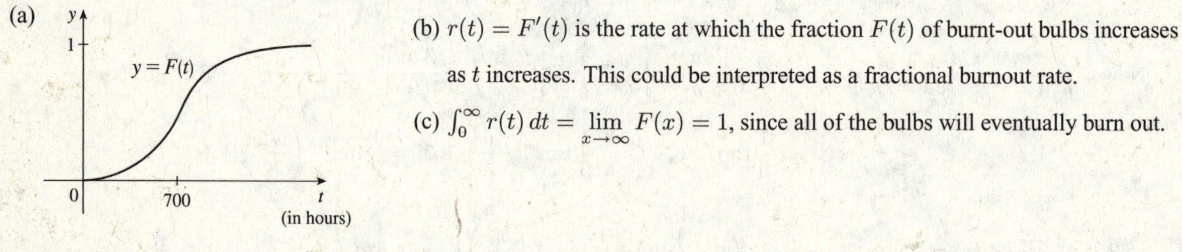

(b) $r(t) = F'(t)$ is the rate at which the fraction $F(t)$ of burnt-out bulbs increases as t increases. This could be interpreted as a fractional burnout rate.

(c) $\int_0^\infty r(t)\,dt = \lim\limits_{x\to\infty} F(x) = 1$, since all of the bulbs will eventually burn out.

31. (a) $I = \int_{-\infty}^{\infty} x\,dx = \int_{-\infty}^{0} x\,dx + \int_0^\infty x\,dx$, and $\int_0^\infty x\,dx = \lim\limits_{t\to\infty}\int_0^t x\,dx = \lim\limits_{t\to\infty}\left[\frac{1}{2}x^2\right]_0^t = \lim\limits_{t\to\infty}\left[\frac{1}{2}t^2 - 0\right] = \infty$, so I is divergent.

(b) $\int_{-t}^{t} x\,dx = \left[\frac{1}{2}x^2\right]_{-t}^{t} = \frac{1}{2}t^2 - \frac{1}{2}t^2 = 0$, so $\lim\limits_{t\to\infty}\int_{-t}^{t} x\,dx = 0$. Therefore, $\int_{-\infty}^{\infty} x\,dx \neq \lim\limits_{t\to\infty}\int_{-t}^{t} x\,dx$.

33. $$\int_0^\infty e^{-x^2/2}\,dx = \lim_{b\to\infty}\int_0^b e^{-x^2/2}\,dx = \sqrt{2}\lim_{b\to\infty}\int_0^{b/\sqrt{2}} e^{-u^2}\,du \qquad \left[\text{Substitution: } u = x/\sqrt{2},\ du = dx/\sqrt{2}\right]$$

$$= \int_0^\infty e^{-u^2}\,du = \sqrt{2}\left(\tfrac{1}{2}\sqrt{\pi}\right) = \tfrac{1}{2}\sqrt{2\pi} = \sqrt{\frac{\pi}{2}}$$

© 2016 Cengage Learning. All Rights Reserved. May not be scanned, copied or duplicated, or posted to a publicly accessible website, in whole or in part.

35. $\displaystyle\int_0^\infty \sqrt{x}e^{-x}\,dx = \lim_{b\to\infty}\int_0^b \sqrt{x}e^{-x}\,dx = \lim_{b\to\infty}\int_0^{\sqrt{b}} ue^{-u^2}\,(2u\,du)$ $\quad\left[\text{Substitution: } u=\sqrt{x},\ du=\tfrac{1}{2}x^{-1/2}dx=\tfrac{1}{2}u^{-1}dx\right]$

$$= 2\lim_{b\to\infty}\int_0^{\sqrt{b}} u^2e^{-u^2}\,du = 2\int_0^\infty u^2e^{-u^2}\,du = 2\left(\tfrac{1}{4}\sqrt{\pi}\right)\ \text{[From Exercise 34]}\ = \tfrac{1}{2}\sqrt{\pi}$$

5 Review

TRUE-FALSE QUIZ

1. True by Property 2 of the Integral in Section 5.2.

3. True by Property 3 of the Integral in Section 5.2.

5. False. For example, let $f(x)=x^2$. Then $\int_0^1 \sqrt{x^2}\,dx = \int_0^1 x\,dx = \frac{1}{2}$, but $\sqrt{\int_0^1 x^2\,dx} = \sqrt{\frac{1}{3}} = \frac{1}{\sqrt{3}}$.

7. True by Comparison Property 7 of the Integral in Section 5.2.

9. True. The integrand is an odd function that is continuous on $[-1,1]$.

11. False. See the paragraph before Note 4 and Figure 4 in Section 4.2, and notice that $y = x - x^3 < 0$ for $1 < x \le 2$.

13. False. For example, the function $y = |x|$ is continuous on $\mathbb{R}$, but has no derivative at $x = 0$.

15. False. If $f(x) = 1/x$, then f is continuous and decreasing on $[1,\infty)$ with $\lim\limits_{x\to\infty} f(x) = 0$, but $\int_1^\infty f(x)\,dx$ is divergent.

17. False. Take $f(x) = 1$ for all x and $g(x) = -1$ for all x. Then $\int_a^\infty f(x)\,dx = \infty$ [divergent] and $\int_a^\infty g(x)\,dx = -\infty$ [divergent], but $\int_a^\infty [f(x)+g(x)]\,dx = 0$ [convergent].

EXERCISES

1. (a)

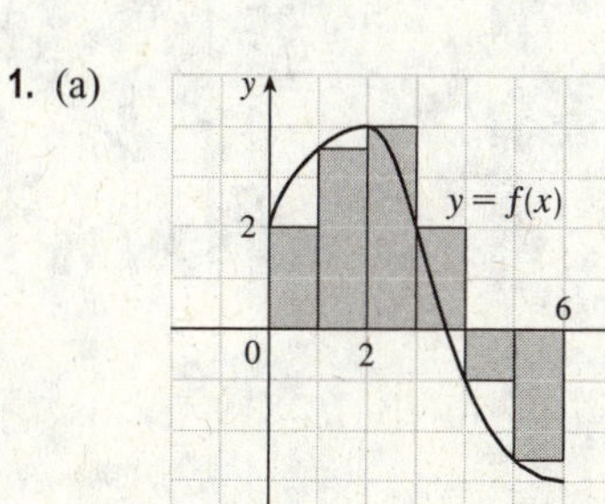

$$L_6 = \sum_{i=1}^{6} f(x_{i-1})\,\Delta x \quad \left[\Delta x = \tfrac{6-0}{6} = 1\right]$$

$$= f(x_0)\cdot 1 + f(x_1)\cdot 1 + f(x_2)\cdot 1 + f(x_3)\cdot 1 + f(x_4)\cdot 1 + f(x_5)\cdot 1$$

$$\approx 2 + 3.5 + 4 + 2 + (-1) + (-2.5) = 8$$

The Riemann sum represents the sum of the areas of the four rectangles above the x-axis minus the sum of the areas of the two rectangles below the x-axis.

(b)

$$M_6 = \sum_{i=1}^{6} f(\overline{x}_i)\,\Delta x \quad \left[\Delta x = \tfrac{6-0}{6} = 1\right]$$

$$= f(\overline{x}_1)\cdot 1 + f(\overline{x}_2)\cdot 1 + f(\overline{x}_3)\cdot 1 + f(\overline{x}_4)\cdot 1 + f(\overline{x}_5)\cdot 1 + f(\overline{x}_6)\cdot 1$$

$$= f(0.5) + f(1.5) + f(2.5) + f(3.5) + f(4.5) + f(5.5)$$

$$\approx 3 + 3.9 + 3.4 + 0.3 + (-2) + (-2.9) = 5.7$$

The Riemann sum represents the sum of the areas of the four rectangles above the x-axis minus the sum of the areas of the two rectangles below the x-axis.

© 2016 Cengage Learning. All Rights Reserved. May not be scanned, copied or duplicated, or posted to a publicly accessible website, in whole or in part.

3. $\int_0^1 (x + \sqrt{1 - x^2})\,dx = \int_0^1 x\,dx + \int_0^1 \sqrt{1 - x^2}\,dx = I_1 + I_2.$

I_1 can be interpreted as the area of the triangle shown in the figure and I_2 can be interpreted as the area of the quarter-circle.

Area $= \frac{1}{2}(1)(1) + \frac{1}{4}(\pi)(1)^2 = \frac{1}{2} + \frac{\pi}{4}.$

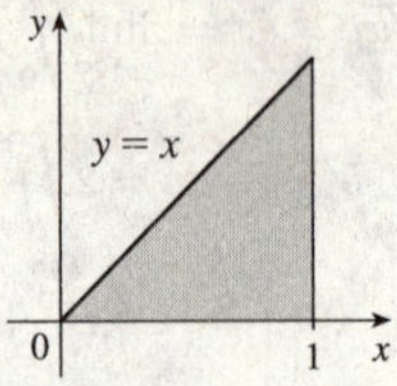

y
y = x
0 1 x

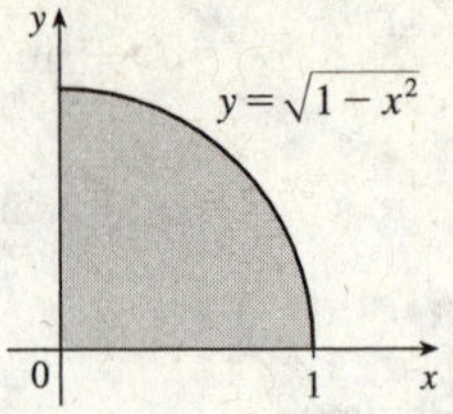

y
$y = \sqrt{1 - x^2}$
0 1 x

5. $\int_0^6 f(x)\,dx = \int_0^4 f(x)\,dx + \int_4^6 f(x)\,dx \quad \Rightarrow \quad 10 = 7 + \int_4^6 f(x)\,dx \quad \Rightarrow \quad \int_4^6 f(x)\,dx = 10 - 7 = 3$

7. First note that either a or b must be the graph of $\int_0^x f(t)\,dt$, since $\int_0^0 f(t)\,dt = 0$, and $c(0) \neq 0$. Now notice that $b > 0$ when c is increasing, and that $c > 0$ when a is increasing. It follows that c is the graph of $f(x)$, b is the graph of $f'(x)$, and a is the graph of $\int_0^x f(t)\,dt$.

9. $\int_1^2 (8x^3 + 3x^2)\,dx = \left[8 \cdot \frac{1}{4}x^4 + 3 \cdot \frac{1}{3}x^3\right]_1^2 = \left[2x^4 + x^3\right]_1^2 = (2 \cdot 2^4 + 2^3) - (2 + 1) = 40 - 3 = 37$

11. $\int_0^1 (1 - x^9)\,dx = \left[x - \frac{1}{10}x^{10}\right]_0^1 = \left(1 - \frac{1}{10}\right) - 0 = \frac{9}{10}$

13. $$\int \left(\frac{1 - x}{x}\right)^2 dx = \int \left(\frac{1}{x} - 1\right)^2 dx = \int \left(\frac{1}{x^2} - \frac{2}{x} + 1\right) dx = -\frac{1}{x} - 2\ln|x| + x + C$$

15. $u = x^2 + 1$, $du = 2x\,dx$, so $\displaystyle\int_0^1 \frac{x}{x^2 + 1}\,dx = \int_1^2 \frac{1}{u}\left(\tfrac{1}{2}\,du\right) = \tfrac{1}{2}\left[\ln u\right]_1^2 = \frac{1}{2}\ln 2.$

17. Let $u = v^3$, so $du = 3v^2\,dv$. When $v = 0$, $u = 0$; when $v = 1$, $u = 1$. Thus,

$\int_0^1 v^2 \cos(v^3)\,dv = \int_0^1 \cos u\left(\frac{1}{3}\,du\right) = \frac{1}{3}\left[\sin u\right]_0^1 = \frac{1}{3}(\sin 1 - 0) = \frac{1}{3}\sin 1.$

19. $\int_0^1 e^{\pi t}\,dt = \left[\frac{1}{\pi}e^{\pi t}\right]_0^1 = \frac{1}{\pi}(e^\pi - 1)$

21. Let $u = x^2 + 4x$. Then $du = (2x + 4)\,dx = 2(x + 2)\,dx$, so

$$\int \frac{x + 2}{\sqrt{x^2 + 4x}}\,dx = \int u^{-1/2}\left(\tfrac{1}{2}\,du\right) = \tfrac{1}{2} \cdot 2u^{1/2} + C = \sqrt{u} + C = \sqrt{x^2 + 4x} + C.$$

23. $$\int_0^5 \frac{x}{x + 10}\,dx = \int_0^5 \left(1 - \frac{10}{x + 10}\right) dx = \left[x - 10\ln(x + 10)\right]_0^5 = 5 - 10\ln 15 + 10\ln 10$$
$$= 5 + 10\ln\tfrac{10}{15} = 5 + 10\ln\tfrac{2}{3}$$

25. $\displaystyle\int_{-\pi/4}^{\pi/4} \frac{t^4 \tan t}{2 + \cos t}\,dt = 0$ by Theorem 4.5.6(b), since $f(t) = \dfrac{t^4 \tan t}{2 + \cos t}$ is an odd function.

27. $$\int_1^4 x^{3/2} \ln x\,dx \quad \begin{bmatrix} u = \ln x, & dv = x^{3/2}\,dx, \\ du = dx/x & v = \frac{2}{5}x^{5/2} \end{bmatrix} = \frac{2}{5}\left[x^{5/2}\ln x\right]_1^4 - \frac{2}{5}\int_1^4 x^{3/2}\,dx = \tfrac{2}{5}(32\ln 4 - \ln 1) - \tfrac{2}{5}\left[\tfrac{2}{5}x^{5/2}\right]_1^4$$
$$= \tfrac{2}{5}(64\ln 2) - \tfrac{4}{25}(32 - 1) = \tfrac{128}{5}\ln 2 - \tfrac{124}{25} \quad \left[\text{or } \tfrac{64}{5}\ln 4 - \tfrac{124}{25}\right]$$

© 2016 Cengage Learning. All Rights Reserved. May not be scanned, copied or duplicated, or posted to a publicly accessible website, in whole or in part.

29. Let $w = \sqrt[3]{x}$. Then $w^3 = x$ and $3w^2\,dw = dx$, so $\int e^{\sqrt[3]{x}}\,dx = \int e^w \cdot 3w^2\,dw = 3I$. To evaluate I, let $u = w^2$, $dv = e^w\,dw \Rightarrow du = 2w\,dw$, $v = e^w$, so $I = \int w^2 e^w\,dw = w^2 e^w - \int 2we^w\,dw$. Now let $U = w$, $dV = e^w\,dw \Rightarrow dU = dw$, $V = e^w$. Thus, $I = w^2e^w - 2\left[we^w - \int e^w\,dw\right] = w^2e^w - 2we^w + 2e^w + C_1$, and hence $3I = 3e^w(w^2 - 2w + 2) + C = 3e^{\sqrt[3]{x}}(x^{2/3} - 2x^{1/3} + 2) + C$.

31. Let $u = 1 + \sec\theta$. Then $du = \sec\theta\,\tan\theta\,d\theta$, so

$$\int \frac{\sec\theta\,\tan\theta}{1+\sec\theta}\,d\theta = \int \frac{1}{1+\sec\theta}\,(\sec\theta\,\tan\theta\,d\theta) = \int \frac{1}{u}\,du = \ln|u| + C = \ln|1+\sec\theta| + C.$$

33. From the graph, it appears that the area under the curve $y = x\sqrt{x}$ between $x = 0$ and $x = 4$ is somewhat less than half the area of an 8×4 rectangle, so perhaps about 13 or 14. To find the exact value, we evaluate

$\int_0^4 x\sqrt{x}\,dx = \int_0^4 x^{3/2}\,dx = \left[\frac{2}{5}x^{5/2}\right]_0^4 = \frac{2}{5}(4)^{5/2} = \frac{64}{5} = 12.8.$

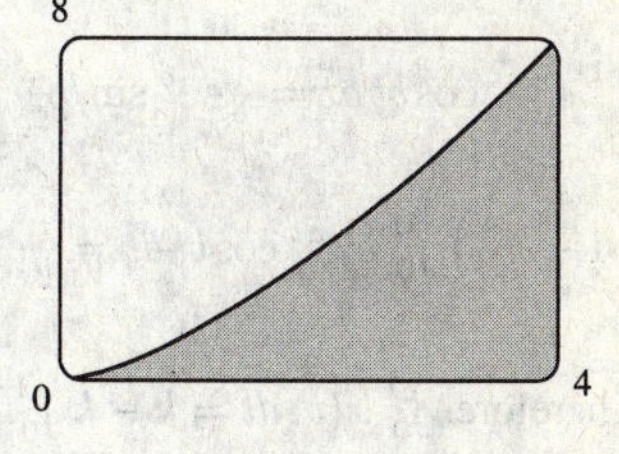

35. Let $u = \sin x$. Then $\dfrac{du}{dx} = \cos x$. Also, $\dfrac{dg}{dx} = \dfrac{dg}{du}\dfrac{du}{dx}$, so

$$g'(x) = \frac{d}{dx}\int_1^{\sin x} \frac{1-t^2}{1+t^4}\,dt = \frac{d}{du}\int_1^{u} \frac{1-t^2}{1+t^4}\,dt \cdot \frac{du}{dx} = \frac{1-u^2}{1+u^4}\cdot\frac{du}{dx} = \frac{1-\sin^2 x}{1+\sin^4 x}\cdot\cos x = \frac{\cos^3 x}{1+\sin^4 x}$$

37. $u = e^x \Rightarrow du = e^x\,dx$, so

$\int e^x\sqrt{1-e^{2x}}\,dx = \int\sqrt{1-u^2}\,du \overset{30}{=} \frac{1}{2}u\sqrt{1-u^2} + \frac{1}{2}\sin^{-1}u + C = \frac{1}{2}\left[e^x\sqrt{1-e^{2x}} + \sin^{-1}(e^x)\right] + C.$

39. If $1 \le x \le 3$, then $\sqrt{1^2+3} \le \sqrt{x^2+3} \le \sqrt{3^2+3} \Rightarrow 2 \le \sqrt{x^2+3} \le 2\sqrt{3}$, so $2(3-1) \le \int_1^3 \sqrt{x^2+3}\,dx \le 2\sqrt{3}(3-1)$; that is, $4 \le \int_1^3 \sqrt{x^2+3}\,dx \le 4\sqrt{3}$.

41.
$$\int_1^\infty \frac{1}{(2x+1)^3}\,dx = \lim_{t\to\infty}\int_1^t \frac{1}{(2x+1)^3}\,dx = \lim_{t\to\infty}\int_1^t \tfrac{1}{2}(2x+1)^{-3}\,2\,dx = \lim_{t\to\infty}\left[-\frac{1}{4(2x+1)^2}\right]_1^t$$
$$= -\frac{1}{4}\lim_{t\to\infty}\left[\frac{1}{(2t+1)^2} - \frac{1}{9}\right] = -\frac{1}{4}\left(0 - \frac{1}{9}\right) = \frac{1}{36}$$

43. $\int_{-\infty}^0 e^{-2x}\,dx = \lim\limits_{t\to-\infty}\int_t^0 e^{-2x}\,dx = \lim\limits_{t\to-\infty}\left[-\frac{1}{2}e^{-2x}\right]_t^0 = \lim\limits_{t\to-\infty}\left(-\frac{1}{2} + \frac{1}{2}e^{-2t}\right) = \infty.$ Divergent

45. Note that $r(t) = b'(t)$, where $b(t) =$ the number of barrels of oil consumed up to time t. So, by the Net Change Theorem, $\int_0^{15} r(t)\,dt = b(15) - b(0)$ represents the number of barrels of oil consumed from Jan. 1, 2000, through Jan. 1, 2015.

47. The total amount of oil that will be spilled is given by:

$$\int_0^\infty r(t)\,dt = \lim_{b\to\infty}\int_0^b 90e^{-0.12t}\,dt = \lim_{b\to\infty}\left[\frac{90}{-0.12}e^{-0.12t}\right]_0^b = 750\lim_{b\to\infty}\left[1 - e^{-0.12b}\right] = 750(1-0) = 750 \text{ gallons}$$

© 2016 Cengage Learning. All Rights Reserved. May not be scanned, copied or duplicated, or posted to a publicly accessible website, in whole or in part.

49. (a) Since $r(t)$ is the net rate of change, the net change in population size between times $t = a$ and $t = b$ is $\int_a^b r(t)\,dt$.

(b) $\int_a^b r(t)\,dt = \int_a^b [b(t) - d(t)]\,dt = \int_a^b b(t)\,dt - \int_a^b d(t)\,dt$ [by Property 4]. Since $b(t)$ and $d(t)$ are the birth and death rates, we have $\int_a^b r(t)\,dt = (\text{Number of births}) - (\text{Number of deaths})$ over the time period from $t = a$ to $t = b$.

51. The measure of impact over the first unit of time is $\int_0^1 x(t)\,dt = \int_0^1 \left[k - ke^{-at}\cos bt\right]\,dt = kt\big]_0^1 - k\int_0^1 e^{-at}\cos bt\,dt$.

Now using parts with $u = e^{-at}$, $dv = \cos bt\,dt$, $du = -ae^{-at}dt$, $v = \frac{1}{b}\sin bt$ gives

$\int_0^1 e^{-at}\cos bt\,dt = \left[\frac{1}{b}e^{-at}\sin bt\right]_0^1 + \frac{a}{b}\int_0^1 e^{-at}\sin bt\,dt$. Using parts again

with $U = e^{-at}$, $dV = \sin bt\,dt$, $dU = -ae^{-at}dt$, $V = -\frac{1}{b}\cos bt$ gives

$\int_0^1 e^{-at}\cos bt\,dt = \frac{1}{b}e^{-a}\sin b + \frac{a}{b}\left(\left[-\frac{1}{b}e^{-at}\cos bt\right]_0^1 - \frac{a}{b}\int_0^1 e^{-at}\cos bt\,dt\right)$. Solving for the integral algebraically gives

$$\left(1 + \frac{a^2}{b^2}\right)\int_0^1 e^{-at}\cos bt\,dt = \frac{1}{b}e^{-a}\sin b - \frac{a}{b^2}e^{-a}\cos b + \frac{a}{b^2} \quad\Rightarrow\quad \int_0^1 e^{-at}\cos bt\,dt = \frac{be^{-a}\sin b - ae^{-a}\cos b + a}{a^2 + b^2}.$$

Therefore, $\int_0^1 x(t)\,dt = k - k\int_0^1 e^{-at}\cos bt\,dt = k - k\left(\dfrac{be^{-a}\sin b - ae^{-a}\cos b + a}{a^2 + b^2}\right)$.

53. Using FTC1, we differentiate both sides of the given equation, $\int_0^x f(t)\,dt = xe^{2x} + \int_0^x e^{-t}f(t)\,dt$, and get

$$f(x) = e^{2x} + 2xe^{2x} + e^{-x}f(x) \quad\Rightarrow\quad f(x)\left(1 - e^{-x}\right) = e^{2x} + 2xe^{2x} \quad\Rightarrow\quad f(x) = \frac{e^{2x}(1 + 2x)}{1 - e^{-x}}.$$

55. Let $u = f(x)$ and $du = f'(x)\,dx$. So $2\int_a^b f(x)f'(x)\,dx = 2\int_{f(a)}^{f(b)} u\,du = \left[u^2\right]_{f(a)}^{f(b)} = [f(b)]^2 - [f(a)]^2$.

© 2016 Cengage Learning. All Rights Reserved. May not be scanned, copied or duplicated, or posted to a publicly accessible website, in whole or in part.

6 □ APPLICATIONS OF INTEGRATION

6.1 Areas Between Curves

1. $A = \int_{x=0}^{x=4} (y_T - y_B)\,dx = \int_0^4 \left[(5x - x^2) - x\right] dx = \int_0^4 (4x - x^2)\,dx = \left[2x^2 - \frac{1}{3}x^3\right]_0^4 = \left(32 - \frac{64}{3}\right) - (0) = \frac{32}{3}$

3. $A = \int_{-1}^{1} \left[e^x - (x^2 - 1)\right] dx = \left[e^x - \frac{1}{3}x^3 + x\right]_{-1}^{1}$

$= \left(e - \frac{1}{3} + 1\right) - \left(e^{-1} + \frac{1}{3} - 1\right) = e - \frac{1}{e} + \frac{4}{3}$

5. $A = \int_0^1 \left(\sqrt{x} - x^2\right) dx$

$= \left[\frac{2}{3}x^{3/2} - \frac{1}{3}x^3\right]_0^1$

$= \frac{2}{3} - \frac{1}{3} = \frac{1}{3}$

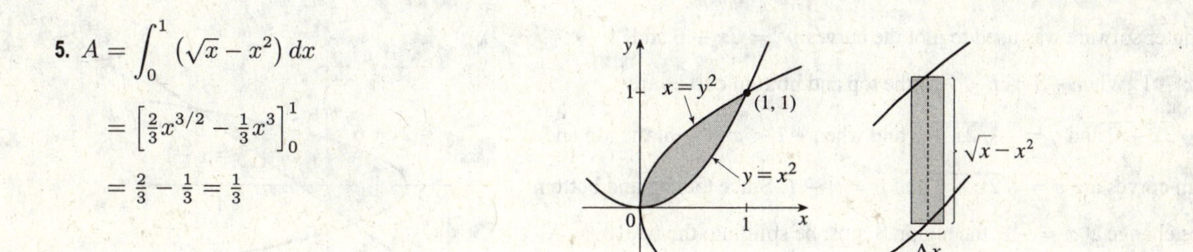

7. $12 - x^2 = x^2 - 6 \;\Leftrightarrow\; 2x^2 = 18 \;\Leftrightarrow$

$x^2 = 9 \;\Leftrightarrow\; x = \pm 3$, so

$A = \int_{-3}^{3} \left[(12 - x^2) - (x^2 - 6)\right] dx$

$= 2\int_0^3 (18 - 2x^2)\,dx$ [by symmetry]

$= 2\left[18x - \frac{2}{3}x^3\right]_0^3 = 2\left[(54 - 18) - 0\right]$

$= 2(36) = 72$

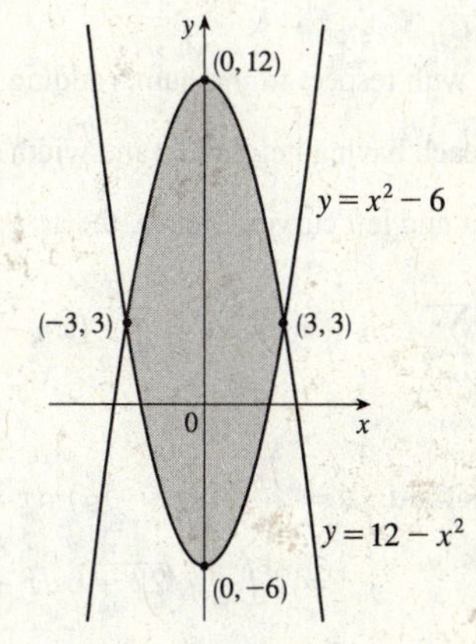

9. $e^x = xe^x \;\Leftrightarrow\; e^x - xe^x = 0 \;\Leftrightarrow\; e^x(1 - x) = 0 \;\Leftrightarrow\; x = 1.$

$A = \int_0^1 (e^x - xe^x)\,dx$

$= \left[e^x - (xe^x - e^x)\right]_0^1$ [use parts with $u = x$ and $dv = e^x\,dx$]

$= \left[2e^x - xe^x\right]_0^1 = (2e - e) - (2 - 0) = e - 2$

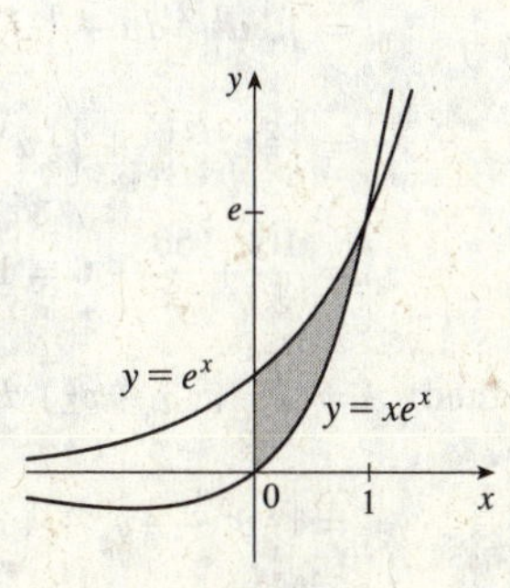

© 2016 Cengage Learning. All Rights Reserved. May not be scanned, copied or duplicated, or posted to a publicly accessible website, in whole or in part.

11.

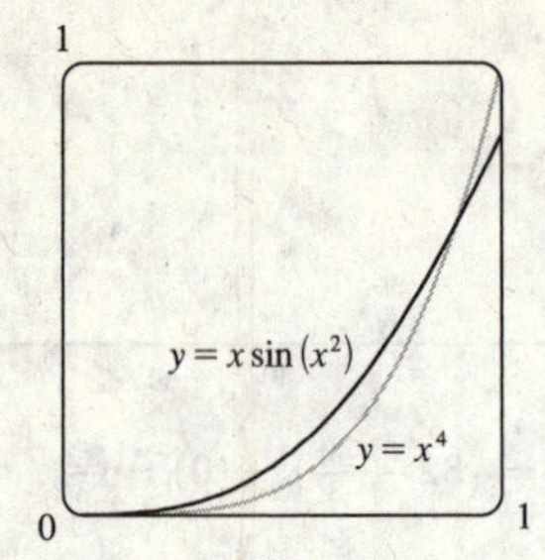

From the graph, we see that the curves intersect at $x = 0$ and $x = a \approx 0.896$, with $x\sin(x^2) > x^4$ on $(0, a)$. So the area A of the region bounded by the curves is

$$A = \int_0^a \left[x\sin(x^2) - x^4\right] dx = \left[-\tfrac{1}{2}\cos(x^2) - \tfrac{1}{5}x^5\right]_0^a$$
$$= -\tfrac{1}{2}\cos(a^2) - \tfrac{1}{5}a^5 + \tfrac{1}{2} \approx 0.037$$

13. $\cos x = \sin 2x = 2\sin x\,\cos x \quad\Leftrightarrow\quad 2\sin x\,\cos x - \cos x = 0 \quad\Leftrightarrow\quad \cos x\,(2\sin x - 1) = 0 \quad\Leftrightarrow$

$2\sin x = 1$ or $\cos x = 0 \quad\Leftrightarrow\quad x = \frac{\pi}{6}$ or $\frac{\pi}{2}$.

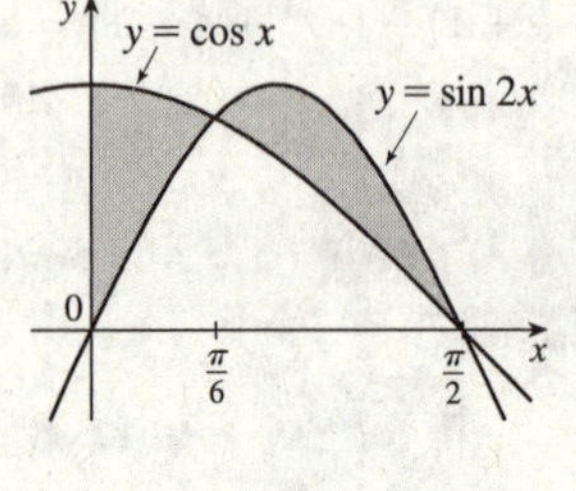

$$A = \int_0^{\pi/6}(\cos x - \sin 2x)\,dx + \int_{\pi/6}^{\pi/2}(\sin 2x - \cos x)\,dx$$
$$= \left[\sin x + \tfrac{1}{2}\cos 2x\right]_0^{\pi/6} + \left[-\tfrac{1}{2}\cos 2x - \sin x\right]_{\pi/6}^{\pi/2}$$
$$= \left(\tfrac{1}{2} + \tfrac{1}{2}\cdot\tfrac{1}{2}\right) - \left(0 + \tfrac{1}{2}\cdot 1\right) + \left[-\tfrac{1}{2}\cdot(-1) - 1\right] - \left(-\tfrac{1}{2}\cdot\tfrac{1}{2} - \tfrac{1}{2}\right)$$
$$= \tfrac{3}{4} - \tfrac{1}{2} - \tfrac{1}{2} + \tfrac{3}{4} = \tfrac{1}{2}$$

15. (a) Computer sofware was used to plot the curves $y^2 = 2x + 6$ and $y = x - 1$. When $-3 < x < -1$, the top and bottom curves are $y = \sqrt{2x+6}$ and $y = -\sqrt{2x+6}$ and when $-1 < x < -5$, the top and bottom curves are $y = \sqrt{2x+6}$ and $y = x - 1$. Since the top and bottom curves change at $x = -1$, the region S must be split into the two areas A_1 and A_2 (shown in the figure) in order to integrate with respect to x.

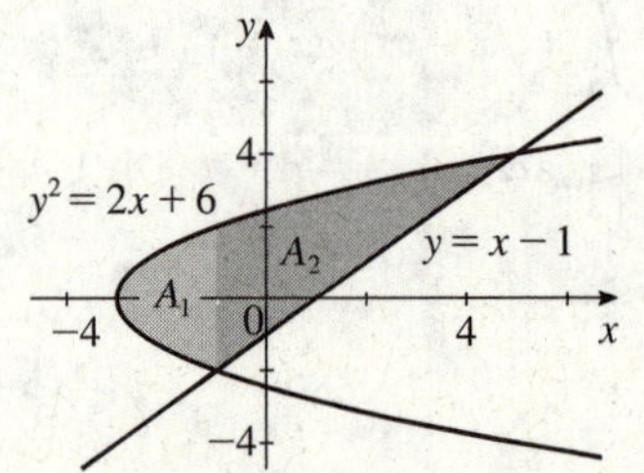

(b) Integrating with respect to y requires adding the areas of approximating rectangles each having height Δy and width $(x_R - x_L)$ where x_R and x_L are the right and left curves. Hence, the area between the curves is

$$A = \lim_{n\to\infty}\sum_{i=1}^{\infty}(x_R - x_L)\,\Delta y = \int_{-2}^{4}(x_R - x_L)\,dy$$

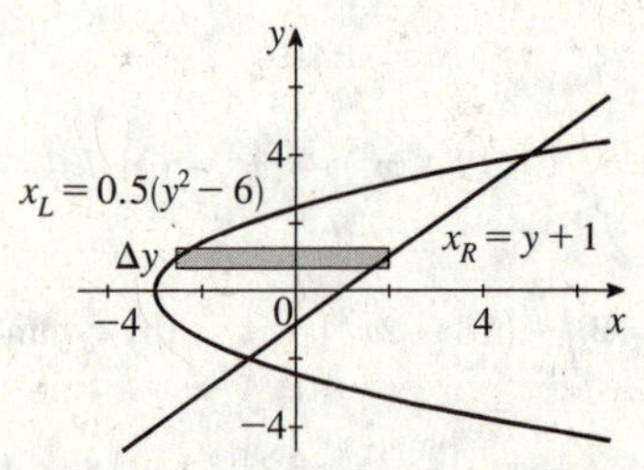

(c) **Part (a) Method:** $A = \int_{-3}^{5}(y_T - y_B)\,dx = \int_{-3}^{-1}\left[\sqrt{2x+6} - (-\sqrt{2x+6})\right]dx + \int_{-1}^{5}\left[\sqrt{2x+6} - (x-1)\right]dx$

$$= 2\int_{-3}^{-1}\sqrt{2x+6}\,dx + \int_{-1}^{5}\sqrt{2x+6}\,dx - \int_{-1}^{5}x\,dx + \int_{-1}^{5}(1)\,dx$$
$$= \int_0^4 u^{1/2}\,du + \tfrac{1}{2}\int_4^{16}u^{1/2}\,du - \tfrac{1}{2}\left[x^2\right]_{-1}^{5} + \left[x\right]_{-1}^{5} \qquad \text{[Substitution: } u = 2x+6,\ du = 2\,dx]$$
$$= \left[\tfrac{2}{3}u^{3/2}\right]_0^4 + \left[\tfrac{1}{3}u^{3/2}\right]_4^{16} - \tfrac{1}{2}(25-1) + (5+1) = \tfrac{2}{3}\left(4^{3/2} - 0\right) + \tfrac{1}{3}\left(16^{3/2} - 4^{3/2}\right) - 6$$
$$= \frac{16}{3} + \frac{56}{3} - 6 = 18$$

Part (b) Method: $A = \displaystyle\int_{-2}^{4}(x_R - x_L)\,dy = \int_{-2}^{4}\left[y + 1 - \tfrac{1}{2}(y^2 - 6)\right]dy = \int_{-2}^{4}\left[y - \tfrac{1}{2}y^2 + 4\right]dy$

$$= \left[\tfrac{1}{2}y^2 - \tfrac{1}{6}y^3 + 4y\right]_{-2}^{4} = \tfrac{1}{2}(4)^2 - \tfrac{1}{6}(4)^3 + 4(4) - \left(\tfrac{1}{2}(-2)^2 - \tfrac{1}{6}(-2)^3 + 4(-2)\right)$$
$$= 8 - \tfrac{32}{3} + 16 - \left(2 + \tfrac{4}{3} + -8\right) = 18$$

© 2016 Cengage Learning. All Rights Reserved. May not be scanned, copied or duplicated, or posted to a publicly accessible website, in whole or in part.

17. Let $W(x)$ denote the width of the leaf at x cm from the left end. Using the Midpoint Rule with $\Delta x = \frac{6-0}{6} = 1$ gives

$$A \approx M_6 = \Delta x\,[W(0.5) + W(1.5) + W(2.5) + W(3.5) + W(4.5) + W(5.5)]$$
$$= 1(0.3 + 2.3 + 3.2 + 3.3 + 2.6 + 1.3) = 1(13) = 13 \text{ cm}^2$$

19. Let $h(x)$ denote the height of the wing at x cm from the left end.

$$A \approx M_5 = \frac{200-0}{5}\,[h(20) + h(60) + h(100) + h(140) + h(180)]$$
$$= 40(20.3 + 29.0 + 27.3 + 20.5 + 8.7) = 40(105.8) = 4232 \text{ cm}^2$$

21. (a) Using the Midpoint Rule with $n = 5$, $\Delta t = \frac{10-0}{5} = 2$ and the midpoints $\{1, 3, 5, 7, 9\}$, we have

$$\int_0^{10} [A(t) - V(t)]\,dt \approx \Delta t\,[(A(1) - V(1)) + (A(3) - V(3)) + (A(5) - V(5)) + (A(7) - V(7)) + (A(9) - V(9))]$$
$$= 2\,[0.023 + 0.012 + 0.007 + 0.002 + 0.001] = 2\,(0.045) = 0.09\,(\text{mL/mL})\cdot\text{min}$$

(b) With $Q_B(10) = 64$ mL, the cerebral blood flow is $F = \dfrac{Q_B(10)}{\int_0^{10}[A(t) - V(t)]\,dt} \approx \dfrac{64\text{ mL}}{0.09\,(\text{mL/mL})\cdot\text{min}} \approx 711$ mL/min.

23. 1 second $= \frac{1}{3600}$ hour, so 10 s $= \frac{1}{360}$ h. With the given data, we can take $n = 5$ to use the Midpoint Rule.

$\Delta t = \frac{1/360 - 0}{5} = \frac{1}{1800}$, so

$$\text{distance}_{\text{Kelly}} - \text{distance}_{\text{Chris}} = \int_0^{1/360} v_K\,dt - \int_0^{1/360} v_C\,dt = \int_0^{1/360} (v_K - v_C)\,dt$$
$$\approx M_5 = \tfrac{1}{1800}\,[(v_K - v_C)(1) + (v_K - v_C)(3) + (v_K - v_C)(5) + (v_K - v_C)(7) + (v_K - v_C)(9)]$$
$$= \tfrac{1}{1800}[(22 - 20) + (52 - 46) + (71 - 62) + (86 - 75) + (98 - 86)]$$
$$= \tfrac{1}{1800}(2 + 6 + 9 + 11 + 12) = \tfrac{1}{1800}(40) = \tfrac{1}{45} \text{ mile, or } 117\tfrac{1}{3} \text{ feet}$$

25. For $0 \le t \le 10$, $b(t) > d(t)$, so the area between the curves is given by

$$\int_0^{10} [b(t) - d(t)]\,dt = \int_0^{10} (2200e^{0.024t} - 1460e^{0.018t})\,dt = \left[\frac{2200}{0.024}e^{0.024t} - \frac{1460}{0.018}e^{0.018t}\right]_0^{10}$$
$$= \left(\frac{275{,}000}{3}e^{0.24} - \frac{730{,}000}{9}e^{0.18}\right) - \left(\frac{275{,}000}{3} - \frac{730{,}000}{9}\right) \approx 8868 \text{ people}$$

This area A represents the increase in population over a 10-year period.

27. We first assume that $c > 0$, since c can be replaced by $-c$ in both equations without changing the graphs, and if $c = 0$ the curves do not enclose a region. We see from the graph that the enclosed area A lies between $x = -c$ and $x = c$, and by symmetry, it is equal to four times the area in the first quadrant. The enclosed area is

$A = 4\int_0^c (c^2 - x^2)\,dx = 4\big[c^2x - \frac{1}{3}x^3\big]_0^c = 4\big(c^3 - \frac{1}{3}c^3\big) = 4\big(\frac{2}{3}c^3\big) = \frac{8}{3}c^3$

So $A = 576 \iff \frac{8}{3}c^3 = 576 \iff c^3 = 216 \iff c = \sqrt[3]{216} = 6$.

Note that $c = -6$ is another solution, since the graphs are the same.

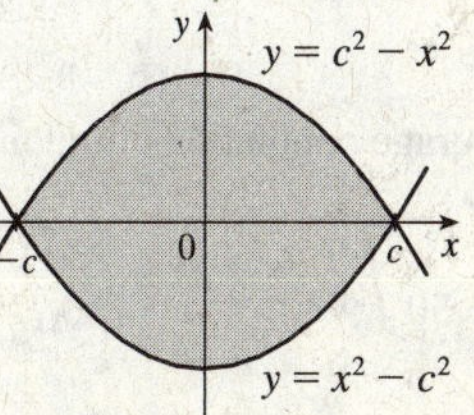

© 2016 Cengage Learning. All Rights Reserved. May not be scanned, copied or duplicated, or posted to a publicly accessible website, in whole or in part.

6.2 Average Values

1. $f_{\text{ave}} = \frac{1}{b-a}\int_a^b f(x)\,dx = \frac{1}{4-0}\int_0^4 (4x - x^2)\,dx = \frac{1}{4}\left[2x^2 - \frac{1}{3}x^3\right]_0^4 = \frac{1}{4}\left[\left(32 - \frac{64}{3}\right) - 0\right] = \frac{1}{4}\left(\frac{32}{3}\right) = \frac{8}{3}$

3. $g_{\text{ave}} = \frac{1}{b-a}\int_a^b g(x)\,dx = \frac{1}{8-1}\int_1^8 \sqrt[3]{x}\,dx = \frac{1}{7}\left[\frac{3}{4}x^{4/3}\right]_1^8 = \frac{3}{28}(16-1) = \frac{45}{28}$

5. $h_{\text{ave}} = \frac{1}{\pi - 0}\int_0^\pi \cos^4 x \sin x\,dx = \frac{1}{\pi}\int_1^{-1} u^4(-du)$ $[u = \cos x,\ du = -\sin x\,dx]$

$= \frac{1}{\pi}\int_{-1}^1 u^4\,du = \frac{1}{\pi}\cdot 2\int_0^1 u^4\,du$ [by Theorem 5.4.6(a)] $= \frac{2}{\pi}\left[\frac{1}{5}u^5\right]_0^1 = \frac{2}{5\pi}$

7. (a) $f_{\text{ave}} = \dfrac{1}{5-2}\displaystyle\int_2^5 (x-3)^2\,dx = \frac{1}{3}\left[\frac{1}{3}(x-3)^3\right]_2^5$

$= \frac{1}{9}\left[2^3 - (-1)^3\right] = \frac{1}{9}(8+1) = 1$

(b) $f(c) = f_{\text{ave}} \Leftrightarrow (c-3)^2 = 1 \Leftrightarrow$ $c - 3 = \pm 1 \Leftrightarrow c = 2$ or 4

(c)

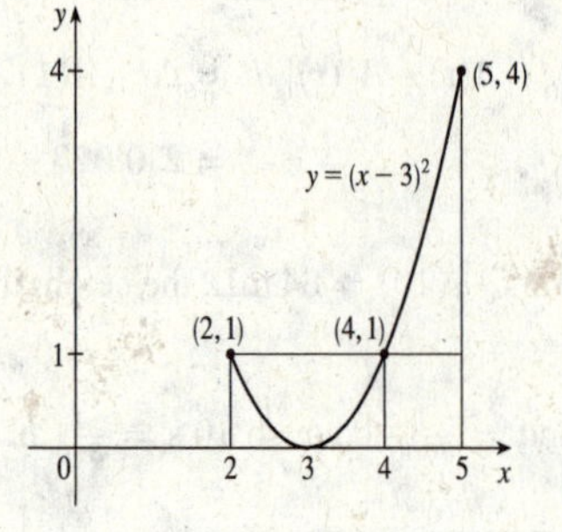

9. (a) $f_{\text{ave}} = \dfrac{1}{\pi - 0}\displaystyle\int_0^\pi (2\sin x - \sin 2x)\,dx$

$= \frac{1}{\pi}\left[-2\cos x + \frac{1}{2}\cos 2x\right]_0^\pi$

$= \frac{1}{\pi}\left[\left(2 + \frac{1}{2}\right) - \left(-2 + \frac{1}{2}\right)\right] = \frac{4}{\pi}$

(b) $f(c) = f_{\text{ave}} \Leftrightarrow 2\sin c - \sin 2c = \frac{4}{\pi} \Leftrightarrow$ $c_1 \approx 1.238$ or $c_2 \approx 2.808$

(c)

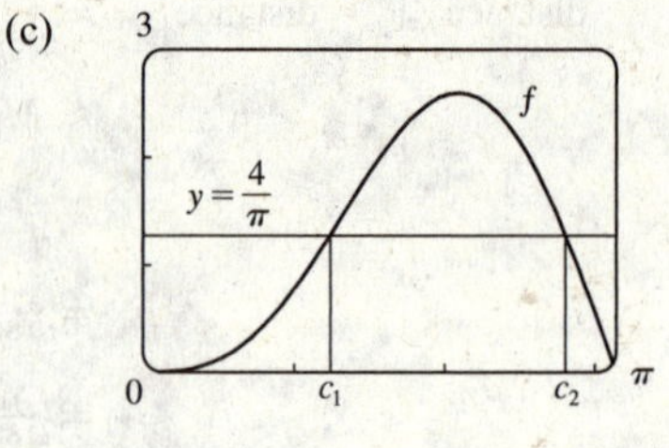

11. Use geometric interpretations to find the values of the integrals.

$\int_0^8 f(x)\,dx = \int_0^1 f(x)\,dx + \int_1^2 f(x)\,dx + \int_2^3 f(x)\,dx + \int_3^4 f(x)\,dx + \int_4^6 f(x)\,dx + \int_6^7 f(x)\,dx + \int_7^8 f(x)\,dx$

$= -\frac{1}{2} + \frac{1}{2} + \frac{1}{2} + 1 + 4 + \frac{3}{2} + 2 = 9$

Thus, the average value of f on $[0,8] = f_{\text{ave}} = \frac{1}{8-0}\int_0^8 f(x)\,dx = \frac{1}{8}(9) = \frac{9}{8}$.

13. Let $t = 0$ and $t = 12$ correspond to 9 AM and 9 PM, respectively.

$$T_{\text{ave}} = \frac{1}{12-0}\int_0^{12}\left[50 + 14\sin\tfrac{1}{12}\pi t\right]dt = \frac{1}{12}\left[50t - 14\cdot\tfrac{12}{\pi}\cos\tfrac{1}{12}\pi t\right]_0^{12}$$

$$= \frac{1}{12}\left[50\cdot 12 + 14\cdot\tfrac{12}{\pi} + 14\cdot\tfrac{12}{\pi}\right] = \left(50 + \tfrac{28}{\pi}\right)\,^\circ\text{F} \approx 59^\circ\text{F}$$

15. The average population of Indonesia in the second half of the century was

$$P_{\text{ave}} = \frac{1}{50}\int_0^{50} P(t)\,dt = \frac{83}{50}\int_0^{50} e^{0.18t}\,dt = \frac{83}{(50)(0.18)}\left[e^{0.18t}\right]_0^{50} = \frac{83}{(50)(0.18)}\left[e^9 - 1\right] \approx 74{,}719 \text{ million people}$$

© 2016 Cengage Learning. All Rights Reserved. May not be scanned, copied or duplicated, or posted to a publicly accessible website, in whole or in part.

17. The average level of infection over the 21 day infection period is

$$f_{\text{ave}} = \frac{1}{21}\int_0^{21} f(t)\,dt = \frac{1}{21}\int_0^{21} [-t(t-21)(t+1)]\,dt = \frac{1}{21}\int_0^{21} (-t^3 + 20t^2 + 21t)\,dt = \frac{1}{21}\left[-\tfrac{1}{4}t^4 + \tfrac{20}{3}t^3 + \tfrac{21}{2}t^2\right]_0^{21}$$

$$= \frac{1}{21}\left[-\tfrac{1}{4}(21)^4 + \tfrac{20}{3}(21)^3 + \tfrac{21}{2}(21)^2\right] - \frac{1}{21}[0] = \frac{17750.25}{21} = 845.25 \text{ cells/mL}$$

19. The average rate of air flow during inhalation is

$$f_{\text{ave}} = \frac{1}{2.5}\int_0^{2.5} \frac{1}{2}\sin(2\pi t/5)\,dt = \frac{1}{5}\left(-\frac{5}{2\pi}\right)[\cos(2\pi t/5)]_0^{2.5} = -\frac{1}{2\pi}[\cos\pi - \cos 0] = -\frac{1}{2\pi}(-1-1) = \frac{1}{\pi} \approx 0.32 \text{ L/s}$$

21. f is continuous on $[1,3]$, so by the Mean Value Theorem for Integrals there exists a number c in $[1,3]$ such that $\int_1^3 f(x)\,dx = f(c)(3-1) \;\Rightarrow\; 8 = 2f(c)$; that is, there is a number c such that $f(c) = \frac{8}{2} = 4$.

23. Let $F(x) = \int_a^x f(t)\,dt$ for x in $[a,b]$. Then F is continuous on $[a,b]$ and differentiable on (a,b), so by the Mean Value Theorem there is a number c in (a,b) such that $F(b) - F(a) = F'(c)(b-a)$. But $F'(x) = f(x)$ by the Fundamental Theorem of Calculus. Therefore, $\int_a^b f(t)\,dt - 0 = f(c)(b-a)$.

6.3 Further Applications to Biology

1. (a) According to the survival function, the proportion of the population that survives is $S(4) = \frac{1}{5}$, so $\frac{1}{5}(7400) = 1480$ of the original members survive.

(b) $\int_0^4 R(t)\,dt = \int_0^4 (2240 + 60t)\,dt = \left[2240t + 30t^2\right]_0^4 = 8960 + 480 = 9440$ members are added.

(c) Not all of the 9440 new members survive.

3. According to Equation 1, the population $T = 12$ weeks from now is

$$P(12) = S(12)\cdot P_0 + \int_0^{12} S(12-t)\,R(t)\,dt = e^{-0.2(12)}(22{,}500) + \int_0^{12} e^{-0.2(12-t)}(1225e^{0.14t})\,dt$$

$$= 22{,}500e^{-2.4} + 1225e^{-2.4}\int_0^{12} e^{0.34t}\,dt = 22{,}500e^{-2.4} + 1225e^{-2.4}\left[\frac{1}{0.34}e^{0.34t}\right]_0^{12}$$

$$= 22{,}500e^{-2.4} + \frac{1225}{0.34}e^{-2.4}(e^{4.08} - e^0) \approx 21{,}046 \text{ insects}$$

5. $P(8) = S(8)\cdot P_0 + \int_0^8 S(8-t)\,R(t)\,dt = e^{-0.25(8)}(50) + \int_0^8 e^{-0.25(8-t)}(12)\,dt$

$$= 50e^{-2} + 12e^{-2}\int_0^8 e^{0.25t}\,dt = 50e^{-2} + 12e^{-2}\left[\frac{1}{0.25}e^{0.25t}\right]_0^8$$

$$= 50e^{-2} + 48e^{-2}(e^2 - e^0) = 50e^{-2} + 48 - 48e^{-2} = 2e^{-2} + 48 \approx 48.3 \text{ mg}$$

7. $P(18) = S(18)\cdot P_0 + \int_0^{18} S(18-t)\,R(t)\,dt = e^{-0.32(18)}(10{,}000) + \int_0^{18} e^{-0.32(18-t)}(1600e^{0.06t})\,dt$

$$= 10{,}000e^{-5.76} + 1600e^{-5.76}\int_0^{18} e^{0.38t}\,dt = 10{,}000e^{-5.76} + 1600e^{-5.76}\left[\frac{1}{0.38}e^{0.38t}\right]_0^{18}$$

$$= 10{,}000e^{-5.76} + \frac{1600}{0.38}e^{-5.76}(e^{6.84} - 1) \approx 12{,}417 \text{ gallons}$$

© 2016 Cengage Learning. All Rights Reserved. May not be scanned, copied or duplicated, or posted to a publicly accessible website, in whole or in part.

9. $F = \dfrac{\pi P R^4}{8\eta l} = \dfrac{\pi(4000)(0.008)^4}{8(0.027)(2)} \approx 0.000119\ \text{cm}^3/\text{s}$

11. From Equation 3, $F = \dfrac{A}{\int_0^T c(t)\,dt} = \dfrac{6}{20I}$, where

$$I = \int_0^{10} te^{-0.6t}\,dt = \left[\frac{1}{(-0.6)^2}(-0.6t-1)\,e^{-0.6t}\right]_0^{10} \quad \left[\begin{array}{c}\text{integrating}\\ \text{by parts}\end{array}\right] = \frac{1}{0.36}(-7e^{-6}+1)$$

Thus, $F = \dfrac{6(0.36)}{20(1-7e^{-6})} = \dfrac{0.108}{1-7e^{-6}} \approx 0.1099\ \text{L/s or } 6.594\ \text{L/min}$.

13. Divide the time interval into 8 subintervals of equal length $\Delta t = (16-0)/8 = 2$. Then the midpoints of the subintervals are $\bar{t}_1 = 1, \bar{t}_2 = 3, \ldots, \bar{t}_8 = 15$. Then

$$\begin{aligned}\int_0^{16} c(t)\,dt \approx M_8 &= c(1)\,\Delta t + c(3)\,\Delta t + c(5)\,\Delta t + c(7)\,\Delta t + c(9)\,\Delta t + c(11)\,\Delta t + c(13)\,\Delta t + c(15)\,\Delta t \\ &\approx \Delta t(4.0 + 7.1 + 7.2 + 6.1 + 4.7 + 3.5 + 2.5 + 1.8) = (2)(36.9) = 73.8\end{aligned}$$

So $F = \dfrac{A}{\int_0^{16} c(t)\,dt} \approx \dfrac{7}{73.8} \approx 0.0949\ \text{L/s} \approx 5.69\ \text{L/min}$.

6.4 Volumes

1. A cross-section is a disk with radius $2 - \frac{1}{2}x$, so its area is $A(x) = \pi\left(2 - \frac{1}{2}x\right)^2$.

$$\begin{aligned}V &= \int_1^2 A(x)\,dx = \int_1^2 \pi\left(2 - \tfrac{1}{2}x\right)^2 dx \\ &= \pi\int_1^2 \left(4 - 2x + \tfrac{1}{4}x^2\right) dx \\ &= \pi\left[4x - x^2 + \tfrac{1}{12}x^3\right]_1^2 \\ &= \pi\left[\left(8 - 4 + \tfrac{8}{12}\right) - \left(4 - 1 + \tfrac{1}{12}\right)\right] \\ &= \pi\left(1 + \tfrac{7}{12}\right) = \tfrac{19}{12}\pi\end{aligned}$$

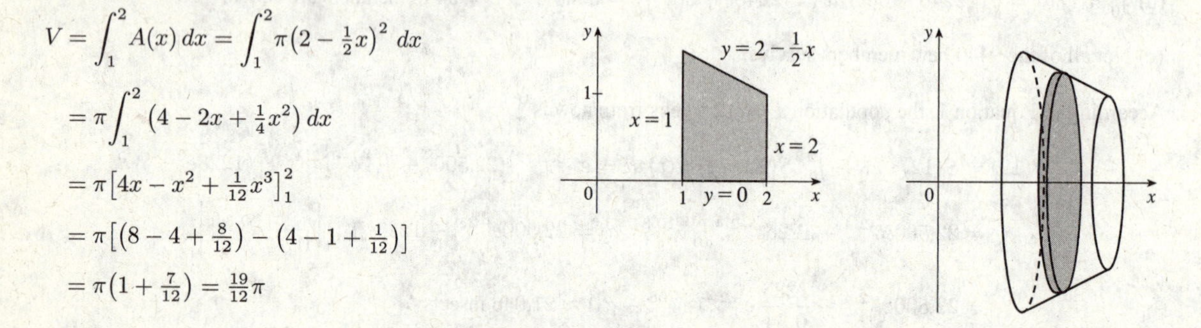

3. A cross-section is a disk with radius $\sqrt{x-1}$, so its area is $A(x) = \pi\left(\sqrt{x-1}\right)^2 = \pi(x-1)$.

$$V = \int_1^5 A(x)\,dx = \int_1^5 \pi(x-1)\,dx = \pi\left[\tfrac{1}{2}x^2 - x\right]_1^5 = \pi\left[\left(\tfrac{25}{2} - 5\right) - \left(\tfrac{1}{2} - 1\right)\right] = 8\pi$$

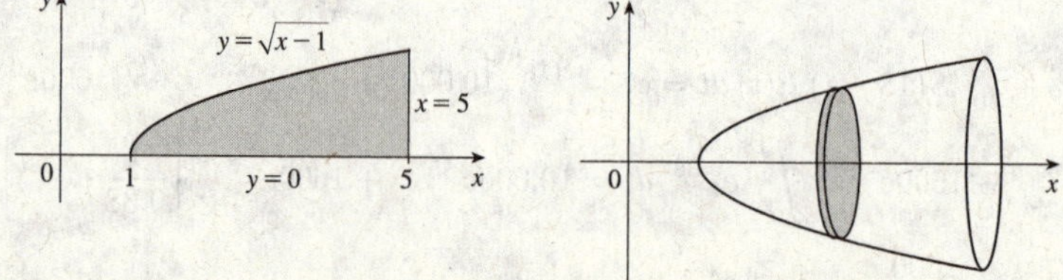

© 2016 Cengage Learning. All Rights Reserved. May not be scanned, copied or duplicated, or posted to a publicly accessible website, in whole or in part.

5. A cross-section is a washer (annulus) with inner radius x^3 and outer radius x, so its area is $A(x) = \pi(x)^2 - \pi(x^3)^2 = \pi(x^2 - x^6)$.

$$V = \int_0^1 A(x)\,dx = \int_0^1 \pi(x^2 - x^6)\,dx$$
$$= \pi\left[\tfrac{1}{3}x^3 - \tfrac{1}{7}x^7\right]_0^1 = \pi\left(\tfrac{1}{3} - \tfrac{1}{7}\right) = \tfrac{4}{21}\pi$$

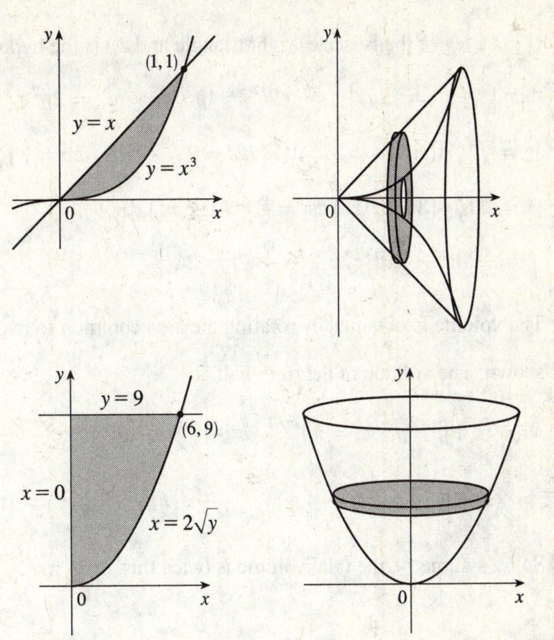

7. A cross-section is a disk with radius $2\sqrt{y}$, so its area is $A(y) = \pi\left(2\sqrt{y}\right)^2$.

$$V = \int_0^9 A(y)\,dy = \int_0^9 \pi\left(2\sqrt{y}\right)^2 dy = 4\pi\int_0^9 y\,dy$$
$$= 4\pi\left[\tfrac{1}{2}y^2\right]_0^9 = 2\pi(81) = 162\pi$$

9. If we use the Midpoint Rule with $n = 6$, the volume of the pancreas is

$$V = \int_0^{12} A(x)\,dx \approx M_6 = \tfrac{12-0}{6}[A(1) + A(3) + A(5) + A(7) + A(9) + A(11)]$$
$$= 2(7.7 + 18.0 + 10.8 + 8.7 + 5.5 + 2.7) = 2(53.4) = 106.8 \text{ cm}^3$$

11. (a) $V = \int_2^{10} \pi\,[f(x)]^2\,dx \approx \pi\,\frac{10-2}{4}\left\{[f(3)]^2 + [f(5)]^2 + [f(7)]^2 + [f(9)]^2\right\}$

$\approx 2\pi\left[(1.5)^2 + (2.2)^2 + (3.8)^2 + (3.1)^2\right] \approx 196 \text{ units}^3$

(b) $V = \int_0^4 \pi\left[(\text{outer radius})^2 - (\text{inner radius})^2\right] dy$

$\approx \pi\,\frac{4-0}{4}\left\{\left[(9.9)^2 - (2.2)^2\right] + \left[(9.7)^2 - (3.0)^2\right] + \left[(9.3)^2 - (5.6)^2\right] + \left[(8.7)^2 - (6.5)^2\right]\right\}$

$\approx 838 \text{ units}^3$

13. We'll form a right circular cone with height h and base radius r by revolving the line $y = \frac{r}{h}x$ about the x-axis.

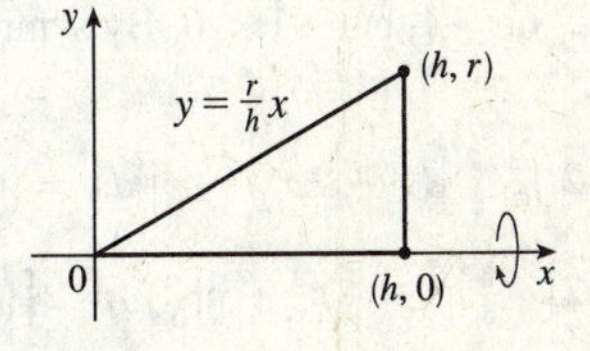

$$V = \pi\int_0^h \left(\frac{r}{h}x\right)^2 dx = \pi\int_0^h \frac{r^2}{h^2}x^2\,dx = \pi\,\frac{r^2}{h^2}\left[\frac{1}{3}x^3\right]_0^h$$
$$= \pi\,\frac{r^2}{h^2}\left(\frac{1}{3}h^3\right) = \frac{1}{3}\pi r^2 h$$

Another solution: Revolve $x = -\dfrac{r}{h}y + r$ about the y-axis.

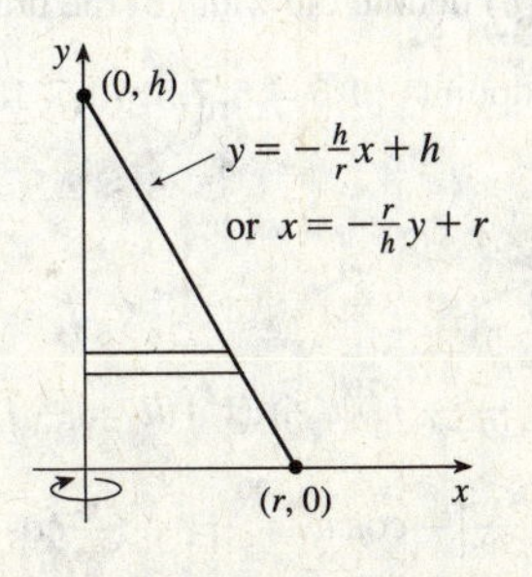

$$V = \pi\int_0^h \left(-\frac{r}{h}y + r\right)^2 dy \overset{*}{=} \pi\int_0^h \left[\frac{r^2}{h^2}y^2 - \frac{2r^2}{h}y + r^2\right] dy$$
$$= \pi\left[\frac{r^2}{3h^2}y^3 - \frac{r^2}{h}y^2 + r^2 y\right]_0^h = \pi\left(\tfrac{1}{3}r^2h - r^2h + r^2h\right) = \tfrac{1}{3}\pi r^2 h$$

* Or use substitution with $u = r - \dfrac{r}{h}y$ and $du = -\dfrac{r}{h}\,dy$ to get

$$\pi\int_r^0 u^2\left(-\frac{h}{r}\,du\right) = -\pi\,\frac{h}{r}\left[\frac{1}{3}u^3\right]_r^0 = -\pi\,\frac{h}{r}\left(-\frac{1}{3}r^3\right) = \frac{1}{3}\pi r^2 h.$$

© 2016 Cengage Learning. All Rights Reserved. May not be scanned, copied or duplicated, or posted to a publicly accessible website, in whole or in part.

15. If l is a leg of the isosceles right triangle and $2y$ is the hypotenuse,

then $l^2 + l^2 = (2y)^2 \;\Rightarrow\; 2l^2 = 4y^2 \;\Rightarrow\; l^2 = 2y^2$.

$V = \int_{-2}^{2} A(x)\,dx = 2\int_0^2 A(x)\,dx = 2\int_0^2 \frac{1}{2}(l)(l)\,dx = 2\int_0^2 y^2\,dx$

$= 2\int_0^2 \frac{1}{4}(36 - 9x^2)\,dx = \frac{9}{2}\int_0^2 (4 - x^2)\,dx$

$= \frac{9}{2}\left[4x - \frac{1}{3}x^3\right]_0^2 = \frac{9}{2}\left(8 - \frac{8}{3}\right) = 24$

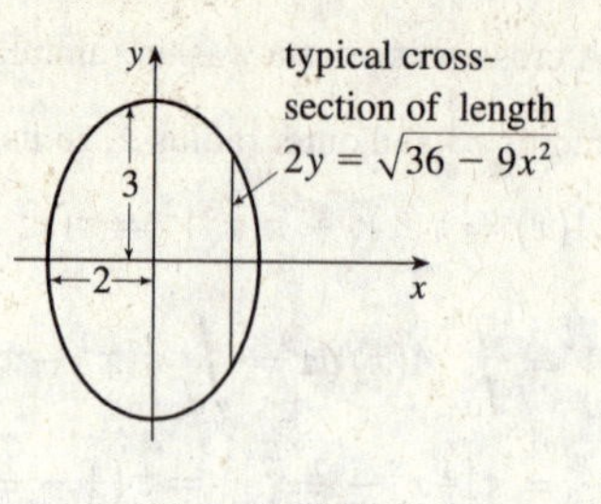

17. The volume is obtained by rotating the area common to two circles of radius r, as shown. The volume of the right half is

$V_{\text{right}} = \pi\int_0^{r/2} y^2\,dx = \pi\int_0^{r/2}\left[r^2 - \left(\frac{1}{2}r + x\right)^2\right]dx$

$= \pi\left[r^2x - \frac{1}{3}\left(\frac{1}{2}r + x\right)^3\right]_0^{r/2} = \pi\left[\left(\frac{1}{2}r^3 - \frac{1}{3}r^3\right) - \left(0 - \frac{1}{24}r^3\right)\right] = \frac{5}{24}\pi r^3$

So by symmetry, the total volume is twice this, or $\frac{5}{12}\pi r^3$.

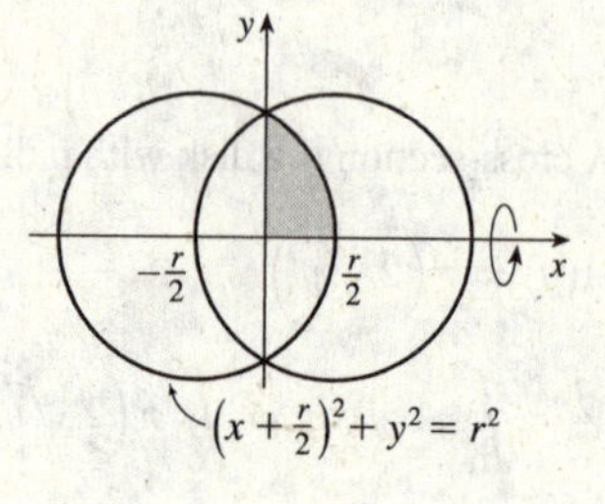

6 Review

EXERCISES

1. The curves intersect when $x^2 = 4x - x^2 \;\Leftrightarrow\; 2x^2 - 4x = 0 \;\Leftrightarrow$ $2x(x - 2) = 0 \;\Leftrightarrow\; x = 0$ or 2.

$A = \int_0^2 \left[(4x - x^2) - x^2\right]dx = \int_0^2 (4x - 2x^2)\,dx$

$= \left[2x^2 - \frac{2}{3}x^3\right]_0^2 = \left[\left(8 - \frac{16}{3}\right) - 0\right] = \frac{8}{3}$

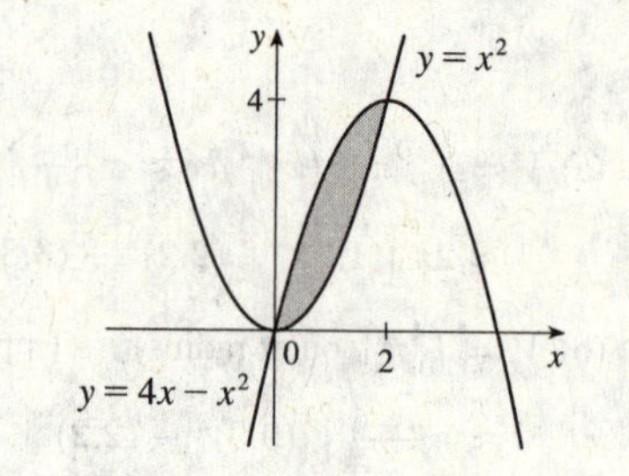

3. If $x \geq 0$, then $|x| = x$, and the graphs intersect when $x = 1 - 2x^2 \;\Leftrightarrow\; 2x^2 + x - 1 = 0 \;\Leftrightarrow\; (2x - 1)(x + 1) = 0 \;\Leftrightarrow$ $x = \frac{1}{2}$ or -1, but $-1 < 0$. By symmetry, we can double the area from $x = 0$ to $x = \frac{1}{2}$.

$A = 2\int_0^{1/2} \left[(1 - 2x^2) - x\right]dx = 2\int_0^{1/2} (-2x^2 - x + 1)\,dx$

$= 2\left[-\frac{2}{3}x^3 - \frac{1}{2}x^2 + x\right]_0^{1/2} = 2\left[\left(-\frac{1}{12} - \frac{1}{8} + \frac{1}{2}\right) - 0\right]$

$= 2\left(\frac{7}{24}\right) = \frac{7}{12}$

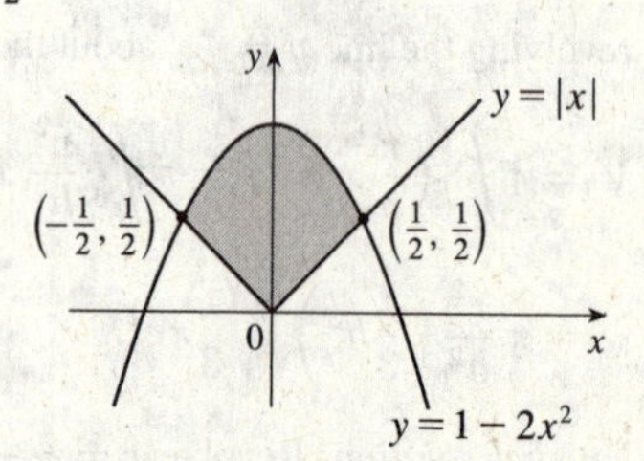

5. Let $T(y)$ denote the width of the brain at y cm from the bottom. Using the Midpoint Rule with $n = 5$, $\Delta y = \frac{15 - 0}{5} = 3$ and the midpoints $\{1.5, 4.5, 7.5, 10.5, 13.5\}$ gives

$$A \approx M_5 = \Delta y\,[T(1.5) + T(4.5) + T(7.5) + T(10.5) + T(13.5)]$$
$$= 3\,[9.8 + 12.9 + 14.3 + 14.0 + 9.2] = 3\,(60.2) = 180.6\ \text{cm}^2$$

7. $f_{\text{ave}} = \frac{1}{10 - 0}\int_0^{10} t\sin(t^2)\,dt = \frac{1}{10}\int_0^{100} \sin u\left(\frac{1}{2}\,du\right) \qquad [u = t^2,\ du = 2t\,dt]$

$= \frac{1}{20}\left[-\cos u\right]_0^{100} = \frac{1}{20}(-\cos 100 + \cos 0) = \frac{1}{20}(1 - \cos 100) \approx 0.007$

© 2016 Cengage Learning. All Rights Reserved. May not be scanned, copied or duplicated, or posted to a publicly accessible website, in whole or in part.

9. The average concentration of the antibiotic during the first two hours is

$$C_{\text{ave}} = \frac{1}{2}\int_0^2 C(t)\,dt = \frac{1}{2}\int_0^2 8(e^{-0.4t} - e^{-0.6t})\,dt = 4\left[-\frac{1}{0.4}e^{-0.4t} + \frac{1}{0.6}e^{-0.6t}\right]_0^2$$

$$= 4\left(\left[-\frac{1}{0.4}e^{-0.8} + \frac{1}{0.6}e^{-1.2}\right] - \left[-\frac{1}{0.4} + \frac{1}{0.6}\right]\right) \approx 0.848\ \mu\text{g/mL}$$

11. The population 10 years from now is

$$P(10) = S(10)\cdot P_0 + \int_0^{10} S(10-t)\,R(t)\,dt = e^{-0.1(10)}(75{,}000) + \int_0^{10} e^{-0.1(10-t)}(3200e^{0.05t})\,dt$$

$$= 75{,}000e^{-1} + 3200e^{-1}\int_0^{10} e^{0.15t}\,dt = 75{,}000e^{-1} + 3200e^{-1}\left[\frac{1}{0.15}e^{0.15t}\right]_0^{10}$$

$$= 75{,}000e^{-1} + \frac{3200}{0.15}e^{-1}(e^{1.5} - 1) \approx 54{,}916$$

13. In Section 6.3, we saw that cardiac output is given by $F = \dfrac{A}{\int_0^T c(t)\,dt}$ where A is the amount of die injected. Estimating the integral using the Midpoint Rule with $n = 6$ and $\Delta t = \dfrac{24-0}{6} = 4$ gives

$$\int_0^T c(t)\,dt \approx M_6 = \Delta t\,[c(2) + c(6) + c(10) + c(14) + c(18) + c(22)]$$

$$= 4\,[1.9 + 5.1 + 7.1 + 4.7 + 2.1 + 0.5] = 4\,(21.4) = 85.6$$

So the cardiac output is $F \approx \dfrac{6}{85.6} \approx 0.070$ L/s.

15. (a) Using the Midpoint Rule on $[0, 1]$ with $f(x) = \tan(x^2)$ and $n = 4$, we estimate

$$A = \int_0^1 \tan(x^2)\,dx \approx \tfrac{1}{4}\left[\tan\left(\left(\tfrac{1}{8}\right)^2\right) + \tan\left(\left(\tfrac{3}{8}\right)^2\right) + \tan\left(\left(\tfrac{5}{8}\right)^2\right) + \tan\left(\left(\tfrac{7}{8}\right)^2\right)\right] \approx \tfrac{1}{4}(1.53) \approx 0.38$$

(b) Using the Midpoint Rule on $[0, 1]$ with $f(x) = \pi\tan^2(x^2)$ (for disks) and $n = 4$, we estimate

$$V = \int_0^1 f(x)\,dx \approx \tfrac{1}{4}\pi\left[\tan^2\left(\left(\tfrac{1}{8}\right)^2\right) + \tan^2\left(\left(\tfrac{3}{8}\right)^2\right) + \tan^2\left(\left(\tfrac{5}{8}\right)^2\right) + \tan^2\left(\left(\tfrac{7}{8}\right)^2\right)\right] \approx \tfrac{\pi}{4}(1.114) \approx 0.87$$

17. (a) A cross-section is a washer with inner radius x^2 and outer radius x.

$$V = \int_0^1 \pi\left[(x)^2 - (x^2)^2\right]dx = \int_0^1 \pi(x^2 - x^4)\,dx = \pi\left[\tfrac{1}{3}x^3 - \tfrac{1}{5}x^5\right]_0^1 = \pi\left[\tfrac{1}{3} - \tfrac{1}{5}\right] = \tfrac{2}{15}\pi$$

(b) A cross-section is a washer with inner radius y and outer radius $\sqrt{y}$.

$$V = \int_0^1 \pi\left[\left(\sqrt{y}\right)^2 - y^2\right]dy = \int_0^1 \pi(y - y^2)\,dy = \pi\left[\tfrac{1}{2}y^2 - \tfrac{1}{3}y^3\right]_0^1 = \pi\left[\tfrac{1}{2} - \tfrac{1}{3}\right] = \tfrac{\pi}{6}$$

19. Take the base to be the disk $x^2 + y^2 \le 9$. Then $V = \int_{-3}^{3} A(x)\,dx$, where $A(x_0)$ is the area of the isosceles right triangle whose hypotenuse lies along the line $x = x_0$ in the xy-plane. The length of the hypotenuse is $2\sqrt{9 - x^2}$ and the length of each leg is $\sqrt{2}\sqrt{9 - x^2}$. $A(x) = \frac{1}{2}\left(\sqrt{2}\sqrt{9 - x^2}\right)^2 = 9 - x^2$, so

$$V = 2\int_0^3 A(x)\,dx = 2\int_0^3 (9 - x^2)\,dx = 2\left[9x - \tfrac{1}{3}x^3\right]_0^3 = 2(27 - 9) = 36$$

© 2016 Cengage Learning. All Rights Reserved. May not be scanned, copied or duplicated, or posted to a publicly accessible website, in whole or in part.

7 □ DIFFERENTIAL EQUATIONS

7.1 Modeling with Differential Equations

1. $y = \frac{2}{3}e^x + e^{-2x} \Rightarrow y' = \frac{2}{3}e^x - 2e^{-2x}$. To show that y is a solution of the differential equation, we will substitute the expressions for y and y' in the left-hand side of the equation and show that the left-hand side is equal to the right-hand side.

$$\text{LHS} = y' + 2y = \tfrac{2}{3}e^x - 2e^{-2x} + 2\left(\tfrac{2}{3}e^x + e^{-2x}\right) = \tfrac{2}{3}e^x - 2e^{-2x} + \tfrac{4}{3}e^x + 2e^{-2x}$$
$$= \tfrac{6}{3}e^x = 2e^x = \text{RHS}$$

The differential equation $y' + 2y = 2e^x$ is a nonautonomous since it involves the function y, its derivative y', and the independent variable x.

3. When $y = e^{-at}\cos t$, the left side of the differential equation is $y' = \frac{d}{dt}\left(e^{-at}\cos t\right) = -ae^{-at}\cos t - e^{-at}\sin t$. This is the same as the right side so y is a solution to the differential equation. The differential equation is pure time since the derivative is expressed as a function of t alone.

5. (a) Since the derivative $y' = -y^2$ is always negative (or 0 if $y = 0$), the function y must be decreasing (or equal to 0) on any interval on which it is defined.

(b) $y = \dfrac{1}{x + C} \Rightarrow y' = -\dfrac{1}{(x + C)^2}$. LHS $= y' = -\dfrac{1}{(x + C)^2} = -\left(\dfrac{1}{x + C}\right)^2 = -y^2 =$ RHS

(c) $y = 0$ is a solution of $y' = -y^2$ that is not a member of the family in part (b).

(d) If $y(x) = \dfrac{1}{x + C}$, then $y(0) = \dfrac{1}{0 + C} = \dfrac{1}{C}$. Since $y(0) = 0.5$, $\dfrac{1}{C} = \dfrac{1}{2} \Rightarrow C = 2$, so $y = \dfrac{1}{x + 2}$.

7. (a) $\dfrac{dP}{dt} = 1.2P\left(1 - \dfrac{P}{4200}\right)$. Now $\dfrac{dP}{dt} > 0 \Rightarrow 1 - \dfrac{P}{4200} > 0$ [assuming that $P > 0$] $\Rightarrow \dfrac{P}{4200} < 1 \Rightarrow$ $P < 4200 \Rightarrow$ the population is increasing for $0 < P < 4200$.

(b) $\dfrac{dP}{dt} < 0 \Rightarrow P > 4200$

(c) $\dfrac{dP}{dt} = 0 \Rightarrow P = 4200$ or $P = 0$

9. (a) This function is increasing *and* also decreasing. But $dy/dt = e^t(y - 1)^2 \geq 0$ for all t, implying that the graph of the solution of the differential equation cannot be decreasing on any interval.

(b) When $y = 1$, $dy/dt = 0$, but the graph does not have a horizontal tangent line.

11. (a) $y' = 1 + x^2 + y^2 \geq 1$ and $y' \to \infty$ as $x \to \infty$. The only curve satisfying these conditions is labeled III.

(b) $y' = xe^{-x^2 - y^2} > 0$ if $x > 0$ and $y' < 0$ if $x < 0$. The only curve with negative tangent slopes when $x < 0$ and positive tangent slopes when $x > 0$ is labeled I.

(c) $y' = \dfrac{1}{1 + e^{x^2 + y^2}} > 0$ and $y' \to 0$ as $x \to \infty$. The only curve satisfying these conditions is labeled IV.

(d) $y' = \sin(xy)\cos(xy) = 0$ if $y = 0$, which is the solution graph labeled II.

© 2016 Cengage Learning. All Rights Reserved. May not be scanned, copied or duplicated, or posted to a publicly accessible website, in whole or in part.

13. The differential equation is pure time since the function $c(t)$ is not present. Note that the differential equation need not be written explicitly in terms of the independent variable t to be classified as pure time. The differential equation states that the rate of change of drug concentration is a positive constant.

$\frac{dc}{dt} = k \;\Rightarrow\; \int \frac{dc}{dt}\,dt = \int k\,dt \;\Rightarrow\; c(t) = kt + C$ and $c(0) = 0 = C$. So $c(t) = kt$. Hence, the drug concentration increases linearly with time.

15. The differential equation is nonautonomous since it contains both the function $c(t)$ and the independent variable t. The differential equation indicates that the rate of change of drug concentration is proportional to the difference between the concentration and c_s, with a constant of proportionality k/t^b, which decreases over time. When $c < c_s$ the concentration increases, when $c = c_s$ the concentration remains unchanged, and when $c > c_s$ the concentration decreases.

Suppose $c(t) = c_s(1 - e^{-\alpha t^{1-b}})$. The left side of the differential equation is

$$\frac{dc}{dt} = \frac{d}{dt}\left(c_s(1 - e^{-\alpha t^{1-b}})\right) = c_s\alpha(1-b)t^{-b}e^{-\alpha t^{1-b}}$$

and the right side is $\frac{k}{t^b}(c_s - c) = \frac{k}{t^b}\left(c_s - c_s(1 - e^{-\alpha t^{1-b}})\right) = c_s k t^{-b} e^{-\alpha t^{1-b}} = c_s\alpha(1-b)t^{-b}e^{-\alpha t^{1-b}}$.

The left and right sides are equal, so $c(t)$ is a solution to the differential equation.

17. With a per capita growth rate of $f(N) = 0.55 - 0.0026N$, the rate of change of population is

$$\frac{dN}{dt} = Nf(N) = N(0.55 - 0.0026N) = 0.55N\left(1 - \frac{0.0026}{0.55}N\right) \approx 0.55N\left(1 - \frac{1}{210}N\right).$$

This is the same as the logistic equation (4) with $r = 0.55$ and $K \approx 210$.

7.2 Phase Plots, Equilibria, and Stability

1. (a) (i) is locally stable and (ii) is unstable. (b) (i) and (ii) are locally stable.

3. (a)

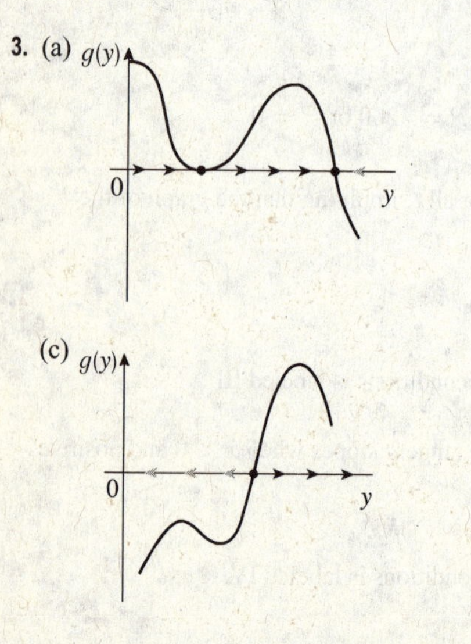

(b)

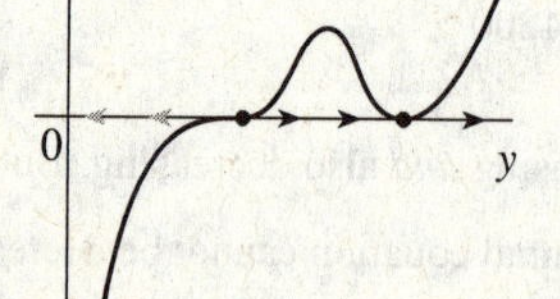

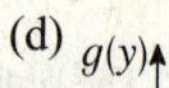

(d)

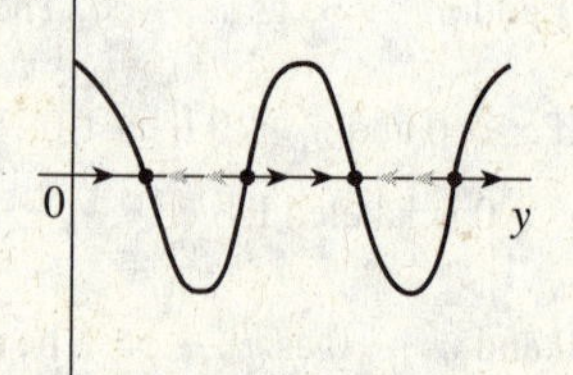

© 2016 Cengage Learning. All Rights Reserved. May not be scanned, copied or duplicated, or posted to a publicly accessible website, in whole or in part.

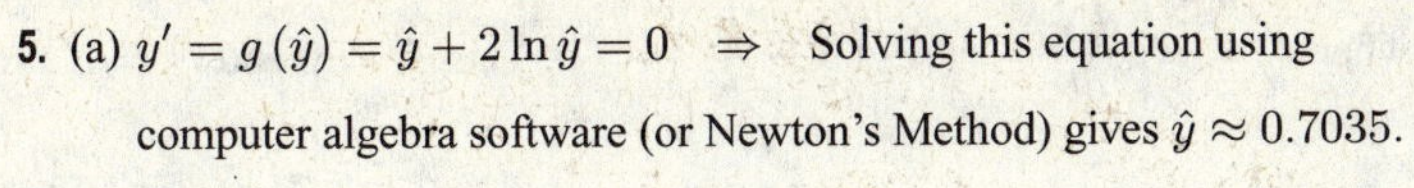

5. (a) $y' = g(\hat{y}) = \hat{y} + 2\ln\hat{y} = 0 \;\Rightarrow\;$ Solving this equation using computer algebra software (or Newton's Method) gives $\hat{y} \approx 0.7035$.

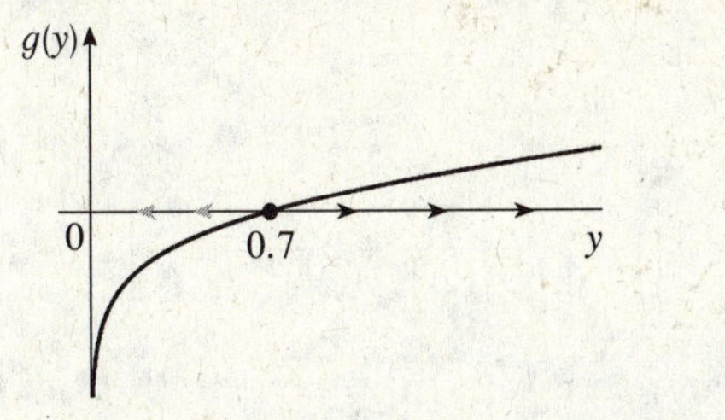

(b) $y' = g(\hat{y}) = \hat{y}^3 - a = 0 \;\Rightarrow\; \hat{y} = \sqrt[3]{a}$

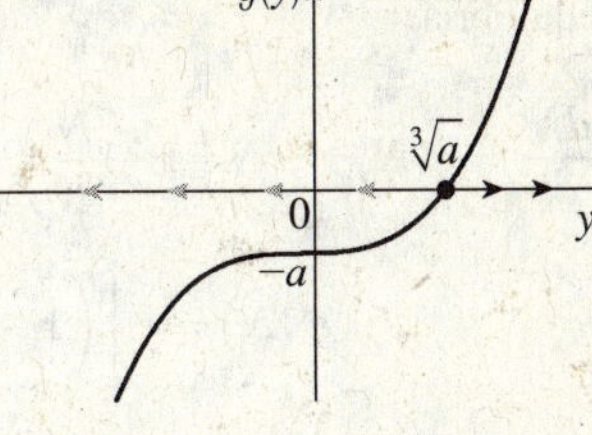

(c) $y' = g(\hat{y}) = \dfrac{5}{2+\hat{y}} = 0 \;\Rightarrow\;$ No solutions/equilibria

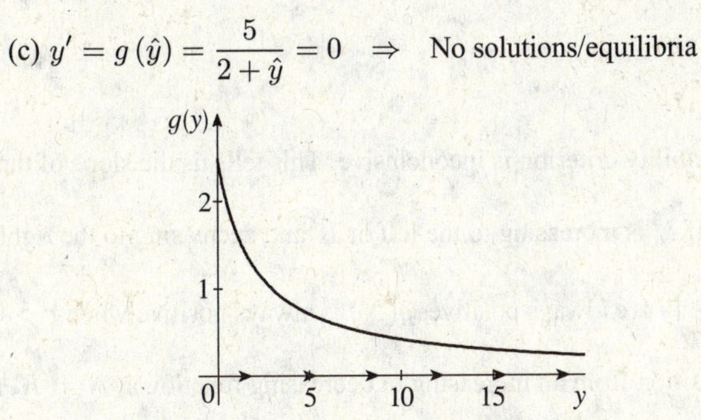

7. (a) $y' = g(\hat{y}) = 5 - 3\hat{y} = 0 \;\Rightarrow\; \hat{y} = \dfrac{5}{3}$

Stability: $g'(y) = -3 \;\Rightarrow\; g'(5/3) = -3 < 0 \;\Rightarrow\; \hat{y} = \dfrac{5}{3}$ is locally stable

(b) $y' = g(\hat{y}) = 2\hat{y} - 3\hat{y}^2 = 0 \;\Rightarrow\; \hat{y}(2 - 3\hat{y}) = 0 \;\Rightarrow\; \hat{y} = 0$ and $\hat{y} = 2/3$

Local Stability Check: $g'(y) = 2 - 6\hat{y}$

So $g'(0) = 2 - 6(0) = 2 > 0 \;\Rightarrow\; \hat{y} = 0$ is unstable

and $g'(2/3) = 2 - 6\left(\frac{2}{3}\right) = -2 < 0 \;\Rightarrow\; \hat{y} = 2/3$ is locally stable

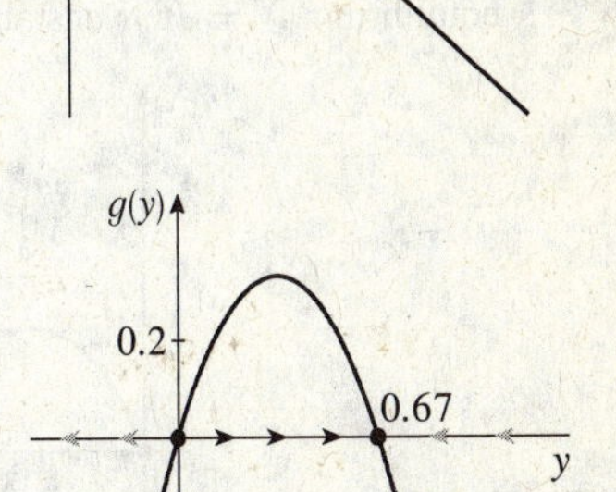

9. (a) $y' = g(\hat{y}) = 1 + a\hat{y} = 0 \;\Rightarrow\; \hat{y} = -\dfrac{1}{a}$. Also, $g'(y) = a$, so $\hat{y} = -\dfrac{1}{a}$ is locally stable when $a < 0$.

(b) $y' = g(\hat{y}) = 1 - e^{-a\hat{y}} = 0 \;\Rightarrow\; -a\hat{y} = \ln 1 = 0 \;\Rightarrow\; \hat{y} = 0$. Also $g'(y) = ae^{-ay} \;\Rightarrow\; g'(0) = a$. So $\hat{y} = 0$ is locally stable when $a < 0$.

(c) $y' = g(\hat{y}) = ae^{\hat{y}}\cos\hat{y} = 0 \;\Rightarrow\; \cos\hat{y} = 0 \;\Rightarrow\; \hat{y} = \cos^{-1}(0) = \pi/2$ (since $0 < y < \pi$). Also,

$g'(y) = a\left(e^y\cos y - e^y\sin y\right) \;\Rightarrow\; g'(\pi/2) = a\left[e^{\pi/2}\cos(\pi/2) - e^{\pi/2}\sin(\pi/2)\right] = -ae^{\pi/2} \approx -4.8a$.

So $\hat{y} = \pi/2$ is locally stable when $g'(\pi/2) < 0 \;\Rightarrow\; a > 0$.

(d) $y' = g(\hat{y}) = \hat{y}(a - \hat{y}) = 0 \;\Rightarrow\; \hat{y} = 0$ and $\hat{y} = a$. Also, $g'(y) = a - 2y \;\Rightarrow\; g'(0) = a$ and $g'(a) = a - 2a = -a$. So $\hat{y} = 0$ is locally stable when $a < 0$ and $\hat{y} = a$ is locally stable when $-a < 0 \;\Rightarrow\; a > 0$.

© 2016 Cengage Learning. All Rights Reserved. May not be scanned, copied or duplicated, or posted to a publicly accessible website, in whole or in part.

11. (a) When $N=0$, $\dfrac{dN}{dt}=r(0)\,(1-0)^2=0$ and when $N=K$, $\dfrac{dN}{dt}=r(K)\left(1-\dfrac{K}{K}\right)^2=0$. Thus $\hat{N}=0$ and $\hat{N}=K$ are equilibria.

(b) $\dfrac{dN}{dt}=g(N)=rN\left(1-\dfrac{N}{K}\right)^2 \;\Rightarrow$

$$g'(N)=r\left[\left(1-\frac{N}{K}\right)^2+2N\left(1-\frac{N}{K}\right)\left(-\frac{1}{K}\right)\right]=r\left[1-2\frac{N}{K}+\frac{N^2}{K^2}+2\frac{N^2}{K^2}-2\frac{N}{K}\right]$$

$$=r\left[1-4\frac{N}{K}+3\frac{N^2}{K^2}\right]=\frac{r}{K^2}\,(K-N)\,(K-3N)$$

Thus $g'(0)=r$, so $\hat{N}=0$ will be unstable when $r>0$.

(c) $g'(K)=r\left[1-4\dfrac{K}{K}+3\dfrac{K^2}{K^2}\right]=r\,[0]=0$, so the local stability criterion is inconclusive. This tells us the slope of the phase plot at $N=K$ is zero (i.e. horizontal tangent line). If N is increasing to the left of K and decreasing to the right of K, then $N=K$ will be stable. Since both N and $\left(1-\dfrac{N}{K}\right)^2$ are always positive, $g(N)$ is always positive when $r>0$ and always negative when $r<0$. Hence, N will never transition from an increasing to decreasing function at $N=K$, so we expect this to be an unstable equilibrium.

(d) We sketch the graph of $g(N)$ noting the zeroes occur at $N=0$ and $N=K$, and the critical points occur when $g'(N)=0$, that is when $N=K$ and $N=K/3$. Also, we have $g(N)\geq 0$ and $\lim\limits_{N\to\infty} g(N)=\infty$ when $r>0$, and $g(N)\leq 0$ and $\lim\limits_{N\to\infty} g(N)=-\infty$ when $r<0$. Combining these properties, we obtain the sketches shown below. The equilibrium $N=K$ is unstable in both cases as predicted in (c).

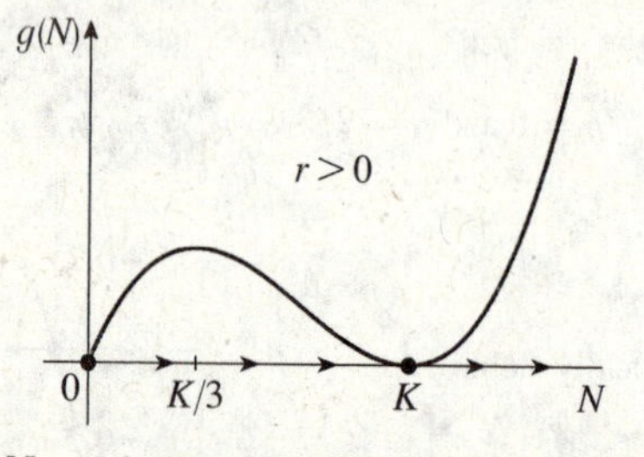

$N=0$ is unstable; $N=K$ is unstable

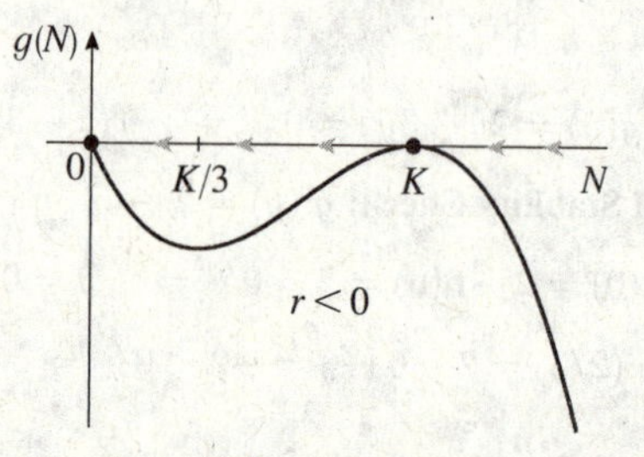

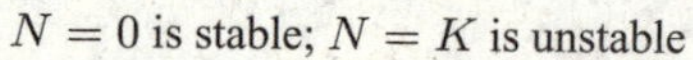
$N=0$ is stable; $N=K$ is unstable

13. (a) With $r=2$ and $K=1000$, we have $\dfrac{dN}{dt}=g(\hat{N})=2\hat{N}\left(1-\dfrac{\hat{N}}{1000}\right)-h\hat{N}=0 \;\Rightarrow\; \hat{N}\left[2-\dfrac{\hat{N}}{500}-h\right]=0 \;\Rightarrow$

$\hat{N}=0$ and $2-\dfrac{\hat{N}}{500}-h=0 \;\Rightarrow\; \hat{N}=500\,(2-h)$.

(b) Extinction occurs when the nonzero equilibrium $\hat{N}=500\,(2-h)$ is less than or equal to zero. So $\hat{N}=500\,(2-h)\leq 0 \;\Rightarrow\; (2-h)\leq 0 \;\Rightarrow\; h\geq 2$. So the population will be extinct when the harvest rate is greater than or equal to 2.

(c) $g'(N)=2\left[\left(1-\dfrac{N}{1000}\right)+N\left(-\dfrac{1}{1000}\right)\right]-h=2\left(1-\dfrac{N}{500}\right)-h \;\Rightarrow$

$g'(500\,(2-h))=2\,(1-(2-h))-h=h-2$. So $\hat{N}$ is stable when $g'(\hat{N})<0 \;\Rightarrow\; h-2<0 \;\Rightarrow\; h<2$.

© 2016 Cengage Learning. All Rights Reserved. May not be scanned, copied or duplicated, or posted to a publicly accessible website, in whole or in part.

15. (a) $\dfrac{dp}{dt} = g(\hat{p}) = c\hat{p}(1-\hat{p}) - m\hat{p} = 0 \Rightarrow \hat{p}[c(1-\hat{p}) - m] = 0 \Rightarrow \hat{p} = 0$ and $c(1-\hat{p}) - m = 0 \Rightarrow$

$\hat{p} = 1 - \dfrac{m}{c}$

(b) The non-zero equilibrium lies between 0 and 1 when $0 < 1 - \dfrac{m}{c} < 1 \Rightarrow -1 < -\dfrac{m}{c} < 0 \Rightarrow 0 < \dfrac{m}{c} < 1 \Rightarrow$

$m < c$ (since m and c are positive constants). This means the colonization rate must be greater than the extinction rate.

(c) $g'(p) = c - 2cp - m \Rightarrow g'(1 - \frac{m}{c}) = c - 2c(1 - \frac{m}{c}) - m = m - c \Rightarrow \hat{p} = 1 - \dfrac{m}{c}$ is locally stable when

$m - c < 0$, that is, when $m < c$.

17. We know that r_1 is a linear function of $1-p$ (frequency of type 2) passing through the points $(1-p, r_1) = (0,0)$ and $(1,\alpha)$, so the growth rate of type 1 is $r_1 = \dfrac{\alpha - 0}{1-0}(1-p-0) + 0 = \alpha(1-p)$. Similarly, r_2 is a linear function of p (frequency of type 1) passing through the points $(p, r_2) = (0,0)$ and $(1,\beta)$, so the growth rate of type 2 is $r_2 = \dfrac{\beta - 0}{1-0}(p-0) + 0 = \beta p$.

Thus, the growth of the two bacterial strains is modelled by $dN_1/dt = r_1N_1 = \alpha(1-p)N_1$ and $dN_2/dt = r_2N_2 = \beta pN_2$.

As in Exercise 16a, we have

$$\frac{dp}{dt} = \frac{N_1'(N_1+N_2) - N_1(N_1'+N_2')}{(N_1+N_2)^2} = \frac{N_1'N_2 - N_1N_2'}{(N_1+N_2)^2} = \frac{(\alpha(1-p)N_1)N_2 - N_1(\beta pN_2)}{(N_1+N_2)^2}$$

$$= \frac{N_1N_2(\alpha(1-p) - \beta p)}{(N_1+N_2)^2} = \left(\frac{N_1}{N_1+N_2}\right)\left(\frac{N_2}{N_1+N_2}\right)[\alpha(1-p) - \beta p]$$

$$= \left(\frac{N_1}{N_1+N_2}\right)\left(1 - \frac{N_1}{N_1+N_2}\right)[\alpha(1-p) - \beta p] = p(1-p)[\alpha(1-p) - \beta p]$$

This is the differential equation for the cross-feeding model used in Example 7.

19. (a) With a type 1 to type 2 per capita mutation rate μ, the rate of change of the two bacterial strains are

$dN_1/dt = r_1N_1 - \mu N_1 = (r_1 - \mu)N_1$ and $dN_2/dt = r_2N_2 + \mu N_1$. So $p(t) = \dfrac{N_1(t)}{N_1(t) + N_2(t)} \Rightarrow$

$$\frac{dp}{dt} = \frac{N_1'(N_1+N_2) - N_1(N_1'+N_2')}{(N_1+N_2)^2} = \frac{N_1'N_2 - N_1N_2'}{(N_1+N_2)^2} = \frac{[(r_1-\mu)N_1]N_2 - N_1(r_2N_2 + \mu N_1)}{(N_1+N_2)^2}$$

$$= \frac{(r_1 - \mu - r_2)N_1N_2 - \mu N_1^2}{(N_1+N_2)^2} = (r_1 - \mu - r_2)\left(\frac{N_1}{N_1+N_2}\right)\left(\frac{N_2}{N_1+N_2}\right) - \mu\left(\frac{N_1}{N_1+N_2}\right)^2$$

$$= (r_1 - \mu - r_2)\left(\frac{N_1}{N_1+N_2}\right)\left(1 - \frac{N_1}{N_1+N_2}\right) - \mu\left(\frac{N_1}{N_1+N_2}\right)^2 = (r_1 - \mu - r_2)p(1-p) - \mu p^2$$

$$= (r_1 - r_2)p(1-p) - \mu p(1-p) - \mu p^2 = (r_1 - r_2)p(1-p) - \mu p$$

$$= sp(1-p) - \mu p \quad \text{where } s = r_1 - r_2$$

(b) $\dfrac{d\hat{p}}{dt} = g(\hat{p}) = s\hat{p}(1-\hat{p}) - \mu\hat{p} = 0 \Rightarrow \hat{p}[s(1-\hat{p}) - \mu] = 0 \Rightarrow \hat{p} = 0$ and $s(1-\hat{p}) - \mu = 0 \Rightarrow$

$\hat{p} = 1 - \dfrac{\mu}{s}$

(c) $g'(p) = s - 2sp - \mu \Rightarrow g'(0) = s - \mu$. So $\hat{p} = 0$ is locally stable when $s < \mu$, that is when $r_1 < r_2 + \mu$. Also,

$g'\left(1 - \frac{\mu}{s}\right) = s - 2s\left(1 - \frac{\mu}{s}\right) - \mu = -s + \mu$, so $\hat{p} = 1 - \dfrac{\mu}{s}$ is locally stable when $s > \mu$, that is when $r_1 > r_2 + \mu$.

© 2016 Cengage Learning. All Rights Reserved. May not be scanned, copied or duplicated, or posted to a publicly accessible website, in whole or in part.

7.3 Direction Fields and Euler's Method

1. (a)

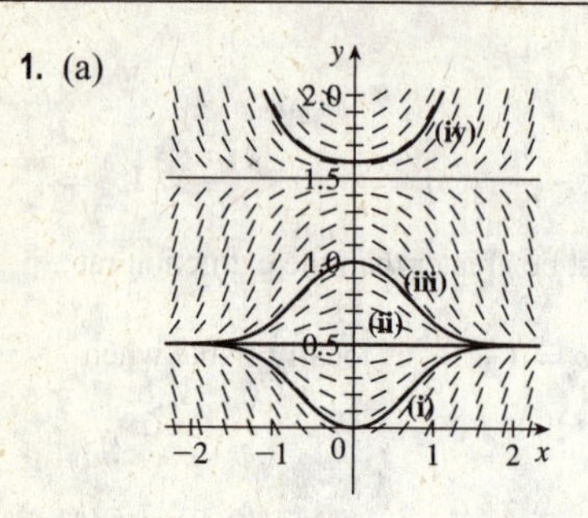

(b) It appears that the constant functions $y = 0.5$ and $y = 1.5$ are equilibrium solutions. Note that these two values of y satisfy the given differential equation $y' = x \cos \pi y$.

3. $y' = 2 - y$. The slopes at each point are independent of x, so the slopes are the same along each line parallel to the x-axis. Thus, III is the direction field for this equation. Note that for $y = 2$, $y' = 0$.

5. $y' = x + y - 1 = 0$ on the line $y = -x + 1$. Direction field IV satisfies this condition. Notice also that on the line $y = -x$ we have $y' = -1$, which is true in IV.

7. (a) $y(0) = 1$

(b) $y(0) = 2$

(c) $y(0) = -1$

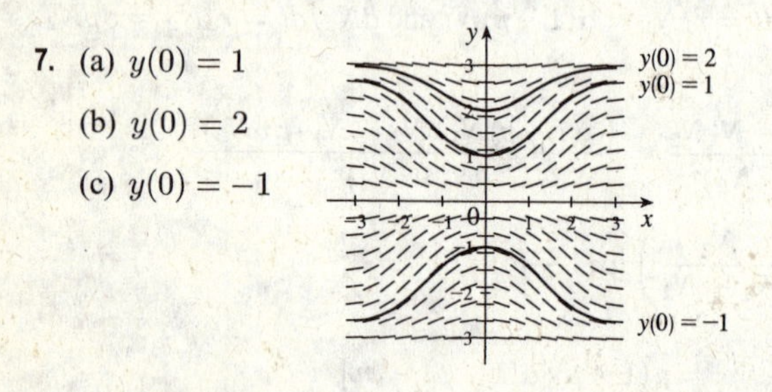

9.

x	y	$y' = \frac{1}{2}y$
0	0	0
0	1	0.5
0	2	1
0	−3	−1.5
0	−2	−1

Note that for $y = 0$, $y' = 0$. The three solution curves sketched go through $(0, 0)$, $(0, 1)$, and $(0, -1)$.

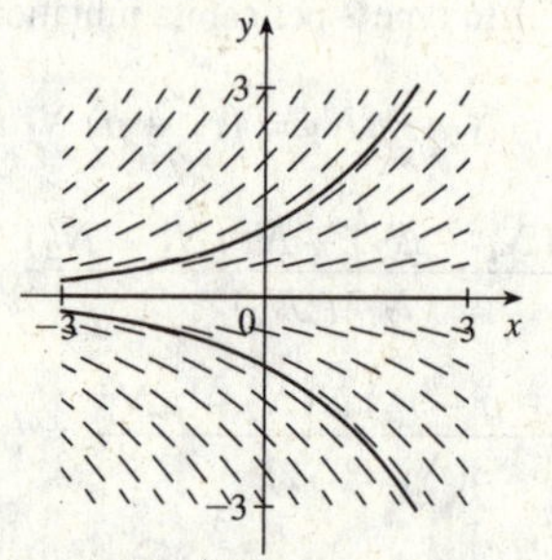

11.

x	y	$y' = y - 2x$
−2	−2	2
−2	2	6
2	2	−2
2	−2	−6

Note that $y' = 0$ for any point on the line $y = 2x$. The slopes are positive to the left of the line and negative to the right of the line. The solution curve in the graph passes through $(1, 0)$.

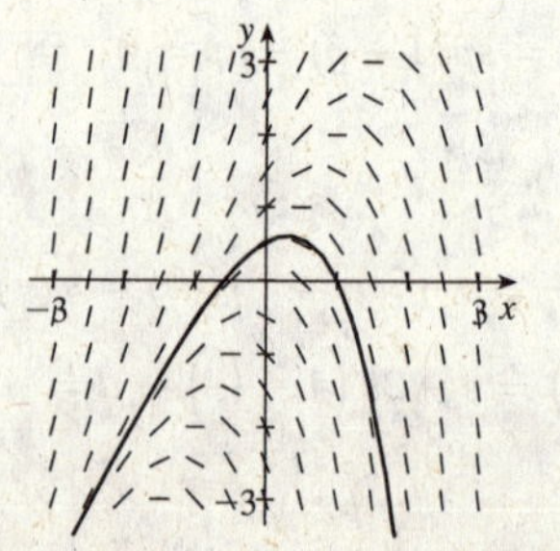

© 2016 Cengage Learning. All Rights Reserved. May not be scanned, copied or duplicated, or posted to a publicly accessible website, in whole or in part.

13.

x	y	$y' = y + xy$
0	± 2	± 2
1	± 2	± 4
-3	± 2	∓ 4

Note that $y' = y(x+1) = 0$ for any point on $y = 0$ or on $x = -1$. The slopes are positive when the factors y and $x+1$ have the same sign and negative when they have opposite signs. The solution curve in the graph passes through $(0, 1)$.

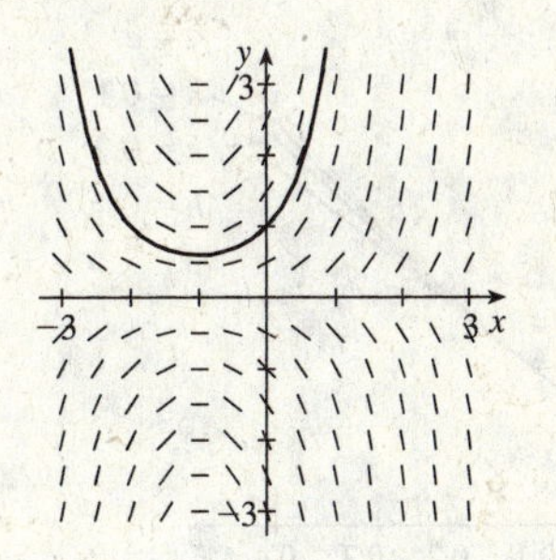

15. In Maple, we can use either `directionfield` (in Maple's share library) or `DEtools[DEplot]` to plot the direction field. To plot the solution, we can either use the initial-value option in `directionfield`, or actually solve the equation.

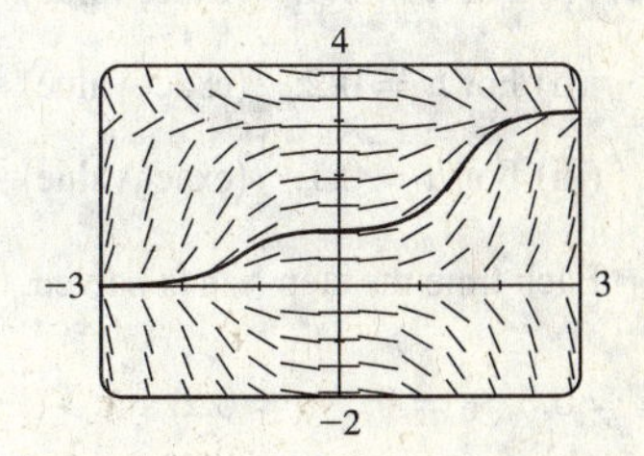

In Mathematica, we use `PlotVectorField` for the direction field, and the `Plot[Evaluate[...]]` construction to plot the solution, which is $y = 2\arctan\left(e^{x^3/3} \cdot \tan\frac{1}{2}\right)$.

In Derive, use `Direction_Field` (in utility file `ODE_APPR`) to plot the direction field. Then use `DSOLVE1(-x^2*SIN(y),1,x,y,0,1)` (in utility file `ODE1`) to solve the equation. Simplify each result.

17. The direction field is for the differential equation $y' = y^3 - 4y$.

$L = \lim_{t\to\infty} y(t)$ exists for $-2 \le c \le 2$;

$L = \pm 2$ for $c = \pm 2$ and $L = 0$ for $-2 < c < 2$.

For other values of c, L does not exist.

19. (a) $y' = F(x, y) = y$ and $y(0) = 1 \;\Rightarrow\; x_0 = 0,\ y_0 = 1$.

(i) $h = 0.4$ and $y_1 = y_0 + hF(x_0, y_0) \;\Rightarrow\; y_1 = 1 + 0.4 \cdot 1 = 1.4$. $x_1 = x_0 + h = 0 + 0.4 = 0.4$, so $y_1 = y(0.4) = 1.4$.

(ii) $h = 0.2 \;\Rightarrow\; x_1 = 0.2$ and $x_2 = 0.4$, so we need to find y_2.

$y_1 = y_0 + hF(x_0, y_0) = 1 + 0.2y_0 = 1 + 0.2 \cdot 1 = 1.2$,

$y_2 = y_1 + hF(x_1, y_1) = 1.2 + 0.2y_1 = 1.2 + 0.2 \cdot 1.2 = 1.44$.

© 2016 Cengage Learning. All Rights Reserved. May not be scanned, copied or duplicated, or posted to a publicly accessible website, in whole or in part.

(iii) $h = 0.1 \Rightarrow x_4 = 0.4$, so we need to find y_4. $y_1 = y_0 + hF(x_0, y_0) = 1 + 0.1y_0 = 1 + 0.1 \cdot 1 = 1.1$,

$y_2 = y_1 + hF(x_1, y_1) = 1.1 + 0.1y_1 = 1.1 + 0.1 \cdot 1.1 = 1.21$,

$y_3 = y_2 + hF(x_2, y_2) = 1.21 + 0.1y_2 = 1.21 + 0.1 \cdot 1.21 = 1.331$,

$y_4 = y_3 + hF(x_3, y_3) = 1.331 + 0.1y_3 = 1.331 + 0.1 \cdot 1.331 = 1.4641$.

(b)

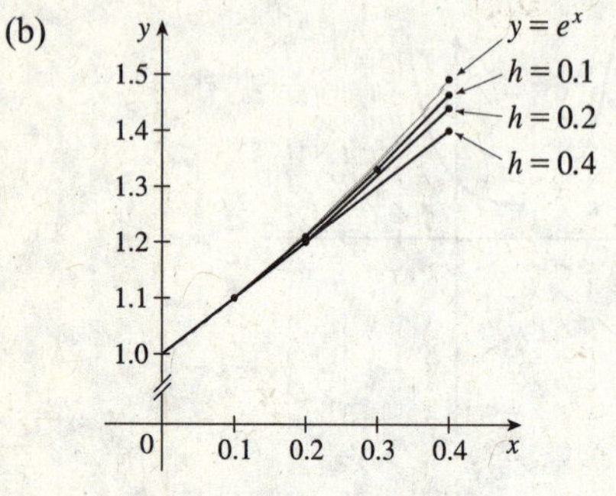

We see that the estimates are underestimates since they are all below the graph of $y = e^x$.

(c) (i) For $h = 0.4$: (exact value) − (approximate value) $= e^{0.4} - 1.4 \approx 0.0918$

(ii) For $h = 0.2$: (exact value) − (approximate value) $= e^{0.4} - 1.44 \approx 0.0518$

(iii) For $h = 0.1$: (exact value) − (approximate value) $= e^{0.4} - 1.4641 \approx 0.0277$

Each time the step size is halved, the error estimate also appears to be halved (approximately).

21. $h = 0.5$, $t_0 = 0$, $N_0 = 0.2$, and $F(t, N) = N\,(0.55 - 0.0026N)$.

Note that $t_1 = t_0 + h = 0 + 0.5 = 0.5$, $t_2 = 1.0$, $t_3 = 1.5$, $t_4 = 2.0$, $t_5 = 2.5$, $t_6 = 3.0$, $t_7 = 3.5$, and $t_8 = 4.0$.

$N_1 = N_0 + hF(t_0, N_0) = 0.2 + 0.5F(0, 0.2) = 0.2 + 0.5\,(0.109896) = 0.254948$.

$N_2 = N_1 + hF(t_1, N_1) = 0.254948 + 0.5F(0.5, 0.254948) = 0.254948 + 0.5\,(0.14005240) = 0.32497420$.

$N_3 = N_2 + hF(t_2, N_2) = 0.32497420 + 0.5F(1.0, 0.0.32497420) = 0.32497420 + 0.5\,(0.17846123) = 0.41420482$.

$N_4 = N_3 + hF(t_3, N_3) = 0.41420482 + 0.5F(1.5, 0.41420482) = 0.41420482 + 0.5\,(0.22736658) = 0.52788811$.

$N_5 = N_4 + hF(t_4, N_4) = 0.52788811 + 0.5F(2.0, 0.52788811) = 0.52788811 + 0.5\,(0.28961393) = 0.67269507$.

$N_6 = N_5 + hF(t_5, N_5) = 0.67269507 + 0.5F(2.5, 0.67269507) = 0.67269507 + 0.5\,(0.36880574) = 0.85709794$

$N_7 = N_6 + hF(t_6, N_6) = 0.85709794 + 0.5F(3.0, 0.85709794) = 0.85709794 + 0.5\,(0.46949386) = 1.09184487$.

$N_8 = N_7 + hF(t_7, N_7) = 1.09184487 + 0.5F(3.5, 1.09184487) = 1.09184487 + 0.5\,(0.59741515) = 1.39055245$.

Thus, $N(4) \approx 1.3906$. This agrees closely with the value of 1.40 found in Table 1 in Section 7.1.

23. $h = 0.5$, $x_0 = 1$, $y_0 = 0$, and $F(x, y) = y - 2x$.

Note that $x_1 = x_0 + h = 1 + 0.5 = 1.5$, $x_2 = 2$, and $x_3 = 2.5$.

$y_1 = y_0 + hF(x_0, y_0) = 0 + 0.5F(1, 0) = 0.5[0 - 2(1)] = -1$.

$y_2 = y_1 + hF(x_1, y_1) = -1 + 0.5F(1.5, -1) = -1 + 0.5[-1 - 2(1.5)] = -3$.

$y_3 = y_2 + hF(x_2, y_2) = -3 + 0.5F(2, -3) = -3 + 0.5[-3 - 2(2)] = -6.5$.

$y_4 = y_3 + hF(x_3, y_3) = -6.5 + 0.5F(2.5, -6.5) = -6.5 + 0.5[-6.5 - 2(2.5)] = -12.25$.

© 2016 Cengage Learning. All Rights Reserved. May not be scanned, copied or duplicated, or posted to a publicly accessible website, in whole or in part.

25. $h = 0.1$, $x_0 = 0$, $y_0 = 1$, and $F(x, y) = y + xy$.

Note that $x_1 = x_0 + h = 0 + 0.1 = 0.1$, $x_2 = 0.2$, $x_3 = 0.3$, and $x_4 = 0.4$.

$y_1 = y_0 + hF(x_0, y_0) = 1 + 0.1F(0, 1) = 1 + 0.1[1 + (0)(1)] = 1.1.$

$y_2 = y_1 + hF(x_1, y_1) = 1.1 + 0.1F(0.1, 1.1) = 1.1 + 0.1[1.1 + (0.1)(1.1)] = 1.221.$

$y_3 = y_2 + hF(x_2, y_2) = 1.221 + 0.1F(0.2, 1.221) = 1.221 + 0.1[1.221 + (0.2)(1.221)] = 1.36752.$

$$\begin{aligned} y_4 = y_3 + hF(x_3, y_3) &= 1.36752 + 0.1F(0.3, 1.36752) = 1.36752 + 0.1[1.36752 + (0.3)(1.36752)] \\ &= 1.5452976. \end{aligned}$$

$$\begin{aligned} y_5 = y_4 + hF(x_4, y_4) &= 1.5452976 + 0.1F(0.4, 1.5452976) \\ &= 1.5452976 + 0.1[1.5452976 + (0.4)(1.5452976)] = 1.761639264. \end{aligned}$$

Thus, $y(0.5) \approx 1.7616$.

27. (a) $dy/dx + 3x^2y = 6x^2 \quad \Rightarrow \quad y' = 6x^2 - 3x^2y$. Store this expression in Y_1 and use the following simple program to evaluate $y(1)$ for each part, using H $= h = 1$ and N $= 1$ for part (i), H $= 0.1$ and N $= 10$ for part (ii), and so forth.

```
h → H: 0 → X: 3 → Y:
For(I, 1, N): Y + H × Y1 → Y: X + H → X:
End(loop):
Display Y.   [To see all iterations, include this statement in the loop.]
```

(i) H $= 1$, N $= 1 \quad \Rightarrow \quad y(1) = 3$

(ii) H $= 0.1$, N $= 10 \quad \Rightarrow \quad y(1) \approx 2.3928$

(iii) H $= 0.01$, N $= 100 \quad \Rightarrow \quad y(1) \approx 2.3701$

(iv) H $= 0.001$, N $= 1000 \quad \Rightarrow \quad y(1) \approx 2.3681$

(b) $y = 2 + e^{-x^3} \quad \Rightarrow \quad y' = -3x^2e^{-x^3}$

$$\text{LHS} = y' + 3x^2y = -3x^2e^{-x^3} + 3x^2\left(2 + e^{-x^3}\right) = -3x^2e^{-x^3} + 6x^2 + 3x^2e^{-x^3} = 6x^2 = \text{RHS}$$

$y(0) = 2 + e^{-0} = 2 + 1 = 3$

(c) The exact value of $y(1)$ is $2 + e^{-1^3} = 2 + e^{-1}$.

(i) For $h = 1$: (exact value) – (approximate value) $= 2 + e^{-1} - 3 \approx -0.6321$

(ii) For $h = 0.1$: (exact value) – (approximate value) $= 2 + e^{-1} - 2.3928 \approx -0.0249$

(iii) For $h = 0.01$: (exact value) – (approximate value) $= 2 + e^{-1} - 2.3701 \approx -0.0022$

(iv) For $h = 0.001$: (exact value) – (approximate value) $= 2 + e^{-1} - 2.3681 \approx -0.0002$

In (ii)–(iv), it seems that when the step size is divided by 10, the error estimate is also divided by 10 (approximately).

© 2016 Cengage Learning. All Rights Reserved. May not be scanned, copied or duplicated, or posted to a publicly accessible website, in whole or in part.

7.4 Separable Equations

1. $\dfrac{dy}{dx} = xy^2 \;\Rightarrow\; \dfrac{dy}{y^2} = x\,dx \;[y \neq 0] \;\Rightarrow\; \displaystyle\int y^{-2}\,dy = \int x\,dx \;\Rightarrow\; -y^{-1} = \tfrac{1}{2}x^2 + C \;\Rightarrow$

$\dfrac{1}{y} = -\tfrac{1}{2}x^2 - C \;\Rightarrow\; y = \dfrac{1}{-\frac{1}{2}x^2 - C} = \dfrac{2}{K - x^2}$, where $K = -2C$. $y = 0$ is also a solution.

3. $(x^2+1)y' = xy \;\Rightarrow\; \dfrac{dy}{dx} = \dfrac{xy}{x^2+1} \;\Rightarrow\; \dfrac{dy}{y} = \dfrac{x\,dx}{x^2+1} \;[y \neq 0] \;\Rightarrow\; \displaystyle\int \frac{dy}{y} = \int \frac{x\,dx}{x^2+1} \;\Rightarrow$

$\ln|y| = \tfrac{1}{2}\ln(x^2+1) + C \;\; [u = x^2+1,\, du = 2x\,dx] \;= \ln(x^2+1)^{1/2} + \ln e^C = \ln\left(e^C\sqrt{x^2+1}\right) \;\Rightarrow$

$|y| = e^C\sqrt{x^2+1} \;\Rightarrow\; y = K\sqrt{x^2+1}$, where $K = \pm e^C$ is a constant. (In our derivation, K was nonzero, but we can restore the excluded case $y = 0$ by allowing K to be zero.)

5. $(y + \sin y)\,y' = x + x^3 \;\Rightarrow\; (y+\sin y)\dfrac{dy}{dx} = x + x^3 \;\Rightarrow\; \displaystyle\int (y + \sin y)\,dy = \int (x + x^3)\,dx \;\Rightarrow$

$\tfrac{1}{2}y^2 - \cos y = \tfrac{1}{2}x^2 + \tfrac{1}{4}x^4 + C$. We cannot solve explicitly for y.

7. $\dfrac{dy}{dt} = \dfrac{te^t}{y\sqrt{1+y^2}} \;\Rightarrow\; y\sqrt{1+y^2}\,dy = te^t\,dt \;\Rightarrow\; \int y\sqrt{1+y^2}\,dy = \int te^t\,dt \;\Rightarrow\; \tfrac{1}{3}\left(1+y^2\right)^{3/2} = te^t - e^t + C$

[where the first integral is evaluated by substitution and the second by parts] $\Rightarrow\; 1 + y^2 = [3(te^t - e^t + C)]^{2/3} \;\Rightarrow$

$y = \pm\sqrt{[3(te^t - e^t + C)]^{2/3} - 1}$

9. $\dfrac{du}{dt} = 2 + 2u + t + tu \;\Rightarrow\; \dfrac{du}{dt} = (1+u)(2+t) \;\Rightarrow\; \displaystyle\int \frac{du}{1+u} = \int (2+t)dt \;\; [u \neq -1] \;\Rightarrow$

$\ln|1+u| = \tfrac{1}{2}t^2 + 2t + C \;\Rightarrow\; |1+u| = e^{t^2/2 + 2t + C} = Ke^{t^2/2+2t}$, where $K = e^C \;\Rightarrow\; 1 + u = \pm Ke^{t^2/2+2t} \;\Rightarrow$

$u = -1 \pm Ke^{t^2/2+2t}$ where $K > 0$. $u = -1$ is also a solution, so $u = -1 + Ae^{t^2/2+2t}$, where A is an arbitrary constant.

11. $\dfrac{dy}{dx} = \dfrac{x}{y} \;\Rightarrow\; y\,dy = x\,dx \;\Rightarrow\; \int y\,dy = \int x\,dx \;\Rightarrow\; \tfrac{1}{2}y^2 = \tfrac{1}{2}x^2 + C.\;\; y(0) = -3 \;\Rightarrow$

$\tfrac{1}{2}(-3)^2 = \tfrac{1}{2}(0)^2 + C \;\Rightarrow\; C = \tfrac{9}{2}$, so $\tfrac{1}{2}y^2 = \tfrac{1}{2}x^2 + \tfrac{9}{2} \;\Rightarrow\; y^2 = x^2 + 9 \;\Rightarrow\; y = -\sqrt{x^2+9}$ since $y(0) = -3 < 0$.

13. $\dfrac{du}{dt} = \dfrac{2t + \sec^2 t}{2u}$, $u(0) = -5$. $\int 2u\,du = \int (2t + \sec^2 t)\,dt \;\Rightarrow\; u^2 = t^2 + \tan t + C$,

where $[u(0)]^2 = 0^2 + \tan 0 + C \;\Rightarrow\; C = (-5)^2 = 25$. Therefore, $u^2 = t^2 + \tan t + 25$, so $u = \pm\sqrt{t^2 + \tan t + 25}$.

Since $u(0) = -5$, we must have $u = -\sqrt{t^2 + \tan t + 25}$.

15. $x\ln x = y\left(1 + \sqrt{3+y^2}\right)y'$, $y(1) = 1$. $\int x\ln x\,dx = \int\left(y + y\sqrt{3+y^2}\right)dy \;\Rightarrow\; \tfrac{1}{2}x^2\ln x - \int \tfrac{1}{2}x\,dx$

[use parts with $u = \ln x$, $dv = x\,dx$] $= \tfrac{1}{2}y^2 + \tfrac{1}{3}(3+y^2)^{3/2} \;\Rightarrow\; \tfrac{1}{2}x^2\ln x - \tfrac{1}{4}x^2 + C = \tfrac{1}{2}y^2 + \tfrac{1}{3}(3+y^2)^{3/2}$.

Now $y(1) = 1 \;\Rightarrow\; 0 - \tfrac{1}{4} + C = \tfrac{1}{2} + \tfrac{1}{3}(4)^{3/2} \;\Rightarrow\; C = \tfrac{1}{2} + \tfrac{8}{3} + \tfrac{1}{4} = \tfrac{41}{12}$, so

$\tfrac{1}{2}x^2\ln x - \tfrac{1}{4}x^2 + \tfrac{41}{12} = \tfrac{1}{2}y^2 + \tfrac{1}{3}(3+y^2)^{3/2}$. We do not solve explicitly for y.

© 2016 Cengage Learning. All Rights Reserved. May not be scanned, copied or duplicated, or posted to a publicly accessible website, in whole or in part.

17. $y' \tan x = a + y,\ 0 < x < \pi/2 \Rightarrow \dfrac{dy}{dx} = \dfrac{a+y}{\tan x} \Rightarrow \dfrac{dy}{a+y} = \cot x\, dx \quad [a + y \neq 0] \Rightarrow$

$\displaystyle\int \frac{dy}{a+y} = \int \frac{\cos x}{\sin x}\, dx \Rightarrow \ln|a+y| = \ln|\sin x| + C \Rightarrow |a+y| = e^{\ln|\sin x|+C} = e^{\ln|\sin x|} \cdot e^C = e^C\,|\sin x| \Rightarrow$

$a + y = K \sin x$, where $K = \pm e^C$. (In our derivation, K was nonzero, but we can restore the excluded case

$y = -a$ by allowing K to be zero.) $y(\pi/3) = a \Rightarrow a + a = K \sin\left(\dfrac{\pi}{3}\right) \Rightarrow 2a = K\dfrac{\sqrt{3}}{2} \Rightarrow K = \dfrac{4a}{\sqrt{3}}$.

Thus, $a + y = \dfrac{4a}{\sqrt{3}} \sin x$ and so $y = \dfrac{4a}{\sqrt{3}} \sin x - a$.

19. If the slope at the point (x, y) is xy, then we have $\dfrac{dy}{dx} = xy \Rightarrow \dfrac{dy}{y} = x\, dx \quad [y \neq 0] \Rightarrow \displaystyle\int \frac{dy}{y} = \int x\, dx \Rightarrow$

$\ln|y| = \frac{1}{2}x^2 + C$. $y(0) = 1 \Rightarrow \ln 1 = 0 + C \Rightarrow C = 0$. Thus, $|y| = e^{x^2/2} \Rightarrow y = \pm e^{x^2/2}$, so $y = e^{x^2/2}$ since $y(0) = 1 > 0$. Note that $y = 0$ is not a solution because it doesn't satisfy the initial condition $y(0) = 1$.

21. $u = x + y \Rightarrow \dfrac{d}{dx}(u) = \dfrac{d}{dx}(x+y) \Rightarrow \dfrac{du}{dx} = 1 + \dfrac{dy}{dx}$, but $\dfrac{dy}{dx} = x + y = u$, so $\dfrac{du}{dx} = 1 + u \Rightarrow$

$\dfrac{du}{1+u} = dx \quad [u \neq -1] \Rightarrow \displaystyle\int \frac{du}{1+u} = \int dx \Rightarrow \ln|1+u| = x + C \Rightarrow |1+u| = e^{x+C} \Rightarrow 1 + u = \pm e^C e^x$

$\Rightarrow u = \pm e^C e^x - 1 \Rightarrow x + y = \pm e^C e^x - 1 \Rightarrow y = Ke^x - x - 1$, where $K = \pm e^C \neq 0$.

If $u = -1$, then $-1 = x + y \Rightarrow y = -x - 1$, which is just $y = Ke^x - x - 1$ with $K = 0$. Thus, the general solution is $y = Ke^x - x - 1$, where $K \in \mathbb{R}$.

23. (a) $y' = 2x\sqrt{1-y^2} \Rightarrow \dfrac{dy}{dx} = 2x\sqrt{1-y^2} \Rightarrow \dfrac{dy}{\sqrt{1-y^2}} = 2x\, dx \quad [\text{if } y \neq 1] \Rightarrow \displaystyle\int \frac{dy}{\sqrt{1-y^2}} = \int 2x\, dx \Rightarrow$

$\sin^{-1} y = x^2 + C$ for $-\frac{\pi}{2} \le x^2 + C \le \frac{\pi}{2}$. $y = 1$ is also a solution which can be verified by substituting into both sides of the differential equation.

(b) $y(0) = 0 \Rightarrow \sin^{-1} 0 = 0^2 + C \Rightarrow C = 0$,

so $\sin^{-1} y = x^2$ and $y = \sin(x^2)$ for $-\sqrt{\pi/2} \le x \le \sqrt{\pi/2}$.

1
$y = \sin(x^2)$
$-\sqrt{\pi/2}$
0
$\sqrt{\pi/2}$

(c) For $\sqrt{1-y^2}$ to be a real number, we must have $-1 \le y \le 1$; that is, $-1 \le y(0) \le 1$. Thus, the initial-value problem $y' = 2x\sqrt{1-y^2}$, $y(0) = 2$ does *not* have a solution.

25. $\dfrac{dy}{dx} = \dfrac{\sin x}{\sin y}$, $y(0) = \dfrac{\pi}{2}$. So $\int \sin y\, dy = \int \sin x\, dx \Leftrightarrow$

$-\cos y = -\cos x + C \Leftrightarrow \cos y = \cos x - C$. From the initial condition, we need $\cos\frac{\pi}{2} = \cos 0 - C \Rightarrow 0 = 1 - C \Rightarrow C = 1$, so the solution is $\cos y = \cos x - 1$. Note that we cannot take $\cos^{-1}$ of both sides, since that would unnecessarily restrict the solution to the case where $-1 \le \cos x - 1 \Leftrightarrow 0 \le \cos x$, as $\cos^{-1}$ is defined only on $[-1, 1]$. Instead we plot the graph using Maple's `plots[implicitplot]` or Mathematica's `Plot[Evaluate[···]]`.

5
−2.5
0
2.5

© 2016 Cengage Learning. All Rights Reserved. May not be scanned, copied or duplicated, or posted to a publicly accessible website, in whole or in part.

27. (a) , (c)

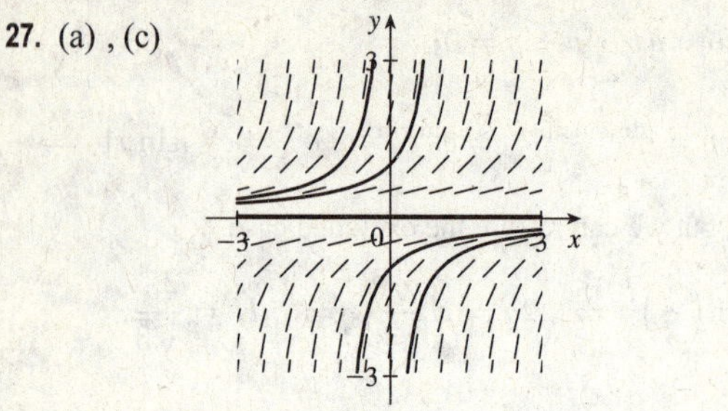

(b) $y' = y^2 \Rightarrow \dfrac{dy}{dx} = y^2 \Rightarrow \displaystyle\int y^{-2}\,dy = \int dx \Rightarrow$

$-y^{-1} = x + C \Rightarrow \dfrac{1}{y} = -x - C \Rightarrow$

$y = \dfrac{1}{K - x}$, where $K = -C$. $y = 0$ is also a solution.

29. $y(x) = 2 + \displaystyle\int_2^x [t - ty(t)]\,dt \Rightarrow y'(x) = x - xy(x)$ [by FTC 1] $\Rightarrow \dfrac{dy}{dx} = x(1-y) \Rightarrow$

$\displaystyle\int \frac{dy}{1-y} = \int x\,dx \Rightarrow -\ln|1-y| = \frac{1}{2}x^2 + C$. Letting $x = 2$ in the original integral equation

gives us $y(2) = 2 + 0 = 2$. Thus, $-\ln|1-2| = \frac{1}{2}(2)^2 + C \Rightarrow 0 = 2 + C \Rightarrow C = -2$.

Thus, $-\ln|1-y| = \frac{1}{2}x^2 - 2 \Rightarrow \ln|1-y| = 2 - \frac{1}{2}x^2 \Rightarrow |1-y| = e^{2-x^2/2} \Rightarrow$

$1 - y = \pm e^{2-x^2/2} \Rightarrow y = 1 + e^{2-x^2/2}$ [$y(2) = 2$].

31. $y(x) = 4 + \displaystyle\int_0^x 2t\sqrt{y(t)}\,dt \Rightarrow y'(x) = 2x\sqrt{y(x)} \Rightarrow \frac{dy}{dx} = 2x\sqrt{y} \Rightarrow \int \frac{dy}{\sqrt{y}} = \int 2x\,dx \Rightarrow$

$2\sqrt{y} = x^2 + C$. Letting $x = 0$ in the original integral equation gives us $y(0) = 4 + 0 = 4$.

Thus, $2\sqrt{4} = 0^2 + C \Rightarrow C = 4$. $2\sqrt{y} = x^2 + 4 \Rightarrow \sqrt{y} = \frac{1}{2}x^2 + 2 \Rightarrow y = \left(\frac{1}{2}x^2 + 2\right)^2$.

33. $\dfrac{dn}{dt} = (e^{-t} - 1)n \Rightarrow \displaystyle\int \frac{dn}{n} = \int (e^{-t} - 1)\,dt$ [if $n \neq 0$] $\Rightarrow \ln|n| = -e^{-t} - t + C \Rightarrow$

$|n| = e^{-(e^{-t}+t)+C} = e^C e^{-(e^{-t}+t)} \Rightarrow n = Ae^{-(e^{-t}+t)}$ where $A = \pm e^C$ is a constant. Note that $n = 0$ is also a

solution and this can be included in the family of solutions by allowing A to be zero. Now $n(0) = n_0 \Rightarrow n_0 = Ae^{-1} \Rightarrow$

$A = en_0$. So $n(t) = en_0 e^{-(e^{-t}+t)} = n_0 e^{1-(e^{-t}+t)}$. The population starts at n_0 and decreases, approaching 0 in the long-run, thus becoming extinct.

35. $\dfrac{dc}{dt} = k(c_s - c) \Rightarrow \displaystyle\int \frac{dc}{(c_s - c)} = \int k\,dt$ [if $c \neq c_s$] $\Rightarrow -\ln|c_s - c| = kt + C \Rightarrow$

$|c_s - c| = e^{-kt-C} = e^{-C}e^{-kt} \Rightarrow c = c_s - Ae^{-kt}$ where $A = \pm e^{-C}$ is a constant. Note that $c = c_s$ is also a solution

and this can be included in the family of solutions by allowing A to be zero. Now $c(0) = 0 \Rightarrow 0 = c_s - A \Rightarrow A = c_s$.

So $c(t) = c_s - c_s e^{-kt}$.

37. $\dfrac{dn}{dt} = kn^{1/2} \Rightarrow \displaystyle\int \frac{dn}{n^{1/2}} = \int k\,dt$ [if $n \neq 0$] $\Rightarrow 2n^{1/2} = kt + C \Rightarrow n = \frac{1}{4}(kt + C)^2$. Now $n(0) = 1 \Rightarrow$

$1 = \frac{1}{4}C^2 \Rightarrow C = \pm 2$. So $n(t) = \frac{1}{4}(kt + 2)^2$. Note: we take the positive value of C in the solution since this ensures

$n' > 0$ in the interval $[0, \infty)$. If $C = -2$, we would have $n(2/k) = 0$ so the population would collapse instead of grow.

© 2016 Cengage Learning. All Rights Reserved. May not be scanned, copied or duplicated, or posted to a publicly accessible website, in whole or in part.

39. (a) With $k = 1$, the differential equation is $\dfrac{dM}{dt} = M^{2/3} - \mu M = M^{2/3}\left(1 - \mu M^{1/3}\right) \Rightarrow$

$\displaystyle\int \frac{dM}{M^{2/3}\left(1 - \mu M^{1/3}\right)} = \int dt$ if $M \neq 0$ and $M \neq \mu^{-3}$. Using the substitution rule for the left side integration with

$v = \left(1 - \mu M^{1/3}\right)$, $dv = -\frac{\mu}{3} M^{-2/3}\, dM$ gives $\displaystyle\int \frac{\left(-\frac{3}{\mu}\, dv\right)}{v} = t + C \Rightarrow$

$-\frac{3}{\mu} \ln|v| = -\frac{3}{\mu} \ln\left|1 - \mu M^{1/3}\right| = t + C \Rightarrow 1 - \mu M^{1/3} = \pm e^{-\mu C/3} e^{-\mu t/3} \Rightarrow M = \mu^{-3}\left(1 - Ae^{-\mu t/3}\right)^3$

where $A = \pm e^{-\mu C/3}$ is a constant. Note that $M = \mu^{-3}$ is also a solution and this can be included in the family of solutions by allowing A to be zero. Now $M(0) = 1 \Rightarrow 1 = \mu^{-3}(1 - A)^3 \Rightarrow A = 1 - \mu$.

So $M(t) = \mu^{-3}\left[1 - (1 - \mu)\, e^{-\mu t/3}\right]^3$.

(b) $\displaystyle\lim_{t\to\infty} M(t) = \lim_{t\to\infty} \mu^{-3}\left[1 - (1 - \mu)\, e^{-\mu t/3}\right]^3 = \mu^{-3}(1 - 0)^3$ [since $\mu > 0$] $= 1/\mu^3$. Thus, the tumor mass approaches μ^{-3} grams as $t \to \infty$.

(c) $D = aM^{1/3} \Rightarrow \dfrac{dD}{dt} = \dfrac{a}{3} M^{-2/3} \dfrac{dM}{dt} = \dfrac{a}{3}\left(\dfrac{a^2}{D^2}\right)\left(kM^{2/3} - \mu M\right) = \dfrac{a^3}{3D^2}\left(k\dfrac{D^2}{a^2} - \mu\dfrac{D^3}{a^3}\right) = \dfrac{\mu}{3}\left(\dfrac{ka}{\mu} - D\right)$.

If we let $A = \dfrac{\mu}{3}$ and $B = \dfrac{ka}{\mu}$, we get $\dfrac{dD}{dt} = A\,(B - D)$ which has the same form as the Bertalanffy equation.

41. $\dfrac{dp}{dt} = sp(1 - p) \Rightarrow \displaystyle\int \frac{dp}{p(1 - p)} = \int s\, dt$ [if $p \neq 0$ and $p \neq 1$] $\Rightarrow \displaystyle\int \left[\frac{1}{p} + \frac{1}{1 - p}\right] dp = st + C \Rightarrow$

$\ln|p| - \ln|1 - p| = \ln\left|\dfrac{p}{1 - p}\right| = st + C \Rightarrow \left|\dfrac{p}{1 - p}\right| = e^C e^{st} \Rightarrow \dfrac{p}{1 - p} = \pm e^C e^{st} \Rightarrow p = \dfrac{Ae^{st}}{1 + Ae^{st}}$ where

$A = \pm e^C$ is a constant. Note that $p = 0$ and $p = 1$ are also solutions to the differential equation. Now $p(0) = p_0 \Rightarrow$

$p_0 = \dfrac{A}{1 + A} \Rightarrow A = \dfrac{p_0}{1 - p_0}$ [if $p_0 \neq 1$]. So $p(t) = \dfrac{\left(\dfrac{p_0}{1 - p_0}\right) e^{st}}{1 + \left(\dfrac{p_0}{1 - p_0}\right) e^{st}} = \dfrac{p_0 e^{st}}{1 - p_0 + p_0 e^{st}}$.

Notice that this also includes the solutions $p = 0$ and $p = 1$, which correspond to the initial conditions $p_0 = 0$ and $p_0 = 1$.

43. (a) $\dfrac{dC}{dt} = r - kC \Rightarrow \dfrac{dC}{dt} = -(kC - r) \Rightarrow \displaystyle\int \frac{dC}{kC - r} = \int -\,dt$ [if $C \neq r/k$] $\Rightarrow$

$(1/k)\ln|kC - r| = -t + M_1 \Rightarrow \ln|kC - r| = -kt + M_2 \Rightarrow |kC - r| = e^{-kt + M_2} \Rightarrow$

$kC - r = M_3 e^{-kt} \Rightarrow kC = M_3 e^{-kt} + r \Rightarrow C(t) = M_4 e^{-kt} + r/k$. Note that $C = r/k$ can be included in this family of solutions by letting $M_4 = 0$. Now $C(0) = C_0 \Rightarrow C_0 = M_4 + r/k \Rightarrow M_4 = C_0 - r/k \Rightarrow$

$C(t) = (C_0 - r/k)e^{-kt} + r/k$.

(b) If $C_0 < r/k$, then $C_0 - r/k < 0$ and the formula for $C(t)$ shows that $C(t)$ increases and $\displaystyle\lim_{t\to\infty} C(t) = r/k$.

As t increases, the formula for $C(t)$ shows how the role of C_0 steadily diminishes as that of r/k increases.

© 2016 Cengage Learning. All Rights Reserved. May not be scanned, copied or duplicated, or posted to a publicly accessible website, in whole or in part.

45. (a) Let $y(t)$ be the amount of salt (in kg) after t minutes. Then $y(0) = 15$. The amount of liquid in the tank is 1000 L at all times, so the concentration at time t (in minutes) is $y(t)/1000$ kg/L and $\frac{dy}{dt} = -\left[\frac{y(t)}{1000}\frac{\text{kg}}{\text{L}}\right]\left(10\,\frac{\text{L}}{\text{min}}\right) = -\frac{y(t)}{100}\frac{\text{kg}}{\text{min}}$.

$\int \frac{dy}{y} = -\frac{1}{100}\int dt$ [if $y \neq 0$] $\Rightarrow$ $\ln y = -\frac{t}{100} + C$. (Note that $y = 0$ is also a solution but it does not satisfy the initial condition.) Now $y(0) = 15 \Rightarrow \ln 15 = C$, so $\ln y = \ln 15 - \frac{t}{100}$.

It follows that $\ln\left(\frac{y}{15}\right) = -\frac{t}{100}$ and $\frac{y}{15} = e^{-t/100}$, so $y = 15e^{-t/100}$ kg.

(b) After 20 minutes, $y = 15e^{-20/100} = 15e^{-0.2} \approx 12.3$ kg.

47. Let $y(t)$ be the amount of alcohol in the vat after t minutes. Then $y(0) = 0.04(500) = 20$ gal. The amount of beer in the vat is 500 gallons at all times, so the percentage at time t (in minutes) is $y(t)/500 \times 100$, and the change in the amount of alcohol with respect to time t is $\frac{dy}{dt} = \text{rate in} - \text{rate out} = 0.06\left(5\,\frac{\text{gal}}{\text{min}}\right) - \frac{y(t)}{500}\left(5\,\frac{\text{gal}}{\text{min}}\right) = 0.3 - \frac{y}{100} = \frac{30-y}{100}\frac{\text{gal}}{\text{min}}$.

Hence, $\int \frac{dy}{30-y} = \int \frac{dt}{100}$ [if $y \neq 30$] $\Rightarrow$ $-\ln|30-y| = \frac{1}{100}t + C$. ($y = 30$ is also a solution but it does not satisfy the initial condition.) Because $y(0) = 20$, we have $-\ln 10 = C$, so

$-\ln|30-y| = \frac{1}{100}t - \ln 10 \Rightarrow \ln|30-y| = -t/100 + \ln 10 \Rightarrow \ln|30-y| = \ln e^{-t/100} + \ln 10 \Rightarrow$

$\ln|30-y| = \ln(10e^{-t/100}) \Rightarrow |30-y| = 10e^{-t/100}$. Since y is continuous, $y(0) = 20$, and the right-hand side is never zero, we deduce that $30 - y$ is always positive. Thus, $30 - y = 10e^{-t/100} \Rightarrow y = 30 - 10e^{-t/100}$. The percentage of alcohol is $p(t) = y(t)/500 \times 100 = y(t)/5 = 6 - 2e^{-t/100}$. The percentage of alcohol after one hour is $p(60) = 6 - 2e^{-60/100} \approx 4.9$.

49. Assume that the raindrop begins at rest, so that $v(0) = 0$. $dm/dt = km$ and $(mv)' = gm \Rightarrow mv' + vm' = gm \Rightarrow$ $mv' + v(km) = gm \Rightarrow v' + vk = g \Rightarrow \frac{dv}{dt} = g - kv \Rightarrow \int \frac{dv}{g-kv} = \int dt$ [if $v \neq g/k$] $\Rightarrow$ $-(1/k)\ln|g - kv| = t + C \Rightarrow \ln|g-kv| = -kt - kC \Rightarrow g - kv = Ae^{-kt}$. (Note that $v = g/k$ is also a solution but it does not satisfy the initial condition.) $v(0) = 0 \Rightarrow A = g$. So $kv = g - ge^{-kt} \Rightarrow v = (g/k)(1 - e^{-kt})$. Since $k > 0$, as $t \to \infty$, $e^{-kt} \to 0$ and therefore, $\lim\limits_{t\to\infty} v(t) = g/k$.

51. (a) The rate of growth of the area is jointly proportional to $\sqrt{A(t)}$ and $M - A(t)$; that is, the rate is proportional to the product of those two quantities. So for some constant k, $dA/dt = k\sqrt{A}\,(M - A)$. We are interested in the maximum of the function dA/dt (when the tissue grows the fastest), so we differentiate, using the Chain Rule and then substituting for dA/dt from the differential equation:

© 2016 Cengage Learning. All Rights Reserved. May not be scanned, copied or duplicated, or posted to a publicly accessible website, in whole or in part.

$$\frac{d}{dt}\left(\frac{dA}{dt}\right) = k\left[\sqrt{A}(-1)\frac{dA}{dt} + (M - A)\cdot\tfrac{1}{2}A^{-1/2}\frac{dA}{dt}\right] = \tfrac{1}{2}kA^{-1/2}\frac{dA}{dt}\left[-2A + (M - A)\right]$$
$$= \tfrac{1}{2}kA^{-1/2}\left[k\sqrt{A}(M - A)\right][M - 3A] = \tfrac{1}{2}k^2(M - A)(M - 3A)$$

This is 0 when $M - A = 0 \Leftrightarrow A(t) = M$ and when $M - 3A = 0 \Leftrightarrow A(t) = M/3$. The first critical point $A(t) = M$ represents a minimum by the First Derivative Test, since $\dfrac{d}{dt}\left(\dfrac{dA}{dt}\right)$ goes from negative to positive when $A(t) = M$. The other critical point $A(t) = M/3$ represents a maximum by the First Derivative Test, since $\dfrac{d}{dt}\left(\dfrac{dA}{dt}\right)$ goes from positive to negative when $A(t) = M/3$.

(b) $dA/dt = k\sqrt{A}(M - A) \Rightarrow \displaystyle\int \frac{dA}{\sqrt{A}(M - A)} = \int k\,dt$ [if $A \neq 0$ and $A \neq M$]

Evaluating the integral using a CAS, we get $\displaystyle\int \frac{dA}{\sqrt{A}(M - A)} = \frac{1}{\sqrt{M}}\ln\left[\frac{\sqrt{M} + \sqrt{A}}{\sqrt{M} - \sqrt{A}}\right]$. (Some CAS's may give an answer in terms of an inverse hyperbolic tangent function. This can be converted to a logarithmic form using the CAS). So $\displaystyle\frac{1}{\sqrt{M}}\ln\left[\frac{\sqrt{M} + \sqrt{A}}{\sqrt{M} - \sqrt{A}}\right] = kt + C_1 \Leftrightarrow \frac{\sqrt{M} + \sqrt{A}}{\sqrt{M} - \sqrt{A}} = e^{\sqrt{M}(kt + C_1)} = Ce^{\sqrt{M}kt}$ where $C = e^{\sqrt{M}C_1}$. Multiplying both sides by $\sqrt{M} - \sqrt{A}$ and then isolating for A gives the solution $A(t) = M\left(\dfrac{Ce^{\sqrt{M}kt} - 1}{Ce^{\sqrt{M}kt} + 1}\right)^2$. To get C in terms of the initial area A_0 and the maximum area M, we substitute $t = 0$ and $A = A_0 = A(0)$: $A_0 = M\left(\dfrac{C - 1}{C + 1}\right)^2 \Leftrightarrow$

$(C + 1)\sqrt{A_0} = (C - 1)\sqrt{M} \Leftrightarrow C\sqrt{A_0} + \sqrt{A_0} = C\sqrt{M} - \sqrt{M} \Leftrightarrow \sqrt{M} + \sqrt{A_0} = C\sqrt{M} - C\sqrt{A_0} \Leftrightarrow$

$\sqrt{M} + \sqrt{A_0} = C\left(\sqrt{M} - \sqrt{A_0}\right) \Leftrightarrow C = \dfrac{\sqrt{M} + \sqrt{A_0}}{\sqrt{M} - \sqrt{A_0}}$. Substituting this into the solution gives

$$A(t) = M\left[\frac{\left(\dfrac{\sqrt{M} + \sqrt{A_0}}{\sqrt{M} - \sqrt{A_0}}\right)e^{\sqrt{M}kt} - 1}{\left(\dfrac{\sqrt{M} + \sqrt{A_0}}{\sqrt{M} - \sqrt{A_0}}\right)e^{\sqrt{M}kt} + 1}\right]^2 = M\left[\frac{\left(\sqrt{M} + \sqrt{A_0}\right)e^{\sqrt{M}kt} - \left(\sqrt{M} - \sqrt{A_0}\right)}{\left(\sqrt{M} + \sqrt{A_0}\right)e^{\sqrt{M}kt} + \left(\sqrt{M} - \sqrt{A_0}\right)}\right]^2$$

Notice that this also includes the solution $A(t) = M$, which corresponds to the initial condition $A_0 = M$. Additionally, $A(t) = 0$ is a solution to the differential equation which can be readily verified.

53. Let $S(A)$ be the number of species on an island having area A. Then $\dfrac{dS}{dA} = k\,(\text{Density}) = k\dfrac{S}{A}$ where k is a constant between 0 and 1. So $\dfrac{dS}{dA} = k\dfrac{S}{A} \Rightarrow \displaystyle\int \frac{dS}{S} = k\int \frac{dA}{A} \Rightarrow \ln|S| = k\ln|A| + C_1 \Rightarrow |S| = e^{C_1}A^k$ [since $A > 0$] $\Rightarrow$ $S = CA^k$ where $C = \pm e^{C_1}$ is a constant. This is the same type of power model used in Example 1.5.14.

© 2016 Cengage Learning. All Rights Reserved. May not be scanned, copied or duplicated, or posted to a publicly accessible website, in whole or in part.

7.5 Systems of Differential Equations

1. $x = t^2 + t,\quad y = t^2 - t,\quad -2 \le t \le 2$

t	-2	-1	0	1	2
x	2	0	0	2	6
y	6	2	0	0	2

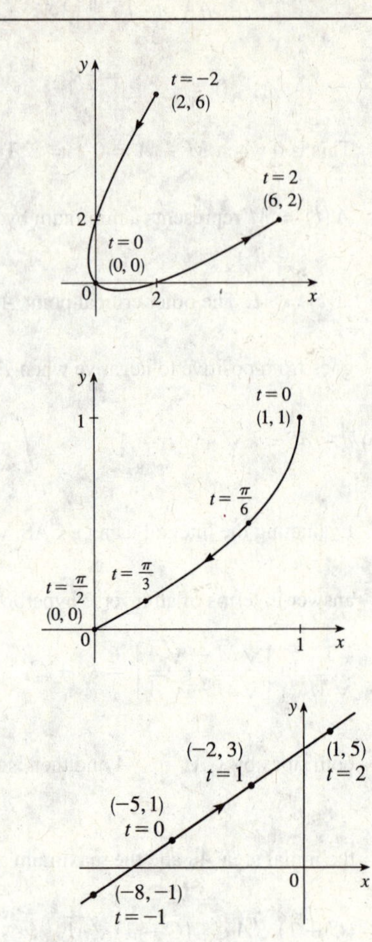

3. $x = \cos^2 t,\quad y = 1 - \sin t,\quad 0 \le t \le \pi/2$

t	0	$\pi/6$	$\pi/3$	$\pi/2$
x	1	$3/4$	$1/4$	0
y	1	$1/2$	$1 - \frac{\sqrt{3}}{2} \approx 0.13$	0

5. $x = 3t - 5,\quad y = 2t + 1$

(a)

t	-2	-1	0	1	2	3	4
x	-11	-8	-5	-2	1	4	7
y	-3	-1	1	3	5	7	9

(b) $x = 3t - 5 \;\Rightarrow\; 3t = x + 5 \;\Rightarrow\; t = \frac{1}{3}(x+5) \;\Rightarrow$

$y = 2 \cdot \frac{1}{3}(x+5) + 1$, so $y = \frac{2}{3}x + \frac{13}{3}$.

7. $x = \sqrt{t},\ y = 1 - t$

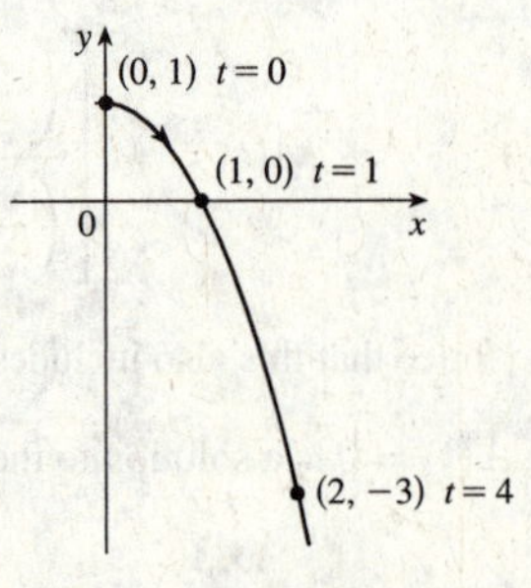

(a)

t	0	1	2	3	4
x	0	1	1.414	1.732	2
y	1	0	-1	-2	-3

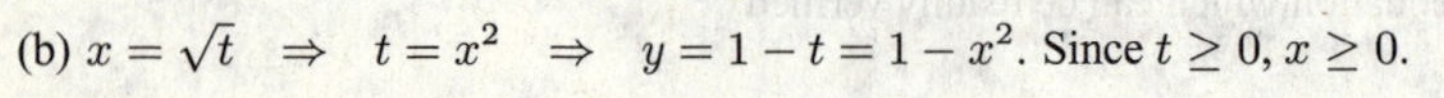

(b) $x = \sqrt{t} \;\Rightarrow\; t = x^2 \;\Rightarrow\; y = 1 - t = 1 - x^2$. Since $t \ge 0$, $x \ge 0$.

So the curve is the right half of the parabola $y = 1 - x^2$.

9. $x = 3 + 2\cos t,\ y = 1 + 2\sin t,\ \pi/2 \le t \le 3\pi/2$. By Example 4 with $r = 2$, $h = 3$, and $k = 1$, the motion of the particle takes place on a circle centered at $(3, 1)$ with a radius of 2. As t goes from $\frac{\pi}{2}$ to $\frac{3\pi}{2}$, the particle starts at the point $(3, 3)$ and moves counterclockwise along the circle $(x-3)^2 + (y-1)^2 = 4$ to $(3, -1)$ [one-half of a circle].

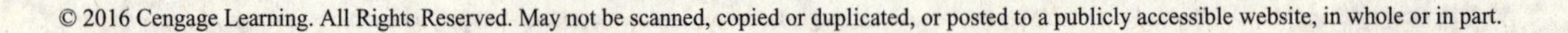
© 2016 Cengage Learning. All Rights Reserved. May not be scanned, copied or duplicated, or posted to a publicly accessible website, in whole or in part.

11. $x = 5\sin t, y = 2\cos t \Rightarrow \sin t = \dfrac{x}{5}, \cos t = \dfrac{y}{2}$. $\sin^2 t + \cos^2 t = 1 \Rightarrow \left(\dfrac{x}{5}\right)^2 + \left(\dfrac{y}{2}\right)^2 = 1$. The motion of the particle takes place on an ellipse centered at $(0, 0)$. As t goes from $-\pi$ to 5π, the particle starts at the point $(0, -2)$ and moves clockwise around the ellipse 3 times.

13. (a) $dx/dt = -0.05x + 0.0001xy$. If $y = 0$, we have $dx/dt = -0.05x$, which indicates that in the absence of y, x declines at a rate proportional to itself. So x represents the predator population and y represents the prey population. The growth of the prey population, $0.1y$ (from $dy/dt = 0.1y - 0.005xy$), is restricted only by encounters with predators (the term $-0.005xy$). The predator population increases only through the term $0.0001xy$; that is, by encounters with the prey and not through additional food sources.

(b) $dy/dt = -0.015y + 0.00008xy$. If $x = 0$, we have $dy/dt = -0.015y$, which indicates that in the absence of x, y would decline at a rate proportional to itself. So y represents the predator population and x represents the prey population. The growth of the prey population, $0.2x$ (from $dx/dt = 0.2x - 0.0002x^2 - 0.006xy = 0.2x(1 - 0.001x) - 0.006xy$), is restricted by a carrying capacity of 1000 [from the term $1 - 0.001x = 1 - x/1000$] and by encounters with predators (the term $-0.006xy$). The predator population increases only through the term $0.00008xy$; that is, by encounters with the prey and not through additional food sources.

15. $dx/dt = 0.5x - 0.004x^2 - 0.001xy = 0.5x(1 - x/125) - 0.001xy$.
$dy/dt = 0.4y - 0.001y^2 - 0.002xy = 0.4y(1 - y/400) - 0.002xy$.

The system shows that x and y have carrying capacities of 125 and 400. An increase in x reduces the growth rate of y due to the negative term $-0.002xy$. An increase in y reduces the growth rate of x due to the negative term $-0.001xy$. Hence the system describes a competition model.

17. (a) At $t = 0$, there are about 300 rabbits and 100 foxes. At $t = t_1$, the number of foxes reaches a minimum of about 20 while the number of rabbits is about 1000. At $t = t_2$, the number of rabbits reaches a maximum of about 2400, while the number of foxes rebounds to 100. At $t = t_3$, the number of rabbits decreases to about 1000 and the number of foxes reaches a maximum of about 315. As t increases, the number of foxes decreases greatly to 100, and the number of rabbits decreases to 300 (the initial populations), and the cycle starts again.

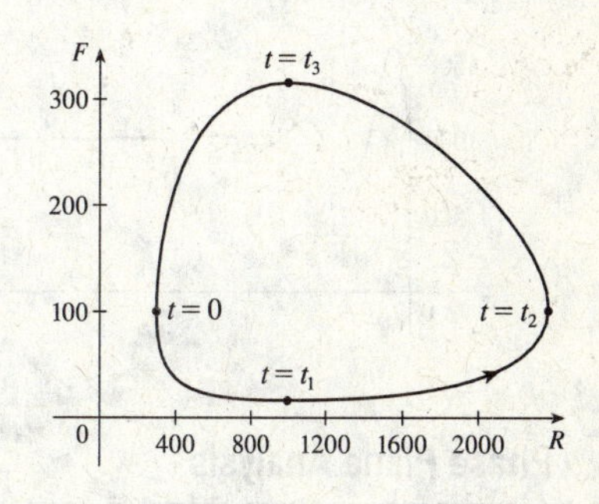

(b)

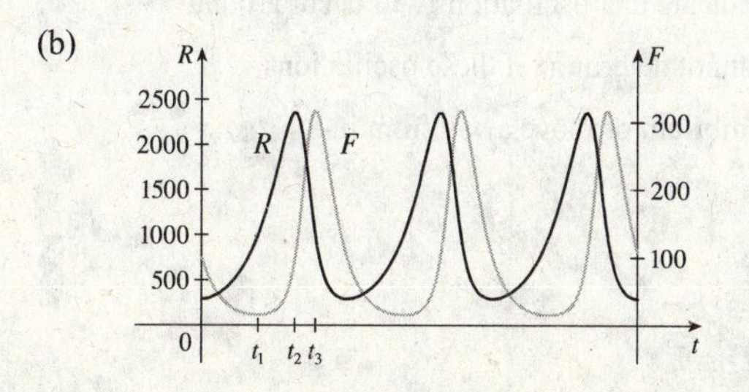

© 2016 Cengage Learning. All Rights Reserved. May not be scanned, copied or duplicated, or posted to a publicly accessible website, in whole or in part.

19.

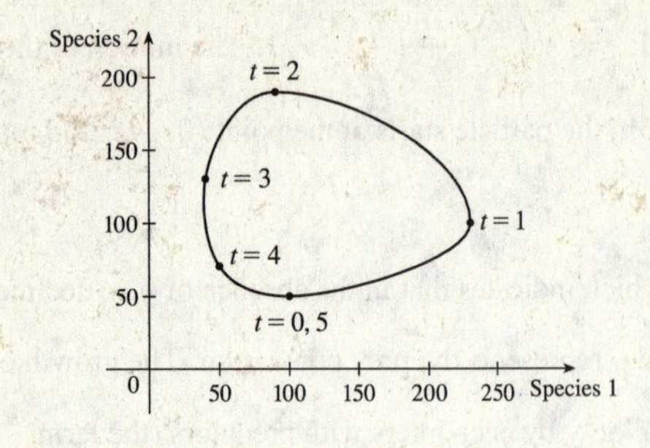

21. $\dfrac{dW}{dR} = \dfrac{-0.02W + 0.00002RW}{0.08R - 0.001RW} \iff (0.08 - 0.001W)R\,dW = (-0.02 + 0.00002R)W\,dR \iff$

$\dfrac{0.08 - 0.001W}{W}\,dW = \dfrac{-0.02 + 0.00002R}{R}\,dR \iff \displaystyle\int\left(\frac{0.08}{W} - 0.001\right)dW = \int\left(-\frac{0.02}{R} + 0.00002\right)dR \iff$

$0.08\ln|W| - 0.001W = -0.02\ln|R| + 0.00002R + K \iff 0.08\ln W + 0.02\ln R = 0.001W + 0.00002R + K \iff$

$\ln\left(W^{0.08}R^{0.02}\right) = 0.00002R + 0.001W + K \iff W^{0.08}R^{0.02} = e^{0.00002R + 0.001W + K} \iff$

$R^{0.02}W^{0.08} = Ce^{0.00002R}e^{0.001W} \iff \dfrac{R^{0.02}W^{0.08}}{e^{0.00002R}e^{0.001W}} = C$. In general, if $\dfrac{dy}{dx} = \dfrac{-ry + bxy}{kx - axy}$, then $C = \dfrac{x^r y^k}{e^{bx}e^{ay}}$.

23. (a) Letting $W = 0$ gives us $dR/dt = 0.08R(1 - 0.0002R)$. $dR/dt = 0 \iff R = 0$ or 5000. Since $dR/dt > 0$ for $0 < R < 5000$, we would expect the rabbit population to *increase* to 5000 for these values of R. Since $dR/dt < 0$ for $R > 5000$, we would expect the rabbit population to *decrease* to 5000 for these values of R. Hence, in the absence of wolves, we would expect the rabbit population to stabilize at 5000.

(b) The populations of wolves and rabbits fluctuate around 64 and 1000, respectively, and eventually stabilize at those values.

(c)

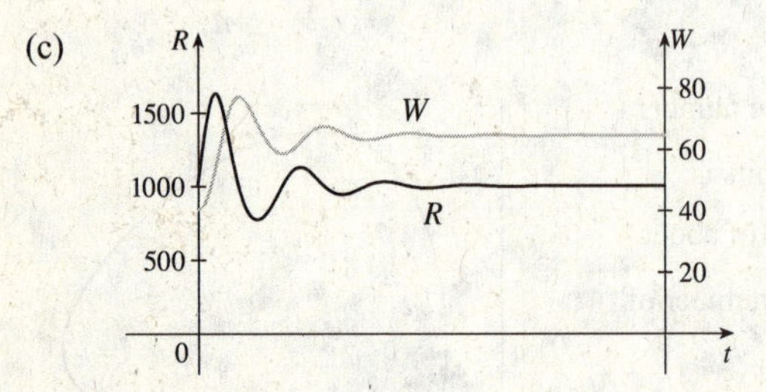

7.6 Phase Plane Analysis

1.

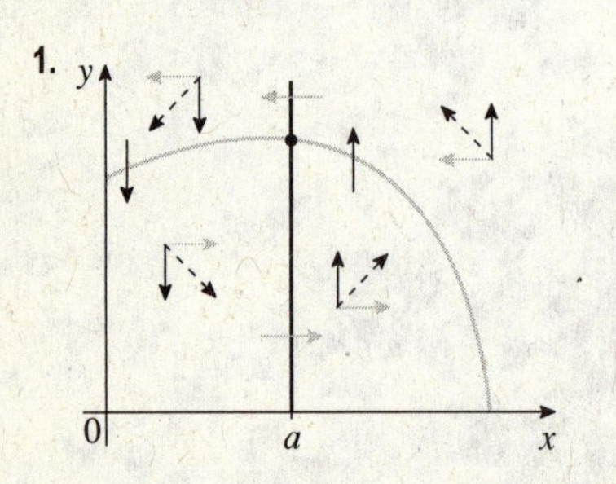

The phase plane arrows indicate that oscillations will occur around the equilibrium, but we cannot determine if these oscillations converge towards the equilibrium or move away from it.

© 2016 Cengage Learning. All Rights Reserved. May not be scanned, copied or duplicated, or posted to a publicly accessible website, in whole or in part.

3.

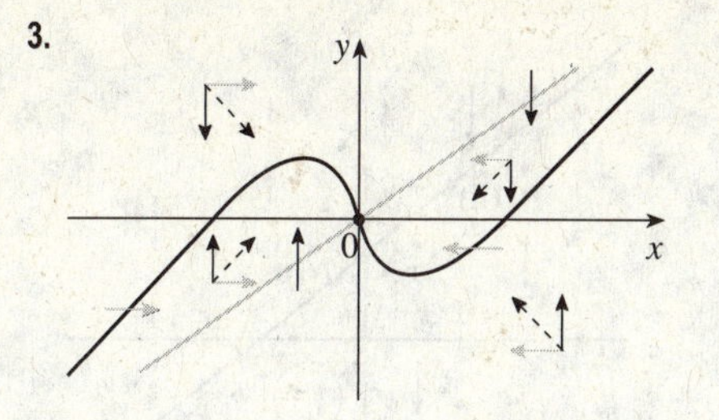

The phase plane arrows indicate that oscillations will occur around the equilibrium, but we cannot determine if these oscillations converge towards the equilibrium or move away from it.

5.

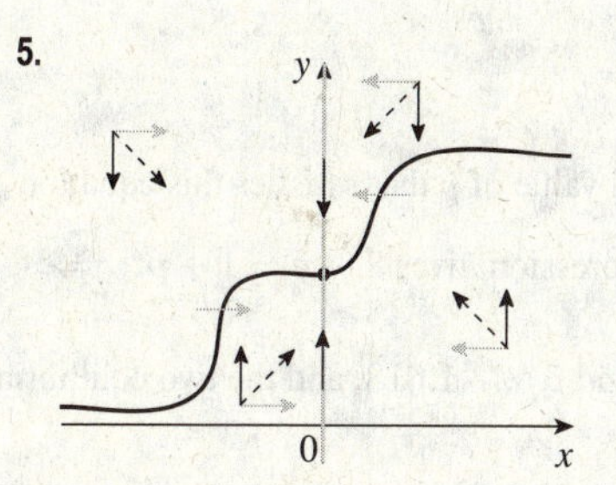

The phase plane arrows indicate that solution curves will follow a path toward the equilibrium for any initial condition. Thus, it is locally stable.

7. (a) $x' = x(3 - x - y),\ y' = y(2 - x - y),\quad x, y \geq 0$

x-nullclines: $x(3 - x - y) = 0 \quad \Rightarrow \quad x = 0$ and $y = 3 - x$

When $y > 3 - x$ (above nullcline), $x' < 0$ and when $y < 3 - x$ (below nullcline), $x' > 0$.

y-nullclines: $y(2 - x - y) = 0 \quad \Rightarrow \quad y = 0$ and $y = 2 - x$

When $y > 2 - x$ (above nullcline), $y' < 0$ and when $y < 2 - x$ (below nullcline), $y' > 0$.

These results are visualized in the phase plane at right.

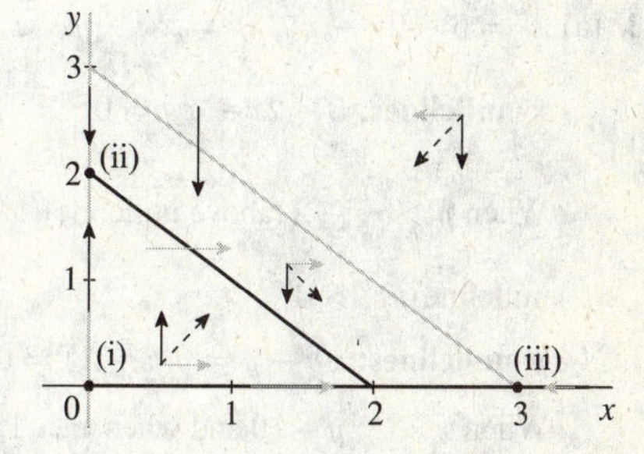

(b) Substituting $\hat{x} = 0$ into the y-nullcline expressions gives $\hat{y} = 0$ and $\hat{y} = 2 - 0 = 2$. So two equilibria are (i) $\hat{x} = 0, \hat{y} = 0$ and (ii) $\hat{x} = 0, \hat{y} = 2$. Substituting $\hat{y} = 0$ into the second x-nullcline expression gives $0 = 3 - \hat{x}$, so $\hat{x} = 3$. Hence, the third equilibrium is (iii) $\hat{x} = 3, \hat{y} = 0$. Lastly, observe that if we try to solve the equations $\hat{y} = 3 - \hat{x}$ and $\hat{y} = 2 - \hat{x}$, we get $3 - \hat{x} = 2 - \hat{x}$. There is no value of $\hat{x}$ that satisfies this equation, so there are only three equilibria as shown in the phase plane.

9. (a) $n' = n(1 - 2m),\ m' = m(2 - 2n - m),\quad n, m \geq 0$

n-nullclines: $n(1 - 2m) = 0 \quad \Rightarrow \quad n = 0$ and $m = 1/2$

When $m > \frac{1}{2}$ (above nullcline), $n' < 0$ and when $m < \frac{1}{2}$ (below nullcline), $n' > 0$.

m-nullclines: $m(2 - 2n - m) = 0 \quad \Rightarrow \quad m = 0$ and $m = 2 - 2\hat{n}$

When $m > 2 - 2n$ (above nullcline), $m' < 0$ and when $m < 2 - 2n$ (below nullcline), $m' > 0$.

These results are visualized in the phase plane at right.

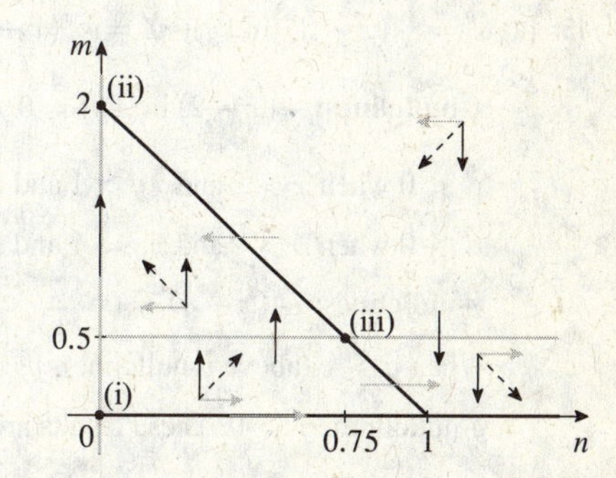

(b) Substituting $\hat{n} = 0$ into the m-nullcline expressions gives $\hat{m} = 0$ and $\hat{m} = 2 - 0 = 2$. So two equilibria are (i) $\hat{n} = 0, \hat{m} = 0$ and (ii) $\hat{n} = 0, \hat{m} = 2$. Substituting $\hat{m} = 1/2$ into the second m-nullcline expression gives $\frac{1}{2} = 2 - 2\hat{n}$, so $\hat{n} = \frac{3}{4}$. Hence, the third equilibrium is (iii) $\hat{n} = \frac{3}{4}, \hat{m} = \frac{1}{2}$. Note that the n and m nullclines $\hat{m} = 1/2$ and $\hat{m} = 0$ do not intersect, so no equilibrium arises from this pair of equations.

© 2016 Cengage Learning. All Rights Reserved. May not be scanned, copied or duplicated, or posted to a publicly accessible website, in whole or in part.

11. (a) $p' = -p^2 + q - 1,\ q' = q(2 - p - q)$

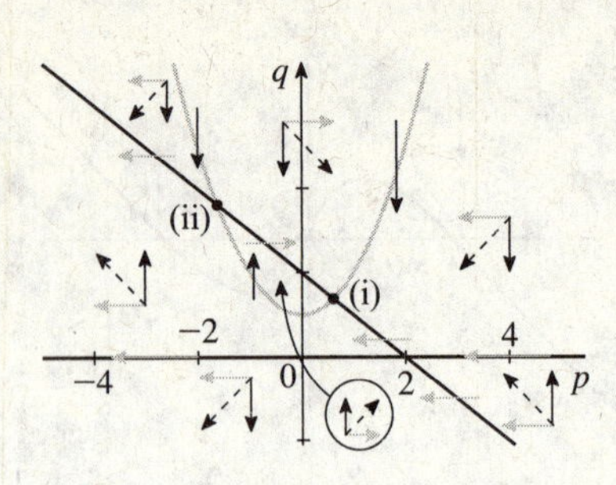

p-nullclines: $-p^2 + q - 1 = 0 \quad \Rightarrow \quad q = 1 + p^2$

When $q > 1 + p^2$ (above nullcline), $p' > 0$ and when $q < 1 + p^2$ (below nullcline), $p' < 0$.

q-nullclines: $q(2 - p - q) = 0 \quad \Rightarrow \quad q = 0$ and $q = 2 - p$

$q' > 0$ when $q > 0$ and $q < 2 - p$ and also when $q < 0$ and $q > 2 - p$.

$q' < 0$ when $q > 0$ and $q > 2 - p$ and also when $q < 0$ and $q < 2 - p$.

These results are graphed in the phase plane at right.

(b) Substituting $\hat{q} = 0$ into the p-nullcline expression gives $0 = 1 + \hat{p}^2$. There is no real value of $\hat{p}$ that satisfies this equation so the two nullclines do not intersect. Substituting $\hat{q} = 2 - \hat{p}$ into the p-nullcline expression gives $2 - \hat{p} = 1 + \hat{p}^2 \quad \Rightarrow$ $\hat{p}^2 + \hat{p} - 1 = 0 \quad \Rightarrow \quad \hat{p} = \dfrac{-1 \pm \sqrt{1^2 - 4(1)(-1)}}{2} = \dfrac{-1 \pm \sqrt{5}}{2}$. So $\hat{p} \approx 0.618$ and $\hat{p} \approx -1.618$, and the two equilibria are (i) $\hat{p} \approx 0.618, \hat{q} \approx 1.382$ and (ii) $\hat{p} \approx -1.618, \hat{q} \approx 3.618$.

13. (a) $x' = 5 - 2x - xy,\ y' = xy - y,\quad x, y \geq 0$

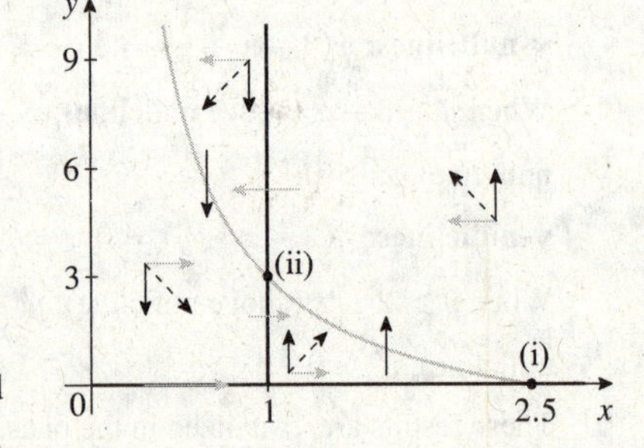

x-nullclines: $5 - 2x - xy = 0 \quad \Rightarrow \quad y = \dfrac{5}{x} - 2$

When $y > \dfrac{5}{x} - 2$ (above nullcline), $x' < 0$ and when $y < \dfrac{5}{x} - 2$ (below nullcline), $x' > 0$

y-nullclines: $xy - y = y(x - 1) = 0 \quad \Rightarrow \quad y = 0$ and $x = 1$

When $x < 1$, $y' < 0$ and when $x > 1$, $y' > 0$. Note the restriction on x and y requires they both be non-negative. These results are graphed in the phase plane at right.

(b) Substituting $\hat{y} = 0$ into the x-nullcline expression gives $0 = \dfrac{5}{\hat{x}} - 2$, so $\hat{x} = 2.5$ and the equilibrium is (i) $\hat{x} = 2.5, \hat{y} = 0$.

Substituting $\hat{x} = 1$ into the x-nullcline expression gives $\hat{y} = \frac{5}{1} - 2 = 3$, so the second equilibrium is (ii) $\hat{x} = 1, \hat{y} = 3$.

15. (a) $x' = -(x - 2)\ln(xy),\ y' = e^x(x - y),\quad x, y > 0$

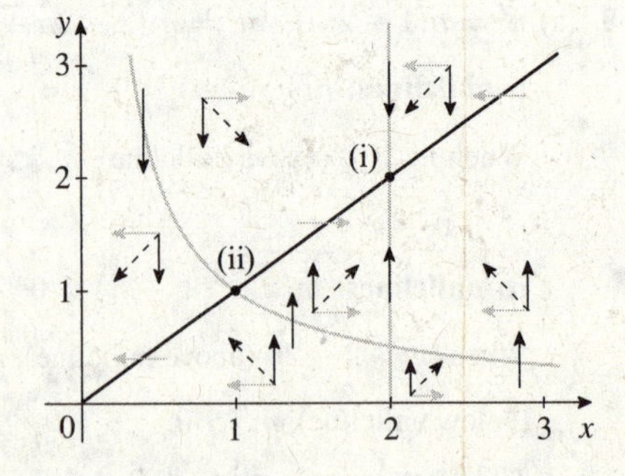

x-nullclines: $-(x - 2)\ln(xy) = 0 \quad \Rightarrow \quad x = 2$ and $y = \dfrac{1}{x}$

$x' < 0$ when $x > 2$ and $xy > 1$ and also when $x < 2$ and $xy < 1$.

$x' > 0$ when $x > 2$ and $xy < 1$ and also when $x < 2$ and $xy > 1$.

y-nullclines: $e^x(x - y) = 0 \quad \Rightarrow \quad y = x$

When $y > x$ (above y-nullcline), $y' < 0$ and when $y < x$ (below y-nullcline), $y' > 0$. These results are graphed in the phase plane at right.

(b) Substituting $\hat{x} = 2$ into the y-nullcline expression gives $\hat{y} = 2$. So the equilibrium is (i) $\hat{x} = 2, \hat{y} = 2$.

Substituting $\hat{y} = 1/\hat{x}$ into the y-nullcline expression gives $\dfrac{1}{\hat{x}} = \hat{x} \quad \Rightarrow \quad \hat{x}^2 = 1 \quad \Rightarrow \quad \hat{x} = 1$ [discard the negative solution since $x > 0$]. So the second equilibrium is (ii) $\hat{x} = 1, \hat{y} = 1$.

© 2016 Cengage Learning. All Rights Reserved. May not be scanned, copied or duplicated, or posted to a publicly accessible website, in whole or in part.

17. (a) $x' = y - ax,\ y' = x - y,\quad a > 0,\quad a \neq 1$

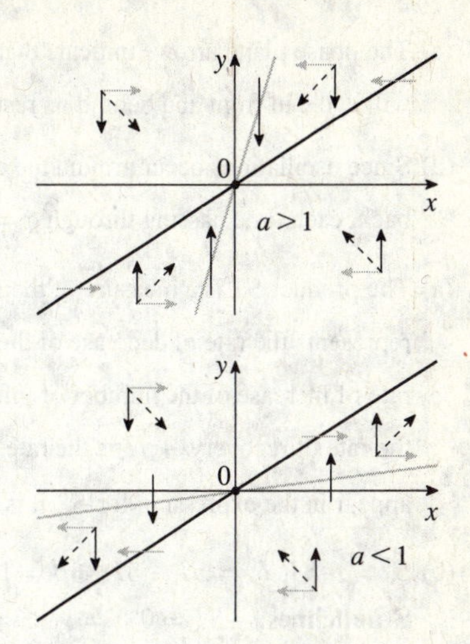

x-nullclines: $y - ax = 0 \quad \Rightarrow \quad y = ax$

When $y < ax$ (below x-nullcline), $x' < 0$ and when $y > ax$ (above x-nullcline), $x' > 0$.

y-nullclines: $x - y = 0 \quad \Rightarrow \quad y = x$

When $y > x$ (above y-nullcline), $y' < 0$ and when $y < x$ (below y-nullcline), $y' > 0$.

Note that when $a > 1$, the x-nullcline has a greater slope than the y-nullcline. The opposite is true when $a < 1$. The phase planes graphed at right show these two situations. We see that the equilibrium will be stable when $a > 1$, since the directional arrows point towards the equilibrium in all regions.

(b) Substituting $\hat{y} = \hat{x}$ into the x-nullcline expression gives $\hat{x} = a\hat{x} \quad \Rightarrow \quad \hat{x}(1 - a) = 0$, so $\hat{x} = 0$. Thus, the equilibrium is $\hat{x} = 0, \hat{y} = 0$.

19. (a) $x' = ay^2 - x + 1,\ y' = 2(1 - y)$

x-nullclines: $ay^2 - x + 1 = 0 \quad \Rightarrow \quad x = 1 \ \ [\text{if } a = 0] \quad$ and $\quad y = \pm\sqrt{(x-1)/a} \ \ [\text{if } a \neq 0]$

If $a = 0$: $x' < 0$ when $x > 1$ and $x' > 0$ when $x < 1$. If $a > 0$: $x' < 0$ when $|y| < \sqrt{(x-1)/a}$ and $x' > 0$ when $|y| > \sqrt{(x-1)/a}$. If $a < 0$: $x' < 0$ when $|y| > \sqrt{(x-1)/a}$ and $x' > 0$ when $|y| < \sqrt{(x-1)/a}$.

y-nullclines: $2(1 - y) = 0 \quad \Rightarrow \quad y = 1 \quad$ When $y > 1$ (above y-nullcline), $y' < 0$ and when $y < 1$ (below y-nullcline), $y' > 0$. The graphs below show the phase planes for different values of a. We see that the equilibrium will be stable for any value of a, since the directional arrows point towards the equilibrium in all regions in all cases.

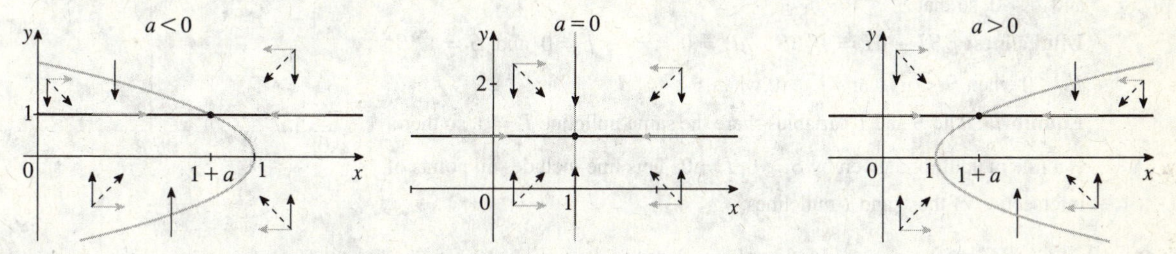

(b) Substituting $\hat{y} = 1$ into the x-nullcline expression gives $a(1) - \hat{x} + 1 = 0 \quad \Rightarrow \quad \hat{x} = a + 1$. Thus, the equilibrium is $\hat{x} = a + 1, \hat{y} = 1$.

21. (a) $m\dfrac{d^2p}{dt^2} = -kp \quad$ Defining the new variable $q = p'$, we have $q' = p'' = \dfrac{d^2p}{dt^2} = -kp/m$. Thus, we have the following system of first order differential equations: $p' = q \quad q' = -kp/m$

(b) $p' = q \ \ q' = -kp/m$

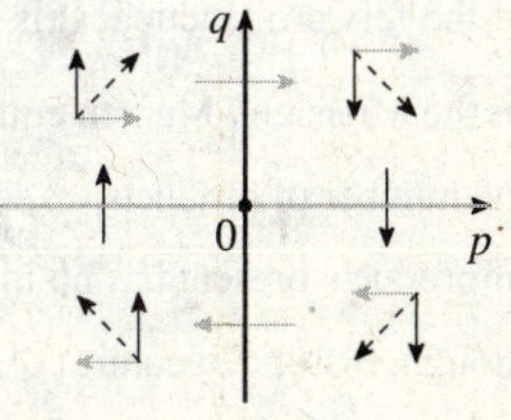

p-nullclines: $q = 0.\quad p' < 0$ when $q < 0$ and $p' > 0$ when $q > 0$.

q-nullclines: $-kp/m = 0 \quad \Rightarrow \quad p = 0$

$q' < 0$ when $p > 0$ and $q' > 0$ when $p < 0$.

The equilibrium is $\hat{p} = 0, \hat{q} = 0$.

© 2016 Cengage Learning. All Rights Reserved. May not be scanned, copied or duplicated, or posted to a publicly accessible website, in whole or in part.

(c) The phase plane arrows indicate that oscillations will occur around the equilibrium. Therefore, the position of the mass will cycle in front and behind its rest position, $p = 0$, as time passes.

(d) Since oscillations occur around the equilibrium, the velocity of the mass, q, will cycle from positive to negative values and back, each time passing through $q = 0$.

23. (a) The product SI is a measure of the number of interactions between susceptible and infected individuals, so $-\beta SI$ represents the rate of decrease of the number of susceptible people due to transmission of the disease. Similarly βSI is the rate of increase of the number of infected people due to transmission of the virus to the susceptible population. Since μ is the rate of recovery, $-\mu I$ is the rate of decrease of the infected population due to recovery. Note that since μI does not appear in the expression for S', it is assumed that recovered individuals are immune from reinfection.

(b) $S' = -SI \quad I' = SI - 5I \quad [\beta = 1, \mu = 5]$

S-nullclines: $-SI = 0 \quad \Rightarrow \quad S = 0$ and $I = 0$

Since S and I both represent a number of individuals, we require that $S \geq 0$ and $I \geq 0$, so that $S' \leq 0$.

I-nullclines: $SI - 5I = I\,(S - 5) = 0 \quad \Rightarrow \quad I = 0$ and $S = 5$

$I' < 0$ when $S < 5$ and $I' > 0$ when $S > 5$.

Equilibria: The S and I variables share the same nullcline $I = 0$, so there is a line of equilibria given by $\hat{S} \geq 0$, $\hat{I} = 0$. This line includes all points of intersection of the S and I nullclines.

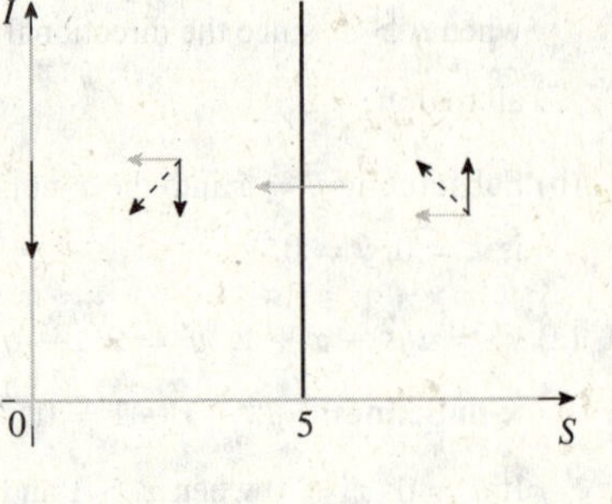

(c) $S' = -\beta SI \qquad I' = \beta SI - \mu I$

S-nullclines: $-\beta SI = 0 \quad \Rightarrow \quad S = 0$ and $I = 0$

Since S and I both represent a number of individuals, we require that $S \geq 0$ and $I \geq 0$, so that $S' \leq 0$.

I-nullclines: $\beta SI - \mu I = I\,(\beta S - \mu) = 0 \quad \Rightarrow \quad I = 0$ and $S = \mu/\beta$

$I' < 0$ when $S < \mu/\beta$ and $I' > 0$ when $S > \mu/\beta$.

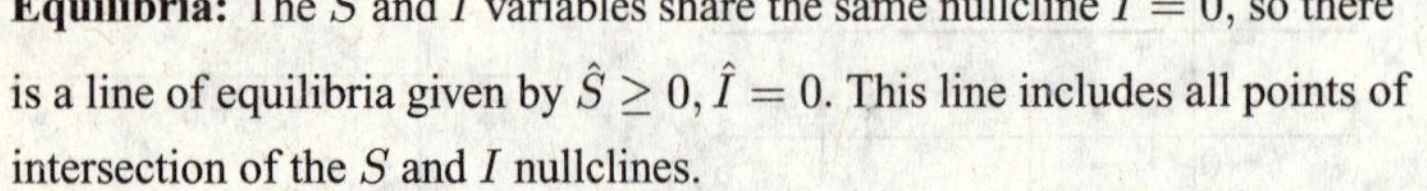

Equilibria: The S and I variables share the same nullcline $I = 0$, so there is a line of equilibria given by $\hat{S} \geq 0$, $\hat{I} = 0$. This line includes all points of intersection of the S and I nullclines.

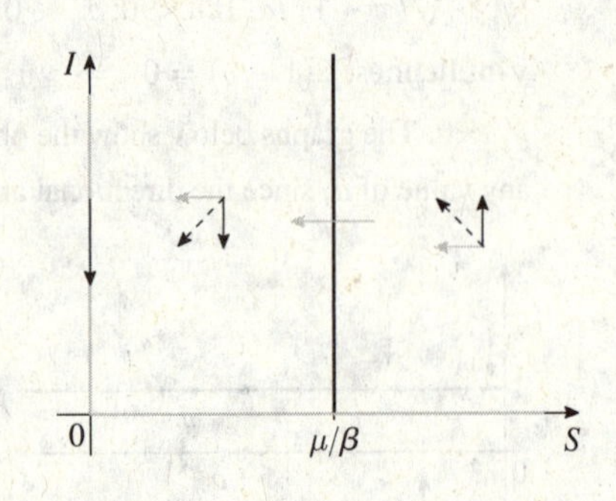

25. (a) yM represents the number of free enzymes, so the product xyM is a measure of the number of interactions, potentially leading to bonds, between substrate molecules and free enzymes. Thus $-k_f xyM$ is the rate of decrease of substrate molecules due to binding with enzymes where k_f is the fractional rate of enzyme binding.

$(1 - y)M$ represents the number of bound enzymes, so the term $k_r(1 - y)M$ is the rate of increase of substrate molecules due to dissociation of the enzyme-substrate complex. k_r is the fractional rate of dissociation.

Lastly, $k_{cat}(1 - y)M$ is the rate of increase of free enzymes due to the catalyzed reaction, where k_{cat} is the fractional rate of the forward reaction. It is also the rate at which product molecules are generated.

(b) In the Michaelis-Menten equations, the dynamics of the substrate molecules, x, and free enzymes, y, are independent of the number of products, z, since the equations for dx/dt and dy/dt are functions of x and y alone. Therefore, the number of products present has no impact on x and y, so a phase plane can be constructed using only these two variables. Note though, that the dynamics of z are affected by y in the dz/dt equation.

© 2016 Cengage Learning. All Rights Reserved. May not be scanned, copied or duplicated, or posted to a publicly accessible website, in whole or in part.

(c) $x' = -k_f xyM + k_r(1-y)M \quad y' = -k_f xyM + k_r(1-y)M + k_{cat}(1-y)M$

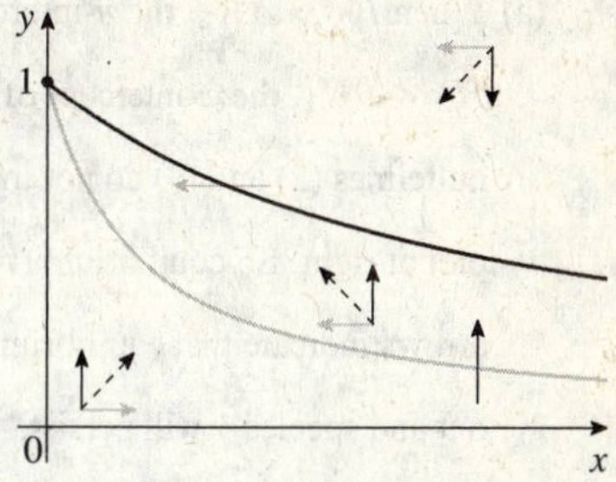

x-nullclines: $-k_f xyM + k_r(1-y)M = 0 \quad \Rightarrow \quad y(k_r + k_f x) = k_r \quad \Rightarrow$

$y = \dfrac{k_r}{k_r + k_f x}$. $x' < 0$ when $y > \dfrac{k_r}{k_r + k_f x}$ (above x-nullcline) and $x' > 0$ when $y < \dfrac{k_r}{k_r + k_f x}$ (below x-nullcline).

y-nullclines: $-k_f xyM + k_r(1-y)M + k_{cat}(1-y)M = 0 \quad \Rightarrow$

$y(k_f x + k_r + k_{cat}) = k_r + k_{cat} \quad \Rightarrow \quad y = \dfrac{k_r + k_{cat}}{k_f x + k_r + k_{cat}}$

$y' < 0$ when $y > \dfrac{k_{cat}}{k_f x + k_r + k_{cat}}$ (above y-nullcline) and $y' > 0$ when $y < \dfrac{k_{cat}}{k_f x + k_r + k_{cat}}$ (below y-nullcline).

Equilibria: Substituting $\hat{y} = \dfrac{k_r + k_{cat}}{k_f \hat{x} + k_r + k_{cat}}$ into the x-nullcline expression gives $\dfrac{k_r + k_{cat}}{k_f \hat{x} + k_r + k_{cat}} = \dfrac{k_r}{k_r + k_f \hat{x}} \quad \Rightarrow$

$(k_r + k_{cat})(k_r + k_f \hat{x}) = k_r(k_f \hat{x} + k_r + k_{cat}) \quad \Rightarrow \quad k_{cat}\hat{x} = 0 \quad \Rightarrow \quad \hat{x} = 0$. So $\hat{y} = \dfrac{k_r + k_{cat}}{0 + k_r + k_{cat}} = 1$ and the equilibrium is $\hat{x} = 0, \hat{y} = 1$.

27. Lotka-Volterra competition equations:

$$\frac{dN_1}{dt} = N_1 r\left(1 - \frac{N_1 + \alpha N_2}{K_1}\right) \qquad \frac{dN_2}{dt} = N_2 r\left(1 - \frac{N_2 + \beta N_1}{K_2}\right) \qquad N_1, N_2 \geq 0$$

N_1-nullclines: $N_1 r\left(1 - \dfrac{N_1 + \alpha N_2}{K_1}\right) = 0 \quad \Rightarrow \quad N_1 = 0$ **(1)** and $N_1 + \alpha N_2 = K_1 \quad \Rightarrow \quad N_2 = -\dfrac{1}{\alpha}N_1 + \dfrac{K_1}{\alpha}$ **(2)**

$dN_1/dt > 0$ when $N_2 < -\dfrac{1}{\alpha}N_1 + \dfrac{K_1}{\alpha}$ (below N_1-nullcline) and $dN_1/dt < 0$ when $N_2 > -\dfrac{1}{\alpha}N_1 + \dfrac{K_1}{\alpha}$ (above N_1-nullcline). Also note that the x and y intercepts of (2) are K_1 and K_1/α respectively.

N_2-nullclines: $N_2 r\left(1 - \dfrac{N_2 + \beta N_1}{K_2}\right) = 0 \quad \Rightarrow \quad N_2 = 0$ **(3)** and $N_2 + \beta N_1 = K_2 \quad \Rightarrow \quad N_2 = -\beta N_1 + K_2$ **(4)**

$dN_2/dt > 0$ when $N_2 < -\beta N_1 + K_2$ (below N_2-nullcline) and $dN_2/dt < 0$ when $N_2 > -\beta N_1 + K_2$ (above N_2-nullcline). Also note that the x and y intercepts of (4) are K_2/β and K_2 respectively.

Equilibria: Equations (1) and (3) give the first equilibrium (i) $\hat{N}_1 = 0, \hat{N}_2 = 0$. Substituting (1) into (4) gives the second equilibrium (ii) $\hat{N}_1 = 0, \hat{N}_2 = K_2$. Substituting (3) into (2) gives $0 = -\dfrac{1}{\alpha}\hat{N}_1 + \dfrac{K_1}{\alpha}$, or $\hat{N}_1 = K_1$, so the third equilibrium is (iii) $\hat{N}_1 = K_1, \hat{N}_2 = 0$. Lastly, substituting (2) into (4) we get $-\dfrac{1}{\alpha}\hat{N}_1 + \dfrac{K_1}{\alpha} = -\beta\hat{N}_1 + K_2 \quad \Rightarrow$

$(\alpha\beta - 1)\hat{N}_1 + K_1 = \alpha K_2 \quad \Rightarrow \quad \hat{N}_1 = \dfrac{\alpha K_2 - K_1}{\alpha\beta - 1}$ and $\hat{N}_2 = -\dfrac{1}{\alpha}\left(\dfrac{\alpha K_2 - K_1}{\alpha\beta - 1}\right) + \dfrac{K_1}{\alpha} = \dfrac{K_1\beta - K_2}{\alpha\beta - 1}$.

Thus, the fourth equilibrium is (iv) $\hat{N}_1 = \dfrac{\alpha K_2 - K_1}{\alpha\beta - 1}, \hat{N}_2 = \dfrac{K_1\beta - K_2}{\alpha\beta - 1}$.

[continued]

© 2016 Cengage Learning. All Rights Reserved. May not be scanned, copied or duplicated, or posted to a publicly accessible website, in whole or in part.

(a) When $K_1 > \alpha K_2$, the y-intercept of (2) is greater than that of (4). Also, when $K_2 < \beta K_1$, the x-intercept of (2) is greater than that of (4). Therefore, the nullclines (2) and (4) do not intersect in the first quadrant as seen in the phase plot at right. So equilibrium (iv) is not biologically feasible. The phase plane arrows indicate that equilibrium (iii) will be locally stable, so species 2 will die out and species 1 will persist.

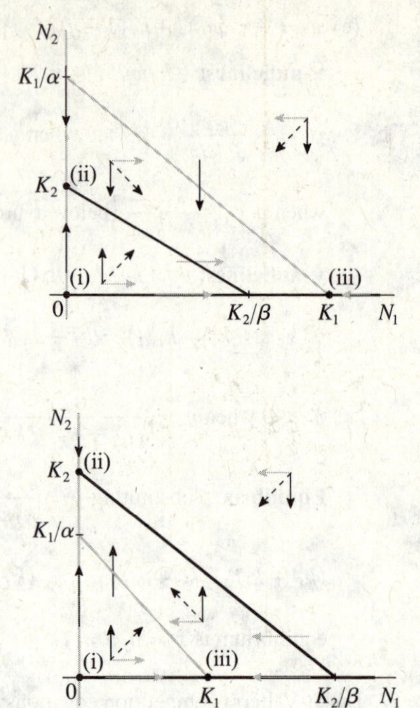

(b) When $K_1 < \alpha K_2$, the y-intercept of (2) is less than that of (4). Also, when $K_2 > \beta K_1$, the x-intercept of (2) is less than that of (4). Therefore, the nullclines (2) and (4) do not intersect in the first quadrant as seen in the phase plot at right. So equilibrium (iv) is not biologically feasible. The phase plane arrows indicate that equilibrium (ii) will be locally stable, so species 1 will die out and species 2 will persist.

(c) When $K_1 < \alpha K_2$, the y-intercept of (2) is less than that of (4). Also, when $K_2 < \beta K_1$, the x-intercept of (2) is greater than that of (4). Therefore, nullclines (2) and (4) intersect in the first quadrant producing equilibrium (iv). This is seen in the phase plot at right. The phase plane arrows indicate that equilibria (i) and (iv) are unstable, whereas equilibria (ii) and (iii) are locally stable. Therefore, species 1 and 2 will not coexist, that is, one of the two species will go extinct.

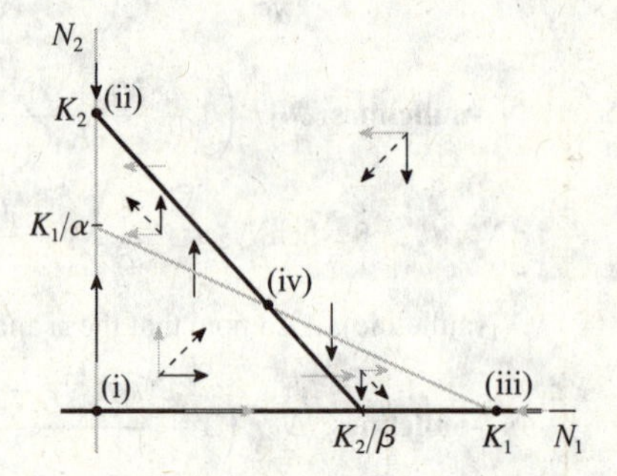

(d) When $K_1 > \alpha K_2$, the y-intercept of (2) is greater than that of (4). Also, when $K_2 > \beta K_1$, the x-intercept of (2) is less than that of (4). Therefore, nullclines (2) and (4) intersect in the first quadrant producing equilibrium (iv). This is seen in the phase plot at right. The phase plane arrows indicate that equilibria (i), (ii) and (iii) are unstable, whereas equilibrium (iv) is locally stable. Therefore, species 1 and 2 will coexist.

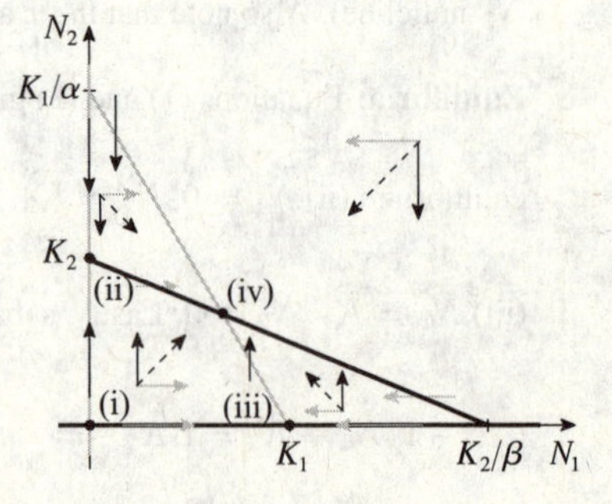

© 2016 Cengage Learning. All Rights Reserved. May not be scanned, copied or duplicated, or posted to a publicly accessible website, in whole or in part.

29. $R' = rR - bRC \qquad C' = \epsilon bRC - \mu C \qquad R, C \geq 0$

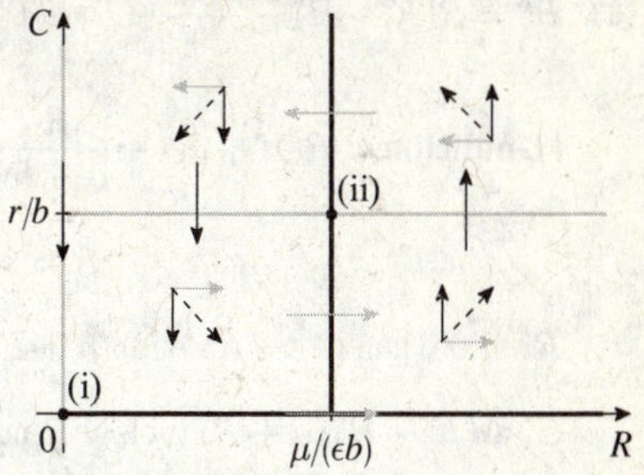

R-nullclines: $rR - bRC = R(r - bC) = 0 \;\Rightarrow\; R = 0$ and $C = \dfrac{r}{b}$

$R' < 0$ when $C > r/b$ (above R-nullcline) and $R' > 0$ when $C < r/b$ (below R-nullcline).

C-nullclines: $\epsilon bRC - \mu C = C(\epsilon bR - \mu) = 0 \;\Rightarrow\; C = 0$ and $R = \dfrac{\mu}{\epsilon b}$

$C' < 0$ when $R < \mu/(\epsilon b)$ (left of C-nullcline) and $C' > 0$ when $R > \mu/(\epsilon b)$ (right of C-nullcline).

C, r/b, (ii), (i), 0, μ/(εb), R

Equilibria: The first R and C nullclines give the equilibrium (i) $\hat{R} = 0, \hat{C} = 0$ and second R and C nullclines give the other equilibrium (ii) $\hat{R} = \dfrac{\mu}{\epsilon b}, \hat{C} = \dfrac{r}{b}$.

The directional arrows in the phase plane indicate that consumer and resource abundance will oscillate over time.

31. (a) The $-\dfrac{K}{V}c$ term represents the rate of decrease of urea concentration in the blood due to dialysis. $\dfrac{c}{V}$ is the fractional amount of concentration per blood volume, so K represents the rate of flow through the dialyzer (see Exercise 7.4.46). The ap term represents the rate of increase of urea concentration in the blood due to urea outflow from the inaccessible pool. The bc term represents the rate of change of urea concentration due to urea flow from the blood back to the inaccessible pool.

(b) $\dfrac{dc}{dt} = -\dfrac{K}{V}c + ap - bc \qquad \dfrac{dp}{dt} = -ap + bc \qquad c, p \geq 0$

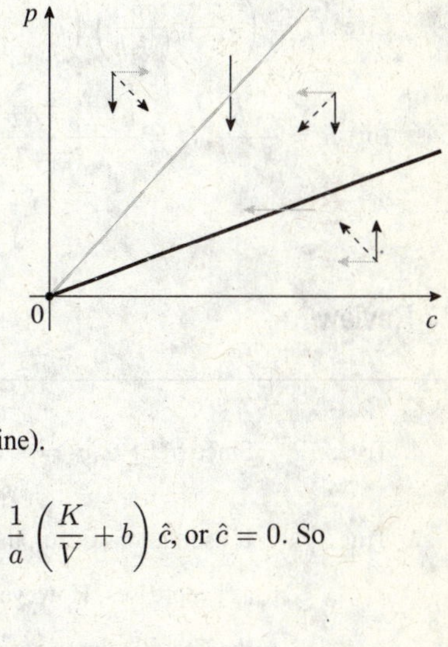

c-nullclines: $-\dfrac{K}{V}c + ap - bc = 0 \;\Rightarrow\; p = \left(\dfrac{K}{aV} + \dfrac{b}{a}\right)c$

$c' < 0$ when $p < \dfrac{1}{a}\left(\dfrac{K}{V} + b\right)c$ (below c-nullcline) and $c' > 0$ when $p > \dfrac{1}{a}\left(\dfrac{K}{V} + b\right)c$ (above c-nullcline).

p, 0, c

p-nullclines: $-ap + bc = 0 \;\Rightarrow\; p = \dfrac{b}{a}c$

$p' < 0$ when $p > \dfrac{b}{a}c$ (above p-nullcline) and $p' > 0$ when $p < \dfrac{b}{a}c$ (below p-nullcline).

Equilibria: Substituting the p-nullcline expression into the c-nullcline gives $\dfrac{b}{a}\hat{c} = \dfrac{1}{a}\left(\dfrac{K}{V} + b\right)\hat{c}$, or $\hat{c} = 0$. So $\hat{p} = \dfrac{b}{a}(0) = 0$ and the equilibrium is (i) $\hat{c} = 0, \hat{p} = 0$.

Since all constants are positive, $\dfrac{K}{aV} + \dfrac{b}{a} > \dfrac{b}{a} > 0$ so the c-nullcline has a steeper slope than the p-nullcline. This is illustrated in the phase plane along with the direction of movement. We see that urea concentration will approach zero in both the blood and pool as $t \to \infty$ for all initial conditions.

© 2016 Cengage Learning. All Rights Reserved. May not be scanned, copied or duplicated, or posted to a publicly accessible website, in whole or in part.

33. $R' = R(K - R) - \dfrac{R}{a + R}C \qquad C' = \dfrac{R}{a + R}C - bC \qquad C > 0$

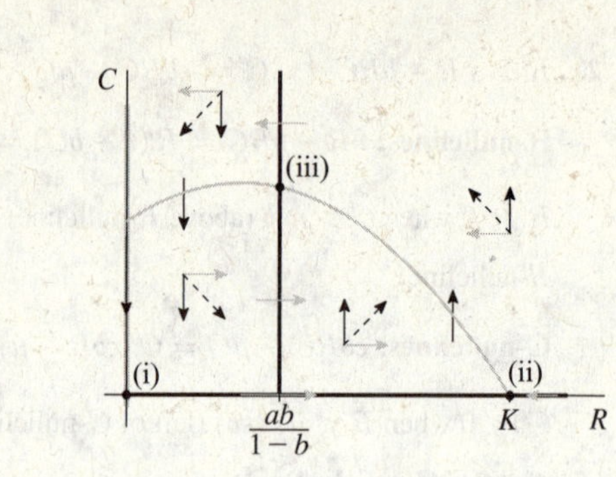

R-nullclines: $R(K - R) - \dfrac{R}{a + R}C = R\left(K - R - \dfrac{1}{a + R}C\right) = 0 \;\Rightarrow$

$$R = 0 \text{ and } C = (K - R)(a + R)$$

$R' < 0$ when $C > (K - R)(a + R)$ (above R-nullcline) and $R' > 0$ when $C < (K - R)(a + R)$ (below R-nullcline).

C-nullclines: $\dfrac{R}{a + R}C - bC = C\left(\dfrac{R}{a + R} - b\right) = 0 \;\Rightarrow\; C = 0$ and $R = b(a + R)$, or $R = \dfrac{ab}{1 - b}$

The restriction $ab/(1 - b) > 0$ requires that $1 - b > 0$ since all constants are positive . This ensures the second C-nullcline has a positive value of R. Under this condition, $C' < 0$ when $R < \dfrac{ab}{1 - b}$ (left of C-nullcline) and $C' > 0$ when $R > \dfrac{ab}{1 - b}$ (right of C-nullcline).

Equilibria: The first two R and C nullclines give the equilibrium (i) $\hat{R} = 0, \hat{C} = 0$. Substituting $\hat{C} = 0$ into the second R-nullcline expression gives $0 = \left(K - \hat{R}\right)(a + \hat{R})$, so $\hat{R} = K$ (reject $\hat{R} = -a$ since R must be non-negative). Hence, the second equilibrium is (ii) $\hat{R} = K, \hat{C} = 0$. Lastly, substituting $\hat{R} = \dfrac{ab}{1 - b}$ into the second R-nullcline expression gives $\hat{C} = \left(K - \dfrac{ab}{1 - b}\right)\left(a + \dfrac{ab}{1 - b}\right) = \dfrac{a(K - Kb - ab)}{(1 - b)^2}$. So the third equilibrium is (iii) $\hat{R} = \dfrac{ab}{1 - b}, \hat{C} = \dfrac{a(K - Kb - ab)}{(1 - b)^2}$. The condition $K > ab/(1 - b)$ ensures both $\hat{R}$ and $\hat{C}$ are positive.

7 Review

TRUE-FALSE QUIZ

1. True. Since $y^4 \geq 0$, $y' = -1 - y^4 < 0$ and the solutions are decreasing functions.

3. True. If oscillations occur, the function $y(t)$ will have repeating maxima and minima so that $dy/dt = 0$ repeatedly as t increases. However, $dy/dt = g(y)$ is not a function of t, so if $g(y) = 0$ at a certain point in time t^*, then dy/dt will remain zero for $t > t^*$. Thus, the function will remain constant if ever $dy/dt = 0$, so it is not possible for oscillations to occur.

5. True. $y' = 3y - 2x + 6xy - 1 = 6xy - 2x + 3y - 1 = 2x(3y - 1) + 1(3y - 1) = (2x + 1)(3y - 1)$, so y' can be written in the form $g(x)f(y)$, and hence, is separable.

© 2016 Cengage Learning. All Rights Reserved. May not be scanned, copied or duplicated, or posted to a publicly accessible website, in whole or in part.

EXERCISES

1. (a) $g(\hat{x}) = a\hat{x} - \hat{x}^2 = 0 \;\Rightarrow\; \hat{x}\,(a - \hat{x}) = 0 \;\Rightarrow\; \hat{x} = 0$ and $\hat{x} = a$.

(b) The graph of $g(x)$ is a parabola that opens down with zeros $x = 0$ and $x = a$. This is illustrated in the phase plots below for three different values of a.

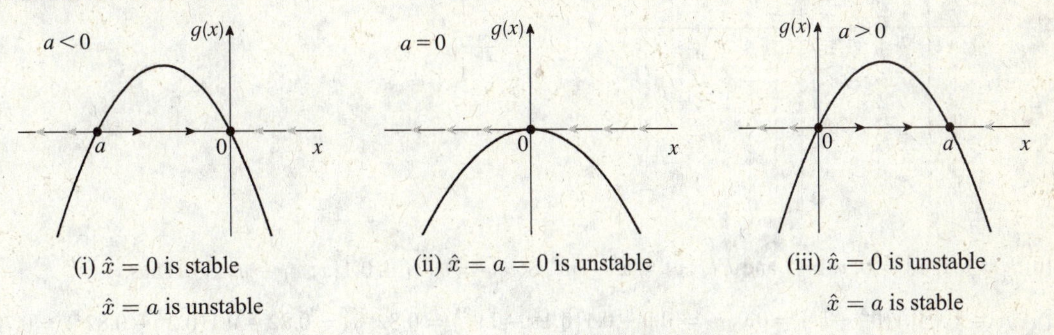

(i) $\hat{x} = 0$ is stable
$\hat{x} = a$ is unstable

(ii) $\hat{x} = a = 0$ is unstable

(iii) $\hat{x} = 0$ is unstable
$\hat{x} = a$ is stable

(c) $g'(x) = a - 2x \;\Rightarrow\; g'(0) = a$ and $g'(a) = a - 2a = -a$. So $\hat{x} = 0$ is locally stable when $a < 0$ and $\hat{x} = a$ is locally stable when $a > 0$. The local stability criterion is inconclusive when $a = 0$ since $g'(0) = 0$.

3. (a) $x' = g(\hat{x}) = a\hat{x} - \hat{x}^3 = 0 \;\Rightarrow\; \hat{x}\left(a - \hat{x}^2\right) = 0 \;\Rightarrow\; \hat{x} = 0$ or $\hat{x}^2 = a$, that is, $\hat{x} = -\sqrt{a}$ or $\hat{x} = \sqrt{a}$ (provided $a > 0$).

(b) The graph of $g(x)$ is a cubic function having zeros at $x = 0$ and $x = \pm\sqrt{a}$ if $a > 0$. As $x \to \infty$, $g(x) \to -\infty$ and as $x \to -\infty$, $g(x) \to \infty$. This is illustrated in the phase plots below for three different values of a.

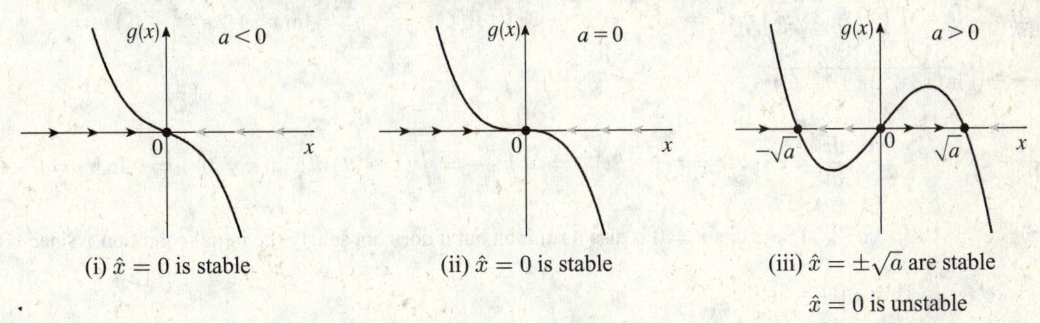

(i) $\hat{x} = 0$ is stable

(ii) $\hat{x} = 0$ is stable

(iii) $\hat{x} = \pm\sqrt{a}$ are stable
$\hat{x} = 0$ is unstable

(c) $g'(x) = a - 3x^2 \;\Rightarrow\; g'(0) = a$ so $\hat{x} = 0$ is locally stable when $a < 0$. Also, $g'(\pm\sqrt{a}) = a - 3\left(\pm\sqrt{a}\right)^2 = -2a$ so the equilibria $\hat{x} = -\sqrt{a}$ and $\hat{x} = \sqrt{a}$ are locally stable when $a > 0$. If $a < 0$ then the equilibria $\hat{x} = \pm\sqrt{a}$ do not exist. If $a = 0$ then $g'(\pm\sqrt{a}) = g'(0) = 0$ so the local stability criterion is inconclusive.

5. (a)

(b) $\lim\limits_{t\to\infty} y(t)$ appears to be finite for $0 \le c \le 4$. In fact $\lim\limits_{t\to\infty} y(t) = 4$ for $c = 4$, $\lim\limits_{t\to\infty} y(t) = 2$ for $0 < c < 4$, and $\lim\limits_{t\to\infty} y(t) = 0$ for $c = 0$. The equilibrium solutions are $y(t) = 0$, $y(t) = 2$, and $y(t) = 4$.

© 2016 Cengage Learning. All Rights Reserved. May not be scanned, copied or duplicated, or posted to a publicly accessible website, in whole or in part.

7. (a)

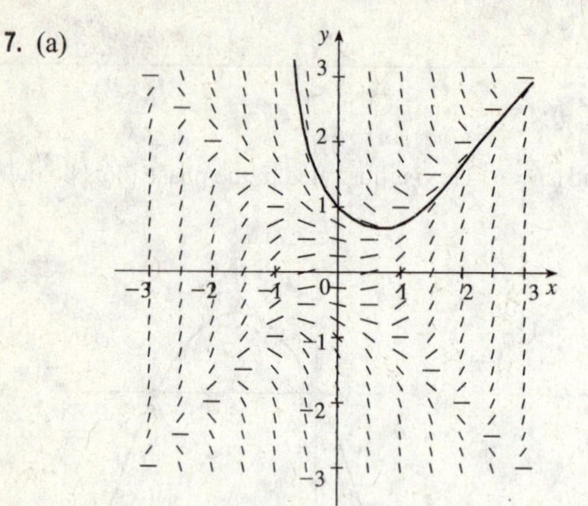

We estimate that when $x = 0.3$, $y = 0.8$, so $y(0.3) \approx 0.8$.

(b) $h = 0.1$, $x_0 = 0$, $y_0 = 1$ and $F(x, y) = x^2 - y^2$. So $y_n = y_{n-1} + 0.1(x_{n-1}^2 - y_{n-1}^2)$. Thus,

$y_1 = 1 + 0.1(0^2 - 1^2) = 0.9$, $y_2 = 0.9 + 0.1(0.1^2 - 0.9^2) = 0.82$, $y_3 = 0.82 + 0.1(0.2^2 - 0.82^2) = 0.75676$.

This is close to our graphical estimate of $y(0.3) \approx 0.8$.

(c) The centers of the horizontal line segments of the direction field are located on the lines $y = x$ and $y = -x$. When a solution curve crosses one of these lines, it has a local maximum or minimum.

9. $2ye^{y^2} y' = 2x + 3\sqrt{x} \;\Rightarrow\; 2ye^{y^2} \dfrac{dy}{dx} = 2x + 3\sqrt{x} \;\Rightarrow\; 2ye^{y^2}\,dy = \left(2x + 3\sqrt{x}\right)dx \;\Rightarrow$

$\int 2ye^{y^2}\,dy = \int \left(2x + 3\sqrt{x}\right)dx \;\Rightarrow\; e^{y^2} = x^2 + 2x^{3/2} + C \;\Rightarrow\; y^2 = \ln(x^2 + 2x^{3/2} + C) \;\Rightarrow$

$y = \pm\sqrt{\ln(x^2 + 2x^{3/2} + C)}$

11. $\dfrac{dr}{dt} + 2tr = r \;\Rightarrow\; \dfrac{dr}{dt} = r - 2tr = r(1 - 2t) \;\Rightarrow\; \displaystyle\int \frac{dr}{r} = \int (1 - 2t)\,dt$ [if $r \neq 0$] $\Rightarrow\; \ln|r| = t - t^2 + C \;\Rightarrow$

$|r| = e^{t - t^2 + C} = ke^{t - t^2}$. (Note that $r = 0$ is also a solution but it does not satisfy the initial condition.) Since $r(0) = 5$, $5 = ke^0 = k$. Thus, $r(t) = 5e^{t - t^2}$.

13. $\dfrac{dn}{dt} = \cos\left(\dfrac{2\pi t}{365}\right) n \;\Rightarrow\; \displaystyle\int \frac{dn}{n} = \int \cos\left(\frac{2\pi t}{365}\right) dt$ [if $n \neq 0$] $\Rightarrow\; \ln|n| = \dfrac{365}{2\pi} \sin\left(\dfrac{2\pi t}{365}\right) + C \;\Rightarrow$

$|n| = e^C e^{(365/2\pi)\sin(2\pi t/365)} \;\Rightarrow\; n = Ae^{(365/2\pi)\sin(2\pi t/365)}$ where $A = \pm e^C$ is a constant. Note that $n = 0$ is also a solution and this can be included in the family of solutions by allowing A to be zero. Now $n(0) = n_0$, so $A = n_0$. Therefore, the population size after t days is $n(t) = n_0 e^{(365/2\pi)\sin(2\pi t/365)}$.

15. (a) $\dfrac{dp}{dt} = cp(1 - p) - mp \;\Rightarrow\; \displaystyle\int \frac{dp}{cp(1 - p) - mp} = \int dt$ $\left[\begin{array}{l}\text{if } p \neq 0 \text{ and} \\ p \neq 1 - m/c\end{array}\right]$ $\Rightarrow\; \displaystyle\int \frac{dp}{p(c - m - cp)} = t + C_1$

We can evaluate the integral by writing the partial fraction decomposition of the integrand, provided $c \neq m$. This gives

$\dfrac{1}{p(c - m - cp)} = \dfrac{A}{p} + \dfrac{B}{c - m - cp} \;\Leftrightarrow\; 1 = A(c - m - cp) + Bp \;\Leftrightarrow\; 1 = (B - Ac)p + A(c - m)$. Setting

$p = 0$ gives $1 = A(c - m)$, so $A = 1/(c - m)$. Equating coefficients of p gives $B - Ac = 0$, so $B = Ac = c/(c - m)$.

© 2016 Cengage Learning. All Rights Reserved. May not be scanned, copied or duplicated, or posted to a publicly accessible website, in whole or in part.

Therefore, $\displaystyle\int \frac{dp}{p\,(c-m-cp)} = \int \left(\frac{1/(c-m)}{p} + \frac{c/(c-m)}{c-m-cp} \right) dp = \frac{1}{c-m} \int \left(\frac{1}{p} + \frac{c}{c-m-cp} \right) dp$

$$= \frac{1}{c-m}\left(\ln|p| - \ln|c-m-cp|\right) = \frac{1}{c-m} \ln\left|\frac{p}{c-m-cp}\right|$$

Continuing to solve the differential equation, we have $\displaystyle\frac{1}{c-m}\ln\left|\frac{p}{c-m-cp}\right| = t + C_1 \;\Leftrightarrow$

$$\ln\left|\frac{p}{c-m-cp}\right| = (c-m)\,t + (c-m)\,C_1 \;\Leftrightarrow\; \left|\frac{p}{c-m-cp}\right| = e^{(c-m)t+(c-m)C_1} \;\Leftrightarrow$$

$$\frac{p}{c-m-cp} = C_2 e^{(c-m)t} \text{ where } C_2 = \pm e^{(c-m)C_1} \;\Leftrightarrow\; p = C_2\,(c-m)\,e^{(c-m)t} - C_2 c e^{(c-m)t} p \;\Leftrightarrow$$

$$p\left(1 + C_2 c e^{(c-m)t}\right) = C_2\,(c-m)\,e^{(c-m)t} \;\Leftrightarrow\; p = \frac{C_2\,(c-m)\,e^{(c-m)t}}{1 + C_2 c e^{(c-m)t}}.$$

Now $p(0) = p_0 \;\Rightarrow\; p_0 = \dfrac{C_2\,(c-m)}{1 + C_2 c} \;\Leftrightarrow\; p_0 + C_2 c p_0 = C_2\,(c-m) \;\Leftrightarrow\; C_2 = \dfrac{p_0}{c-m-cp_0}.$

So $p(t) = \dfrac{\left(\dfrac{p_0}{c-m-cp_0}\right)(c-m)\,e^{(c-m)t}}{1 + \left(\dfrac{p_0}{c-m-cp_0}\right) c e^{(c-m)t}} = \dfrac{p_0\,(c-m)\,e^{(c-m)t}}{c-m-cp_0 + p_0 c e^{(c-m)t}}$ when $c \neq m$.

Note that the solutions $p\,(t) = 0$ and $p\,(t) = 1 - m/c$ are obtained when $p_0 = 0$ and $p_0 = 1 - m/c$ respectively.

If $c = m$, then the differential equation is $\dfrac{dp}{dt} = -cp^2 \;\Rightarrow\; \displaystyle\int \frac{dp}{p^2} = -c \int dt \;\Rightarrow\; -\frac{1}{p} = -ct + C \;\Rightarrow$

$p = \dfrac{1}{ct - C}$. Now, $p(0) = p_0 = \dfrac{1}{-C}$, so $C = -\dfrac{1}{p_0}$. Thus, $p(t) = \dfrac{1}{ct + 1/p_0} = \dfrac{p_0}{cp_0 t + 1}$ when $c = m$.

Note that the solution $p\,(t) = 0$ is obtained when $p_0 = 0$.

Another method: We can get the expression for $p(t)$ when $c = m$ by evaluating the following $\frac{0}{0}$ limit

$$\lim_{c\to m} \frac{p_0\,(c-m)\,e^{(c-m)t}}{c-m-cp_0 + p_0 c e^{(c-m)t}} \overset{\text{H}}{=} \lim_{c\to m} \frac{p_0 e^{(c-m)t} + p_0\,(c-m)\,t e^{(c-m)t}}{1 - p_0 + p_0\left(e^{(c-m)t} + cte^{(c-m)t}\right)} = \frac{p_0 + 0}{1 - p_0 + p_0\,(1+ct)} = \frac{p_0}{cp_0 t + 1}.$$

(b) When $c = m$, $\displaystyle\lim_{t\to\infty} p(t) = \lim_{t\to\infty} \frac{p_0}{cp_0 t + 1} = 0$. Also, when $c \neq m$, the frequency of occupied patches

is $p(t) = \dfrac{p_0\,(c-m)\,e^{(c-m)t}}{c-m-cp_0 + p_0 c e^{(c-m)t}}$. So $p(t)$ will approach zero as $t \to \infty$ if the exponential term $e^{(c-m)t}$ approaches

zero as $t \to \infty$. This occurs when the coefficient of t is negative, that is when $c < m$. Thus, the frequency of occupied patches goes to zero in the long-run if $c \leq m$.

17. (a) Let $V(t)$ represent the volume of oxygen in the lungs (in mL) after t seconds of oxygenation.

$V(0) = 0.2(3000) = 600$ mL since initially 20% of the 3L lung volume is oxygen. The rate at which V increases is equal to the rate at which oxygen flows into the lungs minus the rate at which it flows out. That rate is

$\dfrac{dV}{dt} = 10\dfrac{\text{mL}}{\text{s}} - \dfrac{V}{3000}\dfrac{\text{mL}}{\text{mL}} \times 10\dfrac{\text{mL}}{\text{s}} = 10 - \dfrac{V}{300}\dfrac{\text{mL}}{\text{s}} \;\Rightarrow\; \displaystyle\int \frac{dV}{10 - V/300} = \int dt$ [if $V \neq 3000$] $\Rightarrow$

$-300 \ln|10 - V/300| = t + C \;\Rightarrow\; 10 - V/300 = Ae^{-\frac{1}{300}t} \;\Rightarrow\; V = 3000 - 300Ae^{-\frac{1}{300}t}$. Note that $V = 3000$ is also a solution and this can be included in the family of solutions by allowing A to be zero. Now, $V(0) = 600 \;\Rightarrow$

$600 = 3000 - 300A \;\Rightarrow\; A = 8$. So the amount of oxygen in the lungs is $V(t) = 3000 - 2400e^{-\frac{1}{300}t}$.

© 2016 Cengage Learning. All Rights Reserved. May not be scanned, copied or duplicated, or posted to a publicly accessible website, in whole or in part.

(b) An 80% oxygen content is achieved when $V(t) = 0.8(3000) \Rightarrow 3000 - 2400e^{-\frac{1}{300}t} = 2400 \Rightarrow$

$e^{-\frac{1}{300}t} = \frac{1}{4} \Rightarrow t = -300 \ln \frac{1}{4} = 300 \ln 4 \approx 415.888$ s. Thus, oxygenation should be run for approximately 416 seconds or 6.9 hours.

19. $\frac{dh}{dt} = -\frac{R}{V}\left(\frac{h}{k+h}\right) \Rightarrow \int \frac{k+h}{h}\,dh = \int\left(-\frac{R}{V}\right)dt \quad [\text{if } h \neq 0] \Rightarrow \int\left(1+\frac{k}{h}\right)dh = -\frac{R}{V}\int 1\,dt \Rightarrow$

$h + k\ln h = -\frac{R}{V}t + C$. This equation gives a relationship between h and t, but it is not possible to isolate h and express it in terms of t.

21. (a) $dx/dt = 0.4x(1 - 0.000005x) - 0.002xy$, $dy/dt = -0.2y + 0.000008xy$. If $y = 0$, then $dx/dt = 0.4x(1 - 0.000005x)$, so $dx/dt = 0 \Leftrightarrow x = 0$ or $x = 200{,}000$, which shows that the insect population increases logistically with a carrying capacity of 200,000. Since $dx/dt > 0$ for $0 < x < 200{,}000$ and $dx/dt < 0$ for $x > 200{,}000$, we expect the insect population to stabilize at 200,000.

(b) x and y are constant $\Rightarrow x' = 0$ and $y' = 0 \Rightarrow$

$$\left\{\begin{array}{l} 0 = 0.4x(1 - 0.000005x) - 0.002xy \\ 0 = -0.2y + 0.000008xy \end{array}\right\} \Rightarrow \left\{\begin{array}{l} 0 = 0.4x[(1 - 0.000005x) - 0.005y] \\ 0 = y(-0.2 + 0.000008x) \end{array}\right.$$

The second equation is true if $y = 0$ or $x = \frac{0.2}{0.000008} = 25{,}000$. If $y = 0$ in the first equation, then either $x = 0$ or $x = \frac{1}{0.000005} = 200{,}000$. If $x = 25{,}000$, then $0 = 0.4(25{,}000)[(1 - 0.000005 \cdot 25{,}000) - 0.005y] \Rightarrow$

$0 = 10{,}000[(1 - 0.125) - 0.005y] \Rightarrow 0 = 8750 - 50y \Rightarrow y = 175$.

Case (i): $y = 0, x = 0$: Zero populations

Case (ii): $y = 0, x = 200{,}000$: In the absence of birds, the insect population is always 200,000.

Case (iii): $x = 25{,}000, y = 175$: The predator/prey interaction balances and the populations are stable.

(c) The populations of the birds and insects fluctuate around 175 and 25,000, respectively, and eventually stabilize at those values.

(d)

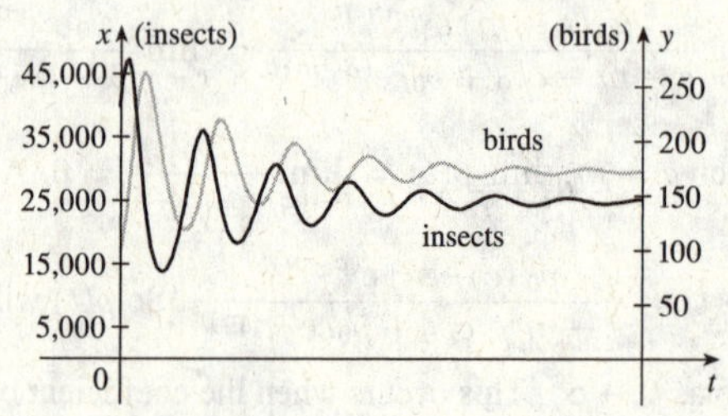

23. (a) The differential equation dp_1/dt has no dependency on p_2, so the growth rate of the species 1 population is unaffected by the presence of species 2. However, an increase in p_1 reduces the growth rate of p_2 due to the negative terms $-c_1p_1p_2$ and $-c_2p_2p_1$ in the expression for dp_2/dt. Hence, species 1 is the superior competitor.

(b) $\frac{dp_1}{dt} = 5p_1(1 - p_1) - 3p_1 \qquad \frac{dp_2}{dt} = 30p_2(1 - p_1 - p_2) - 3p_2 - 5p_1p_2 \qquad [m_1 = m_2 = 3, c_1 = 5, c_2 = 30]$

p_1-nullclines: $5p_1(1 - p_1) - 3p_1 = 0 \Rightarrow p_1(2 - 5p_1) = 0 \Rightarrow p_1 = 0$ and $p_1 = \frac{2}{5} = 0.4$

p_1 is increasing when $p_1' > 0 \Rightarrow p_1(2 - 5p_1) > 0 \Rightarrow 0 < p_1 < \frac{2}{5}$ (since p_1 and $2 - 5p_1$ must have the same sign in order for their product to be positive). Conversely, p_1 is decreasing when $p_1 > \frac{2}{5}$.

© 2016 Cengage Learning. All Rights Reserved. May not be scanned, copied or duplicated, or posted to a publicly accessible website, in whole or in part.

p_2-nullclines: $30p_2(1 - p_1 - p_2) - 3p_2 - 5p_1p_2 = 0 \;\Rightarrow$

$p_2\,(27 - 35p_1 - 30p_2) = 0 \;\Rightarrow\; p_2 = 0$ and $p_2 = -\frac{7}{6}p_1 + \frac{9}{10}$

p_2 is increasing when $p_2' > 0 \;\Rightarrow\; p_2\,(27 - 35p_1 - 30p_2) > 0 \;\Rightarrow$

$27 - 35p_1 - 30p_2 > 0$ (since p_2 is a positive quantity) $\Rightarrow$

$p_2 < -\frac{7}{6}p_1 + \frac{9}{10}$ (below nullcline). Conversely, p_2 is decreasing when

$p_2 > -\frac{7}{6}p_1 + \frac{9}{10}$ (above nullcline).

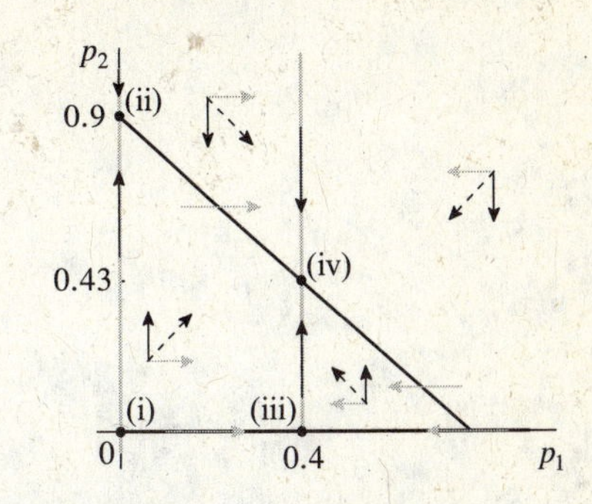

Equilibria: The first p_1 and p_2 nullclines give the equilibrium (i) $\hat{p}_1 = 0, \hat{p}_2 = 0$. Substituting $\hat{p}_1 = 0$ into the second p_2-nullcline gives $\hat{p}_2 = \frac{9}{10}$, so the second equilibrium is (ii) $\hat{p}_1 = 0, \hat{p}_2 = \frac{9}{10}$. The second p_1-nullcline and first p_2-nullcline give (iii) $\hat{p}_1 = \frac{2}{5}, \hat{p}_2 = 0$. Substituting $\hat{p}_1 = \frac{2}{5}$ into the second p_2-nullcline gives $\hat{p}_2 = -\frac{7}{6}\left(\frac{2}{5}\right) + \frac{9}{10} = \frac{13}{30}$. So the fourth equilibrium is (iv) $\hat{p}_1 = \frac{2}{5}, \hat{p}_2 = \frac{13}{30}$. The directional arrows in the phase plane diagram indicate that equilibrium (iv) is locally stable so the two species will coexist.

25. (a) $\dfrac{dM}{dt} = 2C + CM^2 - \dfrac{10M}{1+M} \qquad \dfrac{dC}{dt} = 1 - M \qquad [\alpha = 2, \beta = 1, \gamma = 10, \delta = 1]$

M-nullclines: $2C + CM^2 - \dfrac{10M}{1+M} = 0 \;\Rightarrow\; C\left(2 + M^2\right) = \dfrac{10M}{1+M} \;\Rightarrow\; C = \dfrac{10M}{(1+M)\,(2+M^2)}$

For very large positive values of C, we have $\dfrac{dM}{dt} \approx 2C + CM^2 > 0$, so M is increasing above its nullcline. Similarly, when C is very small, we have $\dfrac{dM}{dt} \approx -\dfrac{10M}{1+M} < 0$, so M is decreasing below its nullcline. We can use a calculator to plot the nullcline in the phase plane.

C-nullclines: $1 - M = 0 \;\Rightarrow\; M = 1$

$C' > 0$ when $M < 1$ (left of nullcline) and $C' < 0$ when $M > 1$ (right of nullcline).

Equilibria: Substituting $\hat{M} = 1$ into the M-nullcline expression gives

$\hat{C} = \dfrac{10(1)}{(1+1)\,(2+1)} = \dfrac{5}{3}$. So the equilibrium is $\hat{M} = 1, \hat{C} = \frac{5}{3}$.

These results are visualized in the phase plane at right.

(b) The directional arrows in the phase plane from part (a) indicate that oscillations will occur around the sole equilibrium $\hat{M} = 1, \hat{C} = \frac{5}{3}$. This means that M will cycle between values above and below 1 mg/mL. Since high concentrations of MPF trigger cell division, we expect cell division to be a periodic process that initiates whenever M reaches a certain threshold concentration.

(c) We saw in part (b) that oscillations will occur around the sole equilibrium, however, we cannot determine, from the phase plane alone, whether these oscillations converge toward the equilibrium.

© 2016 Cengage Learning. All Rights Reserved. May not be scanned, copied or duplicated, or posted to a publicly accessible website, in whole or in part.

8 □ VECTORS AND MATRIX MODELS

8.1 Coordinate Systems

1. We start at the origin, which has coordinates $(0, 0, 0)$. First we move 4 units along the positive x-axis, affecting only the x-coordinate, bringing us to the point $(4, 0, 0)$. We then move 3 units straight downward, in the negative z-direction. Thus only the z-coordinate is affected, and we arrive at $(4, 0, -3)$.

3. The distance from a point to the xz-plane is the absolute value of the y-coordinate of the point. $Q(-5, -1, 4)$ has the y-coordinate with the smallest absolute value, so Q is the point closest to the xz-plane. $R(0, 3, 8)$ must lie in the yz-plane since the distance from R to the yz-plane, given by the x-coordinate of R, is 0.

5. The equation $x + y = 2$ represents the set of all points in $\mathbb{R}^3$ whose x- and y-coordinates have a sum of 2, or equivalently where $y = 2 - x$. This is the set $\{(x, 2 - x, z) \mid x \in \mathbb{R}, z \in \mathbb{R}\}$ which is a vertical plane that intersects the xy-plane in the line $y = 2 - x$, $z = 0$.

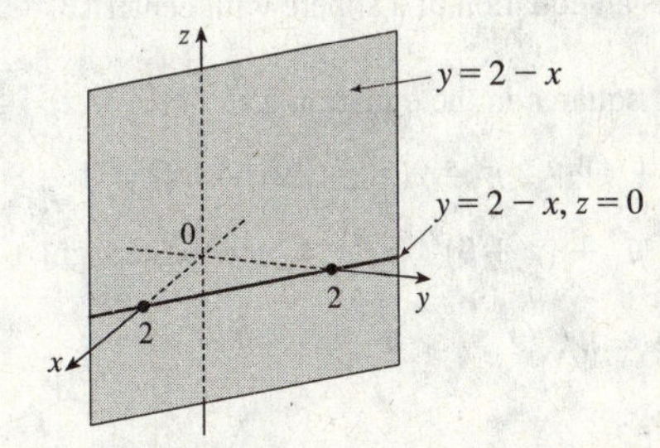

7. (a) We can find the lengths of the sides of the triangle by using the distance formula between pairs of vertices:

$$|PQ| = \sqrt{(7-3)^2 + [0-(-2)]^2 + [1-(-3)]^2} = \sqrt{16+4+16} = 6$$

$$|QR| = \sqrt{(1-7)^2 + (2-0)^2 + (1-1)^2} = \sqrt{36+4+0} = \sqrt{40} = 2\sqrt{10}$$

$$|RP| = \sqrt{(3-1)^2 + (-2-2)^2 + (-3-1)^2} = \sqrt{4+16+16} = 6$$

The longest side is QR, but the Pythagorean Theorem is not satisfied: $|PQ|^2 + |RP|^2 \neq |QR|^2$. Thus PQR is not a right triangle. PQR is isosceles, as two sides have the same length.

(b) Compute the lengths of the sides of the triangle by using the distance formula between pairs of vertices:

$$|PQ| = \sqrt{(4-2)^2 + [1-(-1)]^2 + (1-0)^2} = \sqrt{4+4+1} = 3$$

$$|QR| = \sqrt{(4-4)^2 + (-5-1)^2 + (4-1)^2} = \sqrt{0+36+9} = \sqrt{45} = 3\sqrt{5}$$

$$|RP| = \sqrt{(2-4)^2 + [-1-(-5)]^2 + (0-4)^2} = \sqrt{4+16+16} = 6$$

Since the Pythagorean Theorem is satisfied by $|PQ|^2 + |RP|^2 = |QR|^2$, PQR is a right triangle. PQR is not isosceles, as no two sides have the same length.

9. (a) First we find the distances between points:

$$|AB| = \sqrt{(3-2)^2 + (7-4)^2 + (-2-2)^2} = \sqrt{26}$$

$$|BC| = \sqrt{(1-3)^2 + (3-7)^2 + [3-(-2)]^2} = \sqrt{45} = 3\sqrt{5}$$

$$|AC| = \sqrt{(1-2)^2 + (3-4)^2 + (3-2)^2} = \sqrt{3}$$

In order for the points to lie on a straight line, the sum of the two shortest distances must be equal to the longest distance. Since $\sqrt{26} + \sqrt{3} \neq 3\sqrt{5}$, the three points do not lie on a straight line.

© 2016 Cengage Learning. All Rights Reserved. May not be scanned, copied or duplicated, or posted to a publicly accessible website, in whole or in part.

(b) First we find the distances between points:

$$|DE| = \sqrt{(1-0)^2 + [-2-(-5)]^2 + (4-5)^2} = \sqrt{11}$$

$$|EF| = \sqrt{(3-1)^2 + [4-(-2)]^2 + (2-4)^2} = \sqrt{44} = 2\sqrt{11}$$

$$|DF| = \sqrt{(3-0)^2 + [4-(-5)]^2 + (2-5)^2} = \sqrt{99} = 3\sqrt{11}$$

Since $|DE| + |EF| = |DF|$, the three points lie on a straight line.

11. The radius of the sphere is the distance between $(4, 3, -1)$ and $(3, 8, 1)$: $r = \sqrt{(3-4)^2 + (8-3)^2 + [1-(-1)]^2} = \sqrt{30}$. Thus, an equation of the sphere is $(x-3)^2 + (y-8)^2 + (z-1)^2 = 30$.

13. Completing squares in the equation $x^2 + y^2 + z^2 - 6x + 4y - 2z = 11$ gives $(x^2 - 6x + 9) + (y^2 + 4y + 4) + (z^2 - 2z + 1) = 11 + 9 + 4 + 1 \quad\Rightarrow\quad (x-3)^2 + (y+2)^2 + (z-1)^2 = 25$, which we recognize as an equation of a sphere with center $(3, -2, 1)$ and radius 5.

15. Completing squares in the equation $2x^2 - 8x + 2y^2 + 2z^2 + 24z = 1$ gives $2(x^2 - 4x + 4) + 2y^2 + 2(z^2 + 12z + 36) = 1 + 8 + 72 \quad\Rightarrow\quad 2(x-2)^2 + 2y^2 + 2(z+6)^2 = 81 \quad\Rightarrow$ $(x-2)^2 + y^2 + (z+6)^2 = \frac{81}{2}$, which we recognize as an equation of a sphere with center $(2, 0, -6)$ and radius $\sqrt{\frac{81}{2}} = 9/\sqrt{2}$.

17. (a) If the midpoint of the line segment from $P_1(x_1, y_1, z_1)$ to $P_2(x_2, y_2, z_2)$ is $Q = \left(\frac{x_1 + x_2}{2}, \frac{y_1 + y_2}{2}, \frac{z_1 + z_2}{2}\right)$, then the distances $|P_1Q|$ and $|QP_2|$ are equal, and each is half of $|P_1P_2|$. We verify that this is the case:

$$|P_1P_2| = \sqrt{(x_2 - x_1)^2 + (y_2 - y_1)^2 + (z_2 - z_1)^2}$$

$$\begin{aligned}
|P_1Q| &= \sqrt{\left[\tfrac{1}{2}(x_1 + x_2) - x_1\right]^2 + \left[\tfrac{1}{2}(y_1 + y_2) - y_1\right]^2 + \left[\tfrac{1}{2}(z_1 + z_2) - z_1\right]^2} \\
&= \sqrt{\left(\tfrac{1}{2}x_2 - \tfrac{1}{2}x_1\right)^2 + \left(\tfrac{1}{2}y_2 - \tfrac{1}{2}y_1\right)^2 + \left(\tfrac{1}{2}z_2 - \tfrac{1}{2}z_1\right)^2} \\
&= \sqrt{\left(\tfrac{1}{2}\right)^2\left[(x_2 - x_1)^2 + (y_2 - y_1)^2 + (z_2 - z_1)^2\right]} = \tfrac{1}{2}\sqrt{(x_2 - x_1)^2 + (y_2 - y_1)^2 + (z_2 - z_1)^2} \\
&= \tfrac{1}{2}\,|P_1P_2|
\end{aligned}$$

$$\begin{aligned}
|QP_2| &= \sqrt{\left[x_2 - \tfrac{1}{2}(x_1 + x_2)\right]^2 + \left[y_2 - \tfrac{1}{2}(y_1 + y_2)\right]^2 + \left[z_2 - \tfrac{1}{2}(z_1 + z_2)\right]^2} \\
&= \sqrt{\left(\tfrac{1}{2}x_2 - \tfrac{1}{2}x_1\right)^2 + \left(\tfrac{1}{2}y_2 - \tfrac{1}{2}y_1\right)^2 + \left(\tfrac{1}{2}z_2 - \tfrac{1}{2}z_1\right)^2} = \sqrt{\left(\tfrac{1}{2}\right)^2\left[(x_2 - x_1)^2 + (y_2 - y_1)^2 + (z_2 - z_1)^2\right]} \\
&= \tfrac{1}{2}\sqrt{(x_2 - x_1)^2 + (y_2 - y_1)^2 + (z_2 - z_1)^2} = \tfrac{1}{2}\,|P_1P_2|
\end{aligned}$$

So Q is indeed the midpoint of P_1P_2.

(b) By part (a), the midpoints of sides AB, BC and CA are $P_1\left(-\frac{1}{2}, 1, 4\right)$, $P_2\left(1, \frac{1}{2}, 5\right)$ and $P_3\left(\frac{5}{2}, \frac{3}{2}, 4\right)$. (Recall that a median of a triangle is a line segment from a vertex to the midpoint of the opposite side.) Then the lengths of the medians are:

$$|AP_2| = \sqrt{0^2 + \left(\tfrac{1}{2} - 2\right)^2 + (5-3)^2} = \sqrt{\tfrac{9}{4} + 4} = \sqrt{\tfrac{25}{4}} = \tfrac{5}{2}$$

$$|BP_3| = \sqrt{\left(\tfrac{5}{2} + 2\right)^2 + \left(\tfrac{3}{2}\right)^2 + (4-5)^2} = \sqrt{\tfrac{81}{4} + \tfrac{9}{4} + 1} = \sqrt{\tfrac{94}{4}} = \tfrac{1}{2}\sqrt{94}$$

$$|CP_1| = \sqrt{\left(-\tfrac{1}{2} - 4\right)^2 + (1-1)^2 + (4-5)^2} = \sqrt{\tfrac{81}{4} + 1} = \tfrac{1}{2}\sqrt{85}$$

© 2016 Cengage Learning. All Rights Reserved. May not be scanned, copied or duplicated, or posted to a publicly accessible website, in whole or in part.

19. (a) Since the sphere touches the xy-plane, its radius is the distance from its center, $(2, -3, 6)$, to the xy-plane, namely 6. Therefore $r = 6$ and an equation of the sphere is $(x-2)^2 + (y+3)^2 + (z-6)^2 = 6^2 = 36$.

(b) The radius of this sphere is the distance from its center $(2, -3, 6)$ to the yz-plane, which is 2. Therefore, an equation is $(x-2)^2 + (y+3)^2 + (z-6)^2 = 4$.

(c) Here the radius is the distance from the center $(2, -3, 6)$ to the xz-plane, which is 3. Therefore, an equation is $(x-2)^2 + (y+3)^2 + (z-6)^2 = 9$.

21. The equation $x = 5$ represents a plane parallel to the yz-plane and 5 units in front of it.

23. The inequality $y < 8$ represents a half-space consisting of all points to the left of the plane $y = 8$.

25. The inequality $0 \leq z \leq 6$ represents all points on or between the horizontal planes $z = 0$ (the xy-plane) and $z = 6$.

27. Because $z = -1$, all points in the region must lie in the horizontal plane $z = -1$. In addition, $x^2 + y^2 = 4$, so the region consists of all points that lie on a circle with radius 2 and center on the z-axis that is contained in the plane $z = -1$.

29. The inequality $x^2 + y^2 + z^2 \leq 3$ is equivalent to $\sqrt{x^2 + y^2 + z^2} \leq \sqrt{3}$, so the region consists of those points whose distance from the origin is at most $\sqrt{3}$. This is the set of all points on or inside the sphere with radius $\sqrt{3}$ and center $(0, 0, 0)$.

31. This describes all points whose x-coordinate is between 0 and 5, that is, $0 < x < 5$.

33. This describes a region all of whose points have a distance to the origin which is greater than r, but smaller than R. So inequalities describing the region are $r < \sqrt{x^2 + y^2 + z^2} < R$, or $r^2 < x^2 + y^2 + z^2 < R^2$.

35. (a) To find the x- and y-coordinates of the point P, we project it onto L_2 and project the resulting point Q onto the x- and y-axes. To find the z-coordinate, we project P onto either the xz-plane or the yz-plane (using our knowledge of its x- or y-coordinate) and then project the resulting point onto the z-axis. (Or, we could draw a line parallel to QO from P to the z-axis.) The coordinates of P are $(2, 1, 4)$.

(b) A is the intersection of L_1 and L_2, B is directly below the y-intercept of L_2, and C is directly above the x-intercept of L_2.

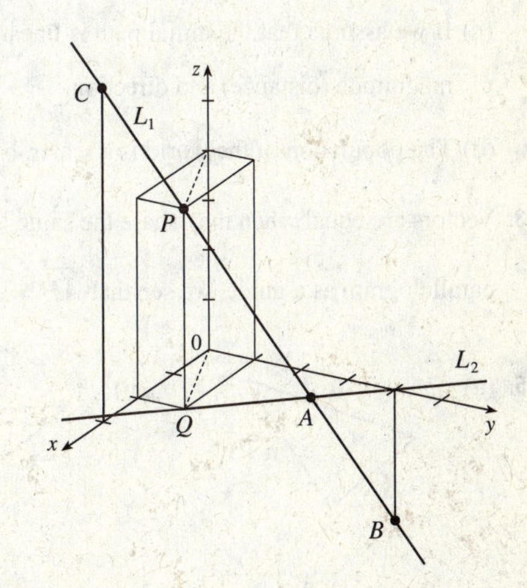

37. At $t = 0$ and $t = 2$, the position coordinates are $P(1.2, 0, 1.5)$ and $Q(1, 2, 0.5)$ respectively.

(a) The magnitude of the net distance traveled in the x direction is $|\Delta x| = |1 - 1.2| = |-0.2| = 0.2$ cm.

(b) The magnitude of the net distance net distance traveled in the z direction is $|\Delta z| = |0.5 - 1.5| = |-1.0| = 1.0$ cm.

(c) The net distance traveled through 3-dimensional space is

$$|PQ| = \sqrt{(1 - 1.2)^2 + (2 - 0)^2 + (0.5 - 1.5)^2} = \sqrt{5.04} \approx 2.24 \text{ cm}.$$

© 2016 Cengage Learning. All Rights Reserved. May not be scanned, copied or duplicated, or posted to a publicly accessible website, in whole or in part.

39. (a) The equation $(x-2)^2+(y-3)^2=1$ defines a circle centered at $C_1(2,3)$ with radius 1 and the equation $(x-3)^2+(y-2)^2=1/4$ defines a circle centered at $C_2(3,2)$ with radius $\frac{1}{2}$. The distance between the centers of each circle is $|C_1C_2|=\sqrt{(3-2)^2+(2-3)^2}=\sqrt{2}\approx 1.4$, so the circles will overlap if the sum of their radii is greater than 1.4. Now, the radii sum is $1+\frac{1}{2}>1.4$, so the circles overlap which suggests that a single vaccine would work for both years.

(b) The equation $(x-2)^2+(y-3)^2+(z-1)^2=1$ defines a sphere centered at $C_1(2,3,1)$ with radius 1 and the equation $(x-3)^2+(y-2)^2+z^2=1/4$ defines a sphere centered at $C_2(3,2,0)$ with radius $\frac{1}{2}$. The distance between the centers of each sphere is $|C_1C_2|=\sqrt{(3-2)^2+(2-3)^2+(0-1)^2}=\sqrt{3}\approx 1.7$, so the spheres will overlap if the sum of their radii is greater than 1.7. But $1+\frac{1}{2}<1.7$, so the spheres do not overlap suggesting that different vaccines would be required for each year.

(c) The plot of the circles in part (a) can be obtained by projecting the plot of the spheres in part (b) onto the xy-plane.

8.2 Vectors

1. (a) The cost of a theater ticket is a scalar, because it has only magnitude.

(b) The current in a river is a vector, because it has both magnitude (the speed of the current) and direction at any given location.

(c) If we assume that the initial path is linear, the initial flight path from Houston to Dallas is a vector, because it has both magnitude (distance) and direction.

(d) The population of the world is a scalar, because it has only magnitude.

3. Vectors are equal when they share the same length and direction (but not necessarily location). Using the symmetry of the parallelogram as a guide, we see that $\overrightarrow{AB}=\overrightarrow{DC}$, $\overrightarrow{DA}=\overrightarrow{CB}$, $\overrightarrow{DE}=\overrightarrow{EB}$, and $\overrightarrow{EA}=\overrightarrow{CE}$.

5.

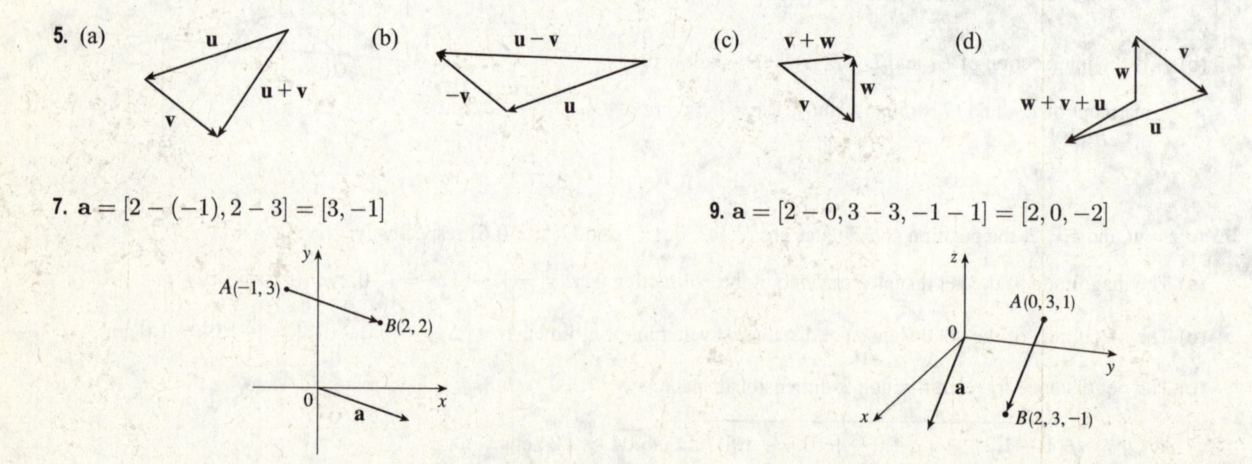

© 2016 Cengage Learning. All Rights Reserved. May not be scanned, copied or duplicated, or posted to a publicly accessible website, in whole or in part.

11. $[-1, 4] + [6, -2] = [-1 + 6, 4 + (-2)] = [5, 2]$

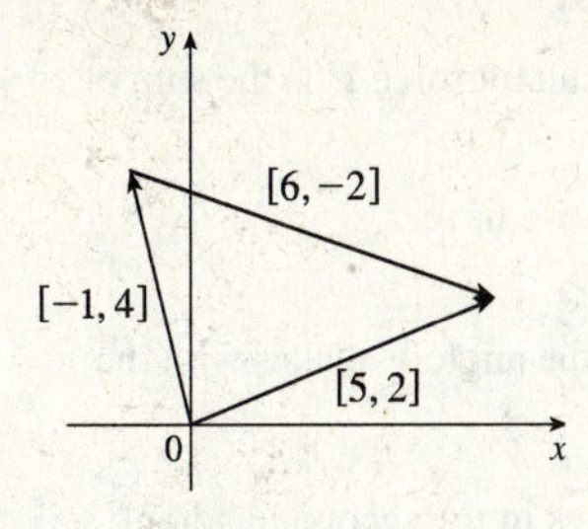

13. $[0, 1, 2] + [0, 0, -3] = [0 + 0, 1 + 0, 2 + (-3)]$

$= [0, 1, -1]$

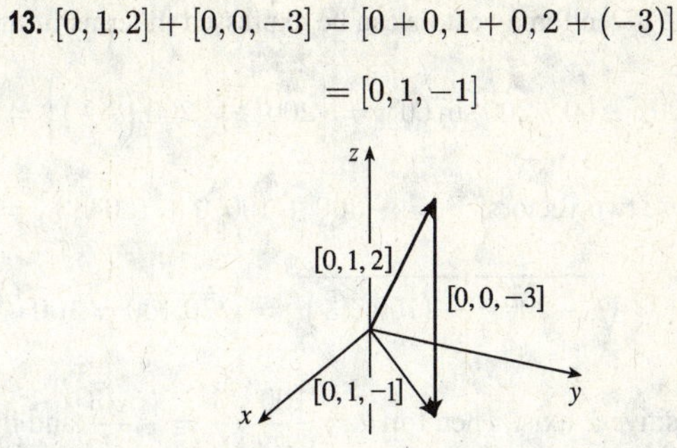

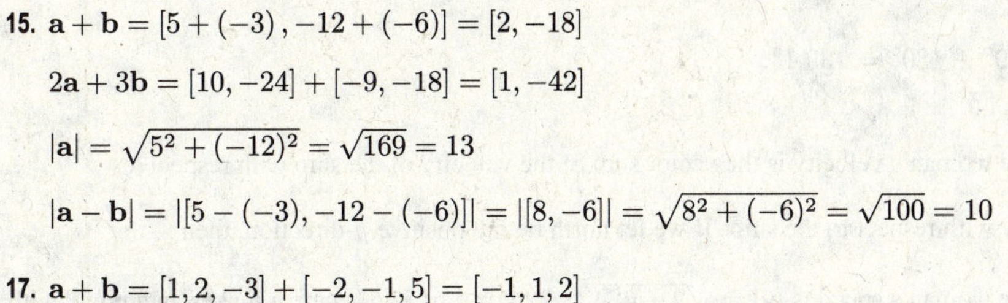

15. $\mathbf{a} + \mathbf{b} = [5 + (-3), -12 + (-6)] = [2, -18]$

$2\mathbf{a} + 3\mathbf{b} = [10, -24] + [-9, -18] = [1, -42]$

$|\mathbf{a}| = \sqrt{5^2 + (-12)^2} = \sqrt{169} = 13$

$|\mathbf{a} - \mathbf{b}| = |[5 - (-3), -12 - (-6)]| = |[8, -6]| = \sqrt{8^2 + (-6)^2} = \sqrt{100} = 10$

17. $\mathbf{a} + \mathbf{b} = [1, 2, -3] + [-2, -1, 5] = [-1, 1, 2]$

$2\mathbf{a} + 3\mathbf{b} = 2\,[1, 2, -3] + 3\,[-2, -1, 5] = [2, 4, -6] + [-6, -3, 15] = [-4, 1, 9]$

$|\mathbf{a}| = \sqrt{1^2 + 2^2 + (-3)^2} = \sqrt{14}$

$|\mathbf{a} - \mathbf{b}| = |[1, 2, -3] - [-2, -1, 5]| = |[3, 3, -8]| = \sqrt{3^2 + 3^2 + (-8)^2} = \sqrt{82}$

19. The vector $[-3, 7]$ has length $|[-3, 7]| = \sqrt{(-3)^2 + 7^2} = \sqrt{58}$, so by Equation 2 the unit vector with the same direction is

$\dfrac{1}{\sqrt{58}}\,([-3, 7]) = \left[-\dfrac{3}{\sqrt{58}}, \dfrac{7}{\sqrt{58}}\right].$

21. The vector $[8, -1, 4]$ has length $|[8, -1, 4]| = \sqrt{8^2 + (-1)^2 + 4^2} = \sqrt{81} = 9$, so by Equation 2 the unit vector with the same direction is $\frac{1}{9}[8, -1, 4] = \left[\frac{8}{9}, -\frac{1}{9}, \frac{4}{9}\right]$.

23. From the figure, we see that the x-component of $\mathbf{v}$ is

$v_1 = |\mathbf{v}| \cos(\pi/3) = 4 \cdot \frac{1}{2} = 2$ and the y-component is

$v_2 = |\mathbf{v}| \sin(\pi/3) = 4 \cdot \frac{\sqrt{3}}{2} = 2\sqrt{3}$. Thus

$\mathbf{v} = [v_1, v_2] = \left[2, 2\sqrt{3}\right]$.

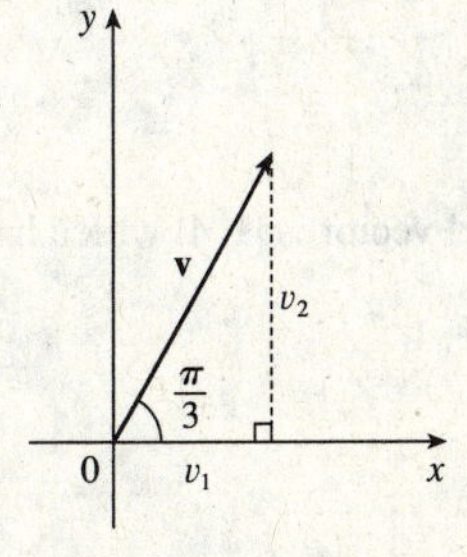

25. The velocity vector $\mathbf{v}$ makes an angle of 40° with the horizontal and has magnitude equal to the speed at which the football was thrown. From the figure, we see that the horizontal component of $\mathbf{v}$ is $|\mathbf{v}| \cos 40^\circ = 60 \cos 40^\circ \approx 45.96$ ft/s and the vertical component is $|\mathbf{v}| \sin 40^\circ = 60 \sin 40^\circ \approx 38.57$ ft/s.

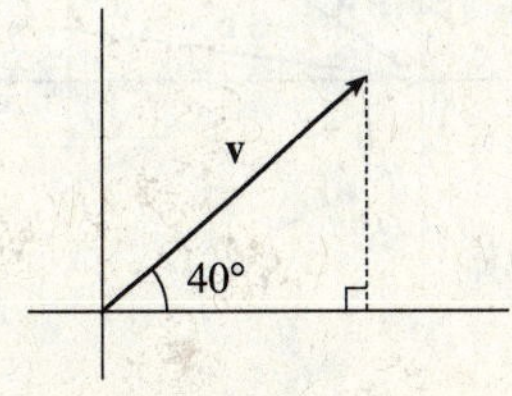

© 2016 Cengage Learning. All Rights Reserved. May not be scanned, copied or duplicated, or posted to a publicly accessible website, in whole or in part.

27. The given force vectors can be expressed in terms of their horizontal and vertical components as $[-300, 0]$ and $[200\cos 60°, 200\sin 60°] = \left[200\left(\frac{1}{2}\right), 200\left(\frac{\sqrt{3}}{2}\right)\right] = \left[100, 100\sqrt{3}\right]$. The resultant force **F** is the sum of these two vectors: $\mathbf{F} = \left[-300+100, 0+100\sqrt{3}\right] = \left[-200, 100\sqrt{3}\right]$. Then we have $|\mathbf{F}| \approx \sqrt{(-200)^2 + \left(100\sqrt{3}\right)^2} = \sqrt{70{,}000} = 100\sqrt{7} \approx 264.6$ N. Let θ be the angle **F** makes with the positive x-axis. Then $\tan\theta = \dfrac{100\sqrt{3}}{-200} = -\dfrac{\sqrt{3}}{2}$ and the terminal point of **F** lies in the second quadrant, so $\theta = \tan^{-1}\left(-\dfrac{\sqrt{3}}{2}\right) + 180° \approx -40.9° + 180° = 139.1°$.

29. With respect to the water's surface, the woman's velocity is the vector sum of the velocity of the ship with respect to the water, and the woman's velocity with respect to the ship. If we let north be the positive y-direction, then $\mathbf{v} = [0, 22] + [-3, 0] = [-3, 22]$. The woman's speed is $|\mathbf{v}| = \sqrt{9+484} \approx 22.2$ mi/h. The vector **v** makes an angle θ with the east, where $\theta = \tan^{-1}\left(\frac{22}{-3}\right) \approx 98°$. Therefore, the woman's direction is about $\text{N}(98-90)°\text{W} = \text{N}8°\text{W}$.

31. $\mathbf{a} = \mathbf{i} + 2\mathbf{j} - 3\mathbf{k} \quad \mathbf{b} = 4\mathbf{i} + 7\mathbf{k}$

(a) $\mathbf{a}+\mathbf{b} = (\mathbf{i}+2\mathbf{j}-3\mathbf{k}) + (4\mathbf{i}+7\mathbf{k}) = (1+4)\,\mathbf{i} + 2\mathbf{j} + (-3+7)\,\mathbf{k} = 5\mathbf{i}+2\mathbf{j}+4\mathbf{k}$

(b) $\mathbf{a}-\mathbf{b} = (\mathbf{i}+2\mathbf{j}-3\mathbf{k}) - (4\mathbf{i}+7\mathbf{k}) = (1-4)\,\mathbf{i} + 2\mathbf{j} + (-3-7)\,\mathbf{k} = -3\mathbf{i}+2\mathbf{j}-10\mathbf{k}$

(c) $2\mathbf{a}+3\mathbf{b} = 2(\mathbf{i}+2\mathbf{j}-3\mathbf{k}) + 3(4\mathbf{i}+7\mathbf{k}) = (2\mathbf{i}+4\mathbf{j}-6\mathbf{k}) + (12\mathbf{i}+21\mathbf{k}) = 14\mathbf{i}+4\mathbf{j}+15\mathbf{k}$

(d) $5\mathbf{a}-7\mathbf{b} = 5(\mathbf{i}+2\mathbf{j}-3\mathbf{k}) - 7(4\mathbf{i}+7\mathbf{k}) = (5\mathbf{i}+10\mathbf{j}-15\mathbf{k}) - (28\mathbf{i}+49\mathbf{k}) = -23\mathbf{i}+10\mathbf{j}-64\mathbf{k}$

33. The slope of the tangent line to the graph of $y = x^2$ at the point $(2, 4)$ is

$$\left.\frac{dy}{dx}\right|_{x=2} = \left.2x\right|_{x=2} = 4$$

and a parallel vector is $[1, 4]$ which has length $|[1,4]| = \sqrt{1^2+4^2} = \sqrt{17}$, so unit vectors parallel to the tangent line are $\pm\frac{1}{\sqrt{17}}\,[1, 4]$.

35. (a), (b)

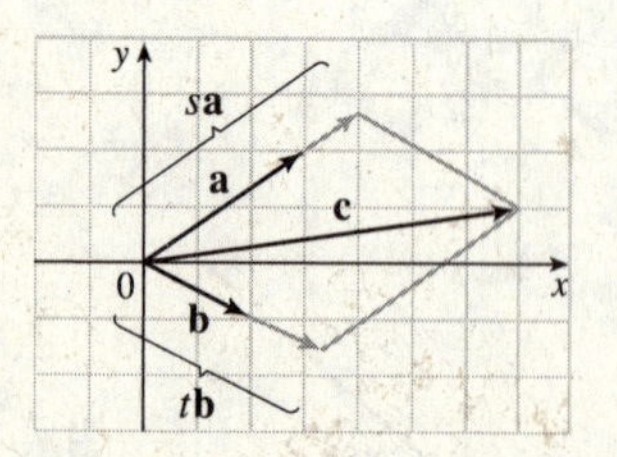

(c) From the sketch, we estimate that $s \approx 1.3$ and $t \approx 1.6$.

(d) $\mathbf{c} = s\,\mathbf{a} + t\,\mathbf{b} \;\Leftrightarrow\; 7 = 3s + 2t$ and $1 = 2s - t$. Solving these equations gives $s = \frac{9}{7}$ and $t = \frac{11}{7}$.

© 2016 Cengage Learning. All Rights Reserved. May not be scanned, copied or duplicated, or posted to a publicly accessible website, in whole or in part.

37.

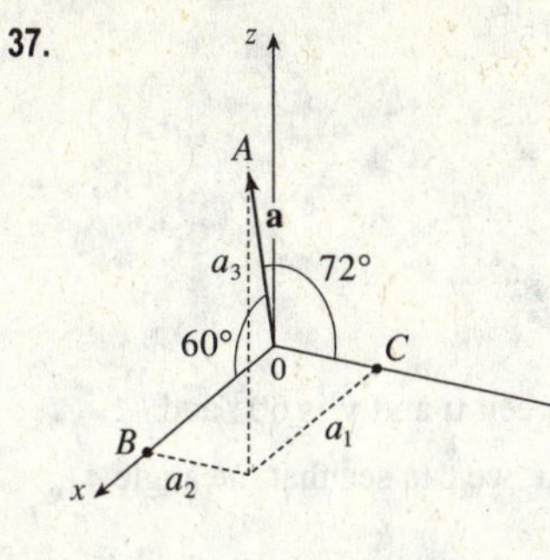

Let $\mathbf{a} = [a_1, a_2, a_3]$, as shown in the figure. Since $|\mathbf{a}| = 1$ and triangle ABO is a right triangle, we have $\cos 60° = \frac{a_1}{1} \Rightarrow a_1 = \cos 60°$. Similarly, triangle ACO is a right triangle, so $a_2 = \cos 72°$. Finally, since $|\mathbf{a}| = 1$ we have

$\sqrt{(\cos 60°)^2 + (\cos 72°)^2 + a_3^2} = 1 \Rightarrow a_3^2 = 1 - (\cos 60°)^2 - (\cos 72°)^2 \Rightarrow$

$a_3 = \sqrt{1 - (\cos 60°)^2 - (\cos 72°)^2}$. Thus

$$\mathbf{a} = \left[\cos 60°, \cos 72°, \sqrt{1 - (\cos 60°)^2 - (\cos 72°)^2}\right] \approx [0.50, 0.31, 0.81].$$

39. (a) A *left anterior hemiblock* can be diagnosed by observing a resultant voltage vector that points upwards and has a smaller (horizontal) magnitude compared to a healthy individual.

(b) Since the direction of the horizontal component of voltage changes, a *left posterior hemiblock* can be diagnosed by observing a resultant voltage vector that points to the right of the patient.

41. $|\mathbf{r} - \mathbf{r}_0|$ is the distance between the points (x, y, z) and (x_0, y_0, z_0), so the set of points is a sphere with radius 1 and center (x_0, y_0, z_0).

Alternate method: $|\mathbf{r} - \mathbf{r}_0| = 1 \Leftrightarrow \sqrt{(x - x_0)^2 + (y - y_0)^2 + (z - z_0)^2} = 1 \Leftrightarrow$

$(x - x_0)^2 + (y - y_0)^2 + (z - z_0)^2 = 1$, which is the equation of a sphere with radius 1 and center (x_0, y_0, z_0).

43.
$$\begin{aligned}\mathbf{a} + (\mathbf{b} + \mathbf{c}) &= [a_1, a_2] + ([b_1, b_2] + [c_1, c_2]) = [a_1, a_2] + [b_1 + c_1, b_2 + c_2] \\ &= [a_1 + b_1 + c_1, a_2 + b_2 + c_2] = [(a_1 + b_1) + c_1, (a_2 + b_2) + c_2] \\ &= [a_1 + b_1, a_2 + b_2] + [c_1, c_2] = ([a_1, a_2] + [b_1, b_2]) + [c_1, c_2] \\ &= (\mathbf{a} + \mathbf{b}) + \mathbf{c}\end{aligned}$$

45. Consider triangle ABC, where D and E are the midpoints of AB and BC. We know that $\overrightarrow{AB} + \overrightarrow{BC} = \overrightarrow{AC}$ **(1)** and $\overrightarrow{DB} + \overrightarrow{BE} = \overrightarrow{DE}$ **(2)**. However, $\overrightarrow{DB} = \frac{1}{2}\overrightarrow{AB}$, and $\overrightarrow{BE} = \frac{1}{2}\overrightarrow{BC}$. Substituting these expressions for $\overrightarrow{DB}$ and $\overrightarrow{BE}$ into **(2)** gives $\frac{1}{2}\overrightarrow{AB} + \frac{1}{2}\overrightarrow{BC} = \overrightarrow{DE}$. Comparing this with **(1)** gives $\overrightarrow{DE} = \frac{1}{2}\overrightarrow{AC}$. Therefore $\overrightarrow{AC}$ and $\overrightarrow{DE}$ are parallel and $\left|\overrightarrow{DE}\right| = \frac{1}{2}\left|\overrightarrow{AC}\right|$.

8.3 The Dot Product

1. (a) $\mathbf{a} \cdot \mathbf{b}$ is a scalar, and the dot product is defined only for vectors, so $(\mathbf{a} \cdot \mathbf{b}) \cdot \mathbf{c}$ has no meaning.

(b) $(\mathbf{a} \cdot \mathbf{b})\,\mathbf{c}$ is a scalar multiple of a vector, so it does have meaning.

(c) Both $|\mathbf{a}|$ and $\mathbf{b} \cdot \mathbf{c}$ are scalars, so $|\mathbf{a}|\,(\mathbf{b} \cdot \mathbf{c})$ is an ordinary product of real numbers, and has meaning.

(d) Both $\mathbf{a}$ and $\mathbf{b} + \mathbf{c}$ are vectors, so the dot product $\mathbf{a} \cdot (\mathbf{b} + \mathbf{c})$ has meaning.

(e) $\mathbf{a} \cdot \mathbf{b}$ is a scalar, but $\mathbf{c}$ is a vector, and so the two quantities cannot be added and $\mathbf{a} \cdot \mathbf{b} + \mathbf{c}$ has no meaning.

(f) $|\mathbf{a}|$ is a scalar, and the dot product is defined only for vectors, so $|\mathbf{a}| \cdot (\mathbf{b} + \mathbf{c})$ has no meaning.

© 2016 Cengage Learning. All Rights Reserved. May not be scanned, copied or duplicated, or posted to a publicly accessible website, in whole or in part.

3. By the definition of the dot product, $\mathbf{a}\cdot\mathbf{b}=|\mathbf{a}|\,|\mathbf{b}|\cos\theta=(6)(5)\cos\frac{2\pi}{3}=30\left(-\frac{1}{2}\right)=-15$.

5. $\mathbf{a}\cdot\mathbf{b}=\left[-2,\frac{1}{3}\right]\cdot[-5,12]=(-2)(-5)+\left(\frac{1}{3}\right)(12)=10+4=14$

7. $\mathbf{a}\cdot\mathbf{b}=\left[4,1,\frac{1}{4}\right]\cdot[6,-3,-8]=(4)(6)+(1)(-3)+\left(\frac{1}{4}\right)(-8)=19$

9. $\mathbf{a}\cdot\mathbf{b}=(2\mathbf{i}+\mathbf{j})\cdot(\mathbf{i}-\mathbf{j}+\mathbf{k})=(2)(1)+(1)(-1)+(0)(1)=1$

11. $\mathbf{u}$, $\mathbf{v}$, and $\mathbf{w}$ are all unit vectors, so the triangle is an equilateral triangle. Thus the angle between $\mathbf{u}$ and $\mathbf{v}$ is 60° and $\mathbf{u}\cdot\mathbf{v}=|\mathbf{u}|\,|\mathbf{v}|\cos 60^\circ=(1)(1)\left(\frac{1}{2}\right)=\frac{1}{2}$. If $\mathbf{w}$ is moved so it has the same initial point as $\mathbf{u}$, we can see that the angle between them is 120° and we have $\mathbf{u}\cdot\mathbf{w}=|\mathbf{u}|\,|\mathbf{w}|\cos 120^\circ=(1)(1)\left(-\frac{1}{2}\right)=-\frac{1}{2}$.

13. (a) $\mathbf{i}\cdot\mathbf{j}=[1,0,0]\cdot[0,1,0]=(1)(0)+(0)(1)+(0)(0)=0$. Similarly, $\mathbf{j}\cdot\mathbf{k}=(0)(0)+(1)(0)+(0)(1)=0$ and $\mathbf{k}\cdot\mathbf{i}=(0)(1)+(0)(0)+(1)(0)=0$.

Another method: Because $\mathbf{i}$, $\mathbf{j}$, and $\mathbf{k}$ are mutually perpendicular, the cosine factor in each dot product is $\cos\frac{\pi}{2}=0$.

(b) By Property 1 of the dot product, $\mathbf{i}\cdot\mathbf{i}=|\mathbf{i}|^2=1^2=1$ since $\mathbf{i}$ is a unit vector. Similarly, $\mathbf{j}\cdot\mathbf{j}=|\mathbf{j}|^2=1$ and $\mathbf{k}\cdot\mathbf{k}=|\mathbf{k}|^2=1$.

15. $|\mathbf{a}|=\sqrt{(-8)^2+6^2}=10$, $|\mathbf{b}|=\sqrt{\left(\sqrt{7}\right)^2+3^2}=4$, and $\mathbf{a}\cdot\mathbf{b}=(-8)\left(\sqrt{7}\right)+(6)(3)=18-8\sqrt{7}$. From the definition of the dot product, we have $\cos\theta=\dfrac{\mathbf{a}\cdot\mathbf{b}}{|\mathbf{a}|\,|\mathbf{b}|}=\dfrac{18-8\sqrt{7}}{10\cdot 4}=\dfrac{9-4\sqrt{7}}{20}$. So the angle between $\mathbf{a}$ and $\mathbf{b}$ is $\theta=\cos^{-1}\left(\dfrac{9-4\sqrt{7}}{20}\right)\approx 95^\circ$.

17. $|\mathbf{a}|=\sqrt{0^2+1^2+1^2}=\sqrt{2}$, $|\mathbf{b}|=\sqrt{1^2+2^2+(-3)^2}=\sqrt{14}$, and $\mathbf{a}\cdot\mathbf{b}=(0)(1)+(1)(2)+(1)(-3)=-1$. Then $\cos\theta=\dfrac{\mathbf{a}\cdot\mathbf{b}}{|\mathbf{a}|\,|\mathbf{b}|}=\dfrac{-1}{\sqrt{2}\cdot\sqrt{14}}=\dfrac{-1}{2\sqrt{7}}$ and $\theta=\cos^{-1}\left(-\dfrac{1}{2\sqrt{7}}\right)\approx 101^\circ$.

19. Let a, b, and c be the angles at vertices A, B, and C respectively. Then a is the angle between vectors $\overrightarrow{AB}$ and $\overrightarrow{AC}$, b is the angle between vectors $\overrightarrow{BA}$ and $\overrightarrow{BC}$, and c is the angle between vectors $\overrightarrow{CA}$ and $\overrightarrow{CB}$.

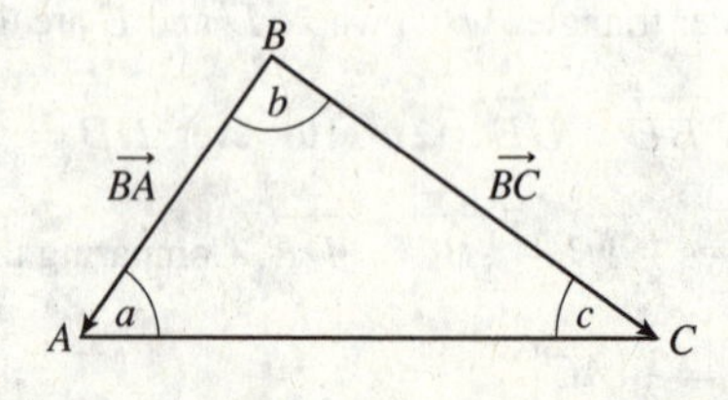

Thus $\cos a=\dfrac{\overrightarrow{AB}\cdot\overrightarrow{AC}}{\left|\overrightarrow{AB}\right|\left|\overrightarrow{AC}\right|}=\dfrac{[2,6]\cdot[-2,4]}{\sqrt{2^2+6^2}\sqrt{(-2)^2+4^2}}=\dfrac{1}{\sqrt{40}\sqrt{20}}(-4+24)=\dfrac{20}{\sqrt{800}}=\dfrac{\sqrt{2}}{2}$ and $a=\cos^{-1}\left(\dfrac{\sqrt{2}}{2}\right)=45^\circ$. Similarly, $\cos b=\dfrac{\overrightarrow{BA}\cdot\overrightarrow{BC}}{\left|\overrightarrow{BA}\right|\left|\overrightarrow{BC}\right|}=\dfrac{[-2,-6]\cdot[-4,-2]}{\sqrt{4+36}\sqrt{16+4}}=\dfrac{1}{\sqrt{40}\sqrt{20}}(8+12)=\dfrac{20}{\sqrt{800}}=\dfrac{\sqrt{2}}{2}$ so $b=\cos^{-1}\left(\dfrac{\sqrt{2}}{2}\right)=45^\circ$ and $c=180^\circ-(45^\circ+45^\circ)=90^\circ$.

Alternate solution: Apply the Law of Cosines three times as follows: $\cos a=\dfrac{\left|\overrightarrow{BC}\right|^2-\left|\overrightarrow{AB}\right|^2-\left|\overrightarrow{AC}\right|^2}{2\left|\overrightarrow{AB}\right|\left|\overrightarrow{AC}\right|}$,

$\cos b=\dfrac{\left|\overrightarrow{AC}\right|^2-\left|\overrightarrow{AB}\right|^2-\left|\overrightarrow{BC}\right|^2}{2\left|\overrightarrow{AB}\right|\left|\overrightarrow{BC}\right|}$, and $\cos c=\dfrac{\left|\overrightarrow{AB}\right|^2-\left|\overrightarrow{AC}\right|^2-\left|\overrightarrow{BC}\right|^2}{2\left|\overrightarrow{AC}\right|\left|\overrightarrow{BC}\right|}$.

© 2016 Cengage Learning. All Rights Reserved. May not be scanned, copied or duplicated, or posted to a publicly accessible website, in whole or in part.

21. (a) $\mathbf{a} \cdot \mathbf{b} = (-5)(6) + (3)(-8) + (7)(2) = -40 \neq 0$, so $\mathbf{a}$ and $\mathbf{b}$ are not orthogonal. Also, since $\mathbf{a}$ is not a scalar multiple of $\mathbf{b}$, $\mathbf{a}$ and $\mathbf{b}$ are not parallel.

(b) $\mathbf{a} \cdot \mathbf{b} = (4)(-3) + (6)(2) = 0$, so $\mathbf{a}$ and $\mathbf{b}$ are orthogonal (and not parallel).

(c) $\mathbf{a} \cdot \mathbf{b} = (-1)(3) + (2)(4) + (5)(-1) = 0$, so $\mathbf{a}$ and $\mathbf{b}$ are orthogonal (and not parallel).

(d) Because $\mathbf{a} = -\frac{2}{3}\,\mathbf{b}$, $\mathbf{a}$ and $\mathbf{b}$ are parallel.

23. $\overrightarrow{QP} = [-1, -3, 2]$, $\overrightarrow{QR} = [4, -2, -1]$, and $\overrightarrow{QP} \cdot \overrightarrow{QR} = -4 + 6 - 2 = 0$. Thus $\overrightarrow{QP}$ and $\overrightarrow{QR}$ are orthogonal, so the angle of the triangle at vertex Q is a right angle.

25. Let $\mathbf{a} = [a_1, a_2, a_3]$ be a vector orthogonal to both $[1, 1, 0]$ and $[1, 0, 1]$. Then $\mathbf{a} \cdot [1, 1, 0] = 0 \quad \Leftrightarrow \quad a_1 + a_2 = 0$ and $\mathbf{a} \cdot [1, 0, 1] = 0 \quad \Leftrightarrow \quad a_1 + a_3 = 0$, so $a_1 = -a_2 = -a_3$. Furthermore $\mathbf{a}$ is to be a unit vector, so $1 = a_1^2 + a_2^2 + a_3^2 = 3a_1^2$ implies $a_1 = \pm\frac{1}{\sqrt{3}}$. Thus $\mathbf{a} = \left[\frac{1}{\sqrt{3}}, -\frac{1}{\sqrt{3}}, -\frac{1}{\sqrt{3}}\right]$ and $\mathbf{a} = \left[-\frac{1}{\sqrt{3}}, \frac{1}{\sqrt{3}}, \frac{1}{\sqrt{3}}\right]$ are two such unit vectors.

27. Since the plane is perpendicular to the vector $[1, -2, 5]$, we can take $[1, -2, 5]$ as a normal vector to the plane. $(0, 0, 0)$ is a point on the plane, so setting $a = 1$, $b = -2$, $c = 5$ and $x_0 = 0$, $y_0 = 0$, $z_0 = 0$ in equation (5) gives $1(x - 0) - 2(y - 0) + 5(z - 0) = 0$ or $x - 2y + 5z = 0$ to be an equation of the plane.

29. $|\mathbf{a}| = \sqrt{3^2 + (-4)^2} = 5$. The scalar projection of $\mathbf{b}$ onto $\mathbf{a}$ is $\text{comp}_{\mathbf{a}}\,\mathbf{b} = \dfrac{\mathbf{a} \cdot \mathbf{b}}{|\mathbf{a}|} = \dfrac{3 \cdot 5 + (-4) \cdot 0}{5} = 3$ and the vector projection of $\mathbf{b}$ onto $\mathbf{a}$ is $\text{proj}_{\mathbf{a}}\,\mathbf{b} = \left(\dfrac{\mathbf{a} \cdot \mathbf{b}}{|\mathbf{a}|}\right)\dfrac{\mathbf{a}}{|\mathbf{a}|} = 3 \cdot \frac{1}{5}[3, -4] = \left[\frac{9}{5}, -\frac{12}{5}\right]$.

31. $|\mathbf{a}| = \sqrt{4 + 1 + 16} = \sqrt{21}$ so the scalar projection of $\mathbf{b}$ onto $\mathbf{a}$ is $\text{comp}_{\mathbf{a}}\,\mathbf{b} = \dfrac{\mathbf{a} \cdot \mathbf{b}}{|\mathbf{a}|} = \dfrac{0 - 1 + 2}{\sqrt{21}} = \dfrac{1}{\sqrt{21}}$ while the vector projection of $\mathbf{b}$ onto $\mathbf{a}$ is $\text{proj}_{\mathbf{a}}\,\mathbf{b} = \dfrac{1}{\sqrt{21}}\dfrac{\mathbf{a}}{|\mathbf{a}|} = \dfrac{1}{\sqrt{21}} \cdot \dfrac{[2, -1, 4]}{\sqrt{21}} = \frac{1}{21}[2, -1, 4] = \left[\frac{2}{21}, -\frac{1}{21}, \frac{4}{21}\right]$.

33. $(\text{orth}_{\mathbf{a}}\,\mathbf{b}) \cdot \mathbf{a} = (\mathbf{b} - \text{proj}_{\mathbf{a}}\,\mathbf{b}) \cdot \mathbf{a} = \mathbf{b} \cdot \mathbf{a} - (\text{proj}_{\mathbf{a}}\,\mathbf{b}) \cdot \mathbf{a} = \mathbf{b} \cdot \mathbf{a} - \dfrac{\mathbf{a} \cdot \mathbf{b}}{|\mathbf{a}|^2}\,\mathbf{a} \cdot \mathbf{a} = \mathbf{b} \cdot \mathbf{a} - \dfrac{\mathbf{a} \cdot \mathbf{b}}{|\mathbf{a}|^2}\,|\mathbf{a}|^2 = \mathbf{b} \cdot \mathbf{a} - \mathbf{a} \cdot \mathbf{b} = 0$.

So they are orthogonal (see discussion after Equation 4).

35. $\text{comp}_{\mathbf{a}}\,\mathbf{b} = \dfrac{\mathbf{a} \cdot \mathbf{b}}{|\mathbf{a}|} = 2 \quad \Leftrightarrow \quad \mathbf{a} \cdot \mathbf{b} = 2\,|\mathbf{a}| = 2\sqrt{10}$. If $\mathbf{b} = [b_1, b_2, b_3]$, then we need $3b_1 + 0b_2 - 1b_3 = 2\sqrt{10}$.

One possible solution is obtained by taking $b_1 = 0$, $b_2 = 0$, $b_3 = -2\sqrt{10}$. In general, $\mathbf{b} = \left[s, t, 3s - 2\sqrt{10}\right]$, $s, t \in \mathbb{R}$.

37. (a) The vector of antigenic change for North American viruses from 2013 to 2014 is $\mathbf{n} = [4 - 2, 3 - 1] = [2, 2]$ and the magnitude of that change is $|\mathbf{n}| = \sqrt{(2)^2 + (2)^2} = \sqrt{8} \approx 2.8$. The vector of antigenic change for Asian viruses from 2013 to 2014 is $\mathbf{a} = [5 - 4, 17 - 18] = [1, -1]$ and the magnitude of that change is $|\mathbf{a}| = \sqrt{(1)^2 + (-1)^2} = \sqrt{2} \approx 1.4$. Thus, the magnitude of antigenic change in North American viruses is approximately twice as large as that in Asian viruses.

(b) The angle θ between the vectors $\mathbf{n}$ and $\mathbf{a}$ from part (a) is a measure of the difference in direction of antigenic change between North American and Asian viruses.

$$\mathbf{n} \cdot \mathbf{a} = |\mathbf{n}|\,|\mathbf{a}| \cos\theta \quad \Rightarrow \quad \cos\theta = \frac{\mathbf{n} \cdot \mathbf{a}}{|\mathbf{n}|\,|\mathbf{a}|} = \frac{[2, 2] \cdot [1, -1]}{(\sqrt{8})\,(\sqrt{2})} = \frac{2(1) + 2(-1)}{(\sqrt{16})} = 0 \quad \Rightarrow \quad \theta = \cos^{-1}(0) = \frac{\pi}{2} \text{ (or } 90^\circ)$$

Thus, the direction of antigenic change in Asian viruses is perpendicular to the direction of antigenic change in North American viruses.

© 2016 Cengage Learning. All Rights Reserved. May not be scanned, copied or duplicated, or posted to a publicly accessible website, in whole or in part.

39. The desired change in expression vector is $\mathbf{D} = [3, 9, -5]$ and the two drug candidates have expression profiles $\mathbf{A} = [2, 4, 1]$ and $\mathbf{B} = [-2, 3, -5]$.

(a) The magnitude of the desired change in expression is $|\mathbf{D}| = |[3, 9, -5]| = \sqrt{(3)^2 + (9)^2 + (-5)^2} = \sqrt{115} \approx 10.7$.

(b) $\text{comp}_{\mathbf{D}}\,\mathbf{A} = \dfrac{\mathbf{D} \cdot \mathbf{A}}{|\mathbf{D}|} = \dfrac{[3, 9, -5] \cdot [2, 4, 1]}{\sqrt{115}} = \dfrac{3(2) + 9\,(4) + (-5)\,(1)}{\sqrt{115}} = \dfrac{37}{\sqrt{115}}$

So the fraction of the desired change induced by drug A is $\dfrac{37/\sqrt{115}}{\sqrt{115}} = \dfrac{37}{115} \approx 0.32$.

$\text{comp}_{\mathbf{D}}\mathbf{B} = \dfrac{\mathbf{D} \cdot \mathbf{B}}{|\mathbf{D}|} = \dfrac{[3, 9, -5] \cdot [-2, 3, -5]}{\sqrt{115}} = \dfrac{3(-2) + 9\,(3) + (-5)\,(-5)}{\sqrt{115}} = \dfrac{46}{\sqrt{115}}$

So the fraction of the desired change induced by drug B is $\dfrac{46/\sqrt{115}}{\sqrt{115}} = \dfrac{46}{115} = 0.4$.

(c) Drug B is closest to achieving the desired effect since the fraction of the desired change it induces is greater than drug A.

41. For convenience, consider the unit cube positioned so that its back left corner is at the origin, and its edges lie along the coordinate axes. The diagonal of the cube that begins at the origin and ends at $(1, 1, 1)$ has vector representation $[1, 1, 1]$. The angle θ between this vector and the vector of the edge which also begins at the origin and runs along the x-axis [that is, $[1, 0, 0]$] is given by $\cos\theta = \dfrac{[1, 1, 1] \cdot [1, 0, 0]}{|[1, 1, 1]|\,|[1, 0, 0]|} = \dfrac{1}{\sqrt{3}} \quad \Rightarrow \quad \theta = \cos^{-1}\left(\frac{1}{\sqrt{3}}\right) \approx 55°$.

43. $(\mathbf{r} - \mathbf{a}) \cdot (\mathbf{r} - \mathbf{b}) = 0$ implies that the vectors $\mathbf{r} - \mathbf{a}$ and $\mathbf{r} - \mathbf{b}$ are orthogonal.

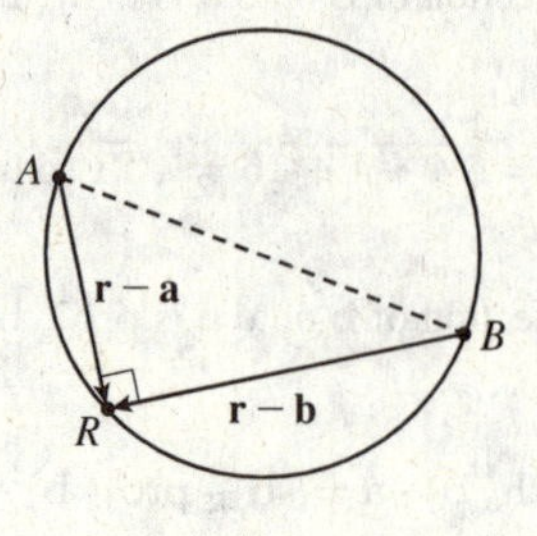

From the diagram (in which A, B and R are the terminal points of the vectors), we see that this implies that R lies on a sphere whose diameter is the line from A to B. The center of this circle is the midpoint of AB, that is, $\frac{1}{2}(\mathbf{a} + \mathbf{b}) = \left[\frac{1}{2}(a_1 + b_1), \frac{1}{2}(a_2 + b_2), \frac{1}{2}(a_3 + b_3)\right]$, and its radius is $\frac{1}{2}\,|\mathbf{a} - \mathbf{b}| = \frac{1}{2}\,\sqrt{(a_1 - b_1)^2 + (a_2 - b_2)^2 + (a_3 - b_3)^2}$.

Or: Expand the given equation, substitute $\mathbf{r} \cdot \mathbf{r} = x^2 + y^2 + z^2$ and complete the squares.

45. If $c = 0$ then $c\mathbf{a} = \mathbf{0}$, so $(c\mathbf{a}) \cdot \mathbf{b} = \mathbf{0} \cdot \mathbf{b} = 0$ by Property 5. Similarly, $\mathbf{a} \cdot (c\mathbf{b}) = \mathbf{a} \cdot \mathbf{0} = 0$, and $c(\mathbf{a} \cdot \mathbf{b}) = 0(|\mathbf{a}|\,|\mathbf{b}|\cos\theta) = 0$, thus $(c\mathbf{a}) \cdot \mathbf{b} = c(\mathbf{a} \cdot \mathbf{b}) = \mathbf{a} \cdot (c\mathbf{b})$. If $c > 0$, the angle θ between $\mathbf{a}$ and $\mathbf{b}$ coincides with the angle between $c\mathbf{a}$ and $\mathbf{b}$, so by definition of the dot product, $(c\mathbf{a}) \cdot \mathbf{b} = |c\mathbf{a}|\,|\mathbf{b}|\cos\theta = |c|\,|\mathbf{a}|\,|\mathbf{b}|\cos\theta = c\,|\mathbf{a}|\,|\mathbf{b}|\cos\theta$. Similarly, $\mathbf{a} \cdot (c\mathbf{b}) = |\mathbf{a}|\,|c\mathbf{b}|\cos\theta = |\mathbf{a}|\,|c|\,|\mathbf{b}|\cos\theta = c\,|\mathbf{a}|\,|\mathbf{b}|\cos\theta$, and $c(\mathbf{a} \cdot \mathbf{b}) = c\,|\mathbf{a}|\,|\mathbf{b}|\cos\theta$. Thus, $(c\mathbf{a}) \cdot \mathbf{b} = c(\mathbf{a} \cdot \mathbf{b}) = \mathbf{a} \cdot (c\mathbf{b})$. The case for $c < 0$ is similar. Using components, let $\mathbf{a} = [a_1, a_2, a_3]$ and $\mathbf{b} = [b_1, b_2, b_3]$. Then

$$\begin{aligned}(c\mathbf{a}) \cdot \mathbf{b} &= [ca_1, ca_2, ca_3] \cdot [b_1, b_2, b_3] = (ca_1)b_1 + (ca_2)b_2 + (ca_3)b_3 \\ &= c(a_1b_1 + a_2b_2 + a_3b_3) = c(\mathbf{a} \cdot \mathbf{b}) \\ &= a_1(cb_1) + a_2(cb_2) + a_3(cb_3) = [a_1, a_2, a_3] \cdot [cb_1, cb_2, cb_3] = \mathbf{a} \cdot (c\mathbf{b})\end{aligned}$$

47. $|\mathbf{a} \cdot \mathbf{b}| = \big|\,|\mathbf{a}|\,|\mathbf{b}|\cos\theta\big| = |\mathbf{a}|\,|\mathbf{b}|\,|\cos\theta|$. Since $|\cos\theta| \leq 1$, $|\mathbf{a} \cdot \mathbf{b}| = |\mathbf{a}|\,|\mathbf{b}|\,|\cos\,\theta| \leq |\mathbf{a}|\,|\mathbf{b}|$.

Note: We have equality in the case of $\cos\theta = \pm 1$, so $\theta = 0$ or $\theta = \pi$, thus equality when $\mathbf{a}$ and $\mathbf{b}$ are parallel.

© 2016 Cengage Learning. All Rights Reserved. May not be scanned, copied or duplicated, or posted to a publicly accessible website, in whole or in part.

49. (a)

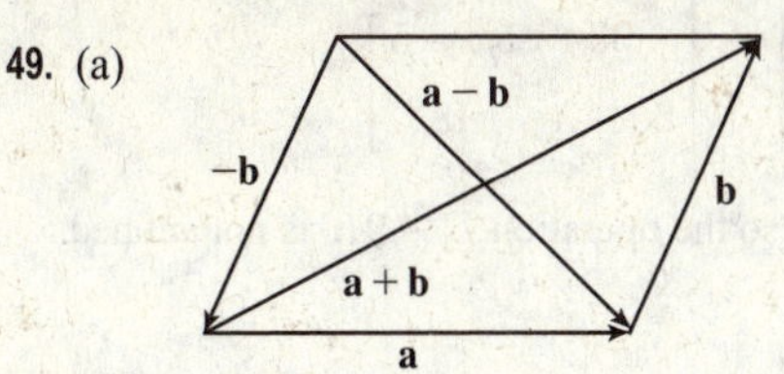

The Parallelogram Law states that the sum of the squares of the lengths of the diagonals of a parallelogram equals the sum of the squares of its (four) sides.

(b) $|\mathbf{a}+\mathbf{b}|^2 = (\mathbf{a}+\mathbf{b})\cdot(\mathbf{a}+\mathbf{b}) = |\mathbf{a}|^2 + 2(\mathbf{a}\cdot\mathbf{b}) + |\mathbf{b}|^2$ and $|\mathbf{a}-\mathbf{b}|^2 = (\mathbf{a}-\mathbf{b})\cdot(\mathbf{a}-\mathbf{b}) = |\mathbf{a}|^2 - 2(\mathbf{a}\cdot\mathbf{b}) + |\mathbf{b}|^2$.

Adding these two equations gives $|\mathbf{a}+\mathbf{b}|^2 + |\mathbf{a}-\mathbf{b}|^2 = 2\,|\mathbf{a}|^2 + 2\,|\mathbf{b}|^2$.

8.4 Matrix Algebra

1. (a) A and $4C$ are both 2×2 matrices, so the operation $A+4C$ results in a 2×2 matrix.

(b) $\frac{1}{2}K$ is a 2×3 matrix and L is a 3×2 matrix, so the operation $\frac{1}{2}K + L$ is not defined.

(c) $5K$ and $3H$ are both 2×3 matrices, so the operation $5K+3H$ results in a 2×3 matrix.

(d) $0G + 3(E+F) = 0G + 3E + 3F$ [by property 5 of matrix addition]

$0G$, $3E$, and $3F$ are all 3×3 matrices, so the operation $0G + 3(E+F)$ results in a 3×3 matrix.

(e) $3A + 6(B+M) = 3A + 6B + 6M$ [by property 5 of matrix addition]

$3A$ and $6B$ are both 2×2 matrices and $6M$ is a 3×2 matrix, so the operation $3A + 6(B+M)$ is not defined.

(f) $12M - L = 12M + (-L)$

$12M$ and $-L$ are both 3×2 matrices, so the operation $12M - L$ results in a 3×2 matrix.

(g) $F + G - 2C = F + G + (-2C)$

F and G are both 3×3 matrices and $-2C$ is a 2×2 matrix, so the operation $F+G-2C$ is not defined.

(h) αF and βG are both 3×3 matrices, so the operation $\alpha F + \beta G$ results in a 3×3 matrix.

3. (a) $A - 3C = \begin{bmatrix} 2 & 5 \\ 1 & 7 \end{bmatrix} - 3\begin{bmatrix} 9 & 2 \\ 7 & 10 \end{bmatrix} = \begin{bmatrix} 2 & 5 \\ 1 & 7 \end{bmatrix} - \begin{bmatrix} 27 & 6 \\ 21 & 30 \end{bmatrix} = \begin{bmatrix} 2-27 & 5-6 \\ 1-21 & 7-30 \end{bmatrix} = \begin{bmatrix} -25 & -1 \\ -20 & -23 \end{bmatrix}$

(b) $$3F + G - E = 3\begin{bmatrix} x & 2 & 9 \\ 6 & y & 13 \\ 0 & 1 & 0 \end{bmatrix} + \begin{bmatrix} 13 & 2 & 0 \\ 5 & 9 & 12 \\ 7 & 0 & 1 \end{bmatrix} - \begin{bmatrix} 0 & 3 & 1 \\ 7 & 6 & 0 \\ 9 & 13 & 5 \end{bmatrix} = \begin{bmatrix} 3x & 6 & 27 \\ 18 & 3y & 39 \\ 0 & 3 & 0 \end{bmatrix} + \begin{bmatrix} 13 & 2 & 0 \\ 5 & 9 & 12 \\ 7 & 0 & 1 \end{bmatrix} - \begin{bmatrix} 0 & 3 & 1 \\ 7 & 6 & 0 \\ 9 & 13 & 5 \end{bmatrix}$$

$$= \begin{bmatrix} 3x+13-0 & 6+2-3 & 27+0-1 \\ 18+5-7 & 3y+9-6 & 39+12-0 \\ 0+7-9 & 3+0-13 & 0+1-5 \end{bmatrix} = \begin{bmatrix} 3x+13 & 5 & 26 \\ 16 & 3y+3 & 51 \\ -2 & -10 & -4 \end{bmatrix}$$

(c) $$5K + 9H = 5\begin{bmatrix} 1 & 0 & 1 \\ 0 & y & 1 \end{bmatrix} + 9\begin{bmatrix} 3 & 1 & 7 \\ 15 & 0 & 2 \end{bmatrix} = \begin{bmatrix} 5 & 0 & 5 \\ 0 & 5y & 5 \end{bmatrix} + \begin{bmatrix} 27 & 9 & 63 \\ 135 & 0 & 18 \end{bmatrix}$$

$$= \begin{bmatrix} 5+27 & 0+9 & 5+63 \\ 0+135 & 5y+0 & 5+18 \end{bmatrix} = \begin{bmatrix} 32 & 9 & 68 \\ 135 & 5y & 23 \end{bmatrix}$$

(d) $H - 12G = H + (-12G)$ H is a (2×3) matrix and $-12G$ is a (3×3) matrix, so the operation $H - 12G$ is not defined.

© 2016 Cengage Learning. All Rights Reserved. May not be scanned, copied or duplicated, or posted to a publicly accessible website, in whole or in part.

(e) $5B - A = 5\begin{bmatrix} 7 & x \\ a & 5 \end{bmatrix} - \begin{bmatrix} 2 & 5 \\ 1 & 7 \end{bmatrix} = \begin{bmatrix} 35 & 5x \\ 5a & 25 \end{bmatrix} - \begin{bmatrix} 2 & 5 \\ 1 & 7 \end{bmatrix} = \begin{bmatrix} 35-2 & 5x-5 \\ 5a-1 & 25-7 \end{bmatrix} = \begin{bmatrix} 33 & 5x-5 \\ 5a-1 & 18 \end{bmatrix}$

(f) $B - 9K = B + (-9K)$ B is a (2×2) matrix and $-9K$ is a (2×3) matrix, so the operation $B - 9K$ is not defined.

(g) $3F - F = 3\begin{bmatrix} x & 2 & 9 \\ 6 & y & 13 \\ 0 & 1 & 0 \end{bmatrix} - \begin{bmatrix} x & 2 & 9 \\ 6 & y & 13 \\ 0 & 1 & 0 \end{bmatrix} = \begin{bmatrix} 3x & 6 & 27 \\ 18 & 3y & 39 \\ 0 & 3 & 0 \end{bmatrix} - \begin{bmatrix} x & 2 & 9 \\ 6 & y & 13 \\ 0 & 1 & 0 \end{bmatrix}$

$= \begin{bmatrix} 3x-x & 6-2 & 27-9 \\ 18-6 & 3y-y & 39-13 \\ 0-0 & 3-1 & 0-0 \end{bmatrix} = \begin{bmatrix} 2x & 4 & 18 \\ 12 & 2y & 26 \\ 0 & 2 & 0 \end{bmatrix}$

Alternative Method: $3F - F = (3-1)F$ [by property 6 of matrix addition] $= 2F = \begin{bmatrix} 2x & 4 & 18 \\ 12 & 2y & 26 \\ 0 & 2 & 0 \end{bmatrix}$

(h) $H - K + 2H = \begin{bmatrix} 3 & 1 & 7 \\ 15 & 0 & 2 \end{bmatrix} - \begin{bmatrix} 1 & 0 & 1 \\ 0 & y & 1 \end{bmatrix} + 2\begin{bmatrix} 3 & 1 & 7 \\ 15 & 0 & 2 \end{bmatrix} = \begin{bmatrix} 3-1 & 1-0 & 7-1 \\ 15-0 & 0-y & 2-1 \end{bmatrix} + \begin{bmatrix} 6 & 2 & 14 \\ 30 & 0 & 4 \end{bmatrix}$

$= \begin{bmatrix} 2+6 & 1+2 & 6+14 \\ 15+30 & -y+0 & 1+4 \end{bmatrix} = \begin{bmatrix} 8 & 3 & 20 \\ 45 & -y & 5 \end{bmatrix}$

Alternative Method: $H - K + 2H = (1+2)H - K$ [by property 1 and 6 of matrix addition]

$= 3H - K = \begin{bmatrix} 9 & 3 & 21 \\ 45 & 0 & 6 \end{bmatrix} - \begin{bmatrix} 1 & 0 & 1 \\ 0 & y & 1 \end{bmatrix} = \begin{bmatrix} 8 & 3 & 20 \\ 45 & -y & 5 \end{bmatrix}$

5. (a) $\begin{bmatrix} 3X \\ 1 \\ 2 \end{bmatrix}^T = \begin{bmatrix} 3X & 1 & 2 \end{bmatrix}$ (b) $\begin{bmatrix} 3 & 3 & 9 \end{bmatrix}^T = \begin{bmatrix} 3 \\ 3 \\ 9 \end{bmatrix}$ (c) $\begin{bmatrix} 2 & 1 & 7 \\ 8 & 3 & 6 \end{bmatrix}^T = \begin{bmatrix} 2 & 8 \\ 1 & 3 \\ 7 & 6 \end{bmatrix}$ (d) $\begin{bmatrix} 2 & 1 \\ 1 & 3 \end{bmatrix}^T = \begin{bmatrix} 2 & 1 \\ 1 & 3 \end{bmatrix}$

7. (a) We can write an $n \times n$ diagonal matrix as $D = \begin{bmatrix} d_{11} & 0 & \cdots & 0 \\ 0 & d_{22} & \cdots & 0 \\ \vdots & \vdots & \ddots & \vdots \\ 0 & 0 & \cdots & d_{nn} \end{bmatrix}$, so that

$$D^2 = \begin{bmatrix} d_{11} & 0 & \cdots & 0 \\ 0 & d_{22} & \cdots & 0 \\ \vdots & \vdots & \ddots & \vdots \\ 0 & 0 & \cdots & d_{nn} \end{bmatrix}\begin{bmatrix} d_{11} & 0 & \cdots & 0 \\ 0 & d_{22} & \cdots & 0 \\ \vdots & \vdots & \ddots & \vdots \\ 0 & 0 & \cdots & d_{nn} \end{bmatrix} = \begin{bmatrix} d_{11}^2+0+\cdots+0 & 0+0+\cdots+0 & \cdots & 0+0+\cdots+0 \\ 0+0+\cdots+0 & 0+d_2^2+\cdots+0 & \cdots & 0+0+\cdots+0 \\ \vdots & \vdots & \ddots & \vdots \\ 0+0+\cdots+0 & 0+0+\cdots+0 & \cdots & 0+0+\cdots+d_{nn}^2 \end{bmatrix}$$

$$= \begin{bmatrix} d_{11}^2 & 0 & \cdots & 0 \\ 0 & d_{22}^2 & \cdots & 0 \\ \vdots & \vdots & \ddots & \vdots \\ 0 & 0 & \cdots & d_{nn}^2 \end{bmatrix}$$

Thus, the matrix D^2 is diagonal with entries d_{ii}^2 (the square of the diagonal entries of D). More generally, if A and B are diagonal matrices, the matrix AB is diagonal with entries $a_{ii}b_{ii}$ (the product of the diagonal entries of A and B).

© 2016 Cengage Learning. All Rights Reserved. May not be scanned, copied or duplicated, or posted to a publicly accessible website, in whole or in part.

(b) We know from (a) that D^2 is a diagonal matrix with entries d_{ii}^2. Let us investigate some additional powers of D:

$D^3 = D\left(D^2\right)$ Matrices D and D^2 are both diagonal, so D^3 must also be diagonal with entries $(d_{ii})\left(d_{ii}^2\right) = d_{ii}^3$ (from part (a)). $D^4 = D\left(D^3\right)$ Matrices D and D^3 are both diagonal, so D^4 must also be diagonal with entries $(d_{ii})\left(d_{ii}^3\right) = d_{ii}^4$ (from part (a)). It appears that for an arbitrary $n \times n$ diagonal matrix D, the matrix D^k is also diagonal with entries d_{ii}^k. (This can be proved using mathematical induction).

9. (a) $C = \begin{bmatrix} 0 & 1 \\ 1 & 0 \end{bmatrix} \Rightarrow C^2 = \begin{bmatrix} 0 & 1 \\ 1 & 0 \end{bmatrix}\begin{bmatrix} 0 & 1 \\ 1 & 0 \end{bmatrix} = \begin{bmatrix} 0+1 & 0+0 \\ 0+0 & 1+0 \end{bmatrix} = \begin{bmatrix} 1 & 0 \\ 0 & 1 \end{bmatrix}$

$$C^3 = CC^2 = \begin{bmatrix} 0 & 1 \\ 1 & 0 \end{bmatrix}\begin{bmatrix} 1 & 0 \\ 0 & 1 \end{bmatrix} = \begin{bmatrix} 0 & 1 \\ 1 & 0 \end{bmatrix} \quad \text{[by property 5 of matrix multiplication]} \quad = C$$

$$C^4 = CC^3 = \begin{bmatrix} 0 & 1 \\ 1 & 0 \end{bmatrix}\begin{bmatrix} 0 & 1 \\ 1 & 0 \end{bmatrix} = \begin{bmatrix} 1 & 0 \\ 0 & 1 \end{bmatrix} = C^2$$

$$C^5 = CC^4 = \begin{bmatrix} 0 & 1 \\ 1 & 0 \end{bmatrix}\begin{bmatrix} 1 & 0 \\ 0 & 1 \end{bmatrix} = \begin{bmatrix} 0 & 1 \\ 1 & 0 \end{bmatrix} \quad \text{[by property 5 of matrix multiplication]} \quad = C$$

(b) Observing the pattern in the even and odd powers of C, we deduce that

$$C^k = \begin{cases} \begin{bmatrix} 0 & 1 \\ 1 & 0 \end{bmatrix} & \text{when } k \text{ is odd} \\ \begin{bmatrix} 1 & 0 \\ 0 & 1 \end{bmatrix} & \text{when } k \text{ is even} \end{cases}$$

11. Let $A = \begin{bmatrix} a_{11} & a_{12} \\ a_{21} & a_{22} \end{bmatrix}$, $B = \begin{bmatrix} b_{11} & b_{12} \\ b_{21} & b_{22} \end{bmatrix}$, and $C = \begin{bmatrix} c_{11} & c_{12} \\ c_{21} & c_{22} \end{bmatrix}$.

(a) $A(BC) = \begin{bmatrix} a_{11} & a_{12} \\ a_{21} & a_{22} \end{bmatrix}\left(\begin{bmatrix} b_{11} & b_{12} \\ b_{21} & b_{22} \end{bmatrix}\begin{bmatrix} c_{11} & c_{12} \\ c_{21} & c_{22} \end{bmatrix}\right) = \begin{bmatrix} a_{11} & a_{12} \\ a_{21} & a_{22} \end{bmatrix}\begin{bmatrix} b_{11}c_{11} + b_{12}c_{21} & b_{11}c_{12} + b_{12}c_{22} \\ b_{21}c_{11} + b_{22}c_{21} & b_{21}c_{12} + b_{22}c_{22} \end{bmatrix}$

$$= \begin{bmatrix} a_{11}b_{11}c_{11} + a_{11}b_{12}c_{21} + a_{12}b_{21}c_{11} + a_{12}b_{22}c_{21} & a_{11}b_{11}c_{12} + a_{11}b_{12}c_{22} + a_{12}b_{21}c_{12} + a_{12}b_{22}c_{22} \\ a_{21}b_{11}c_{11} + a_{21}b_{12}c_{21} + a_{22}b_{21}c_{11} + a_{22}b_{22}c_{21} & a_{21}b_{11}c_{12} + a_{21}b_{12}c_{22} + a_{22}b_{21}c_{12} + a_{22}b_{22}c_{22} \end{bmatrix}$$

$(AB)C = \left(\begin{bmatrix} a_{11} & a_{12} \\ a_{21} & a_{22} \end{bmatrix}\begin{bmatrix} b_{11} & b_{12} \\ b_{21} & b_{22} \end{bmatrix}\right)\begin{bmatrix} c_{11} & c_{12} \\ c_{21} & c_{22} \end{bmatrix} = \begin{bmatrix} a_{11}b_{11} + a_{12}b_{21} & a_{11}b_{12} + a_{12}b_{22} \\ a_{21}b_{11} + a_{22}b_{21} & a_{21}b_{12} + a_{22}b_{22} \end{bmatrix}\begin{bmatrix} c_{11} & c_{12} \\ c_{21} & c_{22} \end{bmatrix}$

$$= \begin{bmatrix} a_{11}b_{11}c_{11} + a_{12}b_{21}c_{11} + a_{11}b_{12}c_{21} + a_{12}b_{22}c_{21} & a_{11}b_{11}c_{12} + a_{12}b_{21}c_{12} + a_{11}b_{12}c_{22} + a_{12}b_{22}c_{22} \\ a_{21}b_{11}c_{11} + a_{22}b_{21}c_{11} + a_{21}b_{12}c_{21} + a_{22}b_{22}c_{21} & a_{21}b_{11}c_{12} + a_{22}b_{21}c_{12} + a_{21}b_{12}c_{22} + a_{22}b_{22}c_{22} \end{bmatrix}$$

$= A(BC)$

(b) $A(B+C) = \begin{bmatrix} a_{11} & a_{12} \\ a_{21} & a_{22} \end{bmatrix}\left(\begin{bmatrix} b_{11} & b_{12} \\ b_{21} & b_{22} \end{bmatrix} + \begin{bmatrix} c_{11} & c_{12} \\ c_{21} & c_{22} \end{bmatrix}\right) = \begin{bmatrix} a_{11} & a_{12} \\ a_{21} & a_{22} \end{bmatrix}\begin{bmatrix} b_{11} + c_{11} & b_{12} + c_{12} \\ b_{21} + c_{21} & b_{22} + c_{22} \end{bmatrix}$

$$= \begin{bmatrix} a_{11}b_{11} + a_{11}c_{11} + a_{12}b_{21} + a_{12}c_{21} & a_{11}b_{12} + a_{11}c_{12} + a_{12}b_{22} + a_{12}c_{22} \\ a_{21}b_{11} + a_{21}c_{11} + a_{22}b_{21} + a_{22}c_{21} & a_{21}b_{12} + a_{21}c_{12} + a_{22}b_{22} + a_{22}c_{22} \end{bmatrix}$$

[continued]

© 2016 Cengage Learning. All Rights Reserved. May not be scanned, copied or duplicated, or posted to a publicly accessible website, in whole or in part.

$$AB+AC = \begin{bmatrix} a_{11} & a_{12} \\ a_{21} & a_{22} \end{bmatrix}\begin{bmatrix} b_{11} & b_{12} \\ b_{21} & b_{22} \end{bmatrix} + \begin{bmatrix} a_{11} & a_{12} \\ a_{21} & a_{22} \end{bmatrix}\begin{bmatrix} c_{11} & c_{12} \\ c_{21} & c_{22} \end{bmatrix}$$

$$= \begin{bmatrix} a_{11}b_{11}+a_{12}b_{21} & a_{11}b_{12}+a_{12}b_{22} \\ a_{21}b_{11}+a_{22}b_{21} & a_{21}b_{12}+a_{22}b_{22} \end{bmatrix} + \begin{bmatrix} a_{11}c_{11}+a_{12}c_{21} & a_{11}c_{12}+a_{12}c_{22} \\ a_{21}c_{11}+a_{22}c_{21} & a_{21}c_{12}+a_{22}c_{22} \end{bmatrix}$$

$$= \begin{bmatrix} a_{11}b_{11}+a_{11}c_{11}+a_{12}b_{21}+a_{12}c_{21} & a_{11}b_{12}+a_{11}c_{12}+a_{12}b_{22}+a_{12}c_{22} \\ a_{21}b_{11}+a_{21}c_{11}+a_{22}b_{21}+a_{22}c_{21} & a_{21}b_{12}+a_{21}c_{12}+a_{22}b_{22}+a_{22}c_{22} \end{bmatrix} = A(B+C)$$

(c) $$(B+C)A = \left(\begin{bmatrix} b_{11} & b_{12} \\ b_{21} & b_{22} \end{bmatrix} + \begin{bmatrix} c_{11} & c_{12} \\ c_{21} & c_{22} \end{bmatrix}\right)\begin{bmatrix} a_{11} & a_{12} \\ a_{21} & a_{22} \end{bmatrix} = \begin{bmatrix} b_{11}+c_{11} & b_{12}+c_{12} \\ b_{21}+c_{21} & b_{22}+c_{22} \end{bmatrix}\begin{bmatrix} a_{11} & a_{12} \\ a_{21} & a_{22} \end{bmatrix}$$

$$= \begin{bmatrix} a_{11}b_{11}+a_{11}c_{11}+a_{21}b_{12}+a_{21}c_{12} & a_{12}b_{11}+a_{12}c_{11}+a_{22}b_{12}+a_{22}c_{12} \\ a_{11}b_{21}+a_{11}c_{21}+a_{21}b_{22}+a_{21}c_{22} & a_{12}b_{21}+a_{12}c_{21}+a_{22}b_{22}+a_{22}c_{22} \end{bmatrix}$$

$$BA+CA = \begin{bmatrix} b_{11} & b_{12} \\ b_{21} & b_{22} \end{bmatrix}\begin{bmatrix} a_{11} & a_{12} \\ a_{21} & a_{22} \end{bmatrix} + \begin{bmatrix} c_{11} & c_{12} \\ c_{21} & c_{22} \end{bmatrix}\begin{bmatrix} a_{11} & a_{12} \\ a_{21} & a_{22} \end{bmatrix}$$

$$= \begin{bmatrix} a_{11}b_{11}+a_{21}b_{12} & a_{12}b_{11}+a_{22}b_{12} \\ a_{11}b_{21}+a_{21}b_{22} & a_{12}b_{21}+a_{22}b_{22} \end{bmatrix} + \begin{bmatrix} a_{11}c_{11}+a_{21}c_{12} & a_{12}c_{11}+a_{22}c_{12} \\ a_{11}c_{21}+a_{21}c_{22} & a_{12}c_{21}+a_{22}c_{22} \end{bmatrix}$$

$$= \begin{bmatrix} a_{11}b_{11}+a_{11}c_{11}+a_{21}b_{12}+a_{21}c_{12} & a_{12}b_{11}+a_{12}c_{11}+a_{22}b_{12}+a_{22}c_{12} \\ a_{11}b_{21}+a_{11}c_{21}+a_{21}b_{22}+a_{21}c_{22} & a_{12}b_{21}+a_{12}c_{21}+a_{22}b_{22}+a_{22}c_{22} \end{bmatrix} = (B+C)A$$

(d) $$AI = \begin{bmatrix} a_{11} & a_{12} \\ a_{21} & a_{22} \end{bmatrix}\begin{bmatrix} 1 & 0 \\ 0 & 1 \end{bmatrix} = \begin{bmatrix} a_{11}+0 & 0+a_{12} \\ a_{21}+0 & 0+a_{22} \end{bmatrix} = \begin{bmatrix} a_{11} & a_{12} \\ a_{21} & a_{22} \end{bmatrix} = A$$

$$IA = \begin{bmatrix} 1 & 0 \\ 0 & 1 \end{bmatrix}\begin{bmatrix} a_{11} & a_{12} \\ a_{21} & a_{22} \end{bmatrix} = \begin{bmatrix} a_{11}+0 & a_{12}+0 \\ 0+a_{21} & 0+a_{22} \end{bmatrix} = \begin{bmatrix} a_{11} & a_{12} \\ a_{21} & a_{22} \end{bmatrix} = AI = A$$

13. If matrix A commutes with matrix X then $AX = XA \Rightarrow \begin{bmatrix} a_{11} & a_{12} \\ a_{21} & a_{22} \end{bmatrix}\begin{bmatrix} 1 & 0 \\ 0 & 0 \end{bmatrix} = \begin{bmatrix} 1 & 0 \\ 0 & 0 \end{bmatrix}\begin{bmatrix} a_{11} & a_{12} \\ a_{21} & a_{22} \end{bmatrix} \Rightarrow$

$\begin{bmatrix} a_{11} & 0 \\ a_{21} & 0 \end{bmatrix} = \begin{bmatrix} a_{11} & a_{12} \\ 0 & 0 \end{bmatrix} \Rightarrow a_{12} = 0$ and $a_{21} = 0$. So A must at least be a diagonal matrix if it is to commute with all possible 2×2 matrices.

Now, if A commutes with Y then $AY = YA \Rightarrow \begin{bmatrix} a_{11} & a_{12} \\ a_{21} & a_{22} \end{bmatrix}\begin{bmatrix} 0 & 1 \\ 0 & 0 \end{bmatrix} = \begin{bmatrix} 0 & 1 \\ 0 & 0 \end{bmatrix}\begin{bmatrix} a_{11} & a_{12} \\ a_{21} & a_{22} \end{bmatrix} \Rightarrow$

$\begin{bmatrix} 0 & a_{11} \\ 0 & a_{21} \end{bmatrix} = \begin{bmatrix} a_{21} & a_{22} \\ 0 & 0 \end{bmatrix} \Rightarrow a_{11} = a_{22}$ and $a_{21} = 0$. Combined with the previous results, we require that $a_{11} = a_{22}$ and $a_{12} = a_{21} = 0$ if A commutes with all possible 2×2 matrices.

© 2016 Cengage Learning. All Rights Reserved. May not be scanned, copied or duplicated, or posted to a publicly accessible website, in whole or in part.

15. (a) $(AB)^T = \left(\begin{bmatrix} a_{11} & a_{12} \\ a_{21} & a_{22} \end{bmatrix} \begin{bmatrix} b_{11} & b_{12} \\ b_{21} & b_{22} \end{bmatrix} \right)^T = \left(\begin{bmatrix} a_{11}b_{11} + a_{12}b_{21} & a_{11}b_{12} + a_{12}b_{22} \\ a_{21}b_{11} + a_{22}b_{21} & a_{21}b_{12} + a_{22}b_{22} \end{bmatrix} \right)^T$

$$= \begin{bmatrix} a_{11}b_{11} + a_{12}b_{21} & a_{21}b_{11} + a_{22}b_{21} \\ a_{11}b_{12} + a_{12}b_{22} & a_{21}b_{12} + a_{22}b_{22} \end{bmatrix}$$

$$B^T A^T = \begin{bmatrix} b_{11} & b_{12} \\ b_{21} & b_{22} \end{bmatrix}^T \begin{bmatrix} a_{11} & a_{12} \\ a_{21} & a_{22} \end{bmatrix}^T = \begin{bmatrix} b_{11} & b_{21} \\ b_{12} & b_{22} \end{bmatrix} \begin{bmatrix} a_{11} & a_{21} \\ a_{12} & a_{22} \end{bmatrix} = \begin{bmatrix} a_{11}b_{11} + a_{12}b_{21} & a_{21}b_{11} + a_{22}b_{21} \\ a_{11}b_{12} + a_{12}b_{22} & a_{21}b_{12} + a_{22}b_{22} \end{bmatrix}$$

Therefore, $(AB)^T = B^T A^T$ for any 2×2 matrices A and B.

(b) Let $\mathbf{a}_i$ be the ith row of the $n \times n$ matrix A and $\mathbf{b}_i$ be the ith column of the $n \times n$ matrix B. Then

$$(AB)^T = \left(\begin{bmatrix} — \mathbf{a}_1 — \\ — \mathbf{a}_2 — \\ \vdots \\ — \mathbf{a}_n — \end{bmatrix} \begin{bmatrix} | & | & & | \\ \mathbf{b}_1 & \mathbf{b}_2 & \cdots & \mathbf{b}_n \\ | & | & & | \end{bmatrix} \right)^T = \begin{bmatrix} \mathbf{a}_1 \cdot \mathbf{b}_1 & \mathbf{a}_1 \cdot \mathbf{b}_2 & \cdots & \mathbf{a}_1 \cdot \mathbf{b}_n \\ \mathbf{a}_2 \cdot \mathbf{b}_1 & \mathbf{a}_2 \cdot \mathbf{b}_2 & \cdots & \mathbf{a}_2 \cdot \mathbf{b}_n \\ \vdots & \vdots & \ddots & \vdots \\ \mathbf{a}_n \cdot \mathbf{b}_1 & \mathbf{a}_n \cdot \mathbf{b}_2 & \cdots & \mathbf{a}_n \cdot \mathbf{b}_n \end{bmatrix}^T$$

$$= \begin{bmatrix} \mathbf{a}_1 \cdot \mathbf{b}_1 & \mathbf{a}_2 \cdot \mathbf{b}_1 & \cdots & \mathbf{a}_n \cdot \mathbf{b}_1 \\ \mathbf{a}_1 \cdot \mathbf{b}_2 & \mathbf{a}_2 \cdot \mathbf{b}_2 & \cdots & \mathbf{a}_n \cdot \mathbf{b}_2 \\ \vdots & \vdots & \ddots & \vdots \\ \mathbf{a}_1 \cdot \mathbf{b}_n & \mathbf{a}_2 \cdot \mathbf{b}_n & \cdots & \mathbf{a}_n \cdot \mathbf{b}_n \end{bmatrix}$$

$$B^T A^T = \begin{bmatrix} | & | & & | \\ \mathbf{b}_1 & \mathbf{b}_2 & \cdots & \mathbf{b}_n \\ | & | & & | \end{bmatrix}^T \begin{bmatrix} — \mathbf{a}_1 — \\ — \mathbf{a}_2 — \\ \vdots \\ — \mathbf{a}_n — \end{bmatrix}^T = \begin{bmatrix} — \mathbf{b}_1 — \\ — \mathbf{b}_2 — \\ \vdots \\ — \mathbf{b}_n — \end{bmatrix} \begin{bmatrix} | & | & & | \\ \mathbf{a}_1 & \mathbf{a}_2 & \cdots & \mathbf{a}_n \\ | & | & & | \end{bmatrix}$$

$$= \begin{bmatrix} \mathbf{b}_1 \cdot \mathbf{a}_1 & \mathbf{b}_2 \cdot \mathbf{a}_1 & \cdots & \mathbf{b}_n \cdot \mathbf{a}_1 \\ \mathbf{b}_1 \cdot \mathbf{a}_2 & \mathbf{b}_2 \cdot \mathbf{a}_2 & \cdots & \mathbf{b}_n \cdot \mathbf{a}_2 \\ \vdots & \vdots & \ddots & \vdots \\ \mathbf{b}_1 \cdot \mathbf{a}_n & \mathbf{b}_2 \cdot \mathbf{a}_n & \cdots & \mathbf{b}_n \cdot \mathbf{a}_n \end{bmatrix} = \begin{bmatrix} \mathbf{a}_1 \cdot \mathbf{b}_1 & \mathbf{a}_2 \cdot \mathbf{b}_1 & \cdots & \mathbf{a}_n \cdot \mathbf{b}_1 \\ \mathbf{a}_1 \cdot \mathbf{b}_2 & \mathbf{a}_2 \cdot \mathbf{b}_2 & \cdots & \mathbf{a}_n \cdot \mathbf{b}_2 \\ \vdots & \vdots & \ddots & \vdots \\ \mathbf{a}_1 \cdot \mathbf{b}_n & \mathbf{a}_2 \cdot \mathbf{b}_n & \cdots & \mathbf{a}_n \cdot \mathbf{b}_n \end{bmatrix} \quad \text{(by property 2 of the dot product)}$$

Thus, $(AB)^T = B^T A^T$ for any $n \times n$ matrices A and B.

Alternative Solution: Let $C = AB$. Then the entries in matrix C are given by $c_{ij} = \sum_{k=1}^{n} a_{ik}b_{kj}$, so the entries in C^T are $c_{ij}^T = \sum_{k=1}^{n} a_{jk}b_{ki}$ (interchange row and column indices i and j).

Let $E = B^T A^T$. Then the entries in matrix E are given by $e_{ij} = \sum_{k=1}^{n} b_{ki}a_{jk}$ (interchange a and b and then switch order of indices on each matrix). So $c_{ij}^T = e_{ij}$ for all i and j, meaning that all entries of matrix C^T are the same as matrix E. Thus $C^T = E$ or $(AB)^T = B^T A^T$.

© 2016 Cengage Learning. All Rights Reserved. May not be scanned, copied or duplicated, or posted to a publicly accessible website, in whole or in part.

17. The sum $A + A^T$ is defined only if A and A^T have the same size, so A must be a square matrix. Suppose A has size $n \times n$. Then

$$A + A^T = \begin{bmatrix} a_{11} & a_{12} & \cdots & a_{1n} \\ a_{21} & a_{22} & \cdots & a_{2n} \\ \vdots & \vdots & \ddots & \vdots \\ a_{n1} & a_{n2} & \cdots & a_{nn} \end{bmatrix} + \begin{bmatrix} a_{11} & a_{12} & \cdots & a_{1n} \\ a_{21} & a_{22} & \cdots & a_{2n} \\ \vdots & \vdots & \ddots & \vdots \\ a_{n1} & a_{n2} & \cdots & a_{nn} \end{bmatrix}^T = \begin{bmatrix} a_{11} & a_{12} & \cdots & a_{1n} \\ a_{21} & a_{22} & \cdots & a_{2n} \\ \vdots & \vdots & \ddots & \vdots \\ a_{n1} & a_{n2} & \cdots & a_{nn} \end{bmatrix} + \begin{bmatrix} a_{11} & a_{21} & \cdots & a_{n1} \\ a_{12} & a_{22} & \cdots & a_{n2} \\ \vdots & \vdots & \ddots & \vdots \\ a_{1n} & a_{2n} & \cdots & a_{nn} \end{bmatrix}$$

$$= \begin{bmatrix} 2a_{11} & a_{12} + a_{21} & \cdots & a_{1n} + a_{n1} \\ a_{21} + a_{12} & 2a_{22} & \cdots & a_{2n} + a_{n2} \\ \vdots & \vdots & \ddots & \vdots \\ a_{n1} + a_{1n} & a_{n2} + a_{2n} & \cdots & 2a_{nn} \end{bmatrix}$$

$A + A^T = 0$ when each entry is zero. This occurs when $a_{11} = a_{22} = \cdots = a_{nn} = 0$ (the diagonal entries of A are zero) and when the the off-diagonal terms satisfy $a_{ij} = -a_{ji}$ for all $1 \le i, j \le n$. In summary, $a_{ij} = \begin{cases} 0 & \text{when } i = j \\ -a_{ji} & \text{when } i \ne j \end{cases}$.

8.5 Matrices and the Dynamics of Vectors

1. Identify variables A, B, C, and D as the first, second, third and fourth variables respectively (when applicable). An arrow drawn from variable j to i is represented with an X in the ijth entry of the corresponding matrix. This rule gives the following matrix representations.

(a) $\begin{bmatrix} \text{X} & 0 \\ \text{X} & 0 \end{bmatrix}$ (b) $\begin{bmatrix} 0 & \text{X} & 0 \\ \text{X} & 0 & \text{X} \\ 0 & \text{X} & 0 \end{bmatrix}$ (c) $\begin{bmatrix} \text{X} & \text{X} & 0 & 0 \\ \text{X} & 0 & \text{X} & 0 \\ 0 & \text{X} & \text{X} & \text{X} \\ 0 & 0 & \text{X} & 0 \end{bmatrix}$ (d) $\begin{bmatrix} 0 & \text{X} & 0 \\ \text{X} & 0 & 0 \\ 0 & 0 & \text{X} \end{bmatrix}$ (e) $\begin{bmatrix} \text{X} & \text{X} \\ \text{X} & \text{X} \end{bmatrix}$

3. Identify variables A, B, C, and D as the first, second, third and fourth variables respectively (when applicable). An arrow drawn from variable j to i is represented with a nonzero entry (corresponding to the number on the arrow) in the ijth position of the corresponding matrix. This rule gives the following matrix representations.

(a) $\begin{bmatrix} 1 & 0 \\ 2 & 3 \end{bmatrix}$ (b) $\begin{bmatrix} 0 & 0 & 4 \\ 1 & 2 & 0 \\ 0 & 3 & 0 \end{bmatrix}$ (c) $\begin{bmatrix} 0 & 0 & 0 & 1 \\ 0 & 0 & 2 & 0 \\ 0 & 3 & 0 & 0 \\ 4 & 0 & 0 & 0 \end{bmatrix}$ (d) $\begin{bmatrix} 0 & 1 \\ 0 & 2 \end{bmatrix}$

5.

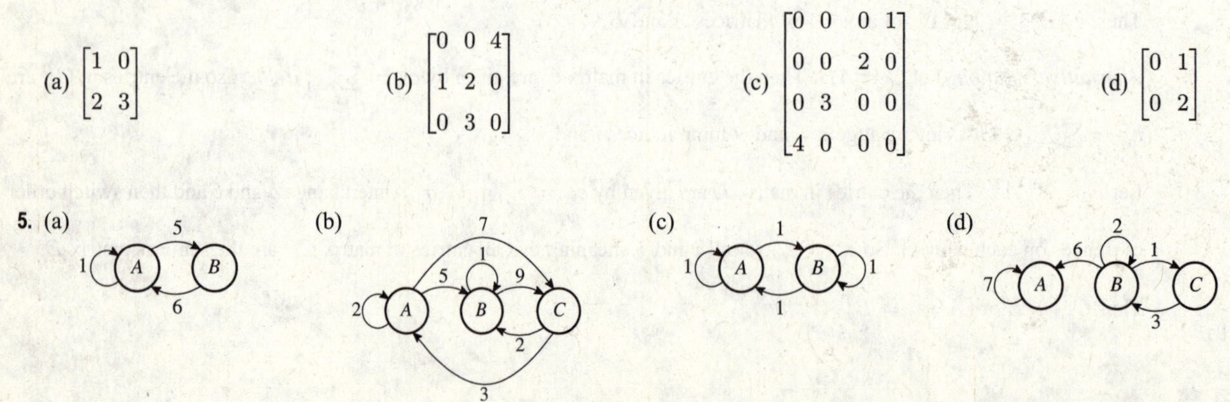

© 2016 Cengage Learning. All Rights Reserved. May not be scanned, copied or duplicated, or posted to a publicly accessible website, in whole or in part.

7. In a matrix model, the ith row of a matrix contains the per capita contributions (inputs) to variable i coming from all variables. Thus, a row of all zeros means there are no contributions to the respective variable. So if the ith variable is X, then an ith row of all zeros implies $X_{t+1} = 0$. E.g. In a matrix diagram, there would be no incoming arrows to the ith circle.

9.
0.2
0.8 A B 0.7
0.3

11.
0.002 0.35
0.998 h e l 0.9
0.65 0.1

13.
0.5
0.5 s b 0.2

15.

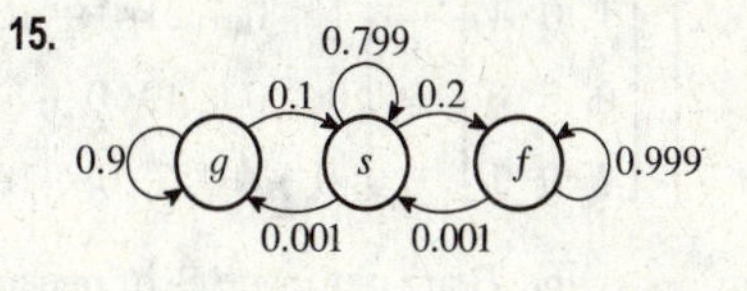

17. $\begin{Bmatrix} x_{t+1} = 2x_t + 3y_t \\ y_{t+1} = 0.9y_t \end{Bmatrix} \Leftrightarrow \begin{bmatrix} x_{t+1} \\ y_{t+1} \end{bmatrix} = \begin{bmatrix} 2 & 3 \\ 0 & 0.9 \end{bmatrix} \begin{bmatrix} x_t \\ y_t \end{bmatrix}$

19. $\begin{Bmatrix} d_{t+1} = 1d_t + \frac{1}{4}h_t \\ h_{t+1} = \frac{1}{2}h_t \\ r_{t+1} = \frac{1}{4}h_t + 1r_t \end{Bmatrix} \Leftrightarrow \begin{bmatrix} d_{t+1} \\ h_{t+1} \\ r_{t+1} \end{bmatrix} = \begin{bmatrix} 1 & \frac{1}{4} & 0 \\ 0 & \frac{1}{2} & 0 \\ 0 & \frac{1}{4} & 1 \end{bmatrix} \begin{bmatrix} d_t \\ h_t \\ r_t \end{bmatrix}$

21. Let X_t, Y_t and Z_t represent the number of individuals carrying allelle X, Y and Z respectively. In each generation: 5% mutate from X to Y and 3% mutate from X to Z, so 92% of X alleles remain unchanged; 0.1% mutate from Y to Z, so 99.9% of Y alleles remain unchanged; 90% mutate from Z to X, so 10% of Z alleles remain unchanged. These generational changes give the following system of equations and matrix model.

$$\begin{Bmatrix} X_{t+1} = 0.92X_t + 0.9Z_t \\ Y_{t+1} = 0.05X_t + 0.999Y_t \\ Z_{t+1} = 0.03X_t + 0.001Y_t + 0.1Z_t \end{Bmatrix} \Leftrightarrow \begin{bmatrix} X_{t+1} \\ Y_{t+1} \\ Z_{t+1} \end{bmatrix} = \begin{bmatrix} 0.92 & 0 & 0.9 \\ 0.05 & 0.999 & 0 \\ 0.03 & 0.001 & 0.1 \end{bmatrix} \begin{bmatrix} X_t \\ Y_t \\ Z_t \end{bmatrix}$$

8.6 The Inverse and Determinant of a Matrix

1. Matrices A and B are inverses of one another if $AB = I$ or $BA = I$.

(a) $AB = \begin{bmatrix} 1 & 5 \\ 2 & 7 \end{bmatrix} \begin{bmatrix} -\frac{7}{3} & \frac{5}{3} \\ \frac{2}{3} & -\frac{1}{3} \end{bmatrix} = \begin{bmatrix} -\frac{7}{3} + \frac{10}{3} & \frac{5}{3} - \frac{5}{3} \\ -\frac{14}{3} + \frac{14}{3} & \frac{10}{3} - \frac{7}{3} \end{bmatrix} = \begin{bmatrix} 1 & 0 \\ 0 & 1 \end{bmatrix} = I$

Therefore, matrices A and B are inverses of one another.

(b) $AB = \begin{bmatrix} 1 & 0 & 3 \\ 2 & 7 & 9 \\ 0 & 2 & 1 \end{bmatrix} \begin{bmatrix} -11 & 6 & -21 \\ -2 & 1 & -3 \\ 4 & -2 & 7 \end{bmatrix} = \begin{bmatrix} -11 + 0 + 12 & 6 + 0 - 6 & -21 + 0 + 21 \\ -22 - 14 + 36 & 12 + 7 - 18 & -42 - 21 + 63 \\ 0 - 4 + 4 & 0 + 2 - 2 & 0 - 6 + 7 \end{bmatrix} = \begin{bmatrix} 1 & 0 & 0 \\ 0 & 1 & 0 \\ 0 & 0 & 1 \end{bmatrix} = I$

Therefore, matrices A and B are inverses of one another.

© 2016 Cengage Learning. All Rights Reserved. May not be scanned, copied or duplicated, or posted to a publicly accessible website, in whole or in part.

(c) $AB = \begin{bmatrix} 0 & 1 \\ 3 & 2 \end{bmatrix} \begin{bmatrix} 2 & 3 \\ 1 & 0 \end{bmatrix} = \begin{bmatrix} 0+1 & 0+0 \\ 6+2 & 9+0 \end{bmatrix} = \begin{bmatrix} 1 & 0 \\ 8 & 9 \end{bmatrix} \neq I$

Therefore, matrices A and B are **not** inverses of one another.

(d) $AB = \begin{bmatrix} 0 & 1 \\ 1 & 1 \end{bmatrix} \begin{bmatrix} -1 & 1 \\ 1 & 0 \end{bmatrix} = \begin{bmatrix} 0+1 & 0+0 \\ -1+1 & 1+0 \end{bmatrix} = \begin{bmatrix} 1 & 0 \\ 0 & 1 \end{bmatrix} = I$

Therefore, matrices A and B are inverses of one another.

(e) $AB = \begin{bmatrix} 1 & 2 & 3 \\ 0 & 1 & 7 \\ 0 & 2 & 1 \end{bmatrix} \begin{bmatrix} 1 & 0 & 1 \\ 0 & 5 & 3 \\ 7 & 0 & 1 \end{bmatrix} = \begin{bmatrix} 1+0+21 & 0+10+0 & 1+6+3 \\ 0+0+49 & 0+5+0 & 0+3+7 \\ 0+0+7 & 0+10+0 & 0+6+1 \end{bmatrix} = \begin{bmatrix} 22 & 10 & 10 \\ 49 & 5 & 10 \\ 7 & 10 & 7 \end{bmatrix} \neq I$

Therefore, matrices A and B are **not** inverses of one another.

(f) $AB = \begin{bmatrix} 9 & 0 \\ 2 & 3 \end{bmatrix} \begin{bmatrix} \frac{1}{9} & 0 \\ -\frac{2}{27} & \frac{1}{3} \end{bmatrix} = \begin{bmatrix} 1+0 & 0+0 \\ \frac{2}{9}-\frac{2}{9} & 0+1 \end{bmatrix} = \begin{bmatrix} 1 & 0 \\ 0 & 1 \end{bmatrix} = I$

Therefore, matrices A and B are inverses of one another.

3. When we multiply D, an $n \times n$ diagonal matrix with entries d_{ii}, by G, an $n \times n$ diagonal matrix with entries $1/d_{ii}$, we get

$$DG = \begin{bmatrix} d_{11} & 0 & \cdots & 0 \\ 0 & d_{22} & \cdots & 0 \\ \vdots & \vdots & \ddots & \vdots \\ 0 & 0 & \cdots & d_{nn} \end{bmatrix} \begin{bmatrix} 1/d_{11} & 0 & \cdots & 0 \\ 0 & 1/d_{22} & \cdots & 0 \\ \vdots & \vdots & \ddots & \vdots \\ 0 & 0 & \cdots & 1/d_{nn} \end{bmatrix} = \begin{bmatrix} 1+0+\cdots+0 & 0+0+\cdots+0 & \cdots & 0+0+\cdots+0 \\ 0+0+\cdots+0 & 0+1+\cdots+0 & \cdots & 0+0+\cdots+0 \\ \vdots & \vdots & \ddots & \vdots \\ 0+0+\cdots+0 & 0+0+\cdots+0 & \cdots & 0+0+\cdots+1 \end{bmatrix}$$

$$= \begin{bmatrix} 1 & 0 & \cdots & 0 \\ 0 & 1 & \cdots & 0 \\ \vdots & \vdots & \ddots & \vdots \\ 0 & 0 & \cdots & 1 \end{bmatrix} = I$$

Thus, G is the inverse of D so $G = D^{-1}$. This means D^{-1} is an $n \times n$ diagonal matrix with entries $1/d_{ii}$.

5. Consider a general 2×2 matrix $A = \begin{bmatrix} a_{11} & a_{12} \\ a_{21} & a_{22} \end{bmatrix}$.

$$A = A^{-1} \Rightarrow \begin{bmatrix} a_{11} & a_{12} \\ a_{21} & a_{22} \end{bmatrix} = \frac{1}{\det A} \begin{bmatrix} a_{22} & -a_{12} \\ -a_{21} & a_{11} \end{bmatrix} \Rightarrow \begin{bmatrix} a_{11} & a_{12} \\ a_{21} & a_{22} \end{bmatrix} = \begin{bmatrix} a_{22} & -a_{12} \\ -a_{21} & a_{11} \end{bmatrix} \quad \text{(since } \det A = 1\text{)}$$

Equating the entries of each matrix, we find that $a_{11} = a_{22}$ (diagonal terms are the same). Also $a_{12} = -a_{12} \Rightarrow a_{12} = 0$ and $a_{21} = -a_{21} \Rightarrow a_{21} = 0$. Lastly, we require $\det A = 0 \Rightarrow a_{11}a_{22} - a_{21}a_{12} = 0 \Rightarrow a_{11}^2 = 0 \Rightarrow$ $a_{11} = \pm 1$ so $a_{22} = \pm 1$. Thus, there are only two 2×2 matrices A with $\det A = 1$ and $A = A^{-1}$ given by $\begin{bmatrix} \pm 1 & 0 \\ 0 & \pm 1 \end{bmatrix}$.

© 2016 Cengage Learning. All Rights Reserved. May not be scanned, copied or duplicated, or posted to a publicly accessible website, in whole or in part.

7. (a) $(AB)^{-1} = \left(\begin{bmatrix} a_{11} & a_{12} \\ a_{21} & a_{22} \end{bmatrix}\begin{bmatrix} b_{11} & b_{12} \\ b_{21} & b_{22} \end{bmatrix}\right)^{-1} = \begin{bmatrix} a_{11}b_{11}+a_{12}b_{21} & a_{11}b_{12}+a_{12}b_{22} \\ a_{21}b_{11}+a_{22}b_{21} & a_{21}b_{12}+a_{22}b_{22} \end{bmatrix}^{-1}$

$$= \frac{\begin{bmatrix} a_{21}b_{12}+a_{22}b_{22} & -a_{11}b_{12}-a_{12}b_{22} \\ -a_{21}b_{11}-a_{22}b_{21} & a_{11}b_{11}+a_{12}b_{21} \end{bmatrix}}{(a_{11}b_{11}+a_{12}b_{21})(a_{21}b_{12}+a_{22}b_{22})-(a_{21}b_{11}+a_{22}b_{21})(a_{11}b_{12}+a_{12}b_{22})}$$

$$= \frac{1}{a_{11}a_{22}b_{11}b_{22}+a_{12}a_{21}b_{12}b_{21}-a_{12}a_{21}b_{11}b_{22}-a_{11}a_{22}b_{12}b_{21}}\begin{bmatrix} a_{21}b_{12}+a_{22}b_{22} & -a_{11}b_{12}-a_{12}b_{22} \\ -a_{21}b_{11}-a_{22}b_{21} & a_{11}b_{11}+a_{12}b_{21} \end{bmatrix}$$

$$B^{-1}A^{-1} = \begin{bmatrix} b_{11} & b_{12} \\ b_{21} & b_{22} \end{bmatrix}^{-1}\begin{bmatrix} a_{11} & a_{12} \\ a_{21} & a_{22} \end{bmatrix}^{-1} = \left(\frac{1}{b_{11}b_{22}-b_{12}b_{21}}\begin{bmatrix} b_{22} & -b_{12} \\ -b_{21} & b_{11} \end{bmatrix}\right)\left(\frac{1}{a_{11}a_{22}-a_{12}a_{21}}\begin{bmatrix} a_{22} & -a_{12} \\ -a_{21} & a_{11} \end{bmatrix}\right)$$

$$= \frac{1}{(a_{11}a_{22}-a_{12}a_{21})(b_{11}b_{22}-b_{12}b_{21})}\begin{bmatrix} b_{22} & -b_{12} \\ -b_{21} & b_{11} \end{bmatrix}\begin{bmatrix} a_{22} & -a_{12} \\ -a_{21} & a_{11} \end{bmatrix}$$

$$= \frac{1}{(a_{11}a_{22}-a_{12}a_{21})(b_{11}b_{22}-b_{12}b_{21})}\begin{bmatrix} a_{22}b_{22}+a_{21}b_{12} & -a_{12}b_{22}-a_{11}b_{12} \\ -a_{22}b_{21}-a_{21}b_{11} & a_{12}b_{21}+a_{11}b_{11} \end{bmatrix} = (AB)^{-1}$$

(b) If AB is nonsingular, the definition of an inverse matrix allows us to write $(AB)^{-1}AB = I \Rightarrow$

$$(AB)^{-1}ABB^{-1} = IB^{-1} \Rightarrow (AB)^{-1}A = B^{-1} \Rightarrow (AB)^{-1}AA^{-1} = B^{-1}A^{-1} \Rightarrow (AB)^{-1} = B^{-1}A^{-1}$$

9. (a) $(A^T)^{-1} = \left(\begin{bmatrix} a_{11} & a_{12} \\ a_{21} & a_{22} \end{bmatrix}^T\right)^{-1} = \begin{bmatrix} a_{11} & a_{21} \\ a_{12} & a_{22} \end{bmatrix}^{-1} = \frac{1}{a_{11}a_{22}-a_{12}a_{21}}\begin{bmatrix} a_{22} & -a_{21} \\ -a_{12} & a_{11} \end{bmatrix}$

$$(A^{-1})^T = \left(\begin{bmatrix} a_{11} & a_{12} \\ a_{21} & a_{22} \end{bmatrix}^{-1}\right)^T = \left(\frac{1}{a_{11}a_{22}-a_{12}a_{21}}\begin{bmatrix} a_{22} & -a_{12} \\ -a_{21} & a_{11} \end{bmatrix}\right)^T = \frac{1}{a_{11}a_{22}-a_{12}a_{21}}\begin{bmatrix} a_{22} & -a_{21} \\ -a_{12} & a_{11} \end{bmatrix}$$

Thus, $(A^T)^{-1} = (A^{-1})^T$ when A is a 2×2 matrix.

(b) If A is nonsingular, the definition of an inverse matrix allows us to write

$$AA^{-1} = I \Rightarrow (AA^{-1})^T = I^T \Rightarrow (A^{-1})^T A^T = I \quad \left(\text{since } (AB)^T = B^T A^T\right)$$

So $(A^{-1})^T$ is the inverse of A^T, that is $(A^T)^{-1} = (A^{-1})^T$.

11. $\det\begin{bmatrix} a & 4 \\ 2 & 8 \end{bmatrix} = (a)(8) - (2)(4) = 8a - 8 = 8(a-1)$

The determinant is zero when $a = 1$, so the matrix is singular when $a = 1$.

13. $\det\begin{bmatrix} 1 & 1 & 1 \\ 1 & 2 & a \\ 1 & 4 & a^2 \end{bmatrix} = (1)(2)(a^2) + (1)(a)(1) + (1)(1)(4) - (1)(2)(1) - (1)(1)(a^2) - (1)(a)(4) = a^2 - 3a + 2$

The determinant is zero when $a^2 - 3a + 2 = 0 \Rightarrow (a-1)(a-2) = 0$, so the matrix is singular when $a = 1$ and $a = 2$.

15. The theorem from Exercise 14 states $\det(AB) = \det A \det B$ so $\det A^2 = \det(AA) = (\det A)^2$, $\det A^3 = \det(AA^2) = (\det A)^3$, and in general $\det A^k = (\det A)^k$ where k is a postive integer.

$$A \text{ is nonsingular} \Leftrightarrow \det A \neq 0 \Rightarrow \det A^k = (\det A)^k \neq 0 \Rightarrow A^k \text{ is nonsingular}$$

© 2016 Cengage Learning. All Rights Reserved. May not be scanned, copied or duplicated, or posted to a publicly accessible website, in whole or in part.

17. $\det D = \det \begin{bmatrix} d_{11} & 0 \\ 0 & d_{22} \end{bmatrix} = (d_{11})(d_{22}) - (0)(0) = d_{11}d_{22}$

19. $\left\{ \begin{matrix} 3x_1 + x_2 = 2 \\ kx_1 + 2x_2 = 4 \end{matrix} \right\} \Rightarrow \underbrace{\begin{bmatrix} 3 & 1 \\ k & 2 \end{bmatrix}}_{A} \underbrace{\begin{bmatrix} x_1 \\ x_2 \end{bmatrix}}_{\mathbf{x}} = \underbrace{\begin{bmatrix} 2 \\ 4 \end{bmatrix}}_{\mathbf{b}} \Rightarrow \det A = (3)(2) - (k)(1) = 6 - k \Rightarrow \det A \neq 0$ when $k \neq 6$

Therefore, there is a unique solution when $k \neq 6$. When $k = 6$, the second equation is identical to the first (within a multiple of 2), that is they represent the same straight line. So there are an infinite number of solutions when $k = 6$.

21. $\left\{ \begin{matrix} a_{11}x_1 + a_{12}x_2 = b_1 \\ a_{21}x_1 + a_{22}x_2 = b_2 \end{matrix} \right\} \Rightarrow \underbrace{\begin{bmatrix} a_{11} & a_{12} \\ a_{21} & a_{22} \end{bmatrix}}_{A} \underbrace{\begin{bmatrix} x_1 \\ x_2 \end{bmatrix}}_{\mathbf{x}} = \underbrace{\begin{bmatrix} b_1 \\ b_2 \end{bmatrix}}_{\mathbf{b}} \Rightarrow$

$$\begin{bmatrix} x_1 \\ x_2 \end{bmatrix} = \begin{bmatrix} a_{11} & a_{12} \\ a_{21} & a_{22} \end{bmatrix}^{-1} \begin{bmatrix} b_1 \\ b_2 \end{bmatrix} = \frac{1}{a_{11}a_{22} - a_{12}a_{21}} \begin{bmatrix} a_{22} & -a_{12} \\ -a_{21} & a_{11} \end{bmatrix} \begin{bmatrix} b_1 \\ b_2 \end{bmatrix}$$

Carrying out the matrix multiplication gives the solution $x_1 = \dfrac{a_{22}b_1 - a_{12}b_2}{a_{11}a_{22} - a_{12}a_{21}}$ and $x_2 = \dfrac{-a_{21}b_1 + a_{11}b_2}{a_{11}a_{22} - a_{12}a_{21}}$.

23. If $\mathbf{x} = \mathbf{p}$ and $\mathbf{x} = \mathbf{q}$ are both solutions to the homogeneous system of equations $A\mathbf{x} = \mathbf{0}$, then $A\mathbf{p} = \mathbf{0}$ and $A\mathbf{q} = \mathbf{0}$. So if $\mathbf{z} = \alpha\mathbf{p} + \beta\mathbf{q}$, then

$$A\mathbf{z} = A(\alpha\mathbf{p} + \beta\mathbf{q}) = \alpha A\mathbf{p} + \beta A\mathbf{q} = \alpha\mathbf{0} + \beta\mathbf{0} = \mathbf{0}$$

Therefore $A\mathbf{z} = \mathbf{0}$, so $\mathbf{z}$ satisfies the system of equations and is also a solution.

25. $\left\{ \begin{matrix} x_1 - 3x_2 = 5 \\ x_1 + x_2 = 2 \end{matrix} \right\} \Rightarrow \underbrace{\begin{bmatrix} 1 & -3 \\ 1 & 1 \end{bmatrix}}_{A} \underbrace{\begin{bmatrix} x_1 \\ x_2 \end{bmatrix}}_{\mathbf{x}} = \underbrace{\begin{bmatrix} 5 \\ 2 \end{bmatrix}}_{\mathbf{b}} \Rightarrow \det A = (1)(1) - (1)(-3) = 4 \neq 0 \Rightarrow A$ is nonsingular

$$\begin{bmatrix} x_1 \\ x_2 \end{bmatrix} = \begin{bmatrix} 1 & -3 \\ 1 & 1 \end{bmatrix}^{-1} \begin{bmatrix} 5 \\ 2 \end{bmatrix} = \frac{1}{4} \begin{bmatrix} 1 & 3 \\ -1 & 1 \end{bmatrix} \begin{bmatrix} 5 \\ 2 \end{bmatrix} = \frac{1}{4} \begin{bmatrix} 5+6 \\ -5+2 \end{bmatrix} = \begin{bmatrix} 11/4 \\ -3/4 \end{bmatrix}$$

Therefore, the solution is $x_1 = \frac{11}{4}$ and $x_2 = -\frac{3}{4}$.

27. $\left\{ \begin{matrix} 3x_1 + 6x_2 = 2 \\ 9x_1 + 12x_2 = 1 \end{matrix} \right\} \Rightarrow \underbrace{\begin{bmatrix} 3 & 6 \\ 9 & 12 \end{bmatrix}}_{A} \underbrace{\begin{bmatrix} x_1 \\ x_2 \end{bmatrix}}_{\mathbf{x}} = \underbrace{\begin{bmatrix} 2 \\ 1 \end{bmatrix}}_{\mathbf{b}} \Rightarrow \det A = (3)(12) - (9)(6) = -18 \neq 0 \Rightarrow A$ is nonsingular

$$\begin{bmatrix} x_1 \\ x_2 \end{bmatrix} = \begin{bmatrix} 3 & 6 \\ 9 & 12 \end{bmatrix}^{-1} \begin{bmatrix} 2 \\ 1 \end{bmatrix} = \frac{1}{-18} \begin{bmatrix} 12 & -6 \\ -9 & 3 \end{bmatrix} \begin{bmatrix} 2 \\ 1 \end{bmatrix} = \frac{1}{-18} \begin{bmatrix} 24 + (-6) \\ -18 + 3 \end{bmatrix} = \begin{bmatrix} -1 \\ 5/6 \end{bmatrix}$$

Therefore, the solution is $x_1 = -1$ and $x_2 = \frac{5}{6}$.

29. $\left\{ \begin{matrix} 8x_1 + 4x_2 = 4 \\ 4x_1 + 2x_2 = 2 \end{matrix} \right\}$ The first equation is identical to the second equation (within a multiple of 2), that is they represent the same straight line. Therefore, there an infinite number of solutions, each of which must satisfy $x_2 = -2x_1 + 1$.

© 2016 Cengage Learning. All Rights Reserved. May not be scanned, copied or duplicated, or posted to a publicly accessible website, in whole or in part.

31. Let the vector in the current season be $\mathbf{x}_{t+1} = \begin{bmatrix} 8 \\ 1.8 \end{bmatrix}$ so $\mathbf{x}_t$ is the vector in the previous season. Using the properties of inverse matrices, we have $\mathbf{x}_{t+1} = \begin{bmatrix} 2 & 3 \\ 0 & 0.9 \end{bmatrix} \mathbf{x}_t \Leftrightarrow \mathbf{x}_t = \begin{bmatrix} 2 & 3 \\ 0 & 0.9 \end{bmatrix}^{-1} \mathbf{x}_{t+1} = \dfrac{1}{1.8 - 0} \begin{bmatrix} 0.9 & -3 \\ 0 & 2 \end{bmatrix} \mathbf{x}_{t+1} \Rightarrow$

$$\mathbf{x}_t = \begin{bmatrix} 0.5 & -5/3 \\ 0 & 10/9 \end{bmatrix} \begin{bmatrix} 8 \\ 1.8 \end{bmatrix} = \begin{bmatrix} 4 + (-3) \\ 0 + 2 \end{bmatrix} = \begin{bmatrix} 1 \\ 2 \end{bmatrix}.$$

33. $\hat{\mathbf{n}} = \begin{bmatrix} b & 2 \\ \frac{1}{2} & 0 \end{bmatrix} \hat{\mathbf{n}} \Rightarrow \hat{\mathbf{n}} - \begin{bmatrix} b & 2 \\ \frac{1}{2} & 0 \end{bmatrix} \hat{\mathbf{n}} = \mathbf{0} \Rightarrow \begin{bmatrix} 1 & 0 \\ 0 & 1 \end{bmatrix} \hat{\mathbf{n}} - \begin{bmatrix} b & 2 \\ \frac{1}{2} & 0 \end{bmatrix} \hat{\mathbf{n}} = \mathbf{0} \Rightarrow \left(\begin{bmatrix} 1 & 0 \\ 0 & 1 \end{bmatrix} - \begin{bmatrix} b & 2 \\ \frac{1}{2} & 0 \end{bmatrix} \right) \hat{\mathbf{n}} = \mathbf{0} \Rightarrow$

$\underbrace{\begin{bmatrix} 1 - b & -2 \\ -\frac{1}{2} & 1 \end{bmatrix}}_{A} \hat{\mathbf{n}} = \begin{bmatrix} 0 \\ 0 \end{bmatrix}$ So finding $\hat{\mathbf{n}}$ requires solving this homogeneous system. The matrix A will be invertible when $\det A \neq 0 \Rightarrow (1 - b)(1) - \left(-\frac{1}{2}\right)(-2) \neq 0 \Rightarrow b \neq 0.$

(a) When $b \neq 0$, A is invertible so by Theorem 5 the homogeneous system has a unique solution given by the trivial solution $\hat{\mathbf{n}} = \begin{bmatrix} 0 \\ 0 \end{bmatrix}$.

(b) When $b = 0$, A is singular so by Theorem 5 the homogeneous system has infinitely many nontrivial solutions which must satisfy $n_1 = 2n_2$ where n_1 and n_2 are the first two components of $\hat{\mathbf{n}}$.

35. $\hat{\mathbf{v}} = \begin{bmatrix} 1 & 0 \\ 0 & -1 \end{bmatrix} \hat{\mathbf{v}} \Rightarrow \hat{\mathbf{v}} - \begin{bmatrix} 1 & 0 \\ 0 & -1 \end{bmatrix} \hat{\mathbf{v}} = \mathbf{0} \Rightarrow \begin{bmatrix} 1 & 0 \\ 0 & 1 \end{bmatrix} \hat{\mathbf{v}} - \begin{bmatrix} 1 & 0 \\ 0 & -1 \end{bmatrix} \hat{\mathbf{v}} = \mathbf{0} \Rightarrow \left(\begin{bmatrix} 1 & 0 \\ 0 & 1 \end{bmatrix} - \begin{bmatrix} 1 & 0 \\ 0 & -1 \end{bmatrix} \right) \hat{\mathbf{v}} = \mathbf{0} \Rightarrow$

$\underbrace{\begin{bmatrix} 0 & 0 \\ 0 & 2 \end{bmatrix}}_{A} \hat{\mathbf{v}} = \begin{bmatrix} 0 \\ 0 \end{bmatrix} \Rightarrow \det A = 0 - 0 = 0 \Rightarrow A$ is singular. Therefore, by Theorem 5 the homogeneous system has infinitely many nontrivial solutions which must satisfy $\hat{\mathbf{v}} = \begin{bmatrix} k \\ 0 \end{bmatrix}$ where k is a real number.

8.7 Eigenvalues and Eigenvectors

1. Let $\mathbf{u} = [0, 2]$ and $\mathbf{v} = [1, 0]$ be the two vectors that make up the letter L.

(a) $A\mathbf{u} = \begin{bmatrix} 1 & 0 \\ 0 & -1 \end{bmatrix} \begin{bmatrix} 0 \\ 2 \end{bmatrix} = \begin{bmatrix} 0 \\ -2 \end{bmatrix}$ $\quad A\mathbf{v} = \begin{bmatrix} 1 & 0 \\ 0 & -1 \end{bmatrix} \begin{bmatrix} 1 \\ 0 \end{bmatrix} = \begin{bmatrix} 1 \\ 0 \end{bmatrix}$

The x components of each vector remain the same but the y components change sign. Therefore, the matrix causes a reflection about the x-axis.

(b) $A\mathbf{u} = \begin{bmatrix} -1 & 0 \\ 0 & 1 \end{bmatrix} \begin{bmatrix} 0 \\ 2 \end{bmatrix} = \begin{bmatrix} 0 \\ 2 \end{bmatrix}$ $\quad A\mathbf{v} = \begin{bmatrix} -1 & 0 \\ 0 & 1 \end{bmatrix} \begin{bmatrix} 1 \\ 0 \end{bmatrix} = \begin{bmatrix} -1 \\ 0 \end{bmatrix}$

The y components of each vector remain the same but the x components change sign. Therefore, the matrix causes a reflection about the y-axis.

© 2016 Cengage Learning. All Rights Reserved. May not be scanned, copied or duplicated, or posted to a publicly accessible website, in whole or in part.

(c) $A\mathbf{u} = \begin{bmatrix} -1 & 0 \\ 0 & -1 \end{bmatrix} \begin{bmatrix} 0 \\ 2 \end{bmatrix} = \begin{bmatrix} 0 \\ -2 \end{bmatrix}$ $\quad A\mathbf{v} = \begin{bmatrix} -1 & 0 \\ 0 & -1 \end{bmatrix} \begin{bmatrix} 1 \\ 0 \end{bmatrix} = \begin{bmatrix} -1 \\ 0 \end{bmatrix}$

The x and y components of each vector change sign. Therefore, the matrix causes a reflection about the x-axis and the y-axis.

(d) $A\mathbf{u} = \begin{bmatrix} 0 & -1 \\ -1 & 0 \end{bmatrix} \begin{bmatrix} 0 \\ 2 \end{bmatrix} = \begin{bmatrix} -2 \\ 0 \end{bmatrix}$ $\quad A\mathbf{v} = \begin{bmatrix} 0 & -1 \\ -1 & 0 \end{bmatrix} \begin{bmatrix} 1 \\ 0 \end{bmatrix} = \begin{bmatrix} 0 \\ -1 \end{bmatrix}$

The new vectors are the same as those from part (c) with the x and y values swapped. Therefore, the matrix causes a reflection about the x-axis and the y-axis and a reflection about the line $y = x$ (swaps x and y values).

3. Let $\mathbf{u} = [0, 2]$ and $\mathbf{v} = [1, 0]$ be the two vectors that make up the letter L.

(a) $A\mathbf{u} = \begin{bmatrix} 1 & 1 \\ 0 & 1 \end{bmatrix} \begin{bmatrix} 0 \\ 2 \end{bmatrix} = \begin{bmatrix} 2 \\ 2 \end{bmatrix}$ $\quad A\mathbf{v} = \begin{bmatrix} 1 & 1 \\ 0 & 1 \end{bmatrix} \begin{bmatrix} 1 \\ 0 \end{bmatrix} = \begin{bmatrix} 1 \\ 0 \end{bmatrix}$

The x components of each vector increase by the size of the y components while the y components remain the same. This causes the vertical segment of the L to rotate clockwise and increase in length.

(b) $A\mathbf{u} = \begin{bmatrix} 1 & 0 \\ -1 & 1 \end{bmatrix} \begin{bmatrix} 0 \\ 2 \end{bmatrix} = \begin{bmatrix} 0 \\ 2 \end{bmatrix}$ $\quad A\mathbf{v} = \begin{bmatrix} 1 & 0 \\ -1 & 1 \end{bmatrix} \begin{bmatrix} 1 \\ 0 \end{bmatrix} = \begin{bmatrix} 1 \\ -1 \end{bmatrix}$

The x components of each vector remain the same and the y components decrease by the size of the x components. This causes the horizontal segment of the L to rotate clockwise and increase in length.

(c) $A\mathbf{u} = \begin{bmatrix} 1 & -2 \\ 0 & 1 \end{bmatrix} \begin{bmatrix} 0 \\ 2 \end{bmatrix} = \begin{bmatrix} -4 \\ 2 \end{bmatrix}$ $\quad A\mathbf{v} = \begin{bmatrix} 1 & -2 \\ 0 & 1 \end{bmatrix} \begin{bmatrix} 1 \\ 0 \end{bmatrix} = \begin{bmatrix} 1 \\ 0 \end{bmatrix}$

The x components of each vector decrease by 2 times the size of the y components and the y components remain the same. This causes the vertical segment of the L to rotate counterclockwise and increase in length.

(d) $A\mathbf{u} = \begin{bmatrix} 1 & 0 \\ 1 & 1 \end{bmatrix} \begin{bmatrix} 0 \\ 2 \end{bmatrix} = \begin{bmatrix} 0 \\ 2 \end{bmatrix}$ $\quad A\mathbf{v} = \begin{bmatrix} 1 & 0 \\ 1 & 1 \end{bmatrix} \begin{bmatrix} 1 \\ 0 \end{bmatrix} = \begin{bmatrix} 1 \\ 1 \end{bmatrix}$

The x components of each vector remain the same and the y components increase by the size of the x components. This causes the horizontal segment of the L to rotate counterclockwise and increase in length.

5. (a) All points on the unit square are reflected about the y-axis, that is, their x values switch sign. This can be achieved using the matrix $\begin{bmatrix} -1 & 0 \\ 0 & 1 \end{bmatrix}$.

(b) The unit square is stretched horizontally by a factor 2, that is, all points have their x value multiplied by 2. This can be achieved using the matrix $\begin{bmatrix} 2 & 0 \\ 0 & 1 \end{bmatrix}$.

(c) This is an example of a shear (see exercise 8.7.3), that is, all points in the unit square have their x values increased by an amount proportional to their y values. Observe that the point $(1, 1)$ in the unit square is transformed to the new point $(1.5, 1)$, so the x values increase by 1.5 times the y values. Therefore, the matrix is $\begin{bmatrix} 1 & 1.5 \\ 0 & 1 \end{bmatrix}$.

© 2016 Cengage Learning. All Rights Reserved. May not be scanned, copied or duplicated, or posted to a publicly accessible website, in whole or in part.

(d) This is an example of a counterclockwise rotation by 45°. The can achieved using the rotation matrix (see exercise 8.3.6)

$$\begin{bmatrix} \cos 45^\circ & -\sin 45^\circ \\ \sin 45^\circ & \cos 45^\circ \end{bmatrix} = \begin{bmatrix} 1/\sqrt{2} & -1/\sqrt{2} \\ 1/\sqrt{2} & 1/\sqrt{2} \end{bmatrix}.$$

When multiplied by this matrix, the vector $[1, 1]$ becomes $\begin{bmatrix} 1/\sqrt{2} & -1/\sqrt{2} \\ 1/\sqrt{2} & 1/\sqrt{2} \end{bmatrix} \begin{bmatrix} 1 \\ 1 \end{bmatrix} = \begin{bmatrix} 0 \\ \sqrt{2} \end{bmatrix}$ which agrees with the y-intercept in the image.

7. Denote the voltage vector of a healthy heart by $\mathbf{v} = [0.3, -0.2]$.

(a) A left anterior hemiblock reflects the heart vector about the x-axis, that is, the y-value switches sign. This can be achieved using the matrix $A = \begin{bmatrix} 1 & 0 \\ 0 & -1 \end{bmatrix}$. Multiplying this matrix by the healthy heart vector gives

$$A\mathbf{v} = \begin{bmatrix} 1 & 0 \\ 0 & -1 \end{bmatrix} \begin{bmatrix} 0.3 \\ -0.2 \end{bmatrix} = \begin{bmatrix} 0.3 \\ 0.2 \end{bmatrix} = \mathbf{h}$$

(b) A left posterior hemiblock reflects the heart vector about the y-axis, that is, the x-value switches sign. This can be achieved using the matrix $A = \begin{bmatrix} -1 & 0 \\ 0 & 1 \end{bmatrix}$. Multiplying this matrix by the healthy heart vector gives

$$A\mathbf{v} = \begin{bmatrix} -1 & 0 \\ 0 & 1 \end{bmatrix} \begin{bmatrix} 0.3 \\ -0.2 \end{bmatrix} = \begin{bmatrix} -0.3 \\ -0.2 \end{bmatrix} = \mathbf{h}$$

(c) An apical ischemia abnormality reflects the heart vector about the x and y axes, that is, the x and y values both switch sign. This can be achieved using the matrix $A = \begin{bmatrix} -1 & 0 \\ 0 & -1 \end{bmatrix}$. Multiplying this matrix by the healthy heart vector gives

$$A\mathbf{v} = \begin{bmatrix} -1 & 0 \\ 0 & -1 \end{bmatrix} \begin{bmatrix} 0.3 \\ -0.2 \end{bmatrix} = \begin{bmatrix} -0.3 \\ 0.2 \end{bmatrix} = \mathbf{h}$$

(d) We want a matrix A that, after multiplication by the healthy heart vector $\mathbf{v}$, gives the chronic obstructive pulmonary disease heart vector $\mathbf{h}$. Thus, we require

$$A\mathbf{v} = \mathbf{h} \quad \Rightarrow \quad \begin{bmatrix} a_{11} & a_{12} \\ a_{21} & a_{22} \end{bmatrix} \begin{bmatrix} 0.3 \\ -0.2 \end{bmatrix} = \begin{bmatrix} 0.1 \\ -0.0667 \end{bmatrix} \quad \Rightarrow \quad \begin{bmatrix} 0.3a_{11} - 0.2a_{12} \\ 0.3a_{21} - 0.2a_{22} \end{bmatrix} = \begin{bmatrix} 0.1 \\ -0.0667 \end{bmatrix} \quad \Rightarrow$$

$\left\{ \begin{array}{l} 0.3a_{11} - 0.2a_{12} = 0.1 \\ 0.3a_{21} - 0.2a_{22} = -0.0667 \end{array} \right\}$ This set of two equations has four unknowns, so we are free to choose values for two unknowns, say a_{12} and a_{21}, and then solve for the remaining two. One possible choice is $a_{11} = a_{12} = 1$, $a_{21} = 0$ and $a_{22} = 0.3335$. This gives the matrix $A = \begin{bmatrix} 1 & 1 \\ 0 & 0.3335 \end{bmatrix}$.

© 2016 Cengage Learning. All Rights Reserved. May not be scanned, copied or duplicated, or posted to a publicly accessible website, in whole or in part.

9. The matrix A has an eigenvalue k if $\det(A - kI) = 0$.

(a) $\det(A - 3I) = \det\begin{bmatrix} 1-3 & 2 \\ 2 & 1-3 \end{bmatrix} = \det\begin{bmatrix} -2 & 2 \\ 2 & -2 \end{bmatrix} = (-2)(-2) - (2)(2) = 0$

Therefore, $k = 3$ is an eigenvalue of the matrix A.

(b) $\det(A - 0I) = \det(A) = \det\begin{bmatrix} 0 & 2 & 1 \\ 2 & 1 & 0 \\ 0 & 2 & 1 \end{bmatrix}$

$= (0)(1)(1) + (2)(0)(0) + (1)(2)(2) - (1)(1)(0) - (2)(2)(1) - (0)(0)(2) = 0$

Therefore, $k = 0$ is an eigenvalue of the matrix A.

(c) $\det(A - 2I) = \det\begin{bmatrix} 5-2 & 2 \\ 0 & 1-2 \end{bmatrix} = \det\begin{bmatrix} 3 & 2 \\ 0 & -1 \end{bmatrix} = (3)(-1) - (0)(2) = -3 \neq 0$

Therefore, $k = 2$ is *not* an eigenvalue of the matrix A.

(d) $\det(A - (1-i)I) = \det\begin{bmatrix} 1-(1-i) & -1 \\ 1 & 1-(1-i) \end{bmatrix} = \det\begin{bmatrix} i & -1 \\ 1 & i \end{bmatrix} = (i)(i) - (1)(-1) = -1 + 1 = 0$

Therefore, $k = 1 - i$ is an eigenvalue of the matrix A.

(e) $\det(A - 0I) = \det A = \det\begin{bmatrix} 1 & 2 & 1 \\ 0 & 2 & 0 \\ 2 & 1 & 0 \end{bmatrix} = (1)(2)(0) + (2)(0)(2) + (1)(0)(1) - (1)(2)(2) - (2)(0)(0) - (1)(0)(1)$

$= -4 \neq 0$

Therefore, $k = 0$ is *not* an eigenvalue of the matrix A.

(f) $\det(A - (1+\sqrt{a})I) = \det\begin{bmatrix} 1-(1+\sqrt{a}) & a \\ 1 & 1-(1+\sqrt{a}) \end{bmatrix} = \det\begin{bmatrix} -\sqrt{a} & a \\ 1 & -\sqrt{a} \end{bmatrix}$

$= (-\sqrt{a})(-\sqrt{a}) - (1)(a) = a - a = 0$

Therefore, $k = 1 + \sqrt{a}$ is an eigenvalue of the matrix A.

11. The eigenvalues λ of a matrix A must satisfy $\det(A - \lambda I) = 0$.

(a) $\det\left(\begin{bmatrix} 2 & 0 \\ 3 & 0 \end{bmatrix} - \lambda I\right) = \det\begin{bmatrix} 2-\lambda & 0 \\ 3 & -\lambda \end{bmatrix} = 0 \Rightarrow (2-\lambda)(-\lambda) - (0)(3) = 0 \Rightarrow \lambda(\lambda - 2) = 0$

The eigenvalues are therefore $\lambda = 0$ and $\lambda = 2$.

(b) $\det\left(\begin{bmatrix} 5 & -4 \\ 6 & -5 \end{bmatrix} - \lambda I\right) = \det\begin{bmatrix} 5-\lambda & -4 \\ 6 & -5-\lambda \end{bmatrix} = 0 \Rightarrow (5-\lambda)(-5-\lambda) - (6)(-4) = 0 \Rightarrow \lambda^2 - 1 = 0$

The eigenvalues are therefore $\lambda = 1$ and $\lambda = -1$.

© 2016 Cengage Learning. All Rights Reserved. May not be scanned, copied or duplicated, or posted to a publicly accessible website, in whole or in part.

(c) $\det\left(\begin{bmatrix} 3 & -1 \\ 0 & 2 \end{bmatrix} - \lambda I\right) = \det\begin{bmatrix} 3-\lambda & -1 \\ 0 & 2-\lambda \end{bmatrix} = 0 \Rightarrow (3-\lambda)(2-\lambda) = 0$

The eigenvalues are therefore $\lambda = 2$ and $\lambda = 3$.

(d) $\det\left(\begin{bmatrix} 0 & 2 \\ -\frac{1}{2} & 0 \end{bmatrix} - \lambda I\right) = \det\begin{bmatrix} 0-\lambda & 2 \\ -\frac{1}{2} & 0-\lambda \end{bmatrix} = 0 \Rightarrow (-\lambda)(-\lambda) - \left(-\frac{1}{2}\right)(2) = \lambda^2 + 1 = 0 \Rightarrow \lambda = \pm\sqrt{-1}$

The eigenvalues are therefore $\lambda = i$ and $\lambda = -i$.

(e) $\det\left(\begin{bmatrix} 6 & -4 & -4 \\ 0 & 0 & 0 \\ 6 & -4 & -4 \end{bmatrix} - \lambda I\right) = \det\begin{bmatrix} 6-\lambda & -4 & -4 \\ 0 & 0-\lambda & 0 \\ 6 & -4 & -4-\lambda \end{bmatrix} = 0 \Rightarrow$

$(6-\lambda)(-\lambda)(-4-\lambda) + 0 + 0 - (-4)(-\lambda)(6) - 0 - 0 = -\lambda\left(\left(-24 - 6\lambda + 4\lambda + \lambda^2\right) + 24\right) = 0 \Rightarrow$

$\lambda^2(\lambda - 2) = 0$

The eigenvalues are therefore $\lambda = 0$ and $\lambda = 2$.

(f) $\det\left(\begin{bmatrix} -1 & 4 & 2 \\ 0 & 3 & 0 \\ -3 & 4 & 4 \end{bmatrix} - \lambda I\right) = \det\begin{bmatrix} -1-\lambda & 4 & 2 \\ 0 & 3-\lambda & 0 \\ -3 & 4 & 4-\lambda \end{bmatrix} = 0 \Rightarrow$

$(-1-\lambda)(3-\lambda)(4-\lambda) + 0 + 0 - (2)(3-\lambda)(-3) - 0 - 0 = (3-\lambda)\left[(-1-\lambda)(4-\lambda) + 6\right] = 0 \Rightarrow$

$(3-\lambda)\left[-4 + \lambda - 4\lambda + \lambda^2 + 6\right] = (3-\lambda)\left[\lambda^2 - 3\lambda + 2\right] = (3-\lambda)(\lambda - 1)(\lambda - 2) = 0$

The eigenvalues are therefore $\lambda = 1$, $\lambda = 2$ and $\lambda = 3$.

13. The eigenvalues λ of the matrix A must satisfy $\det(A - \lambda I) = 0$ and the eigenvectors $\mathbf{v}$ must satisfy $(A - \lambda I)\mathbf{v} = \mathbf{0}$.

(a) **Eigenvalues:** $\det\begin{bmatrix} 1-\lambda & 0 \\ 0 & -1-\lambda \end{bmatrix} = (1-\lambda)(-1-\lambda) = -(1-\lambda)(1+\lambda) = 0$

The eigenvalues are therefore $\lambda = 1$ and $\lambda = -1$.

Eigenvectors: Starting with the eigenvalue $\lambda = 1$ we have

$$\begin{bmatrix} 1-\lambda & 0 \\ 0 & -1-\lambda \end{bmatrix}\begin{bmatrix} v_1 \\ v_2 \end{bmatrix} = \begin{bmatrix} 0 & 0 \\ 0 & -2 \end{bmatrix}\begin{bmatrix} v_1 \\ v_2 \end{bmatrix} = \begin{bmatrix} 0 \\ 0 \end{bmatrix} \Rightarrow \left\{\begin{matrix} 0 = 0 \\ -2v_2 = 0 \end{matrix}\right\}$$

The two equations are satisfied when $v_2 = 0$ and v_1 can be any value. Choosing $v_1 = 1$ gives the eigenvector $\mathbf{v} = \begin{bmatrix} 1 \\ 0 \end{bmatrix}$.

For the second eigenvalue $\lambda = -1$ we have

$$\begin{bmatrix} 1-\lambda & 0 \\ 0 & -1-\lambda \end{bmatrix}\begin{bmatrix} v_1 \\ v_2 \end{bmatrix} = \begin{bmatrix} 2 & 0 \\ 0 & 0 \end{bmatrix}\begin{bmatrix} v_1 \\ v_2 \end{bmatrix} = \begin{bmatrix} 0 \\ 0 \end{bmatrix} \Rightarrow \left\{\begin{matrix} 2v_1 = 0 \\ 0 = 0 \end{matrix}\right\}$$

The two equations are satisfied when $v_1 = 0$ and v_2 can be any value. Choosing $v_2 = 1$ gives the eigenvector $\mathbf{v} = \begin{bmatrix} 0 \\ 1 \end{bmatrix}$.

© 2016 Cengage Learning. All Rights Reserved. May not be scanned, copied or duplicated, or posted to a publicly accessible website, in whole or in part.

(b) **Eigenvalues:** $\det\begin{bmatrix}1-\lambda & 2\\ 2 & 1-\lambda\end{bmatrix} = (1-\lambda)^2 - 4 = 0 \;\Rightarrow\; (1-\lambda) = \pm 2 \;\Rightarrow\; \lambda = 1 \pm 2$

The eigenvalues are therefore $\lambda = 3$ and $\lambda = -1$.

Eigenvectors: Starting with the eigenvalue $\lambda = 3$ we have

$$\begin{bmatrix}1-\lambda & 2\\ 2 & 1-\lambda\end{bmatrix}\begin{bmatrix}v_1\\ v_2\end{bmatrix} = \begin{bmatrix}-2 & 2\\ 2 & -2\end{bmatrix}\begin{bmatrix}v_1\\ v_2\end{bmatrix} = \begin{bmatrix}0\\ 0\end{bmatrix} \;\Rightarrow\; \left\{\begin{matrix}-2v_1 + 2v_2 = 0\\ 2v_1 - 2v_2 = 0\end{matrix}\right\}$$

Both equations specify that $v_2 = v_1$. Choosing $v_1 = 1$ gives the eigenvector $\mathbf{v} = \begin{bmatrix}1\\ 1\end{bmatrix}$.

For the second eigenvalue $\lambda = -1$ we have

$$\begin{bmatrix}1-\lambda & 2\\ 2 & 1-\lambda\end{bmatrix}\begin{bmatrix}v_1\\ v_2\end{bmatrix} = \begin{bmatrix}2 & 2\\ 2 & 2\end{bmatrix}\begin{bmatrix}v_1\\ v_2\end{bmatrix} = \begin{bmatrix}0\\ 0\end{bmatrix} \;\Rightarrow\; \left\{\begin{matrix}2v_1 + 2v_2 = 0\\ 2v_1 + 2v_2 = 0\end{matrix}\right\}$$

Both equations specify that $v_2 = -v_1$. Choosing $v_1 = 1$ gives the eigenvector $\mathbf{v} = \begin{bmatrix}1\\ -1\end{bmatrix}$.

(c) **Eigenvalues:** $\det\begin{bmatrix}1-\lambda & -2\\ 2 & 1-\lambda\end{bmatrix} = (1-\lambda)^2 + 4 = 0 \;\Rightarrow\; (1-\lambda) = \pm 2i \;\Rightarrow\; \lambda = 1 \pm 2i$

The eigenvalues are therefore $\lambda = 1 + 2i$ and $\lambda = 1 - 2i$.

Eigenvectors: Starting with the eigenvalue $\lambda = 1 + 2i$ we have

$$\begin{bmatrix}1-\lambda & -2\\ 2 & 1-\lambda\end{bmatrix}\begin{bmatrix}v_1\\ v_2\end{bmatrix} = \begin{bmatrix}-2i & -2\\ 2 & -2i\end{bmatrix}\begin{bmatrix}v_1\\ v_2\end{bmatrix} = \begin{bmatrix}0\\ 0\end{bmatrix} \;\Rightarrow\; \left\{\begin{matrix}-2iv_1 - 2v_2 = 0\\ 2v_1 - 2iv_2 = 0\end{matrix}\right\}$$

Both equations specify that $v_2 = -iv_1$. Choosing $v_1 = i$ gives the eigenvector $\mathbf{v} = \begin{bmatrix}i\\ 1\end{bmatrix}$.

For the second eigenvalue $\lambda = 1 - 2i$ we have

$$\begin{bmatrix}1-\lambda & -2\\ 2 & 1-\lambda\end{bmatrix}\begin{bmatrix}v_1\\ v_2\end{bmatrix} = \begin{bmatrix}2i & -2\\ 2 & 2i\end{bmatrix}\begin{bmatrix}v_1\\ v_2\end{bmatrix} = \begin{bmatrix}0\\ 0\end{bmatrix} \;\Rightarrow\; \left\{\begin{matrix}2iv_1 - 2v_2 = 0\\ 2v_1 + 2iv_2 = 0\end{matrix}\right\}$$

Both equations specify that $v_2 = iv_1$. Choosing $v_1 = 1$ gives the eigenvector $\mathbf{v} = \begin{bmatrix}1\\ i\end{bmatrix}$.

(d) **Eigenvalues:** $\det\begin{bmatrix}2-\lambda & 7\\ 0 & 5-\lambda\end{bmatrix} = (2-\lambda)(5-\lambda) = 0$

The eigenvalues are therefore $\lambda = 2$ and $\lambda = 5$.

Eigenvectors: Starting with the eigenvalue $\lambda = 2$ we have

$$\begin{bmatrix}2-\lambda & 7\\ 0 & 5-\lambda\end{bmatrix}\begin{bmatrix}v_1\\ v_2\end{bmatrix} = \begin{bmatrix}0 & 7\\ 0 & 3\end{bmatrix}\begin{bmatrix}v_1\\ v_2\end{bmatrix} = \begin{bmatrix}0\\ 0\end{bmatrix} \;\Rightarrow\; \left\{\begin{matrix}7v_2 = 0\\ 3v_2 = 0\end{matrix}\right\}$$

Both equations specify that $v_2 = 0$. Choosing $v_1 = 1$ gives the eigenvector $\mathbf{v} = \begin{bmatrix}1\\ 0\end{bmatrix}$.

© 2016 Cengage Learning. All Rights Reserved. May not be scanned, copied or duplicated, or posted to a publicly accessible website, in whole or in part.

For the second eigenvalue $\lambda = 5$ we have

$$\begin{bmatrix} 2-\lambda & 7 \\ 0 & 5-\lambda \end{bmatrix}\begin{bmatrix} v_1 \\ v_2 \end{bmatrix} = \begin{bmatrix} -3 & 7 \\ 0 & 0 \end{bmatrix}\begin{bmatrix} v_1 \\ v_2 \end{bmatrix} = \begin{bmatrix} 0 \\ 0 \end{bmatrix} \Rightarrow \begin{Bmatrix} -3v_1 + 7v_2 = 0 \\ 0 = 0 \end{Bmatrix}$$

The second equation is always satisfied and the first equation is satisfied when $v_2 = \frac{3}{7}v_1$. Choosing $v_1 = 7$ gives the eigenvector $\mathbf{v} = \begin{bmatrix} 7 \\ 3 \end{bmatrix}$.

(e) **Eigenvalues:** $\det \begin{bmatrix} 1-\lambda & 2 \\ 3 & -3-\lambda \end{bmatrix} = (1-\lambda)(-3-\lambda) - 6 = \lambda^2 + 2\lambda - 9 = 0 \Rightarrow$

$$\lambda = \frac{-2 \pm \sqrt{(2)^2 - 4(1)(-9)}}{2(1)} = -1 \pm \frac{\sqrt{40}}{2} = -1 \pm \sqrt{10}$$

The eigenvalues are therefore $\lambda = -1 + \sqrt{10}$ and $\lambda = -1 - \sqrt{10}$.

Eigenvectors: Starting with the eigenvalue $\lambda = -1 + \sqrt{10}$ we have

$$\begin{bmatrix} 1-\lambda & 2 \\ 3 & -3-\lambda \end{bmatrix}\begin{bmatrix} v_1 \\ v_2 \end{bmatrix} = \begin{bmatrix} 2-\sqrt{10} & 2 \\ 3 & -2-\sqrt{10} \end{bmatrix}\begin{bmatrix} v_1 \\ v_2 \end{bmatrix} = \begin{bmatrix} 0 \\ 0 \end{bmatrix} \Rightarrow \begin{Bmatrix} (2-\sqrt{10})\,v_1 + 2v_2 = 0 \\ 3v_1 - (2+\sqrt{10})\,v_2 = 0 \end{Bmatrix}$$

Both equations specify that $v_2 = -\left(\dfrac{2-\sqrt{10}}{2}\right)v_1$ (try multiplying the second equation by the appropriate radical conjugate if this is unclear). Choosing $v_1 = -2$ gives the eigenvector $\mathbf{v} = \begin{bmatrix} -2 \\ 2-\sqrt{10} \end{bmatrix}$.

For the second eigenvalue $\lambda = -1 - \sqrt{10}$ we have

$$\begin{bmatrix} 1-\lambda & 2 \\ 3 & -3-\lambda \end{bmatrix}\begin{bmatrix} v_1 \\ v_2 \end{bmatrix} = \begin{bmatrix} 2+\sqrt{10} & 2 \\ 3 & -2+\sqrt{10} \end{bmatrix}\begin{bmatrix} v_1 \\ v_2 \end{bmatrix} = \begin{bmatrix} 0 \\ 0 \end{bmatrix} \Rightarrow \begin{Bmatrix} (2+\sqrt{10})\,v_1 + 2v_2 = 0 \\ 3v_1 - (2-\sqrt{10})\,v_2 = 0 \end{Bmatrix}$$

Both equations specify that $v_2 = -\left(\dfrac{2+\sqrt{10}}{2}\right)v_1$ (try multiplying the second equation by the appropriate radical conjugate if this is unclear). Choosing $v_1 = -2$ gives the eigenvector $\mathbf{v} = \begin{bmatrix} -2 \\ 2+\sqrt{10} \end{bmatrix}$.

(f) **Eigenvalues:** $\det \begin{bmatrix} 2-\lambda & 6 \\ 5 & 0-\lambda \end{bmatrix} = -\lambda(2-\lambda) - 30 = \lambda^2 - 2\lambda - 30 = 0 \Rightarrow$

$$\lambda = \frac{2 \pm \sqrt{(-2)^2 - 4(1)(-30)}}{2(1)} = 1 \pm \frac{\sqrt{124}}{2} = 1 \pm \sqrt{31}$$

The eigenvalues are therefore $\lambda = 1 + \sqrt{31}$ and $\lambda = 1 - \sqrt{31}$.

Eigenvectors: Starting with the eigenvalue $\lambda = 1 + \sqrt{31}$ we have

$$\begin{bmatrix} 2-\lambda & 6 \\ 5 & 0-\lambda \end{bmatrix}\begin{bmatrix} v_1 \\ v_2 \end{bmatrix} = \begin{bmatrix} 1-\sqrt{31} & 6 \\ 5 & -1-\sqrt{31} \end{bmatrix}\begin{bmatrix} v_1 \\ v_2 \end{bmatrix} = \begin{bmatrix} 0 \\ 0 \end{bmatrix} \Rightarrow \begin{Bmatrix} (1-\sqrt{31})\,v_1 + 6v_2 = 0 \\ 5v_1 - (1+\sqrt{31})\,v_2 = 0 \end{Bmatrix}$$

[continued]

© 2016 Cengage Learning. All Rights Reserved. May not be scanned, copied or duplicated, or posted to a publicly accessible website, in whole or in part.

Both equations specify that $v_2 = -\left(\frac{1-\sqrt{31}}{6}\right) v_1$ (try multiplying the second equation by the appropriate radical conjugate if this is unclear). Choosing $v_1 = -6$ gives the eigenvector $\mathbf{v} = \begin{bmatrix} -6 \\ 1-\sqrt{31} \end{bmatrix}$.

For the second eigenvalue $\lambda = 1 - \sqrt{31}$ we have

$$\begin{bmatrix} 2-\lambda & 6 \\ 5 & 0-\lambda \end{bmatrix} \begin{bmatrix} v_1 \\ v_2 \end{bmatrix} = \begin{bmatrix} 1+\sqrt{31} & 6 \\ 5 & -1+\sqrt{31} \end{bmatrix} \begin{bmatrix} v_1 \\ v_2 \end{bmatrix} = \begin{bmatrix} 0 \\ 0 \end{bmatrix} \Rightarrow \left\{ \begin{array}{c} (1+\sqrt{31})\, v_1 + 6v_2 = 0 \\ 5v_1 - (1-\sqrt{31})\, v_2 = 0 \end{array} \right\}$$

Both equations specify that $v_2 = -\left(\frac{1+\sqrt{31}}{6}\right) v_1$ (try multiplying the second equation by the appropriate radical conjugate if this is unclear). Choosing $v_1 = -6$ gives the eigenvector $\mathbf{v} = \begin{bmatrix} -6 \\ 1+\sqrt{31} \end{bmatrix}$.

(g) **Eigenvalues:** $\det \begin{bmatrix} 1-\lambda & 0 & 1 \\ 2 & 1-\lambda & 0 \\ 3 & 0 & 1-\lambda \end{bmatrix} = (1-\lambda)^3 - 3(1-\lambda) = (1-\lambda)\left[(1-\lambda)^2 - 3\right]$

$$= (1-\lambda)(1-\lambda+\sqrt{3})(1-\lambda-\sqrt{3}) = 0$$

The eigenvalues are therefore $\lambda = 1$, $\lambda = 1+\sqrt{3}$ and $\lambda = 1-\sqrt{3}$.

Eigenvectors: Starting with the eigenvalue $\lambda = 1$ we have

$$\begin{bmatrix} 1-\lambda & 0 & 1 \\ 2 & 1-\lambda & 0 \\ 3 & 0 & 1-\lambda \end{bmatrix} \begin{bmatrix} v_1 \\ v_2 \\ v_3 \end{bmatrix} = \begin{bmatrix} 0 & 0 & 1 \\ 2 & 0 & 0 \\ 3 & 0 & 0 \end{bmatrix} \begin{bmatrix} v_1 \\ v_2 \\ v_3 \end{bmatrix} = \begin{bmatrix} 0 \\ 0 \\ 0 \end{bmatrix} \Rightarrow \left\{ \begin{array}{c} v_3 = 0 \\ 2v_1 = 0 \\ 3v_1 = 0 \end{array} \right\}$$

The equations specify that $v_1 = v_3 = 0$. Choosing $v_2 = 1$ gives the eigenvector $\mathbf{v} = \begin{bmatrix} 0 \\ 1 \\ 0 \end{bmatrix}$.

For the second eigenvalue $\lambda = 1+\sqrt{3}$ we have

$$\begin{bmatrix} 1-\lambda & 0 & 1 \\ 2 & 1-\lambda & 0 \\ 3 & 0 & 1-\lambda \end{bmatrix} \begin{bmatrix} v_1 \\ v_2 \\ v_3 \end{bmatrix} = \begin{bmatrix} -\sqrt{3} & 0 & 1 \\ 2 & -\sqrt{3} & 0 \\ 3 & 0 & -\sqrt{3} \end{bmatrix} \begin{bmatrix} v_1 \\ v_2 \\ v_3 \end{bmatrix} = \begin{bmatrix} 0 \\ 0 \\ 0 \end{bmatrix} \Rightarrow \left\{ \begin{array}{c} -\sqrt{3}v_1 + v_3 = 0 \\ 2v_1 - \sqrt{3}v_2 = 0 \\ 3v_1 - \sqrt{3}v_3 = 0 \end{array} \right\}$$

The first and last equations specify that $v_3 = \sqrt{3}v_1$. We choose $v_1 = \sqrt{3}$ so that $v_3 = 3$ and the second equation gives $v_2 = \frac{2}{\sqrt{3}}v_1 = 2$. This gives the eigenvector $\mathbf{v} = \begin{bmatrix} \sqrt{3} \\ 2 \\ 3 \end{bmatrix}$.

For the third eigenvalue $\lambda = 1-\sqrt{3}$ we have

$$\begin{bmatrix} 1-\lambda & 0 & 1 \\ 2 & 1-\lambda & 0 \\ 3 & 0 & 1-\lambda \end{bmatrix} \begin{bmatrix} v_1 \\ v_2 \\ v_3 \end{bmatrix} = \begin{bmatrix} \sqrt{3} & 0 & 1 \\ 2 & \sqrt{3} & 0 \\ 3 & 0 & \sqrt{3} \end{bmatrix} \begin{bmatrix} v_1 \\ v_2 \\ v_3 \end{bmatrix} = \begin{bmatrix} 0 \\ 0 \\ 0 \end{bmatrix} \Rightarrow \left\{ \begin{array}{c} \sqrt{3}v_1 + v_3 = 0 \\ 2v_1 + \sqrt{3}v_2 = 0 \\ 3v_1 + \sqrt{3}v_3 = 0 \end{array} \right\}$$

© 2016 Cengage Learning. All Rights Reserved. May not be scanned, copied or duplicated, or posted to a publicly accessible website, in whole or in part.

The first and last equations specify that $v_3 = -\sqrt{3}v_1$. We choose $v_1 = -\sqrt{3}$ so that $v_3 = 3$ and the second equation gives

$v_2 = -\frac{2}{\sqrt{3}}v_1 = 2$. This gives the eigenvector $\mathbf{v} = \begin{bmatrix} -\sqrt{3} \\ 2 \\ 3 \end{bmatrix}$.

(h) **Eigenvalues:** $\det \begin{bmatrix} 1-\lambda & 2 & 3 \\ 0 & 1-\lambda & 7 \\ 0 & 2 & 1-\lambda \end{bmatrix} = (1-\lambda)^3 - 14(1-\lambda) = (1-\lambda)\left[(1-\lambda)^2 - 14\right]$

$$= (1-\lambda)\left(1-\lambda+\sqrt{14}\right)\left(1-\lambda-\sqrt{14}\right)$$

The eigenvalues are therefore $\lambda = 1$, $\lambda = 1+\sqrt{14}$ and $\lambda = 1-\sqrt{14}$.

Eigenvectors: Starting with the eigenvalue $\lambda = 1$ we have

$$\begin{bmatrix} 1-\lambda & 2 & 3 \\ 0 & 1-\lambda & 7 \\ 0 & 2 & 1-\lambda \end{bmatrix}\begin{bmatrix} v_1 \\ v_2 \\ v_3 \end{bmatrix} = \begin{bmatrix} 0 & 2 & 3 \\ 0 & 0 & 7 \\ 0 & 2 & 0 \end{bmatrix}\begin{bmatrix} v_1 \\ v_2 \\ v_3 \end{bmatrix} = \begin{bmatrix} 0 \\ 0 \\ 0 \end{bmatrix} \Rightarrow \left\{\begin{aligned} 2v_2 + 3v_3 &= 0 \\ 7v_3 &= 0 \\ 2v_2 &= 0 \end{aligned}\right\}$$

The equations specify that $v_2 = v_3 = 0$. Choosing $v_1 = 1$ gives the eigenvector $\mathbf{v} = \begin{bmatrix} 1 \\ 0 \\ 0 \end{bmatrix}$.

For the second eigenvalue $\lambda = 1+\sqrt{14}$ we have

$$\begin{bmatrix} 1-\lambda & 2 & 3 \\ 0 & 1-\lambda & 7 \\ 0 & 2 & 1-\lambda \end{bmatrix}\begin{bmatrix} v_1 \\ v_2 \\ v_3 \end{bmatrix} = \begin{bmatrix} -\sqrt{14} & 2 & 3 \\ 0 & -\sqrt{14} & 7 \\ 0 & 2 & -\sqrt{14} \end{bmatrix}\begin{bmatrix} v_1 \\ v_2 \\ v_3 \end{bmatrix} = \begin{bmatrix} 0 \\ 0 \\ 0 \end{bmatrix} \Rightarrow \left\{\begin{aligned} -\sqrt{14}v_1 + 2v_2 + 3v_3 &= 0 \\ -\sqrt{14}v_2 + 7v_3 &= 0 \\ 2v_2 - \sqrt{14}v_3 &= 0 \end{aligned}\right\}$$

The second and third equations specify that $v_3 = \frac{\sqrt{14}}{7}v_2$. We choose $v_2 = 7$ so that $v_3 = \sqrt{14}$ and the first equation gives

$v_1 = \dfrac{2v_2 + 3v_3}{\sqrt{14}} = 3 + \sqrt{14}$. This gives the eigenvector $\mathbf{v} = \begin{bmatrix} 3+\sqrt{14} \\ 7 \\ \sqrt{14} \end{bmatrix}$.

For the third eigenvalue $\lambda = 1-\sqrt{14}$ we have

$$\begin{bmatrix} 1-\lambda & 2 & 3 \\ 0 & 1-\lambda & 7 \\ 0 & 2 & 1-\lambda \end{bmatrix}\begin{bmatrix} v_1 \\ v_2 \\ v_3 \end{bmatrix} = \begin{bmatrix} \sqrt{14} & 2 & 3 \\ 0 & \sqrt{14} & 7 \\ 0 & 2 & \sqrt{14} \end{bmatrix}\begin{bmatrix} v_1 \\ v_2 \\ v_3 \end{bmatrix} = \begin{bmatrix} 0 \\ 0 \\ 0 \end{bmatrix} \Rightarrow \left\{\begin{aligned} \sqrt{14}v_1 + 2v_2 + 3v_3 &= 0 \\ \sqrt{14}v_2 + 7v_3 &= 0 \\ 2v_2 + \sqrt{14}v_3 &= 0 \end{aligned}\right\}$$

The second and third equations specify that $v_3 = -\frac{\sqrt{14}}{7}v_2$. We choose $v_2 = -7$ so that $v_3 = \sqrt{14}$ and the first equation

gives $v_1 = \dfrac{-2v_2 - 3v_3}{\sqrt{14}} = -3 + \sqrt{14}$. This gives the eigenvector $\mathbf{v} = \begin{bmatrix} -3+\sqrt{14} \\ -7 \\ \sqrt{14} \end{bmatrix}$.

© 2016 Cengage Learning. All Rights Reserved. May not be scanned, copied or duplicated, or posted to a publicly accessible website, in whole or in part.

15. In this exercise, there are *repeated eigenvalues* for each matrix. In these cases, there may be more than one distinct (non-scalar multiple) eigenvector associated with the repeated eigenvalue.

(a) **Eigenvalues:** $\det\begin{bmatrix} 1-\lambda & 0 \\ 0 & 1-\lambda \end{bmatrix} = (1-\lambda)^2 = 0$

Therefore, the only eigenvalue is $\lambda = 1$. Note that $\lambda = 1$ is repeated twice (power of 2) in the characteristic polynomial.

Eigenvectors: Using the eigenvalue $\lambda = 1$ we have

$$\begin{bmatrix} 1-\lambda & 0 \\ 0 & 1-\lambda \end{bmatrix}\begin{bmatrix} v_1 \\ v_2 \end{bmatrix} = \begin{bmatrix} 0 & 0 \\ 0 & 0 \end{bmatrix}\begin{bmatrix} v_1 \\ v_2 \end{bmatrix} = \begin{bmatrix} 0 \\ 0 \end{bmatrix} \quad\Rightarrow\quad \begin{Bmatrix} 0 = 0 \\ 0 = 0 \end{Bmatrix}$$

Both equations are always satisfied. So we are free to choose values for v_1 and v_2. Choosing $v_1 = 1$ and $v_2 = 0$ gives one eigenvector $\mathbf{v} = \begin{bmatrix} 1 \\ 0 \end{bmatrix}$ and choosing $v_1 = 0$ and $v_2 = 1$ gives another eigenvector $\mathbf{v} = \begin{bmatrix} 0 \\ 1 \end{bmatrix}$.

Note: When one of the eigenvalues is repeated, there may be more than one eigenvector associated with that eigenvalue. In this exercise, there were two free variables, v_1 and v_2, to which we could assign values, so there were two distinct eigenvectors (not scalar multiples).

(b) **Eigenvalues:** $\det\begin{bmatrix} 1-\lambda & 1 \\ 0 & 1-\lambda \end{bmatrix} = (1-\lambda)^2 - 0 = (1-\lambda)^2 = 0$

Therefore, the only eigenvalue is $\lambda = 1$. Note that $\lambda = 1$ is repeated twice (power of 2) in the characteristic polynomial.

Eigenvectors: Using the eigenvalue $\lambda = 1$ we have

$$\begin{bmatrix} 1-\lambda & 1 \\ 0 & 1-\lambda \end{bmatrix}\begin{bmatrix} v_1 \\ v_2 \end{bmatrix} = \begin{bmatrix} 0 & 1 \\ 0 & 0 \end{bmatrix}\begin{bmatrix} v_1 \\ v_2 \end{bmatrix} = \begin{bmatrix} 0 \\ 0 \end{bmatrix} \quad\Rightarrow\quad \begin{Bmatrix} v_2 = 0 \\ 0 = 0 \end{Bmatrix}$$

The second equation is always satisfied and the first equation is satisfied when $v_2 = 0$. Choosing $v_1 = 1$ gives the eigenvector $\mathbf{v} = \begin{bmatrix} 1 \\ 0 \end{bmatrix}$.

Note: In this exercise, there was one free variable to which we could assign a value for the repeated eigenvalue $\lambda = 1$, so there was only one eigenvector.

(c) **Eigenvalues:** $\det\begin{bmatrix} 1-\lambda & 0 & 0 \\ 0 & 1-\lambda & 0 \\ 0 & 0 & 2-\lambda \end{bmatrix} = (1-\lambda)^2(2-\lambda) = 0$

The eigenvalues are therefore $\lambda = 1$ and $\lambda = 2$. Note that $\lambda = 1$ is repeated twice (power of 2) in the characteristic polynomial.

Eigenvectors: Starting with the eigenvalue $\lambda = 1$ we have

$$\begin{bmatrix} 1-\lambda & 0 & 0 \\ 0 & 1-\lambda & 0 \\ 0 & 0 & 2-\lambda \end{bmatrix}\begin{bmatrix} v_1 \\ v_2 \\ v_3 \end{bmatrix} = \begin{bmatrix} 0 & 0 & 0 \\ 0 & 0 & 0 \\ 0 & 0 & 1 \end{bmatrix}\begin{bmatrix} v_1 \\ v_2 \\ v_3 \end{bmatrix} = \begin{bmatrix} 0 \\ 0 \\ 0 \end{bmatrix} \quad\Rightarrow\quad \begin{Bmatrix} 0 = 0 \\ 0 = 0 \\ v_3 = 0 \end{Bmatrix}$$

The first two equations are always satisfied and the third is satisfied when $v_3 = 0$. So we are free to choose values for v_1

© 2016 Cengage Learning. All Rights Reserved. May not be scanned, copied or duplicated, or posted to a publicly accessible website, in whole or in part.

and v_2. Choosing $v_1 = 1$ and $v_2 = 0$ gives one eigenvector $\mathbf{v} = \begin{bmatrix} 1 \\ 0 \\ 0 \end{bmatrix}$ and choosing $v_1 = 0$ and $v_2 = 1$ gives another

eigenvector $\mathbf{v} = \begin{bmatrix} 0 \\ 1 \\ 0 \end{bmatrix}$. For the second eigenvalue $\lambda = 2$ we have

$$\begin{bmatrix} 1-\lambda & 0 & 0 \\ 0 & 1-\lambda & 0 \\ 0 & 0 & 2-\lambda \end{bmatrix} \begin{bmatrix} v_1 \\ v_2 \\ v_3 \end{bmatrix} = \begin{bmatrix} -1 & 0 & 0 \\ 0 & -1 & 0 \\ 0 & 0 & 0 \end{bmatrix} \begin{bmatrix} v_1 \\ v_2 \\ v_3 \end{bmatrix} = \begin{bmatrix} 0 \\ 0 \\ 0 \end{bmatrix} \Rightarrow \begin{Bmatrix} -v_1 = 0 \\ -v_2 = 0 \\ 0 = 0 \end{Bmatrix}$$

The last equation is always satisfied and the first two equations are satisfied when $v_1 = v_2 = 0$. Choosing $v_3 = 1$ gives the

eigenvector $\mathbf{v} = \begin{bmatrix} 0 \\ 0 \\ 1 \end{bmatrix}$.

Note: In this exercise, there were two free variables, v_1 and v_2, to which we could assign values when finding the eigenvector associated with the repeated eigenvalue $\lambda = 1$. Therefore, there were two distinct eigenvectors (not scalar multiples) associated with this eigenvalue.

17. We determine the eigenvalues of 2×2 upper and lower triangular matrices U and L by calculating the characteristic polynomials

$$\det \begin{bmatrix} u_{11} - \lambda & u_{12} \\ 0 & u_{22} - \lambda \end{bmatrix} = (u_{11} - \lambda)(u_{22} - \lambda) = 0 \Rightarrow \lambda = u_{11} \text{ and } \lambda = u_{22}$$

$$\det \begin{bmatrix} l_{11} - \lambda & 0 \\ l_{21} & l_{22} - \lambda \end{bmatrix} = (l_{11} - \lambda)(l_{22} - \lambda) = 0 \Rightarrow \lambda = l_{11} \text{ and } \lambda = l_{22}$$

Similarly, for 3×3 upper and lower triangular matrices, calculating the characteristic polynomials (using the algorithm on page 531) gives

$$\det \begin{bmatrix} u_{11} - \lambda & u_{12} & u_{13} \\ 0 & u_{22} - \lambda & u_{23} \\ 0 & 0 & u_{33} - \lambda \end{bmatrix} = (u_{11} - \lambda)(u_{22} - \lambda)(u_{33} - \lambda) = 0 \Rightarrow \lambda = u_{11}, u_{22}, \text{ and } u_{33}$$

$$\det \begin{bmatrix} l_{11} - \lambda & 0 & 0 \\ l_{21} & l_{22} - \lambda & 0 \\ l_{31} & l_{32} & l_{33} - \lambda \end{bmatrix} = (l_{11} - \lambda)(l_{22} - \lambda)(l_{33} - \lambda) = 0 \Rightarrow \lambda = l_{11}, l_{22}, \text{ and } l_{33}$$

Therefore, the eigenvalues of 2×2 and 3×3 upper and lower triangular matrices are given by the entries on the diagonal.

19. Let A be a matrix such that $A^2 = 0$. The eigenvalues and eigenvectors of A satisfy $A\mathbf{v} = \lambda\mathbf{v} \Leftrightarrow AA\mathbf{v} = \lambda A\mathbf{v} \Leftrightarrow$ $A^2\mathbf{v} = \lambda(A\mathbf{v}) \Rightarrow 0\mathbf{v} = \lambda^2\mathbf{v} \Rightarrow \lambda^2\mathbf{v} = \mathbf{0}$.

$\mathbf{v}$ is nonzero since it is an eigenvector (Definition 2), so we must have $\lambda = 0$.

© 2016 Cengage Learning. All Rights Reserved. May not be scanned, copied or duplicated, or posted to a publicly accessible website, in whole or in part.

21. The characteristic polynomial of A^T is given by

$$\begin{aligned} \det\left(A^T-\lambda I\right) &= \det\left(A^T-\lambda I^T\right) \quad (\text{since } I=I^T) \\ &= \det\left(\left(A-\lambda I\right)^T\right) \quad \left[\text{since } (A+B)^T = A^T + B^T \text{ (see exercise 8.4.14b)}\right] \\ &= \det\left(\left(A-\lambda I\right)\right) \quad \left[\text{since } \det B = \det B^T \text{ (see exercise 8.6.16)}\right] \end{aligned}$$

Therefore, the characteristic polynomial of A^T is the same as that of A.

23. $A\mathbf{v}=\lambda\mathbf{v} \Rightarrow 2A\mathbf{v}=2\lambda\mathbf{v} \Rightarrow (2A)\,\mathbf{v}=(2\lambda)\,\mathbf{v}$ Therefore, by Definition (2) 2λ is an eigenvalue of the matrix $2A$.

25. $\mathbf{v}$ is an eigenvector of matrices A and B so that $A\mathbf{v}=\lambda_A\mathbf{v}$ and $B\mathbf{v}=\lambda_B\mathbf{v}$

(a) $(A+B)\,\mathbf{v}=A\mathbf{v}+B\mathbf{v}$ [by property 3 of matrix multiplication] $=\lambda_A\mathbf{v}+\lambda_B\mathbf{v}=(\lambda_A+\lambda_B)\,\mathbf{v}$

Therefore, by Definition (2) $\mathbf{v}$ is an eigenvector of $A+B$ with an associated eigenvalue $\lambda=\lambda_A+\lambda_B$.

(b) $AB\mathbf{v}=A\,(B\mathbf{v})=A\,(\lambda_B\mathbf{v})=\lambda_B\,(A\mathbf{v})=\lambda_B\lambda_A\mathbf{v}$

Therefore, by Definition (2) $\mathbf{v}$ is an eigenvector of AB with an associated eigenvalue $\lambda=\lambda_A\lambda_B$.

27. Eigenvalues: $\det\begin{bmatrix}2-\lambda & 3\\ 0 & \frac{9}{10}-\lambda\end{bmatrix}=(2-\lambda)\left(\frac{9}{10}-\lambda\right)-0=(2-\lambda)\left(\frac{9}{10}-\lambda\right)=0$

The eigenvalues are therefore $\lambda=2$ and $\lambda=\frac{9}{10}$.

Eigenvectors: Starting with the eigenvalue $\lambda=2$ we have

$$\begin{bmatrix}2-\lambda & 3\\ 0 & \frac{9}{10}-\lambda\end{bmatrix}\begin{bmatrix}v_1\\ v_2\end{bmatrix}=\begin{bmatrix}0 & 3\\ 0 & -\frac{11}{10}\end{bmatrix}\begin{bmatrix}v_1\\ v_2\end{bmatrix}=\begin{bmatrix}0\\ 0\end{bmatrix} \Rightarrow \left\{\begin{array}{r}3v_2=0\\ -\frac{11}{10}v_2=0\end{array}\right\}$$

Both equations specify that $v_2=0$. Choosing $v_1=1$ gives the eigenvector $\mathbf{v}=\begin{bmatrix}1\\ 0\end{bmatrix}$.

For the second eigenvalue $\lambda=\frac{9}{10}$ we have

$$\begin{bmatrix}2-\lambda & 3\\ 0 & \frac{9}{10}-\lambda\end{bmatrix}\begin{bmatrix}v_1\\ v_2\end{bmatrix}=\begin{bmatrix}\frac{11}{10} & 3\\ 0 & 0\end{bmatrix}\begin{bmatrix}v_1\\ v_2\end{bmatrix}=\begin{bmatrix}0\\ 0\end{bmatrix} \Rightarrow \left\{\begin{array}{r}\frac{11}{10}v_1+3v_2=0\\ 0=0\end{array}\right\}$$

The second equation is always satisfied and the first equation is satisfied when $v_2=-\frac{11}{30}v_1$. Choosing $v_1=30$ gives the eigenvector $\mathbf{v}=\begin{bmatrix}30\\ -11\end{bmatrix}$.

29. Eigenvalues: $\det\begin{bmatrix}b-\lambda & 2\\ \frac{1}{2} & 0-\lambda\end{bmatrix}=-\lambda\,(b-\lambda)-1=\lambda^2-b\lambda-1=0 \Rightarrow \lambda=\dfrac{b\pm\sqrt{b^2+4}}{2}$

The eigenvalues are therefore $\lambda=\frac{1}{2}b+\frac{1}{2}\sqrt{b^2+4}$ and $\lambda=\frac{1}{2}b-\frac{1}{2}\sqrt{b^2+4}$.

Eigenvectors: Starting with the eigenvalue $\lambda=\frac{1}{2}b+\frac{1}{2}\sqrt{b^2+4}$ we have

$$\begin{bmatrix}b-\lambda & 2\\ \frac{1}{2} & 0-\lambda\end{bmatrix}\begin{bmatrix}v_1\\ v_2\end{bmatrix}=\begin{bmatrix}\frac{1}{2}b-\frac{1}{2}\sqrt{b^2+4} & 2\\ \frac{1}{2} & -\frac{1}{2}b-\frac{1}{2}\sqrt{b^2+4}\end{bmatrix}\begin{bmatrix}v_1\\ v_2\end{bmatrix}=\begin{bmatrix}0\\ 0\end{bmatrix} \Rightarrow \left\{\begin{array}{r}\left(\frac{1}{2}b-\frac{1}{2}\sqrt{b^2+4}\right)v_1+2v_2=0\\ \frac{1}{2}v_1-\left(\frac{1}{2}b+\frac{1}{2}\sqrt{b^2+4}\right)v_2=0\end{array}\right\}$$

Both equations specify that $v_2=\left(-\frac{1}{4}b+\frac{1}{4}\sqrt{b^2+4}\right)v_1$ (if this is unclear, try multiplying the second equation by

© 2016 Cengage Learning. All Rights Reserved. May not be scanned, copied or duplicated, or posted to a publicly accessible website, in whole or in part.

$\frac{1}{2}b - \frac{1}{2}\sqrt{b^2+4}$). Choosing $v_1 = 4$ gives the eigenvector $\mathbf{v} = \begin{bmatrix} 4 \\ -b+\sqrt{b^2+4} \end{bmatrix}$.

For the second eigenvalue $\lambda = \frac{1}{2}b - \frac{1}{2}\sqrt{b^2+4}$ we have

$$\begin{bmatrix} b-\lambda & 2 \\ \frac{1}{2} & 0-\lambda \end{bmatrix}\begin{bmatrix} v_1 \\ v_2 \end{bmatrix} = \begin{bmatrix} \frac{1}{2}b+\frac{1}{2}\sqrt{b^2+4} & 2 \\ \frac{1}{2} & -\frac{1}{2}b+\frac{1}{2}\sqrt{b^2+4} \end{bmatrix}\begin{bmatrix} v_1 \\ v_2 \end{bmatrix} = \begin{bmatrix} 0 \\ 0 \end{bmatrix} \Rightarrow \left\{ \begin{array}{l} \left(\frac{1}{2}b+\frac{1}{2}\sqrt{b^2+4}\right)v_1 + 2v_2 = 0 \\ \frac{1}{2}v_1 - \left(\frac{1}{2}b - \frac{1}{2}\sqrt{b^2+4}\right)v_2 = 0 \end{array} \right\}$$

Both equations specify that $v_2 = \left(-\frac{1}{4}b - \frac{1}{4}\sqrt{b^2+4}\right)v_1$ (if this is unclear, try multiplying the second equation by $\frac{1}{2}b + \frac{1}{2}\sqrt{b^2+4}$). Choosing $v_1 = -4$ gives the eigenvector $\mathbf{v} = \begin{bmatrix} -4 \\ b+\sqrt{b^2+4} \end{bmatrix}$.

31. The characteristic polynomial of matrix A is given by

$$\det(A-\lambda I) = \det\begin{bmatrix} 1-\lambda & 2 & 4 \\ \frac{1}{2} & 0-\lambda & 0 \\ 0 & \frac{1}{3} & 0-\lambda \end{bmatrix} = (1-\lambda)(-\lambda)(-\lambda) + 0 + 4\left(\tfrac{1}{2}\right)\left(\tfrac{1}{3}\right) - 0 - 2\left(\tfrac{1}{2}\right)(-\lambda) - 0$$
$$= \lambda^2(1-\lambda) + \tfrac{2}{3} + \lambda = -\lambda^3 + \lambda^2 + \lambda + \tfrac{2}{3}$$

8.8 Iterated Matrix Models

1. The eigenvalues of $A = \begin{bmatrix} 2 & 0 \\ 0 & 1 \end{bmatrix}$ satisfy the characteristic polynomial $(2-\lambda)(1-\lambda) = 0$, which has solutions $\lambda_1 = 1$ and $\lambda_2 = 2$. Solving the system $A\mathbf{v} = \lambda\mathbf{v}$ for each eigenvalue gives the associated eigenvectors $\mathbf{v}_1 = [0,1]$ and $\mathbf{v}_2 = [1,0]$. So

$D = \begin{bmatrix} 1 & 0 \\ 0 & 2 \end{bmatrix}$ and $P = \begin{bmatrix} 0 & 1 \\ 1 & 0 \end{bmatrix} \Rightarrow P^{-1} = \frac{1}{-1}\begin{bmatrix} 0 & -1 \\ -1 & 0 \end{bmatrix} = \begin{bmatrix} 0 & 1 \\ 1 & 0 \end{bmatrix}$. Therefore,

$$PDP^{-1} = \begin{bmatrix} 0 & 1 \\ 1 & 0 \end{bmatrix}\begin{bmatrix} 1 & 0 \\ 0 & 2 \end{bmatrix}\begin{bmatrix} 0 & 1 \\ 1 & 0 \end{bmatrix} = \begin{bmatrix} 0 & 1 \\ 1 & 0 \end{bmatrix}\begin{bmatrix} 0 & 1 \\ 2 & 0 \end{bmatrix} = \begin{bmatrix} 2 & 0 \\ 0 & 1 \end{bmatrix} = A$$

3. The eigenvalues of $A = \begin{bmatrix} 1 & 2 \\ 2 & 1 \end{bmatrix}$ satisfy the characteristic polynomial $(1-\lambda)^2 - 4 = 0$, which has solutions $\lambda_1 = 3$ and $\lambda_2 = -1$. Solving the system $A\mathbf{v} = \lambda\mathbf{v}$ for each eigenvalue gives the associated eigenvectors $\mathbf{v}_1 = [1,1]$ and $\mathbf{v}_2 = [1,-1]$.

So $D = \begin{bmatrix} 3 & 0 \\ 0 & -1 \end{bmatrix}$ and $P = \begin{bmatrix} 1 & 1 \\ 1 & -1 \end{bmatrix} \Rightarrow P^{-1} = \frac{1}{-2}\begin{bmatrix} -1 & -1 \\ -1 & 1 \end{bmatrix} = \begin{bmatrix} \frac{1}{2} & \frac{1}{2} \\ \frac{1}{2} & -\frac{1}{2} \end{bmatrix}$. Therefore,

$$PDP^{-1} = \begin{bmatrix} 1 & 1 \\ 1 & -1 \end{bmatrix}\begin{bmatrix} 3 & 0 \\ 0 & -1 \end{bmatrix}\begin{bmatrix} \frac{1}{2} & \frac{1}{2} \\ \frac{1}{2} & -\frac{1}{2} \end{bmatrix} = \begin{bmatrix} 1 & 1 \\ 1 & -1 \end{bmatrix}\begin{bmatrix} \frac{3}{2} & \frac{3}{2} \\ -\frac{1}{2} & \frac{1}{2} \end{bmatrix} = \begin{bmatrix} 1 & 2 \\ 2 & 1 \end{bmatrix} = A$$

© 2016 Cengage Learning. All Rights Reserved. May not be scanned, copied or duplicated, or posted to a publicly accessible website, in whole or in part.

5. The eigenvalues of $A = \begin{bmatrix} 1 & 2 \\ -3 & 3 \end{bmatrix}$ satisfy the characteristic polynomial $(1-\lambda)(3-\lambda)+6=0 \;\Rightarrow\; \lambda^2 - 4\lambda + 9 = 0$, which has solutions $\lambda_1 = 2+\sqrt{5}i$ and $\lambda_2 = 2-\sqrt{5}i$. Solving the system $A\mathbf{v} = \lambda\mathbf{v}$ for each eigenvalue gives the associated eigenvectors $\mathbf{v}_1 = [2, 1+\sqrt{5}i]$ and $\mathbf{v}_2 = [2, 1-\sqrt{5}i]$. So $D = \begin{bmatrix} 2+\sqrt{5}i & 0 \\ 0 & 2-\sqrt{5}i \end{bmatrix}$ and

$P = \begin{bmatrix} 2 & 2 \\ 1+\sqrt{5}i & 1-\sqrt{5}i \end{bmatrix} \Rightarrow P^{-1} = \dfrac{1}{-4\sqrt{5}i}\begin{bmatrix} 1-\sqrt{5}i & -2 \\ -1-\sqrt{5}i & 2 \end{bmatrix} = \begin{bmatrix} \frac{1}{4\sqrt{5}}i + \frac{1}{4} & -\frac{1}{2\sqrt{5}}i \\ -\frac{1}{4\sqrt{5}}i + \frac{1}{4} & \frac{1}{2\sqrt{5}}i \end{bmatrix}$. Therefore,

$$
\begin{aligned}
PDP^{-1} &= \begin{bmatrix} 2 & 2 \\ 1+\sqrt{5}i & 1-\sqrt{5}i \end{bmatrix}\begin{bmatrix} 2+\sqrt{5}i & 0 \\ 0 & 2-\sqrt{5}i \end{bmatrix}\begin{bmatrix} \frac{1}{4\sqrt{5}}i + \frac{1}{4} & -\frac{1}{2\sqrt{5}}i \\ -\frac{1}{4\sqrt{5}}i + \frac{1}{4} & \frac{1}{2\sqrt{5}}i \end{bmatrix} \\
&= \begin{bmatrix} 2 & 2 \\ 1+\sqrt{5}i & 1-\sqrt{5}i \end{bmatrix}\begin{bmatrix} (2+\sqrt{5}i)\left(\frac{1}{4\sqrt{5}}i + \frac{1}{4}\right) & \left(-\frac{1}{2\sqrt{5}}i\right)(2+\sqrt{5}i) \\ (2-\sqrt{5}i)\left(-\frac{1}{4\sqrt{5}}i + \frac{1}{4}\right) & \left(\frac{1}{2\sqrt{5}}i\right)(2-\sqrt{5}i) \end{bmatrix} \\
&= \begin{bmatrix} 2 & 2 \\ 1+\sqrt{5}i & 1-\sqrt{5}i \end{bmatrix}\begin{bmatrix} \frac{1}{4} + \frac{7\sqrt{5}}{20}i & \frac{1}{2} - \frac{1}{\sqrt{5}}i \\ \frac{1}{4} - \frac{7\sqrt{5}}{20}i & \frac{1}{2} + \frac{1}{\sqrt{5}}i \end{bmatrix} \\
&= \begin{bmatrix} 2\left(\frac{1}{4} + \frac{7\sqrt{5}}{20}i\right) + 2\left(\frac{1}{4} - \frac{7\sqrt{5}}{20}i\right) & 2\left(\frac{1}{2} - \frac{1}{\sqrt{5}}i\right) + 2\left(\frac{1}{2} + \frac{1}{\sqrt{5}}i\right) \\ (1+\sqrt{5}i)\left(\frac{1}{4} + \frac{7\sqrt{5}}{20}i\right) + (1-\sqrt{5}i)\left(\frac{1}{4} - \frac{7\sqrt{5}}{20}i\right) & (1+\sqrt{5}i)\left(\frac{1}{2} - \frac{1}{\sqrt{5}}i\right) + (1-\sqrt{5}i)\left(\frac{1}{2} + \frac{1}{\sqrt{5}}i\right) \end{bmatrix} \\
&= \begin{bmatrix} 1 & 2 \\ -3 & 3 \end{bmatrix} = A
\end{aligned}
$$

7. We use a CAS such as Maple or Mathematica to compute the eigenvalue and eigenvector matrices of $A = \begin{bmatrix} 0 & 0 & 1 \\ 0 & 1 & 0 \\ 0 & 0 & 2 \end{bmatrix}$ and to perform matrix inversion. This gives

$$
D = \begin{bmatrix} 0 & 0 & 0 \\ 0 & 1 & 0 \\ 0 & 0 & 2 \end{bmatrix} \quad P = \begin{bmatrix} 1 & 0 & 1 \\ 0 & 1 & 0 \\ 0 & 0 & 2 \end{bmatrix} \quad P^{-1} = \begin{bmatrix} 1 & 0 & -\frac{1}{2} \\ 0 & 1 & 0 \\ 0 & 0 & \frac{1}{2} \end{bmatrix}
$$

Therefore, $PDP^{-1} = \begin{bmatrix} 1 & 0 & 1 \\ 0 & 1 & 0 \\ 0 & 0 & 2 \end{bmatrix}\begin{bmatrix} 0 & 0 & 0 \\ 0 & 1 & 0 \\ 0 & 0 & 2 \end{bmatrix}\begin{bmatrix} 1 & 0 & -\frac{1}{2} \\ 0 & 1 & 0 \\ 0 & 0 & \frac{1}{2} \end{bmatrix} = \begin{bmatrix} 1 & 0 & 1 \\ 0 & 1 & 0 \\ 0 & 0 & 2 \end{bmatrix}\begin{bmatrix} 0 & 0 & 0 \\ 0 & 1 & 0 \\ 0 & 0 & 1 \end{bmatrix} = \begin{bmatrix} 0 & 0 & 1 \\ 0 & 1 & 0 \\ 0 & 0 & 2 \end{bmatrix} = A.$

9. First, observe that $A^2 = \begin{bmatrix} 1 & 0 \\ 0 & 4 \end{bmatrix}\begin{bmatrix} 1 & 0 \\ 0 & 4 \end{bmatrix} = \begin{bmatrix} 1 & 0 \\ 0 & 16 \end{bmatrix}$. The eigenvalues of A satisfy the characteristic polynomial $(1-\lambda)(4-\lambda) = 0$, which has solutions $\lambda_1 = 1$ and $\lambda_2 = 4$. Solving the system $A\mathbf{v} = \lambda\mathbf{v}$ for each eigenvalue gives the associated eigenvectors $\mathbf{v}_1 = [1, 0]$ and $\mathbf{v}_2 = [0, 1]$. So $D = \begin{bmatrix} 1 & 0 \\ 0 & 4 \end{bmatrix}$, $D^2 = \begin{bmatrix} 1 & 0 \\ 0 & 16 \end{bmatrix}$ and $P = \begin{bmatrix} 1 & 0 \\ 0 & 1 \end{bmatrix} \Rightarrow$

$P^{-1} = \begin{bmatrix} 1 & 0 \\ 0 & 1 \end{bmatrix}$. Therefore, $PD^2P^{-1} = \begin{bmatrix} 1 & 0 \\ 0 & 1 \end{bmatrix}\begin{bmatrix} 1 & 0 \\ 0 & 16 \end{bmatrix}\begin{bmatrix} 1 & 0 \\ 0 & 1 \end{bmatrix} = \begin{bmatrix} 1 & 0 \\ 0 & 1 \end{bmatrix}\begin{bmatrix} 1 & 0 \\ 0 & 16 \end{bmatrix} = \begin{bmatrix} 1 & 0 \\ 0 & 16 \end{bmatrix} = A^2.$

© 2016 Cengage Learning. All Rights Reserved. May not be scanned, copied or duplicated, or posted to a publicly accessible website, in whole or in part.

11. First, observe that $A^2 = \begin{bmatrix} 2 & 2 \\ 0 & 1 \end{bmatrix}\begin{bmatrix} 2 & 2 \\ 0 & 1 \end{bmatrix} = \begin{bmatrix} 4 & 6 \\ 0 & 1 \end{bmatrix}$. The eigenvalues of A satisfy the characteristic polynomial $(2 - \lambda)(1 - \lambda) = 0$, which has solutions $\lambda_1 = 1$ and $\lambda_2 = 2$. Solving the system $A\mathbf{v} = \lambda\mathbf{v}$ for each eigenvalue gives the associated eigenvectors $\mathbf{v}_1 = [2, -1]$ and $\mathbf{v}_2 = [1, 0]$. So $D = \begin{bmatrix} 1 & 0 \\ 0 & 2 \end{bmatrix}$, $D^2 = \begin{bmatrix} 1 & 0 \\ 0 & 4 \end{bmatrix}$ and $P = \begin{bmatrix} 2 & 1 \\ -1 & 0 \end{bmatrix}$ $\Rightarrow$ $P^{-1} = \frac{1}{1}\begin{bmatrix} 0 & -1 \\ 1 & 2 \end{bmatrix}$. Therefore, $PD^2P^{-1} = \begin{bmatrix} 2 & 1 \\ -1 & 0 \end{bmatrix}\begin{bmatrix} 1 & 0 \\ 0 & 4 \end{bmatrix}\begin{bmatrix} 0 & -1 \\ 1 & 2 \end{bmatrix} = \begin{bmatrix} 2 & 1 \\ -1 & 0 \end{bmatrix}\begin{bmatrix} 0 & -1 \\ 4 & 8 \end{bmatrix} = \begin{bmatrix} 4 & 6 \\ 0 & 1 \end{bmatrix} = A^2$.

13. First, observe that $A^2 = \begin{bmatrix} a & 0 \\ 0 & b \end{bmatrix}\begin{bmatrix} a & 0 \\ 0 & b \end{bmatrix} = \begin{bmatrix} a^2 & 0 \\ 0 & b^2 \end{bmatrix}$. The eigenvalues of A satisfy the characteristic polynomial $(a - \lambda)(b - \lambda) = 0$, which has solutions $\lambda_1 = a$ and $\lambda_2 = b$. These are distinct solutions since $a \neq b$. Solving the system $A\mathbf{v} = \lambda\mathbf{v}$ for each eigenvalue gives the associated eigenvectors $\mathbf{v}_1 = [1, 0]$ and $\mathbf{v}_2 = [0, 1]$. So $D = \begin{bmatrix} a & 0 \\ 0 & b \end{bmatrix}$, $D^2 = \begin{bmatrix} a^2 & 0 \\ 0 & b^2 \end{bmatrix}$ and $P = \begin{bmatrix} 1 & 0 \\ 0 & 1 \end{bmatrix}$ $\Rightarrow$ $P^{-1} = \frac{1}{1}\begin{bmatrix} 1 & 0 \\ 0 & 1 \end{bmatrix}$. Therefore,

$$PD^2P^{-1} = \begin{bmatrix} 1 & 0 \\ 0 & 1 \end{bmatrix}\begin{bmatrix} a^2 & 0 \\ 0 & b^2 \end{bmatrix}\begin{bmatrix} 1 & 0 \\ 0 & 1 \end{bmatrix} = \begin{bmatrix} 1 & 0 \\ 0 & 1 \end{bmatrix}\begin{bmatrix} a^2 & 0 \\ 0 & b^2 \end{bmatrix} = \begin{bmatrix} a^2 & 0 \\ 0 & b^2 \end{bmatrix} = A^2$$

15. In Exercise 8.8.1, it was shown that the eigenvalues of $A = \begin{bmatrix} 2 & 0 \\ 0 & 1 \end{bmatrix}$ are $\lambda_1 = 1$ and $\lambda_2 = 2$ and the associated eigenvectors are $\mathbf{v}_1 = [0, 1]$ and $\mathbf{v}_2 = [1, 0]$. So $D = \begin{bmatrix} 1 & 0 \\ 0 & 2 \end{bmatrix}$ and $P = \begin{bmatrix} 0 & 1 \\ 1 & 0 \end{bmatrix}$ $\Rightarrow$ $P^{-1} = \frac{1}{-1}\begin{bmatrix} 0 & -1 \\ -1 & 0 \end{bmatrix} = \begin{bmatrix} 0 & 1 \\ 1 & 0 \end{bmatrix}$.

Also $\begin{bmatrix} c_1 \\ c_2 \end{bmatrix} = P^{-1}\mathbf{n}_0 = \begin{bmatrix} 0 & 1 \\ 1 & 0 \end{bmatrix}\begin{bmatrix} 1 \\ 1 \end{bmatrix} = \begin{bmatrix} 1 \\ 1 \end{bmatrix}$. Therefore, the solution is given by

$$\mathbf{n}_t = c_1 \begin{bmatrix} | \\ \mathbf{v}_1 \\ | \end{bmatrix} \lambda_1^t + c_2 \begin{bmatrix} | \\ \mathbf{v}_2 \\ | \end{bmatrix} \lambda_2^t = (1)\begin{bmatrix} 0 \\ 1 \end{bmatrix}(1)^t + (1)\begin{bmatrix} 1 \\ 0 \end{bmatrix}(2)^t = \begin{bmatrix} 0 \\ 1 \end{bmatrix} + \begin{bmatrix} 1 \\ 0 \end{bmatrix} 2^t$$

17. The eigenvalues of $A = \begin{bmatrix} 1 & 1 \\ 0 & 2 \end{bmatrix}$ satisfy the characteristic polynomial $(1 - \lambda)(2 - \lambda) = 0$, which has solutions $\lambda_1 = 1$ and $\lambda_2 = 2$. Solving the system $A\mathbf{v} = \lambda\mathbf{v}$ for each eigenvalue gives the associated eigenvectors $\mathbf{v}_1 = [1, 0]$ and $\mathbf{v}_2 = [1, 1]$. So $D = \begin{bmatrix} 1 & 0 \\ 0 & 2 \end{bmatrix}$ and $P = \begin{bmatrix} 1 & 1 \\ 0 & 1 \end{bmatrix}$ $\Rightarrow$ $P^{-1} = \frac{1}{1}\begin{bmatrix} 1 & -1 \\ 0 & 1 \end{bmatrix}$. Also $\begin{bmatrix} c_1 \\ c_2 \end{bmatrix} = P^{-1}\mathbf{n}_0 = \begin{bmatrix} 1 & -1 \\ 0 & 1 \end{bmatrix}\begin{bmatrix} 1 \\ 1 \end{bmatrix} = \begin{bmatrix} 0 \\ 1 \end{bmatrix}$.

Therefore, the solution is given by

$$\mathbf{n}_t = c_1 \begin{bmatrix} | \\ \mathbf{v}_1 \\ | \end{bmatrix} \lambda_1^t + c_2 \begin{bmatrix} | \\ \mathbf{v}_2 \\ | \end{bmatrix} \lambda_2^t = (0)\begin{bmatrix} 1 \\ 0 \end{bmatrix} 1^t + (1)\begin{bmatrix} 1 \\ 1 \end{bmatrix} 2^t = \begin{bmatrix} 1 \\ 1 \end{bmatrix} 2^t$$

© 2016 Cengage Learning. All Rights Reserved. May not be scanned, copied or duplicated, or posted to a publicly accessible website, in whole or in part.

19. The eigenvalues of $A = \begin{bmatrix} 1 & a \\ 0 & 2 \end{bmatrix}$ satisfy the characteristic polynomial $(1 - \lambda)(2 - \lambda) = 0$, which has solutions $\lambda_1 = 1$ and $\lambda_2 = 2$. Solving the system $A\mathbf{v} = \lambda\mathbf{v}$ for each eigenvalue gives the associated eigenvectors $\mathbf{v}_1 = [1, 0]$ and $\mathbf{v}_2 = [a, 1]$ with $a \neq 0$. So $D = \begin{bmatrix} 1 & 0 \\ 0 & 2 \end{bmatrix}$ and $P = \begin{bmatrix} 1 & a \\ 0 & 1 \end{bmatrix}$ $\Rightarrow$ $P^{-1} = \frac{1}{1}\begin{bmatrix} 1 & -a \\ 0 & 1 \end{bmatrix}$. Also $\begin{bmatrix} c_1 \\ c_2 \end{bmatrix} = P^{-1}\mathbf{n}_0 = \begin{bmatrix} 1 & -a \\ 0 & 1 \end{bmatrix}\begin{bmatrix} 1 \\ 1 \end{bmatrix} = \begin{bmatrix} 1-a \\ 1 \end{bmatrix}$.

Therefore, the solution is given by

$$\mathbf{n}_t = c_1 \begin{bmatrix} | \\ \mathbf{v}_1 \\ | \end{bmatrix} \lambda_1^t + c_2 \begin{bmatrix} | \\ \mathbf{v}_2 \\ | \end{bmatrix} \lambda_2^t = (1-a)\begin{bmatrix} 1 \\ 0 \end{bmatrix} 1^t + (1)\begin{bmatrix} a \\ 1 \end{bmatrix} 2^t = \begin{bmatrix} 1-a \\ 0 \end{bmatrix} + \begin{bmatrix} a \\ 1 \end{bmatrix} 2^t$$

21-25 Denote the arbitrary initial condition by $\mathbf{n}_0 = \begin{bmatrix} x_0 & y_0 \end{bmatrix}^T$.

21. In exercise 8.8.13, it was shown that the eigenvalues of $A = \begin{bmatrix} a & 0 \\ 0 & b \end{bmatrix}$ are $\lambda_1 = a$ and $\lambda_2 = b$ and the associated eigenvectors are $\mathbf{v}_1 = [1, 0]$ and $\mathbf{v}_2 = [0, 1]$. So $D = \begin{bmatrix} a & 0 \\ 0 & b \end{bmatrix}$ and $P = \begin{bmatrix} 1 & 0 \\ 0 & 1 \end{bmatrix}$ $\Rightarrow$ $P^{-1} = \frac{1}{1}\begin{bmatrix} 1 & 0 \\ 0 & 1 \end{bmatrix}$.

Also $\begin{bmatrix} c_1 \\ c_2 \end{bmatrix} = P^{-1}\mathbf{n}_0 = \begin{bmatrix} 1 & 0 \\ 0 & 1 \end{bmatrix}\begin{bmatrix} x_0 \\ y_0 \end{bmatrix} = \begin{bmatrix} x_0 \\ y_0 \end{bmatrix}$. Therefore, the solution is given by

$$\mathbf{n}_t = c_1 \begin{bmatrix} | \\ \mathbf{v}_1 \\ | \end{bmatrix} \lambda_1^t + c_2 \begin{bmatrix} | \\ \mathbf{v}_2 \\ | \end{bmatrix} \lambda_2^t = x_0 \begin{bmatrix} 1 \\ 0 \end{bmatrix} a^t + y_0 \begin{bmatrix} 0 \\ 1 \end{bmatrix} b^t$$

23. The eigenvalues of $A = \begin{bmatrix} 2 & 1 \\ 0 & 1 \end{bmatrix}$ satisfy the characteristic polynomial $(2 - \lambda)(1 - \lambda) = 0$, which has solutions $\lambda_1 = 1$ and $\lambda_2 = 2$. Solving the system $A\mathbf{v} = \lambda\mathbf{v}$ for each eigenvalue gives the associated eigenvectors $\mathbf{v}_1 = [1, -1]$ and $\mathbf{v}_2 = [1, 0]$.

So $D = \begin{bmatrix} 1 & 0 \\ 0 & 2 \end{bmatrix}$ and $P = \begin{bmatrix} 1 & 1 \\ -1 & 0 \end{bmatrix}$ $\Rightarrow$ $P^{-1} = \frac{1}{1}\begin{bmatrix} 0 & -1 \\ 1 & 1 \end{bmatrix}$. Also $\begin{bmatrix} c_1 \\ c_2 \end{bmatrix} = P^{-1}\mathbf{n}_0 = \begin{bmatrix} 0 & -1 \\ 1 & 1 \end{bmatrix}\begin{bmatrix} x_0 \\ y_0 \end{bmatrix} = \begin{bmatrix} -y_0 \\ x_0 + y_0 \end{bmatrix}$.

Therefore, the solution is given by

$$\mathbf{n}_t = c_1 \begin{bmatrix} | \\ \mathbf{v}_1 \\ | \end{bmatrix} \lambda_1^t + c_2 \begin{bmatrix} | \\ \mathbf{v}_2 \\ | \end{bmatrix} \lambda_2^t = -y_0 \begin{bmatrix} 1 \\ -1 \end{bmatrix} 1^t + (x_0 + y_0)\begin{bmatrix} 1 \\ 0 \end{bmatrix} 2^t = -y_0 \begin{bmatrix} 1 \\ -1 \end{bmatrix} + (x_0 + y_0)\begin{bmatrix} 1 \\ 0 \end{bmatrix} 2^t$$

25. The eigenvalues of $A = \begin{bmatrix} a & b \\ b & a \end{bmatrix}$ satisfy the characteristic polynomial $(a - \lambda)^2 - b^2 = 0$, which has solutions $\lambda_1 = a + b$ and $\lambda_2 = a - b$. The condition $b \neq 0$ ensures the eigenvalues are distinct. Solving the system $A\mathbf{v} = \lambda\mathbf{v}$ for each eigenvalue gives the associated eigenvectors $\mathbf{v}_1 = [1, 1]$ and $\mathbf{v}_2 = [1, -1]$. So $D = \begin{bmatrix} a+b & 0 \\ 0 & a-b \end{bmatrix}$ and $P = \begin{bmatrix} 1 & 1 \\ 1 & -1 \end{bmatrix}$ $\Rightarrow$

$P^{-1} = \frac{1}{-2}\begin{bmatrix} -1 & -1 \\ -1 & 1 \end{bmatrix} = \begin{bmatrix} \frac{1}{2} & \frac{1}{2} \\ \frac{1}{2} & -\frac{1}{2} \end{bmatrix}$. Also $\begin{bmatrix} c_1 \\ c_2 \end{bmatrix} = P^{-1}\mathbf{n}_0 = \begin{bmatrix} \frac{1}{2} & \frac{1}{2} \\ \frac{1}{2} & -\frac{1}{2} \end{bmatrix}\begin{bmatrix} x_0 \\ y_0 \end{bmatrix} = \begin{bmatrix} \frac{1}{2}(x_0 + y_0) \\ \frac{1}{2}(x_0 - y_0) \end{bmatrix}$.

Therefore, the solution is given by

$$\mathbf{n}_t = c_1 \begin{bmatrix} | \\ \mathbf{v}_1 \\ | \end{bmatrix} \lambda_1^t + c_2 \begin{bmatrix} | \\ \mathbf{v}_2 \\ | \end{bmatrix} \lambda_2^t = \tfrac{1}{2}(x_0 + y_0)\begin{bmatrix} 1 \\ 1 \end{bmatrix} (a+b)^t + \tfrac{1}{2}(x_0 - y_0)\begin{bmatrix} 1 \\ -1 \end{bmatrix} (a-b)^t$$

© 2016 Cengage Learning. All Rights Reserved. May not be scanned, copied or duplicated, or posted to a publicly accessible website, in whole or in part.

27. $A\mathbf{x} = \lambda\mathbf{x} \;\Rightarrow\; \overline{A\mathbf{x}} = \overline{\lambda\mathbf{x}} \;\Rightarrow\; \bar{A}\bar{\mathbf{x}} = \bar{\lambda}\bar{\mathbf{x}} \;\Rightarrow\; A\bar{\mathbf{x}} = \bar{\lambda}\bar{\mathbf{x}}$ $(A = \bar{A}$ since A is real$)$

Therefore, if λ is a complex eigenvalue of A with associated eigenvector $\mathbf{x}$, then $\bar{\lambda}$ is also an eigenvector of A with associated eigenvector $\bar{\mathbf{x}}$.

29. (a) If the complex eigenvalues of A are $\lambda = a \pm bi$, then we have $D = \begin{bmatrix} a+bi & 0 \\ 0 & a-bi \end{bmatrix}$. When $T = \begin{bmatrix} a & -b \\ b & a \end{bmatrix}$,

$$R = \begin{bmatrix} i & -i \\ 1 & 1 \end{bmatrix} \;\Rightarrow\; R^{-1} = \frac{1}{2i}\begin{bmatrix} 1 & i \\ -1 & i \end{bmatrix} = \begin{bmatrix} -\frac{1}{2}i & \frac{1}{2} \\ \frac{1}{2}i & \frac{1}{2} \end{bmatrix}, \text{ then}$$

$$R^{-1}TR = \begin{bmatrix} -\frac{1}{2}i & \frac{1}{2} \\ \frac{1}{2}i & \frac{1}{2} \end{bmatrix}\begin{bmatrix} a & -b \\ b & a \end{bmatrix}\begin{bmatrix} i & -i \\ 1 & 1 \end{bmatrix} = \begin{bmatrix} -\frac{1}{2}i & \frac{1}{2} \\ \frac{1}{2}i & \frac{1}{2} \end{bmatrix}\begin{bmatrix} ai-b & -ai-b \\ bi+a & -bi+a \end{bmatrix}$$

$$= \begin{bmatrix} \left(\frac{1}{2}a + \frac{1}{2}bi\right) + \left(\frac{1}{2}bi + \frac{1}{2}a\right) & \left(-\frac{1}{2}a + \frac{1}{2}bi\right) + \left(-\frac{1}{2}bi + \frac{1}{2}a\right) \\ \left(-\frac{1}{2}a - \frac{1}{2}bi\right) + \left(\frac{1}{2}bi + \frac{1}{2}a\right) & \left(\frac{1}{2}a - \frac{1}{2}bi\right) + \left(-\frac{1}{2}bi + \frac{1}{2}a\right) \end{bmatrix} = \begin{bmatrix} a+bi & 0 \\ 0 & a-bi \end{bmatrix} = D$$

(b) Recall from exercise 8.8.27, that the eigenvalues and eigenvectors of real matrices always come in complex conjugate pairs. So if $\mathbf{v} = \begin{bmatrix} v_1 \\ v_2 \end{bmatrix}$ is one eigenvector of A, then the other eigenvector is $\bar{\mathbf{v}} = \begin{bmatrix} \bar{v}_1 \\ \bar{v}_2 \end{bmatrix}$. Using these eigenvectors as the columns of P, we can calculate S

$$S = PR^{-1} = \begin{bmatrix} v_1 & \bar{v}_1 \\ v_2 & \bar{v}_2 \end{bmatrix}\begin{bmatrix} \frac{1}{2i} & \frac{1}{2} \\ -\frac{1}{2i} & \frac{1}{2} \end{bmatrix} = \begin{bmatrix} \frac{1}{2i}(v_1 - \bar{v}_1) & \frac{1}{2}(v_1 + \bar{v}_1) \\ \frac{1}{2i}(v_2 - \bar{v}_2) & \frac{1}{2}(v_2 + \bar{v}_2) \end{bmatrix} = \begin{bmatrix} \operatorname{Im} v_1 & \operatorname{Re} v_1 \\ \operatorname{Im} v_2 & \operatorname{Re} v_2 \end{bmatrix}$$

The last step follows from the fact that for an arbitrary complex number $z = c + di$, the real part can be written as $\operatorname{Re} z = \frac{1}{2}(z + \bar{z}) = c$ and the imaginary part can be expressed as $\operatorname{Im} z = \frac{1}{2i}(z - \bar{z}) = d$.

(c) Using the result of exercise 8.8.28, we can write $T = \begin{bmatrix} a & -b \\ b & a \end{bmatrix} = r\begin{bmatrix} \cos\theta & -\sin\theta \\ \sin\theta & \cos\theta \end{bmatrix}$ where $r = \sqrt{a^2 + b^2}$ is the modulus of the eigenvalues and θ is the argument. So

$$A = STS^{-1} = S\left(r\begin{bmatrix} \cos\theta & -\sin\theta \\ \sin\theta & \cos\theta \end{bmatrix}\right)S^{-1} = rS\begin{bmatrix} \cos\theta & -\sin\theta \\ \sin\theta & \cos\theta \end{bmatrix}S^{-1}$$

(d) $A = STS^{-1} \;\Rightarrow\; A^2 = \left(STS^{-1}\right)\left(STS^{-1}\right) = ST^2S^{-1} \;\Rightarrow\; A^3 = \left(ST^2S^{-1}\right)\left(STS^{-1}\right) = ST^3S^{-1}$

Observing the pattern in the powers of A, we infer that $A^t = ST^tS^{-1} = r^tS\begin{bmatrix} \cos\theta & -\sin\theta \\ \sin\theta & \cos\theta \end{bmatrix}^t S^{-1}$. Now the matrix $\begin{bmatrix} \cos\theta & -\sin\theta \\ \sin\theta & \cos\theta \end{bmatrix}$ rotates vectors by an angle θ, so the matrix $\begin{bmatrix} \cos\theta & -\sin\theta \\ \sin\theta & \cos\theta \end{bmatrix}^t$ applies t of these rotations in sequence, that is, it rotates vectors by an angle θt. Therefore $\begin{bmatrix} \cos\theta & -\sin\theta \\ \sin\theta & \cos\theta \end{bmatrix}^t = \begin{bmatrix} \cos\theta t & -\sin\theta t \\ \sin\theta t & \cos\theta t \end{bmatrix}$ which gives

$$A^t = r^tS\begin{bmatrix} \cos\theta t & -\sin\theta t \\ \sin\theta t & \cos\theta t \end{bmatrix}S^{-1}.$$

© 2016 Cengage Learning. All Rights Reserved. May not be scanned, copied or duplicated, or posted to a publicly accessible website, in whole or in part.

31. In exercise 8.8.4, it was shown that obtaining the eigenvalues and eigenvectors of $A = \begin{bmatrix} 1 & -1 \\ 1 & 1 \end{bmatrix}$ gives $D = \begin{bmatrix} 1+i & 0 \\ 0 & 1-i \end{bmatrix}$,

$P = \begin{bmatrix} i & 1 \\ 1 & i \end{bmatrix}$ and $P^{-1} = \begin{bmatrix} -\frac{1}{2}i & \frac{1}{2} \\ \frac{1}{2} & -\frac{1}{2}i \end{bmatrix}$. So $\begin{bmatrix} c_1 \\ c_2 \end{bmatrix} = P^{-1}\mathbf{n}_0 = \begin{bmatrix} -\frac{1}{2}i & \frac{1}{2} \\ \frac{1}{2} & -\frac{1}{2}i \end{bmatrix} \begin{bmatrix} 1 \\ 1 \end{bmatrix} = \begin{bmatrix} \frac{1}{2} - \frac{1}{2}i \\ \frac{1}{2} - \frac{1}{2}i \end{bmatrix}$.

Therefore, the solution is given by

$$\begin{aligned}
\mathbf{n}_t &= c_1 \begin{bmatrix} | \\ \mathbf{v}_1 \\ | \end{bmatrix} \lambda_1^t + c_2 \begin{bmatrix} | \\ \mathbf{v}_2 \\ | \end{bmatrix} \lambda_2^t = \left(\tfrac{1}{2} - \tfrac{1}{2}i\right) \begin{bmatrix} i \\ 1 \end{bmatrix} (1+i)^t + \left(\tfrac{1}{2} - \tfrac{1}{2}i\right) \begin{bmatrix} 1 \\ i \end{bmatrix} (1-i)^t \\
&= \tfrac{1}{2}(1-i) \left(\begin{bmatrix} i \\ 1 \end{bmatrix} (1+i)^t + \begin{bmatrix} 1 \\ i \end{bmatrix} (1-i)^t \right)
\end{aligned}$$

Using De Moivre's Theorem, we find $(1 \pm i)^t = (\sqrt{2})^t \left(\cos\left(\pm\frac{\pi}{4}t\right) + i\sin\left(\pm\frac{\pi}{4}t\right)\right)$, so the solution becomes

$$\mathbf{n}_t = \tfrac{1}{2}(1-i) \left(\begin{bmatrix} i \\ 1 \end{bmatrix} (\sqrt{2})^t \left(\cos\left(\tfrac{\pi}{4}t\right) + i\sin\left(\tfrac{\pi}{4}t\right)\right) + \begin{bmatrix} 1 \\ i \end{bmatrix} (\sqrt{2})^t \left(\cos\left(-\tfrac{\pi}{4}t\right) + i\sin\left(-\tfrac{\pi}{4}t\right)\right) \right)$$

$$= \tfrac{1}{2}(\sqrt{2})^t (1-i) \left(\begin{bmatrix} i\cos\left(\frac{\pi}{4}t\right) - \sin\left(\frac{\pi}{4}t\right) \\ \cos\left(\frac{\pi}{4}t\right) + i\sin\left(\frac{\pi}{4}t\right) \end{bmatrix} + \begin{bmatrix} \cos\left(-\frac{\pi}{4}t\right) + i\sin\left(-\frac{\pi}{4}t\right) \\ i\cos\left(-\frac{\pi}{4}t\right) - \sin\left(-\frac{\pi}{4}t\right) \end{bmatrix} \right)$$

$$= \tfrac{1}{2}(\sqrt{2})^t (1-i) \begin{bmatrix} i\cos\left(\frac{\pi}{4}t\right) - \sin\left(\frac{\pi}{4}t\right) + \cos\left(\frac{\pi}{4}t\right) - i\sin\left(\frac{\pi}{4}t\right) \\ \cos\left(\frac{\pi}{4}t\right) + i\sin\left(\frac{\pi}{4}t\right) + i\cos\left(\frac{\pi}{4}t\right) + \sin\left(\frac{\pi}{4}t\right) \end{bmatrix} \quad \text{since } \sin(-x) = \sin x \text{ and } \cos(-x) = \cos x$$

$$= \tfrac{1}{2}(\sqrt{2})^t \times$$

$$\begin{bmatrix} i\cos\left(\frac{\pi}{4}t\right) - \sin\left(\frac{\pi}{4}t\right) + \cos\left(\frac{\pi}{4}t\right) - i\sin\left(\frac{\pi}{4}t\right) - i\left(i\cos\left(\frac{\pi}{4}t\right) - \sin\left(\frac{\pi}{4}t\right) + \cos\left(\frac{\pi}{4}t\right) - i\sin\left(\frac{\pi}{4}t\right)\right) \\ \cos\left(\frac{\pi}{4}t\right) + i\sin\left(\frac{\pi}{4}t\right) + i\cos\left(\frac{\pi}{4}t\right) + \sin\left(\frac{\pi}{4}t\right) - i\left(\cos\left(\frac{\pi}{4}t\right) + i\sin\left(\frac{\pi}{4}t\right) + i\cos\left(\frac{\pi}{4}t\right) + \sin\left(\frac{\pi}{4}t\right)\right) \end{bmatrix}$$

$$= \tfrac{1}{2}(\sqrt{2})^t \begin{bmatrix} 2\cos\left(\frac{\pi}{4}t\right) - 2\sin\left(\frac{\pi}{4}t\right) \\ 2\sin\left(\frac{\pi}{4}t\right) + 2\cos\left(\frac{\pi}{4}t\right) \end{bmatrix} = (\sqrt{2})^t \begin{bmatrix} \cos\left(\frac{\pi}{4}t\right) - \sin\left(\frac{\pi}{4}t\right) \\ \sin\left(\frac{\pi}{4}t\right) + \cos\left(\frac{\pi}{4}t\right) \end{bmatrix} = (\sqrt{2})^t \begin{bmatrix} \cos\left(\frac{\pi}{4}t\right) & -\sin\left(\frac{\pi}{4}t\right) \\ \sin\left(\frac{\pi}{4}t\right) & \cos\left(\frac{\pi}{4}t\right) \end{bmatrix} \begin{bmatrix} 1 \\ 1 \end{bmatrix}$$

Note: Another way of completing the exercise is to make use of the matrix factorization derived in Exercise 29d.

33. The eigenvalues of $A = \begin{bmatrix} 1 & 5 \\ -1 & 1 \end{bmatrix}$ satisfy the characteristic polynomial $(1-\lambda)^2 + 5 = 0$, which has solutions $\lambda_1 = 1 + \sqrt{5}i$

and $\lambda_2 = 1 - \sqrt{5}i$. Solving the system $A\mathbf{v} = \lambda\mathbf{v}$ for each eigenvalue gives the associated eigenvectors

$\mathbf{v}_1 = \left[\sqrt{5}, i\right]$ and $\mathbf{v}_2 = \left[\sqrt{5}i, 1\right]$. So $D = \begin{bmatrix} 1+\sqrt{5}i & 0 \\ 0 & 1-\sqrt{5}i \end{bmatrix}$ and $P = \begin{bmatrix} \sqrt{5} & \sqrt{5}i \\ i & 1 \end{bmatrix} \Rightarrow$

$P^{-1} = \dfrac{1}{2\sqrt{5}} \begin{bmatrix} 1 & -\sqrt{5}i \\ -i & \sqrt{5} \end{bmatrix} = \begin{bmatrix} \frac{1}{2\sqrt{5}} & -\frac{1}{2}i \\ -\frac{1}{2\sqrt{5}}i & \frac{1}{2} \end{bmatrix}$. Also $\begin{bmatrix} c_1 \\ c_2 \end{bmatrix} = P^{-1}\mathbf{n}_0 = \begin{bmatrix} \frac{1}{2\sqrt{5}} & -\frac{1}{2}i \\ -\frac{1}{2\sqrt{5}}i & \frac{1}{2} \end{bmatrix} \begin{bmatrix} 0 \\ 1 \end{bmatrix} = \begin{bmatrix} -\frac{1}{2}i \\ \frac{1}{2} \end{bmatrix}$.

© 2016 Cengage Learning. All Rights Reserved. May not be scanned, copied or duplicated, or posted to a publicly accessible website, in whole or in part.

Therefore, the solution is given by

$$\mathbf{n}_t = c_1 \begin{bmatrix} | \\ \mathbf{v}_1 \\ | \end{bmatrix} \lambda_1^t + c_2 \begin{bmatrix} | \\ \mathbf{v}_2 \\ | \end{bmatrix} \lambda_2^t = -\tfrac{1}{2}i \begin{bmatrix} \sqrt{5} \\ i \end{bmatrix} \left(1+\sqrt{5}i\right)^t + \tfrac{1}{2} \begin{bmatrix} \sqrt{5}i \\ 1 \end{bmatrix} \left(1-\sqrt{5}i\right)^t$$

$$= \tfrac{1}{2}\left(\begin{bmatrix} -\sqrt{5}i \\ 1 \end{bmatrix} \left(1+\sqrt{5}i\right)^t + \begin{bmatrix} \sqrt{5}i \\ 1 \end{bmatrix} \left(1-\sqrt{5}i\right)^t \right)$$

Using De Moivre's Theorem, we find $\left(1 \pm \sqrt{5}i\right)^t = \left(\sqrt{6}\right)^t \left(\cos(\pm\theta t) + i\sin(\pm\theta t)\right)$ where $\theta = \tan^{-1}\sqrt{5}$. So the solution becomes

$$\mathbf{n}_t = \tfrac{1}{2}\left(\begin{bmatrix} -\sqrt{5}i \\ 1 \end{bmatrix} \left(\sqrt{6}\right)^t \left(\cos(\theta t) + i\sin(\theta t)\right) + \begin{bmatrix} \sqrt{5}i \\ 1 \end{bmatrix} \left(\sqrt{6}\right)^t \left(\cos(-\theta t) + i\sin(-\theta t)\right) \right)$$

$$= \tfrac{1}{2}\left(\sqrt{6}\right)^t \left(\begin{bmatrix} -\sqrt{5}i\cos(\theta t) + \sqrt{5}\sin(\theta t) \\ \cos(\theta t) + i\sin(\theta t) \end{bmatrix} + \begin{bmatrix} \sqrt{5}i\cos(-\theta t) - \sqrt{5}\sin(-\theta t) \\ \cos(-\theta t) + i\sin(-\theta t) \end{bmatrix} \right)$$

$$= \tfrac{1}{2}\left(\sqrt{6}\right)^t \begin{bmatrix} -\sqrt{5}i\cos(\theta t) + \sqrt{5}\sin(\theta t) + \sqrt{5}i\cos(\theta t) + \sqrt{5}\sin(\theta t) \\ \cos(\theta t) + i\sin(\theta t) + \cos(\theta t) - i\sin(\theta t) \end{bmatrix} \quad \text{since } \sin(-x) = \sin x \text{ and } \cos(-x) = \cos x$$

$$= \tfrac{1}{2}\left(\sqrt{6}\right)^t \begin{bmatrix} 2\sqrt{5}\sin(\theta t) \\ 2\cos(\theta t) \end{bmatrix} = \left(\sqrt{6}\right)^t \begin{bmatrix} \sqrt{5}\sin(\theta t) \\ \cos(\theta t) \end{bmatrix} \quad \text{where } \theta = \tan^{-1}\sqrt{5}$$

Note: Another way of completing the exercise is to make use of the matrix factorization derived in Exercise 29d.

35. (a) $A = \begin{bmatrix} 1 & 0 \\ 0 & -1 \end{bmatrix} \Rightarrow A^2 = \begin{bmatrix} 1 & 0 \\ 0 & -1 \end{bmatrix}\begin{bmatrix} 1 & 0 \\ 0 & -1 \end{bmatrix} = \begin{bmatrix} 1 & 0 \\ 0 & 1 \end{bmatrix} \Rightarrow A^3 = \begin{bmatrix} 1 & 0 \\ 0 & 1 \end{bmatrix}\begin{bmatrix} 1 & 0 \\ 0 & -1 \end{bmatrix} = \begin{bmatrix} 1 & 0 \\ 0 & -1 \end{bmatrix}$

Observe from the pattern of the powers of A that A^k always has zero entries in the off-diagonal terms. Thus, A is *not* primitive so the Perron-Frobenius Theorem cannot be applied to this model.

(b) The eigenvalues of A are $\lambda_1 = 1$ and $\lambda_2 = -1$ and the corresponding eigenvectors are $\mathbf{u}_1 = \begin{bmatrix} 1 \\ 0 \end{bmatrix}$ and $\mathbf{u}_2 = \begin{bmatrix} 0 \\ 1 \end{bmatrix}$ (see exercise 8.7.26). So $D = \begin{bmatrix} 1 & 0 \\ 0 & -1 \end{bmatrix}$ and $P = \begin{bmatrix} 1 & 0 \\ 0 & 1 \end{bmatrix} \Rightarrow P^{-1} = \frac{1}{1}\begin{bmatrix} 1 & 0 \\ 0 & 1 \end{bmatrix}$.

Also $\begin{bmatrix} c_1 \\ c_2 \end{bmatrix} = P^{-1}\mathbf{v}_0 = \begin{bmatrix} 1 & 0 \\ 0 & 1 \end{bmatrix}\begin{bmatrix} 0.3 \\ -0.2 \end{bmatrix} = \begin{bmatrix} 0.3 \\ -0.2 \end{bmatrix}$. Therefore, the solution is given by

$$\mathbf{v}_t = c_1 \begin{bmatrix} | \\ \mathbf{u}_1 \\ | \end{bmatrix} \lambda_1^t + c_2 \begin{bmatrix} | \\ \mathbf{u}_2 \\ | \end{bmatrix} \lambda_2^t = 0.3 \begin{bmatrix} 1 \\ 0 \end{bmatrix} (1)^t + (-0.2) \begin{bmatrix} 0 \\ 1 \end{bmatrix} (-1)^t = \begin{bmatrix} 0.3 \\ 0 \end{bmatrix} + \begin{bmatrix} 0 \\ -0.2 \end{bmatrix} (-1)^t$$

(c) When t is an even integer $(-1)^t = 1$ so that $\mathbf{v}_t = \begin{bmatrix} 0.3 \\ -0.2 \end{bmatrix}$ and when t is an odd integer $(-1)^t = -1$ so that $\mathbf{v}_t = \begin{bmatrix} 0.3 \\ 0.2 \end{bmatrix}$.

Thus, the voltage vector $\mathbf{v}_t$ cycles between $\begin{bmatrix} 0.3 \\ -0.2 \end{bmatrix}$ and $\begin{bmatrix} 0.3 \\ 0.2 \end{bmatrix}$.

© 2016 Cengage Learning. All Rights Reserved. May not be scanned, copied or duplicated, or posted to a publicly accessible website, in whole or in part.

37. (a) The eigenvalues of $A = \begin{bmatrix} \frac{9}{10} & \frac{1}{5} \\ \frac{1}{10} & \frac{4}{5} \end{bmatrix}$ must satisfy $\det(A - \lambda I) = 0 \;\Rightarrow\; \det \begin{bmatrix} \frac{9}{10} - \lambda & \frac{1}{5} \\ \frac{1}{10} & \frac{4}{5} - \lambda \end{bmatrix} = 0 \;\Rightarrow$

$$\left(\tfrac{9}{10} - \lambda\right)\left(\tfrac{4}{5} - \lambda\right) - \tfrac{1}{50} = 0 \;\Rightarrow\; \lambda^2 - \tfrac{17}{10}\lambda + \tfrac{7}{10} = 0 \;\Rightarrow\; \lambda = \frac{17 \pm \sqrt{(-17)^2 - 4\,(10)\,(7)}}{2\,(10)} = \frac{17 \pm 3}{20}.$$

Therefore, the eigenvalues are $\lambda_1 = 1$ and $\lambda_2 = \frac{7}{10}$.

(b) The matrix A is primitive since it contains all positive values, so the Perron-Frobenius theorem says that $\lambda_1 = 1$ is greater in magnitude than all other eigenvalues ($|\lambda_1| > |\lambda_2|$) and the components of the eigenvector corresponding to λ_1 are all positive. Hence, in the long-term, the components of $\mathbf{y}_t$ grow by a factor λ_1 each timestep and asymptotically approach the line defined by the associated eigenvector $\mathbf{v}_1$. But because $\lambda_1 = 1$, the components of $\mathbf{y}_t$ do not actually grow and instead approach a steady state in the long-term.

(c) Solving the system $A\mathbf{v} = \lambda\mathbf{v}$ for each eigenvalue gives the associated eigenvectors $\mathbf{v}_1 = [2, 1]$ and $\mathbf{v}_2 = [1, -1]$.

So $D = \begin{bmatrix} 1 & 0 \\ 0 & \frac{7}{10} \end{bmatrix}$ and $P = \begin{bmatrix} 2 & 1 \\ 1 & -1 \end{bmatrix} \;\Rightarrow\; P^{-1} = \dfrac{1}{-3}\begin{bmatrix} -1 & -1 \\ -1 & 2 \end{bmatrix} = \begin{bmatrix} \frac{1}{3} & \frac{1}{3} \\ \frac{1}{3} & -\frac{2}{3} \end{bmatrix}$.

Also $\begin{bmatrix} c_1 \\ c_2 \end{bmatrix} = P^{-1}\mathbf{y}_0 = \begin{bmatrix} \frac{1}{3} & \frac{1}{3} \\ \frac{1}{3} & -\frac{2}{3} \end{bmatrix}\begin{bmatrix} 1 \\ 0 \end{bmatrix} = \begin{bmatrix} \frac{1}{3} \\ \frac{1}{3} \end{bmatrix}$. Therefore, the solution is given by

$$\mathbf{y}_t = c_1 \begin{bmatrix} | \\ \mathbf{v}_1 \\ | \end{bmatrix} \lambda_1^t + c_2 \begin{bmatrix} | \\ \mathbf{v}_2 \\ | \end{bmatrix} \lambda_2^t = \tfrac{1}{3}\begin{bmatrix} 2 \\ 1 \end{bmatrix}(1)^t + \tfrac{1}{3}\begin{bmatrix} 1 \\ -1 \end{bmatrix}\left(\tfrac{7}{10}\right)^t = \tfrac{1}{3}\left(\begin{bmatrix} 2 \\ 1 \end{bmatrix} + \begin{bmatrix} 1 \\ -1 \end{bmatrix}\left(\tfrac{7}{10}\right)^t\right)$$

(d) $\displaystyle\lim_{t\to\infty} \mathbf{y}_t = \tfrac{1}{3}\begin{bmatrix} 2 \\ 1 \end{bmatrix} + \tfrac{1}{3}\begin{bmatrix} 1 \\ -1 \end{bmatrix}\lim_{t\to\infty}\left(\tfrac{7}{10}\right)^t = \tfrac{1}{3}\begin{bmatrix} 2 \\ 1 \end{bmatrix}$

In the long-term, the solution vector approaches a steady state. As expected, the steady state vector is a scalar multiple of the eigenvector associated with the largest eigenvalue $\lambda_1 = 1$.

39. (a) The eigenvalues of $A = \begin{bmatrix} \frac{8}{10} & \frac{1}{2} \\ \frac{2}{10} & \frac{1}{2} \end{bmatrix}$ must satisfy $\det(A - \lambda I) = 0 \;\Rightarrow\; \det \begin{bmatrix} \frac{8}{10} - \lambda & \frac{1}{2} \\ \frac{2}{10} & \frac{1}{2} - \lambda \end{bmatrix} = 0 \;\Rightarrow$

$$\left(\tfrac{8}{10} - \lambda\right)\left(\tfrac{1}{2} - \lambda\right) - \tfrac{1}{10} = 0 \;\Rightarrow\; \lambda^2 - \tfrac{13}{10}\lambda + \tfrac{3}{10} = 0 \;\Rightarrow\; \lambda = \frac{13 \pm \sqrt{(-13)^2 - 4\,(10)\,(3)}}{2\,(10)} = \frac{13 \pm 7}{20}.$$ Therefore,

the eigenvalues are $\lambda_1 = 1$ and $\lambda_2 = \frac{3}{10}$.

(b) The matrix A is primitive since it contains all positive values, so the Perron-Frobenius theorem says that $\lambda_1 = 1$ is greater in magnitude than all other eigenvalues ($|\lambda_1| > |\lambda_2|$) and the components of the eigenvector corresponding to λ_1 are all positive. Hence, in the long-term, the components of $\mathbf{x}_t$ grow by a factor λ_1 each timestep and asymptotically approach the line defined by the associated eigenvector $\mathbf{v}_1$. But because $\lambda_1 = 1$, the components of $\mathbf{x}_t$ do not actually grow and instead approach a steady state in the long-term.

(c) Solving the system $A\mathbf{v} = \lambda\mathbf{v}$ for each eigenvalue gives the associated eigenvectors $\mathbf{v}_1 = [5, 2]$ and $\mathbf{v}_2 = [1, -1]$.

So $D = \begin{bmatrix} 1 & 0 \\ 0 & \frac{3}{10} \end{bmatrix}$ and $P = \begin{bmatrix} 5 & 1 \\ 2 & -1 \end{bmatrix} \;\Rightarrow\; P^{-1} = \dfrac{1}{-7}\begin{bmatrix} -1 & -1 \\ -2 & 5 \end{bmatrix} = \begin{bmatrix} \frac{1}{7} & \frac{1}{7} \\ \frac{2}{7} & -\frac{5}{7} \end{bmatrix}$.

Also $\begin{bmatrix} c_1 \\ c_2 \end{bmatrix} = P^{-1}\mathbf{x}_0 = \begin{bmatrix} \frac{1}{7} & \frac{1}{7} \\ \frac{2}{7} & -\frac{5}{7} \end{bmatrix}\begin{bmatrix} a \\ 1 - a \end{bmatrix} = \begin{bmatrix} \frac{1}{7} \\ a - \frac{5}{7} \end{bmatrix}$. Therefore, the solution is given by

$$\mathbf{x}_t = c_1 \begin{bmatrix} | \\ \mathbf{v}_1 \\ | \end{bmatrix} \lambda_1^t + c_2 \begin{bmatrix} | \\ \mathbf{v}_2 \\ | \end{bmatrix} \lambda_2^t = \tfrac{1}{7}\begin{bmatrix} 5 \\ 2 \end{bmatrix}(1)^t + \left(a - \tfrac{5}{7}\right)\begin{bmatrix} 1 \\ -1 \end{bmatrix}\left(\tfrac{3}{10}\right)^t = \tfrac{1}{7}\begin{bmatrix} 5 \\ 2 \end{bmatrix} + \left(a - \tfrac{5}{7}\right)\begin{bmatrix} 1 \\ -1 \end{bmatrix}\left(\tfrac{3}{10}\right)^t$$

© 2016 Cengage Learning. All Rights Reserved. May not be scanned, copied or duplicated, or posted to a publicly accessible website, in whole or in part.

(d) $\lim_{t\to\infty} \mathbf{x}_t = \frac{1}{7}\begin{bmatrix}5\\2\end{bmatrix} + \left(a - \frac{5}{7}\right)\begin{bmatrix}1\\-1\end{bmatrix}\lim_{t\to\infty}\left(\frac{3}{10}\right)^t = \frac{1}{7}\begin{bmatrix}5\\2\end{bmatrix}$

In the long-term, the solution vector approaches a steady state. As expected, the steady state vector is a scalar multiple of the eigenvector associated with the largest eigenvalue $\lambda_1 = 1$.

41. (a) $\mathbf{z}_n = \begin{bmatrix}x_n\\y_n\end{bmatrix} = \begin{bmatrix}F_n\\F_{n-1}\end{bmatrix} = \begin{bmatrix}F_{n-1}+F_{n-2}\\F_{n-1}\end{bmatrix} = \begin{bmatrix}1&1\\1&0\end{bmatrix}\begin{bmatrix}F_{n-1}\\F_{n-2}\end{bmatrix} = \begin{bmatrix}1&1\\1&0\end{bmatrix}\begin{bmatrix}x_{n-1}\\y_{n-1}\end{bmatrix} = A\mathbf{z}_{n-1}$

(b) The eigenvalues of $A = \begin{bmatrix}1&1\\1&0\end{bmatrix}$ satisfy the equation $\det(A - \lambda I) = 0 \;\Rightarrow\; \det\begin{bmatrix}1-\lambda&1\\1&-\lambda\end{bmatrix} = 0 \;\Rightarrow$

$-\lambda(1-\lambda) - 1 = 0 \;\Rightarrow\; \lambda^2 - \lambda - 1 = 0 \;\Rightarrow\; \lambda = \dfrac{1 \pm \sqrt{(-1)^2 - 4(1)(-1)}}{2} = \dfrac{1\pm\sqrt{5}}{2}$. Therefore, the eigenvalues are $\lambda_1 = \dfrac{1+\sqrt{5}}{2} \approx 1.6$ and $\lambda_2 = \dfrac{1-\sqrt{5}}{2} \approx -0.6$.

(c) Observe that $A^2 = \begin{bmatrix}1&1\\1&0\end{bmatrix}\begin{bmatrix}1&1\\1&0\end{bmatrix} = \begin{bmatrix}2&1\\1&1\end{bmatrix}$ contains all positive entries so A is primitive. The Perron-Frobenius theorem then says that $\lambda_1 \approx 1.6$ is greater in magnitude than all other eigenvalues ($|\lambda_1| > |\lambda_2|$) and the components of the eigenvector corresponding to λ_1 are all positive. Hence, in the long-term, the components of $\mathbf{z}_n$ grow by a factor λ_1 each timestep and asymptotically approach the line defined by the associated eigenvector $\mathbf{v}_1$.

(d) The eigenvectors $\mathbf{v}$ must satisfy $(A - \lambda I)\,\mathbf{v} = \mathbf{0}$. Starting with the eigenvalue $\lambda_1 = \dfrac{1+\sqrt{5}}{2}$ we have

$$\begin{bmatrix}1-\lambda&1\\1&0-\lambda\end{bmatrix}\begin{bmatrix}v_1\\v_2\end{bmatrix} = \begin{bmatrix}\frac{1-\sqrt{5}}{2}&1\\1&-\frac{1+\sqrt{5}}{2}\end{bmatrix}\begin{bmatrix}v_1\\v_2\end{bmatrix} = \begin{bmatrix}0\\0\end{bmatrix} \;\Rightarrow\; \left\{\begin{aligned}\frac{1-\sqrt{5}}{2}v_1 + v_2 &= 0\\ v_1 - \frac{1+\sqrt{5}}{2}v_2 &= 0\end{aligned}\right\}$$

Both equations specify that $v_2 = -\dfrac{1-\sqrt{5}}{2}v_1$. Choosing $v_1 = 2$ gives the eigenvector $\mathbf{v}_1 = \begin{bmatrix}2\\-1+\sqrt{5}\end{bmatrix}$.

For the second eigenvalue $\lambda_2 = \dfrac{1-\sqrt{5}}{2}$ we have

$$\begin{bmatrix}1-\lambda&1\\1&0-\lambda\end{bmatrix}\begin{bmatrix}v_1\\v_2\end{bmatrix} = \begin{bmatrix}\frac{1+\sqrt{5}}{2}&1\\1&-\frac{1-\sqrt{5}}{2}\end{bmatrix}\begin{bmatrix}v_1\\v_2\end{bmatrix} = \begin{bmatrix}0\\0\end{bmatrix} \;\Rightarrow\; \left\{\begin{aligned}\frac{1+\sqrt{5}}{2}v_1 + v_2 &= 0\\ v_1 - \frac{1-\sqrt{5}}{2}v_2 &= 0\end{aligned}\right\}$$

Both equations specify that $v_2 = -\dfrac{1+\sqrt{5}}{2}v_1$. Choosing $v_1 = 2$ gives the eigenvector $\mathbf{v}_2 = \begin{bmatrix}2\\-1-\sqrt{5}\end{bmatrix}$.

So $D = \begin{bmatrix}\frac{1+\sqrt{5}}{2}&0\\0&\frac{1-\sqrt{5}}{2}\end{bmatrix}$ and $P = \begin{bmatrix}2&2\\-1+\sqrt{5}&-1-\sqrt{5}\end{bmatrix} \;\Rightarrow\; P^{-1} = \dfrac{1}{-4\sqrt{5}}\begin{bmatrix}-1-\sqrt{5}&-2\\1-\sqrt{5}&2\end{bmatrix}$.

[continued]

© 2016 Cengage Learning. All Rights Reserved. May not be scanned, copied or duplicated, or posted to a publicly accessible website, in whole or in part.

Also note that $\mathbf{z}_2 = \begin{bmatrix} x_2 \\ y_2 \end{bmatrix} = \begin{bmatrix} F_2 \\ F_1 \end{bmatrix} = \begin{bmatrix} 1 \\ 1 \end{bmatrix}$ can be used as an initial condition. Therefore, the solution is given by

$$\mathbf{z}_n = PD^{n-2}P^{-1}\mathbf{z}_2 = \frac{1}{-4\sqrt{5}}\begin{bmatrix} 2 & 2 \\ -1+\sqrt{5} & -1-\sqrt{5} \end{bmatrix}\begin{bmatrix} \left(\frac{1+\sqrt{5}}{2}\right)^{n-2} & 0 \\ 0 & \left(\frac{1-\sqrt{5}}{2}\right)^{n-2} \end{bmatrix}\begin{bmatrix} -1-\sqrt{5} & -2 \\ 1-\sqrt{5} & 2 \end{bmatrix}\begin{bmatrix} 1 \\ 1 \end{bmatrix}$$

$$= \frac{1}{-4\sqrt{5}}\begin{bmatrix} 2 & 2 \\ -1+\sqrt{5} & -1-\sqrt{5} \end{bmatrix}\begin{bmatrix} \left(\frac{1+\sqrt{5}}{2}\right)^{n-2} & 0 \\ 0 & \left(\frac{1-\sqrt{5}}{2}\right)^{n-2} \end{bmatrix}\begin{bmatrix} -3-\sqrt{5} \\ 3-\sqrt{5} \end{bmatrix}$$

$$= \frac{1}{-4\sqrt{5}}\begin{bmatrix} 2 & 2 \\ -1+\sqrt{5} & -1-\sqrt{5} \end{bmatrix}\begin{bmatrix} (-3-\sqrt{5})\left(\frac{1+\sqrt{5}}{2}\right)^{n-2} \\ (3-\sqrt{5})\left(\frac{1-\sqrt{5}}{2}\right)^{n-2} \end{bmatrix}$$

$$= \frac{1}{-4\sqrt{5}}\begin{bmatrix} 2\,(-3-\sqrt{5})\left(\frac{1+\sqrt{5}}{2}\right)^{n-2} + 2\,(3-\sqrt{5})\left(\frac{1-\sqrt{5}}{2}\right)^{n-2} \\ (-1+\sqrt{5})\,(-3-\sqrt{5})\left(\frac{1+\sqrt{5}}{2}\right)^{n-2} + (-1-\sqrt{5})\,(3-\sqrt{5})\left(\frac{1-\sqrt{5}}{2}\right)^{n-2} \end{bmatrix}$$

$$= \frac{1}{2\sqrt{5}}\begin{bmatrix} (3+\sqrt{5})\left(\frac{1+\sqrt{5}}{2}\right)^{n-2} + (-3+\sqrt{5})\left(\frac{1-\sqrt{5}}{2}\right)^{n-2} \\ (1+\sqrt{5})\left(\frac{1+\sqrt{5}}{2}\right)^{n-2} + (-1+\sqrt{5})\left(\frac{1-\sqrt{5}}{2}\right)^{n-2} \end{bmatrix} = \begin{bmatrix} F_n \\ F_{n-1} \end{bmatrix}$$

where $n \geq 2$. Note that D was raised to the power $n-2$ instead of n since the initial condition started at $n = 2$. Therefore, the nth Fibonacci number is

$$F_n = \frac{1}{2\sqrt{5}}\,(3+\sqrt{5})\left(\frac{1+\sqrt{5}}{2}\right)^{n-2} + \frac{1}{2\sqrt{5}}\,(-3+\sqrt{5})\left(\frac{1-\sqrt{5}}{2}\right)^{n-2} = \frac{1}{\sqrt{5}}\left[\left(\frac{1+\sqrt{5}}{2}\right)^{n} - \left(\frac{1-\sqrt{5}}{2}\right)^{n}\right]$$

8 Review

TRUE-FALSE QUIZ

1. True, by Property 2 of the dot product. (See Section 8.3.)

3. False. For example, if $\mathbf{u} = [1, 0]$ and $\mathbf{v} = [0, 1]$ then $[1, 0] \cdot [0, 1] = 0$ but $\mathbf{u} \neq \mathbf{0}$ and $\mathbf{v} \neq \mathbf{0}$.

5. This is false. In $\mathbb{R}^2$, $x^2 + y^2 = 1$ represents a circle, but $\{(x, y, z) \mid x^2 + y^2 = 1\}$ represents a *three-dimensional surface*, namely, a circular cylinder aligned with the z-axis.

7. False. In general, matrices do not commute.

For example, if $A = \begin{bmatrix} 1 & 0 \\ 0 & -1 \end{bmatrix}$ and $B = \begin{bmatrix} 0 & 1 \\ 1 & 0 \end{bmatrix}$, then $AB = \begin{bmatrix} 0 & 1 \\ -1 & 0 \end{bmatrix}$ and $BA = \begin{bmatrix} 0 & -1 \\ 1 & 0 \end{bmatrix}$ so $AB \neq BA$.

© 2016 Cengage Learning. All Rights Reserved. May not be scanned, copied or duplicated, or posted to a publicly accessible website, in whole or in part.

9. True. The determinant of a diagonal matrix is given by the product of all diagonal terms (Exercise 8.6.17). Since the identity matrix is diagonal with ones in every non-zero entry, we have $\det I = 1$.

11. True, by the definition of an eigenvector (Definition 8.7.2). The scalar k is the eigenvalue associated with the eigenvector $\mathbf{y}$.

13. False. For example, the matrix $A = \begin{bmatrix} 2 & 1 \\ 1 & 2 \end{bmatrix}$ has eigenvalues $\lambda = 3$ and $\lambda = 1$. These do not lie on the diagonal of A, so in general the statement is false.

15. True. If A has n distinct eigenvalues, it will also have n distinct eigenvectors. In this case, the matrix of eigenvectors P is invertible so P^{-1} can be calculated. The matrix A can then be written as $A = PDP^{-1}$ where D is an $n \times n$ matrix whose diagonal entries are the distinct eigenvalues.

17. True, by Theorem 8.6.5.

EXERCISES

1. (a) The radius of the sphere is the distance between the points $(-1, 2, 1)$ and $(6, -2, 3)$, namely, $\sqrt{[6 - (-1)]^2 + (-2 - 2)^2 + (3 - 1)^2} = \sqrt{69}$. By the formula for an equation of a sphere (see page 522), an equation of the sphere with center $(-1, 2, 1)$ and radius $\sqrt{69}$ is $(x + 1)^2 + (y - 2)^2 + (z - 1)^2 = 69$.

(b) The intersection of this sphere with the yz-plane is the set of points on the sphere whose x-coordinate is 0. Putting $x = 0$ into the equation, we have $(y - 2)^2 + (z - 1)^2 = 68, x = 0$ which represents a circle in the yz-plane with center $(0, 2, 1)$ and radius $\sqrt{68}$.

(c) Completing squares gives $(x - 4)^2 + (y + 1)^2 + (z + 3)^2 = -1 + 16 + 1 + 9 = 25$. Thus the sphere is centered at $(4, -1, -3)$ and has radius 5.

3. Let the center of the hypersphere be $P\,(c_1, c_2, \ldots, c_n)$. If a point $T\,(x_1, x_2, \ldots, x_n)$ lies on the hypersphere, it must satisfy $|PT| = r$. Using the Distance Formula in n Dimension from Section 8.1, we have $|PT| = r \;\;\Rightarrow$ $\sqrt{(x_1 - c_1)^2 + (x_2 - c_2)^2 + \cdots + (x_n - c_n)^2} = r \;\;\Rightarrow\;\; (x_1 - c_1)^2 + (x_2 - c_2)^2 + \cdots + (x_n - c_n)^2 = r^2$. This is the equation of a hypersphere in n dimensions.

5. For the two vectors to be orthogonal, we need $[3, 2, x] \cdot [2x, 4, x] = 0 \;\;\Leftrightarrow\;\; (3)(2x) + (2)(4) + (x)(x) = 0 \;\;\Leftrightarrow$ $x^2 + 6x + 8 = 0 \;\;\Leftrightarrow\;\; (x + 2)(x + 4) = 0 \;\;\Leftrightarrow\;\; x = -2$ or $x = -4$.

7. (a) The line $2x - y = 3 \;\;\Leftrightarrow\;\; y = 2x - 3$ has slope 2, so a vector parallel to the line is $\mathbf{a} = [1, 2]$. The line $3x + y = 7 \;\;\Leftrightarrow$ $y = -3x + 7$ has slope -3, so a vector parallel to the line is $\mathbf{b} = [1, -3]$. The angle between the lines is the same as the angle θ between the vectors. Here we have $\mathbf{a} \cdot \mathbf{b} = (1)(1) + (2)(-3) = -5$, $|\mathbf{a}| = \sqrt{1^2 + 2^2} = \sqrt{5}$, and $|\mathbf{b}| = \sqrt{1^2 + (-3)^2} = \sqrt{10}$, so $\cos\theta = \dfrac{\mathbf{a} \cdot \mathbf{b}}{|\mathbf{a}|\,|\mathbf{b}|} = \dfrac{-5}{\sqrt{5} \cdot \sqrt{10}} = \dfrac{-5}{5\sqrt{2}} = -\dfrac{1}{\sqrt{2}}$ or $-\dfrac{\sqrt{2}}{2}$. Thus $\theta = 135^\circ$, and the acute angle between the lines is $180^\circ - 135^\circ = 45^\circ$.

© 2016 Cengage Learning. All Rights Reserved. May not be scanned, copied or duplicated, or posted to a publicly accessible website, in whole or in part.

(b) The line $x + 2y = 7 \quad \Leftrightarrow \quad y = -\frac{1}{2}x + \frac{7}{2}$ has slope $-\frac{1}{2}$, so a vector parallel to the line is $\mathbf{a} = [2, -1]$. The line $5x - y = 2 \quad \Leftrightarrow \quad y = 5x - 2$ has slope 5, so a vector parallel to the line is $\mathbf{b} = [1, 5]$. The lines meet at the same angle θ that the vectors meet at. Here we have $\mathbf{a} \cdot \mathbf{b} = (2)(1) + (-1)(5) = -3$, $|\mathbf{a}| = \sqrt{2^2 + (-1)^2} = \sqrt{5}$, and $|\mathbf{b}| = \sqrt{1^2 + 5^2} = \sqrt{26}$, so $\cos\theta = \dfrac{\mathbf{a} \cdot \mathbf{b}}{|\mathbf{a}|\,|\mathbf{b}|} = \dfrac{-3}{\sqrt{5} \cdot \sqrt{26}} = \dfrac{-3}{\sqrt{130}}$ and $\theta = \cos^{-1}\left(\dfrac{-3}{\sqrt{130}}\right) \approx 105.3°$. The acute angle between the lines is approximately $180° - 105.3° = 74.7°$.

9. First note that $\mathbf{n} = [a, b]$ is perpendicular to the line, because if $Q_1 = (a_1, b_1)$ and $Q_2 = (a_2, b_2)$ lie on the line, then $\mathbf{n} \cdot \overrightarrow{Q_1Q_2} = aa_2 - aa_1 + bb_2 - bb_1 = 0$, since $aa_2 + bb_2 = -c = aa_1 + bb_1$ from the equation of the line.
Let $P_2 = (x_2, y_2)$ lie on the line. Then the distance from P_1 to the line is the absolute value of the scalar projection of $\overrightarrow{P_1P_2}$ onto $\mathbf{n}$. $\text{comp}_{\mathbf{n}}\left(\overrightarrow{P_1P_2}\right) = \dfrac{|\mathbf{n} \cdot [x_2 - x_1, y_2 - y_1]|}{|\mathbf{n}|} = \dfrac{|ax_2 - ax_1 + by_2 - by_1|}{\sqrt{a^2 + b^2}} = \dfrac{|ax_1 + by_1 + c|}{\sqrt{a^2 + b^2}}$

since $ax_2 + by_2 = -c$. The required distance is $\dfrac{|(3)(-2) + (-4)(3) + 5|}{\sqrt{3^2 + (-4)^2}} = \dfrac{13}{5}$.

11. (a) The distance between the points $(-1, 3)$ and $(1, 1)$ in 'expression space' is

$$\sqrt{(-1-1)^2 + (3-1)^2} = \sqrt{4+4} = \sqrt{8} = 2\sqrt{2} \approx 2.83$$

(b) Treating the points in expression space as the tips of position vectors gives us the vectors $\mathbf{a} = [-1, 3]$ and $\mathbf{b} = [1, 1]$. The angle θ between $\mathbf{a}$ and $\mathbf{b}$ can be found using the dot product (Equation 8.3.4)

$$\mathbf{a} \cdot \mathbf{b} = |\mathbf{a}|\,|\mathbf{b}| \cos\theta \quad \Leftrightarrow \quad \cos\theta = \frac{\mathbf{a} \cdot \mathbf{b}}{|\mathbf{a}|\,|\mathbf{b}|} = \frac{[-1, 3] \cdot [1, 1]}{|[-1, 3]|\,|[1, 1]|} = \frac{(-1)(1) + (3)(1)}{\sqrt{(-1)^2 + 3^2}\sqrt{1^2 + 1^2}} = \frac{2}{\sqrt{10}\sqrt{2}} = \frac{1}{\sqrt{5}} \quad \Rightarrow$$

$$\theta = \cos^{-1}\left(\frac{1}{\sqrt{5}}\right) \approx 1.11 \text{ (or } 63.4°)$$

13. An arrow drawn from variable j to i is represented with a nonzero entry (the number on the arrow) in the ijth position of the corresponding matrix. This rule gives the following matrix representation.

$$\begin{bmatrix} 0 & 0 & 0 \\ 0.5 & 1 & 2 \\ 0 & 0.5 & 0 \end{bmatrix}$$

15. (a) $A - D = A + (-D)$ $\quad A$ is a 2×2 matrix and $-D$ is a 2×3 matrix, so the operation $A - D$ is not defined.

(b) $2C + A = 2\begin{bmatrix} 1 & 2 \\ 9 & 6 \end{bmatrix} + \begin{bmatrix} 7 & 0 \\ 1 & 9 \end{bmatrix} = \begin{bmatrix} 2 & 4 \\ 18 & 12 \end{bmatrix} + \begin{bmatrix} 7 & 0 \\ 1 & 9 \end{bmatrix} = \begin{bmatrix} 2+7 & 4+0 \\ 18+1 & 12+9 \end{bmatrix} = \begin{bmatrix} 9 & 4 \\ 19 & 21 \end{bmatrix}$

(c) $A + 2B + C = A + (2B + C)$

$2B$ is a 3×2 matrix and C is a 2×2 matrix, so the operation $A + 2B + C$ is not defined.

(d) $D - B = D + (-B)$ $\quad D$ is a 2×3 matrix and $-B$ is a 3×2 matrix, so the operation $D - B$ is not defined.

(e) $B^T + D = \begin{bmatrix} 1 & 4 \\ 1 & 0 \\ 0 & 9 \end{bmatrix}^T + \begin{bmatrix} 15 & 0 & 7 \\ 0 & 5 & 2 \end{bmatrix} = \begin{bmatrix} 1 & 1 & 0 \\ 4 & 0 & 9 \end{bmatrix} + \begin{bmatrix} 15 & 0 & 7 \\ 0 & 5 & 2 \end{bmatrix} = \begin{bmatrix} 1+15 & 1+0 & 0+7 \\ 4+0 & 0+5 & 9+2 \end{bmatrix} = \begin{bmatrix} 16 & 1 & 7 \\ 4 & 5 & 11 \end{bmatrix}$

© 2016 Cengage Learning. All Rights Reserved. May not be scanned, copied or duplicated, or posted to a publicly accessible website, in whole or in part.

(f) $2C - A^{-1} = 2\begin{bmatrix}1 & 2\\ 9 & 6\end{bmatrix} - \begin{bmatrix}7 & 0\\ 1 & 9\end{bmatrix}^{-1} = \begin{bmatrix}2 & 4\\ 18 & 12\end{bmatrix} - \frac{1}{(7)(9)-(1)(0)}\begin{bmatrix}9 & 0\\ -1 & 7\end{bmatrix} = \begin{bmatrix}2 & 4\\ 18 & 12\end{bmatrix} - \frac{1}{63}\begin{bmatrix}9 & 0\\ -1 & 7\end{bmatrix}$

$$= \begin{bmatrix}2 & 4\\ 18 & 12\end{bmatrix} - \begin{bmatrix}\frac{1}{7} & 0\\ -\frac{1}{63} & \frac{1}{9}\end{bmatrix} = \begin{bmatrix}2-\frac{1}{7} & 4-0\\ 18-(-\frac{1}{63}) & 12-\frac{1}{9}\end{bmatrix} = \begin{bmatrix}\frac{13}{7} & 4\\ \frac{1135}{63} & \frac{107}{9}\end{bmatrix}$$

17. If $A = A^{-1}$, then A must be invertible and $\det A \neq 0$. Equating the matrices A and A^{-1}, we have

$$A = A^{-1} \quad\Leftrightarrow\quad \begin{bmatrix}a_{11} & a_{12}\\ a_{21} & a_{22}\end{bmatrix} = \begin{bmatrix}a_{11} & a_{12}\\ a_{21} & a_{22}\end{bmatrix}^{-1} \quad\Leftrightarrow\quad \begin{bmatrix}a_{11} & a_{12}\\ a_{21} & a_{22}\end{bmatrix} = \frac{1}{\det A}\begin{bmatrix}a_{22} & -a_{12}\\ -a_{21} & a_{11}\end{bmatrix}$$

Equating each entry of A and A^{-1} gives a system of four equations

$$\left\{\begin{aligned} a_{11} &= \frac{a_{22}}{\det A}\\ a_{21} &= -\frac{a_{21}}{\det A}\\ a_{12} &= -\frac{a_{12}}{\det A}\\ a_{22} &= \frac{a_{11}}{\det A}\end{aligned}\right\} \quad\Leftrightarrow\quad \left\{\begin{aligned} a_{22} &= a_{11}\det A\\ a_{21} &= -a_{21}\det A\\ a_{12} &= -a_{12}\det A\\ a_{11} &= a_{22}\det A\end{aligned}\right\}$$

The second and third equations are satisfied only when $\det A = -1$ (since a_{ij} are nonzero). Substituting this result in the first (or fourth) equation gives $a_{11} = -a_{22}$.

19. $\det I = \det\begin{bmatrix}1 & 0\\ 0 & 1\end{bmatrix} = (1)(1) - (0)(0) = 1$

21. $\left\{\begin{aligned} 3x - y &= 2\\ x + 7y &= 4\end{aligned}\right\} \quad\Rightarrow\quad \begin{bmatrix}3 & -1\\ 1 & 7\end{bmatrix}\begin{bmatrix}x\\ y\end{bmatrix} = \begin{bmatrix}2\\ 4\end{bmatrix} \quad\Rightarrow$

$$\begin{bmatrix}x\\ y\end{bmatrix} = \begin{bmatrix}3 & -1\\ 1 & 7\end{bmatrix}^{-1}\begin{bmatrix}2\\ 4\end{bmatrix} = \frac{1}{(3)(7)-(1)(-1)}\begin{bmatrix}7 & 1\\ -1 & 3\end{bmatrix}\begin{bmatrix}2\\ 4\end{bmatrix} = \frac{1}{22}\begin{bmatrix}14+4\\ -2+12\end{bmatrix} = \begin{bmatrix}9/11\\ 5/11\end{bmatrix}$$

Therefore, the solution to the system of equations is $x = \frac{9}{11}$ and $y = \frac{5}{11}$.

23. $\left\{\begin{aligned} 7x + y &= 0\\ 14x + 2y &= 0\end{aligned}\right\}$

The first equation is identical to the second equation (within a multiple of 2), that is they represent the same straight line. Therefore, there are an infinite number of solutions, each of which must satisfy $y = -7x$.

25. $\hat{\mathbf{n}} = \begin{bmatrix}1 & 2 & 4\\ \frac{1}{2} & 0 & 0\\ 0 & \frac{1}{3} & 0\end{bmatrix}\hat{\mathbf{n}} \;\Rightarrow\; \hat{\mathbf{n}} - \begin{bmatrix}1 & 2 & 4\\ \frac{1}{2} & 0 & 0\\ 0 & \frac{1}{3} & 0\end{bmatrix}\hat{\mathbf{n}} = \mathbf{0} \;\Rightarrow\; \left(\begin{bmatrix}1 & 0 & 0\\ 0 & 1 & 0\\ 0 & 0 & 1\end{bmatrix} - \begin{bmatrix}1 & 2 & 4\\ \frac{1}{2} & 0 & 0\\ 0 & \frac{1}{3} & 0\end{bmatrix}\right)\hat{\mathbf{n}} = \mathbf{0} \;\Rightarrow\; \underbrace{\begin{bmatrix}0 & -2 & -4\\ -\frac{1}{2} & 1 & 0\\ 0 & -\frac{1}{3} & 1\end{bmatrix}}_{A}\hat{\mathbf{n}} = \begin{bmatrix}0\\ 0\\ 0\end{bmatrix}$

So finding $\hat{\mathbf{n}}$ requires solving the homogeneous system above. If $\det A \neq 0$, there is only one solution given by $\hat{\mathbf{n}} = \mathbf{0}$, otherwise there are infinitely many solutions.

$$\det A = 0 + 0 + (-\tfrac{2}{3}) - 0 - (1) - 0 = -\tfrac{5}{3} \neq 0$$

Therefore, the unique solution is $\hat{\mathbf{n}} = \begin{bmatrix}0\\ 0\\ 0\end{bmatrix}$. This indicates that at equilibrium, the age-structured population is extinct.

© 2016 Cengage Learning. All Rights Reserved. May not be scanned, copied or duplicated, or posted to a publicly accessible website, in whole or in part.

27. Eigenvalues: $\det \begin{bmatrix} 3-\lambda & 1 \\ 1 & 3-\lambda \end{bmatrix} = (3-\lambda)(3-\lambda) - 1 = \lambda^2 - 6\lambda + 8 = (\lambda-4)(\lambda-2) = 0$

The eigenvalues are therefore $\lambda = 2$ and $\lambda = 4$.

Eigenvectors: Starting with the eigenvalue $\lambda = 2$ we have

$$\begin{bmatrix} 3-\lambda & 1 \\ 1 & 3-\lambda \end{bmatrix}\begin{bmatrix} v_1 \\ v_2 \end{bmatrix} = \begin{bmatrix} 1 & 1 \\ 1 & 1 \end{bmatrix}\begin{bmatrix} v_1 \\ v_2 \end{bmatrix} = \begin{bmatrix} 0 \\ 0 \end{bmatrix} \quad \Rightarrow \quad \begin{Bmatrix} v_1 + v_2 = 0 \\ v_1 + v_2 = 0 \end{Bmatrix}$$

Both equations specify that $v_2 = -v_1$. Choosing $v_1 = 1$ gives the eigenvector $\mathbf{v} = \begin{bmatrix} 1 \\ -1 \end{bmatrix}$.

For the second eigenvalue $\lambda = 4$ we have

$$\begin{bmatrix} 3-\lambda & 1 \\ 1 & 3-\lambda \end{bmatrix}\begin{bmatrix} v_1 \\ v_2 \end{bmatrix} = \begin{bmatrix} -1 & 1 \\ 1 & -1 \end{bmatrix}\begin{bmatrix} v_1 \\ v_2 \end{bmatrix} = \begin{bmatrix} 0 \\ 0 \end{bmatrix} \quad \Rightarrow \quad \begin{Bmatrix} -v_1 + v_2 = 0 \\ v_1 - v_2 = 0 \end{Bmatrix}$$

Both equations specify that $v_2 = v_1$. Choosing $v_1 = 1$ gives the eigenvector $\mathbf{v} = \begin{bmatrix} 1 \\ 1 \end{bmatrix}$.

29. Eigenvalues: $\det \begin{bmatrix} 2-\lambda & 1 \\ -1 & 2-\lambda \end{bmatrix} = (2-\lambda)^2 + 1 = \lambda^2 - 4\lambda + 5 = 0 \quad \Rightarrow$

$\lambda = \dfrac{4 \pm \sqrt{(-4)^2 - 4(1)(5)}}{2(1)} = \dfrac{4 \pm \sqrt{-4}}{2} = 2 \pm i$ The eigenvalues are therefore $\lambda = 2 + i$ and $\lambda = 2 - i$.

Eigenvectors: Starting with the eigenvalue $\lambda = 2 + i$ we have

$$\begin{bmatrix} 2-\lambda & 1 \\ -1 & 2-\lambda \end{bmatrix}\begin{bmatrix} v_1 \\ v_2 \end{bmatrix} = \begin{bmatrix} -i & 1 \\ -1 & -i \end{bmatrix}\begin{bmatrix} v_1 \\ v_2 \end{bmatrix} = \begin{bmatrix} 0 \\ 0 \end{bmatrix} \quad \Rightarrow \quad \begin{Bmatrix} -iv_1 + v_2 = 0 \\ -v_1 - iv_2 = 0 \end{Bmatrix}$$

Both equations specify that $v_2 = iv_1$. Choosing $v_1 = 1$ gives the complex eigenvector $\mathbf{v} = \begin{bmatrix} 1 \\ i \end{bmatrix}$.

For the second eigenvalue $\lambda = 2 - i$ we have

$$\begin{bmatrix} 2-\lambda & 1 \\ -1 & 2-\lambda \end{bmatrix}\begin{bmatrix} v_1 \\ v_2 \end{bmatrix} = \begin{bmatrix} i & 1 \\ -1 & i \end{bmatrix}\begin{bmatrix} v_1 \\ v_2 \end{bmatrix} = \begin{bmatrix} 0 \\ 0 \end{bmatrix} \quad \Rightarrow \quad \begin{Bmatrix} iv_1 + v_2 = 0 \\ -v_1 + iv_2 = 0 \end{Bmatrix}$$

Both equations specify that $v_2 = -iv_1$. Choosing $v_1 = i$ gives the eigenvector $\mathbf{v} = \begin{bmatrix} i \\ 1 \end{bmatrix}$.

31. Let $A = \begin{bmatrix} a & b \\ 1-a & 1-b \end{bmatrix}$ where a and b are real numbers. Observe that each column of A sums to one. We verify $\lambda = 1$ is an eigenvalue by evaluating

$$\det(A - \lambda I) = \det(A - I) = \det \begin{bmatrix} a-1 & b \\ 1-a & 1-b-1 \end{bmatrix} = (a-1)(-b) - (1-a)(b) = -ab + b - b + ab = 0$$

Therefore, $\lambda = 1$ is an eigenvalue of a 2×2 matrix whose columns sum to one.

© 2016 Cengage Learning. All Rights Reserved. May not be scanned, copied or duplicated, or posted to a publicly accessible website, in whole or in part.

33. The eigenvalues of $A = \begin{bmatrix} 2 & 3 \\ -4 & -6 \end{bmatrix}$ satisfy the characteristic polynomial $(2-\lambda)(-6-\lambda)+12 = \lambda(\lambda+4) = 0$, which has solutions $\lambda_1 = 0$ and $\lambda_2 = -4$. Solving the system $A\mathbf{v} = \lambda\mathbf{v}$ for each eigenvalue gives the associated eigenvectors $\mathbf{v}_1 = [3, -2]$ and $\mathbf{v}_2 = [1, -2]$. So $D = \begin{bmatrix} 0 & 0 \\ 0 & -4 \end{bmatrix}$ and $P = \begin{bmatrix} 3 & 1 \\ -2 & -2 \end{bmatrix} \Rightarrow P^{-1} = \frac{1}{-4}\begin{bmatrix} -2 & -1 \\ 2 & 3 \end{bmatrix} = \begin{bmatrix} \frac{1}{2} & \frac{1}{4} \\ -\frac{1}{2} & -\frac{3}{4} \end{bmatrix}$.

Also $\begin{bmatrix} c_1 \\ c_2 \end{bmatrix} = P^{-1}\mathbf{n}_0 = \begin{bmatrix} \frac{1}{2} & \frac{1}{4} \\ -\frac{1}{2} & -\frac{3}{4} \end{bmatrix}\begin{bmatrix} 1 \\ 0 \end{bmatrix} = \begin{bmatrix} \frac{1}{2} \\ -\frac{1}{2} \end{bmatrix}$. Therefore, the solution is given by

$$\mathbf{n}_t = c_1 \begin{bmatrix} | \\ \mathbf{v}_1 \\ | \end{bmatrix} \lambda_1^t + c_2 \begin{bmatrix} | \\ \mathbf{v}_2 \\ | \end{bmatrix} \lambda_2^t = \tfrac{1}{2}\begin{bmatrix} 3 \\ -2 \end{bmatrix} 0^t - \tfrac{1}{2}\begin{bmatrix} 1 \\ -2 \end{bmatrix}(-4)^t = -\tfrac{1}{2}\begin{bmatrix} 1 \\ -2 \end{bmatrix}(-4)^t$$

where $t >= 1$ since 0^t is undefined when $t = 0$ in the second last step.

35. The eigenvalues of $A = \begin{bmatrix} a & -1 \\ 1 & a \end{bmatrix}$ satisfy the characteristic polynomial $(a-\lambda)^2 + 1 = 0$, which has solutions $\lambda_1 = a + i$ and $\lambda_2 = a - i$. Solving the system $A\mathbf{v} = \lambda\mathbf{v}$ for each eigenvalue gives the associated eigenvectors $\mathbf{v}_1 = [i, 1]$ and $\mathbf{v}_2 = [1, i]$. So $D = \begin{bmatrix} a+i & 0 \\ 0 & a-i \end{bmatrix}$ and $P = \begin{bmatrix} i & 1 \\ 1 & i \end{bmatrix} \Rightarrow P^{-1} = \frac{1}{-2}\begin{bmatrix} i & -1 \\ -1 & i \end{bmatrix} = \begin{bmatrix} -\frac{1}{2}i & \frac{1}{2} \\ \frac{1}{2} & -\frac{1}{2}i \end{bmatrix}$.

Also $\begin{bmatrix} c_1 \\ c_2 \end{bmatrix} = P^{-1}\mathbf{n}_0 = \begin{bmatrix} -\frac{1}{2}i & \frac{1}{2} \\ \frac{1}{2} & -\frac{1}{2}i \end{bmatrix}\begin{bmatrix} 1 \\ 1 \end{bmatrix} = \begin{bmatrix} \frac{1}{2} - \frac{1}{2}i \\ \frac{1}{2} - \frac{1}{2}i \end{bmatrix}$. Therefore, the solution is given by

$$\mathbf{n}_t = c_1 \begin{bmatrix} | \\ \mathbf{v}_1 \\ | \end{bmatrix} \lambda_1^t + c_2 \begin{bmatrix} | \\ \mathbf{v}_2 \\ | \end{bmatrix} \lambda_2^t = \left(\tfrac{1}{2} - \tfrac{1}{2}i\right)\begin{bmatrix} i \\ 1 \end{bmatrix}(a+i)^t + \left(\tfrac{1}{2} - \tfrac{1}{2}i\right)\begin{bmatrix} 1 \\ i \end{bmatrix}(a-i)^t$$

Using De Moivre's Theorem, we find $(a \pm i)^t = r^t(\cos(\pm\theta t) + i\sin(\pm\theta t))$ where $r = \sqrt{1+a^2}$ and $\theta = \tan^{-1}\left(\frac{1}{a}\right)$. So the solution becomes

$$\begin{aligned}
\mathbf{n}_t &= \left(\tfrac{1}{2} - \tfrac{1}{2}i\right)\begin{bmatrix} i \\ 1 \end{bmatrix} r^t(\cos(\theta t) + i\sin(\theta t)) + \left(\tfrac{1}{2} - \tfrac{1}{2}i\right)\begin{bmatrix} 1 \\ i \end{bmatrix} r^t(\cos(-\theta t) + i\sin(-\theta t)) \\
&= \begin{bmatrix} \frac{1}{2} + \frac{1}{2}i \\ \frac{1}{2} - \frac{1}{2}i \end{bmatrix} r^t(\cos(\theta t) + i\sin(\theta t)) + \begin{bmatrix} \frac{1}{2} - \frac{1}{2}i \\ \frac{1}{2} + \frac{1}{2}i \end{bmatrix} r^t(\cos(\theta t) - i\sin(\theta t)) \qquad \begin{bmatrix} \text{since } \sin(-x) = \sin x \\ \text{and } \cos(-x) = \cos x \end{bmatrix} \\
&= \tfrac{1}{2}r^t\left(\begin{bmatrix} \cos(\theta t) + i\sin(\theta t) + i\cos(\theta t) - \sin(\theta t) \\ \cos(\theta t) + i\sin(\theta t) - i\cos(\theta t) + \sin(\theta t) \end{bmatrix} + \begin{bmatrix} \cos(\theta t) - i\sin(\theta t) - i\cos(\theta t) - \sin(\theta t) \\ \cos(\theta t) - i\sin(\theta t) + i\cos(\theta t) + \sin(\theta t) \end{bmatrix}\right) \\
&= \tfrac{1}{2}r^t\left(\begin{bmatrix} 2\cos(\theta t) - 2\sin(\theta t) \\ 2\sin(\theta t) + 2\cos(\theta t) \end{bmatrix}\right) = r^t\left(\begin{bmatrix} \cos(\theta t) - \sin(\theta t) \\ \sin(\theta t) + \cos(\theta t) \end{bmatrix}\right)
\end{aligned}$$

© 2016 Cengage Learning. All Rights Reserved. May not be scanned, copied or duplicated, or posted to a publicly accessible website, in whole or in part.

37. The Perron-Frobenius Theorem can be used when A is primitive, that is when A^k contains all positive entries for some positive integer k.

$$A = \begin{bmatrix} 0.9 & 0.001 & 0 \\ 0.1 & 0.799 & 0.001 \\ 0 & 0.2 & 0.999 \end{bmatrix} \Rightarrow A^2 = \begin{bmatrix} 0.9 & 0.001 & 0 \\ 0.1 & 0.799 & 0.001 \\ 0 & 0.2 & 0.999 \end{bmatrix} \begin{bmatrix} 0.9 & 0.001 & 0 \\ 0.1 & 0.799 & 0.001 \\ 0 & 0.2 & 0.999 \end{bmatrix}$$

$$= \begin{bmatrix} 0.8101 & 0.001699 & 0.000001 \\ 0.1699 & 0.638701 & 0.001798 \\ 0.02 & 0.3596 & 0.998201 \end{bmatrix}$$

A^2 contains all positive entries so the Perron-Frobenius Theorem can be applied in this case.

39. (a) $\mathbf{z}_n = \begin{bmatrix} x_n \\ y_n \end{bmatrix} = \begin{bmatrix} a_n \\ a_{n-1} \end{bmatrix} = \begin{bmatrix} \frac{1}{2}a_{n-1} + \frac{1}{2}a_{n-2} \\ a_{n-1} \end{bmatrix} = \begin{bmatrix} \frac{1}{2} & \frac{1}{2} \\ 1 & 0 \end{bmatrix} \begin{bmatrix} a_{n-1} \\ a_{n-2} \end{bmatrix} = \underbrace{\begin{bmatrix} \frac{1}{2} & \frac{1}{2} \\ 1 & 0 \end{bmatrix}}_{A} \underbrace{\begin{bmatrix} x_{n-1} \\ y_{n-1} \end{bmatrix}}_{\mathbf{z}_{n-1}} = A\mathbf{z}_{n-1}$

(b) Since $A^2 = \begin{bmatrix} \frac{1}{2} & \frac{1}{2} \\ 1 & 0 \end{bmatrix} \begin{bmatrix} \frac{1}{2} & \frac{1}{2} \\ 1 & 0 \end{bmatrix} = \begin{bmatrix} \frac{3}{4} & \frac{1}{4} \\ \frac{1}{2} & \frac{1}{2} \end{bmatrix}$ contains all positive entries, the matrix A is primitive and the Perron-Frobenius Theorem is applicable.

(c) The eigenvalues of A must satisfy $\det(A - \lambda I) = 0 \Rightarrow \det \begin{bmatrix} \frac{1}{2} - \lambda & \frac{1}{2} \\ 1 & -\lambda \end{bmatrix} = 0 \Rightarrow -\lambda\left(\frac{1}{2} - \lambda\right) - \frac{1}{2} = 0 \Rightarrow$ $2\lambda^2 - \lambda - 1 = 0 \Rightarrow (2\lambda + 1)(\lambda - 1) = 0$. Therefore, the eigenvalues are $\lambda_1 = 1$ and $\lambda_2 = -\frac{1}{2}$.

(d) The Perron-Frobenius Theorem says that the components of the eigenvector corresponding to $\lambda_1 = 1$ are all positive. Also, in the long-term, the components of $\mathbf{z}_n$ grow by a factor 1 each timestep, that is they do not vary, and asymptotically approach the line defined by the associated eigenvector $\mathbf{v}_1$.

(e) Because $|\lambda_1| = 1$ the solution vector will approach a constant multiple of the eigenvector $\mathbf{v}_1$ in the long-term, that is,

$\lim_{n\to\infty} \mathbf{z}_n = \lim_{n\to\infty} \begin{bmatrix} a_n \\ a_{n-1} \end{bmatrix} = c\mathbf{v}_1$ where c is a constant. Hence, $\lim_{n\to\infty} a_n = c\,(\text{Component 1 of } \mathbf{v}_1)$.

© 2016 Cengage Learning. All Rights Reserved. May not be scanned, copied or duplicated, or posted to a publicly accessible website, in whole or in part.

9 □ MULTIVARIABLE CALCULUS

9.1 Functions of Several Variables

1. (a) From Table 1, $f(-15, 40) = -27$, which means that if the temperature is $-15°$C and the wind speed is 40 km/h, then the air would feel equivalent to approximately $-27°$C without wind.

(b) The question is asking: when the temperature is $-20°$C, what wind speed gives a wind-chill index of $-30°$C? From Table 1, the speed is 20 km/h.

(c) The question is asking: when the wind speed is 20 km/h, what temperature gives a wind-chill index of $-49°$C? From Table 1, the temperature is $-35°$C.

(d) The function $W = f(-5, v)$ means that we fix T at -5 and allow v to vary, resulting in a function of one variable. In other words, the function gives wind-chill index values for different wind speeds when the temperature is $-5°$C. From Table 1 (look at the row corresponding to $T = -5$), the function decreases and appears to approach a constant value as v increases.

(e) The function $W = f(T, 50)$ means that we fix v at 50 and allow T to vary, again giving a function of one variable. In other words, the function gives wind-chill index values for different temperatures when the wind speed is 50 km/h . From Table 1 (look at the column corresponding to $v = 50$), the function increases almost linearly as T increases.

3. (a) $f(160, 70) = 0.1091(160)^{0.425}(70)^{0.725} \approx 20.5$, which means that the surface area of a person 70 inches (5 feet 10 inches) tall who weighs 160 pounds is approximately 20.5 square feet.

(b) Answers will vary depending on the height and weight of the reader.

5. (a) $P(L, K) = 1.47L^{0.65}K^{0.35} \quad \Rightarrow \quad P(120, 20) = 1.47(120)^{0.65}(20)^{0.35} \approx 94.2$. This means that when the manufacturer invests \$20 million and 120,000 labor hours are spent, its yearly production is about \$94.2 million.

(b) If the amounts of labor and capital are both doubled, we replace L, K in the function with $2L, 2K$, giving

$$P(2L, 2K) = 1.47(2L)^{0.65}(2K)^{0.35} = 1.47(2^{0.65})(2^{0.35})L^{0.65}K^{0.35} = (2^1)1.47L^{0.65}K^{0.35} = 2P(L, K)$$

Thus, the production is doubled.

7. $F(R, S) = 4.2 + 0.008R + 0.102S + 0.017R^2 - 0.034S^2 - 0.268RS \quad \Rightarrow$

$F(3, 1) = 4.2 + 0.008(3) + 0.102(1) + 0.017(3)^2 - 0.034(1)^2 - 0.268(3)(1) = 3.641$ and

$F(1, 3) = 4.2 + 0.008(1) + 0.102(3) + 0.017(1)^2 - 0.034(3)^2 - 0.268(1)(3) = 3.421$

The snake with $R = 3$ and $S = 1$ has a greater measure of fitness so it is likely to survive longer.

9. (a) $g(2, -1) = \cos(2 + 2(-1)) = \cos(0) = 1$

(b) $x + 2y$ is defined for all choices of values for x and y and the cosine function is defined for all input values, so the domain of g is $\mathbb{R}^2$.

(c) The range of the cosine function is $[-1, 1]$ and $x + 2y$ generates all possible input values for the cosine function, so the range of $\cos(x + 2y)$ is $[-1, 1]$.

© 2016 Cengage Learning. All Rights Reserved. May not be scanned, copied or duplicated, or posted to a publicly accessible website, in whole or in part.

11. (a) $f(1,1,1) = \sqrt{1} + \sqrt{1} + \sqrt{1} + \ln(4 - 1^2 - 1^2 - 1^2) = 3 + \ln 1 = 3$

(b) $\sqrt{x}$, $\sqrt{y}$, $\sqrt{z}$ are defined only when $x \geq 0$, $y \geq 0$, $z \geq 0$, and $\ln(4 - x^2 - y^2 - z^2)$ is defined when $4 - x^2 - y^2 - z^2 > 0 \quad \Leftrightarrow \quad x^2 + y^2 + z^2 < 4$, thus the domain is $\{(x, y, z) \mid x^2 + y^2 + z^2 < 4,\ x \geq 0,\ y \geq 0,\ z \geq 0\}$, the portion of the interior of a sphere of radius 2, centered at the origin, that is in the first octant.

13. $\sqrt{2x - y}$ is defined only when $2x - y \geq 0$, or $y \leq 2x$.

So the domain of f is $\{(x, y) \mid y \leq 2x\}$.

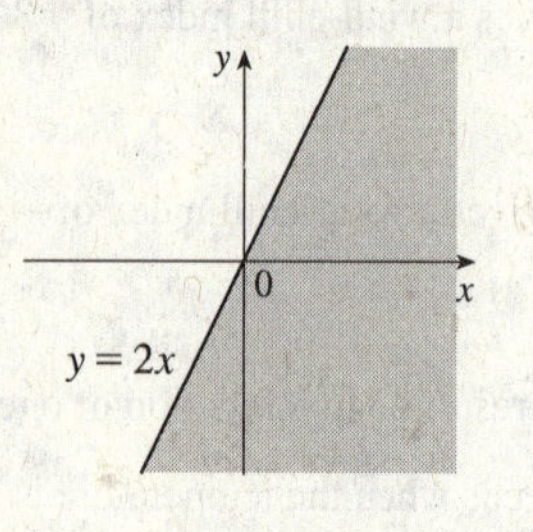

15. $\sqrt{1 - x^2}$ is defined only when $1 - x^2 \geq 0$, or $x^2 \leq 1 \quad \Leftrightarrow \quad -1 \leq x \leq 1$, and $\sqrt{1 - y^2}$ is defined only when $1 - y^2 \geq 0$, or $y^2 \leq 1 \quad \Leftrightarrow \quad -1 \leq y \leq 1$.

Thus the domain of f is $\{(x, y) \mid -1 \leq x \leq 1,\ -1 \leq y \leq 1\}$.

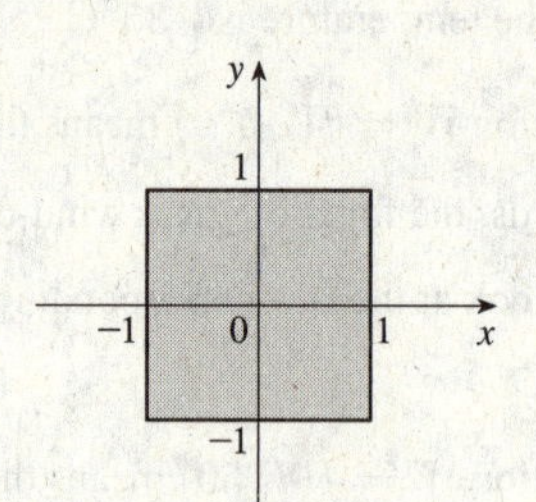

17. $\sqrt{y - x^2}$ is defined only when $y - x^2 \geq 0$, or $y \geq x^2$.

In addition, f is not defined if $1 - x^2 = 0 \quad \Leftrightarrow \quad x = \pm 1$. Thus the domain of f is $\{(x, y) \mid y \geq x^2,\ x \neq \pm 1\}$.

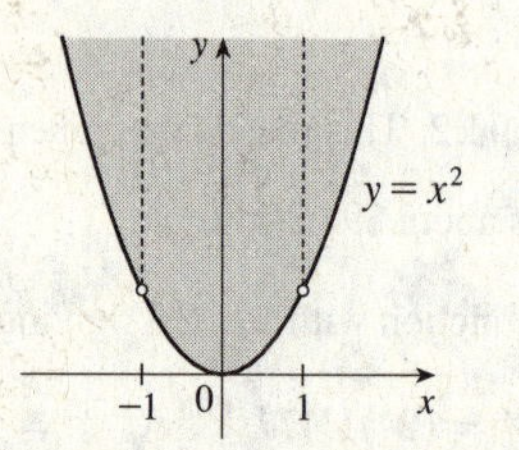

19. We need $1 - x^2 - y^2 - z^2 \geq 0$ or $x^2 + y^2 + z^2 \leq 1$, so $D = \{(x, y, z) \mid x^2 + y^2 + z^2 \leq 1\}$ (the points inside or on the sphere of radius 1, center the origin).

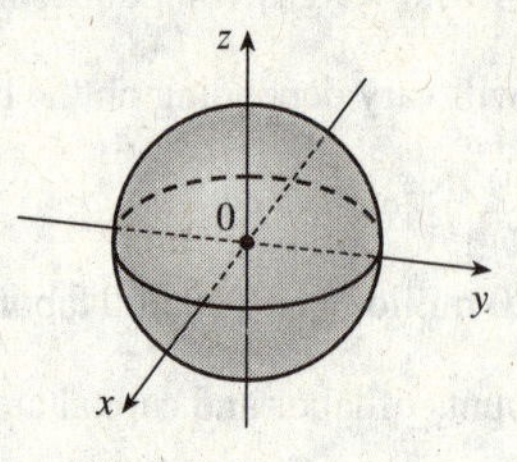

21. $z = 3$, a horizontal plane through the point $(0, 0, 3)$.

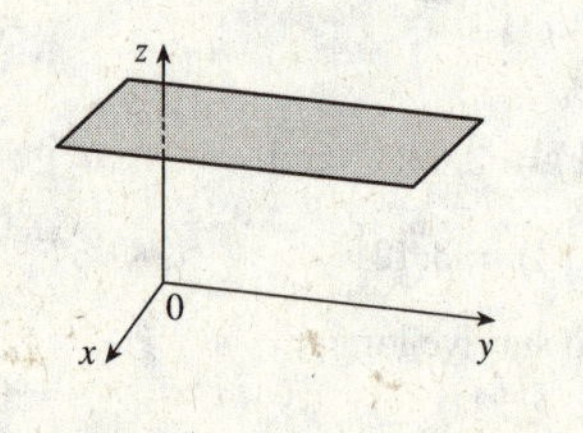

23. $z = 10 - 4x - 5y$ or $4x + 5y + z = 10$, a plane with intercepts 2.5, 2, and 10.

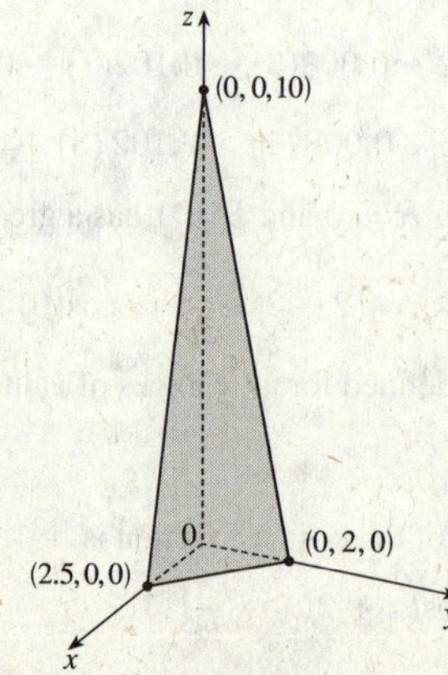

© 2016 Cengage Learning. All Rights Reserved. May not be scanned, copied or duplicated, or posted to a publicly accessible website, in whole or in part.

25. $z = y^2 + 1$, a parabolic cylinder

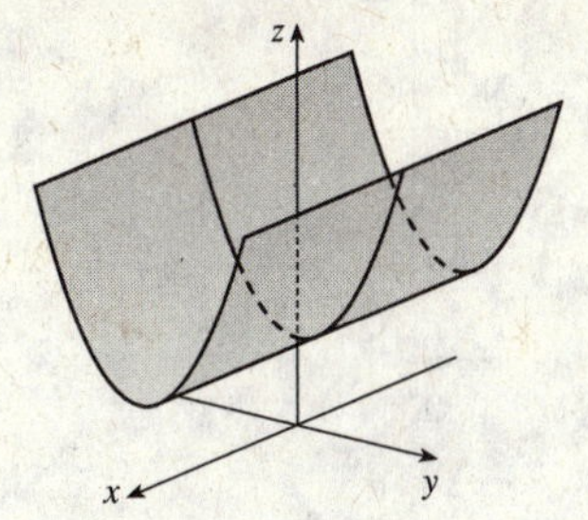

27. The point $(-3, 3)$ lies between the level curves with z-values 50 and 60. Since the point is a little closer to the level curve with $z = 60$, we estimate that $f(-3, 3) \approx 56$. The point $(3, -2)$ appears to be just about halfway between the level curves with z-values 30 and 40, so we estimate $f(3, -2) \approx 35$. The graph rises as we approach the origin, gradually from above, steeply from below.

29. The point $(160, 10)$, corresponding to day 160 and a depth of 10 m, lies between the isothermals with temperature values of 8 and 12°C. Since the point appears to be located about three-fourths the distance from the 8°C isothermal to the 12°C isothermal, we estimate the temperature at that point to be approximately 11°C. The point $(180, 5)$ lies between the 16 and 20°C isothermals, very close to the 20°C level curve, so we estimate the temperature there to be about 19.5°C.

31. Near A, the level curves are very close together, indicating that the terrain is quite steep. At B, the level curves are much farther apart, so we would expect the terrain to be much less steep than near A, perhaps almost flat.

33. The level curves of $f(x, y) = 2x - y$ are $2x - y = k$ or $y = 2x - k$, a family of lines with slope 2 and y-intercept $-k$.

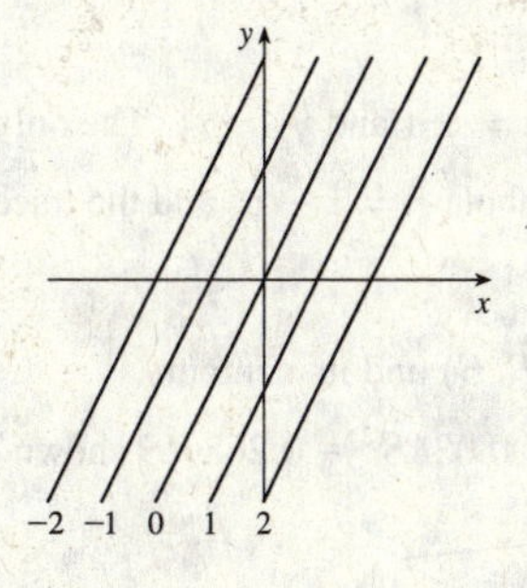

35. The level curves are $xy = k$ or $y = k/x$. For $k = 0$ the curves are the coordinate axes; if $k > 0$, they are hyperbolas in the first and third quadrants (multiples of the reciprocal function); if $k < 0$, they are hyperbolas in the second and fourth quadrants.

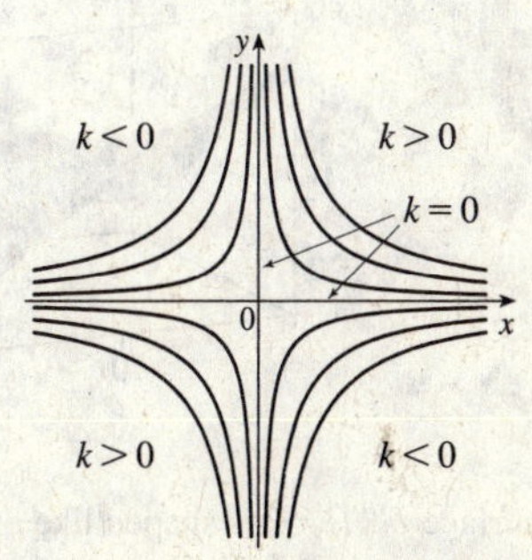

© 2016 Cengage Learning. All Rights Reserved. May not be scanned, copied or duplicated, or posted to a publicly accessible website, in whole or in part.

37. The level curves are $ye^x = k$ or $y = ke^{-x}$, a family of exponential curves.

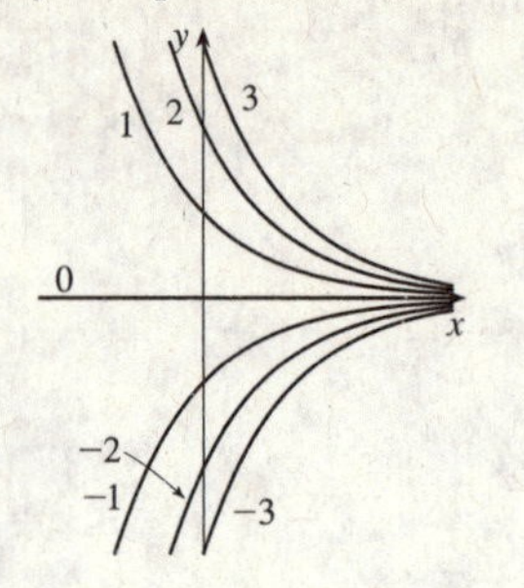

39. The level curves of the BMI function $B(m,h) = \dfrac{m}{h^2}$ are given by $m = kh^2$ ($k \geq 0$ since m and h are positive), a family of quadratic functions passing through the origin. The level curves are plotted for $k = 18.5$, 25, 30, and 40 with the optimal BMI region ($18.5 \leq B \leq 25$) shaded. The point $(m,h) = (62, 1.52)$ falls outside of the shaded region, so a 1.52 m tall individual weighing 62 kg does *not* have an optimal BMI.

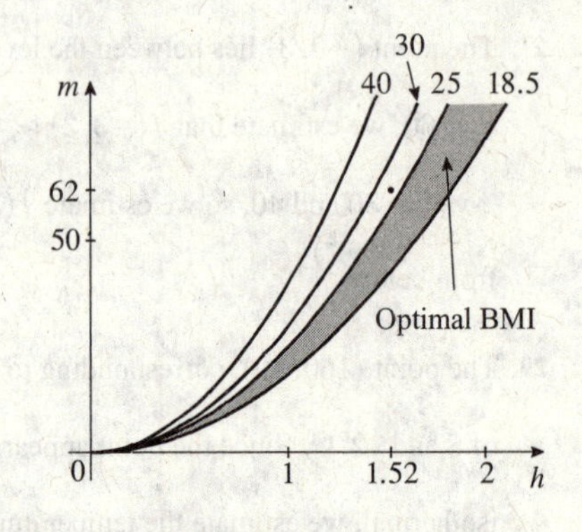

41. $z = \sin(xy)$ (a) C (b) II

Reasons: This function is periodic in both x and y, and the function is the same when x is interchanged with y, so its graph is symmetric about the plane $y = x$. In addition, the function is 0 along the x- and y-axes. These conditions are satisfied only by C and II.

43. $z = \sin(x - y)$ (a) F (b) I

Reasons: This function is periodic in both x and y but is constant along the lines $y = x + k$, a condition satisfied only by F and I.

45. $z = (1 - x^2)(1 - y^2)$ (a) B (b) VI

Reasons: This function is 0 along the lines $x = \pm 1$ and $y = \pm 1$. The only contour map in which this could occur is VI. Also note that the trace in the xz-plane is the parabola $z = 1 - x^2$ and the trace in the yz-plane is the parabola $z = 1 - y^2$, so the graph is B.

47. A CAS was used to graph the function $F(R,S)$ and its contours $k = 4.2 + 0.008R + 0.102S + 0.017R^2 - 0.034S^2 - 0.268RS$ shown below.

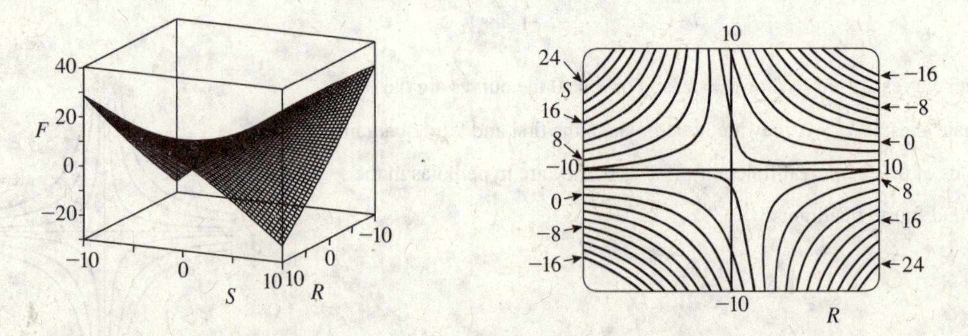

The surface $F(R,S)$ is shaped like a saddle with corners at opposite combinations of R and S giving the highest values of F. This means that snakes with a high value of reversals and low value of stripedness (and vice versa) will likely survive longer.

© 2016 Cengage Learning. All Rights Reserved. May not be scanned, copied or duplicated, or posted to a publicly accessible website, in whole or in part.

49. $f(x,y) = 5x^3 - x^2y^2$ is a polynomial, and hence continuous, so $\lim\limits_{(x,y)\to(1,2)} f(x,y) = f(1,2) = 5(1)^3 - (1)^2(2)^2 = 1$.

51. $f(x,y) = \dfrac{4-xy}{x^2+3y^2}$ is a rational function and hence continuous on its domain.

$(2,1)$ is in the domain of f, so f is continuous there and $\lim\limits_{(x,y)\to(2,1)} f(x,y) = f(2,1) = \dfrac{4-(2)(1)}{(2)^2+3(1)^2} = \dfrac{2}{7}$.

53. $f(x,y) = y^4/(x^4+3y^4)$. First approach $(0,0)$ along the x-axis. Then $f(x,0) = 0/x^4 = 0$ for $x \neq 0$, so $f(x,y) \to 0$. Now approach $(0,0)$ along the y-axis. Then for $y \neq 0$, $f(0,y) = y^4/3y^4 = 1/3$, so $f(x,y) \to 1/3$. Since f has two different limits along two different lines, the limit does not exist.

55. $f(x,y) = (xy\cos y)/(3x^2+y^2)$. On the x-axis, $f(x,0) = 0$ for $x \neq 0$, so $f(x,y) \to 0$ as $(x,y) \to (0,0)$ along the x-axis. Approaching $(0,0)$ along the line $y = x$, $f(x,x) = (x^2\cos x)/4x^2 = \frac{1}{4}\cos x$ for $x \neq 0$, so $f(x,y) \to \frac{1}{4}$ along this line. Thus the limit does not exist.

57. $f(x,y) = \dfrac{xy}{\sqrt{x^2+y^2}}$. We can see that the limit along any line through $(0,0)$ is 0, as well as along other paths through $(0,0)$ such as $x = y^2$ and $y = x^2$. So we suspect that the limit exists and equals 0; we use the Squeeze Theorem to prove our assertion. $0 \leq \left|\dfrac{xy}{\sqrt{x^2+y^2}}\right| \leq |x|$ since $|y| \leq \sqrt{x^2+y^2}$, and $|x| \to 0$ as $(x,y) \to (0,0)$. So $\lim\limits_{(x,y)\to(0,0)} f(x,y) = 0$.

59. $F(x,y) = \arctan\left(x+\sqrt{y}\right) = g(f(x,y))$ where $f(x,y) = x+\sqrt{y}$, continuous on its domain $\{(x,y) \mid y \geq 0\}$, and $g(t) = \arctan t$ is continuous everywhere. Thus F is continuous on its domain $\{(x,y) \mid y \geq 0\}$.

61. $G(x,y) = \ln(x^2+y^2-4) = g(f(x,y))$ where $f(x,y) = x^2+y^2-4$, continuous on $\mathbb{R}^2$, and $g(t) = \ln t$, continuous on its domain $\{t \mid t > 0\}$. Thus G is continuous on its domain $\{(x,y) \mid x^2+y^2-4 > 0\} = \{(x,y) \mid x^2+y^2 > 4\}$, the exterior of the circle $x^2+y^2 = 4$.

63. $\sqrt{y}$ is continuous on its domain $\{y \mid y \geq 0\}$ and $x^2-y^2+z^2$ is continuous everywhere, so $f(x,y,z) = \dfrac{\sqrt{y}}{x^2-y^2+z^2}$ is continuous for $y \geq 0$ and $x^2-y^2+z^2 \neq 0 \;\Rightarrow\; y^2 \neq x^2+z^2$, that is, $\left\{(x,y,z) \mid y \geq 0, y \neq \sqrt{x^2+z^2}\,\right\}$.

65. $f(x,y) = \begin{cases} \dfrac{x^2y^3}{2x^2+y^2} & \text{if } (x,y) \neq (0,0) \\ 1 & \text{if } (x,y) = (0,0) \end{cases}$ The first piece of f is a rational function defined everywhere except at the origin, so f is continuous on $\mathbb{R}^2$ except possibly at the origin. Since $x^2 \leq 2x^2+y^2$, we have $\left|x^2y^3/(2x^2+y^2)\right| \leq |y^3|$. We know that $|y^3| \to 0$ as $(x,y) \to (0,0)$. So, by the Squeeze Theorem, $\lim\limits_{(x,y)\to(0,0)} f(x,y) = \lim\limits_{(x,y)\to(0,0)} \dfrac{x^2y^3}{2x^2+y^2} = 0$.

But $f(0,0) = 1$, so f is discontinuous at $(0,0)$. Therefore, f is continuous on the set $\{(x,y) \mid (x,y) \neq (0,0)\}$.

© 2016 Cengage Learning. All Rights Reserved. May not be scanned, copied or duplicated, or posted to a publicly accessible website, in whole or in part.

9.2 Partial Derivatives

1. (a) $\partial T/\partial x$ represents the rate of change of T when we fix y and t and consider T as a function of the single variable x, which describes how quickly the temperature changes when longitude changes but latitude and time are constant. $\partial T/\partial y$ represents the rate of change of T when we fix x and t and consider T as a function of y, which describes how quickly the temperature changes when latitude changes but longitude and time are constant. $\partial T/\partial t$ represents the rate of change of T when we fix x and y and consider T as a function of t, which describes how quickly the temperature changes over time for a constant longitude and latitude.

(b) $f_x(158, 21, 9)$ represents the rate of change of temperature at longitude 158°W, latitude 21°N at 9:00 AM when only longitude varies. Since the air is warmer to the west than to the east, increasing longitude results in an increased air temperature, so we would expect $f_x(158, 21, 9)$ to be positive. $f_y(158, 21, 9)$ represents the rate of change of temperature at the same time and location when only latitude varies. Since the air is warmer to the south and cooler to the north, increasing latitude results in a decreased air temperature, so we would expect $f_y(158, 21, 9)$ to be negative. $f_t(158, 21, 9)$ represents the rate of change of temperature at the same time and location when only time varies. Since typically air temperature increases from the morning to the afternoon as the sun warms it, we would expect $f_t(158, 21, 9)$ to be positive.

3. (a) By Definition 4, $f_T(-15, 30) = \lim\limits_{h\to 0} \dfrac{f(-15+h, 30) - f(-15, 30)}{h}$, which we can approximate by considering $h = 5$ and $h = -5$ and using the values given in the table:

$f_T(-15, 30) \approx \dfrac{f(-10, 30) - f(-15, 30)}{5} = \dfrac{-20 - (-26)}{5} = \dfrac{6}{5} = 1.2$,

$f_T(-15, 30) \approx \dfrac{f(-20, 30) - f(-15, 30)}{-5} = \dfrac{-33 - (-26)}{-5} = \dfrac{-7}{-5} = 1.4$. Averaging these values, we estimate $f_T(-15, 30)$ to be approximately 1.3. Thus, when the actual temperature is $-15°$C and the wind speed is 30 km/h, the apparent temperature rises by about 1.3°C for every degree that the actual temperature rises.

Similarly, $f_v(-15, 30) = \lim\limits_{h\to 0} \dfrac{f(-15, 30+h) - f(-15, 30)}{h}$ which we can approximate by considering $h = 10$ and $h = -10$: $f_v(-15, 30) \approx \dfrac{f(-15, 40) - f(-15, 30)}{10} = \dfrac{-27 - (-26)}{10} = \dfrac{-1}{10} = -0.1$,

$f_v(-15, 30) \approx \dfrac{f(-15, 20) - f(-15, 30)}{-10} = \dfrac{-24 - (-26)}{-10} = \dfrac{2}{-10} = -0.2$. Averaging these values, we estimate $f_v(-15, 30)$ to be approximately -0.15. Thus, when the actual temperature is $-15°$C and the wind speed is 30 km/h, the apparent temperature decreases by about 0.15°C for every km/h that the wind speed increases.

(b) For a fixed wind speed v, the values of the wind-chill index W increase as temperature T increases (look at a column of the table), so $\dfrac{\partial W}{\partial T}$ is positive. For a fixed temperature T, the values of W decrease (or remain constant) as v increases (look at a row of the table), so $\dfrac{\partial W}{\partial v}$ is negative (or perhaps 0).

(c) For fixed values of T, the function values $f(T, v)$ appear to become constant (or nearly constant) as v increases, so the corresponding rate of change is 0 or near 0 as v increases. This suggests that $\lim\limits_{v\to\infty} (\partial W/\partial v) = 0$.

© 2016 Cengage Learning. All Rights Reserved. May not be scanned, copied or duplicated, or posted to a publicly accessible website, in whole or in part.

5. (a) If we start at $(1,2)$ and move in the positive x-direction, the graph of f increases. Thus $f_x(1,2)$ is positive.

(b) If we start at $(1,2)$ and move in the positive y-direction, the graph of f decreases. Thus $f_y(1,2)$ is negative.

7. $f(x,y) = 16 - 4x^2 - y^2 \Rightarrow f_x(x,y) = -8x$ and $f_y(x,y) = -2y \Rightarrow f_x(1,2) = -8$ and $f_y(1,2) = -4$. The graph of f is the paraboloid $z = 16 - 4x^2 - y^2$ and the vertical plane $y = 2$ intersects it in the parabola $z = 12 - 4x^2$, $y = 2$ (the curve C_1 in the first figure). The slope of the tangent line to this parabola at $(1,2,8)$ is $f_x(1,2) = -8$. Similarly the plane $x = 1$ intersects the paraboloid in the parabola $z = 12 - y^2$, $x = 1$ (the curve C_2 in the second figure) and the slope of the tangent line at $(1,2,8)$ is $f_y(1,2) = -4$.

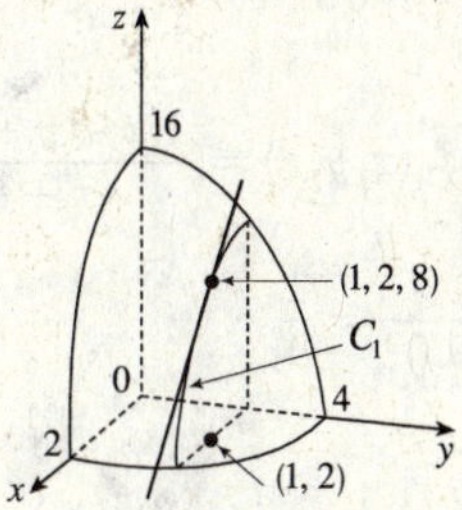

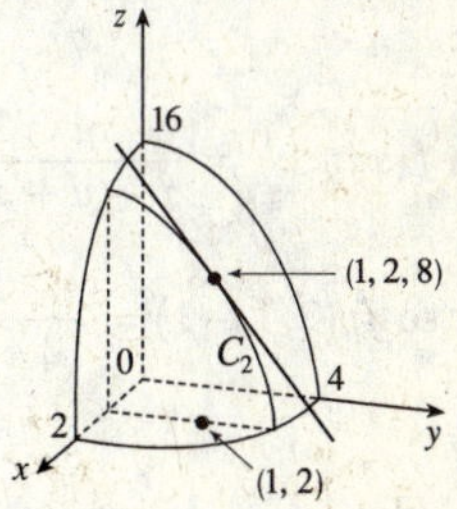

9. $f(x,y) = y^5 - 3xy \Rightarrow f_x(x,y) = 0 - 3y = -3y$, $f_y(x,y) = 5y^4 - 3x$

11. $f(x,t) = e^{-t}\cos\pi x \Rightarrow f_x(x,t) = e^{-t}(-\sin\pi x)(\pi) = -\pi e^{-t}\sin\pi x$, $f_t(x,t) = e^{-t}(-1)\cos\pi x = -e^{-t}\cos\pi x$

13. $z = (2x+3y)^{10} \Rightarrow \dfrac{\partial z}{\partial x} = 10(2x+3y)^9 \cdot 2 = 20(2x+3y)^9$, $\dfrac{\partial z}{\partial y} = 10(2x+3y)^9 \cdot 3 = 30(2x+3y)^9$

15. $f(x,y) = \dfrac{x-y}{x+y} \Rightarrow f_x(x,y) = \dfrac{(1)(x+y)-(x-y)(1)}{(x+y)^2} = \dfrac{2y}{(x+y)^2}$,

$f_y(x,y) = \dfrac{(-1)(x+y)-(x-y)(1)}{(x+y)^2} = -\dfrac{2x}{(x+y)^2}$

17. $w = \sin\alpha\cos\beta \Rightarrow \dfrac{\partial w}{\partial\alpha} = \cos\alpha\cos\beta$, $\dfrac{\partial w}{\partial\beta} = -\sin\alpha\sin\beta$

19. $f(r,s) = r\ln(r^2+s^2) \Rightarrow f_r(r,s) = r \cdot \dfrac{2r}{r^2+s^2} + \ln(r^2+s^2)\cdot 1 = \dfrac{2r^2}{r^2+s^2} + \ln(r^2+s^2)$,

$f_s(r,s) = r \cdot \dfrac{2s}{r^2+s^2} + 0 = \dfrac{2rs}{r^2+s^2}$

21. $u = te^{w/t} \Rightarrow \dfrac{\partial u}{\partial t} = t\cdot e^{w/t}(-wt^{-2}) + e^{w/t}\cdot 1 = e^{w/t} - \dfrac{w}{t}e^{w/t} = e^{w/t}\left(1 - \dfrac{w}{t}\right)$, $\dfrac{\partial u}{\partial w} = te^{w/t}\cdot\dfrac{1}{t} = e^{w/t}$

23. $f(x,y,z) = xz - 5x^2y^3z^4 \Rightarrow f_x(x,y,z) = z - 10xy^3z^4$, $f_y(x,y,z) = -15x^2y^2z^4$, $f_z(x,y,z) = x - 20x^2y^3z^3$

25. $w = \ln(x+2y+3z) \Rightarrow \dfrac{\partial w}{\partial x} = \dfrac{1}{x+2y+3z}$, $\dfrac{\partial w}{\partial y} = \dfrac{2}{x+2y+3z}$, $\dfrac{\partial w}{\partial z} = \dfrac{3}{x+2y+3z}$

27. $u = xe^{-t}\sin\theta \Rightarrow \dfrac{\partial u}{\partial x} = e^{-t}\sin\theta$, $\dfrac{\partial u}{\partial t} = -xe^{-t}\sin\theta$, $\dfrac{\partial u}{\partial\theta} = xe^{-t}\cos\theta$

29. $f(x,y,z,t) = xyz^2\tan(yt) \Rightarrow f_x(x,y,z,t) = yz^2\tan(yt)$,

$f_y(x,y,z,t) = xyz^2\cdot\sec^2(yt)\cdot t + xz^2\tan(yt) = xyz^2t\sec^2(yt) + xz^2\tan(yt)$,

$f_z(x,y,z,t) = 2xyz\tan(yt)$, $f_t(x,y,z,t) = xyz^2\sec^2(yt)\cdot y = xy^2z^2\sec^2(yt)$

31. $u = \sqrt{x_1^2 + x_2^2 + \cdots + x_n^2}$. For each $i = 1, \ldots, n$, $u_{x_i} = \frac{1}{2}\left(x_1^2 + x_2^2 + \cdots + x_n^2\right)^{-1/2}(2x_i) = \dfrac{x_i}{\sqrt{x_1^2 + x_2^2 + \cdots + x_n^2}}$.

© 2016 Cengage Learning. All Rights Reserved. May not be scanned, copied or duplicated, or posted to a publicly accessible website, in whole or in part.

33. $f(x,y) = \ln\left(x + \sqrt{x^2+y^2}\right) \Rightarrow$

$$f_x(x,y) = \frac{1}{x+\sqrt{x^2+y^2}}\left[1 + \tfrac{1}{2}(x^2+y^2)^{-1/2}(2x)\right] = \frac{1}{x+\sqrt{x^2+y^2}}\left(1 + \frac{x}{\sqrt{x^2+y^2}}\right),$$

so $f_x(3,4) = \dfrac{1}{3+\sqrt{3^2+4^2}}\left(1 + \dfrac{3}{\sqrt{3^2+4^2}}\right) = \frac{1}{8}\left(1+\frac{3}{5}\right) = \frac{1}{5}$.

35. $f(x,y,z) = \dfrac{y}{x+y+z} \quad \Rightarrow \quad f_y(x,y,z) = \dfrac{1(x+y+z) - y(1)}{(x+y+z)^2} = \dfrac{x+z}{(x+y+z)^2}$,

so $f_y(2,1,-1) = \dfrac{2+(-1)}{(2+1+(-1))^2} = \dfrac{1}{4}$.

37. $x^2+y^2+z^2 = 3xyz \quad \Rightarrow \quad \dfrac{\partial}{\partial x}(x^2+y^2+z^2) = \dfrac{\partial}{\partial x}(3xyz) \quad \Rightarrow \quad 2x + 0 + 2z\,\dfrac{\partial z}{\partial x} = 3y\left(x\,\dfrac{\partial z}{\partial x} + z\cdot 1\right) \quad \Leftrightarrow$

$2z\,\dfrac{\partial z}{\partial x} - 3xy\,\dfrac{\partial z}{\partial x} = 3yz - 2x \quad \Leftrightarrow \quad (2z - 3xy)\,\dfrac{\partial z}{\partial x} = 3yz - 2x$, so $\dfrac{\partial z}{\partial x} = \dfrac{3yz-2x}{2z-3xy}$.

$\dfrac{\partial}{\partial y}(x^2+y^2+z^2) = \dfrac{\partial}{\partial y}(3xyz) \quad \Rightarrow \quad 0 + 2y + 2z\,\dfrac{\partial z}{\partial y} = 3x\left(y\,\dfrac{\partial z}{\partial y} + z\cdot 1\right) \quad \Leftrightarrow \quad 2z\,\dfrac{\partial z}{\partial y} - 3xy\,\dfrac{\partial z}{\partial y} = 3xz - 2y \quad \Leftrightarrow$

$(2z-3xy)\,\dfrac{\partial z}{\partial y} = 3xz - 2y$, so $\dfrac{\partial z}{\partial y} = \dfrac{3xz-2y}{2z-3xy}$.

39. $x - z = \arctan(yz) \quad \Rightarrow \quad \dfrac{\partial}{\partial x}(x-z) = \dfrac{\partial}{\partial x}(\arctan(yz)) \quad \Rightarrow \quad 1 - \dfrac{\partial z}{\partial x} = \dfrac{1}{1+(yz)^2}\cdot y\,\dfrac{\partial z}{\partial x} \quad \Leftrightarrow$

$1 = \left(\dfrac{y}{1+y^2z^2} + 1\right)\dfrac{\partial z}{\partial x} \quad \Leftrightarrow \quad 1 = \left(\dfrac{y+1+y^2z^2}{1+y^2z^2}\right)\dfrac{\partial z}{\partial x}$, so $\dfrac{\partial z}{\partial x} = \dfrac{1+y^2z^2}{1+y+y^2z^2}$.

$\dfrac{\partial}{\partial y}(x-z) = \dfrac{\partial}{\partial y}(\arctan(yz)) \quad \Rightarrow \quad 0 - \dfrac{\partial z}{\partial y} = \dfrac{1}{1+(yz)^2}\cdot\left(y\,\dfrac{\partial z}{\partial y} + z\cdot 1\right) \quad \Leftrightarrow$

$-\dfrac{z}{1+y^2z^2} = \left(\dfrac{y}{1+y^2z^2} + 1\right)\dfrac{\partial z}{\partial y} \quad \Leftrightarrow \quad -\dfrac{z}{1+y^2z^2} = \left(\dfrac{y+1+y^2z^2}{1+y^2z^2}\right)\dfrac{\partial z}{\partial y} \quad \Leftrightarrow \quad \dfrac{\partial z}{\partial y} = -\dfrac{z}{1+y+y^2z^2}$.

41. $S = f(w,h) = 0.1091w^{0.425}h^{0.725}$

(a) $\dfrac{\partial S}{\partial w}(w,h) = 0.1091(0.425)w^{0.425-1}h^{0.725} = 0.0463675w^{-0.575}h^{0.725} \quad \Rightarrow$

$\dfrac{\partial S}{\partial w}(160,70) = 0.0463675\,(160)^{-0.575}\,(70)^{0.725} \approx 0.0545\ \text{ft}^2/\text{lb}$ This is the rate at which body surface area increases with respect to weight when an individual weighs 160 lb and has a height 70 in.

(b) $\dfrac{\partial S}{\partial h}(w,h) = 0.1091\,(0.725)\,w^{0.425}h^{0.725-1} = 0.0790975w^{0.425}h^{-0.275} \quad \Rightarrow$

$\dfrac{\partial S}{\partial h}(160,70) = 0.0790975\,(160)^{0.425}\,(70)^{-0.275} \approx 0.213\ \text{ft}^2/\text{in}$ This is the rate at which body surface area increases with respect to height when an individual weighs 160 lb and has a height 70 in.

© 2016 Cengage Learning. All Rights Reserved. May not be scanned, copied or duplicated, or posted to a publicly accessible website, in whole or in part.

43. $R = C\dfrac{L}{r^4} \;\Rightarrow\; \dfrac{\partial R}{\partial L} = \dfrac{C}{r^4}$ and $\dfrac{\partial R}{\partial r} = C\left(-4\dfrac{L}{r^5}\right) = -4C\dfrac{L}{r^5}$

$\partial R/\partial L$ represents the rate of change in resistance of blood that occurs from an increase in artery length while holding artery radius constant. It is positive, meaning that blood resistance increases as the artery length increases.

$\partial R/\partial r$ represents the rate of change in resistance of blood that occurs from an increase in artery radius while holding artery length constant. It is negative, meaning that the blood resistance decreases as the artery radius increases.

45. $P(v,x,m) = Av^3 + \dfrac{B(mg/x)^2}{v} = Av^3 + \dfrac{Bm^2g^2}{vx^2} \;\Rightarrow\; \dfrac{\partial P}{\partial v} = 3Av^2 - \dfrac{B(mg/x)^2}{v^2}$ and $\dfrac{\partial P}{\partial x} = -2\dfrac{Bm^2g^2}{vx^3}$ and $\dfrac{\partial P}{\partial m} = 2\dfrac{Bmg^2}{vx^2}$

$\partial P/\partial v$ represents the rate of change in required power that occurs from an increase in the bird's velocity.

$\partial P/\partial x$ represents the rate of change in required power that occurs from an increase in the fraction of time the bird spends flapping. It is negative, meaning that the required power decreases as the fraction of time spent flapping increases.

$\partial P/\partial m$ represents the rate of change in required power that occurs from an increase in the bird's mass. It is positive, meaning that the required power increases as the bird's mass increases.

47. $E(m,v) = 2.65m^{0.66} + \dfrac{3.5m^{0.75}}{v} \;\Rightarrow\; E_m(m,v) = 1.749m^{-0.34} + \dfrac{2.625m^{-0.25}}{v}$ and $E_v(m,v) = -\dfrac{3.5m^{0.75}}{v^2}$

$E_m(400,8) = 1.749(400)^{-0.34} + \dfrac{2.625(400)^{-0.25}}{8} \approx 0.301$ kcal/ g, so an increase in mass of 1 g will require approximately 0.3 kcal of additional energy when $m = 400$ and $v = 8$.

$E_v(400,8) = -\dfrac{3.5(400)^{0.75}}{(8)^2} \approx -4.89$ kcal/(km/h), so an increase in speed of 1km/h will result in approximately 4.9 kcal of energy saved when $m = 400$ and $v = 8$.

49. $\left(P + \dfrac{n^2a}{V^2}\right)(V - nb) = nRT \;\Rightarrow\; T = \dfrac{1}{nR}\left(P + \dfrac{n^2a}{V^2}\right)(V - nb)$, so $\dfrac{\partial T}{\partial P} = \dfrac{1}{nR}(1)(V - nb) = \dfrac{V - nb}{nR}$.

We can also write $P + \dfrac{n^2a}{V^2} = \dfrac{nRT}{V - nb} \;\Rightarrow\; P = \dfrac{nRT}{V - nb} - \dfrac{n^2a}{V^2} = nRT(V - nb)^{-1} - n^2aV^{-2}$, so

$$\frac{\partial P}{\partial V} = -nRT(V - nb)^{-2}(1) + 2n^2aV^{-3} = \frac{2n^2a}{V^3} - \frac{nRT}{(V - nb)^2}.$$

51. $f(x,y) = x^3y^5 + 2x^4y \;\Rightarrow\; f_x(x,y) = 3x^2y^5 + 8x^3y$, $f_y(x,y) = 5x^3y^4 + 2x^4$. Then $f_{xx}(x,y) = 6xy^5 + 24x^2y$, $f_{xy}(x,y) = 15x^2y^4 + 8x^3$, $f_{yx}(x,y) = 15x^2y^4 + 8x^3$, and $f_{yy}(x,y) = 20x^3y^3$.

53. $w = \sqrt{u^2 + v^2} \;\Rightarrow\; w_u = \frac{1}{2}(u^2 + v^2)^{-1/2} \cdot 2u = \dfrac{u}{\sqrt{u^2 + v^2}}$, $w_v = \frac{1}{2}(u^2 + v^2)^{-1/2} \cdot 2v = \dfrac{v}{\sqrt{u^2 + v^2}}$. Then

$$w_{uu} = \frac{1 \cdot \sqrt{u^2 + v^2} - u \cdot \frac{1}{2}(u^2 + v^2)^{-1/2}(2u)}{\left(\sqrt{u^2 + v^2}\right)^2} = \frac{\sqrt{u^2 + v^2} - u^2/\sqrt{u^2 + v^2}}{u^2 + v^2} = \frac{u^2 + v^2 - u^2}{(u^2 + v^2)^{3/2}} = \frac{v^2}{(u^2 + v^2)^{3/2}},$$

$$w_{uv} = u\left(-\tfrac{1}{2}\right)\left(u^2 + v^2\right)^{-3/2}(2v) = -\frac{uv}{(u^2 + v^2)^{3/2}},\quad w_{vu} = v\left(-\tfrac{1}{2}\right)\left(u^2 + v^2\right)^{-3/2}(2u) = -\frac{uv}{(u^2 + v^2)^{3/2}},$$

$$w_{vv} = \frac{1 \cdot \sqrt{u^2 + v^2} - v \cdot \frac{1}{2}(u^2 + v^2)^{-1/2}(2v)}{\left(\sqrt{u^2 + v^2}\right)^2} = \frac{\sqrt{u^2 + v^2} - v^2/\sqrt{u^2 + v^2}}{u^2 + v^2} = \frac{u^2 + v^2 - v^2}{(u^2 + v^2)^{3/2}} = \frac{u^2}{(u^2 + v^2)^{3/2}}.$$

© 2016 Cengage Learning. All Rights Reserved. May not be scanned, copied or duplicated, or posted to a publicly accessible website, in whole or in part.

55. $z = \arctan \dfrac{x+y}{1-xy} \quad \Rightarrow$

$$z_x = \frac{1}{1 + \left(\frac{x+y}{1-xy}\right)^2} \cdot \frac{(1)(1-xy) - (x+y)(-y)}{(1-xy)^2} = \frac{1+y^2}{(1-xy)^2 + (x+y)^2} = \frac{1+y^2}{1+x^2+y^2+x^2y^2}$$

$$= \frac{1+y^2}{(1+x^2)(1+y^2)} = \frac{1}{1+x^2},$$

$$z_y = \frac{1}{1 + \left(\frac{x+y}{1-xy}\right)^2} \cdot \frac{(1)(1-xy) - (x+y)(-x)}{(1-xy)^2} = \frac{1+x^2}{(1-xy)^2 + (x+y)^2} = \frac{1+x^2}{(1+x^2)(1+y^2)} = \frac{1}{1+y^2}.$$

Then $z_{xx} = -(1+x^2)^{-2} \cdot 2x = -\dfrac{2x}{(1+x^2)^2}$, $z_{xy} = 0$, $z_{yx} = 0$, $z_{yy} = -(1+y^2)^{-2} \cdot 2y = -\dfrac{2y}{(1+y^2)^2}$.

57. $u = xe^{xy} \quad \Rightarrow \quad u_x = x \cdot ye^{xy} + e^{xy} \cdot 1 = (xy+1)e^{xy}$, $u_{xy} = (xy+1) \cdot xe^{xy} + e^{xy} \cdot x = (x^2y + 2x)e^{xy}$ and $u_y = x(xe^{xy}) = x^2e^{xy}$, $u_{yx} = x^2 \cdot ye^{xy} + e^{xy} \cdot 2x = (x^2y + 2x)e^{xy}$. Thus $u_{xy} = u_{yx}$.

59. $f(x,y) = 3xy^4 + x^3y^2 \quad \Rightarrow \quad f_x = 3y^4 + 3x^2y^2$, $f_{xx} = 6xy^2$, $f_{xxy} = 12xy$ and $f_y = 12xy^3 + 2x^3y$, $f_{yy} = 36xy^2 + 2x^3$, $f_{yyy} = 72xy$.

61. $f(x,y,z) = \cos(4x + 3y + 2z) \quad \Rightarrow$

$f_x = -\sin(4x+3y+2z)(4) = -4\sin(4x+3y+2z)$, $f_{xy} = -4\cos(4x+3y+2z)(3) = -12\cos(4x+3y+2z)$,

$f_{xyz} = -12(-\sin(4x+3y+2z))(2) = 24\sin(4x+3y+2z)$ and

$f_y = -\sin(4x+3y+2z)(3) = -3\sin(4x+3y+2z)$,

$f_{yz} = -3\cos(4x+3y+2z)(2) = -6\cos(4x+3y+2z)$, $f_{yzz} = -6(-\sin(4x+3y+2z))(2) = 12\sin(4x+3y+2z)$.

63. $u = e^{r\theta}\sin\theta \quad \Rightarrow \quad \dfrac{\partial u}{\partial \theta} = e^{r\theta}\cos\theta + \sin\theta \cdot e^{r\theta}\,(r) = e^{r\theta}\,(\cos\theta + r\sin\theta)$,

$$\frac{\partial^2 u}{\partial r\,\partial\theta} = e^{r\theta}\,(\sin\theta) + (\cos\theta + r\sin\theta)\,e^{r\theta}\,(\theta) = e^{r\theta}\,(\sin\theta + \theta\cos\theta + r\theta\sin\theta),$$

$$\frac{\partial^3 u}{\partial r^2\,\partial\theta} = e^{r\theta}\,(\theta\sin\theta) + (\sin\theta + \theta\cos\theta + r\theta\sin\theta) \cdot e^{r\theta}\,(\theta) = \theta e^{r\theta}\,(2\sin\theta + \theta\cos\theta + r\theta\sin\theta).$$

65. Assuming that the third partial derivatives of f are continuous (easily verified), we can write $f_{xzy} = f_{yxz}$. Then $f(x,y,z) = xy^2z^3 + \sec^2(x\sqrt{z}) \quad \Rightarrow \quad f_y = 2xyz^3 \quad \Rightarrow \quad f_{yx} = 2yz^3 \quad \Rightarrow \quad f_{yxz} = 6yz^2 = f_{xzy}$.

67. $u = e^{-\alpha^2k^2t}\sin kx \quad \Rightarrow \quad u_x = ke^{-\alpha^2k^2t}\cos kx$, $u_{xx} = -k^2e^{-\alpha^2k^2t}\sin kx$, and $u_t = -\alpha^2k^2e^{-\alpha^2k^2t}\sin kx$. Thus $\alpha^2u_{xx} = u_t$.

69. (a) $u = \sin(kx)\,\sin(akt) \quad \Rightarrow \quad u_t = ak\sin(kx)\,\cos(akt)$, $u_{tt} = -a^2k^2\sin(kx)\,\sin(akt)$, $u_x = k\cos(kx)\,\sin(akt)$, $u_{xx} = -k^2\sin(kx)\,\sin(akt)$. Thus $u_{tt} = a^2u_{xx}$.

(b) $u = \dfrac{t}{a^2t^2 - x^2} \quad \Rightarrow \quad u_t = \dfrac{(a^2t^2 - x^2) - t(2a^2t)}{(a^2t^2 - x^2)^2} = -\dfrac{a^2t^2 + x^2}{(a^2t^2 - x^2)^2}$,

$$u_{tt} = \frac{-2a^2t(a^2t^2 - x^2)^2 + (a^2t^2 + x^2)(2)(a^2t^2 - x^2)(2a^2t)}{(a^2t^2 - x^2)^4} = \frac{2a^4t^3 + 6a^2tx^2}{(a^2t^2 - x^2)^3},$$

$$u_x = t(-1)(a^2t^2 - x^2)^{-2}(2x) = \frac{2tx}{(a^2t^2 - x^2)^2},$$

$$u_{xx} = \frac{2t(a^2t^2 - x^2)^2 - 2tx\,(2)(a^2t^2 - x^2)(-2x)}{(a^2t^2 - x^2)^4} = \frac{2a^2t^3 - 2tx^2 + 8tx^2}{(a^2t^2 - x^2)^3} = \frac{2a^2t^3 + 6tx^2}{(a^2t^2 - x^2)^3}.$$

Thus $u_{tt} = a^2u_{xx}$.

© 2016 Cengage Learning. All Rights Reserved. May not be scanned, copied or duplicated, or posted to a publicly accessible website, in whole or in part.

(c) $u = (x - at)^6 + (x + at)^6 \Rightarrow u_t = -6a(x - at)^5 + 6a(x + at)^5$, $u_{tt} = 30a^2(x - at)^4 + 30a^2(x + at)^4$,

$u_x = 6(x - at)^5 + 6(x + at)^5$, $u_{xx} = 30(x - at)^4 + 30(x + at)^4$. Thus $u_{tt} = a^2 u_{xx}$.

(d) $u = \sin(x - at) + \ln(x + at) \Rightarrow u_t = -a\cos(x - at) + \dfrac{a}{x + at}$, $u_{tt} = -a^2 \sin(x - at) - \dfrac{a^2}{(x + at)^2}$,

$u_x = \cos(x - at) + \dfrac{1}{x + at}$, $u_{xx} = -\sin(x - at) - \dfrac{1}{(x + at)^2}$. Thus $u_{tt} = a^2 u_{xx}$.

71. $u = xe^y + ye^x \Rightarrow \dfrac{\partial u}{\partial x} = e^y + ye^x$, $\dfrac{\partial^2 u}{\partial x^2} = ye^x$, $\dfrac{\partial^3 u}{\partial x^3} = ye^x$, $\dfrac{\partial u}{\partial y} = xe^y + e^x$, $\dfrac{\partial^2 u}{\partial y^2} = xe^y$, $\dfrac{\partial^3 u}{\partial y^3} = xe^y$,

$\dfrac{\partial^3 u}{\partial x\,\partial y^2} = \dfrac{\partial}{\partial x}\left(\dfrac{\partial^2 u}{\partial y^2}\right) = \dfrac{\partial}{\partial x}(xe^y) = e^y$, $\dfrac{\partial^2 u}{\partial x\,\partial y} = e^y + e^x \Rightarrow \dfrac{\partial^3 u}{\partial x^2\,\partial y} = e^x$. Then

$$\frac{\partial^3 u}{\partial x^3} + \frac{\partial^3 u}{\partial y^3} = ye^x + xe^y = x\,(e^y) + y\,(e^x) = x\,\frac{\partial^3 u}{\partial x\,\partial y^2} + y\,\frac{\partial^3 u}{\partial x^2\,\partial y}$$

73. $f_x(x, y) = x + 4y \Rightarrow f_{xy}(x, y) = 4$ and $f_y(x, y) = 3x - y \Rightarrow f_{yx}(x, y) = 3$. Since f_{xy} and f_{yx} are continuous everywhere but $f_{xy}(x, y) \neq f_{yx}(x, y)$, Clairaut's Theorem implies that such a function $f(x, y)$ does not exist.

9.3 Tangent Planes and Linear Approximations

1. $z = f(x, y) = 3y^2 - 2x^2 + x \Rightarrow f_x(x, y) = -4x + 1$, $f_y(x, y) = 6y$, so $f_x(2, -1) = -7$, $f_y(2, -1) = -6$.

By Equation 2, an equation of the tangent plane is $z - (-3) = f_x(2, -1)(x - 2) + f_y(2, -1)[y - (-1)] \Rightarrow$

$z + 3 = -7(x - 2) - 6(y + 1)$ or $z = -7x - 6y + 5$.

3. $z = f(x, y) = \sqrt{xy} \Rightarrow f_x(x, y) = \frac{1}{2}(xy)^{-1/2} \cdot y = \frac{1}{2}\sqrt{y/x}$, $f_y(x, y) = \frac{1}{2}(xy)^{-1/2} \cdot x = \frac{1}{2}\sqrt{x/y}$, so $f_x(1, 1) = \frac{1}{2}$ and $f_y(1, 1) = \frac{1}{2}$. Thus an equation of the tangent plane is $z - 1 = f_x(1, 1)(x - 1) + f_y(1, 1)(y - 1) \Rightarrow$

$z - 1 = \frac{1}{2}(x - 1) + \frac{1}{2}(y - 1)$ or $x + y - 2z = 0$.

5. $z = f(x, y) = y\cos(x - y) \Rightarrow f_x = y(-\sin(x - y)(1)) = -y\sin(x - y)$,

$f_y = y(-\sin(x - y)(-1)) + \cos(x - y) = y\sin(x - y) + \cos(x - y)$, so $f_x(2, 2) = -2\sin(0) = 0$,

$f_y(2, 2) = 2\sin(0) + \cos(0) = 1$ and an equation of the tangent plane is $z - 2 = 0(x - 2) + 1(y - 2)$ or $z = y$.

7. $f(x, y) = x\sqrt{y}$. The partial derivatives are $f_x(x, y) = \sqrt{y}$ and $f_y(x, y) = \dfrac{x}{2\sqrt{y}}$, so $f_x(1, 4) = 2$ and $f_y(1, 4) = \frac{1}{4}$. Both f_x and f_y are continuous functions for $y > 0$, so by Theorem 8, f is differentiable at $(1, 4)$. By Equation 3, the linearization of f at $(1, 4)$ is given by $L(x, y) = f(1, 4) + f_x(1, 4)(x - 1) + f_y(1, 4)(y - 4) = 2 + 2(x - 1) + \frac{1}{4}(y - 4) = 2x + \frac{1}{4}y - 1$.

9. $f(x, y) = \dfrac{x}{x + y}$. The partial derivatives are $f_x(x, y) = \dfrac{1(x + y) - x(1)}{(x + y)^2} = y/(x + y)^2$ and

$f_y(x, y) = x(-1)(x + y)^{-2} \cdot 1 = -x/(x + y)^2$, so $f_x(2, 1) = \frac{1}{9}$ and $f_y(2, 1) = -\frac{2}{9}$. Both f_x and f_y are continuous functions for $y \neq -x$, so f is differentiable at $(2, 1)$ by Theorem 8. The linearization of f at $(2, 1)$ is given by

$L(x, y) = f(2, 1) + f_x(2, 1)(x - 2) + f_y(2, 1)(y - 1) = \frac{2}{3} + \frac{1}{9}(x - 2) - \frac{2}{9}(y - 1) = \frac{1}{9}x - \frac{2}{9}y + \frac{2}{3}$.

© 2016 Cengage Learning. All Rights Reserved. May not be scanned, copied or duplicated, or posted to a publicly accessible website, in whole or in part.

11. $f(x,y,z) = x^2y + y^2z$. The partial derivatives are $f_x(x,y,z) = 2xy$, $f_y(x,y,z) = x^2 + 2yz$ and $f_z(x,y,z) = y^2$, so $f_x(1,2,3) = 2(1)(2) = 4$, $f_y(1,2,3) = (1)^2 + 2(2)(3) = 13$ and $f_z(1,2,3) = 2^2 = 4$. f_x, f_y and f_z are continuous functions near $(1,2,3)$, so f is differentiable at $(1,2,3)$. The linearization of f at $(1,2,3)$ is

$$\begin{aligned} L(x,y,z) &= f(1,2,3) + f_x(1,2,3)(x-1) + f_y(1,2,3)(y-2) + f_z(1,2,3)(z-3) \\ &= 14 + 4(x-1) + 13(y-2) + 4(z-3) = 4x + 13y + 4z - 28 \end{aligned}$$

13. Let $f(x,y) = \dfrac{2x+3}{4y+1}$. Then $f_x(x,y) = \dfrac{2}{4y+1}$ and $f_y(x,y) = (2x+3)(-1)(4y+1)^{-2}(4) = \dfrac{-8x-12}{(4y+1)^2}$. Both f_x and f_y are continuous functions for $y \neq -\frac{1}{4}$, so by Theorem 8, f is differentiable at $(0,0)$. We have $f_x(0,0) = 2$, $f_y(0,0) = -12$ and the linear approximation of f at $(0,0)$ is $f(x,y) \approx f(0,0) + f_x(0,0)(x-0) + f_y(0,0)(y-0) = 3 + 2x - 12y$.

15. We can estimate $f(2.2, 4.9)$ using a linear approximation of f at $(2,5)$, given by $f(x,y) \approx f(2,5) + f_x(2,5)(x-2) + f_y(2,5)(y-5) = 6 + 1(x-2) + (-1)(y-5) = x - y + 9$. Thus $f(2.2, 4.9) \approx 2.2 - 4.9 + 9 = 6.3$.

17. $f(x,y,z) = \sqrt{x^2+y^2+z^2} \quad \Rightarrow \quad f_x(x,y,z) = \dfrac{x}{\sqrt{x^2+y^2+z^2}}$, $f_y(x,y,z) = \dfrac{y}{\sqrt{x^2+y^2+z^2}}$, and

$f_z(x,y,z) = \dfrac{z}{\sqrt{x^2+y^2+z^2}}$, so $f_x(3,2,6) = \frac{3}{7}$, $f_y(3,2,6) = \frac{2}{7}$, $f_z(3,2,6) = \frac{6}{7}$. Then the linear approximation of f at $(3,2,6)$ is given by

$$\begin{aligned} f(x,y,z) &\approx f(3,2,6) + f_x(3,2,6)(x-3) + f_y(3,2,6)(y-2) + f_z(3,2,6)(z-6) \\ &= 7 + \tfrac{3}{7}(x-3) + \tfrac{2}{7}(y-2) + \tfrac{6}{7}(z-6) = \tfrac{3}{7}x + \tfrac{2}{7}y + \tfrac{6}{7}z \end{aligned}$$

Thus $\sqrt{(3.02)^2 + (1.97)^2 + (5.99)^2} = f(3.02, 1.97, 5.99) \approx \frac{3}{7}(3.02) + \frac{2}{7}(1.97) + \frac{6}{7}(5.99) \approx 6.9914$.

19. From the table, $f(94,80) = 127$. To estimate $f_T(94,80)$ and $f_H(94,80)$ we follow the procedure used in 9.2.4. Since $f_T(94,80) = \lim\limits_{h\to 0} \dfrac{f(94+h,80) - f(94,80)}{h}$, we approximate this quantity with $h = \pm 2$ and use the values given in the table:

$$f_T(94,80) \approx \frac{f(96,80) - f(94,80)}{2} = \frac{135-127}{2} = 4, \quad f_T(94,80) \approx \frac{f(92,80) - f(94,80)}{-2} = \frac{119-127}{-2} = 4$$

Averaging these values gives $f_T(94,80) \approx 4$. Similarly, $f_H(94,80) = \lim\limits_{h\to 0} \dfrac{f(94,80+h) - f(94,80)}{h}$, so we use $h = \pm 5$:

$$f_H(94,80) \approx \frac{f(94,85) - f(94,80)}{5} = \frac{132-127}{5} = 1, \quad f_H(94,80) \approx \frac{f(94,75) - f(94,80)}{-5} = \frac{122-127}{-5} = 1$$

Averaging these values gives $f_H(94,80) \approx 1$. The linear approximation, then, is

$$\begin{aligned} f(T,H) &\approx f(94,80) + f_T(94,80)(T-94) + f_H(94,80)(H-80) \\ &\approx 127 + 4(T-94) + 1(H-80) \qquad [\text{or } 4T + H - 329] \end{aligned}$$

Thus when $T = 95$ and $H = 78$, $f(95,78) \approx 127 + 4(95-94) + 1(78-80) = 129$, so we estimate the heat index to be approximately 129°F.

© 2016 Cengage Learning. All Rights Reserved. May not be scanned, copied or duplicated, or posted to a publicly accessible website, in whole or in part.

21. $E(m,v) = 2.65m^{0.66} + \dfrac{3.5m^{0.75}}{v} \;\Rightarrow\; E_m(m,v) = 1.749m^{-0.34} + \dfrac{2.625m^{-0.25}}{v}$ and

$E_v(m,v) = -\dfrac{3.5m^{0.75}}{v^2}$, so $E_m(400,8) = 1.749(400)^{-0.34} + \dfrac{2.625(400)^{-0.25}}{8} \approx 0.301$ kcal/ g and

$E_v(400,8) = -\dfrac{3.5(400)^{0.75}}{(8)^2} \approx -4.89$ kcal/(km/h). Then the linearization of E at $(400,8)$ is given by

$$\begin{aligned} L(m,v) &= E(400,8) + E_m(400,8)\,(m-400) + E_v(400,8)\,(v-400) \\ &\approx 177.36 + 0.301\,(m-400) - 4.89\,(v-8) \approx 0.301m - 4.89v + 96 \end{aligned}$$

9.4 The Chain Rule

1. $z = x^2 + y^2 + xy,\; x = \sin t,\; y = e^t \;\Rightarrow\; \dfrac{dz}{dt} = \dfrac{\partial z}{\partial x}\dfrac{dx}{dt} + \dfrac{\partial z}{\partial y}\dfrac{dy}{dt} = (2x+y)\cos t + (2y+x)e^t$

3. $z = \sqrt{1+x^2+y^2},\; x = \ln t,\; y = \cos t \;\Rightarrow$

$$\frac{dz}{dt} = \frac{\partial z}{\partial x}\frac{dx}{dt} + \frac{\partial z}{\partial y}\frac{dy}{dt} = \tfrac{1}{2}(1+x^2+y^2)^{-1/2}(2x)\cdot\frac{1}{t} + \tfrac{1}{2}(1+x^2+y^2)^{-1/2}(2y)(-\sin t) = \frac{1}{\sqrt{1+x^2+y^2}}\left(\frac{x}{t} - y\sin t\right)$$

5. $w = xe^{y/z},\; x = t^2,\; y = 1-t,\; z = 1+2t \;\Rightarrow$

$$\frac{dw}{dt} = \frac{\partial w}{\partial x}\frac{dx}{dt} + \frac{\partial w}{\partial y}\frac{dy}{dt} + \frac{\partial w}{\partial z}\frac{dz}{dt} = e^{y/z}\cdot 2t + xe^{y/z}\left(\frac{1}{z}\right)\cdot(-1) + xe^{y/z}\left(-\frac{y}{z^2}\right)\cdot 2 = e^{y/z}\left(2t - \frac{x}{z} - \frac{2xy}{z^2}\right)$$

7. To calculate $\partial z/\partial s$ we hold t fixed and compute the ordinary derivative of z with respect to s as given by Theorem 2.

Thus $\dfrac{\partial z}{\partial s} = \dfrac{\partial z}{\partial x}\dfrac{\partial x}{\partial s} + \dfrac{\partial z}{\partial y}\dfrac{\partial y}{\partial s}$. Similarly, holding s fixed and applying Theorem 2 gives $\dfrac{\partial z}{\partial t} = \dfrac{\partial z}{\partial x}\dfrac{\partial x}{\partial t} + \dfrac{\partial z}{\partial y}\dfrac{\partial y}{\partial t}$.

9. $z = x^2y^3,\; x = s\cos t,\; y = s\sin t \;\Rightarrow$

$$\frac{\partial z}{\partial s} = \frac{\partial z}{\partial x}\frac{\partial x}{\partial s} + \frac{\partial z}{\partial y}\frac{\partial y}{\partial s} = 2xy^3\cos t + 3x^2y^2\sin t$$

$$\frac{\partial z}{\partial t} = \frac{\partial z}{\partial x}\frac{\partial x}{\partial t} + \frac{\partial z}{\partial y}\frac{\partial y}{\partial t} = (2xy^3)(-s\sin t) + (3x^2y^2)(s\cos t) = -2sxy^3\sin t + 3sx^2y^2\cos t$$

11. When $t = 3$, $x = g(3) = 2$ and $y = h(3) = 7$. By the Chain Rule (2),

$$\frac{dz}{dt} = \frac{\partial f}{\partial x}\frac{dx}{dt} + \frac{\partial f}{\partial y}\frac{dy}{dt} = f_x(2,7)g'(3) + f_y(2,7)\,h'(3) = (6)(5) + (-8)(-4) = 62.$$

13. $y\cos x = x^2 + y^2$, so let $F(x,y) = y\cos x - x^2 - y^2 = 0$. Then by Equation 6

$$\frac{dy}{dx} = -\frac{F_x}{F_y} = -\frac{-y\sin x - 2x}{\cos x - 2y} = \frac{2x + y\sin x}{\cos x - 2y}.$$

15. $\cos(x-y) = xe^y$, so let $F(x,y) = \cos(x-y) - xe^y = 0$.

Then $\dfrac{dy}{dx} = -\dfrac{F_x}{F_y} = -\dfrac{-\sin(x-y) - e^y}{-\sin(x-y)(-1) - xe^y} = \dfrac{\sin(x-y) + e^y}{\sin(x-y) - xe^y}$.

© 2016 Cengage Learning. All Rights Reserved. May not be scanned, copied or duplicated, or posted to a publicly accessible website, in whole or in part.

17. $x^2 + y^2 + z^2 = 3xyz$, so let $F(x, y, z) = x^2 + y^2 + z^2 - 3xyz = 0$. Then by Equations 7

$$\frac{\partial z}{\partial x} = -\frac{F_x}{F_z} = -\frac{2x - 3yz}{2z - 3xy} = \frac{3yz - 2x}{2z - 3xy} \quad \text{and} \quad \frac{\partial z}{\partial y} = -\frac{F_y}{F_z} = -\frac{2y - 3xz}{2z - 3xy} = \frac{3xz - 2y}{2z - 3xy}.$$

19. $x - z = \arctan(yz)$, so let $F(x, y, z) = x - z - \arctan(yz) = 0$. Then

$$\frac{\partial z}{\partial x} = -\frac{F_x}{F_z} = -\frac{1}{-1 - \dfrac{1}{1 + (yz)^2}(y)} = \frac{1 + y^2z^2}{1 + y + y^2z^2} \text{ and}$$

$$\frac{\partial z}{\partial y} = -\frac{F_y}{F_z} = -\frac{-\dfrac{1}{1 + (yz)^2}(z)}{-1 - \dfrac{1}{1 + (yz)^2}(y)} = -\frac{\dfrac{z}{1 + y^2z^2}}{\dfrac{1 + y^2z^2 + y}{1 + y^2z^2}} = -\frac{z}{1 + y + y^2z^2}.$$

21. (a) $W = f(F(t), C(t)) \Rightarrow \dfrac{dW}{dt} = \dfrac{\partial W}{\partial F}\dfrac{dF}{dt} + \dfrac{\partial W}{\partial C}\dfrac{dC}{dt}$

(b) We expect the wolf population to increase as the size of the food supply increases so $\partial W/\partial F$ is positive. Also, the size of the wolf population will decline as the number of competitors increases so $\partial W/\partial C$ is negative.

(c) When the food supply increases dF/dt is positive and when the competition decreases dC/dt is negative. So both terms in dW/dt are positive. Therefore, the wolf population will increase.

(d) If dF/dt and dC/dt are both positive, then dW/dt is the sum of a positive term and a negative term. So we are unable to determine whether dW/dt is positive or negative.

23. $B(m, h) = \dfrac{m}{h^2} \Rightarrow \dfrac{dB}{da} = \dfrac{\partial B}{\partial m}\dfrac{dm}{da} + \dfrac{\partial B}{\partial h}\dfrac{dh}{da} = \dfrac{1}{h^2}\dfrac{dm}{da} - 2\dfrac{m}{h^3}\dfrac{dh}{da}$

25. Let $F(\rho, q, A) = \rho e^{-qA} - 1 + A$. Then, from equation (5), the rate of change of A with respect to ρ is

$$\frac{\partial A}{\partial \rho} = -\frac{F_\rho}{F_A} = -\frac{e^{-qA}}{-\rho q e^{-qA} + 1} = \frac{1}{\rho q - e^{qA}}$$

27. $\dfrac{dP}{dt} = 0.05$, $\dfrac{dT}{dt} = 0.15$, $V = 8.31\dfrac{T}{P}$ and $\dfrac{dV}{dt} = \dfrac{8.31}{P}\dfrac{dT}{dt} - 8.31\dfrac{T}{P^2}\dfrac{dP}{dt}$. Thus when $P = 20$ and $T = 320$,

$$\frac{dV}{dt} = 8.31\left[\frac{0.15}{20} - \frac{(0.05)(320)}{400}\right] \approx -0.27 \text{ L/s}.$$

29. $F(x, y, z) = 0$ is assumed to define z as a function of x and y, that is, $z = f(x, y)$. So by (7), $\dfrac{\partial z}{\partial x} = -\dfrac{F_x}{F_z}$ since $F_z \neq 0$. Similarly, it is assumed that $F(x, y, z) = 0$ defines x as a function of y and z, that is $x = h(x, z)$. Then $F(h(y, z), y, z) = 0$ and by the Chain Rule, $F_x\dfrac{\partial x}{\partial y} + F_y\dfrac{\partial y}{\partial y} + F_z\dfrac{\partial z}{\partial y} = 0$. But $\dfrac{\partial z}{\partial y} = 0$ and $\dfrac{\partial y}{\partial y} = 1$, so $F_x\dfrac{\partial x}{\partial y} + F_y = 0 \Rightarrow \dfrac{\partial x}{\partial y} = -\dfrac{F_y}{F_x}$.

A similar calculation shows that $\dfrac{\partial y}{\partial z} = -\dfrac{F_z}{F_y}$. Thus $\dfrac{\partial z}{\partial x}\dfrac{\partial x}{\partial y}\dfrac{\partial y}{\partial z} = \left(-\dfrac{F_x}{F_z}\right)\left(-\dfrac{F_y}{F_x}\right)\left(-\dfrac{F_z}{F_y}\right) = -1$.

© 2016 Cengage Learning. All Rights Reserved. May not be scanned, copied or duplicated, or posted to a publicly accessible website, in whole or in part.

9.5 Directional Derivatives and the Gradient Vector

1. We can approximate the directional derivative of the pressure function at K in the direction of S by the average rate of change of pressure between the points where the red line intersects the contour lines closest to K (extend the red line slightly at the left). In the direction of S, the pressure changes from 1000 millibars to 996 millibars and we estimate the distance between these two points to be approximately 50 km (using the fact that the distance from K to S is 300 km). Then the rate of change of pressure in the direction given is approximately $\frac{996-1000}{50} = -0.08$ millibar/km.

3. $D_{\mathbf{u}}\, f(-20,30) = \nabla f\,(-20,30)\cdot \mathbf{u} = f_T(-20,30)\left(\frac{1}{\sqrt{2}}\right) + f_v(-20,30)\left(\frac{1}{\sqrt{2}}\right)$.

$f_T(-20,30) = \lim\limits_{h\to 0} \dfrac{f(-20+h,30)-f(-20,30)}{h}$, so we can approximate $f_T(-20,30)$ by considering $h=\pm 5$ and using the values given in the table: $f_T(-20,30) \approx \dfrac{f(-15,30)-f(-20,30)}{5} = \dfrac{-26-(-33)}{5} = 1.4$,

$f_T(-20,30) \approx \dfrac{f(-25,30)-f(-20,30)}{-5} = \dfrac{-39-(-33)}{-5} = 1.2$. Averaging these values gives $f_T(-20,30)\approx 1.3$.

Similarly, $f_v(-20,30) = \lim\limits_{h\to 0} \dfrac{f(-20,30+h)-f(-20,30)}{h}$, so we can approximate $f_v(-20,30)$ with $h=\pm 10$:

$f_v(-20,30) \approx \dfrac{f(-20,40)-f(-20,30)}{10} = \dfrac{-34-(-33)}{10} = -0.1$,

$f_v(-20,30) \approx \dfrac{f(-20,20)-f(-20,30)}{-10} = \dfrac{-30-(-33)}{-10} = -0.3$. Averaging these values gives $f_v(-20,30)\approx -0.2$.

Then $D_{\mathbf{u}}f(-20,30) \approx 1.3\left(\frac{1}{\sqrt{2}}\right) + (-0.2)\left(\frac{1}{\sqrt{2}}\right) \approx 0.778$.

5. $f(x,y) = ye^{-x} \;\Rightarrow\; f_x(x,y) = -ye^{-x}$ and $f_y(x,y) = e^{-x}$. If $\mathbf{u}$ is a unit vector in the direction of $\theta = 2\pi/3$, then from Equation 6, $D_{\mathbf{u}}\, f(0,4) = f_x(0,4)\cos\left(\frac{2\pi}{3}\right) + f_y(0,4)\sin\left(\frac{2\pi}{3}\right) = -4\cdot\left(-\frac{1}{2}\right) + 1\cdot\frac{\sqrt{3}}{2} = 2+\frac{\sqrt{3}}{2}$.

7. $f(x,y) = 5xy^2 - 4x^3y$

(a) $\nabla f(x,y) = [f_x(x,y), f_y(x,y)] = \left[5y^2 - 12x^2y, 10xy - 4x^3\right]$

(b) $\nabla f(1,2) = \left[5(2)^2 - 12(1)^2(2), 10(1)(2) - 4(1)^3\right] = [-4,16]$

(c) By Equation 9, $D_{\mathbf{u}}\, f(1,2) = \nabla f(1,2)\cdot\mathbf{u} = [-4,16]\cdot\left[\frac{5}{13},\frac{12}{13}\right] = (-4)\left(\frac{5}{13}\right) + (16)\left(\frac{12}{13}\right) = \frac{172}{13}$.

9. $f(x,y) = \sin(2x+3y)$

(a) $\nabla f(x,y) = \left[\dfrac{\partial f}{\partial x}, \dfrac{\partial f}{\partial y}\right] = [\cos(2x+3y)\cdot 2, \cos(2x+3y)\cdot 3] = [2\cos(2x+3y), 3\cos(2x+3y)]$

(b) $\nabla f(-6,4) = [2\cos 0, 3\cos 0] = [2,3]$

(c) By Equation 9, $D_{\mathbf{u}}\, f(-6,4) = \nabla f(-6,4)\cdot\mathbf{u} = [2,3]\cdot\left[\frac{1}{2}\sqrt{3}, -\frac{1}{2}\right] = \frac{1}{2}\left(2\sqrt{3}-3\right) = \sqrt{3} - \frac{3}{2}$.

11. $f(x,y) = 1 + 2x\sqrt{y} \;\Rightarrow\; \nabla f(x,y) = \left[2\sqrt{y}, 2x\cdot\frac{1}{2}y^{-1/2}\right] = \left[2\sqrt{y}, x/\sqrt{y}\right]$, $\nabla f(3,4) = \left[4,\frac{3}{2}\right]$, and a unit vector in the direction of $\mathbf{v}$ is $\mathbf{u} = \dfrac{1}{\sqrt{4^2+(-3)^2}}[4,-3] = \left[\frac{4}{5},-\frac{3}{5}\right]$, so $D_{\mathbf{u}}\, f(3,4) = \nabla f(3,4)\cdot\mathbf{u} = \left[4,\frac{3}{2}\right]\cdot\left[\frac{4}{5},-\frac{3}{5}\right] = \frac{23}{10}$.

© 2016 Cengage Learning. All Rights Reserved. May not be scanned, copied or duplicated, or posted to a publicly accessible website, in whole or in part.

13. $g(p,q) = p^4 - p^2q^3 \Rightarrow \nabla g(p,q) = [4p^3 - 2pq^3, -3p^2q^2]$, $\nabla g(2,1) = [28,12]$, and a unit vector in the direction of $\mathbf{v}$ is

$\mathbf{u} = \frac{1}{\sqrt{1^2+3^2}}[1,3] = \frac{1}{\sqrt{10}}[1,3]$, so $D_{\mathbf{u}}\, g(2,1) = \nabla g(2,1) \cdot \mathbf{u} = [28,12] \cdot \frac{1}{\sqrt{10}}[1,3] = \frac{1}{\sqrt{10}}(28-36) = -\frac{8}{\sqrt{10}}$ or $-\frac{4\sqrt{10}}{5}$.

15. $V(u,t) = e^{-ut} \Rightarrow \nabla V(u,t) = [e^{-ut} \cdot (-t), e^{-ut} \cdot (-u)] = [-te^{-ut}, -ue^{-ut}]$,

$\nabla V(0,3) = [-3,0]$, and a unit vector in the direction of $\mathbf{v}$ is $\mathbf{u} = \frac{1}{\sqrt{2^2+1^2}}[2,-1] = \left[\frac{2}{\sqrt{5}}, -\frac{1}{\sqrt{5}}\right]$, so

$D_{\mathbf{u}}\, V(0,3) = \nabla V(0,3) \cdot \mathbf{u} = [-3,0] \cdot \left[\frac{2}{\sqrt{5}}, -\frac{1}{\sqrt{5}}\right] = \frac{-6}{\sqrt{5}} + 0 = \frac{-6}{\sqrt{5}}$ or $\frac{-6\sqrt{5}}{5}$.

17. $f(x,y) = \sqrt{xy} \Rightarrow \nabla f(x,y) = \left[\frac{1}{2}(xy)^{-1/2}(y), \frac{1}{2}(xy)^{-1/2}(x)\right] = \left[\frac{y}{2\sqrt{xy}}, \frac{x}{2\sqrt{xy}}\right]$, so

$\nabla f(2,8) = \left[1, \frac{1}{4}\right]$. The unit vector in the direction of $\overrightarrow{PQ} = [5-2, 4-8] = [3,-4]$ is $\mathbf{u} = \left[\frac{3}{5}, -\frac{4}{5}\right]$, so

$D_{\mathbf{u}}\, f(2,8) = \nabla f(2,8) \cdot \mathbf{u} = \left[1, \frac{1}{4}\right] \cdot \left[\frac{3}{5}, -\frac{4}{5}\right] = \frac{2}{5}$.

19. $f(x,y) = y^2/x = y^2x^{-1} \Rightarrow \nabla f(x,y) = [-y^2x^{-2}, 2yx^{-1}] = [-y^2/x^2, 2y/x]$.

$\nabla f(2,4) = [-4,4]$, or equivalently $[-1,1]$, is the direction of maximum rate of change, and the maximum rate is $|\nabla f(2,4)| = \sqrt{16+16} = 4\sqrt{2}$.

21. $f(x,y) = \sin(xy) \Rightarrow \nabla f(x,y) = [y\cos(xy), x\cos(xy)]$, $\nabla f(1,0) = [0,1]$. Thus the maximum rate of change is $|\nabla f(1,0)| = 1$ in the direction $[0,1]$.

23. (a) The rate of change of f at (a,b) in the direction of the unit vector $\mathbf{u}$ is given by the directional derivative $D_{\mathbf{u}}\, f(a,b) = |\nabla f(a,b)|\,|\mathbf{u}|\cos\theta = |\nabla f(a,b)|\cos\theta$. Since the minimum value of $\cos\theta$ is -1 occurring when $\theta = \pi$, the minimum value of $D_{\mathbf{u}}\, f(a,b)$ is $-|\nabla f|$ occurring when $\theta = \pi$, that is when $\mathbf{u}$ is in the opposite direction of $\nabla f(a,b)$ (assuming $\nabla f(a,b) \neq \mathbf{0}$).

(b) $f(x,y) = x^4y - x^2y^3 \Rightarrow \nabla f(x,y) = [4x^3y - 2xy^3, x^4 - 3x^2y^2]$, so f decreases fastest at the point $(2,-3)$ in the direction $-\nabla f(2,-3) = -[12,-92] = [-12,92]$.

25. The direction of fastest change is $\nabla f(x,y) = [2x-2, 2y-4]$, so we need to find all points (x,y) where $\nabla f(x,y)$ is parallel to $[1,1] \Leftrightarrow [2x-2, 2y-4] = k[1,1] \Leftrightarrow k = 2x-2$ and $k = 2y-4$. Then $2x-2 = 2y-4 \Rightarrow y = x+1$, so the direction of fastest change is $[1,1]$ at all points on the line $y = x+1$.

27. The fisherman is traveling in the direction $[-80,-60]$. A unit vector in this direction is $\mathbf{u} = \frac{1}{100}[-80,-60] = \left[-\frac{4}{5}, -\frac{3}{5}\right]$, and if the depth of the lake is given by $f(x,y) = 200 + 0.02x^2 - 0.001y^3$, then $\nabla f(x,y) = [0.04x, -0.003y^2]$.

$D_{\mathbf{u}}\, f(80,60) = \nabla f(80,60) \cdot \mathbf{u} = [3.2, -10.8] \cdot \left[-\frac{4}{5}, -\frac{3}{5}\right] = 3.92$. Since $D_{\mathbf{u}}\, f(80,60)$ is positive, the depth of the lake is increasing near $(80,60)$ in the direction toward the buoy.

© 2016 Cengage Learning. All Rights Reserved. May not be scanned, copied or duplicated, or posted to a publicly accessible website, in whole or in part.

29. $f(x,y) = \dfrac{1}{1+x^2+y^2} \Rightarrow \nabla f(x,y) = \left[-\dfrac{2x}{(1+x^2+y^2)^2}, -\dfrac{2y}{(1+x^2+y^2)^2}\right]$, $\nabla f(1,2) = \left[-\frac{1}{18}, -\frac{1}{9}\right]$. Thus, the bacterium will initially move in the direction $\left[-\frac{1}{18}, -\frac{1}{9}\right]$, or equivalently $[-1,-2]$, which is the direction of maximum change at the point $(1,2)$.

31. (a) $R_0(d,v) = 5(1-v)\dfrac{d}{1+d} \Rightarrow \nabla R_0(d,v) = \left[\dfrac{5(1-v)}{(1+d)^2}, -\dfrac{5d}{1+d}\right]$, so $\nabla R_0(2,0.1) = \left[\frac{1}{2}, -\frac{10}{3}\right]$. Since d increases twice as much as v, the direction of change is $(2,1)$ which gives the unit vector $\mathbf{u} = \left[\frac{2}{\sqrt{5}}, \frac{1}{\sqrt{5}}\right]$. So

$D_{\mathbf{u}} R_0(2,0.1) = \nabla R_0(2,0.1)\cdot\mathbf{u} = \left[\frac{1}{2}, -\frac{10}{3}\right]\cdot\left[\frac{2}{\sqrt{5}}, \frac{1}{\sqrt{5}}\right] = \dfrac{1}{\sqrt{5}} - \dfrac{10}{3\sqrt{5}} = -\dfrac{7}{3\sqrt{5}}$. The directional derivative is negative, so a small reallocation of resources will be beneficial leading to a smaller value of R_0.

(b) The level curves are $5(1-v)\dfrac{d}{1+d} = k$ or

$v = 1 - \dfrac{k}{5}\left(\dfrac{1}{d}+1\right) = -\dfrac{k}{5}\dfrac{1}{d} + \left(1-\dfrac{k}{5}\right)$, a family of reciprocal functions.

Given the biological context, the domain includes $d \geq 0$ and $0 \leq v \leq 1$. These curves are plotted in the contour map for several values of k. Observe that starting at $(2,0.1)$ and following a path in the direction of $\mathbf{u}$ leads to a decrease in R_0 as found in part (a).

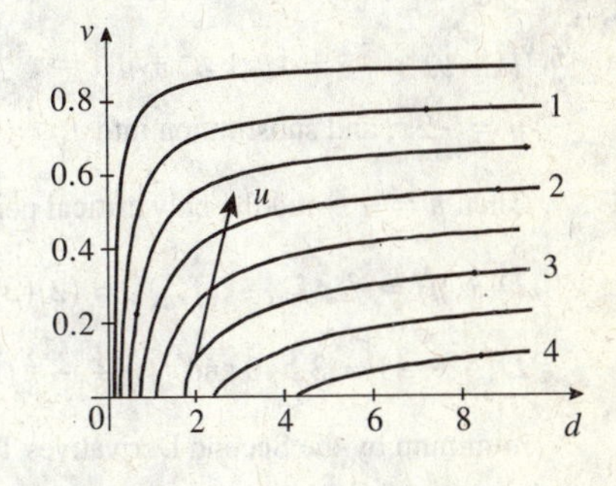

33. $f(x,y) = xy \Rightarrow \nabla f(x,y) = \langle y, x\rangle$, $\nabla f(3,2) = [2,3]$. $\nabla f(3,2)$ is perpendicular to the tangent line, so the tangent line has equation

$\nabla f(3,2)\cdot[x-3, y-2] = 0 \Rightarrow [2,3]\cdot[x-3, x-2] = 0 \Rightarrow$

$2(x-3) + 3(y-2) = 0$ or $2x + 3y = 12$.

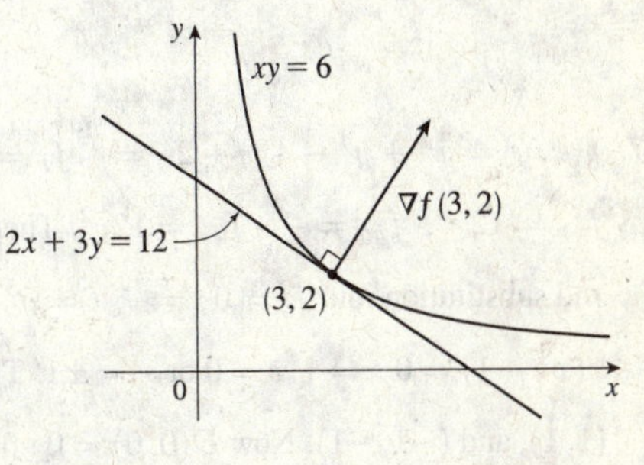

35. $f(x,y) = x^3 + 5x^2y + y^3 \Rightarrow$

$D_{\mathbf{u}} f(x,y) = \nabla f(x,y)\cdot\mathbf{u} = \left[3x^2 + 10xy, 5x^2 + 3y^2\right]\cdot\left[\frac{3}{5}, \frac{4}{5}\right] = \frac{9}{5}x^2 + 6xy + 4x^2 + \frac{12}{5}y^2 = \frac{29}{5}x^2 + 6xy + \frac{12}{5}y^2$. Then

$D^2_{\mathbf{u}} f(x,y) = D_{\mathbf{u}}\left[D_{\mathbf{u}} f(x,y)\right] = \nabla\left[D_{\mathbf{u}} f(x,y)\right]\cdot\mathbf{u} = \left[\frac{58}{5}x + 6y, 6x + \frac{24}{5}y\right]\cdot\left[\frac{3}{5}, \frac{4}{5}\right]$

$= \frac{174}{25}x + \frac{18}{5}y + \frac{24}{5}x + \frac{96}{25}y = \frac{294}{25}x + \frac{186}{25}y$

and $D^2_{\mathbf{u}} f(2,1) = \frac{294}{25}(2) + \frac{186}{25}(1) = \frac{774}{25}$.

9.6 Maximum and Minimum Values

1. (a) First we compute $D(1,1) = f_{xx}(1,1)\,f_{yy}(1,1) - [f_{xy}(1,1)]^2 = (4)(2) - (1)^2 = 7$. Since $D(1,1) > 0$ and $f_{xx}(1,1) > 0$, f has a local minimum at $(1,1)$ by the Second Derivatives Test.

(b) $D(1,1) = f_{xx}(1,1)\,f_{yy}(1,1) - [f_{xy}(1,1)]^2 = (4)(2) - (3)^2 = -1$. Since $D(1,1) < 0$, f has a saddle point at $(1,1)$ by the Second Derivatives Test.

© 2016 Cengage Learning. All Rights Reserved. May not be scanned, copied or duplicated, or posted to a publicly accessible website, in whole or in part.

3. In the figure, a point at approximately $(1,1)$ is enclosed by level curves which are oval in shape and indicate that as we move away from the point in any direction the values of f are increasing. Hence we would expect a local minimum at or near $(1,1)$. The level curves near $(0,0)$ resemble hyperbolas, and as we move away from the origin, the values of f increase in some directions and decrease in others, so we would expect to find a saddle point there.

To verify our predictions, we have $f(x,y)=4+x^3+y^3-3xy \;\Rightarrow\; f_x(x,y)=3x^2-3y$, $f_y(x,y)=3y^2-3x$. We have critical points where these partial derivatives are equal to 0: $3x^2-3y=0$, $3y^2-3x=0$. Substituting $y=x^2$ from the first equation into the second equation gives $3(x^2)^2-3x=0 \;\Rightarrow\; 3x(x^3-1)=0 \;\Rightarrow\; x=0$ or $x=1$. Then we have two critical points, $(0,0)$ and $(1,1)$. The second partial derivatives are $f_{xx}(x,y)=6x$, $f_{xy}(x,y)=-3$, and $f_{yy}(x,y)=6y$, so $D(x,y)=f_{xx}(x,y)\,f_{yy}(x,y)-[f_{xy}(x,y)]^2=(6x)(6y)-(-3)^2=36xy-9$. Then $D(0,0)=36(0)(0)-9=-9$, and $D(1,1)=36(1)(1)-9=27$. Since $D(0,0)<0$, f has a saddle point at $(0,0)$ by the Second Derivatives Test. Since $D(1,1)>0$ and $f_{xx}(1,1)>0$, f has a local minimum at $(1,1)$.

5. $f(x,y)=x^2+xy+y^2+y \;\Rightarrow\; f_x=2x+y,\; f_y=x+2y+1,\; f_{xx}=2,\; f_{xy}=1,\; f_{yy}=2$. Then $f_x=0$ implies $y=-2x$, and substitution into $f_y=x+2y+1=0$ gives $x+2(-2x)+1=0 \;\Rightarrow\; -3x=-1 \;\Rightarrow\; x=\frac{1}{3}$. Then $y=-\frac{2}{3}$ and the only critical point is $\left(\frac{1}{3},-\frac{2}{3}\right)$.

$D(x,y)=f_{xx}f_{yy}-(f_{xy})^2=(2)(2)-(1)^2=3$, and since $D\left(\frac{1}{3},-\frac{2}{3}\right)=3>0$ and $f_{xx}\left(\frac{1}{3},-\frac{2}{3}\right)=2>0$, $f\left(\frac{1}{3},-\frac{2}{3}\right)=-\frac{1}{3}$ is a local minimum by the Second Derivatives Test.

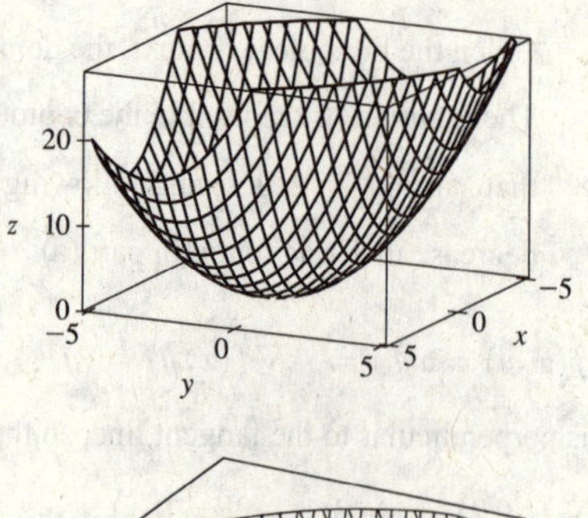

7. $f(x,y)=x^4+y^4-4xy+2 \;\Rightarrow\; f_x=4x^3-4y$, $f_y=4y^3-4x$, $f_{xx}=12x^2$, $f_{xy}=-4$, $f_{yy}=12y^2$. Then $f_x=0$ implies $y=x^3$, and substitution into $f_y=0 \;\Rightarrow\; x=y^3$ gives $x^9-x=0 \;\Rightarrow\; x(x^8-1)=0 \;\Rightarrow\; x=0$ or $x=\pm1$. Thus the critical points are $(0,0)$, $(1,1)$, and $(-1,-1)$. Now $D(0,0)=0\cdot0-(-4)^2=-16<0$, so $(0,0)$ is a saddle point. $D(1,1)=(12)(12)-(-4)^2>0$ and $f_{xx}(1,1)=12>0$, so $f(1,1)=0$ is a local minimum. $D(-1,-1)=(12)(12)-(-4)^2>0$ and $f_{xx}=(-1,-1)=12>0$, so $f(-1,-1)=0$ is also a local minimum.

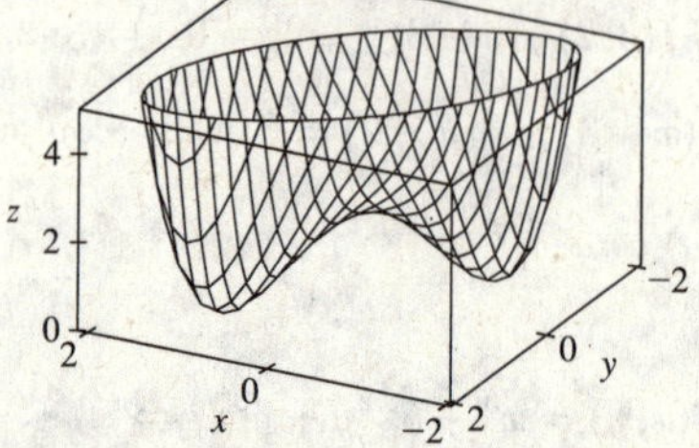

9. $f(x,y)=x^3-12xy+8y^3 \;\Rightarrow\; f_x=3x^2-12y$, $f_y=-12x+24y^2$, $f_{xx}=6x$, $f_{xy}=-12$, $f_{yy}=48y$. Then $f_x=0$ implies $x^2=4y$ and $f_y=0$ implies $x=2y^2$. Substituting the second equation into the first gives $(2y^2)^2=4y \;\Rightarrow\; 4y^4=4y \;\Rightarrow\; 4y(y^3-1)=0 \;\Rightarrow\; y=0$ or $y=1$. If $y=0$ then $x=0$ and if $y=1$ then $x=2$, so the critical points are $(0,0)$ and $(2,1)$.

$D(0,0)=(0)(0)-(-12)^2=-144<0$, so $(0,0)$ is a saddle point.

$D(2,1)=(12)(48)-(-12)^2=432>0$ and $f_{xx}(2,1)=12>0$ so $f(2,1)=-8$ is a local minimum.

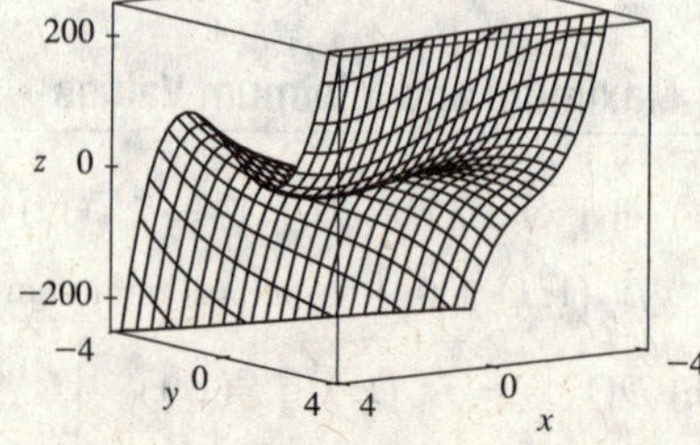

© 2016 Cengage Learning. All Rights Reserved. May not be scanned, copied or duplicated, or posted to a publicly accessible website, in whole or in part.

11. $f(x,y) = e^x \cos y \;\Rightarrow\; f_x = e^x \cos y,\ f_y = -e^x \sin y.$

Now $f_x = 0$ implies $\cos y = 0$ or $y = \frac{\pi}{2} + n\pi$ for n an integer.

But $\sin\left(\frac{\pi}{2} + n\pi\right) \neq 0$, so there are no critical points.

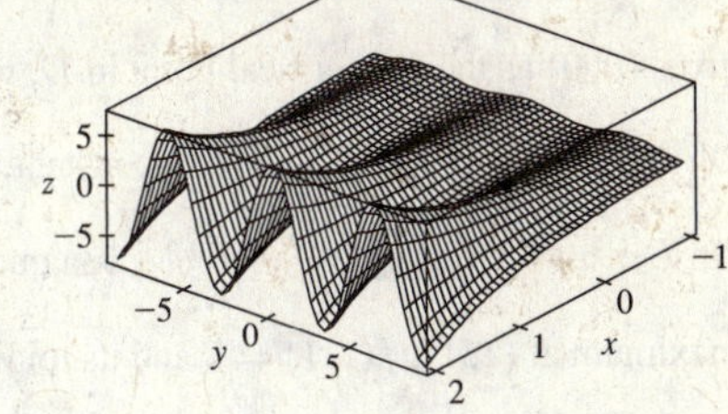

13. $f(x,y) = (x^2 + y^2)e^{y^2 - x^2} \;\Rightarrow$

$$f_x = (x^2+y^2)e^{y^2-x^2}(-2x) + 2xe^{y^2-x^2} = 2xe^{y^2-x^2}(1 - x^2 - y^2),$$

$$f_y = (x^2+y^2)e^{y^2-x^2}(2y) + 2ye^{y^2-x^2} = 2ye^{y^2-x^2}(1 + x^2 + y^2),$$

$$f_{xx} = 2xe^{y^2-x^2}(-2x) + (1 - x^2 - y^2)\left(2x\left(-2xe^{y^2-x^2}\right) + 2e^{y^2-x^2}\right) = 2e^{y^2-x^2}((1-x^2-y^2)(1-2x^2) - 2x^2),$$

$$f_{xy} = 2xe^{y^2-x^2}(-2y) + 2x(2y)e^{y^2-x^2}(1 - x^2 - y^2) = -4xye^{y^2-x^2}(x^2+y^2),$$

$$f_{yy} = 2ye^{y^2-x^2}(2y) + (1 + x^2 + y^2)\left(2y\left(2ye^{y^2-x^2}\right) + 2e^{y^2-x^2}\right) = 2e^{y^2-x^2}((1+x^2+y^2)(1+2y^2) + 2y^2).$$

$f_y = 0$ implies $y = 0$, and substituting into $f_x = 0$ gives $2xe^{-x^2}(1 - x^2) = 0 \;\Rightarrow\; x = 0$ or $x = \pm 1$. Thus the critical points are $(0,0)$ and $(\pm 1, 0)$. Now $D(0,0) = (2)(2) - 0 > 0$ and $f_{xx}(0,0) = 2 > 0$, so $f(0,0) = 0$ is a local minimum. $D(\pm 1, 0) = (-4e^{-1})(4e^{-1}) - 0 < 0$ so $(\pm 1, 0)$ are saddle points.

15. $f(x,y) = y^2 - 2y\cos x \;\Rightarrow\; f_x = 2y\sin x,\ f_y = 2y - 2\cos x,$ $f_{xx} = 2y\cos x,\ f_{xy} = 2\sin x,\ f_{yy} = 2$. Then $f_x = 0$ implies $y = 0$ or $\sin x = 0 \;\Rightarrow\; x = 0, \pi$, or 2π for $-1 \le x \le 7$. Substituting $y = 0$ into $f_y = 0$ gives $\cos x = 0 \;\Rightarrow\; x = \frac{\pi}{2}$ or $\frac{3\pi}{2}$, substituting $x = 0$ or $x = 2\pi$ into $f_y = 0$ gives $y = 1$, and substituting $x = \pi$ into $f_y = 0$ gives $y = -1$.

Thus the critical points are $(0,1)$, $\left(\frac{\pi}{2}, 0\right)$, $(\pi, -1)$, $\left(\frac{3\pi}{2}, 0\right)$, and $(2\pi, 1)$.

$D\left(\frac{\pi}{2}, 0\right) = D\left(\frac{3\pi}{2}, 0\right) = -4 < 0$ so $\left(\frac{\pi}{2}, 0\right)$ and $\left(\frac{3\pi}{2}, 0\right)$ are saddle points. $D(0,1) = D(\pi,-1) = D(2\pi,1) = 4 > 0$ and $f_{xx}(0,1) = f_{xx}(\pi,-1) = f_{xx}(2\pi,1) = 2 > 0$, so $f(0,1) = f(\pi,-1) = f(2\pi,1) = -1$ are local minima.

17. Since f is a polynomial it is continuous on D, so an absolute maximum and minimum exist. Here $f_x = 4$, $f_y = -5$ so there are no critical points inside D. Thus the absolute extrema must both occur on the boundary. Along L_1: $x = 0$ and $f(0,y) = 1 - 5y$ for $0 \le y \le 3$, a decreasing function in y, so the maximum value is $f(0,0) = 1$ and the minimum value is $f(0,3) = -14$. Along L_2: $y = 0$ and $f(x,0) = 1 + 4x$ for $0 \le x \le 2$, an increasing function in x, so the minimum value is $f(0,0) = 1$ and the maximum value is $f(2,0) = 9$. Along L_3: $y = -\frac{3}{2}x + 3$ and $f\left(x, -\frac{3}{2}x + 3\right) = \frac{23}{2}x - 14$ for $0 \le x \le 2$, an increasing function in x, so the minimum value is $f(0,3) = -14$ and the maximum value is $f(2,0) = 9$. Thus the absolute maximum of f on D is $f(2,0) = 9$ and the absolute minimum is $f(0,3) = -14$.

© 2016 Cengage Learning. All Rights Reserved. May not be scanned, copied or duplicated, or posted to a publicly accessible website, in whole or in part.

19. $f_x(x,y) = 2x + 2xy$, $f_y(x,y) = 2y + x^2$, and setting $f_x = f_y = 0$ gives $(0,0)$ as the only critical point in D, with $f(0,0) = 4$.

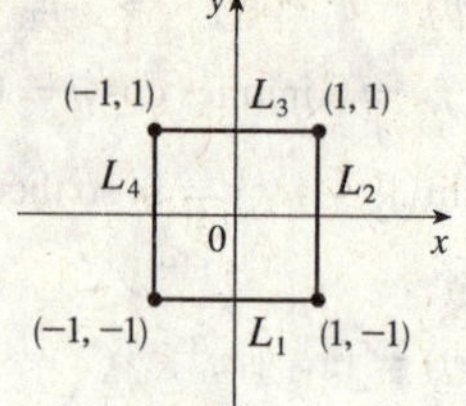

On L_1: $y = -1$, $f(x,-1) = 5$, a constant.

On L_2: $x = 1$, $f(1,y) = y^2 + y + 5$, a quadratic in y which attains its maximum at $(1,1)$, $f(1,1) = 7$ and its minimum at $\left(1,-\frac{1}{2}\right)$, $f\left(1,-\frac{1}{2}\right) = \frac{19}{4}$.

On L_3: $f(x,1) = 2x^2 + 5$ which attains its maximum at $(-1,1)$ and $(1,1)$ with $f(\pm 1,1) = 7$ and its minimum at $(0,1)$, $f(0,1) = 5$.

On L_4: $f(-1,y) = y^2 + y + 5$ with maximum at $(-1,1)$, $f(-1,1) = 7$ and minimum at $\left(-1,-\frac{1}{2}\right)$, $f\left(-1,-\frac{1}{2}\right) = \frac{19}{4}$.

Thus the absolute maximum is attained at both $(\pm 1,1)$ with $f(\pm 1,1) = 7$ and the absolute minimum on D is attained at $(0,0)$ with $f(0,0) = 4$.

21. $f(x,y) = -(x^2-1)^2 - (x^2y - x - 1)^2 \quad\Rightarrow\quad f_x(x,y) = -2(x^2-1)(2x) - 2(x^2y - x - 1)(2xy - 1)$ and $f_y(x,y) = -2(x^2y - x - 1)x^2$. Setting $f_y(x,y) = 0$ gives either $x = 0$ or $x^2y - x - 1 = 0$.

There are no critical points for $x = 0$, since $f_x(0,y) = -2$, so we set $x^2y - x - 1 = 0 \quad\Leftrightarrow\quad y = \dfrac{x+1}{x^2} \quad [x \neq 0]$,

so $f_x\left(x, \dfrac{x+1}{x^2}\right) = -2(x^2-1)(2x) - 2\left(x^2\,\dfrac{x+1}{x^2} - x - 1\right)\left(2x\,\dfrac{x+1}{x^2} - 1\right) = -4x(x^2-1)$. Therefore $f_x(x,y) = f_y(x,y) = 0$ at the points $(1,2)$ and $(-1,0)$. To classify these critical points, we calculate

$f_{xx}(x,y) = -12x^2 - 12x^2y^2 + 12xy + 4y + 2$, $f_{yy}(x,y) = -2x^4$, and $f_{xy}(x,y) = -8x^3y + 6x^2 + 4x$. In order to use the Second Derivatives Test we calculate

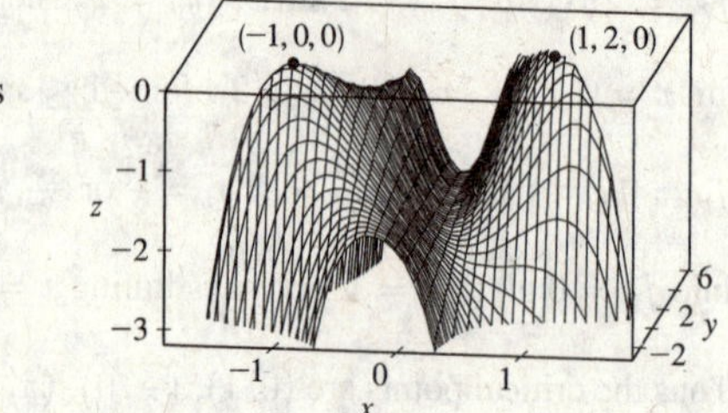

$D(-1,0) = f_{xx}(-1,0)\,f_{yy}(-1,0) - [f_{xy}(-1,0)]^2 = 16 > 0$,

$f_{xx}(-1,0) = -10 < 0$, $D(1,2) = 16 > 0$, and $f_{xx}(1,2) = -26 < 0$, so both $(-1,0)$ and $(1,2)$ give local maxima.

23. Let d be the distance from $(2,1,-1)$ to any point (x,y,z) on the plane $x + y - z = 1$, so $d = \sqrt{(x-2)^2 + (y-1)^2 + (z+1)^2}$ where $z = x + y - 1$, and we minimize $d^2 = f(x,y) = (x-2)^2 + (y-1)^2 + (x+y)^2$. Then $f_x(x,y) = 2(x-2) + 2(x+y) = 4x + 2y - 4$, $f_y(x,y) = 2(y-1) + 2(x+y) = 2x + 4y - 2$. Solving $4x + 2y - 4 = 0$ and $2x + 4y - 2 = 0$ simultaneously gives $x = 1$, $y = 0$. An absolute minimum exists (since there is a minimum distance from the point to the plane) and it must occur at a critical point, so the shortest distance occurs for $x = 1$, $y = 0$ for which $d = \sqrt{(1-2)^2 + (0-1)^2 + (0+1)^2} = \sqrt{3}$.

25. Let d be the distance from the point $(4,2,0)$ to any point (x,y,z) on the cone, so $d = \sqrt{(x-4)^2 + (y-2)^2 + z^2}$ where $z^2 = x^2 + y^2$, and we minimize $d^2 = (x-4)^2 + (y-2)^2 + x^2 + y^2 = f(x,y)$. Then $f_x(x,y) = 2\,(x-4) + 2x = 4x - 8$, $f_y(x,y) = 2\,(y-2) + 2y = 4y - 4$, and the critical points occur when

© 2016 Cengage Learning. All Rights Reserved. May not be scanned, copied or duplicated, or posted to a publicly accessible website, in whole or in part.

$f_x = 0 \;\Rightarrow\; x = 2,\; f_y = 0 \;\Rightarrow\; y = 1$. Thus the only critical point is $(2, 1)$. An absolute minimum exists (since there is a minimum distance from the cone to the point) which must occur at a critical point, so the points on the cone closest to $(4, 2, 0)$ are $\left(2, 1, \pm\sqrt{5}\right)$.

27. $x + y + z = 100$, so maximize $f(x, y) = xy(100 - x - y)$. $f_x = 100y - 2xy - y^2$, $f_y = 100x - x^2 - 2xy$, $f_{xx} = -2y$, $f_{yy} = -2x$, $f_{xy} = 100 - 2x - 2y$. Then $f_x = 0$ implies $y = 0$ or $y = 100 - 2x$. Substituting $y = 0$ into $f_y = 0$ gives $x = 0$ or $x = 100$ and substituting $y = 100 - 2x$ into $f_y = 0$ gives $3x^2 - 100x = 0$ so $x = 0$ or $\frac{100}{3}$.

Thus the critical points are $(0, 0)$, $(100, 0)$, $(0, 100)$ and $\left(\frac{100}{3}, \frac{100}{3}\right)$.

$D(0, 0) = D(100, 0) = D(0, 100) = -10{,}000$ while $D\left(\frac{100}{3}, \frac{100}{3}\right) = \frac{10{,}000}{3}$ and $f_{xx}\left(\frac{100}{3}, \frac{100}{3}\right) = -\frac{200}{3} < 0$. Thus $(0, 0)$, $(100, 0)$ and $(0, 100)$ are saddle points whereas $f\left(\frac{100}{3}, \frac{100}{3}\right)$ is a local maximum. Thus the numbers are $x = y = z = \frac{100}{3}$.

29. Center the sphere at the origin so that its equation is $x^2 + y^2 + z^2 = r^2$, and orient the inscribed rectangular box so that its edges are parallel to the coordinate axes. Any vertex of the box satisfies $x^2 + y^2 + z^2 = r^2$, so take (x, y, z) to be the vertex in the first octant. Then the box has length $2x$, width $2y$, and height $2z = 2\sqrt{r^2 - x^2 - y^2}$ with volume given by

$V(x, y) = (2x)(2y)\left(2\sqrt{r^2 - x^2 - y^2}\right) = 8xy\sqrt{r^2 - x^2 - y^2}$ for $0 < x < r$, $0 < y < r$. Then

$$V_x = (8xy)\cdot\tfrac{1}{2}(r^2 - x^2 - y^2)^{-1/2}(-2x) + \sqrt{r^2 - x^2 - y^2}\cdot 8y = \frac{8y(r^2 - 2x^2 - y^2)}{\sqrt{r^2 - x^2 - y^2}} \text{ and } V_y = \frac{8x(r^2 - x^2 - 2y^2)}{\sqrt{r^2 - x^2 - y^2}}.$$

Setting $V_x = 0$ gives $y = 0$ or $2x^2 + y^2 = r^2$, but $y > 0$ so only the latter solution applies. Similarly, $V_y = 0$ with $x > 0$ implies $x^2 + 2y^2 = r^2$. Substituting, we have $2x^2 + y^2 = x^2 + 2y^2 \;\Rightarrow\; x^2 = y^2 \;\Rightarrow\; y = x$. Then $x^2 + 2y^2 = r^2 \;\Rightarrow\; 3x^2 = r^2 \;\Rightarrow\; x = \sqrt{r^2/3} = r/\sqrt{3} = y$. Thus the only critical point is $\left(r/\sqrt{3}, r/\sqrt{3}\right)$. There must be a maximum volume and here it must occur at a critical point, so the maximum volume occurs when $x = y = r/\sqrt{3}$ and the maximum volume is $V\left(\frac{r}{\sqrt{3}}, \frac{r}{\sqrt{3}}\right) = 8\left(\frac{r}{\sqrt{3}}\right)\left(\frac{r}{\sqrt{3}}\right)\sqrt{r^2 - \left(\frac{r}{\sqrt{3}}\right)^2 - \left(\frac{r}{\sqrt{3}}\right)^2} = \dfrac{8}{3\sqrt{3}}\,r^3$.

31. Maximize $f(x, y) = \dfrac{xy}{3}(6 - x - 2y)$, then the maximum volume is $V = xyz$.

$f_x = \frac{1}{3}(6y - 2xy - 2y^2) = \frac{1}{3}y(6 - 2x - 2y)$ and $f_y = \frac{1}{3}x(6 - x - 4y)$. Setting $f_x = 0$ and $f_y = 0$ gives the critical point $(2, 1)$ which geometrically must give a maximum. Thus the volume of the largest such box is $V = (2)(1)\left(\frac{2}{3}\right) = \frac{4}{3}$.

33. Let the dimensions be x, y, and z; then $4x + 4y + 4z = c$ and the volume is

$V = xyz = xy\left(\frac{1}{4}c - x - y\right) = \frac{1}{4}cxy - x^2y - xy^2$, $x > 0$, $y > 0$. Then $V_x = \frac{1}{4}cy - 2xy - y^2$ and $V_y = \frac{1}{4}cx - x^2 - 2xy$, so $V_x = 0 = V_y$ when $2x + y = \frac{1}{4}c$ and $x + 2y = \frac{1}{4}c$. Solving, we get $x = \frac{1}{12}c$, $y = \frac{1}{12}c$ and $z = \frac{1}{4}c - x - y = \frac{1}{12}c$. From the geometrical nature of the problem, this critical point must give an absolute maximum. Thus the box is a cube with edge length $\frac{1}{12}c$.

© 2016 Cengage Learning. All Rights Reserved. May not be scanned, copied or duplicated, or posted to a publicly accessible website, in whole or in part.

35. Let the dimensions be x, y and z, then minimize $xy + 2(xz + yz)$ if $xyz = 32{,}000\text{ cm}^3$. Then

$f(x,y) = xy + [64{,}000(x+y)/xy] = xy + 64{,}000(x^{-1} + y^{-1})$, $f_x = y - 64{,}000x^{-2}$, $f_y = x - 64{,}000y^{-2}$.

And $f_x = 0$ implies $y = 64{,}000/x^2$; substituting into $f_y = 0$ implies $x^3 = 64{,}000$ or $x = 40$ and then $y = 40$. Now

$D(x,y) = [(2)(64{,}000)]^2 x^{-3} y^{-3} - 1 > 0$ for $(40, 40)$ and $f_{xx}(40, 40) > 0$ so this is indeed a minimum. Thus the dimensions of the box are $x = y = 40$ cm, $z = 20$ cm.

37. (a) Substituting $p_3 = 1 - p_1 - p_2$ into the Shannon index equation gives

$$\begin{aligned} H(p_1,p_2) &= -p_1 \ln p_1 - p_2 \ln p_2 - (1 - p_1 - p_2)\ln(1 - p_1 - p_2) \\ &= -p_1(\ln p_1 - \ln(1 - p_1 - p_2)) - p_2(\ln p_2 - \ln(1 - p_1 - p_2)) - \ln(1 - p_1 - p_2) \\ &= -p_1 \ln\frac{p_1}{1 - p_1 - p_2} - p_2 \ln\frac{p_2}{1 - p_1 - p_2} - \ln(1 - p_1 - p_2) \\ &= \ln\left[\frac{(1 - p_1 - p_2)^{p_1 + p_2 - 1}}{p_1^{p_1} p_2^{p_2}}\right] \end{aligned}$$

(b) The quantities p_1 and p_2 represent proportions so they must each be positive and their sum must be no larger than 1. Hence the domain of H is $\{(p_1,p_2) \mid p_1 \geq 0, p_2 \geq 0, p_1 + p_2 \leq 1\}$ as depicted in the figure in part (c).

(c) $H_{p_1} = -\ln\dfrac{p_1}{1 - p_1 - p_2} - p_1\left(\dfrac{1 - p_1 - p_2}{p_1}\right)\left(\dfrac{1 - p_2}{(1 - p_1 - p_2)^2}\right) - \dfrac{p_2}{1 - p_1 - p_2} + \dfrac{1}{1 - p_1 - p_2} = -\ln\dfrac{p_1}{1 - p_1 - p_2}$,

$H_{p2} = -\dfrac{p_1}{1 - p_1 - p_2} - \ln p_2 - 1 + \ln(1 - p_1 - p_2) - p_2\dfrac{1}{1 - p_1 - p_2} + \dfrac{1}{1 - p_1 - p_2} = -\ln\dfrac{p_2}{1 - p_1 - p_2}$. Then

$H_{p_1} = 0$ implies $\dfrac{p_1}{1 - p_1 - p_2} = 1 \Rightarrow p_2 = 1 - 2p_1$. Substituting $p_2 = 1 - 2p_1$ into $H_{p_2} = 0$ gives

$-\ln\dfrac{1 - 2p_1}{p_1} = 0 \Rightarrow p_1 = \frac{1}{3}$ so $p_2 = 1 - 2\left(\frac{1}{3}\right) = \frac{1}{3}$. Thus the only critical point is $\left(\frac{1}{3}, \frac{1}{3}\right)$ with

$H\left(\frac{1}{3}, \frac{1}{3}\right) = -\ln\left(\frac{1}{3}\right) \approx 1.1$.

On L_1: $p_1 = 0$, $H(0,p_2) = -p_2 \ln p_2 + p_2 \ln(1 - p_2) - \ln(1 - p_2) = f(p_2)$, which has a maximum value when $f'(p_2) = 0 \Rightarrow \ln(1 - p_2) - \ln p_2 = 0 \Rightarrow$ $p_2 = \frac{1}{2}$. So $H\left(0, \frac{1}{2}\right) = \ln 2 \approx 0.69$ is the maximum on L_1 and the minima occur at the domain boundaries $(0,0)$ and $(0,1)$ with $H(0,0) = H(0,1) = 0$.

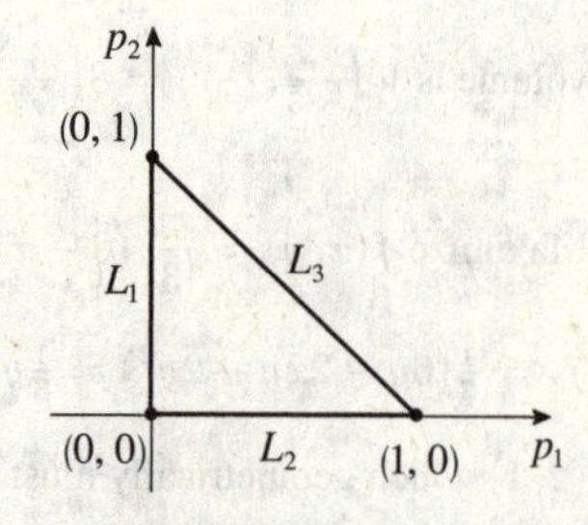

On L_2: $p_2 = 0$, $H(p_1,0) = -p_1 \ln p_1 + p_1 \ln(1 - p_1) - \ln(1 - p_1)$, which attains its maximum at $\left(\frac{1}{2}, 0\right)$, $H\left(\frac{1}{2}, 0\right) = \ln 2 \approx 0.69$ and its minima at $(0,0)$ and $(1,0)$ with $H(0,0) = H(1,0) = 0$.

On L_3: $p_2 = 1 - p_1$, $H(p_1, 1 - p_1) = -\ln\left[p_1^{p_1}(1 - p_1)^{1 - p_1}\right]$, which attains its maximum at $\left(\frac{1}{2}, \frac{1}{2}\right)$, $H\left(\frac{1}{2}, \frac{1}{2}\right) = \ln 2 \approx 0.69$ and its minima at $(0,1)$ and $(1,0)$ with $H(0,1) = H(1,0) = 0$.

Thus the absolute maximum value is $\ln 3 \approx 1.1$ which occurs when $p_1 = p_2 = p_3 = \frac{1}{3}$.

© 2016 Cengage Learning. All Rights Reserved. May not be scanned, copied or duplicated, or posted to a publicly accessible website, in whole or in part.

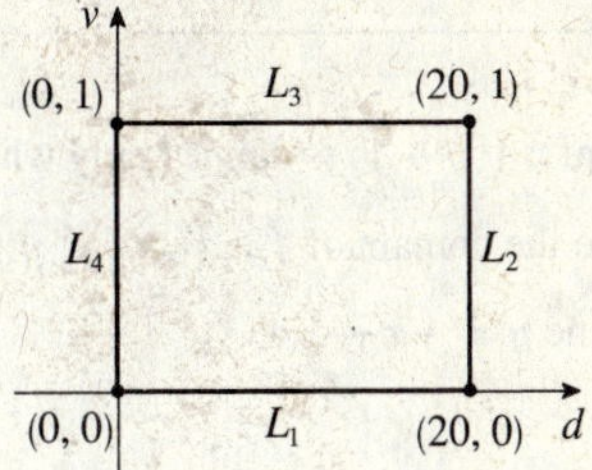

39. Since v represents a fraction of the population, the domain of $R_0(d,v)$ is

$D = \{(d,v) \mid 0 \le d \le 20, 0 \le v \le 1\}$.

$R_0(d,v) = 5(1-v)\dfrac{d}{1+d} \quad \Rightarrow \quad \dfrac{\partial R_0}{\partial d} = \dfrac{5(1-v)}{(1+d)^2}$ and $\dfrac{\partial R_0}{\partial v} = -\dfrac{5d}{1+d}$ so the

only critical point in D is $(0,1)$ where $R_0(0,1) = 0$. Along L_1: $v = 0$, so

$R_0(d,0) = \dfrac{5d}{1+d}$, for $0 \le d \le 20$, an increasing function in d, so the minimum

value is $R_0(0,0) = 0$ and the maximum value is $R_0(20,0) = \frac{100}{21} \approx 4.76$.

Along L_2: $d = 20$, so $R_0(20,v) = \frac{100}{21}(1-v)$, for $0 \le v \le 1$, an decreasing function in v, so the minimum value is

$R_0(20,1) = 0$ and the maximum value is $R_0(20,0) = \frac{100}{21} \approx 4.76$.

Along L_3: $v = 1$, so $R_0(d,1) = 0$, for $0 \le d \le 20$, a constant function. Along L_4: $d = 0$, so $R_0(0,v) = 0$, for $0 \le v \le 1$, a constant function in v. Thus, the absolute maximum value is $R_0\,(20,0) = \frac{100}{21} \approx 4.76$.

9 Review

TRUE-FALSE QUIZ

1. True. $f_y(a,b) = \lim\limits_{h\to 0} \dfrac{f(a,b+h) - f(a,b)}{h}$ from Equation 9.2.3. Let $h = y - b$. As $h \to 0$, $y \to b$. Then by substituting, we get $f_y(a,b) = \lim\limits_{y\to b} \dfrac{f(a,y) - f(a,b)}{y - b}$.

3. False. $f_{xy} = \dfrac{\partial^2 f}{\partial y\,\partial x}$.

5. False. As an example, consider the function $f\,(x,y) = \dfrac{xy^2}{x^2+y^4}$. Approaching $(0,0)$ along any line through the origin $y = mx$, we have $f\,(x,mx) = \dfrac{x\,(mx)^2}{x^2+(mx)^4} = \dfrac{m^2x}{1+m^4x^2}$. So $f\,(x,y) \to 0$ as $(x,y) \to (0,0)$ along $y = mx$. However, if we approach $(0,0)$ along the parabola $x = y^2$, we have $f\,(y^2,y) = \dfrac{(y^2)\,y^2}{(y^2)^2+y^4} = \dfrac{y^4}{2y^4} = \dfrac{1}{2}$, so $f\,(x,y) \to \frac{1}{2}$ as $(x,y) \to (0,0)$ along $x = y^2$. Thus, we see that it is not always sufficient to check the limiting behaviour along linear paths.

7. True. If f has a local minimum and f is differentiable at (a,b) then by Theorem 9.6.2, $f_x(a,b) = 0$ and $f_y(a,b) = 0$, so $\nabla f(a,b) = [f_x(a,b), f_y(a,b)] = [0,0] = \mathbf{0}$.

9. False. $\nabla f(x,y) = [0, 1/y]$.

11. True. $\nabla f = [\cos x, \cos y]$, so $|\nabla f| = \sqrt{\cos^2 x + \cos^2 y}$. But $|\cos\theta| \le 1$, so $|\nabla f| \le \sqrt{2}$. Now $D_{\mathbf{u}}\,f(x,y) = \nabla f \cdot \mathbf{u} = |\nabla f|\,|\mathbf{u}|\cos\theta$, but $\mathbf{u}$ is a unit vector, so $|D_{\mathbf{u}}\,f(x,y)| \le \sqrt{2}\cdot 1 \cdot 1 = \sqrt{2}$.

© 2016 Cengage Learning. All Rights Reserved. May not be scanned, copied or duplicated, or posted to a publicly accessible website, in whole or in part.

EXERCISES

1. $\ln(x + y + 1)$ is defined only when $x + y + 1 > 0 \quad \Leftrightarrow \quad y > -x - 1$, so the domain of f is $\{(x, y) \mid y > -x - 1\}$, all those points above the line $y = -x - 1$.

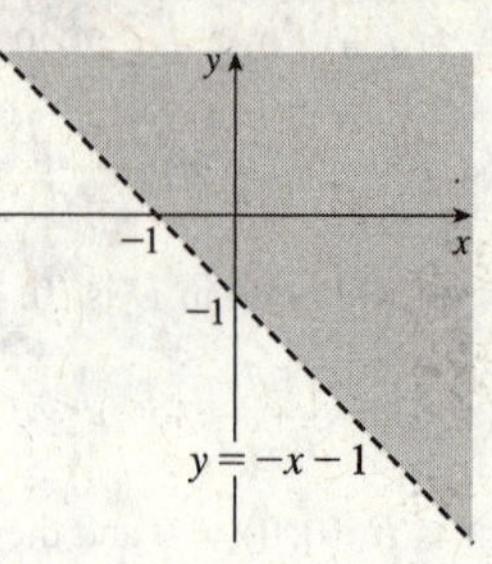

3. $z = f(x, y) = 1 - y^2$, a parabolic cylinder

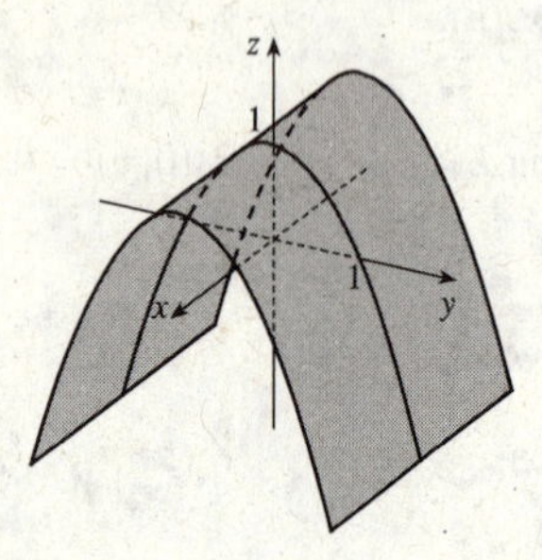

5. The level curves are $\ln(1 + x^2 + y^2) = k$ or $x^2 + y^2 = e^k - 1$ $(k \geq 0)$. This is the equation of a circle centered at the origin with radius $\sqrt{e^k - 1}$. If $k = 0$, the graph consists of just the origin.

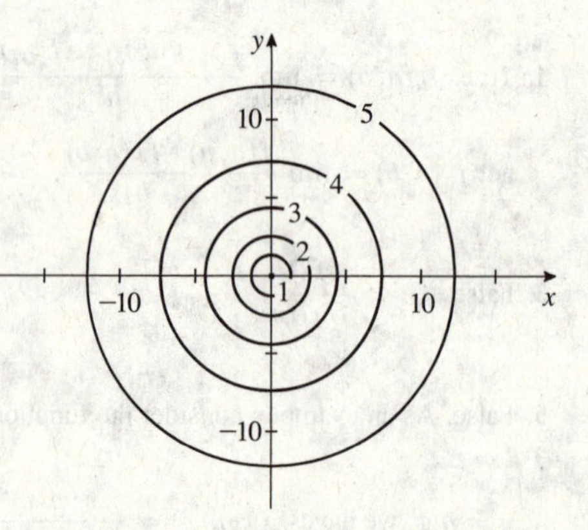

7. f is a rational function, so it is continuous on its domain. Since f is defined at $(1, 1)$, we use direct substitution to evaluate the limit: $\displaystyle\lim_{(x,y)\to(1,1)} \frac{2xy}{x^2 + 2y^2} = \frac{2(1)(1)}{1^2 + 2(1)^2} = \frac{2}{3}$.

9. (a) $T_x(6, 4) = \displaystyle\lim_{h\to 0} \frac{T(6 + h, 4) - T(6, 4)}{h}$, so we can approximate $T_x(6, 4)$ by considering $h = \pm 2$ and using the values given in the table: $T_x(6, 4) \approx \dfrac{T(8, 4) - T(6, 4)}{2} = \dfrac{86 - 80}{2} = 3$,

$T_x(6, 4) \approx \dfrac{T(4, 4) - T(6, 4)}{-2} = \dfrac{72 - 80}{-2} = 4$. Averaging these values, we estimate $T_x(6, 4)$ to be approximately $3.5°\text{C/m}$. Similarly, $T_y(6, 4) = \displaystyle\lim_{h\to 0} \frac{T(6, 4 + h) - T(6, 4)}{h}$, which we can approximate with $h = \pm 2$:

$T_y(6, 4) \approx \dfrac{T(6, 6) - T(6, 4)}{2} = \dfrac{75 - 80}{2} = -2.5$, $T_y(6, 4) \approx \dfrac{T(6, 2) - T(6, 4)}{-2} = \dfrac{87 - 80}{-2} = -3.5$. Averaging these values, we estimate $T_y(6, 4)$ to be approximately $-3.0°\text{C/m}$.

© 2016 Cengage Learning. All Rights Reserved. May not be scanned, copied or duplicated, or posted to a publicly accessible website, in whole or in part.

(b) Here $\mathbf{u} = \left[\frac{1}{\sqrt{2}}, \frac{1}{\sqrt{2}}\right]$, so by Equation 9.5.9, $D_{\mathbf{u}}\,T(6,4) = \nabla T(6,4) \cdot \mathbf{u} = T_x(6,4)\,\frac{1}{\sqrt{2}} + T_y(6,4)\,\frac{1}{\sqrt{2}}$. Using our estimates from part (a), we have $D_{\mathbf{u}}\,T(6,4) \approx (3.5)\,\frac{1}{\sqrt{2}} + (-3.0)\,\frac{1}{\sqrt{2}} = \frac{1}{2\sqrt{2}} \approx 0.35$. This means that as we move through the point $(6,4)$ in the direction of $\mathbf{u}$, the temperature increases at a rate of approximately 0.35°C/m.

Alternatively, we can use Definition 9.5.2: $D_{\mathbf{u}}\,T(6,4) = \lim\limits_{h\to 0} \dfrac{T\left(6 + h\,\frac{1}{\sqrt{2}}, 4 + h\,\frac{1}{\sqrt{2}}\right) - T(6,4)}{h}$,

which we can estimate with $h = \pm 2\sqrt{2}$. Then $D_{\mathbf{u}}\,T(6,4) \approx \dfrac{T(8,6) - T(6,4)}{2\sqrt{2}} = \dfrac{80 - 80}{2\sqrt{2}} = 0$,

$D_{\mathbf{u}}\,T(6,4) \approx \dfrac{T(4,2) - T(6,4)}{-2\sqrt{2}} = \dfrac{74 - 80}{-2\sqrt{2}} = \dfrac{3}{\sqrt{2}}$. Averaging these values, we have $D_{\mathbf{u}}\,T(6,4) \approx \frac{3}{2\sqrt{2}} \approx 1.1^\circ\text{C/m}$.

(c) $T_{xy}(x,y) = \dfrac{\partial}{\partial y}\left[T_x(x,y)\right] = \lim\limits_{h\to 0} \dfrac{T_x(x,y+h) - T_x(x,y)}{h}$, so $T_{xy}(6,4) = \lim\limits_{h\to 0} \dfrac{T_x(6,4+h) - T_x(6,4)}{h}$ which we can estimate with $h = \pm 2$. We have $T_x(6,4) \approx 3.5$ from part (a), but we will also need values for $T_x(6,6)$ and $T_x(6,2)$. If we use $h = \pm 2$ and the values given in the table, we have

$T_x(6,6) \approx \dfrac{T(8,6) - T(6,6)}{2} = \dfrac{80 - 75}{2} = 2.5$, $T_x(6,6) \approx \dfrac{T(4,6) - T(6,6)}{-2} = \dfrac{68 - 75}{-2} = 3.5$.

Averaging these values, we estimate $T_x(6,6) \approx 3.0$. Similarly,

$T_x(6,2) \approx \dfrac{T(8,2) - T_x(6,2)}{2} = \dfrac{90 - 87}{2} = 1.5$, $T_x(6,2) \approx \dfrac{T(4,2) - T(6,2)}{-2} = \dfrac{74 - 87}{-2} = 6.5$.

Averaging these values, we estimate $T_x(6,2) \approx 4.0$. Finally, we estimate $T_{xy}(6,4)$:

$T_{xy}(6,4) \approx \dfrac{T_x(6,6) - T_x(6,4)}{2} = \dfrac{3.0 - 3.5}{2} = -0.25$, $T_{xy}(6,4) \approx \dfrac{T_x(6,2) - T_x(6,4)}{-2} = \dfrac{4.0 - 3.5}{-2} = -0.25$.

Averaging these values, we have $T_{xy}\,(6,4) \approx -0.25$.

11. $f(x,y) = \sqrt{2x + y^2} \quad \Rightarrow \quad f_x = \frac{1}{2}(2x + y^2)^{-1/2}(2) = \dfrac{1}{\sqrt{2x + y^2}}$, $f_y = \frac{1}{2}(2x + y^2)^{-1/2}(2y) = \dfrac{y}{\sqrt{2x + y^2}}$

13. $g(u,v) = u\tan^{-1} v \quad \Rightarrow \quad g_u = \tan^{-1} v$, $g_v = \dfrac{u}{1 + v^2}$

15. $T(p,q,r) = p\ln(q + e^r) \quad \Rightarrow \quad T_p = \ln(q + e^r)$, $T_q = \dfrac{p}{q + e^r}$, $T_r = \dfrac{pe^r}{q + e^r}$

17. (a) $P(m,T) = \dfrac{1}{1 + e^{3.222 - 31.669m + 0.083T}} \quad \Rightarrow \quad \dfrac{\partial P}{\partial m}\,(m,T) = \dfrac{31.669e^{3.222 - 31.669m + 0.083T}}{\left(1 + e^{3.222 - 31.669m + 0.083T}\right)^2}$ and

$\dfrac{\partial P}{\partial T}\,(m,T) = -\dfrac{0.083e^{3.222 - 31.669m + 0.083T}}{\left(1 + e^{3.222 - 31.669m + 0.083T}\right)^2}$. Then $\dfrac{\partial P}{\partial m}\,(0.2, 20) = \dfrac{31.669e^{3.222 - 31.669(0.2) + 0.083(20)}}{\left(1 + e^{3.222 - 31.669(0.2) + 0.083(20)}\right)^2} \approx 4.87$ and

$\dfrac{\partial P}{\partial T}\,(0.2, 20) = -\dfrac{0.083e^{3.222 - 31.669(0.2) + 0.083(20)}}{\left(1 + e^{3.222 - 31.669(0.2) + 0.083(20)}\right)^2} \approx -0.013$.

(b) $\dfrac{\partial P}{\partial T}$ is negative so an increase in water temperature will decrease the probability that the tadpole escapes, that is, worsen its chance of escape (assuming the tadpole's mass remains the same).

(c) $\dfrac{\partial P}{\partial m}$ is positive so an increase in mass will increase the probability that the tadpole escapes, that is, improve its chance of escape (assuming the water temperature remains the same as the tadpole grows).

© 2016 Cengage Learning. All Rights Reserved. May not be scanned, copied or duplicated, or posted to a publicly accessible website, in whole or in part.

19. $f(x,y) = 4x^3 - xy^2 \quad \Rightarrow \quad f_x = 12x^2 - y^2,\ f_y = -2xy,\ f_{xx} = 24x,\ f_{yy} = -2x,\ f_{xy} = f_{yx} = -2y$

21. $f(x,y,z) = x^k y^l z^m \quad \Rightarrow \quad f_x = kx^{k-1}y^l z^m,\ f_y = lx^k y^{l-1} z^m,\ f_z = mx^k y^l z^{m-1},\ f_{xx} = k(k-1)x^{k-2}y^l z^m,$

$f_{yy} = l(l-1)x^k y^{l-2} z^m,\ f_{zz} = m(m-1)x^k y^l z^{m-2},\ f_{xy} = f_{yx} = klx^{k-1}y^{l-1}z^m,\ f_{xz} = f_{zx} = kmx^{k-1}y^l z^{m-1},$

$f_{yz} = f_{zy} = lmx^k y^{l-1} z^{m-1}$

23. $z = xy + xe^{y/x} \quad \Rightarrow \quad \dfrac{\partial z}{\partial x} = y - \dfrac{y}{x}e^{y/x} + e^{y/x},\ \dfrac{\partial z}{\partial y} = x + e^{y/x}$ and

$$x\,\frac{\partial z}{\partial x} + y\,\frac{\partial z}{\partial y} = x\left(y - \frac{y}{x}e^{y/x} + e^{y/x}\right) + y\left(x + e^{y/x}\right) = xy - ye^{y/x} + xe^{y/x} + xy + ye^{y/x} = xy + xy + xe^{y/x} = xy + z.$$

25. $z = f(x,y) = 4x^2 - y^2 + 2y \quad \Rightarrow \quad f_x(x,y) = 8x,\ f_y(x,y) = -2y + 2$, so $f_x(-1,2) = -8,\ f_y(-1,2) = -2$. By Equation 2, an equation of the tangent plane is $z - 4 = f_x(-1,2)[x - (-1)] + f_y(-1,2)(y-2) \quad \Rightarrow$ $z - 4 = -8(x+1) - 2(y-2)$ or $z = -8x - 2y$.

27. $z = f(x,y) = \sqrt{4 - x^2 - 2y^2} \quad \Rightarrow \quad f_x(x,y) = \frac{1}{2}(4 - x^2 - 2y^2)^{-1/2}(-2x) = -\dfrac{x}{\sqrt{4 - x^2 - 2y^2}},$

$f_y(x,y) = \frac{1}{2}(4 - x^2 - 2y^2)^{-1/2}(-4y) = -\dfrac{2y}{\sqrt{4 - x^2 - 2y^2}}$, so $f_x(1,-1) = -1$ and $f_y(1,-1) = 2$. Thus, an equation of the tangent plane is $z - 1 = f_x(1,-1)(x-1) + f_y(1,-1)[y - (-1)] \quad \Rightarrow \quad z - 1 = -1(x-1) + 2(y+1)$ or $x - 2y + z = 4$.

29. $f(x,y) = xy + y^3$. The partial derivatives are $f_x(x,y) = y$, and $f_y(x,y) = x + 3y^2$, so $f_x(2,1) = 1$, and $f_y(2,1) = 5$. f_x and f_y are continuous functions near $(2,1)$, so f is differentiable at $(2,1)$. The linearization of f at $(2,1)$ is $L(x,y) = f(2,1) + f_x(2,1)(x-2) + f_y(2,1)(y-1) = 3 + (x-2) + 5(y-1) = x + 5y - 4$.

31. $\dfrac{du}{dp} = \dfrac{\partial u}{\partial x}\dfrac{dx}{dp} + \dfrac{\partial u}{\partial y}\dfrac{dy}{dp} + \dfrac{\partial u}{\partial z}\dfrac{dz}{dp} = 2xy^3(1 + 6p) + 3x^2y^2(pe^p + e^p) + 4z^3(p\cos p + \sin p)$

33. $\dfrac{\partial z}{\partial x} = \dfrac{\partial z}{\partial u}\,y + \dfrac{\partial z}{\partial v}\,\dfrac{-y}{x^2}$ and

$$\frac{\partial^2 z}{\partial x^2} = y\,\frac{\partial}{\partial x}\left(\frac{\partial z}{\partial u}\right) + \frac{2y}{x^3}\frac{\partial z}{\partial v} + \frac{-y}{x^2}\frac{\partial}{\partial x}\left(\frac{\partial z}{\partial v}\right) = \frac{2y}{x^3}\frac{\partial z}{\partial v} + y\left(\frac{\partial^2 z}{\partial u^2}\,y + \frac{\partial^2 z}{\partial v\,\partial u}\,\frac{-y}{x^2}\right) + \frac{-y}{x^2}\left(\frac{\partial^2 z}{\partial v^2}\,\frac{-y}{x^2} + \frac{\partial^2 z}{\partial u\,\partial v}\,y\right)$$

$$= \frac{2y}{x^3}\frac{\partial z}{\partial v} + y^2\,\frac{\partial^2 z}{\partial u^2} - \frac{2y^2}{x^2}\,\frac{\partial^2 z}{\partial u\,\partial v} + \frac{y^2}{x^4}\,\frac{\partial^2 z}{\partial v^2}$$

Also $\dfrac{\partial z}{\partial y} = x\,\dfrac{\partial z}{\partial u} + \dfrac{1}{x}\dfrac{\partial z}{\partial v}$ and

$$\frac{\partial^2 z}{\partial y^2} = x\,\frac{\partial}{\partial y}\left(\frac{\partial z}{\partial u}\right) + \frac{1}{x}\frac{\partial}{\partial y}\left(\frac{\partial z}{\partial v}\right) = x\left(\frac{\partial^2 z}{\partial u^2}\,x + \frac{\partial^2 z}{\partial v\,\partial u}\,\frac{1}{x}\right) + \frac{1}{x}\left(\frac{\partial^2 z}{\partial v^2}\,\frac{1}{x} + \frac{\partial^2 z}{\partial u\,\partial v}\,x\right) = x^2\,\frac{\partial^2 z}{\partial u^2} + 2\,\frac{\partial^2 z}{\partial u\,\partial v} + \frac{1}{x^2}\,\frac{\partial^2 z}{\partial v^2}$$

Thus

$$x^2\,\frac{\partial^2 z}{\partial x^2} - y^2\,\frac{\partial^2 z}{\partial y^2} = \frac{2y}{x}\frac{\partial z}{\partial v} + x^2y^2\,\frac{\partial^2 z}{\partial u^2} - 2y^2\,\frac{\partial^2 z}{\partial u\,\partial v} + \frac{y^2}{x^2}\frac{\partial^2 z}{\partial v^2} - x^2y^2\,\frac{\partial^2 z}{\partial u^2} - 2y^2\,\frac{\partial^2 z}{\partial u\,\partial v} - \frac{y^2}{x^2}\frac{\partial^2 z}{\partial v^2}$$

$$= \frac{2y}{x}\frac{\partial z}{\partial v} - 4y^2\,\frac{\partial^2 z}{\partial u\,\partial v} = 2v\,\frac{\partial z}{\partial v} - 4uv\,\frac{\partial^2 z}{\partial u\,\partial v}$$

since $y = xv = \dfrac{uv}{y}$ or $y^2 = uv$.

© 2016 Cengage Learning. All Rights Reserved. May not be scanned, copied or duplicated, or posted to a publicly accessible website, in whole or in part.

35. $g(p,q) = pq^2e^{-pq} \;\Rightarrow\; \nabla g(p,q) = \left[\frac{\partial g}{\partial p}, \frac{\partial g}{\partial q}\right] = \left[q^2e^{-pq} - pq^3e^{-pq}, 2pqe^{-pq} - p^2q^2e^{-pq}\right]$

37. $f(x,y) = x^2e^{-y} \;\Rightarrow\; \nabla f = \left[2xe^{-y}, -x^2e^{-y}\right]$, $\nabla f(-2,0) = [-4,-4]$. The direction is given by $[4,-3]$, so $\mathbf{u} = \frac{1}{\sqrt{4^2+(-3)^2}}[4,-3] = \frac{1}{5}[4,-3]$ and $D_{\mathbf{u}}\, f(-2,0) = \nabla f(-2,0) \cdot \mathbf{u} = [-4,-4] \cdot \frac{1}{5}[4,-3] = \frac{1}{5}(-16+12) = -\frac{4}{5}$.

39. First we draw a line passing through Homestead and the eye of the hurricane. We can approximate the directional derivative at Homestead in the direction of the eye of the hurricane by the average rate of change of wind speed between the points where this line intersects the contour lines closest to Homestead. In the direction of the eye of the hurricane, the wind speed changes from 45 to 50 knots. We estimate the distance between these two points to be approximately 8 miles, so the rate of change of wind speed in the direction given is approximately $\frac{50-45}{8} = \frac{5}{8} = 0.625$ knot/mi.

41. $f(x,y) = 3xy - x^2y - xy^2 \;\Rightarrow\; f_x = 3y - 2xy - y^2$, $f_y = 3x - x^2 - 2xy$, $f_{xx} = -2y$, $f_{yy} = -2x$, $f_{xy} = 3 - 2x - 2y$. Then $f_x = 0$ implies $y(3-2x-y) = 0$ so $y = 0$ or $y = 3 - 2x$. Substituting into $f_y = 0$ implies $x(3-x) = 0$ or $3x(-1+x) = 0$. Hence the critical points are $(0,0)$, $(3,0)$, $(0,3)$ and $(1,1)$. $D(0,0) = D(3,0) = D(0,3) = -9 < 0$ so $(0,0)$, $(3,0)$, and $(0,3)$ are saddle points. $D(1,1) = 3 > 0$ and $f_{xx}(1,1) = -2 < 0$, so $f(1,1) = 1$ is a local maximum.

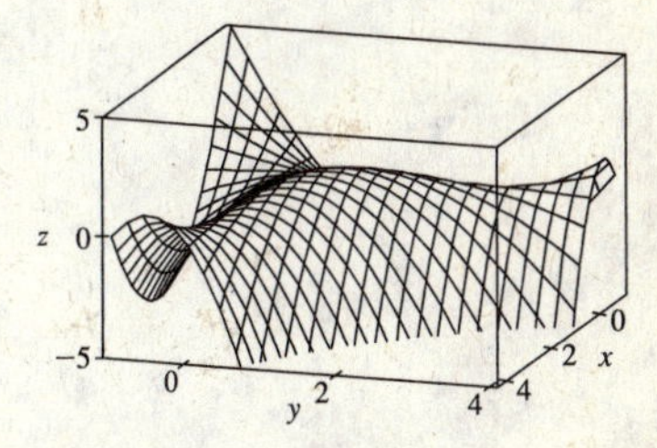

43. First solve inside D. Here $f_x = 4y^2 - 2xy^2 - y^3$, $f_y = 8xy - 2x^2y - 3xy^2$. Then $f_x = 0$ implies $y = 0$ or $y = 4 - 2x$, but $y = 0$ isn't inside D. Substituting $y = 4 - 2x$ into $f_y = 0$ implies $x = 0$, $x = 2$ or $x = 1$, but $x = 0$ isn't inside D, and when $x = 2$, $y = 0$ but $(2,0)$ isn't inside D. Thus the only critical point inside D is $(1,2)$ and $f(1,2) = 4$. Secondly we consider the boundary of D. On L_1: $f(x,0) = 0$ and so $f = 0$ on L_1. On L_2: $x = -y + 6$ and $f(-y+6, y) = y^2(6-y)(-2) = -2(6y^2 - y^3)$ which has critical points at $y = 0$ and $y = 4$. Then $f(6,0) = 0$ while $f(2,4) = -64$. On L_3: $f(0,y) = 0$, so $f = 0$ on L_3. Thus on D the absolute maximum of f is $f(1,2) = 4$ while the absolute minimum is $f(2,4) = -64$.

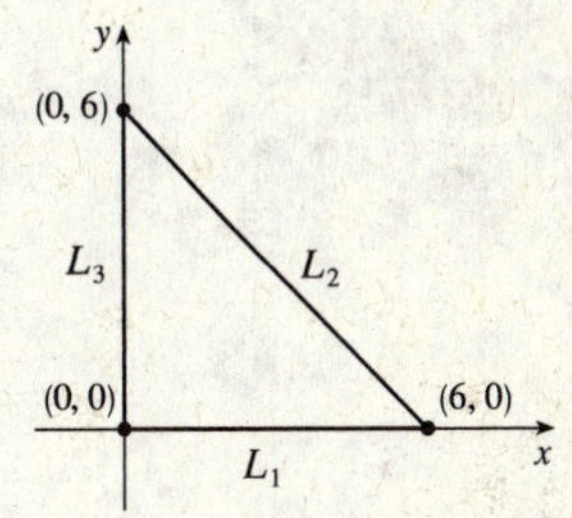

© 2016 Cengage Learning. All Rights Reserved. May not be scanned, copied or duplicated, or posted to a publicly accessible website, in whole or in part.

10 □ SYSTEMS OF LINEAR DIFFERENTIAL EQUATIONS

10.1 Qualitative Analysis of Linear Systems

1. $\left\{\begin{array}{l} dy/dt = -3ty + x \\ dx/dt = x - y \end{array}\right\} \Rightarrow \left\{\begin{array}{l} dy/dt = a_{11}(t)\,y + a_{12}x \\ dx/dt = a_{21}x + a_{22}y \end{array}\right\}$ where $a_{11}(t) = -3t$, $a_{12} = a_{21} = 1$, and $a_{22} = -1$.

This is a linear, homogeneous, nonautonomous system since the coefficient a_{11} contains the independent variable t.

3. $dy/dt = 3yz - 2z$, $\quad dz/dt = 2z + 5y$ This is a nonlinear, autonomous system since the differential equations are independent of t and there is a nonlinear term $3yz$ in dy/dt.

5. $\left\{\begin{array}{l} dy/dz = 2z + 3y \\ dx/dz = 3x - 2y \end{array}\right\} \Rightarrow \left\{\begin{array}{l} dy/dz = a_{11}y + a_{12}x + g_1(z) \\ dx/dz = a_{21}y + a_{22}x + g_2 \end{array}\right\}$ where $\begin{array}{l} a_{11} = 3, a_{21} = -2, a_{22} = 3, g_1(z) = 2z \\ a_{12} = g_2 = 0 \end{array}$

This is a linear, nonhomogeneous, nonautonomous system since g_1 contains the independent variable z.

7. $\left\{\begin{array}{l} dx/dt = 5x - 3y \\ dy/dt = 2y - x \end{array}\right\} \Rightarrow \begin{bmatrix} dx/dt \\ dy/dt \end{bmatrix} = \begin{bmatrix} 5 & -3 \\ -1 & 2 \end{bmatrix} \begin{bmatrix} x \\ y \end{bmatrix}$

9. $\left\{\begin{array}{l} dx/dt = 3ty - 7 \\ dy/dt = 2x - 3y \end{array}\right\} \Rightarrow \begin{bmatrix} dx/dt \\ dy/dt \end{bmatrix} = \begin{bmatrix} 0 & 3t \\ 2 & -3 \end{bmatrix} \begin{bmatrix} x \\ y \end{bmatrix} + \begin{bmatrix} -7 \\ 0 \end{bmatrix}$

11. $\left\{\begin{array}{l} dx/dt = 2x - 5 \\ dy/dt = 3x + 7y \end{array}\right\} \Rightarrow \begin{bmatrix} dx/dt \\ dy/dt \end{bmatrix} = \begin{bmatrix} 2 & 0 \\ 3 & 7 \end{bmatrix} \begin{bmatrix} x \\ y \end{bmatrix} + \begin{bmatrix} -5 \\ 0 \end{bmatrix}$

13. $\left\{\begin{array}{l} dx/dt = x + 4y - 3t \\ dy/dt = y - x \end{array}\right\} \Rightarrow \begin{bmatrix} dx/dt \\ dy/dt \end{bmatrix} = \begin{bmatrix} 1 & 4 \\ -1 & 1 \end{bmatrix} \begin{bmatrix} x \\ y \end{bmatrix} + \begin{bmatrix} -3t \\ 0 \end{bmatrix}$

15. $A = \begin{bmatrix} -3 & 1 \\ 2 & -2 \end{bmatrix} \Rightarrow \begin{array}{l} x_1 \text{ nullcline: } -3x_1 + x_2 = 0 \\ x_2 \text{ nullcline: } 2x_1 - 2x_2 = 0 \end{array} \Rightarrow \begin{array}{l} x_2 = 3x_1 \\ x_2 = x_1 \end{array}$

x_1 is increasing when $x_2 > 3x_1$ (above x_1 nullcline) and decreasing below its nullcline. x_2 is increasing when $x_2 < x_1$ (below x_2 nullcline) and decreasing above its nullcline. Using these properties, we sketch the solution curves on the phase plane shown at right. The equilibrium is a stable node.

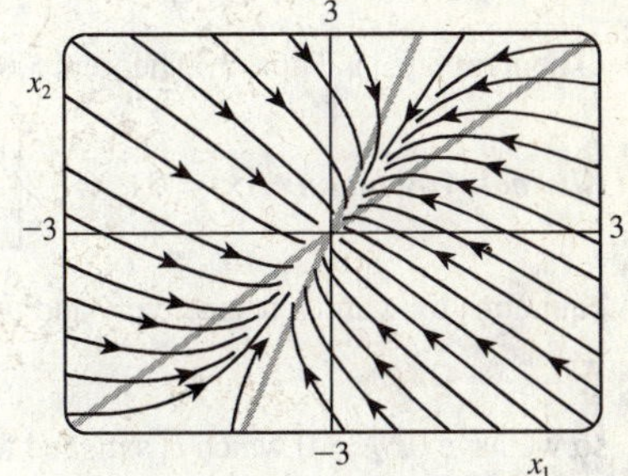

17. $A = \begin{bmatrix} 1 & 2 \\ -2 & 1 \end{bmatrix} \Rightarrow \begin{array}{l} x_1 \text{ nullcline: } \quad x_1 + 2x_2 = 0 \\ x_2 \text{ nullcline: } -2x_1 + x_2 = 0 \end{array} \Rightarrow \begin{array}{l} x_2 = -\frac{1}{2}x_1 \\ x_2 = 2x_1 \end{array}$

x_1 is increasing when $x_2 > -\frac{1}{2}x_1$ (above x_1 nullcline) and decreasing below its nullcline. x_2 is increasing when $x_2 > 2x_1$ (above x_2 nullcline) and decreasing below its nullcline. Using these properties, we sketch the solution curves on the phase plane shown at right. The equilibrium is a spiral.

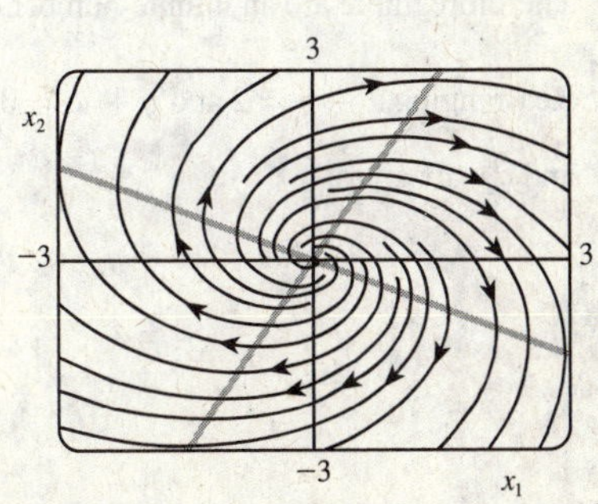

© 2016 Cengage Learning. All Rights Reserved. May not be scanned, copied or duplicated, or posted to a publicly accessible website, in whole or in part.

19. $A = \begin{bmatrix} -1 & 2 \\ -3 & 0 \end{bmatrix} \Rightarrow \begin{matrix} x_1 \text{ nullcline: } -x_1 + 2x_2 = 0 \\ x_2 \text{ nullcline: } \quad -3x_1 = 0 \end{matrix} \Rightarrow \begin{matrix} x_2 = \frac{1}{2}x_1 \\ x_1 = 0 \end{matrix}$

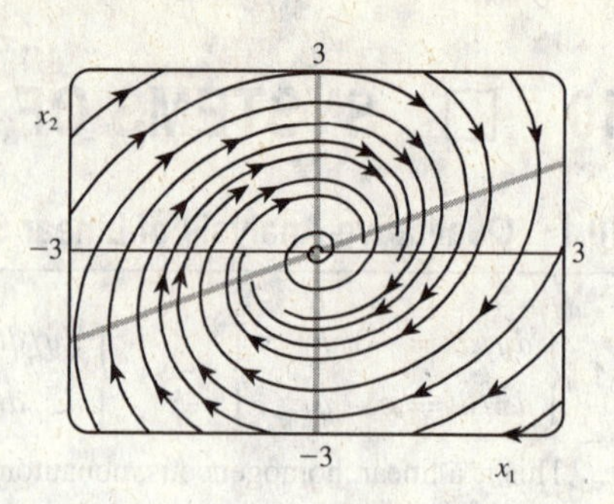

x_1 is increasing when $x_2 > \frac{1}{2}x_1$ (above x_1 nullcline) and decreasing below its nullcline. x_2 is increasing when $x_1 < 0$ (left of x_2 nullcline) and decreasing to the right of its nullcline. Using these properties, we sketch the solution curves on the phase plane shown at right. The equilibrium is a spiral.

21. $A = \begin{bmatrix} 2 & -1 \\ -1 & 2 \end{bmatrix} \Rightarrow \begin{matrix} x_1 \text{ nullcline: } \quad 2x_1 - x_2 = 0 \\ x_2 \text{ nullcline: } -x_1 + 2x_2 = 0 \end{matrix} \Rightarrow \begin{matrix} x_2 = 2x_1 \\ x_2 = \frac{1}{2}x_1 \end{matrix}$

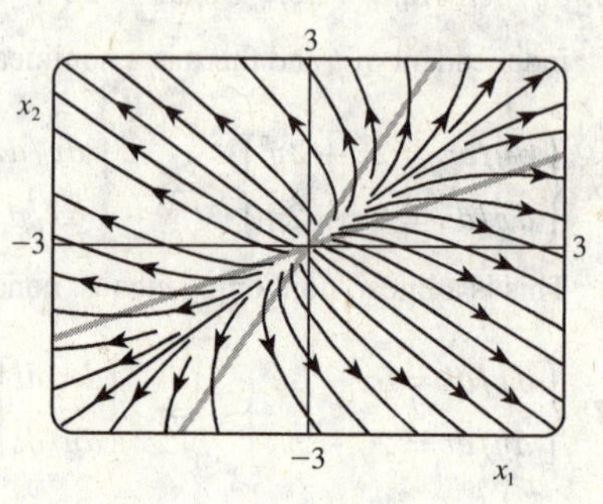

x_1 is increasing when $x_2 < -x_1$ (below x_1 nullcline) and decreasing above its nullcline. x_2 is increasing when $x_2 > \frac{1}{2}x_1$ (above x_2 nullcline) and decreasing below its nullcline. Using these properties, we sketch the solution curves on the phase plane shown at right. The equilibrium is an unstable node.

23. A is nonsingular $\Leftrightarrow$ A is invertible

(a) At equilibrium $d\hat{\mathbf{x}}/dt = \mathbf{0} \Rightarrow A\hat{\mathbf{x}} + \mathbf{g} = \mathbf{0} \Rightarrow A\hat{\mathbf{x}} = -\mathbf{g} \Rightarrow \hat{\mathbf{x}} = -A^{-1}\mathbf{g}$

(b) The components of $\mathbf{y} = \mathbf{x} - \hat{\mathbf{x}}$ represent the distance between the vector $\mathbf{x}$ and the equilibrium vector $\hat{\mathbf{x}}$ along each coordinate dimension.

(c) $\dfrac{d\mathbf{y}}{dt} = \dfrac{d}{dt}(\mathbf{x} - \hat{\mathbf{x}}) = \dfrac{d\mathbf{x}}{dt} - \dfrac{d\hat{\mathbf{x}}}{dt} = (A\mathbf{x} + \mathbf{g}) - (A\hat{\mathbf{x}} + \mathbf{g}) = A\mathbf{x} - A\hat{\mathbf{x}} = A(\mathbf{x} - \hat{\mathbf{x}}) = Ay$

25. The homogeneous system $\dfrac{d\mathbf{x}}{dt} = A\mathbf{x}$ is at equilibrium when $\dfrac{d\mathbf{x}}{dt} = \mathbf{0} \Rightarrow A\mathbf{x} = \mathbf{0}$. Now $\det A = 0$ so A is not invertible (Theorem 8.6.1). Thus, by Theorem 8.6.5, there are an infinite number of solutions.

Alternative Solution: $A\mathbf{x} = \mathbf{0} \Rightarrow \begin{bmatrix} a & b \\ c & d \end{bmatrix}\begin{bmatrix} x_1 \\ x_2 \end{bmatrix} = \begin{bmatrix} 0 \\ 0 \end{bmatrix} \Rightarrow \left\{ \begin{matrix} ax_1 + bx_2 = 0 \\ cx_1 + dx_2 = 0 \end{matrix} \right\}$ Dividing the first and second equations by a and c respectively and then subtracting them gives $\frac{b}{a}x_2 - \frac{d}{c}x_2 = 0 \Rightarrow \underbrace{(bc - ad)}_{\det A}x_2 = 0$. Now $\det A = 0$ so we have $0x_2 = 0$ which is satisfied for any choice of x_2. Similarly, we can solve the equations for x_1 to find that $0x_1 = 0$. Therefore, there are an infinite number of choices of x_1 and x_2 implying that there are an infinite number of equilibria.

27. Rearranging $x + z = 2$ and $y + w = 3$, we get $z = 2 - x$ and $y = 3 - w$. Substituting these into the expressions for dw/dt and dx/dt gives

$$\begin{matrix} dw/dt = 2x + y - z \\ dx/dt = 3x + z \end{matrix} \Rightarrow \begin{matrix} dw/dt = 2x + (3 - w) - (2 - x) \\ dx/dt = 3x + (2 - x) \end{matrix} \Rightarrow \begin{matrix} dw/dt = 3x - w + 1 \\ dx/dt = 2x + 2 \end{matrix}$$

© 2016 Cengage Learning. All Rights Reserved. May not be scanned, copied or duplicated, or posted to a publicly accessible website, in whole or in part.

29. Let $x_1(t) = y(t)$ and $x_2(t) = y'(t)$ so that $x_1' = x_2$. Also $y''(t) + p(t)\,y'(t) + q(t)\,y(t) = g(t)$ becomes

$x_2' + p(t)x_2 + q(t)x_1 = g(t) \quad \Rightarrow \quad x_2' = -p(t)x_2 - q(t)x_1 + g(t)$. The system of differential equations can then be expressed in matrix form as follows:

$$\left\{\begin{aligned} x_1' &= x_2 \\ x_2' &= -p(t)x_2 - q(t)x_1 + g(t) \end{aligned}\right\} \Rightarrow \underbrace{\begin{bmatrix} x_1' \\ x_2' \end{bmatrix}}_{d\mathbf{x}/dt} = \begin{bmatrix} 0 & 1 \\ -q(t) & -p(t) \end{bmatrix} \underbrace{\begin{bmatrix} x_1 \\ x_2 \end{bmatrix}}_{\mathbf{x}} = \begin{bmatrix} 0 \\ g(t) \end{bmatrix}$$

and $\mathbf{x}(0) = \begin{bmatrix} x_1(0) \\ x_2(0) \end{bmatrix} = \begin{bmatrix} y(0) \\ y'(0) \end{bmatrix} = \begin{bmatrix} a \\ b \end{bmatrix}$.

31. $$\begin{aligned} \frac{dm}{dt} &= 1 - p - m \\ \frac{dp}{dt} &= m - p \end{aligned} \quad \Rightarrow \quad \begin{aligned} &m \text{ nullcline: } 1 - p - m = 0 \\ &p \text{ nullcline: } \quad m - p = 0 \end{aligned} \quad \Rightarrow \quad \begin{aligned} p &= 1 - m \\ p &= m \end{aligned}$$

m is increasing when $p < 1 - m$ (below m nullcline) and decreasing above its nullcline. p is increasing when $p < m$ (below p nullcline) and decreasing above its nullcline. Using these properties, we sketch the solution curves on the phase plane shown at right. Nullclines appear as grey lines.

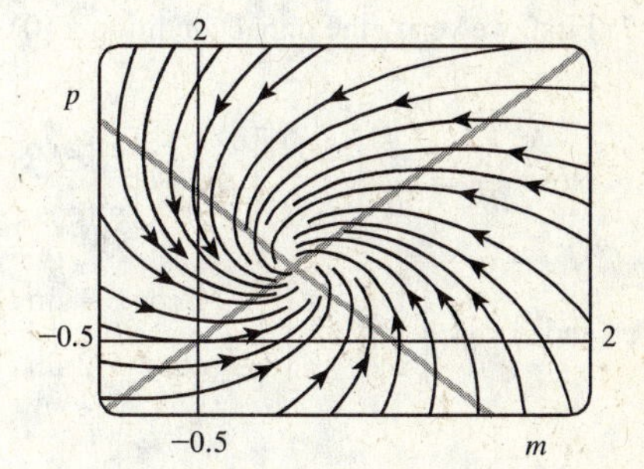

33. (a) $$\begin{aligned} \frac{dx_1}{dt} &= -ax_1 - bx_1 \\ \frac{dx_2}{dt} &= bx_1 - cx_2 \end{aligned} \quad \Rightarrow \quad \begin{aligned} &x_1 \text{ nullcline: } -(a+b)\,x_1 = 0 \\ &x_2 \text{ nullcline: } \quad bx_1 - cx_2 = 0 \end{aligned} \quad \Rightarrow \quad \begin{aligned} x_1 &= 0 \\ x_2 &= \frac{b}{c}x_1 \end{aligned}$$

x_1 is increasing when $x_1 < 0$ (left of x_1 nullcline) and decreasing to the right of its nullcline. x_2 is increasing when $x_2 < \frac{b}{c}x_1$ (below x_2 nullcline) and decreasing above its nullcline. Using these properties, we sketch the solution curves on the phase plane shown at right.

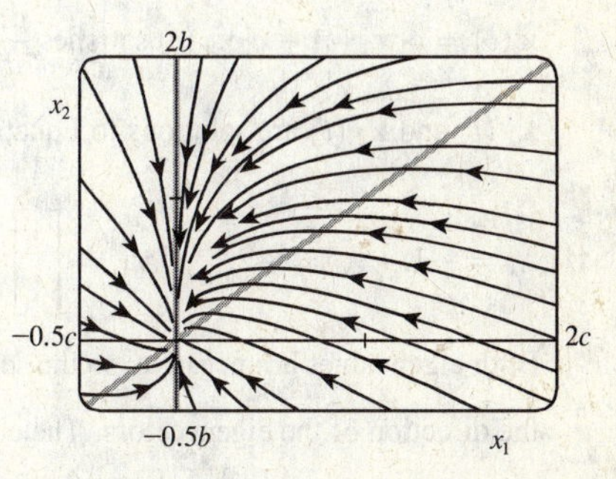

(b) The initial condition will be located in the first quadrant since x_1 and x_2 represent amounts of the antibody and must be nonnegative. The solution curves move toward the origin in the phase plane indicating the equilibrium is stable. Thus, the amount of antibody in the blood and tumor will approach zero as $t \to \infty$.

10.2 Solving Systems of Linear Differential Systems

1. $\dfrac{d\mathbf{x}}{dt} = \dfrac{d}{dt}\begin{bmatrix} x_1(t) \\ x_2(t) \end{bmatrix} = \dfrac{d}{dt}\begin{bmatrix} \frac{1}{3}(4e^{2t} - e^{-t}) \\ \frac{2}{3}(e^{2t} - e^{-t}) \end{bmatrix} = \begin{bmatrix} \frac{1}{3}(8e^{2t} + e^{-t}) \\ \frac{2}{3}(2e^{2t} + e^{-t}) \end{bmatrix}$

$$A\mathbf{x} = \begin{bmatrix} 3 & -2 \\ 2 & -2 \end{bmatrix} \begin{bmatrix} \frac{1}{3}(4e^{2t} - e^{-t}) \\ \frac{2}{3}(e^{2t} - e^{-t}) \end{bmatrix} = \begin{bmatrix} (4e^{2t} - e^{-t}) - \frac{4}{3}(e^{2t} - e^{-t}) \\ \frac{2}{3}(4e^{2t} - e^{-t}) - \frac{4}{3}(e^{2t} - e^{-t}) \end{bmatrix} = \begin{bmatrix} \frac{8}{3}e^{2t} + \frac{1}{3}e^{-t} \\ \frac{4}{3}e^{2t} + \frac{2}{3}e^{-t} \end{bmatrix} = \frac{d\mathbf{x}}{dt}$$

© 2016 Cengage Learning. All Rights Reserved. May not be scanned, copied or duplicated, or posted to a publicly accessible website, in whole or in part.

3. $\dfrac{d\mathbf{x}}{dt} = \dfrac{d}{dt}\begin{bmatrix} x_1(t) \\ x_2(t) \end{bmatrix} = \dfrac{d}{dt}\begin{bmatrix} e^{-t}\cos 2t \\ \frac{1}{2}e^{-t}\sin 2t \end{bmatrix} = \begin{bmatrix} -e^{-t}\cos 2t - 2e^{-t}\sin 2t \\ -\frac{1}{2}e^{-t}\sin 2t + e^{-t}\cos 2t \end{bmatrix}$

$A\mathbf{x} = \begin{bmatrix} -1 & -4 \\ 1 & -1 \end{bmatrix}\begin{bmatrix} e^{-t}\cos 2t \\ \frac{1}{2}e^{-t}\sin 2t \end{bmatrix} = \begin{bmatrix} -e^{-t}\cos 2t - 2e^{-t}\sin 2t \\ e^{-t}\cos 2t - \frac{1}{2}e^{-t}\sin 2t \end{bmatrix} = \dfrac{d\mathbf{x}}{dt}$

5. First, we verify the initial condition $\mathbf{x}(0) = \begin{bmatrix} x_1(0) \\ x_2(0) \end{bmatrix} = \begin{bmatrix} e^0 \\ 2e^0 \end{bmatrix} = \begin{bmatrix} 1 \\ 2 \end{bmatrix} = \mathbf{x}_0$.

Now $\dfrac{d\mathbf{x}}{dt} = \dfrac{d}{dt}\begin{bmatrix} x_1(t) \\ x_2(t) \end{bmatrix} = \dfrac{d}{dt}\begin{bmatrix} e^{-t} \\ 2e^{-t} \end{bmatrix} = \begin{bmatrix} -e^{-t} \\ -2e^{-t} \end{bmatrix}$ and $A\mathbf{x} = \begin{bmatrix} 3 & -2 \\ 6 & -4 \end{bmatrix}\begin{bmatrix} e^{-t} \\ 2e^{-t} \end{bmatrix} = \begin{bmatrix} 3e^{-t} - 4e^{-t} \\ 6e^{-t} - 8e^{-t} \end{bmatrix} = \begin{bmatrix} -e^{-t} \\ -2e^{-t} \end{bmatrix} = \dfrac{d\mathbf{x}}{dt}$.

7. First, we verify the initial condition $\mathbf{x}(0) = \begin{bmatrix} x_1(0) \\ x_2(0) \end{bmatrix} = \begin{bmatrix} e^0(2\cos 0 - \sin 0) \\ e^0(\cos 0 + 2\sin 0) \end{bmatrix} = \begin{bmatrix} 2 \\ 1 \end{bmatrix} = \mathbf{x}_0$.

Now $\dfrac{d\mathbf{x}}{dt} = \dfrac{d}{dt}\begin{bmatrix} x_1(t) \\ x_2(t) \end{bmatrix} = \dfrac{d}{dt}\begin{bmatrix} e^t(2\cos t - \sin t) \\ e^t(\cos t + 2\sin t) \end{bmatrix} = \begin{bmatrix} e^t(2\cos t - \sin t) + e^t(-2\sin t - \cos t) \\ e^t(\cos t + 2\sin t) + e^t(-\sin t + 2\cos t) \end{bmatrix} = \begin{bmatrix} e^t(\cos t - 3\sin t) \\ e^t(3\cos t + \sin t) \end{bmatrix}$

and $A\mathbf{x} = \begin{bmatrix} 1 & -1 \\ 1 & 1 \end{bmatrix}\begin{bmatrix} e^t(2\cos t - \sin t) \\ e^t(\cos t + 2\sin t) \end{bmatrix} = \begin{bmatrix} e^t(2\cos t - \sin t) - e^t(\cos t + 2\sin t) \\ e^t(2\cos t - \sin t) + e^t(\cos t + 2\sin t) \end{bmatrix} = \begin{bmatrix} e^t(\cos t - 3\sin t) \\ e^t(3\cos t + \sin t) \end{bmatrix} = \dfrac{d\mathbf{x}}{dt}$.

9. Suppose $\mathbf{x}_1(t)$ and $\mathbf{x}_2(t)$ are solutions to the system of linear differential equations $\dfrac{d\mathbf{x}}{dt} = A\mathbf{x}$. The function

$\mathbf{x}(t) = c_1\mathbf{x}_1(t) + c_2\mathbf{x}_2(t)$ satisfies $\dfrac{d\mathbf{x}}{dt} = c_1\dfrac{d\mathbf{x}_1}{dt} + c_2\dfrac{d\mathbf{x}_2}{dt} = c_1A\mathbf{x}_1 + c_2A\mathbf{x}_2 = A(c_1\mathbf{x}_1 + c_2\mathbf{x}_2) = A\mathbf{x}$. Therefore, if

$\mathbf{x}_1(t)$ and $\mathbf{x}_2(t)$ are solutions to Equation (1), then so is $\mathbf{x}(t) = c_1\mathbf{x}_1(t) + c_2\mathbf{x}_2(t)$.

11. $\lambda_1 = -1, \quad \lambda_2 = -2; \qquad \mathbf{v}_1 = \begin{bmatrix} 1 \\ 1 \end{bmatrix} \quad \mathbf{v}_2 = \begin{bmatrix} -1 \\ 1 \end{bmatrix}$

Both eigenvalues are negative, so the solution curves move toward the origin in the direction of the eigenvectors. The eigenvectors are plotted as grey lines in the phase plane along with a sketch of the solution curves at right.

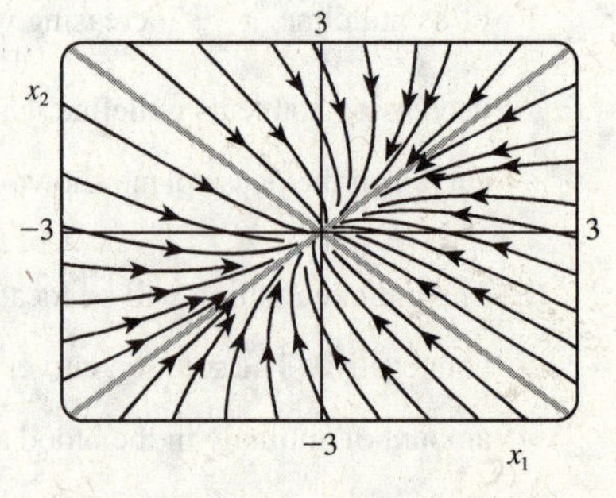

13. $\lambda_1 = 2, \quad \lambda_2 = -2; \qquad \mathbf{v}_1 = \begin{bmatrix} 3 \\ 1 \end{bmatrix} \quad \mathbf{v}_2 = \begin{bmatrix} 1 \\ 1 \end{bmatrix}$

λ_1 is positive and λ_2 is negative, so the solution curves move away from the origin along the direction of $\pm\mathbf{v}_1$ and toward the origin along the direction of $\pm\mathbf{v}_2$. The eigenvectors are plotted as grey lines in the phase plane along with a sketch of the solution curves at right.

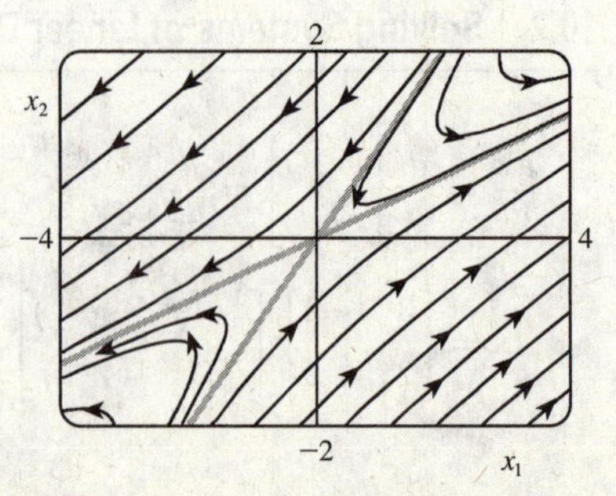

© 2016 Cengage Learning. All Rights Reserved. May not be scanned, copied or duplicated, or posted to a publicly accessible website, in whole or in part.

15. $\lambda_1 = 5, \quad \lambda_2 = 1; \qquad \mathbf{v}_1 = \begin{bmatrix} 2 \\ 2 \end{bmatrix} \quad \mathbf{v}_2 = \begin{bmatrix} -2 \\ 1 \end{bmatrix}$

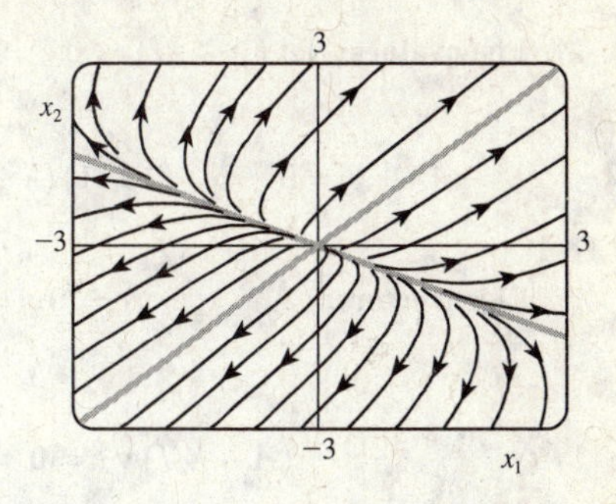

Both eigenvalues are positive, so the solution curves move away from the origin along the direction of the eigenvectors. The eigenvectors are plotted as grey lines in the phase plane along with a sketch of the solution curves at right. Since $\lambda_1 > \lambda_2$, the solution curves grow at a faster rate in the direction of $\pm\mathbf{v}_1$ as compared to $\pm\mathbf{v}_2$.

17. Eigenvalues: $\det(A - \lambda I) = 0 \;\Rightarrow\; \det\begin{bmatrix} -\frac{3}{2} - \lambda & \frac{1}{2} \\ \frac{1}{2} & -\frac{3}{2} - \lambda \end{bmatrix} = 0 \;\Rightarrow\; \left(-\frac{3}{2} - \lambda\right)^2 - \frac{1}{4} = 0 \;\Rightarrow\; \lambda = -\frac{3}{2} \pm \frac{1}{2}.$

The eigenvalues are therefore $\lambda_1 = -1$ and $\lambda_2 = -2$.

Eigenvectors: $(A - \lambda_1 I)\,\mathbf{v}_1 = \mathbf{0} \;\Rightarrow\; \begin{bmatrix} -\frac{1}{2} & \frac{1}{2} \\ \frac{1}{2} & -\frac{1}{2} \end{bmatrix} \mathbf{v}_1 = \begin{bmatrix} 0 \\ 0 \end{bmatrix} \;\Rightarrow\; \mathbf{v}_1 = \begin{bmatrix} 1 \\ 1 \end{bmatrix}$ and

$$(A - \lambda_2 I)\,\mathbf{v}_2 = \mathbf{0} \;\Rightarrow\; \begin{bmatrix} \frac{1}{2} & \frac{1}{2} \\ \frac{1}{2} & \frac{1}{2} \end{bmatrix} \mathbf{v}_2 = \begin{bmatrix} 0 \\ 0 \end{bmatrix} \;\Rightarrow\; \mathbf{v}_2 = \begin{bmatrix} 1 \\ -1 \end{bmatrix}$$

The general solution given by equation (8) is $\mathbf{x}(t) = c_1 e^{\lambda_1 t} \begin{bmatrix} | \\ \mathbf{v}_1 \\ | \end{bmatrix} + c_2 e^{\lambda_2 t} \begin{bmatrix} | \\ \mathbf{v}_2 \\ | \end{bmatrix} = c_1 e^{-t} \begin{bmatrix} 1 \\ 1 \end{bmatrix} + c_2 e^{-2t} \begin{bmatrix} 1 \\ -1 \end{bmatrix}.$

Initial Condition: Using equation (9) to determine the constants c_1 and c_2 in the general solution, we find

$$\begin{bmatrix} | & | \\ \mathbf{v}_1 & \mathbf{v}_2 \\ | & | \end{bmatrix} \begin{bmatrix} c_1 \\ c_2 \end{bmatrix} = \begin{bmatrix} x_1(0) \\ x_2(0) \end{bmatrix} \;\Rightarrow\; \begin{bmatrix} 1 & 1 \\ 1 & -1 \end{bmatrix} \begin{bmatrix} c_1 \\ c_2 \end{bmatrix} = \begin{bmatrix} 1 \\ 2 \end{bmatrix} \;\Rightarrow\; \left\{ \begin{matrix} c_1 + c_2 = 1 \\ c_1 - c_2 = 2 \end{matrix} \right\}$$

which is satisfied when $c_1 = \frac{3}{2}$ and $c_2 = -\frac{1}{2}$. Thus, the solution is $\mathbf{x}(t) = \frac{3}{2} e^{-t} \begin{bmatrix} 1 \\ 1 \end{bmatrix} - \frac{1}{2} e^{-2t} \begin{bmatrix} 1 \\ -1 \end{bmatrix}.$

19. Eigenvalues: $\det(A - \lambda I) = 0 \;\Rightarrow\; \det\begin{bmatrix} 1 - \lambda & 0 \\ 4 & -1 - \lambda \end{bmatrix} = 0 \;\Rightarrow\; (1 - \lambda)(-1 - \lambda) = 0$. The eigenvalues are therefore $\lambda_1 = 1$ and $\lambda_2 = -1$.

Eigenvectors: $(A - \lambda_1 I)\,\mathbf{v}_1 = \mathbf{0} \;\Rightarrow\; \begin{bmatrix} 0 & 0 \\ 4 & -2 \end{bmatrix} \mathbf{v}_1 = \begin{bmatrix} 0 \\ 0 \end{bmatrix} \;\Rightarrow\; \mathbf{v}_1 = \begin{bmatrix} 1 \\ 2 \end{bmatrix}$ and

$$(A - \lambda_2 I)\,\mathbf{v}_2 = \mathbf{0} \;\Rightarrow\; \begin{bmatrix} 2 & 0 \\ 4 & 0 \end{bmatrix} \mathbf{v}_2 = \begin{bmatrix} 0 \\ 0 \end{bmatrix} \;\Rightarrow\; \mathbf{v}_2 = \begin{bmatrix} 0 \\ 1 \end{bmatrix}$$

The general solution given by equation (8) is $\mathbf{x}(t) = c_1 e^{\lambda_1 t} \begin{bmatrix} | \\ \mathbf{v}_1 \\ | \end{bmatrix} + c_2 e^{\lambda_2 t} \begin{bmatrix} | \\ \mathbf{v}_2 \\ | \end{bmatrix} = c_1 e^{t} \begin{bmatrix} 1 \\ 2 \end{bmatrix} + c_2 e^{-t} \begin{bmatrix} 0 \\ 1 \end{bmatrix}.$

Initial Condition: Using equation (9) to determine the constants c_1 and c_2 in the general solution, we find

$$\begin{bmatrix} | & | \\ \mathbf{v}_1 & \mathbf{v}_2 \\ | & | \end{bmatrix} \begin{bmatrix} c_1 \\ c_2 \end{bmatrix} = \begin{bmatrix} x_1(0) \\ x_2(0) \end{bmatrix} \;\Rightarrow\; \begin{bmatrix} 1 & 0 \\ 2 & 1 \end{bmatrix} \begin{bmatrix} c_1 \\ c_2 \end{bmatrix} = \begin{bmatrix} 3 \\ 2 \end{bmatrix} \;\Rightarrow\; \left\{ \begin{matrix} c_1 = 3 \\ 2c_1 + c_2 = 2 \end{matrix} \right\}$$

which is satisfied when $c_1 = 3$ and $c_2 = -4$. Thus, the solution is $\mathbf{x}(t) = 3e^{t} \begin{bmatrix} 1 \\ 2 \end{bmatrix} - 4e^{-t} \begin{bmatrix} 0 \\ 1 \end{bmatrix}.$

© 2016 Cengage Learning. All Rights Reserved. May not be scanned, copied or duplicated, or posted to a publicly accessible website, in whole or in part.

21. Eigenvalues: $\det(A-\lambda I)=0 \;\Rightarrow\; \det\begin{bmatrix}-3-\lambda & 4\\ -6 & 7-\lambda\end{bmatrix}=0 \;\Rightarrow\; (-3-\lambda)(7-\lambda)+24=0 \;\Rightarrow$

$\lambda^2-4\lambda+3=0 \;\Rightarrow\; (\lambda-1)(\lambda-3)=0$. The eigenvalues are therefore $\lambda_1=1$ and $\lambda_2=3$.

Eigenvectors: $(A-\lambda_1 I)\,\mathbf{v}_1=\mathbf{0} \;\Rightarrow\; \begin{bmatrix}-4 & 4\\ -6 & 6\end{bmatrix}\mathbf{v}_1=\begin{bmatrix}0\\0\end{bmatrix} \;\Rightarrow\; \mathbf{v}_1=\begin{bmatrix}1\\1\end{bmatrix}$ and

$$(A-\lambda_2 I)\,\mathbf{v}_2=\mathbf{0} \;\Rightarrow\; \begin{bmatrix}-6 & 4\\ -6 & 4\end{bmatrix}\mathbf{v}_2=\begin{bmatrix}0\\0\end{bmatrix} \;\Rightarrow\; \mathbf{v}_2=\begin{bmatrix}2\\3\end{bmatrix}$$

The general solution given by equation (8) is $\mathbf{x}(t)=c_1e^{\lambda_1 t}\begin{bmatrix}|\\ \mathbf{v}_1\\ |\end{bmatrix}+c_2e^{\lambda_2 t}\begin{bmatrix}|\\ \mathbf{v}_2\\ |\end{bmatrix}=c_1e^{t}\begin{bmatrix}1\\1\end{bmatrix}+c_2e^{3t}\begin{bmatrix}2\\3\end{bmatrix}$.

Initial Condition: Using equation (9) to determine the constants c_1 and c_2 in the general solution, we find

$$\begin{bmatrix}| & |\\ \mathbf{v}_1 & \mathbf{v}_2\\ | & |\end{bmatrix}\begin{bmatrix}c_1\\c_2\end{bmatrix}=\begin{bmatrix}x_1(0)\\x_2(0)\end{bmatrix} \;\Rightarrow\; \begin{bmatrix}1 & 2\\ 1 & 3\end{bmatrix}\begin{bmatrix}c_1\\c_2\end{bmatrix}=\begin{bmatrix}-1\\3\end{bmatrix} \;\Rightarrow\; \begin{Bmatrix}c_1+2c_2=-1\\ c_1+3c_2=3\end{Bmatrix}$$

which is satisfied when $c_1=-9$ and $c_2=4$. Thus, the solution is $\mathbf{x}(t)=-9e^{t}\begin{bmatrix}1\\1\end{bmatrix}+4e^{3t}\begin{bmatrix}2\\3\end{bmatrix}$.

23. Eigenvalues: $\det(A-\lambda I)=0 \;\Rightarrow\; \det\begin{bmatrix}-1-\lambda & 2\\ -3 & -1-\lambda\end{bmatrix}=0 \;\Rightarrow\; (-1-\lambda)^2+6=0 \;\Rightarrow\; \lambda=\underbrace{-1}_{a}\pm i\underbrace{\sqrt{6}}_{b}$.

The eigenvalues are therefore $\lambda_1=-1+i\sqrt{6}$ and $\lambda_2=-1-i\sqrt{6}$.

Eigenvectors: $(A-\lambda_1 I)\,\mathbf{v}_1=\mathbf{0} \;\Rightarrow\; \begin{bmatrix}-i\sqrt{6} & 2\\ -3 & -i\sqrt{6}\end{bmatrix}\mathbf{v}_1=\begin{bmatrix}0\\0\end{bmatrix} \;\Rightarrow\; \mathbf{v}_1=\begin{bmatrix}\sqrt{6}\\3i\end{bmatrix}=\underbrace{\begin{bmatrix}\sqrt{6}\\0\end{bmatrix}}_{\mathbf{u}}+i\underbrace{\begin{bmatrix}0\\3\end{bmatrix}}_{\mathbf{w}}$ and since

complex eigenvectors come in conjugate pairs, we know that $\mathbf{v}_2=\begin{bmatrix}\sqrt{6}\\0\end{bmatrix}-i\begin{bmatrix}0\\3\end{bmatrix}$.

Initial Condition: Using equation (14) to determine the constants k_1 and k_2 in the general solution (13), we find

$$\begin{bmatrix}| & |\\ \mathbf{u} & \mathbf{w}\\ | & |\end{bmatrix}\begin{bmatrix}k_1\\k_2\end{bmatrix}=\begin{bmatrix}x_1(0)\\x_2(0)\end{bmatrix} \;\Rightarrow\; \begin{bmatrix}\sqrt{6} & 0\\ 0 & 3\end{bmatrix}\begin{bmatrix}k_1\\k_2\end{bmatrix}=\begin{bmatrix}2\\0\end{bmatrix} \;\Rightarrow\; \begin{Bmatrix}\sqrt{6}k_1=2\\ 3k_2=0\end{Bmatrix}$$

which is satisfied when $k_1=\frac{\sqrt{6}}{3}$ and $k_2=0$. Thus, the solution as given by equation (13) is

$$\begin{aligned}\mathbf{x}(t)&=k_1e^{at}(\mathbf{u}\cos bt-\mathbf{w}\sin bt)+k_2e^{at}(\mathbf{w}\cos bt+\mathbf{u}\sin bt)\\ &=\tfrac{\sqrt{6}}{3}e^{-t}\left(\begin{bmatrix}\sqrt{6}\\0\end{bmatrix}\cos\sqrt{6}t-\begin{bmatrix}0\\3\end{bmatrix}\sin\sqrt{6}t\right)+0=e^{-t}\begin{bmatrix}2\cos\sqrt{6}t\\ -\sqrt{6}\sin\sqrt{6}t\end{bmatrix}\end{aligned}$$

© 2016 Cengage Learning. All Rights Reserved. May not be scanned, copied or duplicated, or posted to a publicly accessible website, in whole or in part.

25. **Eigenvalues:** $\det(A-\lambda I)=0 \Rightarrow \det\begin{bmatrix}0-\lambda & -1\\ -1 & 0-\lambda\end{bmatrix}=0 \Rightarrow \lambda^2-1=0 \Rightarrow \lambda=\pm 1$. The eigenvalues are therefore $\lambda_1=1$ and $\lambda_2=-1$.

Eigenvectors: $(A-\lambda_1 I)\,\mathbf{v}_1=\mathbf{0} \Rightarrow \begin{bmatrix}-1 & -1\\ -1 & -1\end{bmatrix}\mathbf{v}_1=\begin{bmatrix}0\\0\end{bmatrix} \Rightarrow \mathbf{v}_1=\begin{bmatrix}1\\-1\end{bmatrix}$ and

$$(A-\lambda_2 I)\,\mathbf{v}_2=\mathbf{0} \Rightarrow \begin{bmatrix}1 & -1\\ -1 & 1\end{bmatrix}\mathbf{v}_2=\begin{bmatrix}0\\0\end{bmatrix} \Rightarrow \mathbf{v}_2=\begin{bmatrix}1\\1\end{bmatrix}$$

The general solution given by equation (8) is $\mathbf{x}(t)=c_1e^{\lambda_1 t}\begin{bmatrix}|\\ \mathbf{v}_1\\ |\end{bmatrix}+c_2e^{\lambda_2 t}\begin{bmatrix}|\\ \mathbf{v}_2\\ |\end{bmatrix}=c_1e^{t}\begin{bmatrix}1\\-1\end{bmatrix}+c_2e^{-t}\begin{bmatrix}1\\1\end{bmatrix}$.

Initial Condition: Using equation (9) to determine the constants c_1 and c_2 in the general solution, we find

$$\begin{bmatrix}| & |\\ \mathbf{v}_1 & \mathbf{v}_2\\ | & |\end{bmatrix}\begin{bmatrix}c_1\\c_2\end{bmatrix}=\begin{bmatrix}x_1(0)\\x_2(0)\end{bmatrix} \Rightarrow \begin{bmatrix}1 & 1\\ -1 & 1\end{bmatrix}\begin{bmatrix}c_1\\c_2\end{bmatrix}=\begin{bmatrix}2\\1\end{bmatrix} \Rightarrow \left\{\begin{array}{r}c_1+c_2=2\\ -c_1+c_2=1\end{array}\right\}$$

which is satisfied when $c_1=\frac{1}{2}$ and $c_2=\frac{3}{2}$. Thus, the solution is $\mathbf{x}(t)=\frac{1}{2}e^{t}\begin{bmatrix}1\\-1\end{bmatrix}+\frac{3}{2}e^{-t}\begin{bmatrix}1\\1\end{bmatrix}$.

27. **Eigenvalues:** $\det(A-\lambda I)=0 \Rightarrow \det\begin{bmatrix}2-\lambda & -5\\ 2 & 1-\lambda\end{bmatrix}=0 \Rightarrow (2-\lambda)(1-\lambda)+10=0 \Rightarrow$

$\lambda^2-3\lambda+12=0 \Rightarrow \lambda=\underbrace{\tfrac{3}{2}}_{a}\pm i\underbrace{\tfrac{\sqrt{39}}{2}}_{b}$. The eigenvalues are therefore $\lambda_1=\frac{3}{2}+i\frac{\sqrt{39}}{2}$ and $\lambda_2=\frac{3}{2}-i\frac{\sqrt{39}}{2}$.

Eigenvectors: $(A-\lambda_1 I)\,\mathbf{v}_1=\mathbf{0} \Rightarrow \begin{bmatrix}\frac{1}{2}-i\frac{\sqrt{39}}{2} & -5\\ 2 & -\frac{1}{2}-i\frac{\sqrt{39}}{2}\end{bmatrix}\mathbf{v}_1=\begin{bmatrix}0\\0\end{bmatrix} \Rightarrow$

$\mathbf{v}_1=\begin{bmatrix}1+i\sqrt{39}\\4\end{bmatrix}=\underbrace{\begin{bmatrix}1\\4\end{bmatrix}}_{\mathbf{u}}+i\underbrace{\begin{bmatrix}\sqrt{39}\\0\end{bmatrix}}_{\mathbf{w}}$ and since complex eigenvectors come in conjugate pairs, we know that

$\mathbf{v}_2=\begin{bmatrix}1\\4\end{bmatrix}-i\begin{bmatrix}\sqrt{39}\\0\end{bmatrix}$.

Initial Condition: Using equation (14) to determine the constants k_1 and k_2 in the general solution (13), we find

$$\begin{bmatrix}| & |\\ \mathbf{u} & \mathbf{w}\\ | & |\end{bmatrix}\begin{bmatrix}k_1\\k_2\end{bmatrix}=\begin{bmatrix}x_1(0)\\x_2(0)\end{bmatrix} \Rightarrow \begin{bmatrix}1 & \sqrt{39}\\ 4 & 0\end{bmatrix}\begin{bmatrix}k_1\\k_2\end{bmatrix}=\begin{bmatrix}1\\1\end{bmatrix} \Rightarrow \left\{\begin{array}{r}k_1+\sqrt{39}k_2=1\\ 4k_1=1\end{array}\right\}$$

which is satisfied when $k_1=\frac{1}{4}$ and $k_2=\frac{\sqrt{39}}{52}$. Thus, the solution as given by equation (13) is

$$\begin{aligned}\mathbf{x}(t)&=k_1e^{at}(\mathbf{u}\cos bt-\mathbf{w}\sin bt)+k_2e^{at}(\mathbf{w}\cos bt+\mathbf{u}\sin bt)\\ &=\tfrac{1}{4}e^{3t/2}\left(\begin{bmatrix}1\\4\end{bmatrix}\cos\tfrac{\sqrt{39}}{2}t-\begin{bmatrix}\sqrt{39}\\0\end{bmatrix}\sin\tfrac{\sqrt{39}}{2}t\right)+\tfrac{\sqrt{39}}{52}e^{3t/2}\left(\begin{bmatrix}\sqrt{39}\\0\end{bmatrix}\cos\tfrac{\sqrt{39}}{2}t+\begin{bmatrix}1\\4\end{bmatrix}\sin\tfrac{\sqrt{39}}{2}t\right)\\ &=e^{3t/2}\begin{bmatrix}\frac{1}{4}\cos\frac{\sqrt{39}}{2}t-\frac{\sqrt{39}}{4}\sin\frac{\sqrt{39}}{2}t+\frac{3}{4}\cos\frac{\sqrt{39}}{2}t+\frac{\sqrt{39}}{52}\sin\frac{\sqrt{39}}{2}t\\ \cos\frac{\sqrt{39}}{2}t+\frac{\sqrt{39}}{13}\sin\frac{\sqrt{39}}{2}t\end{bmatrix}=e^{3t/2}\begin{bmatrix}\cos\frac{\sqrt{39}}{2}t-\frac{3\sqrt{39}}{13}\sin\frac{\sqrt{39}}{2}t\\ \cos\frac{\sqrt{39}}{2}t+\frac{\sqrt{39}}{13}\sin\frac{\sqrt{39}}{2}t\end{bmatrix}\end{aligned}$$

© 2016 Cengage Learning. All Rights Reserved. May not be scanned, copied or duplicated, or posted to a publicly accessible website, in whole or in part.

29. Eigenvalues: $\det(A-\lambda I)=0 \;\Rightarrow\; \det\begin{bmatrix}-2-\lambda & -1\\ 2 & 1-\lambda\end{bmatrix}=0 \;\Rightarrow\; (-2-\lambda)(1-\lambda)+2=0 \;\Rightarrow$

$\lambda(\lambda+1)=0$. The eigenvalues are therefore $\lambda_1=0$ and $\lambda_2=-1$.

Eigenvectors: $(A-\lambda_1 I)\,\mathbf{v}_1=\mathbf{0} \;\Rightarrow\; \begin{bmatrix}-2 & -1\\ 2 & 1\end{bmatrix}\mathbf{v}_1=\begin{bmatrix}0\\0\end{bmatrix} \;\Rightarrow\; \mathbf{v}_1=\begin{bmatrix}1\\-2\end{bmatrix}$ and

$(A-\lambda_2 I)\,\mathbf{v}_2=\mathbf{0} \;\Rightarrow\; \begin{bmatrix}-1 & -1\\ 2 & 2\end{bmatrix}\mathbf{v}_2=\begin{bmatrix}0\\0\end{bmatrix} \;\Rightarrow\; \mathbf{v}_2=\begin{bmatrix}1\\-1\end{bmatrix}$

The general solution given by equation (8) is $\mathbf{x}(t)=c_1e^{\lambda_1 t}\begin{bmatrix}|\\ \mathbf{v}_1\\ |\end{bmatrix}+c_2e^{\lambda_2 t}\begin{bmatrix}|\\ \mathbf{v}_2\\ |\end{bmatrix}=c_1\begin{bmatrix}1\\-2\end{bmatrix}+c_2e^{-t}\begin{bmatrix}1\\-1\end{bmatrix}$.

31. (a) $\mathbf{x}(t)=\begin{bmatrix}c_1e^{-t}\\ c_2e^{-t}\end{bmatrix} \;\Rightarrow\; \dfrac{d\mathbf{x}}{dt}=\begin{bmatrix}-c_1e^{-t}\\ -c_2e^{-t}\end{bmatrix}$ and $\begin{bmatrix}-1 & 0\\ 0 & -1\end{bmatrix}\mathbf{x}=\begin{bmatrix}-1 & 0\\ 0 & -1\end{bmatrix}\begin{bmatrix}c_1e^{-t}\\ c_2e^{-t}\end{bmatrix}=\begin{bmatrix}-c_1e^{-t}\\ -c_2e^{-t}\end{bmatrix}=\dfrac{d\mathbf{x}}{dt}$

(b) **Eigenvalues:** $\det(A-\lambda I)=0 \;\Rightarrow\; \det\begin{bmatrix}-1-\lambda & 0\\ 0 & -1-\lambda\end{bmatrix}=0 \;\Rightarrow\; (-1-\lambda)^2=0$. The eigenvalues are therefore $\lambda=-1$ (repeated twice).

Eigenvectors: $(A-\lambda I)\,\mathbf{v}=\mathbf{0} \;\Rightarrow\; \begin{bmatrix}0 & 0\\ 0 & 0\end{bmatrix}\mathbf{v}=\begin{bmatrix}0\\0\end{bmatrix}$

Any choice of $\mathbf{v}$ will satisfy this matrix system, that is, there are no restrictions on the components of $\mathbf{v}$. In the case of distinct eigenvalues, there was always a restriction on the components of $\mathbf{v}$. In this case (with repeated eigenvalues), we say there are 2 free parameters since we are free to choose both components of $\mathbf{v}$. We make two different choices for $\mathbf{v}_1$ and $\mathbf{v}_2$ ensuring that the eigenvectors are not scalar multiples (termed linearly independent vectors in the 2×2 case in linear algebra textbooks). So we choose $\mathbf{v}_1=\begin{bmatrix}1\\0\end{bmatrix}$ and $\mathbf{v}_2=\begin{bmatrix}0\\1\end{bmatrix}$ and the general solution given by equation (8) is

$$\mathbf{x}(t)=c_1e^{\lambda_1 t}\begin{bmatrix}|\\ \mathbf{v}_1\\ |\end{bmatrix}+c_2e^{\lambda_2 t}\begin{bmatrix}|\\ \mathbf{v}_2\\ |\end{bmatrix}=c_1e^{-t}\begin{bmatrix}1\\0\end{bmatrix}+c_2e^{-t}\begin{bmatrix}0\\1\end{bmatrix}.$$

33. The eigenvalues satisfy $\det(A-\lambda I)=0 \;\Rightarrow\; \det\begin{bmatrix}-1-\lambda & 1\\ a & -1-\lambda\end{bmatrix}=0 \;\Rightarrow\; (-1-\lambda)^2-a=0 \;\Rightarrow$

$\lambda=-1\pm\sqrt{a}$. The eigenvalues are therefore $\lambda_1=-1+\sqrt{a}$ and $\lambda_2=-1-\sqrt{a}$.

- If $a<0$: The eigenvalues are complex with negative real parts, so the origin is a stable spiral.
- If $a=0$: The eigenvalue $\lambda=-1$ (repeated twice) is real and negative, so the origin is a stable node (possibly an improper node [see exercise 10.2.32]).
- If $0<a<1$: Both eigenvalues are real and negative, so the origin is a stable node.
- If $a=1$: The eigenvalues are $\lambda_1=0$ and $\lambda_2=-2$ and the x_1 and x_2 nullclines are both $x_2=x_1$. Therefore, there are an infinite number of equilibrium points so the origin is *not* stable.
- If $a>1$: One eigenvalue is real positive and the other eigenvalue is real negative, so the origin is a saddle.

© 2016 Cengage Learning. All Rights Reserved. May not be scanned, copied or duplicated, or posted to a publicly accessible website, in whole or in part.

35. The eigenvalues satisfy $\det(A - \lambda I) = 0 \;\Rightarrow\; \det\begin{bmatrix} a-\lambda & 1 \\ 1 & a-\lambda \end{bmatrix} = 0 \;\Rightarrow\; (a-\lambda)^2 - 1 = 0 \;\Rightarrow\; \lambda = a \pm 1.$

The eigenvalues are therefore $\lambda_1 = a + 1$ and $\lambda_2 = a - 1$.

- If $a < -1$: Both eigenvalues are real and negative, so the origin is a stable node.
- If $a = -1$: The eigenvalues are $\lambda_1 = 0$ and $\lambda_2 = -2$ and the x_1 and x_2 nullclines are both $x_2 = x_1$. Therefore, there are an infinite number of equilibrium points so the origin is *not* stable.
- If $-1 < a < 1$: One eigenvalue is real positive and the other eigenvalue is real negative, so the origin is a saddle.
- If $a = 1$: The eigenvalues are $\lambda_1 = 2$ and $\lambda_2 = 0$ so the solution curves will grow in the direction of the eigenvector $\mathbf{v}_2$. Therefore, the origin is an unstable equilibrium.
- If $a > 1$: Both eigenvalues are real and positive, so the origin is an unstable node.

37. The eigenvalues of a general 2×2 system $A = \begin{bmatrix} a_{11} & a_{12} \\ a_{21} & a_{22} \end{bmatrix}$ satisfy the characteristic polynomial

$\lambda^2 - (a_{11} + a_{22})\,\lambda + (a_{11}a_{22} - a_{12}a_{21}) = 0 \;\Rightarrow\; \lambda^2 - (\text{trace } A)\,\lambda + (\det A) = 0 \;\Rightarrow$

$\lambda = \dfrac{(\text{trace } A) \pm \sqrt{(\text{trace } A)^2 - 4\det A}}{2}$. If the origin is stable, then the real parts of both eigenvalues of A are negative by Theorem 15. Thus, if the eigenvalues are

i) Complex: trace $A < 0$ ensures the real part of λ is negative, and $\det A > 0$ is a necessary condition for λ to be complex.

ii) Real: Since both eigenvalues are negative, their product is positive so $\lambda_1\lambda_2 > 0 \;\Rightarrow$ $\frac{1}{4}(\text{trace } A)^2 - \frac{1}{4}\left((\text{trace } A)^2 - 4\det A\right) = \det A > 0$. Also, trace $A < 0$ otherwise one of the eigenvalues would be positive.

To prove the converse statement, we assume that trace $A < 0$ and $\det A > 0$. It follows from Theorem 15 that if the eigenvalues are

i) Complex: The real parts of both eigenvalues are negative, so the origin is stable.

ii) Real: $\lambda_1\lambda_2 = \det A > 0 \;\Rightarrow$ The eigenvalues are either both positive or both negative, but since trace $A < 0$, they must both be negative. Thus, the origin is stable.

39. Consider the 2-dimensional, autonomous, system of linear differential equations $\dfrac{d\mathbf{x}}{dt} = A\mathbf{x}$ with two different initial conditions $\mathbf{x}_0$ and $\mathbf{y}_0$. Theorem 2 states that each initial condition results in a unique solution $\mathbf{x}(t)$ and $\mathbf{y}(t)$ for $t \in \mathbb{R}$. Suppose the solution curves cross at some time $t = t^*$ so that $\mathbf{x}(t^*) = \mathbf{y}(t^*) \;\Rightarrow\; A\mathbf{x}(t^*) = A\mathbf{y}(t^*) \;\Rightarrow$ $\dfrac{d\mathbf{x}}{dt}(t^*) = \dfrac{d\mathbf{y}}{dt}(t^*)$. This means that at $t = t^*$, the components of each solution curve vary at the same rate, and since the two curves intersect at that time, they will follow the same path in the phase plane for all $t > t^*$. But this contradicts the uniqueness theorem (2) which states that the solution $\mathbf{x}(t)$ is unique and defined for *all* t. Therefore, the solution curves cannot cross in the phase plane.

© 2016 Cengage Learning. All Rights Reserved. May not be scanned, copied or duplicated, or posted to a publicly accessible website, in whole or in part.

10.3 Applications

1. $\left\{\begin{array}{l} z_1' = z_2 \\ z_2' = -\frac{k}{m} z_1 - \frac{b}{m} z_2 = -0.01 z_1 - 0.001 z_2 \end{array}\right\} \Rightarrow \frac{d}{dt}\begin{bmatrix} z_1 \\ z_2 \end{bmatrix} = \underbrace{\begin{bmatrix} 0 & 1 \\ -0.01 & -0.001 \end{bmatrix}}_{A} \underbrace{\begin{bmatrix} z_1 \\ z_2 \end{bmatrix}}_{\mathbf{z}}$

Eigenvalues: $\det(A - \lambda I) = 0 \;\Rightarrow\; \det\begin{bmatrix} -\lambda & 1 \\ -0.01 & -0.001 - \lambda \end{bmatrix} = 0 \;\Rightarrow\; -\lambda(-0.001 - \lambda) + 0.01 = 0 \;\Rightarrow$

$\lambda^2 + 0.001\lambda + 0.01 = 0 \;\Rightarrow\; \lambda = -\frac{0.001}{2} \pm \frac{1}{2}\sqrt{10^{-6} - 0.04} \approx \underbrace{-5 \times 10^{-4}}_{a} \pm i\,\underbrace{0.1}_{b}$. The eigenvalues are therefore

$\lambda_1 \approx -5 \times 10^{-4} + 0.1i$ and $\lambda_2 \approx -5 \times 10^{-4} - 0.1i$.

Eigenvectors: $(A - \lambda_1 I)\,\mathbf{v}_1 = \mathbf{0} \;\Rightarrow\; \begin{bmatrix} 5 \times 10^{-4} - 0.1i & 1 \\ -0.01 & -5 \times 10^{-4} - 0.1i \end{bmatrix} \mathbf{v}_1 = \begin{bmatrix} 0 \\ 0 \end{bmatrix} \;\Rightarrow$

$\mathbf{v}_1 \approx \begin{bmatrix} 1 \\ -5 \times 10^{-4} + 0.1i \end{bmatrix} = \underbrace{\begin{bmatrix} 1 \\ -5 \times 10^{-4} \end{bmatrix}}_{\mathbf{u}} + i\,\underbrace{\begin{bmatrix} 0 \\ 0.1 \end{bmatrix}}_{\mathbf{w}}$ and since complex eigenvectors come in conjugate pairs, we know

that $\mathbf{v}_2 \approx \begin{bmatrix} 1 \\ -5 \times 10^{-4} \end{bmatrix} - i\begin{bmatrix} 0 \\ 0.1 \end{bmatrix}$.

In the case of complex eigenvalues, the general solution given by equation (13) is

$$\begin{aligned} \mathbf{z}(t) &= k_1 e^{at}(\mathbf{u}\cos bt - \mathbf{w}\sin bt) + k_2 e^{at}(\mathbf{w}\cos bt + \mathbf{u}\sin bt) \\ &\approx k_1 e^{-5\times10^{-4}t}\left(\begin{bmatrix} 1 \\ -5 \times 10^{-4} \end{bmatrix}\cos 0.1t - \begin{bmatrix} 0 \\ 0.1 \end{bmatrix}\sin 0.1t\right) \\ &\qquad + k_2 e^{-5\times10^{-4}t}\left(\begin{bmatrix} 0 \\ 0.1 \end{bmatrix}\cos 0.1t + \begin{bmatrix} 1 \\ -5 \times 10^{-4} \end{bmatrix}\sin 0.1t\right) \\ &= e^{-5\times10^{-4}t}\begin{bmatrix} k_1\cos 0.1t + k_2 \sin 0.1t \\ (-5 \times 10^{-4}k_1 + 0.1k_2)\cos 0.1t - (0.1k_1 + 5 \times 10^{-4}k_2)\sin 0.1t \end{bmatrix} \end{aligned}$$

3. $\left\{\begin{array}{l} y_1' = -4y_1 \\ y_2' = y_1 - 2y_2 + d \end{array}\right\} \Rightarrow \begin{bmatrix} y_1' \\ y_2' \end{bmatrix} = \underbrace{\begin{bmatrix} -4 & 0 \\ 1 & -2 \end{bmatrix}}_{A}\begin{bmatrix} y_1 \\ y_2 \end{bmatrix} + \begin{bmatrix} 0 \\ d \end{bmatrix}$

(a) First, we find the equilibrium by setting $y_1' = y_2' = 0$ which results in the system $\begin{bmatrix} -4 & 0 \\ 1 & -2 \end{bmatrix}\begin{bmatrix} y_1 \\ y_2 \end{bmatrix} = -\begin{bmatrix} 0 \\ d \end{bmatrix}$. The solution of this system is $\hat{\mathbf{y}} = \begin{bmatrix} 0 \\ d/2 \end{bmatrix}$. Now, let $x_1(t) = y_1(t) - \hat{y}_1 = y_1(t)$ and $x_2(t) = y_2(t) - \hat{y}_2 = y_2(t) - d/2$. From Exercise 10.1.23, we know that $\mathbf{x} = \begin{bmatrix} x_1(t) \\ x_2(t) \end{bmatrix}$ satisfies the homogeneous system $\frac{d\mathbf{x}}{dt} = A\mathbf{x}$.

(b) **Eigenvalues:** $\det(A - \lambda I) = 0 \;\Rightarrow\; \det\begin{bmatrix} -4 - \lambda & 0 \\ 1 & -2 - \lambda \end{bmatrix} = 0 \;\Rightarrow\; (-4 - \lambda)(-2 - \lambda) = 0$. The eigenvalues are therefore $\lambda_1 = -2$ and $\lambda_2 = -4$.

© 2016 Cengage Learning. All Rights Reserved. May not be scanned, copied or duplicated, or posted to a publicly accessible website, in whole or in part.

Eigenvectors: $(A - \lambda_1 I)\,\mathbf{v}_1 = \mathbf{0} \;\Rightarrow\; \begin{bmatrix} -2 & 0 \\ 1 & 0 \end{bmatrix} \mathbf{v}_1 = \begin{bmatrix} 0 \\ 0 \end{bmatrix} \;\Rightarrow\; \mathbf{v}_1 = \begin{bmatrix} 0 \\ 1 \end{bmatrix}$ and

$$(A - \lambda_2 I)\,\mathbf{v}_2 = \mathbf{0} \;\Rightarrow\; \begin{bmatrix} 0 & 0 \\ 1 & 2 \end{bmatrix} \mathbf{v}_2 = \begin{bmatrix} 0 \\ 0 \end{bmatrix} \;\Rightarrow\; \mathbf{v}_2 = \begin{bmatrix} -2 \\ 1 \end{bmatrix}$$

The general solution given by equation 10.2.8 is $\mathbf{x}(t) = c_1 e^{\lambda_1 t} \begin{bmatrix} | \\ \mathbf{v}_1 \\ | \end{bmatrix} + c_2 e^{\lambda_2 t} \begin{bmatrix} | \\ \mathbf{v}_2 \\ | \end{bmatrix} = c_1 e^{-2t} \begin{bmatrix} 0 \\ 1 \end{bmatrix} + c_2 e^{-4t} \begin{bmatrix} -2 \\ 1 \end{bmatrix}$.

(c) If $\mathbf{y}(0) = \begin{bmatrix} y_1(0) \\ y_2(0) \end{bmatrix} = \begin{bmatrix} 1 \\ 0 \end{bmatrix}$, then $\mathbf{x}(0) = \begin{bmatrix} y_1(0) \\ y_2(0) - d/2 \end{bmatrix} = \begin{bmatrix} 1 \\ -d/2 \end{bmatrix}$. Setting $t = 0$ in the general solution gives

$\mathbf{x}(0) = c_1 \begin{bmatrix} 0 \\ 1 \end{bmatrix} + c_2 \begin{bmatrix} -2 \\ 1 \end{bmatrix} = \begin{bmatrix} 1 \\ -d/2 \end{bmatrix} \;\Rightarrow\; \begin{bmatrix} 0 & -2 \\ 1 & 1 \end{bmatrix} \begin{bmatrix} c_1 \\ c_2 \end{bmatrix} = \begin{bmatrix} 1 \\ -d/2 \end{bmatrix}$ which is satisfied when $c_1 = \frac{1}{2} - \frac{d}{2}$ and

$c_2 = -\frac{1}{2}$. Thus the solution to the initial value problem for $\mathbf{x}(t)$ is $\mathbf{x}(t) = \left(\frac{1}{2} - \frac{d}{2}\right) e^{-2t} \begin{bmatrix} 0 \\ 1 \end{bmatrix} - \frac{1}{2} e^{-4t} \begin{bmatrix} -2 \\ 1 \end{bmatrix}$.

(d) Expressing the solution from part (c) in terms of the original variables, we have

$$\mathbf{y}(t) = \begin{bmatrix} y_1(t) \\ y_2(t) \end{bmatrix} = \mathbf{x}(t) + \hat{\mathbf{y}} = \left(\tfrac{1}{2} - \tfrac{d}{2}\right) e^{-2t} \begin{bmatrix} 0 \\ 1 \end{bmatrix} - \tfrac{1}{2} e^{-4t} \begin{bmatrix} -2 \\ 1 \end{bmatrix} + \begin{bmatrix} 0 \\ d/2 \end{bmatrix}$$

Therefore, the fraction of cells that are irreversibly resistant at time t is given by the function $y_2(t) = \left(\frac{1}{2} - \frac{d}{2}\right) e^{-2t} - \frac{1}{2} e^{-4t} + \frac{d}{2}$.

(e) 50% of the cells are irreversibly resistant when $y_2(t) = \frac{1}{2} \;\Rightarrow\; \left(\frac{1}{2} - \frac{d}{2}\right) e^{-2t} - \frac{1}{2} e^{-4t} + \frac{d}{2} = \frac{1}{2}$. If we let $z = e^{-2t}$, the equation can be expressed as the following quadratic function $\left(\frac{1}{2} - \frac{d}{2}\right) z - \frac{1}{2} z^2 + \frac{d}{2} = \frac{1}{2} \;\Rightarrow$

$$z^2 + (d-1)\,z + (1-d) = 0 \;\Rightarrow\; z = \frac{-(d-1) \pm \sqrt{(d-1)^2 - 4(1-d)}}{2} = \frac{1 - d \pm \sqrt{(d-1)(d+3)}}{2}.$$

Assuming $d > 1$, we take the positive root to ensure z is positive. Since $z = e^{-2t}$, the time taken before 50% of the cells are irreversibly resistant is $t = -\frac{1}{2} \ln z = -\frac{1}{2} \ln\left(\frac{1}{2}\left(1 - d + \sqrt{(d-1)(d+3)}\right)\right)$. Note that if $d \le 1$, there is no time at which 50% of the cells are irreversibly resistant.

5. $\left\{ \begin{array}{l} C' = -\alpha C - \beta C \\ I' = \alpha C - \delta I + \rho I \end{array} \right\} \;\Rightarrow\; \dfrac{d}{dt} \begin{bmatrix} C \\ I \end{bmatrix} = \underbrace{\begin{bmatrix} -\alpha - \beta & 0 \\ \alpha & \rho - \delta \end{bmatrix}}_{A} \underbrace{\begin{bmatrix} C \\ I \end{bmatrix}}_{\mathbf{x}}$

(a) **Eigenvalues:** $\det(A - \lambda I) = 0 \;\Rightarrow\; \det \begin{bmatrix} -\alpha - \beta - \lambda & 0 \\ \alpha & \rho - \delta - \lambda \end{bmatrix} = 0 \;\Rightarrow\; (-\alpha - \beta - \lambda)(\rho - \beta - \lambda) = 0.$

The eigenvalues are therefore $\lambda_1 = \rho - \delta$ and $\lambda_2 = -\alpha - \beta$.

Eigenvectors: $(A - \lambda_1 I)\,\mathbf{v}_1 = \mathbf{0} \;\Rightarrow\; \begin{bmatrix} -\alpha - \beta - \rho + \delta & 0 \\ \alpha & 0 \end{bmatrix} \mathbf{v}_1 = \begin{bmatrix} 0 \\ 0 \end{bmatrix} \;\Rightarrow\; \mathbf{v}_1 = \begin{bmatrix} 0 \\ 1 \end{bmatrix}$ and

$$(A - \lambda_2 I)\,\mathbf{v}_2 = \mathbf{0} \;\Rightarrow\; \begin{bmatrix} 0 & 0 \\ \alpha & \rho - \delta + \alpha + \beta \end{bmatrix} \mathbf{v}_2 = \begin{bmatrix} 0 \\ 0 \end{bmatrix} \;\Rightarrow\; \mathbf{v}_2 = \begin{bmatrix} \rho - \delta + \alpha + \beta \\ -\alpha \end{bmatrix}$$

The general solution given by equation 10.2.8 is

$$\mathbf{x}(t) = c_1 e^{\lambda_1 t} \begin{bmatrix} | \\ \mathbf{v}_1 \\ | \end{bmatrix} + c_2 e^{\lambda_2 t} \begin{bmatrix} | \\ \mathbf{v}_2 \\ | \end{bmatrix} = c_1 e^{(\rho - \delta)t} \begin{bmatrix} 0 \\ 1 \end{bmatrix} + c_2 e^{-(\alpha + \beta)t} \begin{bmatrix} \rho - \delta + \alpha + \beta \\ -\alpha \end{bmatrix}.$$

© 2016 Cengage Learning. All Rights Reserved. May not be scanned, copied or duplicated, or posted to a publicly accessible website, in whole or in part.

(b) $\det A = -(\alpha+\beta)(\rho-\delta)$, trace $A = -(\alpha+\beta)+\rho-\delta$ and after some simplification

$\det A - \frac{1}{4}(\text{trace } A)^2 = -\frac{1}{4}(\alpha+\beta+\rho-\delta)^2 < 0.$

If $\rho > \delta$, then $\det A < 0$ so Figure 10.2.8 indicates the origin is a saddle in this case. If $\rho < \delta$, then $\det A > 0$, trace $A < 0$ and $\det A < \frac{1}{4}(\text{trace } A)^2$ so the origin will be a stable node.

Alternative Solution: Referring back to the general solution in part (a), if $\rho > \delta$, then λ_1 is real positive and λ_2 is real negative so the origin is a saddle. If $\rho < \delta$, then λ_1 and λ_2 are both real negative so the origin is a stable node.

(c) Setting $t = 0$ in the general solution gives $\mathbf{x}(0) = c_1\begin{bmatrix}0\\1\end{bmatrix} + c_2\begin{bmatrix}\rho-\delta+\alpha+\beta\\-\alpha\end{bmatrix} = \begin{bmatrix}C_0\\0\end{bmatrix} \Rightarrow$

$\begin{bmatrix}0 & \rho-\delta+\alpha+\beta\\1 & -\alpha\end{bmatrix}\begin{bmatrix}c_1\\c_2\end{bmatrix} = \begin{bmatrix}C_0\\0\end{bmatrix}$ which is satisfied when $c_1 = \dfrac{\alpha C_0}{\rho-\delta+\alpha+\beta}$ and $c_2 = \dfrac{C_0}{\rho-\delta+\alpha+\beta}$.

Thus the solution to the initial value problem for $\mathbf{x}(t)$ is

$$\mathbf{x}(t) = \frac{\alpha C_0}{\rho-\delta+\alpha+\beta}e^{(\rho-\delta)t}\begin{bmatrix}0\\1\end{bmatrix} + \frac{C_0}{\rho-\delta+\alpha+\beta}e^{-(\alpha+\beta)t}\begin{bmatrix}\rho-\delta+\alpha+\beta\\-\alpha\end{bmatrix}.$$

(d) From the result of part (c), the number of cancer cells that have invaded the organ is

$I(t) = \dfrac{\alpha C_0}{\rho-\delta+\alpha+\beta}e^{(\rho-\delta)t} - \dfrac{\alpha C_0}{\rho-\delta+\alpha+\beta}e^{-(\alpha+\beta)t}$. If $\rho > \delta$, then the exponential term $e^{(\rho-\delta)t}$ will grow indefinitely as t increases while the term $e^{-(\alpha+\beta)t}$ approaches zero, so the tumor will grow in the long term. Conversely, if the tumor grows in the long term then $I(t)$ increases as t increases (and does not converge). This is possible only when the coefficient of t in the exponential term $e^{(\rho-\delta)t}$ is positive, that is when $\rho > \delta$.

7. (a) With $n = 2$, the system of differential equations for the three stages is

$$x_0' = -u_0x_0 \qquad x_1' = u_0x_0 - u_1x_1 \qquad x_2' = u_1x_1$$

(b) $\left\{\begin{aligned}x_0' &= -u_0x_0\\ x_1' &= u_0x_0 - u_1x_1\end{aligned}\right\} \Rightarrow \dfrac{d}{dt}\begin{bmatrix}x_0\\x_1\end{bmatrix} = \underbrace{\begin{bmatrix}-u_0 & 0\\u_0 & -u_1\end{bmatrix}}_{A}\underbrace{\begin{bmatrix}x_0\\x_1\end{bmatrix}}_{\mathbf{x}}$

Eigenvalues: $\det(A-\lambda I) = 0 \Rightarrow \det\begin{bmatrix}-u_0-\lambda & 0\\u_0 & -u_1-\lambda\end{bmatrix} = 0 \Rightarrow (-u_0-\lambda)(-u_1-\lambda) = 0$. The eigenvalues are therefore $\lambda_1 = -u_0$ and $\lambda_2 = -u_1$.

Eigenvectors: $(A-\lambda_1 I)\mathbf{v}_1 = \mathbf{0} \Rightarrow \begin{bmatrix}0 & 0\\u_0 & -u_1+u_0\end{bmatrix}\mathbf{v}_1 = \begin{bmatrix}0\\0\end{bmatrix} \Rightarrow \mathbf{v}_1 = \begin{bmatrix}u_0-u_1\\-u_0\end{bmatrix}$ and

$(A-\lambda_2 I)\mathbf{v}_2 = \mathbf{0} \Rightarrow \begin{bmatrix}-u_0+u_1 & 0\\u_0 & 0\end{bmatrix}\mathbf{v}_2 = \begin{bmatrix}0\\0\end{bmatrix} \Rightarrow \mathbf{v}_2 = \begin{bmatrix}0\\1\end{bmatrix}$

The general solution given by equation 10.2.8 is

$$\mathbf{x}(t) = c_1e^{\lambda_1 t}\begin{bmatrix}|\\\mathbf{v}_1\\|\end{bmatrix} + c_2e^{\lambda_2 t}\begin{bmatrix}|\\\mathbf{v}_2\\|\end{bmatrix} = c_1e^{-u_0t}\begin{bmatrix}u_0-u_1\\-u_0\end{bmatrix} + c_2e^{-u_1t}\begin{bmatrix}0\\1\end{bmatrix}.$$

Initial Condition: Using equation 10.2.9 to determine the constants c_1 and c_2 in the general solution, we find

$$\begin{bmatrix}| & |\\\mathbf{v}_1 & \mathbf{v}_2\\| & |\end{bmatrix}\begin{bmatrix}c_1\\c_2\end{bmatrix} = \begin{bmatrix}x_0(0)\\x_1(0)\end{bmatrix} \Rightarrow \begin{bmatrix}u_0-u_1 & 0\\-u_0 & 1\end{bmatrix}\begin{bmatrix}c_1\\c_2\end{bmatrix} = \begin{bmatrix}k\\0\end{bmatrix} \Rightarrow \left\{\begin{aligned}(u_0-u_1)c_1 &= k\\-u_0c_1+c_2 &= 0\end{aligned}\right\}$$

© 2016 Cengage Learning. All Rights Reserved. May not be scanned, copied or duplicated, or posted to a publicly accessible website, in whole or in part.

which is satisfied when $c_1 = \dfrac{k}{u_0 - u_1}$ and $c_2 = \dfrac{u_0 k}{u_0 - u_1}$. Thus, the solution of the initial value problem is

$$\mathbf{x}(t) = \frac{k}{u_0 - u_1} e^{-u_0 t} \begin{bmatrix} u_0 - u_1 \\ -u_0 \end{bmatrix} + \frac{u_0 k}{u_0 - u_1} e^{-u_1 t} \begin{bmatrix} 0 \\ 1 \end{bmatrix}.$$

(c) From the result of part (b), we have $x_1(t) = -\dfrac{u_0 k}{u_0 - u_1} e^{-u_0 t} + \dfrac{u_0 k}{u_0 - u_1} e^{-u_1 t}$. From part (a), x_2 must satisfy

$\dfrac{dx_2}{dt} = u_1 x_1 = -\dfrac{u_0 u_1 k}{u_0 - u_1} e^{-u_0 t} + \dfrac{u_0 u_1 k}{u_0 - u_1} e^{-u_1 t}$.

(d) Using a separation of variables to solve the differential equation, we find

$\displaystyle\int dx_2 = \int \left[-\frac{u_0 u_1 k}{u_0 - u_1} e^{-u_0 t} + \frac{u_0 u_1 k}{u_0 - u_1} e^{-u_1 t} \right] dt \;\Rightarrow\; x_2 = \frac{u_1 k}{u_0 - u_1} e^{-u_0 t} - \frac{u_0 k}{u_0 - u_1} e^{-u_1 t} + C$. Now

$x_2(0) = 0 \;\Rightarrow\; \dfrac{u_1 k}{u_0 - u_1} - \dfrac{u_0 k}{u_0 - u_1} + C = 0 \;\Rightarrow\; C = \dfrac{u_0 k}{u_0 - u_1} - \dfrac{u_1 k}{u_0 - u_1}$. Therefore,

$x_2(t) = \dfrac{u_1 k}{u_0 - u_1} e^{-u_0 t} - \dfrac{u_0 k}{u_0 - u_1} e^{-u_1 t} + \dfrac{u_0 k}{u_0 - u_1} - \dfrac{u_1 k}{u_0 - u_1} = \dfrac{u_1 k}{u_0 - u_1}\left(e^{-u_0 t} - 1\right) - \dfrac{u_0 k}{u_0 - u_1}\left(e^{-u_1 t} - 1\right)$.

9. $$\left.\begin{cases} \dfrac{dw}{dt} = -k_w(w - R) \\ \dfrac{dp}{dt} = -k_p(p - w) \end{cases}\right\} \Rightarrow \frac{d}{dt}\begin{bmatrix} w \\ p \end{bmatrix} = \underbrace{\begin{bmatrix} -k_w & 0 \\ k_p & -k_p \end{bmatrix}}_{A}\underbrace{\begin{bmatrix} w \\ p \end{bmatrix}}_{\mathbf{y}} + \begin{bmatrix} k_w R \\ 0 \end{bmatrix}$$

(a) The differential equation dw/dt indicates that the rate of change of water temperature is proportional to the temperature difference between the water and the room, with a negative proportionality constant $-k_w$. If $w > R$ then $dw/dt < 0$ and if $w < R$ then $dw/dt > 0$, so the water temperature will approach room temperature in the long run. Since dw/dt does not depend on p, it has been assumed that, relative to the room's temperature, the coin will have no effect on the water's temperature. The differential equation dp/dt indicates that the rate of change of the coin's temperature is proportional to the temperature difference between the coin and the water, with a negative proportionality constant $-k_p$. So the coin's temperature will approach the water's temperature in the long run. Also, since dp/dt does not depend on the room temperature R, it has been assumed that the coin's temperature is affected directly by the water's temperature.

(b) First, we find the equilibrium by setting $dw/dt = dp/dt = 0$ which results in the system

$\begin{bmatrix} -k_w & 0 \\ k_p & -k_p \end{bmatrix}\begin{bmatrix} \hat{w} \\ \hat{p} \end{bmatrix} = -\begin{bmatrix} k_w R \\ 0 \end{bmatrix}$. The solution of this system is $\begin{bmatrix} \hat{w} \\ \hat{p} \end{bmatrix} = \begin{bmatrix} R \\ R \end{bmatrix}$. Now, let

$\mathbf{x} = \begin{bmatrix} w(t) - \hat{w} \\ p(t) - \hat{p} \end{bmatrix} = \begin{bmatrix} w(t) - R \\ p(t) - R \end{bmatrix}$. From Exercise 10.1.23, we know that $\mathbf{x}$ satisfies the homogeneous system $\dfrac{d\mathbf{x}}{dt} = A\mathbf{x}$.

(c) **Eigenvalues:** $\det(A - \lambda I) = 0 \;\Rightarrow\; \det\begin{bmatrix} -k_w - \lambda & 0 \\ k_p & -k_p - \lambda \end{bmatrix} = 0 \;\Rightarrow\; (-k_w - \lambda)(-k_p - \lambda) = 0$. The eigenvalues are therefore $\lambda_1 = -k_w$ and $\lambda_2 = -k_p$.

Eigenvectors: $(A - \lambda_1 I)\mathbf{v}_1 = \mathbf{0} \;\Rightarrow\; \begin{bmatrix} 0 & 0 \\ k_p & -k_p + k_w \end{bmatrix}\mathbf{v}_1 = \begin{bmatrix} 0 \\ 0 \end{bmatrix} \;\Rightarrow\; \mathbf{v}_1 = \begin{bmatrix} k_w - k_p \\ -k_p \end{bmatrix}$ and

$$(A - \lambda_2 I)\mathbf{v}_2 = \mathbf{0} \;\Rightarrow\; \begin{bmatrix} -k_w + k_p & 0 \\ k_p & 0 \end{bmatrix}\mathbf{v}_2 = \begin{bmatrix} 0 \\ 0 \end{bmatrix} \;\Rightarrow\; \mathbf{v}_2 = \begin{bmatrix} 0 \\ 1 \end{bmatrix}$$

The general solution given by equation 10.2.8 is

$$\mathbf{x}(t) = c_1 e^{\lambda_1 t}\begin{bmatrix} | \\ \mathbf{v}_1 \\ | \end{bmatrix} + c_2 e^{\lambda_2 t}\begin{bmatrix} | \\ \mathbf{v}_2 \\ | \end{bmatrix} = c_1 e^{-k_w t}\begin{bmatrix} k_w - k_p \\ -k_p \end{bmatrix} + c_2 e^{-k_p t}\begin{bmatrix} 0 \\ 1 \end{bmatrix}.$$

© 2016 Cengage Learning. All Rights Reserved. May not be scanned, copied or duplicated, or posted to a publicly accessible website, in whole or in part.

(d) Using the result of part (b), the general solution expressed in terms of the original variables w and p is

$\begin{bmatrix} w(t) \\ p(t) \end{bmatrix} = \mathbf{x} + \begin{bmatrix} R \\ R \end{bmatrix} = c_1 e^{-k_w t} \begin{bmatrix} k_w - k_p \\ -k_p \end{bmatrix} + c_2 e^{-k_p t} \begin{bmatrix} 0 \\ 1 \end{bmatrix} + \begin{bmatrix} R \\ R \end{bmatrix}$. Now setting $t = 0$, we have

$\begin{bmatrix} w(0) \\ p(0) \end{bmatrix} = c_1 \begin{bmatrix} k_w - k_p \\ -k_p \end{bmatrix} + c_2 \begin{bmatrix} 0 \\ 1 \end{bmatrix} + \begin{bmatrix} R \\ R \end{bmatrix} \Rightarrow \begin{bmatrix} k_w - k_p & 0 \\ -k_p & 1 \end{bmatrix} \begin{bmatrix} c_1 \\ c_2 \end{bmatrix} = \begin{bmatrix} w_0 - R \\ p_0 - R \end{bmatrix}$ which is satisfied when

$c_1 = \dfrac{w_0 - R}{k_w - k_p}$ and $c_2 = k_p \dfrac{w_0 - R}{k_w - k_p} + p_0 - R$. Therefore the solution to the initial value problem is

$$\begin{bmatrix} w(t) \\ p(t) \end{bmatrix} = \frac{w_0 - R}{k_w - k_p} e^{-k_w t} \begin{bmatrix} k_w - k_p \\ -k_p \end{bmatrix} + \left(k_p \frac{w_0 - R}{k_w - k_p} + p_0 - R \right) e^{-k_p t} \begin{bmatrix} 0 \\ 1 \end{bmatrix} + \begin{bmatrix} R \\ R \end{bmatrix}.$$

11. (a) To find the equilibrium, we set $dx_1/dt = dx_2/dt = 0$ which gives the linear system $\left\{ \begin{array}{r} ac + a\dfrac{\hat{x}_2}{V} - 2a\dfrac{\hat{x}_1}{V} = 0 \\ ac - a\dfrac{\hat{x}_2}{V} = 0 \end{array} \right\}$. This system is satisfied when $\hat{x}_1 = \hat{x}_2 = cV$. Thus, at equilibrium, the amount of CO_2 in both the shallow and deep compartments during normal lung functioning is given by cV.

(b) $\left\{ \begin{array}{l} \dfrac{dx_1}{dt} = ac + a\dfrac{x_2}{V} - 2a\dfrac{x_1}{V} \\ \dfrac{dx_2}{dt} = -a\dfrac{x_2}{V} \end{array} \right\} \Rightarrow \dfrac{d}{dt}\begin{bmatrix} x_1 \\ x_2 \end{bmatrix} = \underbrace{\begin{bmatrix} -2a/V & a/V \\ 0 & -a/V \end{bmatrix}}_{A} \underbrace{\begin{bmatrix} x_1 \\ x_2 \end{bmatrix}}_{\mathbf{x}} + \begin{bmatrix} ac \\ 0 \end{bmatrix}$

First, we find the equilibrium by setting $x_1' = x_2' = 0$ which results in the system $\begin{bmatrix} -2a/V & a/V \\ 0 & -a/V \end{bmatrix} \begin{bmatrix} \hat{x}_1 \\ \hat{x}_2 \end{bmatrix} = -\begin{bmatrix} ac \\ 0 \end{bmatrix}$.

The solution of this system is $\hat{\mathbf{x}} = \begin{bmatrix} cV/2 \\ 0 \end{bmatrix}$. Thus, during an embolism, there is half the amount of CO_2 in the first compartment as during normal lung functioning at equilibrium. Now, let $\mathbf{y} = \mathbf{x} - \hat{\mathbf{x}} = \mathbf{x} - \begin{bmatrix} cV/2 \\ 0 \end{bmatrix}$. From Exercise 10.1.23, we know that $\mathbf{y}$ satisfies the homogeneous system $\dfrac{d\mathbf{y}}{dt} = A\mathbf{y}$.

(c) **Eigenvalues:** $\det(A - \lambda I) = 0 \Rightarrow \det \begin{bmatrix} -2a/V - \lambda & a/V \\ 0 & -a/V - \lambda \end{bmatrix} = 0 \Rightarrow (-2a/V - \lambda)(-a/V - \lambda) = 0.$

The eigenvalues are therefore $\lambda_1 = -\dfrac{a}{V}$ and $\lambda_2 = -2\dfrac{a}{V}$.

Eigenvectors: $(A - \lambda_1 I)\mathbf{v}_1 = \mathbf{0} \Rightarrow \begin{bmatrix} -a/V & a/V \\ 0 & 0 \end{bmatrix} \mathbf{v}_1 = \begin{bmatrix} 0 \\ 0 \end{bmatrix} \Rightarrow \mathbf{v}_1 = \begin{bmatrix} 1 \\ 1 \end{bmatrix}$ and

$(A - \lambda_2 I)\mathbf{v}_2 = \mathbf{0} \Rightarrow \begin{bmatrix} 0 & a/V \\ 0 & a/V \end{bmatrix} \mathbf{v}_2 = \begin{bmatrix} 0 \\ 0 \end{bmatrix} \Rightarrow \mathbf{v}_2 = \begin{bmatrix} 1 \\ 0 \end{bmatrix}$

The general solution given by equation 10.2.8 is $\mathbf{y}(t) = c_1 e^{\lambda_1 t} \begin{bmatrix} | \\ \mathbf{v}_1 \\ | \end{bmatrix} + c_2 e^{\lambda_2 t} \begin{bmatrix} | \\ \mathbf{v}_2 \\ | \end{bmatrix} = c_1 e^{-at/V} \begin{bmatrix} 1 \\ 1 \end{bmatrix} + c_2 e^{-2at/V} \begin{bmatrix} 1 \\ 0 \end{bmatrix}$.

© 2016 Cengage Learning. All Rights Reserved. May not be scanned, copied or duplicated, or posted to a publicly accessible website, in whole or in part.

Initial Condition: Using the results of parts (a) and (b), the initial condition is

$\mathbf{y}(0) = \begin{bmatrix} y_1(0) \\ y_2(0) \end{bmatrix} = \begin{bmatrix} x_1(0) - cV/2 \\ x_2(0) \end{bmatrix} = \begin{bmatrix} cV - cV/2 \\ cV \end{bmatrix} = \begin{bmatrix} cV/2 \\ cV \end{bmatrix}$. Using Equation 10.2.9 to determine the constants c_1 and c_2 in the general solution, we find

$$\begin{bmatrix} | & | \\ \mathbf{v}_1 & \mathbf{v}_2 \\ | & | \end{bmatrix} \begin{bmatrix} c_1 \\ c_2 \end{bmatrix} = \begin{bmatrix} y_1(0) \\ y_2(0) \end{bmatrix} \Rightarrow \begin{bmatrix} 1 & 1 \\ 1 & 0 \end{bmatrix} \begin{bmatrix} c_1 \\ c_2 \end{bmatrix} = \begin{bmatrix} cV/2 \\ cV \end{bmatrix} \Rightarrow \left\{ \begin{array}{c} c_1 + c_2 = cV/2 \\ c_1 = cV \end{array} \right\}$$

which is satisfied when $c_1 = cV$ and $c_2 = -\dfrac{cV}{2}$. Thus, the solution of the initial value problem is

$$\mathbf{y}(t) = cVe^{-at/V} \begin{bmatrix} 1 \\ 1 \end{bmatrix} - \frac{cV}{2} e^{-2at/V} \begin{bmatrix} 1 \\ 0 \end{bmatrix}.$$

(d) $\mathbf{x}(t) = \mathbf{y}(t) + \hat{\mathbf{x}} = cVe^{-at/V} \begin{bmatrix} 1 \\ 1 \end{bmatrix} - \dfrac{cV}{2} e^{-2at/V} \begin{bmatrix} 1 \\ 0 \end{bmatrix} + \begin{bmatrix} cV/2 \\ 0 \end{bmatrix}$

10.4 Systems of Nonlinear Differential Equations

1. $\begin{array}{l} \dfrac{dx_1}{dt} = f_1(x_1, x_2) = 4x_1 - 2x_1x_2 \\ \dfrac{dx_2}{dt} = f_2(x_1, x_2) = -2x_2 + 8x_1x_2 \end{array} \Rightarrow \begin{array}{ll} \dfrac{\partial f_1(x_1,x_2)}{\partial x_1} = 4 - 2x_2 & \dfrac{\partial f_1(x_1,x_2)}{\partial x_2} = -2x_1 \\ \dfrac{\partial f_2(x_1,x_2)}{\partial x_1} = 8x_2 & \dfrac{\partial f_2(x_1,x_2)}{\partial x_2} = -2 + 8x_1 \end{array}$

When $\hat{x}_1 = \hat{x}_2 = 0$, the linearization of the nonlinear system as given by equation (12) is

$$\frac{d\boldsymbol{\epsilon}}{dt} = \begin{bmatrix} \dfrac{\partial f_1(0,0)}{\partial x_1} & \dfrac{\partial f_1(0,0)}{\partial x_2} \\ \dfrac{\partial f_2(0,0)}{\partial x_1} & \dfrac{\partial f_2(0,0)}{\partial x_2} \end{bmatrix} \boldsymbol{\epsilon} = \begin{bmatrix} 4 & 0 \\ 0 & -2 \end{bmatrix} \boldsymbol{\epsilon} \quad \text{where } \boldsymbol{\epsilon} = \begin{bmatrix} x_1 \\ x_2 \end{bmatrix} - \begin{bmatrix} \hat{x}_1 \\ \hat{x}_2 \end{bmatrix} = \begin{bmatrix} x_1 \\ x_2 \end{bmatrix}$$

3. $\begin{array}{l} \dfrac{dx_1}{dt} = f_1(x_1, x_2) = \sin x_1 + x_1x_2 + 3x_2^2 \\ \dfrac{dx_2}{dt} = f_2(x_1, x_2) = \cos x_2 - 1 + x_1(x_1 - 1) + 7x_2 \end{array} \Rightarrow \begin{array}{ll} \dfrac{\partial f_1(x_1,x_2)}{\partial x_1} = \cos x_1 + x_2 & \dfrac{\partial f_1(x_1,x_2)}{\partial x_2} = x_1 + 6x_2 \\ \dfrac{\partial f_2(x_1,x_2)}{\partial x_1} = 2x_1 - 1 & \dfrac{\partial f_2(x_1,x_2)}{\partial x_2} = -\sin x_2 + 7 \end{array}$

When $\hat{x}_1 = \hat{x}_2 = 0$, the linearization of the nonlinear system as given by equation (12) is

$$\frac{d\boldsymbol{\epsilon}}{dt} = \begin{bmatrix} \dfrac{\partial f_1(0,0)}{\partial x_1} & \dfrac{\partial f_1(0,0)}{\partial x_2} \\ \dfrac{\partial f_2(0,0)}{\partial x_1} & \dfrac{\partial f_2(0,0)}{\partial x_2} \end{bmatrix} \boldsymbol{\epsilon} = \begin{bmatrix} 1 & 0 \\ -1 & 7 \end{bmatrix} \boldsymbol{\epsilon} \quad \text{where } \boldsymbol{\epsilon} = \begin{bmatrix} x_1 \\ x_2 \end{bmatrix} - \begin{bmatrix} \hat{x}_1 \\ \hat{x}_2 \end{bmatrix} = \begin{bmatrix} x_1 \\ x_2 \end{bmatrix}$$

5. $\begin{array}{l} \dfrac{dx_1}{dt} = f_1(x_1, x_2) = 1 + x_1^3 - \dfrac{1 + x_1}{1 + x_2} \\ \dfrac{dx_2}{dt} = f_2(x_1, x_2) = 2x_2 + x_1^2 \end{array} \Rightarrow \begin{array}{ll} \dfrac{\partial f_1(x_1,x_2)}{\partial x_1} = 3x_1^2 - \dfrac{1}{1 + x_2} & \dfrac{\partial f_1(x_1,x_2)}{\partial x_2} = \dfrac{1 + x_1}{(1 + x_2)^2} \\ \dfrac{\partial f_2(x_1,x_2)}{\partial x_1} = 2x_1 & \dfrac{\partial f_2(x_1,x_2)}{\partial x_2} = 2 \end{array}$

When $\hat{x}_1 = \hat{x}_2 = 0$, the linearization of the nonlinear system as given by equation (12) is

$$\frac{d\boldsymbol{\epsilon}}{dt} = \begin{bmatrix} \dfrac{\partial f_1(0,0)}{\partial x_1} & \dfrac{\partial f_1(0,0)}{\partial x_2} \\ \dfrac{\partial f_2(0,0)}{\partial x_1} & \dfrac{\partial f_2(0,0)}{\partial x_2} \end{bmatrix} \boldsymbol{\epsilon} = \begin{bmatrix} -1 & 1 \\ 0 & 2 \end{bmatrix} \boldsymbol{\epsilon} \quad \text{where } \boldsymbol{\epsilon} = \begin{bmatrix} x_1 \\ x_2 \end{bmatrix} - \begin{bmatrix} \hat{x}_1 \\ \hat{x}_2 \end{bmatrix} = \begin{bmatrix} x_1 \\ x_2 \end{bmatrix}$$

© 2016 Cengage Learning. All Rights Reserved. May not be scanned, copied or duplicated, or posted to a publicly accessible website, in whole or in part.

7. The equilibria of the system $\left\{\begin{array}{l}\dfrac{dx_1}{dt}=f_1(x_1,x_2)=-5x_1+x_1x_2\\ \dfrac{dx_2}{dt}=f_2(x_1,x_2)=x_2-5x_1x_2\end{array}\right\}$ must satisfy $\left\{\begin{array}{l}0=-5x_1+x_1x_2\\ 0=x_2-5x_1x_2\end{array}\right\}\Rightarrow$

$\left\{\begin{array}{l}0=x_1(-5+x_2)\\ 0=x_2(1-5x_1)\end{array}\right\}$. Thus, the equilibria are (i) $\hat{x}_1=0,\ \hat{x}_2=0$ and (ii) $\hat{x}_1=\frac{1}{5},\ \hat{x}_2=5$. The Jacobian matrix of the

differential system is $J(x_1,x_2)=\begin{bmatrix}\dfrac{\partial f_1(x_1,x_2)}{\partial x_1} & \dfrac{\partial f_1(x_1,x_2)}{\partial x_2}\\ \dfrac{\partial f_2(x_1,x_2)}{\partial x_1} & \dfrac{\partial f_2(x_1,x_2)}{\partial x_2}\end{bmatrix}=\begin{bmatrix}-5+x_2 & x_1\\ -5x_2 & 1-5x_1\end{bmatrix}\Rightarrow J(0,0)=\begin{bmatrix}-5 & 0\\ 0 & 1\end{bmatrix}$

and $J\left(\frac{1}{5},5\right)=\begin{bmatrix}0 & \frac{1}{5}\\ -25 & 0\end{bmatrix}$. Hence, the linearization around equilibrium (i) is $\dfrac{d\epsilon}{dt}=\begin{bmatrix}-5 & 0\\ 0 & 1\end{bmatrix}\epsilon$ where $\epsilon=\begin{bmatrix}x_1\\ x_2\end{bmatrix}$ and the

linearization around equilibrium (ii) is $\dfrac{d\epsilon}{dt}=\begin{bmatrix}0 & \frac{1}{5}\\ -25 & 0\end{bmatrix}\epsilon$ where $\epsilon=\begin{bmatrix}x_1\\ x_2\end{bmatrix}-\begin{bmatrix}\frac{1}{5}\\ 5\end{bmatrix}$.

9. The equilibria of the system $\left\{\begin{array}{l}\dfrac{dx_1}{dt}=f_1(x_1,x_2)=x_1-6x_2^2+x_1x_2\\ \dfrac{dx_2}{dt}=f_2(x_1,x_2)=8x_1+4x_1x_2\end{array}\right\}$ must satisfy $\left\{\begin{array}{l}0=x_1-6x_2^2+x_1x_2\\ 0=8x_1+4x_1x_2\end{array}\right\}\Rightarrow$

$\left\{\begin{array}{l}0=x_1-6x_2^2+x_1x_2\\ 0=x_1(8+4x_2)\end{array}\right\}$. Thus, the equilibria are (i) $\hat{x}_1=0,\ \hat{x}_2=0$ and (ii) $\hat{x}_1=-24,\ \hat{x}_2=-2$. The Jacobian matrix of

the differential system is $J(x_1,x_2)=\begin{bmatrix}\dfrac{\partial f_1(x_1,x_2)}{\partial x_1} & \dfrac{\partial f_1(x_1,x_2)}{\partial x_2}\\ \dfrac{\partial f_2(x_1,x_2)}{\partial x_1} & \dfrac{\partial f_2(x_1,x_2)}{\partial x_2}\end{bmatrix}=\begin{bmatrix}1+x_2 & -12x_2+x_1\\ 8+4x_2 & 4x_1\end{bmatrix}\Rightarrow$

$J(0,0)=\begin{bmatrix}1 & 0\\ 8 & 0\end{bmatrix}$ and $J(-24,-2)=\begin{bmatrix}-1 & 0\\ 0 & -96\end{bmatrix}$. Hence, the linearization around equilibrium (i) is $\dfrac{d\epsilon}{dt}=\begin{bmatrix}1 & 0\\ 8 & 0\end{bmatrix}\epsilon$

where $\epsilon=\begin{bmatrix}x_1\\ x_2\end{bmatrix}$ and the linearization around equilibrium (ii) is $\dfrac{d\epsilon}{dt}=\begin{bmatrix}-1 & 0\\ 0 & -96\end{bmatrix}\epsilon$ where $\epsilon=\begin{bmatrix}x_1\\ x_2\end{bmatrix}-\begin{bmatrix}-24\\ -2\end{bmatrix}$.

11. The equilibria of the system $\left\{\begin{array}{l}\dfrac{dx_1}{dt}=f_1(x_1,x_2)=e^{-x_1}(x_1-x_2)\\ \dfrac{dx_2}{dt}=f_2(x_1,x_2)=x_1-x_2^2+2x_1x_2\end{array}\right\}$ must satisfy $\left\{\begin{array}{l}0=e^{-x_1}(x_1-x_2)\\ 0=x_1-x_2^2+2x_1x_2\end{array}\right\}$. Thus, the

equilibria are (i) $\hat{x}_1=0,\ \hat{x}_2=0$ and (ii) $\hat{x}_1=-1,\ \hat{x}_2=-1$. The Jacobian matrix of the differential system is

$J(x_1,x_2)=\begin{bmatrix}\dfrac{\partial f_1(x_1,x_2)}{\partial x_1} & \dfrac{\partial f_1(x_1,x_2)}{\partial x_2}\\ \dfrac{\partial f_2(x_1,x_2)}{\partial x_1} & \dfrac{\partial f_2(x_1,x_2)}{\partial x_2}\end{bmatrix}=\begin{bmatrix}e^{-x_1}-e^{-x_1}x_1+e^{-x_1}x_2 & -e^{-x_1}\\ 1+2x_2 & -2x_2+2x_1\end{bmatrix}\Rightarrow$

$J(0,0)=\begin{bmatrix}1 & -1\\ 1 & 0\end{bmatrix}$, and $J(-1,-1)=\begin{bmatrix}e & -e\\ -1 & 0\end{bmatrix}$. Hence, the linearization around each equilibrium is:

(i) $\dfrac{d\epsilon}{dt}=\begin{bmatrix}1 & -1\\ 1 & 0\end{bmatrix}\epsilon$ where $\epsilon=\begin{bmatrix}x_1\\ x_2\end{bmatrix}$, and (ii) $\dfrac{d\epsilon}{dt}=\begin{bmatrix}e & -e\\ -1 & 0\end{bmatrix}\epsilon$ where $\epsilon=\begin{bmatrix}x_1\\ x_2\end{bmatrix}-\begin{bmatrix}-1\\ -1\end{bmatrix}$.

© 2016 Cengage Learning. All Rights Reserved. May not be scanned, copied or duplicated, or posted to a publicly accessible website, in whole or in part.

13. When $\widehat{x}_1 = 0$ and $\widehat{x}_2 = 2$, the Jacobian matrix is $J = \begin{bmatrix} -4 & 0 \\ 0 & 3 \end{bmatrix} \Rightarrow \det J = -12 < 0$ and $\operatorname{trace} J = -1 < 0$.

Also, when $\widehat{x}_1 = 2$ and $\widehat{x}_2 = -1$, the Jacobian matrix is $J = \begin{bmatrix} -2 & 0 \\ 0 & -3 \end{bmatrix} \Rightarrow \det J = 6 > 0$ and $\operatorname{trace} J = -5 < 0$.

Thus, by Theorem 16, equilibrium (i) is unstable and equilibrium (ii) is locally stable.

15. When $\widehat{x}_1 = 0$ and $\widehat{x}_2 = 0$, the Jacobian matrix is $J = \begin{bmatrix} 0 & 0 \\ 1 & -\sin 1 \end{bmatrix} \Rightarrow \lambda_1 = 0$ and $\lambda_2 = -\sin 1 \approx -0.84$. Also,

when $\widehat{x}_1 = 1$ and $\widehat{x}_2 = 1$, the Jacobian matrix is $J = \begin{bmatrix} 1 - \cos 1 & 0 \\ \cos 1 & -\sin 1 \end{bmatrix} \Rightarrow \lambda_1 = 1 - \cos 1 \approx 0.46$ and

$\lambda_2 = -\sin 1 \approx -0.84$. Thus, by Theorem 14, the analysis is inconclusive for equilibrium (i) and equilibrium (ii) is unstable.

17. When $\widehat{x}_1 = 0$ and $\widehat{x}_2 = 0$, the Jacobian matrix is $J = \begin{bmatrix} -\frac{1}{2} & -1 \\ 0 & -2 \end{bmatrix} \Rightarrow \det J = 1 > 0$ and $\operatorname{trace} J = -\frac{5}{2} < 0$.

Also, when $\widehat{x}_1 = -2$ and $\widehat{x}_2 = -1 - \sqrt{3}$, the Jacobian matrix is $J = \begin{bmatrix} -\dfrac{1}{1-\sqrt{3}} & -1 - \dfrac{2}{(1-\sqrt{3})^2} \\ -1-\sqrt{3} & 0 \end{bmatrix} \Rightarrow$

$\det J = -\left(1 + \sqrt{3}\right)\left[1 + 2/\left(1 - \sqrt{3}\right)^2\right] < 0$. Thus, by Theorem 16, equilibrium (i) is locally stable and (ii) is unstable.

19. The equilibria of the system $\left\{ \begin{array}{l} x' = f_1(x,y) = x(3 - x - y) \\ y' = f_2(x,y) = y(2 - x - y) \end{array} \right\}$ must satisfy $\left\{ \begin{array}{l} 0 = x(3 - x - y) \\ 0 = y(2 - x - y) \end{array} \right\}$. Thus, the equilibria are (i) $\hat{x} = 0$, $\hat{y} = 0$, (ii) $\hat{x} = 0$, $\hat{y} = 2$, and (iii) $\hat{x} = 3$, $\hat{y} = 0$. The Jacobian matrix of the differential system is

$$J(x,y) = \begin{bmatrix} \dfrac{\partial f_1(x,y)}{\partial x} & \dfrac{\partial f_1(x,y)}{\partial y} \\ \dfrac{\partial f_2(x,y)}{\partial x} & \dfrac{\partial f_2(x,y)}{\partial y} \end{bmatrix} = \begin{bmatrix} 3 - 2x - y & -x \\ -y & 2 - x - 2y \end{bmatrix} \Rightarrow J(0,0) = \begin{bmatrix} 3 & 0 \\ 0 & 2 \end{bmatrix}, J(0,2) = \begin{bmatrix} 1 & 0 \\ -2 & -2 \end{bmatrix}$$

and $J(3,0) = \begin{bmatrix} -3 & -3 \\ 0 & -1 \end{bmatrix}$. Now $\det J(0,0) = 6 > 0$, $\operatorname{trace} J(0,0) = 5 > 0$, $\det J(0,2) = -2 < 0$, $\det J(3,0) = 3 > 0$, $\operatorname{trace} J(3,0) = -4 < 0$. So equilibria (i) and (ii) are unstable and equilibrium (iii) is locally stable.

21. The equilibria of the system $\left\{ \begin{array}{l} n' = f_1(n,m) = n(1 - 2m) \\ m' = f_2(n,m) = m(2 - 2n - m) \end{array} \right\}$ must satisfy $\left\{ \begin{array}{l} 0 = n(1 - 2m) \\ 0 = m(2 - 2n - m) \end{array} \right\}$. Thus, the equilibria are (i) $\hat{n} = 0$, $\hat{m} = 0$, (ii) $\hat{n} = 0$, $\hat{m} = 2$ and (iii) $n = \frac{3}{4}$, $m = \frac{1}{2}$. The Jacobian matrix of the differential system is

$$J(n,m) = \begin{bmatrix} \dfrac{\partial f_1(n,m)}{\partial n} & \dfrac{\partial f_1(n,m)}{\partial m} \\ \dfrac{\partial f_2(n,m)}{\partial n} & \dfrac{\partial f_2(n,m)}{\partial m} \end{bmatrix} = \begin{bmatrix} 1 - 2m & -2n \\ -2m & 2 - 2n - 2m \end{bmatrix} \Rightarrow J(0,0) = \begin{bmatrix} 1 & 0 \\ 0 & 2 \end{bmatrix}, J(0,2) = \begin{bmatrix} -3 & 0 \\ -4 & -2 \end{bmatrix}$$

and $J\left(\frac{3}{4}, \frac{1}{2}\right) = \begin{bmatrix} 0 & -\frac{3}{2} \\ -1 & -\frac{1}{2} \end{bmatrix}$.

Now $\det J(0,0) = 2 > 0$, $\operatorname{trace} J(0,0) = 3 > 0$; $\det J(0,2) = 6 > 0$, $\operatorname{trace} J(0,2) = -5 < 0$; $\det J\left(\frac{3}{4}, \frac{1}{2}\right) = -\frac{3}{2} < 0$. Thus, equilibria (i) and (iii) are unstable and equilibrium (ii) is locally stable.

© 2016 Cengage Learning. All Rights Reserved. May not be scanned, copied or duplicated, or posted to a publicly accessible website, in whole or in part.

23. The equilibria of the system $\left\{\begin{array}{l} p' = f_1(p,q) = -p^2 + q - 1 \\ q' = f_2(p,q) = q(2-p-q) \end{array}\right\}$ must satisfy $\left\{\begin{array}{l} 0 = -p^2 + q - 1 \\ 0 = q(2-p-q) \end{array}\right\}$. The second equation is satisfied when $q = 0$, however, the first equation becomes $0 = -p^2 - 1$ which has no real solution. The second equation also specifies that $q = 2 - p$ which when substituted into the first equation gives $-p^2 - p + 1 = 0 \Rightarrow p = \dfrac{-1 \pm \sqrt{5}}{2}$. Thus, the equilibria are (i) $\hat{p} = -\frac{1}{2} + \frac{\sqrt{5}}{2}, \hat{q} = \frac{5}{2} - \frac{\sqrt{5}}{2}$ and (ii) $\hat{p} = -\frac{1}{2} - \frac{\sqrt{5}}{2}, \hat{q} = \frac{5}{2} + \frac{\sqrt{5}}{2}$. The Jacobian matrix of the differential system is $J(p,q) = \begin{bmatrix} \dfrac{\partial f_1(p,q)}{\partial p} & \dfrac{\partial f_1(p,q)}{\partial q} \\ \dfrac{\partial f_2(p,q)}{\partial p} & \dfrac{\partial f_2(p,q)}{\partial q} \end{bmatrix} = \begin{bmatrix} -2p & 1 \\ -q & 2-p-2q \end{bmatrix} \Rightarrow$

$J\left(-\frac{1}{2} + \frac{\sqrt{5}}{2}, \frac{5}{2} - \frac{\sqrt{5}}{2}\right) = \begin{bmatrix} 1-\sqrt{5} & 1 \\ -\frac{5}{2} + \frac{\sqrt{5}}{2} & -\frac{5}{2} + \frac{\sqrt{5}}{2} \end{bmatrix}$ and $J\left(-\frac{1}{2} - \frac{\sqrt{5}}{2}, \frac{5}{2} + \frac{\sqrt{5}}{2}\right) = \begin{bmatrix} 1+\sqrt{5} & 1 \\ -\frac{5}{2} - \frac{\sqrt{5}}{2} & -\frac{5}{2} - \frac{\sqrt{5}}{2} \end{bmatrix}$.

Now $\det J\left(-\frac{1}{2} + \frac{\sqrt{5}}{2}, \frac{5}{2} - \frac{\sqrt{5}}{2}\right) = \frac{5}{2}\left(\sqrt{5} - 1\right) > 0$, trace $J\left(-\frac{1}{2} + \frac{\sqrt{5}}{2}, \frac{5}{2} - \frac{\sqrt{5}}{2}\right) = -\frac{3}{2} - \frac{\sqrt{5}}{2} < 0$;

$\det J\left(-\frac{1}{2} - \frac{\sqrt{5}}{2}, \frac{5}{2} + \frac{\sqrt{5}}{2}\right) = -\frac{5}{2}\left(1 + \sqrt{5}\right) < 0$. Thus, equilibrium (i) is locally stable and equilibrium (ii) is unstable.

25. The equilibria of the system $\left\{\begin{array}{l} x' = f_1(x,y) = ax^2 + ay - x \\ y' = f_2(x,y) = x - y \end{array}\right\}$ must satisfy $\left\{\begin{array}{l} 0 = ax^2 + ay - x \\ 0 = x - y \end{array}\right\}$. The second equation specifies that $y = x$. Substituting this into the first equation gives $ax^2 + ax - x = 0 \Rightarrow x(ax + a - 1) = 0$. Thus, the equilibria are (i) $\hat{x} = 0, \hat{y} = 0$ and (ii) $\hat{x} = \dfrac{1-a}{a}, \hat{y} = \dfrac{1-a}{a}$. The Jacobian matrix of the differential system is

$J(x,y) = \begin{bmatrix} \dfrac{\partial f_1(x,y)}{\partial x} & \dfrac{\partial f_1(x,y)}{\partial y} \\ \dfrac{\partial f_2(x,y)}{\partial x} & \dfrac{\partial f_2(x,y)}{\partial y} \end{bmatrix} = \begin{bmatrix} 2ax-1 & a \\ 1 & -1 \end{bmatrix} \Rightarrow J(0,0) = \begin{bmatrix} -1 & a \\ 1 & -1 \end{bmatrix}$ and

$J\left(\dfrac{1-a}{a}, \dfrac{1-a}{a}\right) = \begin{bmatrix} 1-2a & a \\ 1 & -1 \end{bmatrix}$. Now $\det J(0,0) = 1 - a$ and trace $J(0,0) = -2 < 0$, so by Theorem 16, equilibrium (i) is locally stable when $a < 1$. Also, $\det J\left(\dfrac{1-a}{a}, \dfrac{1-a}{a}\right) = a - 1$ and trace $J\left(\dfrac{1-a}{a}, \dfrac{1-a}{a}\right) = -2a$, so equilibrium (ii) is locally stable when $a > 1$. Note that the stability analysis is inconclusive for both equilibria when $a = 1$.

27. (a) $\left\{\begin{array}{l} \dfrac{dp_1}{dt} = f_1(p_1,p_2) = 5p_1(1-p_1) - 3p_1 = p_1(2-5p_1) \\ \dfrac{dp_2}{dt} = f_2(p_1,p_2) = 30p_2(1-p_1-p_2) - 3p_2 - 5p_1p_2 = p_2(27 - 35p_1 - 30p_2) \end{array}\right\}$ $\begin{bmatrix} m_1 = m_2 = 3, \\ c_1 = 5,\ c_2 = 30 \end{bmatrix}$

The equilibria must satisfy $\left\{\begin{array}{l} 0 = p_1(2-5p_1) \\ 0 = p_2(27 - 35p_1 - 30p_2) \end{array}\right\}$ which has solutions (i) $\hat{p}_1 = 0, \hat{p}_2 = 0$, (ii) $\hat{p}_1 = 0, \hat{p}_2 = \frac{9}{10}$, (iii) $\hat{p}_1 = \frac{2}{5}, \hat{p}_2 = 0$ and (iv) $\hat{p}_1 = \frac{2}{5}, \hat{p}_2 = \frac{13}{30}$.

(b) $J(p_1,p_2) = \begin{bmatrix} \dfrac{\partial f_1(p_1,p_2)}{\partial p_1} & \dfrac{\partial f_1(p_1,p_2)}{\partial p_2} \\ \dfrac{\partial f_2(p_1,p_2)}{\partial p_1} & \dfrac{\partial f_2(p_1,p_2)}{\partial p_2} \end{bmatrix} = \begin{bmatrix} 2-10p_1 & 0 \\ -35p_2 & 27 - 35p_1 - 60p_2 \end{bmatrix}$

© 2016 Cengage Learning. All Rights Reserved. May not be scanned, copied or duplicated, or posted to a publicly accessible website, in whole or in part.

(c) $J(0,0) = \begin{bmatrix} 2 & 0 \\ 0 & 27 \end{bmatrix} \Rightarrow \det J(0,0) = 54 > 0$ and trace $J(0,0) = 29 > 0$. Therefore $(0,0)$ is an unstable equilibrium.

$J\left(0, \frac{9}{10}\right) = \begin{bmatrix} 2 & 0 \\ -\frac{63}{2} & -27 \end{bmatrix} \Rightarrow \det J\left(0, \frac{9}{10}\right) = -54 < 0$. Therefore $\left(0, \frac{9}{10}\right)$ is an unstable equilibrium.

$J\left(\frac{2}{5}, 0\right) = \begin{bmatrix} -2 & 0 \\ 0 & 13 \end{bmatrix} \Rightarrow \det J\left(\frac{2}{5}, 0\right) = -26 < 0$. Therefore $\left(\frac{2}{5}, 0\right)$ is an unstable equilibrium.

$J\left(\frac{2}{5}, \frac{13}{30}\right) = \begin{bmatrix} -2 & 0 \\ -\frac{91}{6} & -13 \end{bmatrix} \Rightarrow \det J\left(\frac{2}{5}, \frac{13}{30}\right) = 26 > 0$ and trace $J\left(\frac{2}{5}, \frac{13}{30}\right) = -15 < 0$. Therefore, $\left(\frac{2}{5}, \frac{13}{30}\right)$ is locally stable.

(d) $\hat{p}_1 = \frac{2}{5}, \hat{p}_2 = \frac{13}{30}$ is the only locally stable equilibrium. Thus, the species are predicted to coexist with species 1 occupying 40% of patches and species 2 occupying approximately 43% of patches.

29. $R' = f_1(R,C) = 2 - RC \qquad C' = f_2(R,C) = RC - C \qquad [f(R) = 2, g(R,C) = RC, h(C) = C, \text{and } \epsilon = 1]$

The equilibria must satisfy $2 - RC = 0$ and $RC - C = 0$. The first equation requires that $RC = 2$ and substituting this into the second equation gives $2 - C = 0 \Rightarrow C = 2$ and $R = \dfrac{2}{C} = 1$. Therefore, the only equilibrium is $\hat{R} = 1$, $\hat{C} = 2$. Now, the Jacobian matrix is $J(R,C) = \begin{bmatrix} \dfrac{\partial f_1(R,C)}{\partial R} & \dfrac{\partial f_1(R,C)}{\partial C} \\ \dfrac{\partial f_2(R,C)}{\partial R} & \dfrac{\partial f_2(R,C)}{\partial C} \end{bmatrix} = \begin{bmatrix} -C & -R \\ C & R-1 \end{bmatrix} \Rightarrow$

$J(1,2) = \begin{bmatrix} -2 & -1 \\ 2 & 0 \end{bmatrix} \Rightarrow \det J(1,2) = 2 > 0$ and trace $J(1,2) = -2 < 0$. Therefore, by Theorem 16, the equilibrium is locally stable.

31. $R' = 2R(1 - R/5) - RC \qquad C' = RC - C \qquad [f(R) = 2R(1 - R/5), g(R,C) = RC, h(C) = C, \text{and } \epsilon = 1]$

The equilibria must satisfy $\left\{\begin{array}{l} 0 = R[2(1 - R/5) - C] \\ 0 = C(R-1) \end{array}\right\}$ which has the obvious solution (i) $\hat{R} = 0, \hat{C} = 0$. Substituting $C = 0$ into the first equation gives $R[2(1 - R/5)] = 0$ which is also satisfied when $R = 5$. Therefore the second equilibrium is (ii) $\hat{R} = 5, \hat{C} = 0$. The second equation is also satisfied when $R = 1$ and substituting this into the first equation gives $2\left(1 - \frac{1}{5}\right) - C = 0 \Rightarrow C = \frac{8}{5}$. Thus, the third equilibrium is (iii) $\hat{R} = 1, \hat{C} = \frac{8}{5}$. Now, the Jacobian matrix is

$$J(R,C) = \begin{bmatrix} \dfrac{\partial f_1(R,C)}{\partial R} & \dfrac{\partial f_1(R,C)}{\partial C} \\ \dfrac{\partial f_2(R,C)}{\partial R} & \dfrac{\partial f_2(R,C)}{\partial C} \end{bmatrix} = \begin{bmatrix} 2(1 - 2R/5) - C & -R \\ C & R-1 \end{bmatrix} \Rightarrow$$

$J(0,0) = \begin{bmatrix} 2 & 0 \\ 0 & -1 \end{bmatrix} \Rightarrow \det J(0,0) = -2 < 0$. Therefore $(0,0)$ is an unstable equilibrium.

$J(5,0) = \begin{bmatrix} -2 & -5 \\ 0 & 4 \end{bmatrix} \Rightarrow \det J(5,0) = -8 < 0$. Therefore $(5,0)$ is an unstable equilibrium.

$J\left(1, \frac{8}{5}\right) = \begin{bmatrix} -\frac{2}{5} & -1 \\ \frac{8}{5} & 0 \end{bmatrix} \Rightarrow \det J\left(1, \frac{8}{5}\right) = \frac{8}{5} > 0$ and trace $J\left(1, \frac{8}{5}\right) = -\frac{2}{5} < 0$. Therefore $\left(1, \frac{8}{5}\right)$ is locally stable.

© 2016 Cengage Learning. All Rights Reserved. May not be scanned, copied or duplicated, or posted to a publicly accessible website, in whole or in part.

33. (a) The differential equations dx/dt and dy/dt do not contain any terms with z. Thus, the rate of change of x and y are unaffected by z, so x and y can analyzed separately.

(b) The equilibria of the system $\left\{\begin{aligned} \frac{dx}{dt} &= f_1(x,y) = -k_f xyM + k_r(1-y)M \\ \frac{dy}{dt} &= f_2(x,y) = -k_f xyM + k_r(1-y)M + k_{cat}(1-y)M \end{aligned}\right\}$ must satisfy

$$\left\{\begin{aligned} 0 &= -k_f xyM + k_r(1-y)M \\ 0 &= -k_f xyM + k_r(1-y)M + k_{cat}(1-y)M \end{aligned}\right\} \Rightarrow \left\{\begin{aligned} 0 &= -k_f xyM + k_r(1-y)M \\ 0 &= -k_f xyM + k_r(1-y)M + k_{cat}(1-y)M \end{aligned}\right\}.$$

Subtracting the first equation from the second gives $k_{cat}(1-y)M = 0 \Rightarrow y = 1$ and substituting this into the first equation we find $x = 0$. Thus, the only equilibrium is (i) $\hat{x} = 0$, $\hat{y} = 1$.

(c) $$J(x,y) = \begin{bmatrix} \dfrac{\partial f_1(x,y)}{\partial x} & \dfrac{\partial f_1(x,y)}{\partial y} \\ \dfrac{\partial f_2(x,y)}{\partial x} & \dfrac{\partial f_2(x,y)}{\partial y} \end{bmatrix} = \begin{bmatrix} -k_f yM & -k_f xM - k_r M \\ -k_f yM & -k_f xM - k_r M - k_{cat}M \end{bmatrix}$$

$$= \begin{bmatrix} -k_f yM & -M(k_f x + k_r) \\ -k_f yM & -M(k_f x + k_r + k_{cat}) \end{bmatrix}$$

(d) $$J(0,1) = \begin{bmatrix} -k_f M & -Mk_r \\ -k_f M & -M(k_r + k_{cat}) \end{bmatrix} \Rightarrow \det J(0,1) = k_f M^2(k_r + k_{cat}) - k_f k_r M^2 = k_{cat} k_f M^2 > 0$$

since all constants are positive. Also, trace $J(0,1) = -k_f M - M(k_r + k_{cat}) = -[k_f + k_r + k_{cat}]M < 0$. Therefore, by Theorem 16, the equilibrium is locally stable.

35. $\dfrac{dv}{dt} = f_1(v,w) = v(v-a)(1-v) - w \qquad \dfrac{dw}{dt} = f_2(v,w) = \epsilon(v-w)$

(a) Substituting the $v = 0$ and $w = 0$ into the differential equations gives $\dfrac{dv}{dt} = 0$ and $\dfrac{dw}{dt} = 0$. Thus, the origin is an equilibrium.

(b) $$J(v,w) = \begin{bmatrix} \dfrac{\partial f_1(v,w)}{\partial v} & \dfrac{\partial f_1(v,w)}{\partial w} \\ \dfrac{\partial f_2(v,w)}{\partial v} & \dfrac{\partial f_2(v,w)}{\partial w} \end{bmatrix} = \begin{bmatrix} (v-a)(1-v) + v(1-v) - v(v-a) & -1 \\ \epsilon & -\epsilon \end{bmatrix}$$

$$= \begin{bmatrix} -3v^2 + 2(1+a)v - a & -1 \\ \epsilon & -\epsilon \end{bmatrix}$$

(c) $$J(0,0) = \begin{bmatrix} -a & -1 \\ \epsilon & -\epsilon \end{bmatrix} \Rightarrow \det J(0,0) = a\epsilon + \epsilon > 0 \quad \text{and} \quad \text{trace } J(0,0) = -(a+\epsilon) < 0$$

Therefore, by Theorem 16, the origin is locally stable for all positive values a and ϵ.

© 2016 Cengage Learning. All Rights Reserved. May not be scanned, copied or duplicated, or posted to a publicly accessible website, in whole or in part.

10 Review

TRUE-FALSE QUIZ

1. False, the term $2xy$ in dy/dt is a product of the two dependent variables, so the system is nonlinear.

3. True, the nullclines of the general two-variable system $\mathbf{x}' = A\mathbf{x}$ are $a_{11}x_1 + a_{12}x_2 = 0$ and $a_{21}x_1 + a_{22}x_2 = 0$. Both are straight lines.

5. True. A saddle equilibrium of the system $\mathbf{x}' = A\mathbf{x}$ must have $\det A < 0$, so the equilibrium is unstable by Theorem 10.2.16. Alternatively, this is true because there is always some initial condition for which the system moves away from the origin.

7. True, in fact, spirals have complex eigenvalues with nonzero real parts. If the eigenvalues are purely imaginary, then the solution curves will be centers.

9. False. In Section 10.4, there were several examples of nonlinear systems of differential equations that have multiple equilibria. For example, the Lotka-Volterra competition equations and the Kermack-McKendrick infectious disease model.

EXERCISES

1. $p' = 2q - 1, \quad q' = q^2 - q - p$ Nonlinear due to the q^2 term in the expression for q'.

3. $$\left\{\begin{array}{l} z' = tz - 2w \\ w' = z - w - 1 \end{array}\right\} \Leftrightarrow \frac{d}{dt}\begin{bmatrix} z \\ w \end{bmatrix} = \underbrace{\begin{bmatrix} t & -2 \\ 1 & -1 \end{bmatrix}}_{A(t)} \begin{bmatrix} z \\ w \end{bmatrix} + \underbrace{\begin{bmatrix} 0 \\ -1 \end{bmatrix}}_{\mathbf{g}}$$

Linear since the differential equations can be expressed as a linear matrix system.

5. First, we verify the initial condition $\mathbf{x}(0) = \begin{bmatrix} x_1(0) \\ x_2(0) \end{bmatrix} = \begin{bmatrix} \cos 0 - 3\sin 0 \\ 2\cos 0 - \sin 0 \end{bmatrix} = \begin{bmatrix} 1 \\ 2 \end{bmatrix} = \mathbf{x}_0$.

Now $\dfrac{d\mathbf{x}}{dt} = \dfrac{d}{dt}\begin{bmatrix} x_1(t) \\ x_2(t) \end{bmatrix} = \dfrac{d}{dt}\begin{bmatrix} \cos t - 3\sin t \\ 2\cos t - \sin t \end{bmatrix} = \begin{bmatrix} -\sin t - 3\cos t \\ -2\sin t - \cos t \end{bmatrix}$ and

$$A\mathbf{x} = \begin{bmatrix} 1 & -2 \\ 1 & -1 \end{bmatrix}\begin{bmatrix} \cos t - 3\sin t \\ 2\cos t - \sin t \end{bmatrix} = \begin{bmatrix} \cos t - 3\sin t - 4\cos t + 2\sin t \\ \cos t - 3\sin t - 2\cos t + \sin t \end{bmatrix} = \begin{bmatrix} -3\cos t - \sin t \\ -\cos t - 2\sin t \end{bmatrix} = \frac{d\mathbf{x}}{dt}.$$

7. First, we verify the initial condition $\mathbf{x}(0) = \begin{bmatrix} x_1(0) \\ x_2(0) \end{bmatrix} = \begin{bmatrix} 2e^0 - 0e^0 \\ e^0 \end{bmatrix} = \begin{bmatrix} 2 \\ 1 \end{bmatrix} = \mathbf{x}_0$.

Now $\dfrac{d\mathbf{x}}{dt} = \dfrac{d}{dt}\begin{bmatrix} x_1(t) \\ x_2(t) \end{bmatrix} = \dfrac{d}{dt}\begin{bmatrix} 2e^t - te^t \\ e^t \end{bmatrix} = \begin{bmatrix} 2e^t - e^t - te^t \\ e^t \end{bmatrix} = \begin{bmatrix} e^t - te^t \\ e^t \end{bmatrix}$ and

$$A\mathbf{x} = \begin{bmatrix} 1 & -1 \\ 0 & 1 \end{bmatrix}\begin{bmatrix} 2e^t - te^t \\ e^t \end{bmatrix} = \begin{bmatrix} 2e^t - te^t - e^t \\ e^t \end{bmatrix} = \begin{bmatrix} e^t - te^t \\ e^t \end{bmatrix} = \frac{d\mathbf{x}}{dt}.$$

© 2016 Cengage Learning. All Rights Reserved. May not be scanned, copied or duplicated, or posted to a publicly accessible website, in whole or in part.

9. Eigenvalues: $\det(A-\lambda I)=0 \;\Rightarrow\; \det\begin{bmatrix}0-\lambda & 1\\ -1 & 0-\lambda\end{bmatrix}=0 \;\Rightarrow\; \lambda^2+1=0 \;\Rightarrow\; \lambda=\underbrace{0}_{a}\pm i\underbrace{1}_{b}.$

The eigenvalues are therefore $\lambda_1=i$ and $\lambda_2=-i$.

Eigenvectors: $(A-\lambda_1 I)\,\mathbf{v}_1=\mathbf{0} \;\Rightarrow\; \begin{bmatrix}-i & 1\\ -1 & -i\end{bmatrix}\mathbf{v}_1=\begin{bmatrix}0\\0\end{bmatrix} \;\Rightarrow\; \mathbf{v}_1=\begin{bmatrix}1\\i\end{bmatrix}=\underbrace{\begin{bmatrix}1\\0\end{bmatrix}}_{\mathbf{u}}+i\underbrace{\begin{bmatrix}0\\1\end{bmatrix}}_{\mathbf{w}}$ and since complex eigenvectors come in conjugate pairs, we know that $\mathbf{v}_2=\begin{bmatrix}1\\0\end{bmatrix}-i\begin{bmatrix}0\\1\end{bmatrix}$.

Initial Condition: Using equation 10.2.14 to determine the constants k_1 and k_2 in the general solution 10.2.13, we find

$$\begin{bmatrix}| & |\\ \mathbf{u} & \mathbf{w}\\ | & |\end{bmatrix}\begin{bmatrix}k_1\\k_2\end{bmatrix}=\begin{bmatrix}x_1(0)\\x_2(0)\end{bmatrix} \;\Rightarrow\; \begin{bmatrix}1 & 0\\ 0 & 1\end{bmatrix}\begin{bmatrix}k_1\\k_2\end{bmatrix}=\begin{bmatrix}1\\1\end{bmatrix} \;\Rightarrow\; \begin{Bmatrix}k_1=1\\k_2=1\end{Bmatrix}$$

Thus, the solution as given by equation 10.2.13 is

$$\begin{aligned}\mathbf{x}(t)&=k_1e^{at}(\mathbf{u}\cos bt-\mathbf{w}\sin bt)+k_2e^{at}(\mathbf{w}\cos bt+\mathbf{u}\sin bt)\\ &=\left(\begin{bmatrix}1\\0\end{bmatrix}\cos t-\begin{bmatrix}0\\1\end{bmatrix}\sin t\right)+\left(\begin{bmatrix}0\\1\end{bmatrix}\cos t+\begin{bmatrix}1\\0\end{bmatrix}\sin t\right)=\begin{bmatrix}\cos t+\sin t\\ \cos t-\sin t\end{bmatrix}\end{aligned}$$

11. Eigenvalues: $\det(A-\lambda I)=0 \;\Rightarrow\; \det\begin{bmatrix}1-\lambda & 1\\ 0 & -2-\lambda\end{bmatrix}=0 \;\Rightarrow\; (1-\lambda)(-2-\lambda)=0.$

The eigenvalues are therefore $\lambda_1=1$ and $\lambda_2=-2$.

Eigenvectors: $(A-\lambda_1 I)\,\mathbf{v}_1=\mathbf{0} \;\Rightarrow\; \begin{bmatrix}0 & 1\\ 0 & -3\end{bmatrix}\mathbf{v}_1=\begin{bmatrix}0\\0\end{bmatrix} \;\Rightarrow\; \mathbf{v}_1=\begin{bmatrix}1\\0\end{bmatrix}$ and

$$(A-\lambda_2 I)\,\mathbf{v}_2=\mathbf{0} \;\Rightarrow\; \begin{bmatrix}3 & 1\\ 0 & 0\end{bmatrix}\mathbf{v}_2=\begin{bmatrix}0\\0\end{bmatrix} \;\Rightarrow\; \mathbf{v}_2=\begin{bmatrix}1\\-3\end{bmatrix}$$

The general solution given by equation 10.2.8 is $\mathbf{x}(t)=c_1e^{\lambda_1 t}\begin{bmatrix}|\\ \mathbf{v}_1\\ |\end{bmatrix}+c_2e^{\lambda_2 t}\begin{bmatrix}|\\ \mathbf{v}_2\\ |\end{bmatrix}=c_1e^{t}\begin{bmatrix}1\\0\end{bmatrix}+c_2e^{-2t}\begin{bmatrix}1\\-3\end{bmatrix}$.

Initial Condition: Using equation 10.2.9 to determine the constants c_1 and c_2 in the general solution, we find

$$\begin{bmatrix}| & |\\ \mathbf{v}_1 & \mathbf{v}_2\\ | & |\end{bmatrix}\begin{bmatrix}c_1\\c_2\end{bmatrix}=\begin{bmatrix}x_1(0)\\x_2(0)\end{bmatrix} \;\Rightarrow\; \begin{bmatrix}1 & 1\\ 0 & -3\end{bmatrix}\begin{bmatrix}c_1\\c_2\end{bmatrix}=\begin{bmatrix}1\\-2\end{bmatrix} \;\Rightarrow\; \begin{Bmatrix}c_1+c_2=1\\ -3c_2=-2\end{Bmatrix}$$

which is satisfied when $c_1=\frac{1}{3}$ and $c_2=\frac{2}{3}$. Thus, the solution is $\mathbf{x}(t)=\frac{1}{3}e^{t}\begin{bmatrix}1\\0\end{bmatrix}+\frac{2}{3}e^{-2t}\begin{bmatrix}1\\-3\end{bmatrix}$.

13. (a) Referring to the general two-compartment mixing model provided in the exercise instructions, we can write the initial value problem as

$$\begin{aligned}\frac{dx_1}{dt}&=rc-r\frac{x_1}{V_1}=(8)(2)-(8)\frac{x_1}{50}=16-\tfrac{4}{25}x_1\\ \frac{dx_2}{dt}&=r\frac{x_1}{V_1}-r\frac{x_2}{V_2}=(8)\frac{x_1}{50}-(8)\frac{x_2}{25}=\tfrac{4}{25}x_1-\tfrac{8}{25}x_2\end{aligned} \;\Rightarrow\; \frac{d}{dt}\begin{bmatrix}x_1\\x_2\end{bmatrix}=\underbrace{\begin{bmatrix}-\frac{4}{25} & 0\\ \frac{4}{25} & -\frac{8}{25}\end{bmatrix}}_{A}\underbrace{\begin{bmatrix}x_1\\x_2\end{bmatrix}}_{\mathbf{x}}+\begin{bmatrix}16\\0\end{bmatrix} \text{ and } \mathbf{x}(0)=\begin{bmatrix}8\\0\end{bmatrix}$$

© 2016 Cengage Learning. All Rights Reserved. May not be scanned, copied or duplicated, or posted to a publicly accessible website, in whole or in part.

(b) Before solving the IVP, we convert the the nonhomogeneous system to a homogeneous system. First, we find the equilibrium by setting $x_1' = x_2' = 0$ which results in the system $\begin{bmatrix} -\frac{4}{25} & 0 \\ \frac{4}{25} & -\frac{8}{25} \end{bmatrix} \begin{bmatrix} \hat{x}_1 \\ \hat{x}_2 \end{bmatrix} = -\begin{bmatrix} 16 \\ 0 \end{bmatrix}$. The solution of this system is $\hat{\mathbf{x}} = \begin{bmatrix} 100 \\ 50 \end{bmatrix}$. Now, let $\mathbf{y} = \mathbf{x} - \hat{\mathbf{x}} = \mathbf{x} - \begin{bmatrix} 100 \\ 50 \end{bmatrix}$. From Exercise 10.1.23, we know that $\mathbf{y}$ satisfies the homogeneous system $\frac{d\mathbf{y}}{dt} = A\mathbf{y}$.

Eigenvalues: $\det(A - \lambda I) = 0 \;\Rightarrow\; \det \begin{bmatrix} -\frac{4}{25} - \lambda & 0 \\ \frac{4}{25} & -\frac{8}{25} - \lambda \end{bmatrix} = 0 \;\Rightarrow\; \left(-\frac{4}{25} - \lambda\right)\left(-\frac{8}{25} - \lambda\right) = 0.$

The eigenvalues are therefore $\lambda_1 = -\frac{4}{25}$ and $\lambda_2 = -\frac{8}{25}$.

Eigenvectors: $(A - \lambda_1 I)\,\mathbf{v}_1 = \mathbf{0} \;\Rightarrow\; \begin{bmatrix} 0 & 0 \\ \frac{4}{25} & -\frac{4}{25} \end{bmatrix} \mathbf{v}_1 = \begin{bmatrix} 0 \\ 0 \end{bmatrix} \;\Rightarrow\; \mathbf{v}_1 = \begin{bmatrix} 1 \\ 1 \end{bmatrix}$ and

$$(A - \lambda_2 I)\,\mathbf{v}_2 = \mathbf{0} \;\Rightarrow\; \begin{bmatrix} \frac{4}{25} & 0 \\ \frac{4}{25} & 0 \end{bmatrix} \mathbf{v}_2 = \begin{bmatrix} 0 \\ 0 \end{bmatrix} \;\Rightarrow\; \mathbf{v}_2 = \begin{bmatrix} 0 \\ 1 \end{bmatrix}$$

The general solution given by equation 10.2.8 is $\mathbf{y}(t) = c_1 e^{\lambda_1 t} \begin{bmatrix} | \\ \mathbf{v}_1 \\ | \end{bmatrix} + c_2 e^{\lambda_2 t} \begin{bmatrix} | \\ \mathbf{v}_2 \\ | \end{bmatrix} = c_1 e^{-4t/25} \begin{bmatrix} 1 \\ 1 \end{bmatrix} + c_2 e^{-8t/25} \begin{bmatrix} 0 \\ 1 \end{bmatrix}$.

Expressed in terms of the original variables, the general solution is

$$\mathbf{x}(t) = \mathbf{y} + \begin{bmatrix} 100 \\ 50 \end{bmatrix} = c_1 e^{-4t/25} \begin{bmatrix} 1 \\ 1 \end{bmatrix} + c_2 e^{-8t/25} \begin{bmatrix} 0 \\ 1 \end{bmatrix} + \begin{bmatrix} 100 \\ 50 \end{bmatrix}.$$

Initial Condition: Setting $t = 0$ in the general solution gives $\mathbf{x}(0) = \begin{bmatrix} 8 \\ 0 \end{bmatrix} = c_1 \begin{bmatrix} 1 \\ 1 \end{bmatrix} + c_2 \begin{bmatrix} 0 \\ 1 \end{bmatrix} + \begin{bmatrix} 100 \\ 50 \end{bmatrix} \;\Rightarrow$

$\begin{bmatrix} 1 & 0 \\ 1 & 1 \end{bmatrix} \begin{bmatrix} c_1 \\ c_2 \end{bmatrix} = \begin{bmatrix} -92 \\ -50 \end{bmatrix}$ which is satisfied when $c_1 = -92$ and $c_2 = 42$. Therefore the solution to the initial value problem is $\begin{bmatrix} x_1(t) \\ x_2(t) \end{bmatrix} = -92 e^{-4t/25} \begin{bmatrix} 1 \\ 1 \end{bmatrix} + 42 e^{-8t/25} \begin{bmatrix} 0 \\ 1 \end{bmatrix} + \begin{bmatrix} 100 \\ 50 \end{bmatrix}$.

(c) The amount of salt in tank 2 at time t is given by $x_2(t) = -92e^{-4t/25} + 42e^{-8t/25} + 50$. Now $x_2' = \frac{368}{25} e^{-4t/25} - \frac{336}{25} e^{-8t/25} = \frac{1}{25} e^{-8t/25} \left(368 e^{4t/25} - 336\right)$ so $x_2' > 0$ for all $t \geq 0$. This means the amount of salt in tank 2 increases with time, however, in the long-run $\lim\limits_{t\to\infty} x_2(t) = \lim\limits_{t\to\infty} \left(-92e^{-4t/25} + 42e^{-8t/25} + 50\right) = 50$. So the maximum amount of salt in tank 2 is 50 g.

15. $\frac{dx_1}{dt} = ac + a\frac{x_2}{V} - 2a\frac{x_1}{V}$ $\quad \frac{dx_2}{dt} = ac - a\frac{x_2}{V}$ $\quad$ The term ac represents the rate at which CO_2 is added to both the shallow and deep compartments. The term $a\frac{x_2}{V}$ represents a flow of CO_2 from the deep to the shallow compartment, and $2a\frac{x_1}{V}$ is the rate of outflow of CO_2 from the shallow compartment. Thus, the equations can be viewed as a two compartment mixing problem in which there is a fixed amount of CO_2 added to each compartment, a flow of CO_2 from compartment 2 to 1, and CO_2 exits the system from compartment 1.

© 2016 Cengage Learning. All Rights Reserved. May not be scanned, copied or duplicated, or posted to a publicly accessible website, in whole or in part.

17. $\dfrac{dc}{dt} = -\dfrac{K}{V}c + ap \qquad \dfrac{dp}{dt} = -ap$ The term ap represents a flow of urea concentration from the pool to the blood and $\dfrac{K}{V}c$ is the rate of outflow of urea concentration from the blood. Thus, the equations can be viewed as a two compartment mixing problem in which urea flows in one direction, from the pool compartment to the blood compartment and urea leaves the system through the blood.

19. $J(p,q) = \begin{bmatrix} 0 & 2 \\ -1 & 2q-1 \end{bmatrix} \Rightarrow J\left(-\frac{1}{4},\frac{1}{2}\right) = \begin{bmatrix} 0 & 2 \\ -1 & 0 \end{bmatrix} \Rightarrow \det J\left(-\frac{1}{4},\frac{1}{2}\right) = 2 > 0$ and $\operatorname{trace} J\left(-\frac{1}{4},\frac{1}{2}\right) = 0$.

Therefore, the local stability analysis is inconclusive and we cannot tell whether the equilibrium is stable.

21. $J(z,w) = \begin{bmatrix} 3z^2 - 8z + 3 & -2 \\ 1 & -1 \end{bmatrix} \Rightarrow J(1,0) = \begin{bmatrix} -2 & -2 \\ 1 & -1 \end{bmatrix} \Rightarrow \det J(1,0) = 4 > 0$ and

$\operatorname{trace} J(1,0) = -3 < 0$. Therefore $\widehat{z} = 1$, $\widehat{w} = 0$ is locally stable.

23. The equilibria of the system $\left\{\begin{array}{l} x' = f_1(x,y) = 5x - xy \\ y' = f_2(x,y) = 4y - y^2 - 2xy \end{array}\right\}$ must satisfy $\left\{\begin{array}{l} 0 = x(5-y) \\ 0 = y(4-y-2x) \end{array}\right\}$. Thus, the equilibria are (i) $\hat{x} = 0, \hat{y} = 0$, (ii) $\hat{x} = 0, \hat{y} = 4$, and (iii) $\hat{x} = -\frac{1}{2}, \hat{y} = 5$. The Jacobian matrix of the differential system is

$$J(x,y) = \begin{bmatrix} \dfrac{\partial f_1(x,y)}{\partial x} & \dfrac{\partial f_1(x,y)}{\partial y} \\ \dfrac{\partial f_2(x,y)}{\partial x} & \dfrac{\partial f_2(x,y)}{\partial y} \end{bmatrix} = \begin{bmatrix} 5-y & -x \\ -2y & 4-2y-2x \end{bmatrix} \Rightarrow J(0,0) = \begin{bmatrix} 5 & 0 \\ 0 & 4 \end{bmatrix}, J(0,4) = \begin{bmatrix} 1 & 0 \\ -8 & -4 \end{bmatrix}$$

and $J\left(-\frac{1}{2},5\right) = \begin{bmatrix} 0 & \frac{1}{2} \\ -10 & -5 \end{bmatrix}$. Now $\det J(0,0) = 20 > 0$, $\operatorname{trace} J(0,0) = 9 > 0$; $\det J(0,4) = -4 < 0$;

$\det J\left(-\frac{1}{2},5\right) = 5 > 0$, $\operatorname{trace} J\left(-\frac{1}{2},5\right) = -5 < 0$. So equilibria (i) and (ii) are unstable and equilibrium (iii) is locally stable.

25. The equilibria of the system $\left\{\begin{array}{l} R' = f_1(R,W) = 8R(1-2R) - RW \\ W' = f_2(R,W) = -2W + 2RW \end{array}\right\}$ must satisfy $\left\{\begin{array}{l} 0 = R(8 - 16R - W) \\ 0 = W(-2+2R) \end{array}\right\}$. Thus, the equilibria are (i) $\hat{R} = 0, \hat{W} = 0$, (ii) $\hat{R} = \frac{1}{2}, \hat{W} = 0$, and (iii) $\hat{R} = 1, \hat{W} = -8$. The Jacobian matrix of the differential

system is $J(R,W) = \begin{bmatrix} \dfrac{\partial f_1(R,W)}{\partial R} & \dfrac{\partial f_1(R,W)}{\partial W} \\ \dfrac{\partial f_2(R,W)}{\partial R} & \dfrac{\partial f_2(R,W)}{\partial W} \end{bmatrix} = \begin{bmatrix} 8-32R-W & -R \\ 2W & -2+2R \end{bmatrix} \Rightarrow J(0,0) = \begin{bmatrix} 8 & 0 \\ 0 & -2 \end{bmatrix}$,

$J\left(\frac{1}{2},0\right) = \begin{bmatrix} -8 & -\frac{1}{2} \\ 0 & -1 \end{bmatrix}$, and $J(1,-8) = \begin{bmatrix} -16 & -1 \\ -16 & 0 \end{bmatrix}$. Now $\det J(0,0) = -16 < 0$; $\det J\left(\frac{1}{2},0\right) = 8 > 0$,

$\operatorname{trace} J\left(\frac{1}{2},0\right) = -9 < 0$; $\det J(1,-8) = -16 < 0$. So equilibria (i) and (iii) are unstable and equilibrium (ii) is locally stable.

27. $\dfrac{dN_1}{dt} = f_1(N_1,N_2) = N_1\left(1 - \dfrac{N_1 + \alpha N_2}{K_1}\right) \qquad \dfrac{dN_2}{dt} = f_2(N_1,N_2) = N_2\left(1 - \dfrac{N_2 + \beta N_1}{K_2}\right)$

(a) The equilibria must satisfy $N_1\left(1 - \dfrac{N_1 + \alpha N_2}{K_1}\right) = 0$ and $N_2\left(1 - \dfrac{N_2 + \beta N_1}{K_2}\right) = 0$. The first equation is satisfied when **(1)** $N_1 = 0$ and **(2)** $N_2 = \dfrac{K_1 - N_1}{\alpha}$, and the second equation is satisfied when **(3)** $N_2 = 0$ and

© 2016 Cengage Learning. All Rights Reserved. May not be scanned, copied or duplicated, or posted to a publicly accessible website, in whole or in part.

(4) $N_2 = K_2 - \beta N_1$. Equations (1) and (3) give the first equilibrium (i) $\hat{N}_1 = 0, \hat{N}_2 = 0$. Substituting (1) into (4) gives the second equilibrium (ii) $\hat{N}_1 = 0, \hat{N}_2 = K_2$. Substituting (3) into (2) gives $0 = -\frac{1}{\alpha}\hat{N}_1 + \frac{K_1}{\alpha}$, or $\hat{N}_1 = K_1$, so the third equilibrium is (iii) $\hat{N}_1 = K_1, \hat{N}_2 = 0$. Lastly, substituting (2) into (4) we get $-\frac{1}{\alpha}\hat{N}_1 + \frac{K_1}{\alpha} = -\beta\hat{N}_1 + K_2 \Rightarrow$ $(\alpha\beta - 1)\hat{N}_1 + K_1 = \alpha K_2 \Rightarrow \hat{N}_1 = \frac{\alpha K_2 - K_1}{\alpha\beta - 1}$ and $\hat{N}_2 = -\frac{1}{\alpha}\left(\frac{\alpha K_2 - K_1}{\alpha\beta - 1}\right) + \frac{K_1}{\alpha} = \frac{K_1\beta - K_2}{\alpha\beta - 1}$. Thus, the fourth equilibrium is (iv) $\hat{N}_1 = \frac{\alpha K_2 - K_1}{\alpha\beta - 1}, \hat{N}_2 = \frac{K_1\beta - K_2}{\alpha\beta - 1}$ assuming $\alpha\beta \neq 1$.

(b) $J(N_1, N_2) = \begin{bmatrix} \frac{\partial f_1(N_1,N_2)}{\partial N_1} & \frac{\partial f_1(N_1,N_2)}{\partial N_2} \\ \frac{\partial f_2(N_1,N_2)}{\partial N_1} & \frac{\partial f_2(N_1,N_2)}{\partial N_2} \end{bmatrix} = \begin{bmatrix} 1 - \frac{2N_1 + \alpha N_2}{K_1} & -\frac{\alpha N_1}{K_1} \\ -\frac{\beta N_2}{K_2} & 1 - \frac{2N_2 + \beta N_1}{K_2} \end{bmatrix}$

(c) Assuming that $K_1 > \alpha K_2$ and $K_2 < \beta K_1$, we can evaluate the Jacobian at each equilibrium to determine its stability.

$J(0,0) = \begin{bmatrix} 1 & 0 \\ 0 & 1 \end{bmatrix} \Rightarrow \det J(0,0) = 1 > 0$ and $\operatorname{trace} J(0,0) = 2 > 0$. Therefore, equilibrium (i) is unstable for all constant values.

$J(0, K_2) = \begin{bmatrix} 1 - \frac{\alpha K_2}{K_1} & 0 \\ -\beta & -1 \end{bmatrix} \Rightarrow \det J(0, K_2) = \frac{\alpha K_2}{K_1} - 1 < 0$. Therefore, equilibrium (ii) is unstable.

$J(K_1, 0) = \begin{bmatrix} -1 & -\alpha \\ 0 & 1 - \frac{\beta K_1}{K_2} \end{bmatrix} \Rightarrow \det J(K_1, 0) = \frac{\beta K_1}{K_2} - 1 > 0$ and $\operatorname{trace} J(K_1, 0) = -\frac{\beta K_1}{K_2} < 0$.

Therefore, equilibrium (iii) is locally stable.

Equilibrium (iv) is not biologically feasible in this case since $\hat{N}_1 < 0$ if $\alpha\beta > 1$ and $\hat{N}_2 < 0$ if $\alpha\beta < 1$.

(d) Assuming that $K_1 < \alpha K_2$ and $K_2 > \beta K_1$, we can determine the stability of each equilibrium as in part (c).

$\det J(0,0) = 1 > 0$ and $\operatorname{trace} J(0,0) = 2 > 0$. Therefore, equilibrium (i) is unstable for all constant values.

$\det J(0, K_2) = \frac{\alpha K_2}{K_1} - 1 > 0$ and $\operatorname{trace} J(0, K_2) = -\frac{\alpha K_2}{K_1} < 0$. Therefore, equilibrium (ii) is locally stable.

$\det J(K_1, 0) = \frac{\beta K_1}{K_2} - 1 < 0$. Therefore, equilibrium (iii) is unstable.

Equilibrium (iv) is not biologically feasible in this case since $\hat{N}_1 < 0$ if $\alpha\beta < 1$ and $\hat{N}_2 < 0$ if $\alpha\beta > 1$.

(e) Assume that $K_1 < \alpha K_2$ and $K_2 < \beta K_1 \Rightarrow \alpha K_2 < \alpha\beta K_1 \Rightarrow K_1 < \alpha\beta K_1 \Rightarrow \alpha\beta > 1$. We can determine the stability of each equilibrium as in part (c).

$\det J(0,0) = 1 > 0$ and $\operatorname{trace} J(0,0) = 2 > 0$. Therefore, equilibrium (i) is unstable for all constant values.

$\det J(0, K_2) = \frac{\alpha K_2}{K_1} - 1 > 0$ and $\operatorname{trace} J(0, K_2) = -\frac{\alpha K_2}{K_1} < 0$. Therefore, equilibrium (ii) is locally stable.

$\det J(K_1, 0) = \frac{\beta K_1}{K_2} - 1 > 0$ and $\operatorname{trace} J(K_1, 0) = -\frac{\beta K_1}{K_2} < 0$. Therefore, equilibrium (iii) is locally stable.

[continued]

© 2016 Cengage Learning. All Rights Reserved. May not be scanned, copied or duplicated, or posted to a publicly accessible website, in whole or in part.

$$J\left(\frac{\alpha K_2 - K_1}{\alpha\beta - 1}, \frac{K_1\beta - K_2}{\alpha\beta - 1}\right) = \begin{bmatrix} \dfrac{K_1 - \alpha K_2}{K_1(\alpha\beta - 1)} & -\dfrac{\alpha(\alpha K_2 - K_1)}{K_1(\alpha\beta - 1)} \\ -\dfrac{\beta(\beta K_1 - K_2)}{K_2(\alpha\beta - 1)} & \dfrac{K_2 - \beta K_1}{K_2(\alpha\beta - 1)} \end{bmatrix} \Rightarrow$$

$$\begin{aligned} \det J\left(\frac{\alpha K_2 - K_1}{\alpha\beta - 1}, \frac{K_1\beta - K_2}{\alpha\beta - 1}\right) &= \frac{(K_2 - \beta K_1)(K_1 - \alpha K_2)}{K_1K_2(\alpha\beta - 1)^2} - \frac{\alpha\beta(\beta K_1 - K_2)(\alpha K_2 - K_1)}{K_1K_2(\alpha\beta - 1)^2} \\ &= \frac{(K_2 - \beta K_1)(K_1 - \alpha K_2)}{K_1K_2(1 - \alpha\beta)} < 0. \end{aligned}$$

Therefore, equilibrium (iv) is unstable.

(f) Assume that $K_1 > \alpha K_2$ and $K_2 > \beta K_1 \Rightarrow \alpha K_2 > \alpha\beta K_1 \Rightarrow K_1 > \alpha\beta K_1 \Rightarrow \alpha\beta < 1$. We can determine the stability of each equilibrium as in part (e).

$\det J(0,0) = 1 > 0$ and trace $J(0,0) = 2 > 0$. Therefore, equilibrium (i) is unstable for all constant values.

$\det J(0, K_2) = \dfrac{\alpha K_2}{K_1} - 1 < 0$ and $\det J(K_1, 0) = \dfrac{\beta K_1}{K_2} - 1 < 0$. Therefore, equilibria (ii) and (iii) are unstable.

$\det J\left(\dfrac{\alpha K_2 - K_1}{\alpha\beta - 1}, \dfrac{K_1\beta - K_2}{\alpha\beta - 1}\right) = \dfrac{(K_2 - \beta K_1)(K_1 - \alpha K_2)}{K_1K_2(1 - \alpha\beta)} > 0$ and

$\operatorname{trace} J\left(\dfrac{\alpha K_2 - K_1}{\alpha\beta - 1}, \dfrac{K_1\beta - K_2}{\alpha\beta - 1}\right) = \dfrac{K_1 - \alpha K_2}{K_1(\alpha\beta - 1)} + \dfrac{K_2 - \beta K_1}{K_2(\alpha\beta - 1)} < 0$. Therefore, equilibrium (iv) is locally stable.

29. $\dfrac{dp_1}{dt} = f_1(p_1, p_2) = c_1p_1(h - p_1) - m_1p_1 \qquad \dfrac{dp_2}{dt} = f_2(p_1, p_2) = c_2p_2(h - p_1 - p_2) - m_2p_2 - c_1p_1p_2$

(a) The equilibria must satisfy $\left\{\begin{array}{l} 0 = p_1[c_1(h - p_1) - m_1] \\ 0 = p_2[c_2(h - p_1 - p_2) - m_2 - c_1p_1] \end{array}\right\}$. The first equation is satisfied when $p_1 = 0$ and $p_1 = h - \dfrac{m_1}{c_1}$. Substituting $p_1 = 0$ into the second equation gives $p_2[c_2(h - p_2) - m_2] = 0$ which has solutions $p_2 = 0$ and $p_2 = h - \dfrac{m_2}{c_2}$. Therefore, two equilibria are (i) $\hat{p}_1 = 0, \hat{p}_2 = 0$ and (ii) $\hat{p}_1 = 0, \hat{p}_2 = h - \dfrac{m_2}{c_2}$. Substituting $p_1 = h - \dfrac{m_1}{c_1}$ into the second equation and simplifying gives $p_2\left[c_2\dfrac{m_1}{c_1} - m_2 - c_1h + m_1 - c_2p_2\right] = 0$ which has solutions $p_2 = 0$ and $p_2 = \dfrac{m_1}{c_1} - \dfrac{m_2}{c_2} - \dfrac{c_1h}{c_2} + \dfrac{m_1}{c_2}$. So the other two equilibria are (iii) $\hat{p}_1 = h - \dfrac{m_1}{c_1}, \hat{p}_2 = 0$ and (iv) $\hat{p}_1 = h - \dfrac{m_1}{c_1}, \hat{p}_2 = \dfrac{m_1}{c_1} - \dfrac{m_2}{c_2} - \dfrac{c_1h}{c_2} + \dfrac{m_1}{c_2}$.

(b) $$J(p_1, p_2) = \begin{bmatrix} \dfrac{\partial f_1(p_1,p_2)}{\partial p_1} & \dfrac{\partial f_1(p_1,p_2)}{\partial p_2} \\ \dfrac{\partial f_2(p_1,p_2)}{\partial p_1} & \dfrac{\partial f_2(p_1,p_2)}{\partial p_2} \end{bmatrix} = \begin{bmatrix} c_1(h - 2p_1) - m_1 & 0 \\ -c_2p_2 - c_1p_2 & c_2(h - p_1 - 2p_2) - m_2 - c_1p_1 \end{bmatrix}$$

(c) The extinction equilibrium (i) $\hat{p}_1 = 0, \hat{p}_2 = 0$ is locally stable when $\det J > 0$ and trace $J > 0$. Evaluating the Jacobian matrix at the equilibrium gives $J(0,0) = \begin{bmatrix} hc_1 - m_1 & 0 \\ 0 & hc_2 - m_2 \end{bmatrix} \Rightarrow \det J(0,0) = (hc_1 - m_1)(hc_2 - m_2)$.

This is positive when both terms have the same sign, that is, when **(1)** $hc_1 > m_1$ and $hc_2 > m_2$ or when **(2)** $hc_1 < m_1$ and $hc_2 < m_2$. Also, trace $J(0,0) = (hc_1 - m_1) + (hc_2 - m_2)$. The trace is negative when conditions (2) are met. Therefore, the extinction equilibrium is stable when $hc_1 < m_1$ and $hc_2 < m_2$.

© 2016 Cengage Learning. All Rights Reserved. May not be scanned, copied or duplicated, or posted to a publicly accessible website, in whole or in part.

31. $\left\{\begin{aligned} \frac{df}{dt} &= f_1(f,s) = af\frac{f}{f+s} - gf(f+s) \\ \frac{ds}{dt} &= f_2(f,s) = r - gs(f+s) \end{aligned}\right\} \Rightarrow$

$$J(f,s) = \begin{bmatrix} \dfrac{\partial f_1(f,s)}{\partial f} & \dfrac{\partial f_1(f,s)}{\partial s} \\ \dfrac{\partial f_2(f,s)}{\partial f} & \dfrac{\partial f_2(f,s)}{\partial s} \end{bmatrix} = \begin{bmatrix} a\dfrac{f^2+2fs}{(f+s)^2} - g(2f+s) & -af\dfrac{f}{(f+s)^2} - gf \\ -gs & -g(f+2s) \end{bmatrix}$$

(a) When $r = 0$, substituting $f = a/g$ and $s = 0$ into the differential equations gives

$\dfrac{df}{dt} = a\left(\dfrac{a}{g}\right)\dfrac{a/g}{a/g+0} - g\left(\dfrac{a}{g}\right)\left(\dfrac{a}{g}+0\right) = 0$ and $\dfrac{ds}{dt} = -g(0)(a/g+0) = 0$. So $\widehat{f} = a/g$ and $\widehat{s} = 0$ is an equilibrium. To determine its stability, we evaluate the Jacobian matrix at the equilibrium values.

$J\left(\dfrac{a}{g},0\right) = \begin{bmatrix} -a & -2a \\ 0 & -a \end{bmatrix} \Rightarrow \det J\left(\dfrac{a}{g},0\right) = a > 0$ and $\operatorname{trace} J\left(\dfrac{a}{g},0\right) = -2a < 0$. Therefore, $\left(\dfrac{a}{g},0\right)$ is a locally stable equilibrium.

(b) When $r > 0$, substituting $f = 0$ and $s = \sqrt{r/g}$ into the differential equations gives

$\dfrac{df}{dt} = a(0)\dfrac{0}{0+\sqrt{r/g}} - g(0)(0+\sqrt{r/g}) = 0$ and $\dfrac{ds}{dt} = r - g\sqrt{r/g}(0+\sqrt{r/g}) = 0$. So $\widehat{f} = 0$ and $\widehat{s} = \sqrt{r/g}$ is an equilibrium. To determine its stability, we evaluate the Jacobian matrix at the equilibrium values.

$J\left(0,\sqrt{r/g}\right) = \begin{bmatrix} -g\sqrt{r/g} & 0 \\ -g\sqrt{r/g} & -2g\sqrt{r/g} \end{bmatrix} \Rightarrow \det J\left(\dfrac{a}{g},0\right) = 2gr > 0$ and $\operatorname{trace} J\left(\dfrac{a}{g},0\right) = -3g\sqrt{r/g} < 0$.

Therefore, $\left(0,\sqrt{r/g}\right)$ is a locally stable equilibrium.

© 2016 Cengage Learning. All Rights Reserved. May not be scanned, copied or duplicated, or posted to a publicly accessible website, in whole or in part.

11 □ DESCRIPTIVE STATISTICS

11.1 Numerical Descriptions of Data

1. Beak width is a length measurement, likely made in units of mm, and can take on a continuum of positive values. Hence it is numerical and continuous.

3. The variable "number of bristles" is measured using nonnegative integer values. Thus, it is numerical and discrete.

5. The variable "health care facility" has three categories which do not have a natural ordering. Hence, it is categorical and nominal.

7. Ordered data: 902, 920, 924, 932, 937, 939, 945, 948, 949, 951, 957, 958, 961, 965, 969, 970, 975, 982, 987, 991

$$\text{Median}=\frac{951+957}{2}=954,\quad Q_1=\frac{937+939}{2}=938,\quad Q_3=\frac{969+970}{2}=969.5\quad \text{IQR}=969.5-938=31.5$$

$$\bar{x}=\left[\begin{array}{l}902+920+924+932+937+939+945+948+949+951\\ \quad +957+958+961+965+969+970+975+982+987+991\end{array}\right]\Big/20=\frac{19062}{20}=953.1\text{ lbs}$$

$$\text{s.d.}=\sqrt{\frac{1}{20}\left[\begin{array}{l}(902-953.1)^2+(920-953.1)^2+(924-953.1)^2+(932-953.1)^2+(937-953.1)^2\\ \quad +(939-953.1)^2+(945-953.1)^2+(948-953.1)^2+(949-953.1)^2+(951-953.1)^2\\ \quad +(957-953.1)^2+(958-953.1)^2+(961-953.1)^2+(965-953.1)^2+(969-953.1)^2\\ \quad +(970-953.1)^2+(975-953.1)^2+(982-953.1)^2+(987-953.1)^2+(991-953.1)^2\end{array}\right]}$$

$$=\sqrt{10251.8/20}\approx 22.6405\text{ lbs}$$

9. Ordered data: 0, 0, 5, 5, 5, 12, 12, 15, 23, 24, 27, 34, 36, 36, 42, 43, 54, 56, 58, 62

$$\text{Median}=\frac{24+27}{2}=25.5,\quad Q_1=\frac{5+12}{2}=8.5,\quad Q_3=\frac{42+43}{2}=42.5\quad \text{IQR}=42.5-8.5=34$$

$$\bar{x}=\left[\begin{array}{l}0+0+5+5+5+12+12+15+23+24\\ \quad +27+34+36+36+42+43+54+56+58+62\end{array}\right]\Big/20=\frac{549}{20}=27.45$$

$$\text{s.d.}=\sqrt{\frac{1}{20}\left[\begin{array}{l}(0-27.45)^2+(0-27.45)^2+(5-27.45)^2+(5-27.45)^2+(5-27.45)^2\\ \quad +(12-27.45)^2+(12-27.45)^2+(15-27.45)^2+(23-27.45)^2+(24-27.45)^2\\ \quad +(27-27.45)^2+(34-27.45)^2+(36-27.45)^2+(36-27.45)^2+(42-27.45)^2\\ \quad +(43-27.45)^2+(54-27.45)^2+(56-27.45)^2+(58-27.45)^2+(62-27.45)^2\end{array}\right]}$$

$$=\sqrt{7972.95/20}\approx 19.966$$

11. Ordered Data: 0.05, 0.06, 0.07, 0.09, 0.12, 0.13, 0.16, 0.16, 0.19, 0.21, 0.21, 0.22, 0.23, 0.26, 0.32, 0.34, 0.35, 0.35, 0.45, 0.48

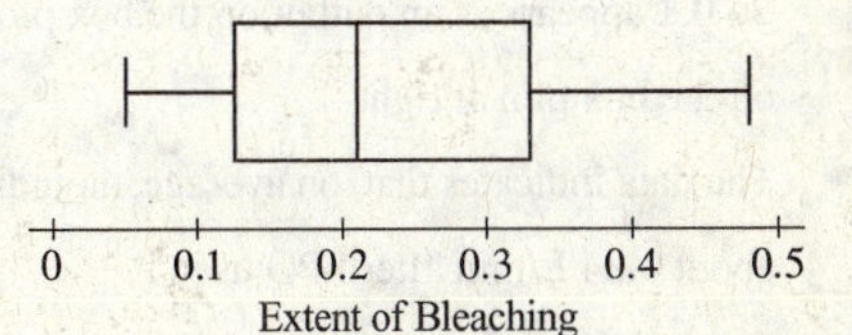

$$\text{Min}=0.05\quad Q_1=\frac{0.12+0.13}{2}=0.125\quad \text{Median}=\frac{0.21+0.21}{2}=0.21$$

$$Q_3=\frac{0.32+0.34}{2}=0.33\quad \text{Max}=0.48$$

© 2016 Cengage Learning. All Rights Reserved. May not be scanned, copied or duplicated, or posted to a publicly accessible website, in whole or in part.

13. Ordered Data: 0.2, 0.3, 0.4, 0.5, 0.9, 1.0, 1.1, 1.1, 1.2, 1.3

Min $= 0.2$ $\quad Q_1 = 0.4$ $\quad$ Median $= \dfrac{0.9 + 1.0}{2} = 0.95$ $\quad Q_3 = 1.1$ $\quad$ Max $= 1.3$

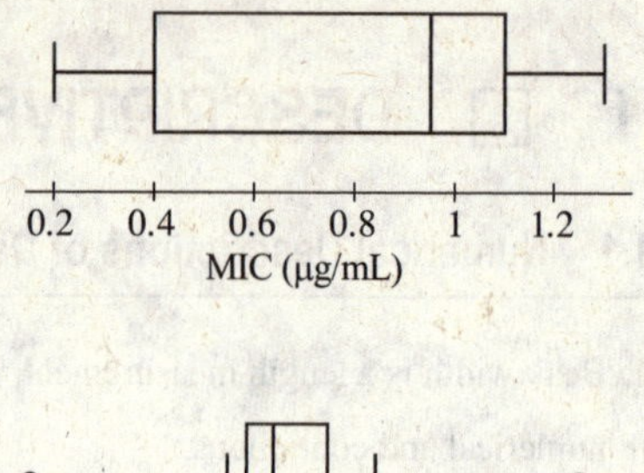

15. Ordered Data: 0.68, 1.13, 1.22, 1.22, 1.25, 1.25, 1.47, 1.94

Min $= 0.68$ $\quad Q_1 = \dfrac{1.13 + 1.22}{2} = 1.175$ $\quad$ Median $= \dfrac{1.22 + 1.25}{2} = 1.235$

$Q_3 = \dfrac{1.25 + 1.47}{2} = 1.36$ $\quad$ Max $= 1.94$

0.6 0.8 1 1.2 1.4 1.6 1.8 2
Birth Weight (kg)

Also, IQR $= 1.36 - 1.175 = 0.185$ and outliers must be outside the range

$[Q_1 - 1.5 \times \text{IQR}, Q_3 + 1.5 \times \text{IQR}] = [0.8975, 1.6375]$. So 0.68 and 1.94 appear as outliers on the box plot.

17. Ordered Data: 2.0, 2.3, 2.5, 2.9, 3.0, 3.1, 3.2, 3.3, 3.3,
3.5, 3.6, 3.6, 3.7, 4.2, 4.5, 4.5, 6.9, 7.3

Min $= 2.0$ $\quad Q_1 = 3.0$ $\quad$ Median $= \dfrac{3.3 + 3.5}{2} = 3.4$ $\quad Q_3 = 4.2$ $\quad$ Max $= 7.3$

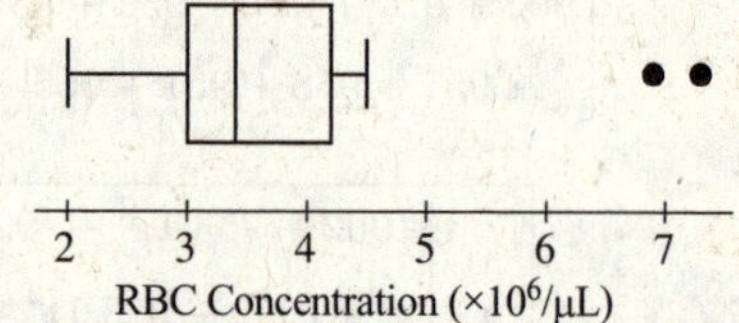

Also, IQR $= 4.2 - 3.0 = 1.2$ and outliers must be outside the range

$[Q_1 - 1.5 \times \text{IQR}, Q_3 + 1.5 \times \text{IQR}] = [1.2, 6.0]$. So 6.9 and 7.3 appear as outliers on the box plot.

19. Ordered data: 0.1, 0.2, 0.3, 0.3, 0.3, 0.3, 0.3, 0.3, 0.4, 0.4, 0.4, 0.5, 0.5, 0.5

$$\bar{x} = \left[\begin{array}{r} 0.1 + 0.2 + 0.3 + 0.3 + 0.3 + 0.3 + 0.3 \\ +0.3 + 0.4 + 0.4 + 0.4 + 0.5 + 0.5 + 0.5 \end{array}\right] \Big/ 14 = \frac{4.8}{14} \approx 0.3429 \text{ L/min}$$

$$\text{s.d.} = \sqrt{\frac{1}{14}\left[\begin{array}{r} \left(0.1 - \frac{4.8}{14}\right)^2 + \left(0.2 - \frac{4.8}{14}\right)^2 + \left(0.3 - \frac{4.8}{14}\right)^2 + \left(0.3 - \frac{4.8}{14}\right)^2 + \left(0.3 - \frac{4.8}{14}\right)^2 + \left(0.3 - \frac{4.8}{14}\right)^2 \\ + \left(0.3 - \frac{4.8}{14}\right)^2 + \left(0.3 - \frac{4.8}{14}\right)^2 + \left(0.4 - \frac{4.8}{14}\right)^2 + \left(0.4 - \frac{4.8}{14}\right)^2 + \left(0.4 - \frac{4.8}{14}\right)^2 + \left(0.5 - \frac{4.8}{14}\right)^2 \\ + \left(0.5 - \frac{4.8}{14}\right)^2 + \left(0.5 - \frac{4.8}{14}\right)^2 \end{array}\right]}$$

$$\approx \sqrt{0.174285714/14} \approx 0.1116 \text{ L/min}$$

Min $= 0.1$ $\quad Q_1 = 0.3$ $\quad$ Median $= \dfrac{0.3 + 0.3}{2} = 0.3$ $\quad Q_3 = 0.4$ $\quad$ Max $= 0.5$

Also, IQR $= 0.4 - 0.3 = 0.1$ and outliers must be outside the range $[Q_1 - 1.5 \times \text{IQR}, Q_3 + 1.5 \times \text{IQR}] = [0.15, 0.55]$.

So 0.1 appears as an outlier on the box plot. Note that the median and Q_1 coincide on the box plot at right.

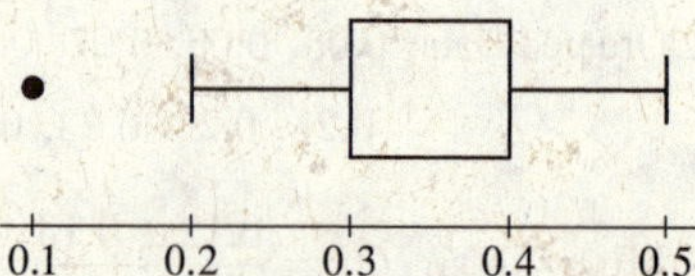

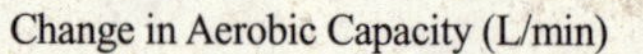

The data indicates that, on average, an individual's aerobic capacity increases by about 0.34 L/min after EPO use.

© 2016 Cengage Learning. All Rights Reserved. May not be scanned, copied or duplicated, or posted to a publicly accessible website, in whole or in part.

11.2 Graphical Descriptions of Data

1. **3.**

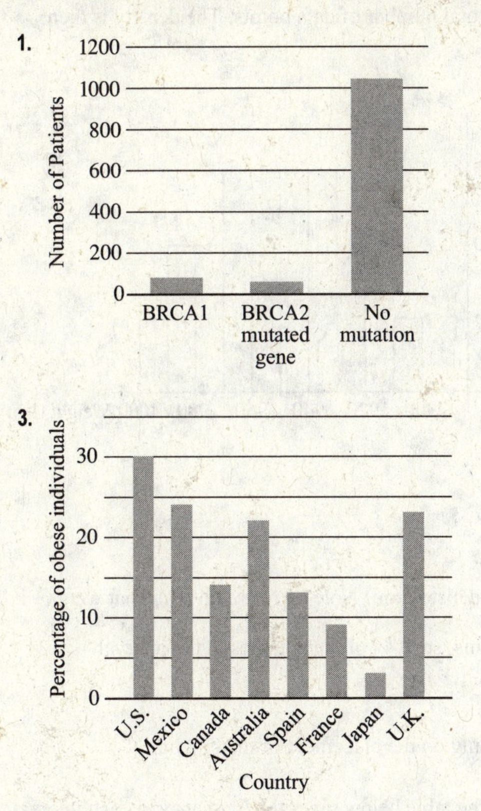

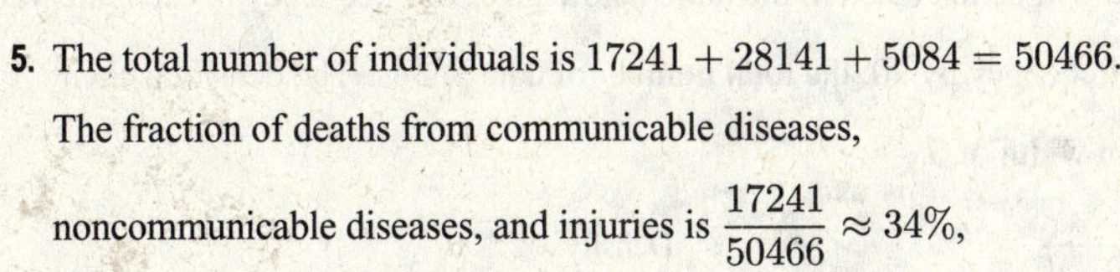

5. The total number of individuals is $17241 + 28141 + 5084 = 50466$. The fraction of deaths from communicable diseases, noncommunicable diseases, and injuries is $\frac{17241}{50466} \approx 34\%$, $\frac{28141}{50466} \approx 56\%$, and $\frac{5084}{50466} \approx 10\%$ respectively. We construct a pie chart with the area of each sector proportional to the percentages of each category.

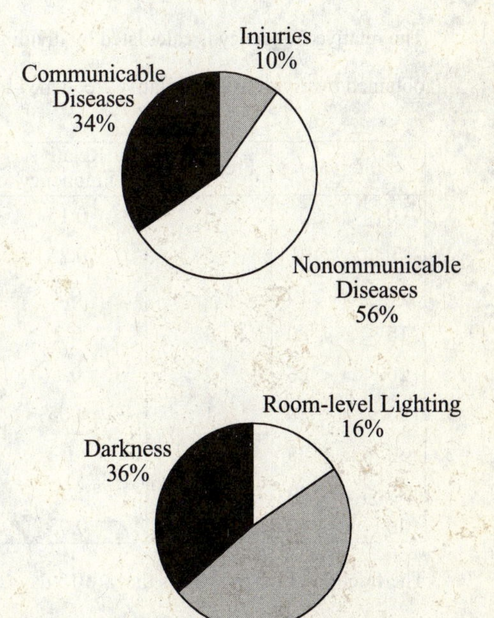

7. The total number of individuals is $172 + 232 + 75 = 479$. The fraction of children who sleep in darkness, sleep with a nightlight, and sleep with room-level lighting is $\frac{172}{479} \approx 36\%$, $\frac{232}{479} \approx 48\%$, and $\frac{75}{479} \approx 16\%$ respectively. We construct a pie chart with the area of each sector proportional to the percentages of each category.

© 2016 Cengage Learning. All Rights Reserved. May not be scanned, copied or duplicated, or posted to a publicly accessible website, in whole or in part.

9. Counting the number of data points that fall between the intervals listed in the table below gives the frequency in each interval. The relative frequency is calculated by dividing each frequency by 20, the total number of data points. The density is then obtained by dividing each relative frequency by the bin width of 10.

Interval	Frequency	Relative Frequency	Density
$900 \leq x < 910$	1	0.05	0.005
$910 \leq x < 920$	0	0	0
$920 \leq x < 930$	2	0.10	0.010
$930 \leq x < 940$	3	0.15	0.015
$940 \leq x < 950$	3	0.15	0.015
$950 \leq x < 960$	3	0.15	0.015
$960 \leq x < 970$	3	0.15	0.015
$970 \leq x < 980$	2	0.10	0.010
$980 \leq x < 990$	2	0.10	0.010
$990 \leq x < 1000$	1	0.05	0.005

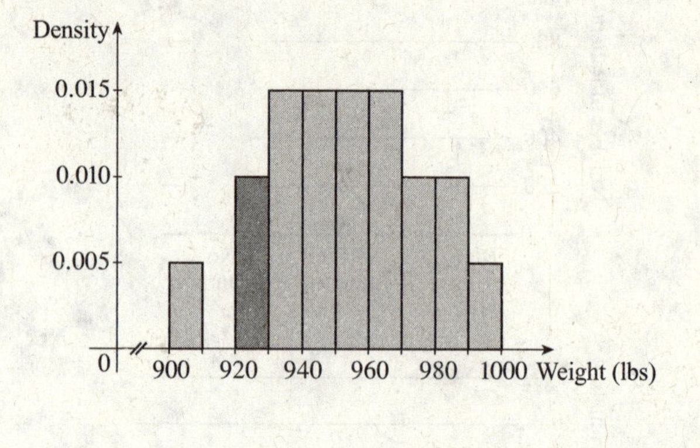

The fraction of seals weighing between 910 and 930 pounds is shaded in the histogram. Note there are no seals that weigh between 910 and 920 pounds. The shaded region amounts to $2/20$ data points, so 10% of the data lies in the interval $[910, 930)$.

Note: A student's histogram may differ slightly from that presented depending on the placement of the first bin.

11. Counting the number of data points that fall between the intervals listed in the table below gives the frequency in each interval. The relative frequency is calculated by dividing each frequency by 20, the total number of data points. The density is then obtained by dividing each relative frequency by the bin width of 5.

Interval	Frequency	Relative Frequency	Density
$0 \leq x < 5$	3	0.15	0.03
$5 \leq x < 10$	5	0.25	0.05
$10 \leq x < 15$	4	0.2	0.04
$15 \leq x < 20$	4	0.2	0.04
$20 \leq x < 25$	2	0.1	0.02
$25 \leq x < 30$	1	0.05	0.01
$30 \leq x < 35$	0	0	0
$35 \leq x < 40$	0	0	0
$40 \leq x < 45$	1	0.05	0.01

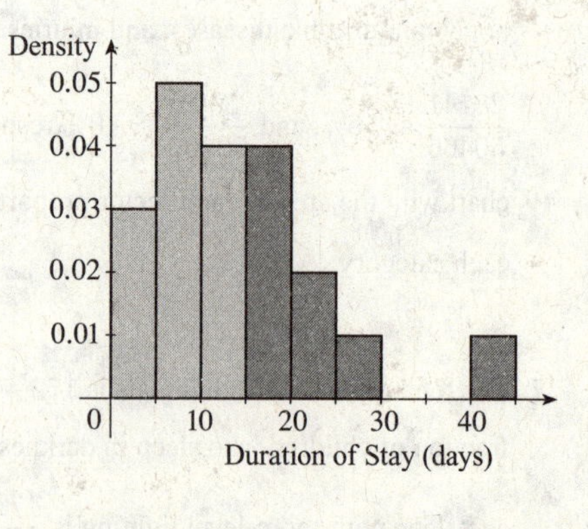

The fraction of hospital stays that are 15 days or more is shaded in the histogram. The shaded region amounts to $8/20$ data points, so 40% of the data lies in the interval $[15, 45)$.

Note: A student's histogram may differ slightly from that presented depending on the placement of the first bin.

© 2016 Cengage Learning. All Rights Reserved. May not be scanned, copied or duplicated, or posted to a publicly accessible website, in whole or in part.

13. Counting the number of data points that fall between the intervals listed in the table below gives the frequency in each interval. The relative frequency is calculated by dividing each frequency by 18, the total number of data points. The density is then obtained by dividing each relative frequency by the bin width of 0.5.

Interval	Frequency	Relative Frequency	Density
$2 \le x < 2.5$	2	0.11	0.22
$2.5 \le x < 3$	2	0.11	0.22
$3 \le x < 3.5$	5	0.28	0.56
$3.5 \le x < 4$	4	0.22	0.44
$4 \le x < 4.5$	1	0.056	0.11
$4.5 \le x < 5$	2	0.11	0.22
$5 \le x < 5.5$	0	0	0
$5.5 \le x < 6$	0	0	0
$6 \le x < 6.5$	0	0	0
$6.5 \le x < 7$	1	0.056	0.11
$7 \le x < 7.5$	1	0.056	0.11

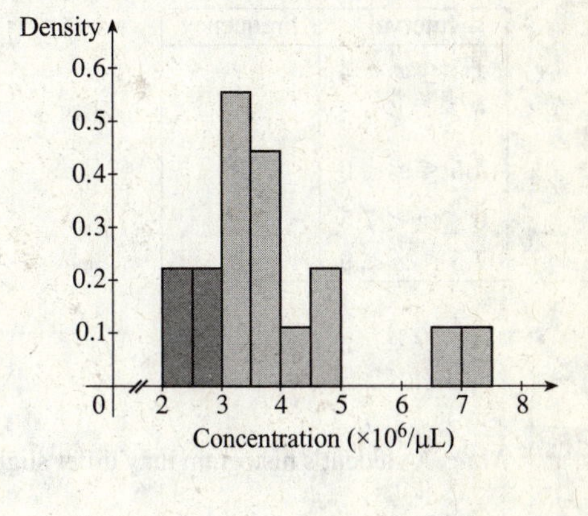

The fraction of mice having a concentration less than $3 \times 10^6/\mu\text{L}$ is shaded in the histogram. The shaded region amounts to $4/18$ data points, so approximately 22% of the data lies in the interval $\left[2 \times 10^6, 3 \times 10^6\right)$.

Note: A student's histogram may differ slightly from that presented depending on the placement of the first bin.

15. With 15 data points, Sturges' formula gives $1 + \ln(15)/\ln 2 \approx 4.9$. This suggests a good starting point is 5 bins which would require a bin width of $(71 - 45)/5 = 5.2$. However, we will opt for an integer bin width of 5 which requires 6 bins to span the data. Note that the choice of bin width is not unique and different bin widths will result in different histograms.

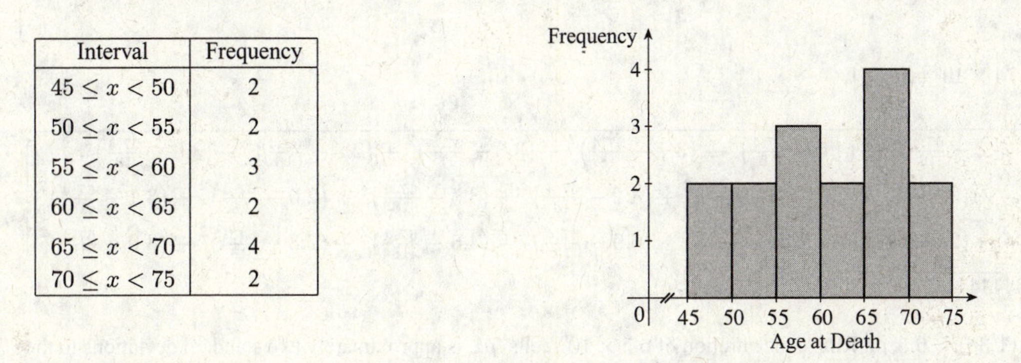

Interval	Frequency
$45 \le x < 50$	2
$50 \le x < 55$	2
$55 \le x < 60$	3
$60 \le x < 65$	2
$65 \le x < 70$	4
$70 \le x < 75$	2

Note: A student's histogram may differ slightly from that presented depending on the placement of the first bin and the choice of bin width.

© 2016 Cengage Learning. All Rights Reserved. May not be scanned, copied or duplicated, or posted to a publicly accessible website, in whole or in part.

17. With 15 data points, Sturges' formula gives $1 + \ln(15) / \ln 2 \approx 4.9$. This suggests a good starting point is 5 bins which would require a bin width of $(7.6 - 1.5)/5 = 1.22$. However, we will opt for a bin width of 1.5. Note that the choice of bin width is not unique and different bin widths will result in histograms with a different appearance.

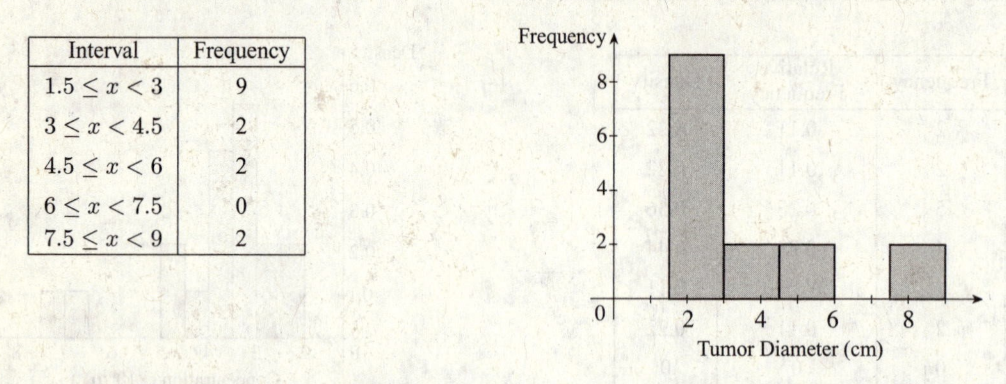

Interval	Frequency
$1.5 \le x < 3$	9
$3 \le x < 4.5$	2
$4.5 \le x < 6$	2
$6 \le x < 7.5$	0
$7.5 \le x < 9$	2

Note: A student's histogram may differ slightly from that presented depending on the placement of the first bin and the choice of bin width.

19. The standard normal curve $f(x) = \frac{1}{\sqrt{2\pi}} e^{-x^2/2}$ has critical points when $f'(x) = 0 \quad\Rightarrow\quad -\frac{1}{\sqrt{2\pi}} x e^{-x^2/2} = 0 \quad\Rightarrow\quad x = 0$. Now, $f'(x) > 0$ when $x < 0$ and $f'(x) < 0$ when $x > 0$, so $x = 0$ is a local maximum. Also, $\lim_{x \to \pm\infty} f(x) = 0$, so we conclude that $f(x)$ has an absolute maximum at $x = 0$.

21. The mean and standard deviation of the Weddel seal data (as calculated in Exercise 11.1.7) is $\bar{x} = 953.1$ lbs and s.d. ≈ 22.6405 lbs. Since $953.1 - 22.6 \approx 930$ and $953.1 + 22.6 \approx 975$, seal weights between 930 and 975 pounds lie approximately within one standard deviation of the mean. Thus, approximately 68.2% of the data lies in the interval $[930, 975]$.

23. $$\bar{x} = \frac{1}{18}\left[3.2 + 4.5 + 3.7 + 3.3 + 2.5 + 3.6 + 3.3 + 3.6 + 4.2 + 3.5 + 2.0 + 3.1 + 2.3 + 2.9 + 6.9 + 4.5 + 7.3 + 3.0\right]$$

$$= \frac{67.4}{18} \approx 3.74 \times 10^6 \text{ cells}/\mu\text{L}$$

$$\text{s.d.} = \sqrt{\frac{1}{18}\left[\begin{array}{l} \left(3.2 - \frac{67.4}{18}\right)^2 + \left(4.5 - \frac{67.4}{18}\right)^2 + \left(3.7 - \frac{67.4}{18}\right)^2 + \left(3.3 - \frac{67.4}{18}\right)^2 + \left(2.5 - \frac{67.4}{18}\right)^2 + \left(3.6 - \frac{67.4}{18}\right)^2 \\ + \left(3.3 - \frac{67.4}{18}\right)^2 + \left(3.6 - \frac{67.4}{18}\right)^2 + \left(4.2 - \frac{67.4}{18}\right)^2 + \left(3.5 - \frac{67.4}{18}\right)^2 + \left(2.0 - \frac{67.4}{18}\right)^2 + \left(3.1 - \frac{67.4}{18}\right)^2 \\ + \left(2.3 - \frac{67.4}{18}\right)^2 + \left(2.9 - \frac{67.4}{18}\right)^2 + \left(6.9 - \frac{67.4}{18}\right)^2 + \left(4.5 - \frac{67.4}{18}\right)^2 + \left(7.3 - \frac{67.4}{18}\right)^2 + \left(3.0 - \frac{67.4}{18}\right)^2 \end{array}\right]}$$

$$= \sqrt{33.10444444/18} \approx 1.356147 \times 10^6 \text{ cells}/\mu\text{L}$$

Since $3.74 + 2(1.36) \approx 6.5$, an RBC concentration of 6.5×10^6 cells $/\mu$L is approximately two standard deviations to the right of the mean. Referring to Figure 10, we see that when $x > \mu + 2\sigma$ the area under the normal curve is $100\% - (50\% + 34.1\% + 13.65\%) = 2.25\%$. Thus, approximately 2.25% of the data lies above 6.5×10^6 cells $/\mu$L.

© 2016 Cengage Learning. All Rights Reserved. May not be scanned, copied or duplicated, or posted to a publicly accessible website, in whole or in part.

25. The mean and standard deviation of the butterfly data (as calculated in Exercise 11.1.8) is

$\bar{x} \approx 10.5857\,\text{cm}$ and s.d. $\approx 0.8778\,\text{cm}$. Since $10.586 - 0.878 \approx 9.7$ and $10.586 + 2\,(0.878) \approx 12.3$, butterfly wingspans between 9.7 and 12.3 cm lie approximately within a band one standard deviation to the left of the mean and two standard deviations to the right of the mean. Referring to Figure 10, we see that when $\mu - \sigma < x < \mu + 2\sigma$ the area under the normal curve is $2\,(34.1\%) + 13.65\% = 81.85\%$. Thus, approximately 81.85% of the data lies between 9.7 and 12.3 cm.

27. $\bar{x} = \dfrac{1}{20}\,[6 + 10 + 5 + 12 + 5 + 3 + 41 + 12 + 6 + 15 + 20 + 4 + 3 + 7 + 13 + 16 + 18 + 29 + 23 + 17]$

$= \dfrac{265}{20} = 13.25$ days

So $\lambda = 1/\bar{x} = \frac{4}{53}$ and $f\,(x) = \frac{4}{53}e^{-4x/53}$. The area under the curve $f\,(x)$ in the interval $[20, \infty)$ is

$$\int_{20}^{\infty} f\,(x)\;dx = \lim_{t\to\infty}\int_{20}^{t}\tfrac{4}{53}e^{-4x/53}\,dx = -\lim_{t\to\infty}\left[e^{-4x/53}\right]_{20}^{t} = -\lim_{t\to\infty}\left[e^{-4t/53} - e^{-80/53}\right] = e^{-80/53} \approx 0.221.$$

Therefore, approximately 22% of stays last longer than 20 days.

29. If a lies between 45 and 75, then the area under the curve $f\,(x)$ in the interval $[45, a]$ is

$$\int_{45}^{a} f\,(x)\;dx = \int_{45}^{a}\tfrac{1}{30}\,dx = \tfrac{1}{30}\,x\Big]_{45}^{a} = \tfrac{1}{30}\,(a - 45)$$

If $a > 75$, then the area is

$$\int_{45}^{a} f\,(x)\;dx = \int_{45}^{75}\tfrac{1}{30}\,dx + \int_{75}^{a}(0)\;dx = \tfrac{1}{30}\,x\Big]_{45}^{75} = \tfrac{1}{30}\,(75 - 45) = 1$$

Therefore, the fraction of patients that died between ages 45 and a is $\begin{cases}\frac{1}{30}\,(a - 45) & \text{if } 45 \le a \le 75\\ 1 & \text{if } a > 75\end{cases}$.

11.3 Relationships Between Variables

1.

	Survived	Died	Row Total
Vaccinated	25	0	25
Unvaccinated	0	25	25
Column Total	25	25	50

It appears that vaccinated individuals always survive whereas unvaccinated individuals always die.

3.

	Developed Myopia	No Myopia	Row Total
Darkness	9	163	172
Night-light	31	201	232
Room lighting	48	27	75
Column Total	88	391	479

64% (48/75) of children who slept with room-level lighting developed myopia whereas children who slept in darkness or with a night-light developed myopia only about 5% (9/172) and 13% (31/232) of the time, respectively.

This suggests there is a relationship between exposure to ambient light and the likelihood of developing myopia.

© 2016 Cengage Learning. All Rights Reserved. May not be scanned, copied or duplicated, or posted to a publicly accessible website, in whole or in part.

5.

	Species X present	Species X not present	Row Total
Herbivore-dominated	9	3	12
Carnivore-dominated	4	11	15
Column Total	13	14	27

The data suggests that 75% (9/12) of herbivore-dominated communities have species X whereas only about 27% (4/15) of carnivore-dominated communities have species X. Thus, it appears that species X is a good indicator of a herbivore-dominated community.

7. Sorted Data (Before): 3.6, 4.0, 4.1, 4.2, 4.2, 4.3, 4.4, 4.6, 4.6, 4.8, 4.8, 5.0, 5.1, 5.2

Sorted Data (After): 4.0, 4.4, 4.5, 4.6, 4.6, 4.7, 4.7, 5.0, 5.1, 5.1, 5.1, 5.2, 5.3, 5.4

Before EPO: Min $= 3.6$ $Q_1 = 4.2$ Median $= \dfrac{4.4 + 4.6}{2} = 4.5$ $Q_3 = 4.8$ Max $= 5.2$

Also, IQR $= 4.8 - 4.2 = 0.6$ and outliers must be outside the range $[Q_1 - 1.5 \times \text{IQR}, Q_3 + 1.5 \times \text{IQR}] = [3.3, 5.7]$.

So there are no outliers in the before EPO data.

After EPO: Min $= 4.0$ $Q_1 = 4.6$ Median $= \dfrac{4.7 + 5.0}{2} = 4.85$ $Q_3 = 5.1$ Max $= 5.4$

Also, IQR $= 5.1 - 4.6 = 0.5$ and outliers must be outside the range $[Q_1 - 1.5 \times \text{IQR}, Q_3 + 1.5 \times \text{IQR}] = [3.85, 5.85]$.

So there are no outliers in the after EPO data.

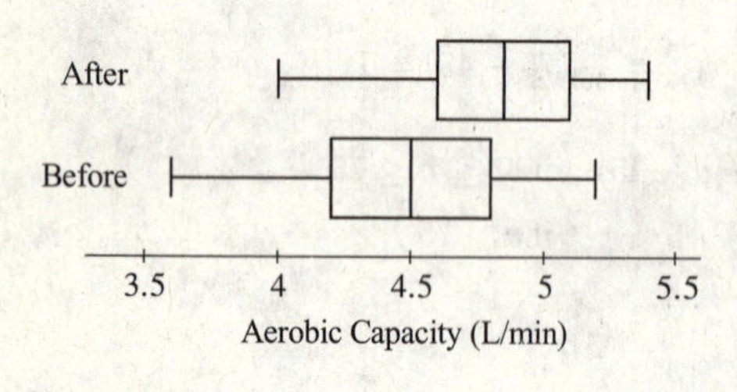

The distribution of the after-injection data is shifted toward slightly larger aerobic capacity values compared to the before-injection data. The median, and the first and third quartiles have all increased indicating EPO may have an impact on aerobic capacity. Note that the shift in the after-injection data is not a very large one, so we cannot conclude from the outlier plots alone that EPO use is linked to higher aerobic capacities. To draw a conclusion of this nature, we would need to use the methods of inferential statistics (see Chapter 13).

9. Sorted Data (Benthic): 1.6, 2.4, 2.6, 3.1, 3.1, 3.2, 3.2, 3.5, 3.5, 3.9, 4.3, 4.5, 4.5, 4.6, 5.5

Sorted Data (Limnetic): 0.4, 1.6, 1.8, 1.9, 2.1, 2.2, 2.2, 2.6, 3.0, 3.3, 3.5, 3.5, 3.6, 3.7, 4.2

Benthic Feeders Five-Number Summary: Min $= 1.6$ $Q_1 = 3.1$ Median $= 3.5$ $Q_3 = 4.5$ Max $= 5.5$

Also, IQR $= 4.5 - 3.1 = 1.4$ and outliers must be outside the range $[Q_1 - 1.5 \times \text{IQR}, Q_3 + 1.5 \times \text{IQR}] = [1.0, 6.6]$.

So there are no outliers in the benthic feeders data.

Limnetic Feeders Five-Number Summary: Min $= 0.4$ $Q_1 = 1.9$ Median $= 2.6$ $Q_3 = 3.5$ Max $= 4.2$

Also, IQR $= 3.5 - 1.9 = 1.6$ and outliers must be outside the range $[Q_1 - 1.5 \times \text{IQR}, Q_3 + 1.5 \times \text{IQR}] = [-0.5, 5.9]$.

So there are no outliers in the limnetic feeders data.

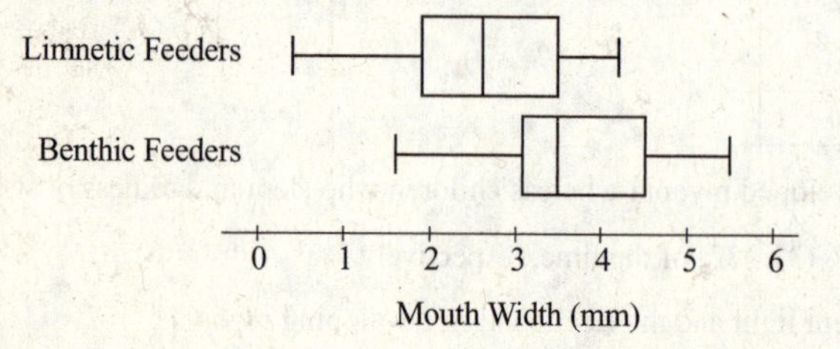

The distribution of the limnetic feeder data is shifted toward smaller mouth widths compared to the benthic feeder data. There is some overlap between the two data sets, though the five-number summary values are all smaller in the limnetic data.

© 2016 Cengage Learning. All Rights Reserved. May not be scanned, copied or duplicated, or posted to a publicly accessible website, in whole or in part.

11. Sorted Data (Females): 46, 48, 49, 50, 51, 53, 53, 54, 54, 55, 56, 58, 60

Sorted Data (Males): 29, 29, 30, 30, 31, 32, 34, 36, 41, 42, 43, 44, 46, 51, 52, 52, 55

Five # Summary (Female): Min $= 46$ $Q_1 = \dfrac{49+50}{2} = 49.5$ Median $= 53$ $Q_3 = \dfrac{55+56}{2} = 55.5$ Max $= 60$

Also, IQR $= 55.5 - 49.5 = 6$ and outliers must be outside the range $[Q_1 - 1.5 \times \text{IQR}, Q_3 + 1.5 \times \text{IQR}] = [40.5, 64.5]$. So there are no outliers in the female data.

Five # Summary (Male): Min $= 29$ $Q_1 = \dfrac{30+31}{2} = 30.5$ Median $= 41$ $Q_3 = \dfrac{46+51}{2} = 48.5$ Max $= 55$

Also, IQR $= 48.5 - 30.5 = 18$ and outliers must be outside the range $[Q_1 - 1.5 \times \text{IQR}, Q_3 + 1.5 \times \text{IQR}] = [3.5, 75.5]$. So there are no outliers in the male data.

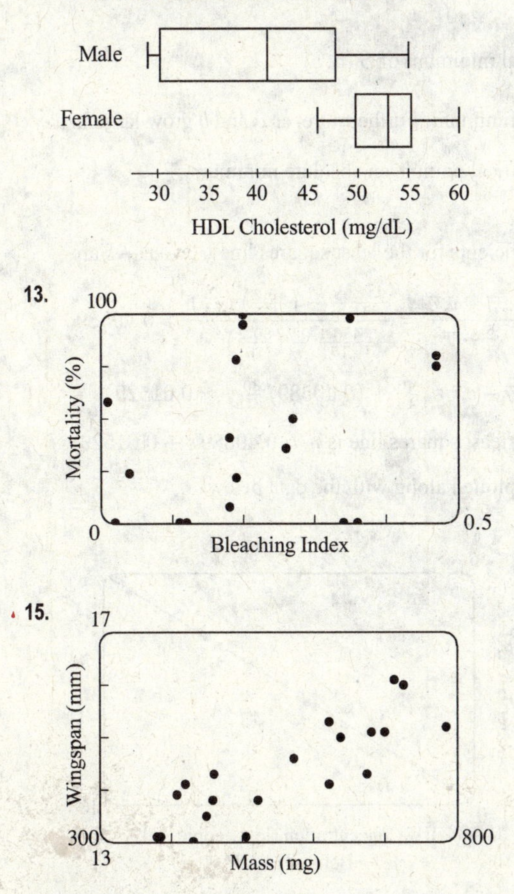

The distribution of the male data is shifted toward smaller HDL cholesterol levels compared to the female data. Thus, it appears that females tend to have higher levels of HDL cholesterol than males.

13. The data points in the scatterplot do not appear to exhibit a distinct pattern. Datapoints with a high bleaching index can have a high or a low mortality, and reefs with a low bleaching index can also have a high or a low mortality. Therefore, there is no apparent relationship between bleaching and mortality.

15. The scatterplot of the data shows a roughly positive linear relationship between mass and wingspan, that is, as mass increases wingspan increases. Thus, damselflies with larger masses tend to have larger wingspans.

17. (a) In deriving Definition 3, it was shown that $\dfrac{\partial S}{\partial a} = 2b\sum_{i=1}^n x_i + 2a\sum_{i=1}^n x_i^2 - 2\sum_{i=1}^n x_i y_i$ and $\dfrac{\partial S}{\partial b} = 2nb + 2a\sum_{i=1}^n x_i - 2\sum_{i=1}^n y_i$. The second partial derivates are $\dfrac{\partial^2 S}{\partial a^2} = 2\sum_{i=1}^n x_i^2$, $\dfrac{\partial^2 S}{\partial b\,\partial a} = 2\sum_{i=1}^n x_i$ and $\dfrac{\partial^2 S}{\partial b^2} = 2n$. Since the partial derivatives do not contain the independent variables a and b, we can immediately apply the second derivative test

[continued]

© 2016 Cengage Learning. All Rights Reserved. May not be scanned, copied or duplicated, or posted to a publicly accessible website, in whole or in part.

$$D = S_{aa}S_{bb} - [S_{ab}]^2 = \left(2\sum_{i=1}^{n} x_i^2\right)(2n) - \left(2\sum_{i=1}^{n} x_i\right)^2$$
$$= 4\left[n\sum_{i=1}^{n} x_i^2 - \left(\sum_{i=1}^{n} x_i\right)^2\right] = 4\left[n\sum_{i=1}^{n} x_i^2 - 2\left(\sum_{i=1}^{n} x_i\right)^2 + \left(\sum_{i=1}^{n} x_i\right)^2\right]$$

To verify that $D > 0$, we factor D using the relation $\bar{x} = \sum_{i=1}^{n} x_i/n$ as follows

$$D = 4\left[n\sum_{i=1}^{n} x_i^2 - 2(n\bar{x})\left(\sum_{i=1}^{n} x_i\right) + (n\bar{x})^2\right]$$
$$= 4n\left[\sum_{i=1}^{n} x_i^2 - 2\bar{x}\left(\sum_{i=1}^{n} x_i\right) + n\bar{x}^2\right] = 4n\left[\sum_{i=1}^{n} x_i^2 - 2\bar{x}\left(\sum_{i=1}^{n} x_i\right) + \sum_{i=1}^{n} \bar{x}^2\right]$$
$$= 4n\left[\sum_{i=1}^{n}\left(x_i^2 - 2\bar{x}x_i + \bar{x}^2\right)\right] = 4n\left[\sum_{i=1}^{n}(x_i - \bar{x})^2\right] > 0$$

Thus, $D > 0$ and $S_{aa} = 2\sum_{i=1}^{n} x_i^2 > 0$, so the critical point is a local minimum of $S(a, b)$.

(b) $S(a, b)$ is a continuous function and the only critical point is a local minimum. Furthermore, as a and b grow large in any direction, $S(a, b)$ also grows large. We therefore conclude that the critical point is an absolute minimum.

19.

Before x	After y	x^2	xy
7.40	3.70	54.76	27.38
5.10	2.60	26.01	13.26
6.90	3.40	47.61	23.46
7.20	3.60	51.84	25.92
1.40	0.70	1.96	0.98
4.30	2.10	18.49	9.03
5.10	2.60	26.01	13.26
2.90	1.50	8.41	4.35
7.20	3.60	51.84	25.92
3.50	1.70	12.25	5.95
4.70	2.30	22.09	10.81
4.70	2.40	22.09	11.28
9.30	4.60	86.49	42.78
4.50	2.20	20.25	9.90
6.00	3.00	36.00	18.00
$\bar{x} = \frac{80.2}{15}$	$\bar{y} = \frac{40}{15}$	$\overline{x^2} = \frac{486.1}{15}$	$\overline{xy} = \frac{242.28}{15}$

The coefficients for the least squares line $y = ax + b$ are

$$a = \frac{\overline{xy} - \bar{x}\,\bar{y}}{\overline{x^2} - \bar{x}^2} = \frac{\frac{242.28}{15} - \left(\frac{80.2}{15}\right)\left(\frac{40}{15}\right)}{\frac{486.1}{15} - \left(\frac{80.2}{15}\right)^2} \approx 0.49589$$

$$b = \bar{y} - a\bar{x} \approx \tfrac{40}{15} - (0.49589)\,\tfrac{80.2}{15} \approx 0.01529$$

Thus, the least squares line is $y = 0.49589x + 0.01529$ which is plotted along with the data below.

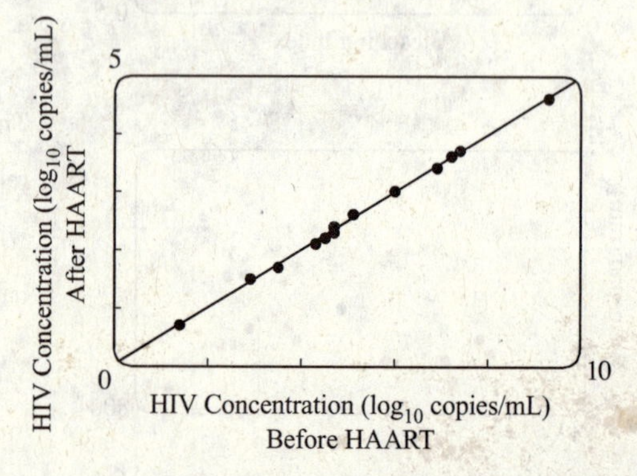

Individuals with high values of HIV concentration before HAART also tend to have high values after HAART and vice versa. An increase of 1 $\log_{10}$ copies/mL in before treatment HIV concentration leads to an increase of about 0.5 $\log_{10}$ copies/mL in after treatment HIV concentration.

© 2016 Cengage Learning. All Rights Reserved. May not be scanned, copied or duplicated, or posted to a publicly accessible website, in whole or in part.

21.

Bleaching Index x	Mortality y	x^2	xy
0.01	58	0.0001	0.58
0.02	0	0.0004	0
0.04	24	0.0016	0.96
0.11	0	0.0121	0
0.12	0	0.0144	0
0.18	8	0.0324	1.44
0.18	41	0.0324	7.38
0.19	78	0.0361	14.82
0.19	22	0.0361	4.18
0.20	95	0.0400	19
0.20	100	0.0400	20
0.26	36	0.0676	9.36
0.27	50	0.0729	13.5
0.34	0	0.1156	0
0.35	98	0.1225	34.3
0.36	0	0.1296	0
0.47	80	0.2209	37.6
0.47	75	0.2209	35.25
$\overline{x} = \frac{3.96}{18}$	$\overline{y} = \frac{765}{18}$	$\overline{x^2} = \frac{1.1956}{18}$	$\overline{xy} = \frac{198.37}{18}$

The coefficients for the least squares line $y = ax + b$ are

$$a = \frac{\overline{xy} - \overline{x}\,\overline{y}}{\overline{x^2} - \overline{x}^2} = \frac{\frac{198.37}{18} - \left(\frac{3.96}{18}\right)\left(\frac{765}{18}\right)}{\frac{1.1956}{18} - \left(\frac{3.96}{18}\right)^2} \approx 92.69420$$

$$b = \overline{y} - a\overline{x} \approx \tfrac{765}{18} - (92.69420)\,\tfrac{3.96}{18} \approx 22.10727$$

Thus, the least squares line is $y = 92.6942x + 22.10727$ which is plotted along with the data below.

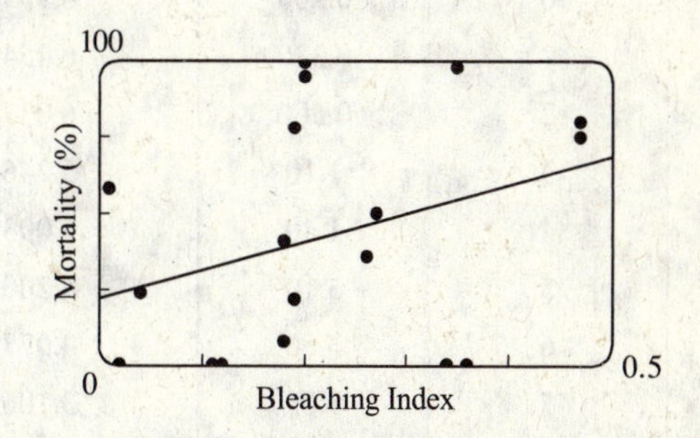

The data points are scattered in the plot and do not appear to exhibit a linear pattern. Thus, the least squares line is a poor approximation to the data from the study.

23.

Year x	Emergence Date y	x^2	xy
1992	112	3968064	223104
1993	115	3972049	229195
1994	111	3976036	221334
1995	116	3980025	231420
1996	115	3984016	229540
1997	111	3988009	221667
1998	112	3992004	223776
1999	112	3996001	223888
2000	116	4000000	232000
2001	116	4004001	232116
2002	121	4008004	242242
2003	119	4012009	238357
2004	108	4016016	216432
2005	116	4020025	232580
2006	115	4024036	230690
2007	118	4028049	236826
2008	120	4032064	240960
2009	122	4036081	245098
2010	111	4040100	223110
2011	125	4044121	251375
$\overline{x} = \frac{40030}{20}$	$\overline{y} = \frac{2311}{20}$	$\overline{x^2} = \frac{80120710}{20}$	$\overline{xy} = \frac{4625710}{20}$

The coefficients for the least squares line $y = ax + b$ are

$$a = \frac{\overline{xy} - \overline{x}\,\overline{y}}{\overline{x^2} - \overline{x}^2} = \frac{\frac{4625710}{20} - \left(\frac{40030}{20}\right)\left(\frac{2311}{20}\right)}{\frac{80120710}{20} - \left(\frac{40030}{20}\right)^2} \approx 0.366$$

$$b = \overline{y} - a\overline{x} \approx \tfrac{2311}{20} - (0.366)\,\tfrac{40030}{20} \approx -617.33$$

Thus, the least squares line is $y = 0.366x - 617.33$ which is plotted along with the data below.

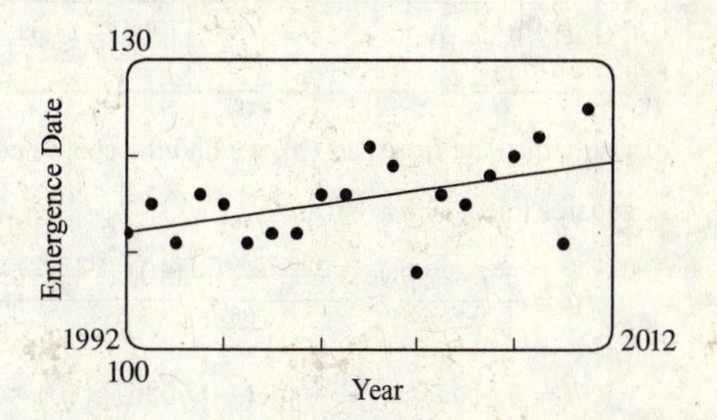

The least squares line appears to be a rough approximation of the data in the sample. The slope is approximately 0.366 which indicates that as each year passes, the emergence date increases. Specifically, after one year the emergence date will increase by about 0.366 days.

© 2016 Cengage Learning. All Rights Reserved. May not be scanned, copied or duplicated, or posted to a publicly accessible website, in whole or in part.

25. (a) $y(t) = \ln\left(\frac{210}{N(t)} - 1\right) = \ln\left(\frac{210}{\frac{210}{1+Ae^{-rt}}} - 1\right) = \ln\left(1 + Ae^{-rt} - 1\right) = \ln\left(Ae^{-rt}\right) = \ln A + \ln e^{-rt} = -rt + \ln A$

Thus, the graph of $y(t)$ is a line having slope $-r$ and y-intercept $\ln A$.

(b)

Time x	Population N	$\ln\left(210N^{-1} - 1\right)$ y	x^2	xy
0	0.200	6.9556	0	0.0000
1	0.330	6.4542	1	6.4542
2	0.500	6.0379	4	12.0757
3	1.10	5.2465	9	15.7396
4	1.40	5.0039	16	20.0158
5	3.10	4.2008	25	21.0042
6	3.50	4.0775	36	24.4652
7	9.00	3.1061	49	21.7426
8	10.0	2.9957	64	23.9659
9	25.4	1.9834	81	17.8510
10	27.0	1.9136	100	19.1365
11	55.0	1.0361	121	11.3970
12	76.0	0.5671	144	6.8053
13	115	-0.1911	169	-2.4837
14	160	-1.1632	196	-16.2841
15	162	-1.2164	225	-18.2459
16	190	-2.2513	256	-36.0207
17	193	-2.4295	289	-41.3011
18	190	-2.2513	324	-40.5233
19	209	-5.3423	361	-101.5044
20	190	-2.2513	400	-45.0258
$\overline{x} = \frac{210}{21} = 10$		$\overline{y} \approx \frac{32.4823}{21}$	$\overline{x^2} = \frac{2870}{21}$	$\overline{xy} \approx \frac{-100.7360}{21}$

(c) Using the data from part (b), we find the coefficients for the least squares line $y = ax + b$ are

$$a = \frac{\overline{xy} - \overline{x}\,\overline{y}}{\overline{x^2} - \overline{x}^2} \approx \frac{\frac{-100.7360}{21} - (10)\left(\frac{32.4823}{21}\right)}{\frac{2870}{21} - (10)^2} \approx -0.55267$$

$$b = \overline{y} - a\overline{x} \approx \frac{32.4823}{21} - (-0.55267)(10) \approx 7.07352$$

Thus, the least squares line is $y = -0.5527x + 7.0735$ which is plotted along with the data at right. Since $y(t) = -rt + \ln A$ from part (a), we have $r \approx 0.55267$ and $\ln A \approx 7.07352 \Rightarrow A \approx e^{7.07352} \approx 1180.30$.

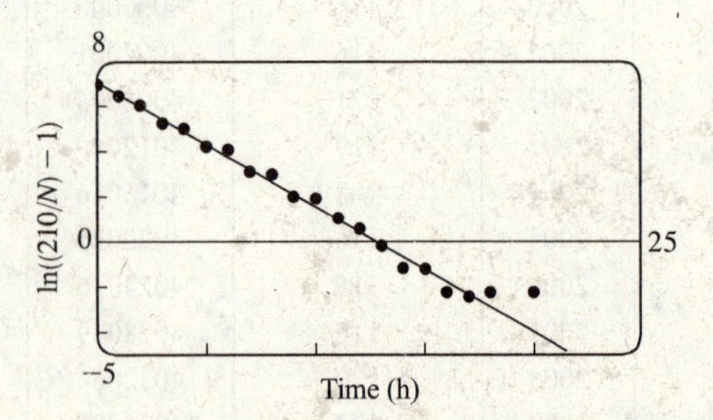

© 2016 Cengage Learning. All Rights Reserved. May not be scanned, copied or duplicated, or posted to a publicly accessible website, in whole or in part.

(d) The population function is

$N(t) = \dfrac{210}{1 + Ae^{-rt}} \approx \dfrac{210}{1 + 1180.30e^{-0.55267t}}$. This function is plotted along with the population data for the entire time period. The population function approximates the data quite well over the entire 36 hour time period.

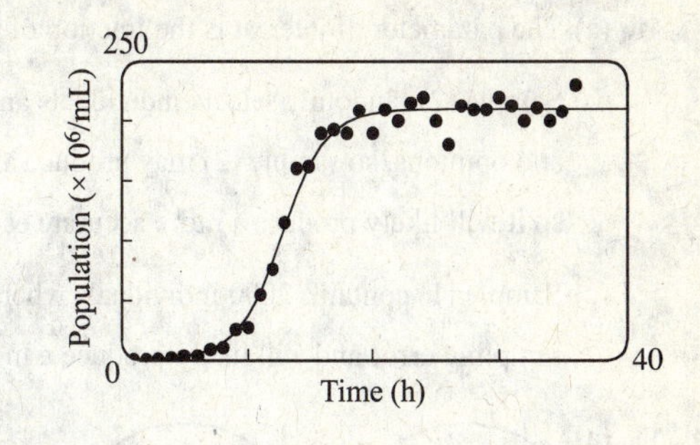

11.4 Populations, Samples, and Inference

1. (a) The parameter of interest is the reproductive success of all yellow perch fish.

(b) The data are experimental because ecologists control which fish are exposed to contaminants.

(c) Provided the yellow perch in the lake are representative of the entire yellow perch population, then the data likely come from a random sample since the fish exposed to the contaminant are chosen at random. However, if the global yellow perch population is not well represented by the yellow perch in the lake, then the estimate of reproductive success may be biased. For example, a nearby factory may pollute the lake ecosystem predisposing the perch to reproductive failure. This would bias the estimate of the "unexposed" sample toward lower reproductive success.

3. (a) The parameter of interest is the fraction of Nigerians with polio.

(b) The data are observational because researchers test if individuals have polio but do not directly alter the patients' likelihood of having polio.

(c) The estimate of the fraction of Nigerians with polio is potentially biased since individuals who regularly visit health clinics are likely ill or prone to illness. This would bias the estimated prevalence of polio toward higher values.

5. (a) The parameter of interest is the fraction of treated individuals with TB.

(b) The data are experimental because researchers determined which patients received the antibiotic and which received the placebo instead.

(c) The data likely come from a random sample since the newly diagnosed TB patients are sampled randomly and patients are randomly chosen from that sample for the antibiotic treatment.

7. (a) The parameter of interest is the reproductive success of all plants.

(b) The data are experimental because botanists exposed selected individuals to increased temperatures.

(c) If the species *Arabidopsis thaliana* is representative of all plant species, then the estimate of reproductive success obtained from the data is likely unbiased. However, if this species is more adaptive to temperature changes than other plants, the estimate of the population parameter may be biased toward higher reproductive success.

9. (a) The parameter of interest is the fraction of salmon larger than 30 cm.

(b) Sample (1) captures fish using a mesh net so very small fish will not be caught. Thus, the sample will produce a biased estimate of the fraction of salmon larger than 30 cm. Sample (2) will likely produce more accurate estimates because all fish are directly observed.

(c) Sample (1) contains 500 fish while sample (2) only has 100 fish. Thus, sample (1) will likely have a smaller sampling error and will produce more precise estimates.

© 2016 Cengage Learning. All Rights Reserved. May not be scanned, copied or duplicated, or posted to a publicly accessible website, in whole or in part.

11. (a) The parameter of interest is the fraction of individuals who support socialized health care.

(b) Sample (2) randomly selects individuals and their housemates. Individuals who share a residence likely share similar views and opinions, so sample (2) may produce a biased estimate. Sample (1) randomly chooses individuals from the census data so it will likely produce a more accurate estimate of the parameter.

(c) Sample (1) contains 2000 individuals whereas sample (2) contains 1500 individuals. Thus, sample (1) has a smaller sampling error and will likely produce a more precise estimate.

13. (a)

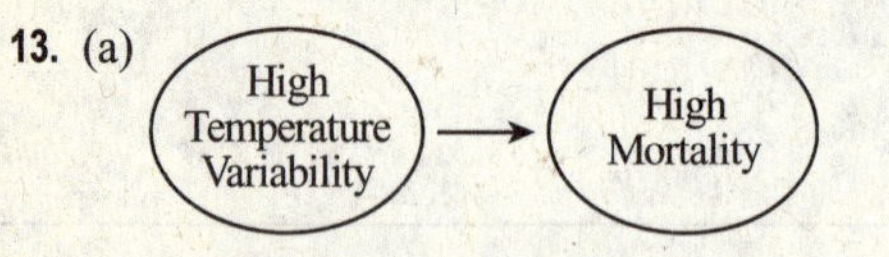

(b) Potential confounding variables include:

(i) Level of urbanization: High levels of urbanization may lead to larger temperature variability due to paved land surfaces and other human activities. Urbanization may also reduce the biodiversity of a region as habitats are destroyed leading to reduced plant and animal populations. This can affect all species in an ecosystem's food web and hence, can lead to increased insect mortality.

(ii) Canopy cover: High amounts of canopy cover can reduce temperature variability and also provides additional leaves on which insects can feed leading to lower insect mortality rates. Conversely, less canopy cover results in more sunlight leading to higher temperature variability and also less food for insects which increases insect mortality.

(c)

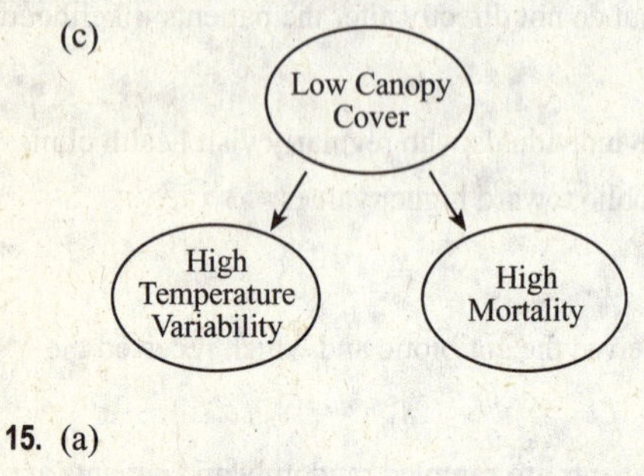

15. (a)

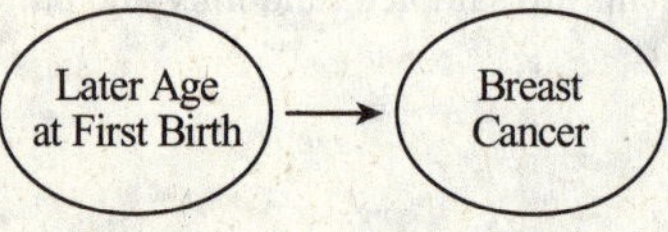

(b) A potential confounding variable is an underlying health or fertility problem. Women with this fertility issue may take several years attempting to conceive eventually using fertility medication and hence having a later in life pregnancy. The underlying health problem could also lead to breast cancer later in life.

(c)

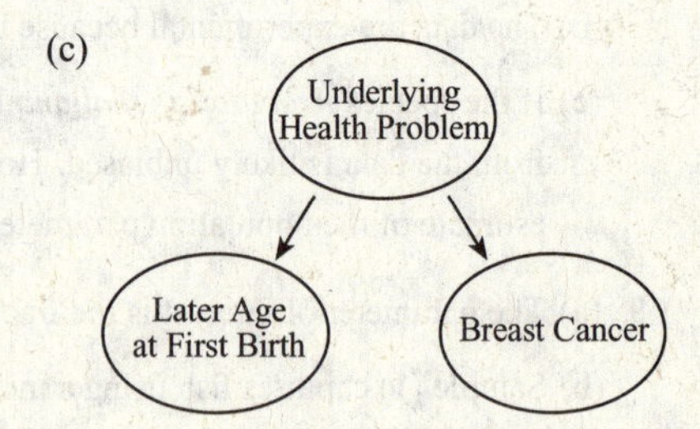

17. (a)

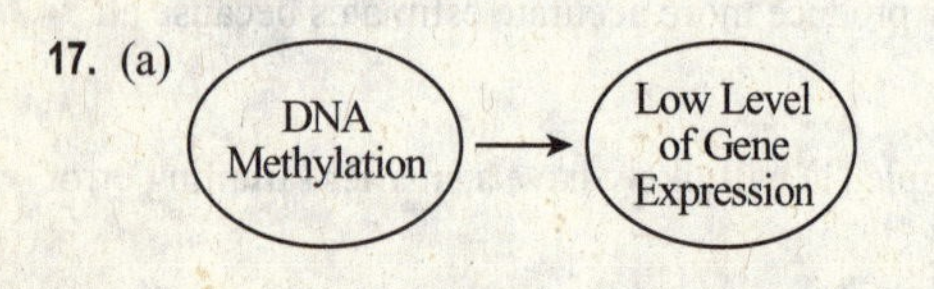

© 2016 Cengage Learning. All Rights Reserved. May not be scanned, copied or duplicated, or posted to a publicly accessible website, in whole or in part.

(b) A potential confounding variable is a specialized inhibitor molecule. This molecule might assist in the attachment of methyl groups to DNA and also limit gene expression. In the presence of this inhibitor, there would be high levels of DNA methylation and low levels of gene expression. If absent, there would be low levels of DNA methylation and higher levels of gene expression.

(c)

Inhibitor Molecule
DNA Methylation
Low Level of Gene Expression

11 Review

TRUE-FALSE QUIZ

1. False. If the data are skewed to the left, the median is greater than the mean (see Figure 11.2.5).

3. True. The interquartile range captures the middle 50% of the data and the range captures all of the data, so the IQR will be less than or equal to the range. Mathematically, Min $\leq Q_1$ and Max $\geq Q_3 \;\Rightarrow\; IQR = Q_3 - Q_1 \leq$ Max $-$ Min=Range.

5. True. The area of a histogram bar is proportional to its height and the height of a bar in a histogram is proportional to the number of data points contained in the corresponding bin. This is true for frequency, relative frequency, and density histograms (see the discussion on Histograms in Section 11.2). Therefore, the area of the bar is proportional to the number of data points.

7. False. Bias refers to the systematic over or under-estimation of a parameter.

9. False, observational data is useful for inferring patterns and associations in the population. Inferring causative relationships is best done by conducting an experiment in which variables can be controlled and manipulated by the researcher.

EXERCISES

1. Ordered data: 8 , 9 , 15 , 16 , 18 , 18 , 18 , 20 , 21 , 22

(a) $\bar{x} = [8+9+15+16+18+18+18+20+21+22]/10 = \dfrac{165}{10} = 16.5\,\text{mg}$

(b) Min $= 8 \quad Q_1 = 15 \quad$ Median $= \dfrac{18+18}{2} = 18 \quad Q_3 = 20 \quad$ Max $= 22$

(c) IQR $= Q_3 - Q_1 = 20 - 15 = 5$. Outliers must be outside the range $[Q_1 - 1.5 \times \text{IQR}, Q_3 + 1.5 \times \text{IQR}] = [7.5, 27.5]$, so there are no outliers in the data set.

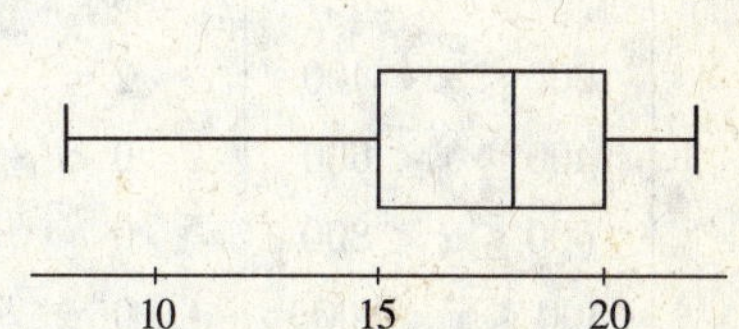

Pollen Amount (mg)

3. Ordered data: 471, 502, 520, 553, 559, 577, 597, 652, 677, 720, 733, 807

(a) $\bar{x} = [471+502+520+553+559+577+597+652+677+720+733+807]/12 = \dfrac{7368}{12} = 614\,\text{mg}$

(b) Min $= 471 \quad Q_1 = \dfrac{520+553}{2} = 536.5 \quad$ Median $= \dfrac{577+597}{2} = 587 \quad Q_3 = \dfrac{677+720}{2} = 698.5 \quad$ Max $= 807$

(c) IQR $= Q_3 - Q_1 = 698.5 - 536.5 = 162$. Outliers must be outside the range $[Q_1 - 1.5 \times \text{IQR}, Q_3 + 1.5 \times \text{IQR}] = [293.5, 941.5]$, so there are no outliers in the data set.

500 600 700 800

Mass (mg)

© 2016 Cengage Learning. All Rights Reserved. May not be scanned, copied or duplicated, or posted to a publicly accessible website, in whole or in part.

5. The data contains two unordered categories so it is categorical and nominal. A bar (or pie) graph can be used to display the data.

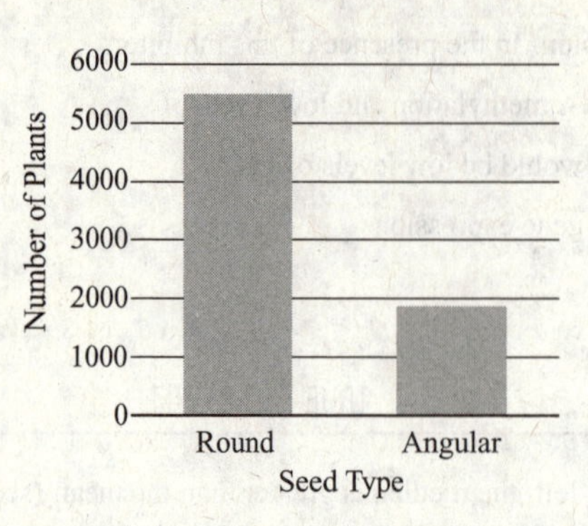

7. The data contains five ordered categories so it is categorical and ordinal. A bar graph can be used to display the data.

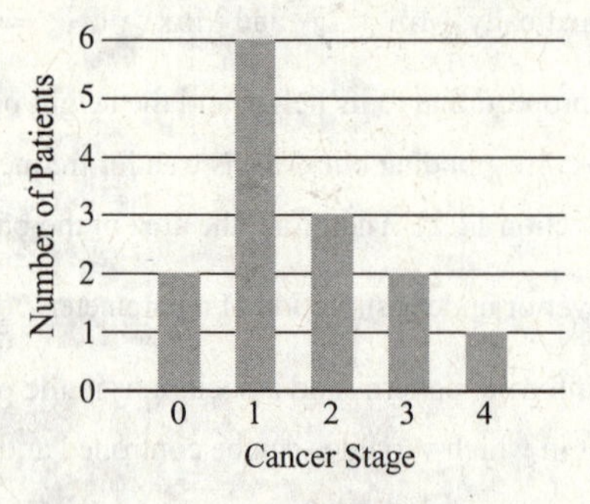

9. Counting the number of data points that fall between the intervals listed in the table below gives the frequency in each interval. The relative frequency is calculated by dividing each frequency by 10, the total number of data points. The density is then obtained by dividing each relative frequency by the bin width of 200.

Interval	Frequency	Relative Frequency	Density
$0 \le x < 200$	6	0.6	0.003
$200 \le x < 400$	2	0.2	0.001
$400 \le x < 600$	0	0	0
$600 \le x < 800$	0	0	0
$800 \le x < 1000$	0	0	0
$1000 \le x < 1200$	0	0	0
$1200 \le x < 1400$	2	0.2	0.001

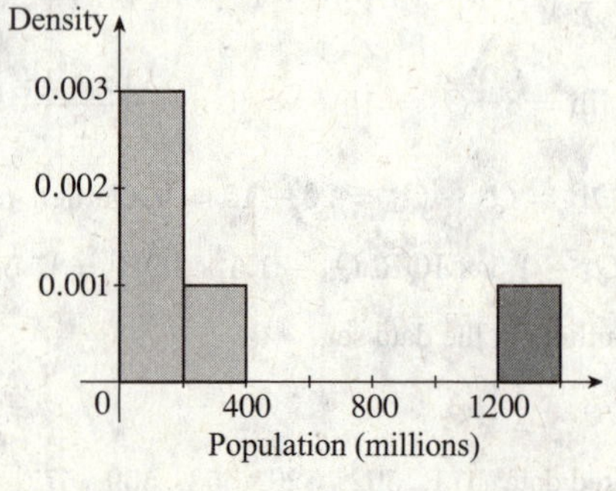

Density
0.003
0.002
0.001
0
400 800 1200
Population (millions)

The fraction of countries with a population of 400 million or more is shaded in the histogram. The shaded region amounts to $1/5$ of the data points, so 20% of the data lies in the interval $[400, 1400)$.

11. Since $500 + 2(25) = 550$ and $500 - 25 = 475$, fish masses between 475 and 550 lie within a band one standard deviation to the left of the mean and two standard deviations to the right of the mean. Referring to Figure 11.2.10, we see that when $\mu - \sigma < x < \mu + 2\sigma$ the area under the normal curve is $2(34.1\%) + 13.65\% = 81.85\%$. Thus, approximately 81.85% of the data lies between 475 and 500 g.

© 2016 Cengage Learning. All Rights Reserved. May not be scanned, copied or duplicated, or posted to a publicly accessible website, in whole or in part.

13. The area under the curve $f(x)$ in the interval $[8, \infty)$ is

$$\int_8^{\infty} f(x)\,dx = \lim_{t\to\infty} \int_8^t \tfrac{1}{5} e^{-x/5}\,dx = -\lim_{t\to\infty} \left[e^{-x/5}\right]_8^t = -\lim_{t\to\infty} \left[e^{-t/5} - e^{-8/5}\right] = e^{-8/5} \approx 20\%.$$ Therefore,

approximately 20% of antibiotics remained useful for at least 8 years.

15.

	Positive Test	Negative Test	Row Total
Cancer	$0.75(490) \approx 368$	122	490
No Cancer	$0.55\,(346) \approx 190$	156	346
Column Total	558	278	836

A large fraction of individuals with cancer tested positive, however, individuals without cancer still tested positive 55% of the time. Also, about 44% (122/278) of the individuals who tested negative had cancer. Therefore, the result of the PSA test may not be a reliable indicator of prostate cancer.

17.

	Positive Test	Negative Test	Row Total
Male	13	8	21
Female	10	5	15
Column Total	23	13	36

The fraction of males who test positive ($13/21 \approx 62\%$) is close to the fraction of females who test positive ($10/15 \approx 67\%$). Similarly, approximately 57% of positive tests come from males and 62% of negative tests come from males. Hence, there does not appear to be a strong relationship between sex and test results.

19. Sorted Data (Rotation): 29, 31, 37, 37, 39, 41, 41, 42, 42, 44, 45, 46

Sorted Data (No Rotation): 27, 29, 29, 31, 31, 34, 35, 36, 36, 37, 40, 41

Rotation: Min = 29 $Q_1 = \dfrac{37+37}{2} = 37$ Median $= \dfrac{41+41}{2} = 41$ $Q_3 = \dfrac{42+44}{2} = 43$ Max = 46

Also, IQR $= 43 - 37 = 6$ and outliers must be outside the range $[Q_1 - 1.5 \times \text{IQR}, Q_3 + 1.5 \times \text{IQR}] = [28, 52]$. So there are no outliers in the crop rotation data.

No Rotation: Min = 27 $Q_1 = \dfrac{29+31}{2} = 30$ Median $= \dfrac{34+35}{2} = 34.5$ $Q_3 = \dfrac{36+37}{2} = 36.5$ Max = 41

Also, IQR $= 36.5 - 30 = 6.5$ and outliers must be outside the range $[Q_1 - 1.5 \times \text{IQR}, Q_3 + 1.5 \times \text{IQR}] = [20.25, 46.25]$. So there are no outliers in the no crop rotation data.

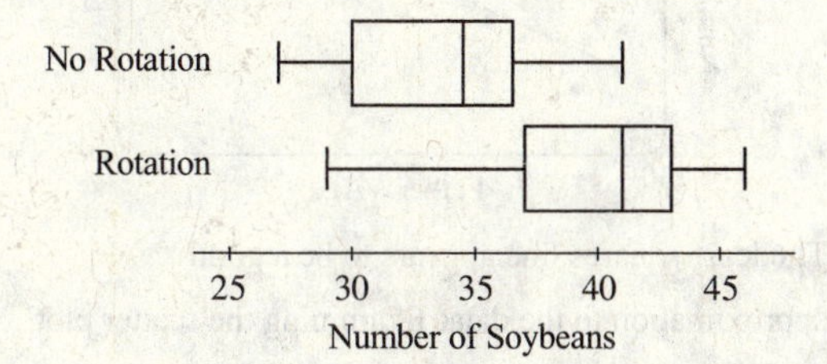

The distribution of the crop rotation data is shifted toward higher soybean harvests compared to the no crop rotation data. The minimum values have not changed significantly; However, the median, and the first and third quartiles have all substantially increased, so it appears that crop rotation has an effect on the soybean harvest in the sample.

© 2016 Cengage Learning. All Rights Reserved. May not be scanned, copied or duplicated, or posted to a publicly accessible website, in whole or in part.

21.

Brain Size x	Intelligence y	x^2	xy
965	52	931225	50180
1050	59	1102500	61950
1060	52	1123600	55120
1090	53	1188100	57770
1090	55	1188100	59950
1130	61	1276900	68930
1060	65	1123600	68900
1100	67	1210000	73700
1055	75	1113025	79125
1050	75	1102500	78750
1055	74	1113025	78070
1060	76	1123600	80560
1100	76	1210000	83600
1190	78	1416100	92820
1210	79	1464100	95590
1220	82	1488400	100040
1300	82	1690000	106600
$\overline{x} = \frac{18785}{17} = 1105$	$\overline{y} = \frac{1161}{17}$	$\overline{x^2} = \frac{20864775}{17}$	$\overline{xy} = \frac{1291655}{17}$

The coefficients for the least squares line $y = ax + b$ are

$$a = \frac{\overline{xy} - \overline{x}\,\overline{y}}{\overline{x^2} - \overline{x}^2} = \frac{\frac{1291655}{17} - (1105)\left(\frac{1161}{17}\right)}{\frac{20864775}{17} - (1105)^2}$$

$$\approx 0.08151$$

$$b = \overline{y} - a\overline{x} \approx \tfrac{1161}{17} - (0.08151)(1105) \approx -21.77342$$

Thus, the least squares line is $y = 0.08151x - 21.77342$ which is plotted along with the data below.

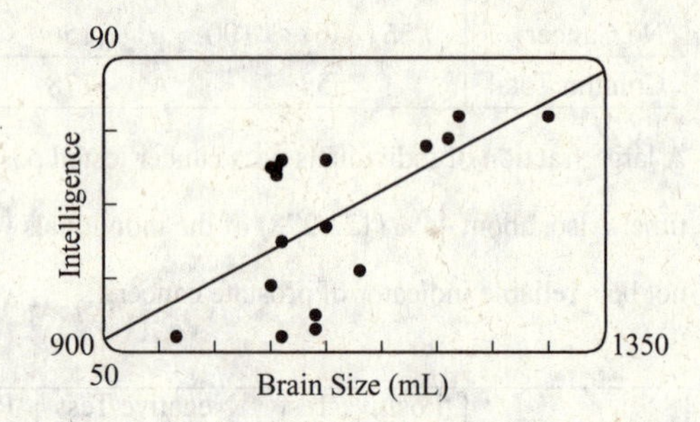

The least squares line indicates that there is a positive relationship in the data, that is, as brain size increases intelligence increases. However, further examining the datapoints, we see that this relationship appears to be weak for moderate sized brains since intelligence ranges across approximately 30 units when brain sizes are close to 1100 mL.

23.

Father's Age x	Number of Mutations y	x^2	xy
16	40	256	640
17	41	289	697
19	39	361	741
18	49	324	882
23	49	529	1127
24	51	576	1224
24	52	576	1248
24	60	576	1440
25	55	625	1375
25	56	625	1400
27	50	729	1350
29	52	841	1508
30	57	900	1710
31	61	961	1891
34	75	1156	2550
36	70	1296	2520
37	65	1369	2405
$\overline{x} = \frac{439}{17}$	$\overline{y} = \frac{922}{17}$	$\overline{x^2} = \frac{11989}{17}$	$\overline{xy} = \frac{24708}{17}$

The coefficients for the least squares line $y = ax + b$ are

$$a = \frac{\overline{xy} - \overline{x}\,\overline{y}}{\overline{x^2} - \overline{x}^2} = \frac{\frac{24708}{17} - \left(\frac{439}{17}\right)\left(\frac{922}{17}\right)}{\frac{11989}{17} - \left(\frac{439}{17}\right)^2} \approx 1.37739$$

$$b = \overline{y} - a\overline{x} \approx \tfrac{922}{17} - (1.37739)\tfrac{439}{17} \approx 18.66625$$

Thus, the least squares line is $y = 1.3774x + 18.6662$ which is plotted along with the data below.

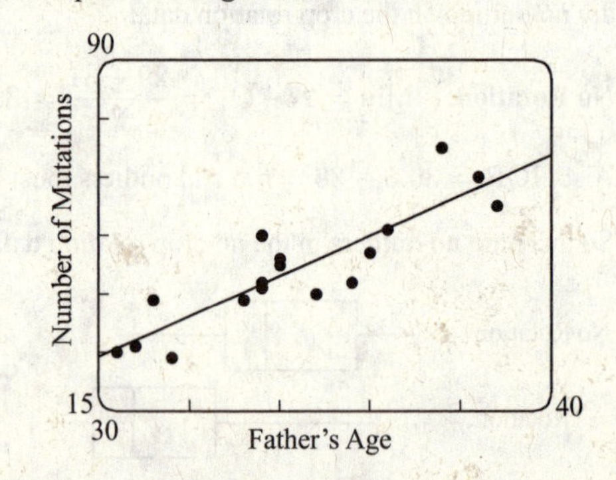

The least-squares line appears to be a good approximation to the data. Examining the scatter plot and least-squares line, we see that there is a positive relationship in the data, that is, as the father's age increases, the number of mutations increases.

© 2016 Cengage Learning. All Rights Reserved. May not be scanned, copied or duplicated, or posted to a publicly accessible website, in whole or in part.

25. (a)

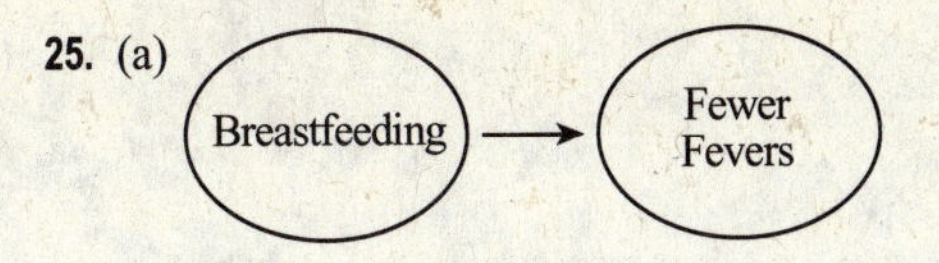

(b) A potential confounding variable is a genetic condition (such as galactosemia) that may make breast feeding unsafe for the infant and may also cause fevers.

(c)

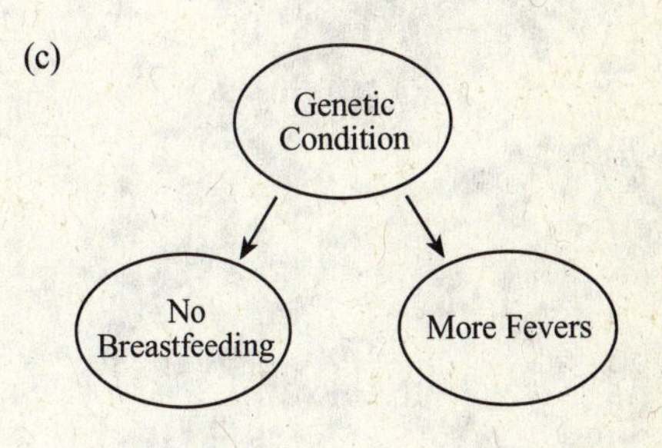

27. (a)

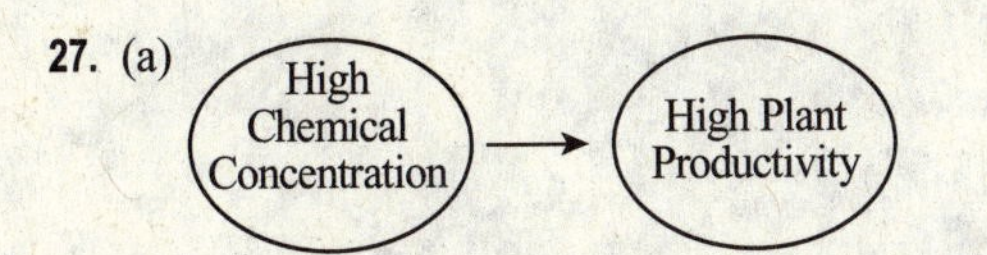

(b) A potential confounding variable is soil type. Soil may vary in different regions, having different density, particle size, etc. There may be a particular soil type that contains the chemical of interest and also has a density and/or particle size that improves plant productivity.

(c)

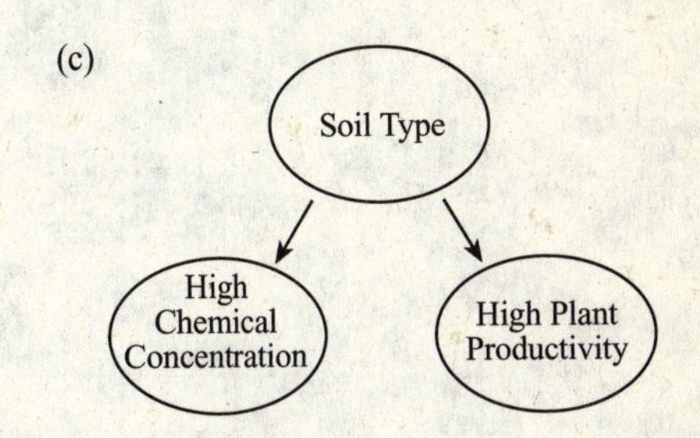

© 2016 Cengage Learning. All Rights Reserved. May not be scanned, copied or duplicated, or posted to a publicly accessible website, in whole or in part.

12 □ PROBABILITY

12.1 Principles of Counting

1. In the experiment, there are $n_1 = 20$ choices of fish, $n_2 = 2$ choices of food, $n_3 = 3$ choices of temperature, and $n_4 = 2$ choices of sex. Therefore, the total number of possible choices is $n_1 \cdot n_2 \cdot n_3 \cdot n_4 = 20 \cdot 2 \cdot 3 \cdot 2 = 240$.

3. The first digit has $n_1 = 8$ possible number choices, and the remaining nine digits have $n_2 = n_3 = \cdots = n_{10} = 10$ choices. Thus, the total number of possible telephone numbers is $n_1 \cdot n_2 \cdot \cdots \cdot n_{10} = 8 \cdot 10 \cdot \cdots \cdot 10 = 8 \cdot 10^9 = 8,000,000,000$.

5. (a) There are 20 different amino acids, so the three mutation positions each have 19 choices for the mutant amino acid. This gives a total number of mutant peptides of $19 \cdot 19 \cdot 19 = 19^3 = 6859$.

(b) The first selection has $n_1 = 10$ choices for the mutation position and the second selection has $n_2 = 19$ choices for the mutant amino acid at the chosen mutant position. This gives a total number of mutant peptides of $n_1 \cdot n_2 = 10 \cdot 19 = 190$.

7. There are 26 letter choices for the first three character selections, so $n_1 = n_2 = n_3 = 26$. Since the palindrome must read the same forwards as backwards, the fourth character must match the second character and the fifth must match the first. Thus, there is only one choice, $n_4 = n_5 = 1$, for the final two letters. The number of possible five-letter palindromes is $n_1 \cdot n_2 \cdot n_3 \cdot n_4 \cdot n_5 = 26 \cdot 26 \cdot 26 \cdot 1 \cdot 1 = 26^3 = 17,576$.

9. Each of the 800 possible nucleotide selections has 4 choices so that $n_1 = n_2 = \cdots = n_{800} = 4$. Thus, the number of different sequences is $n_1 \cdot n_2 \cdot \cdots \cdot n_{800} = 4 \cdot 4 \cdot \cdots \cdot 4 = 4^{800}$.

11. There are $n_1 = 4$ choices for the ABO grouping and $n_2 = 2$ choices for the Rh grouping. Thus, the number of different blood types is $n_1 \cdot n_2 = 4 \cdot 2 = 8$.

13. Let n_i be the number of choices for segment i.

(a) There are 2 different choices (one from each strain) for each of the 8 DNA segments, so $n_1 = n_2 = \cdots = n_8 = 2$. Using the Fundamental Principle of Counting, the number of different possible viral types is $n_1 \cdot n_2 \cdot \cdots \cdot n_8 = 2 \cdot 2 \cdot \cdots \cdot 2 = 2^8 = 256$.

(b) Now there are 2 different choices (one from each strain) for 5 of the DNA segments so $n_1 = n_2 = \cdots = n_5 = 2$. The remaining 3 DNA segments are identical in both strains, so there is, effectively, only 1 choice for the final 3 segments. Thus, $n_6 = n_7 = n_8 = 1$. Using the Fundamental Principle of Counting, the number of different possible viral types is $n_1 \cdot n_2 \cdot \cdots \cdot n_8 = 2^5 \cdot 1^3 = 32$.

15. Let n_i be the number of choices for 'letter' i. Each 'letter' has four possible choices, so if there are k 'letters' in a group then $n_1 = n_2 = \cdots = n_k = 4$. Using the Fundamental Principle of Counting, the number of possible protein subunits is $n_1 \cdot n_2 \cdot \cdots \cdot n_k = \underbrace{4 \cdot 4 \cdot \cdots \cdot 4}_{k \text{ products}} = 4^k$. If $k = 2$, there are $4^2 = 16$ possible subunits, too few to code for the 20 different amino acids. If $k = 3$, there are $4^3 = 64$ possible subunits, more than enough to represent 20 different amino acids. Thus, $k = 3$ is the smallest value that still allows coding for all the amino acids.

© 2016 Cengage Learning. All Rights Reserved. May not be scanned, copied or duplicated, or posted to a publicly accessible website, in whole or in part.

17. There are 7 possible choices for the first ball, and then 6 possible choices for the second ball, 5 possible choices for the third ball, 4 possible choices for the fourth ball, and 3 possible choices for the fifth ball. Therefore, using the Fundamental Principle of Counting, there are a total of $7 \cdot 6 \cdot 5 \cdot 4 \cdot 3$ different possible color sequences. This is just the number of permutations of 7 balls when taken 5 at a time. Specifically, using the formula for $P_{n,k}$, we obtain

$P_{7,5} = \dfrac{7!}{(7-5)!} = \dfrac{7 \cdot 6 \cdot 5 \cdot 4 \cdot 3 \cdot 2 \cdot 1}{2 \cdot 1} = 7 \cdot 6 \cdot 5 \cdot 4 \cdot 3 = 2520$. The number of possible color combinations is given by

$C_{7,5} = \dfrac{7!}{5!\,(7-5)!} = \dfrac{7 \cdot 6}{2 \cdot 1} = 21$.

19. We want the number of ordered arrangements of a set of 4 tires since each tire location is unique. Put another way, there are 4 possible choices of tire for the first location, and then 3 possible choices for the second location, 2 possible choices for the third location, and 1 possible choice for the fourth location. Using the Fundamental Principle of Counting we obtain $4 \cdot 3 \cdot 2 \cdot 1 = 4!$. This is just the number of permutations of the 4 tires which we write as $P_4 = 4! = 24$.

21. (a) The number of ordered arrangements of the 5 finishers is given by $P_5 = 5! = 120$. Put another way, there are 5 possible choices for first place, and then 4 possible choices for second place, 3 possible choices for third place, 2 possible choices for fourth place, and 1 possible choice for fifth place. From the Fundamental Principle of Counting we obtain $5 \cdot 4 \cdot 3 \cdot 2 \cdot 1 = 5! = 120$. So there are 120 possible ways the race can be completed.

(b) The number of ordered arrangements of the 5 finishers when taken 3 at a time is given by

$P_{5,3} = \dfrac{5!}{(5-3)!} = \dfrac{5!}{2!} = 5 \cdot 4 \cdot 3 = 60$. Put another way, there are 5 possible choices for first place, and then 4 possible choices for second place, and 3 possible choices for third place. From the Fundamental Principle of Counting we obtain $5 \cdot 4 \cdot 3 = 60$. So there are 60 possible arrangements for the podium positions.

23. If a particular object must be excluded, then we really have $n - 1$ objects from which to make permutations of k objects. Therefore, the number of permutations is $P_{n-1,k} = \dfrac{(n-1)!}{(n-1-k)!}$.

25. The number of ordered arrangements of the 6 individuals is given by $P_6 = 6! = 720$. Put another way, there are 6 possible choices for the most preferred male, and then 5 possible choices for the second best, 4 possible choices for the third best, and so on. From the Fundamental Principle of Counting we obtain $6 \cdot 5 \cdot 4 \cdot 3 \cdot 2 \cdot 1 = 720$. So there are 720 possible dominance hierarchies.

27. The number of ordered arrangements of the 4 genes is given by $P_4 = 4! = 24$. Put another way, there are 4 possible choices for the first gene on the chromosome, and then 3 possible choices for the second, 2 possible choices for the third, and 1 possible choice for the fourth. From the Fundamental Principle of Counting we obtain $4 \cdot 3 \cdot 2 \cdot 1 = 24$. So there are 24 possible gene arrangements.

29. The number of arrangements of the genes within operon A is $P_3 = 3!$ and similarly for operons B and C we have $P_5 = 5!$ and $P_6 = 6!$. Now, the number of arrangements of the 3 operons along the DNA is $P_3 = 3!$. Therefore, using the Fundamental Principle of Counting, the total number of different genomic arrangements is $3! \cdot 5! \cdot 6! \cdot 3! = 3,110,400$.

© 2016 Cengage Learning. All Rights Reserved. May not be scanned, copied or duplicated, or posted to a publicly accessible website, in whole or in part.

31. We can calculate the total number of possible arrangements and subtract the number of arrangements in which species 1 and 2 are next to one another. The total number of arrangements including adjacent placements of species 1 and 2 is $P_7 = 7! = 5040$. Now, if species 1 and 2 are adjacent to each other, there are $P_2 = 2!$ choices for the order of species 1 and 2 (either 1 then 2 or 2 then 1), 6 choices for the location of the species 1 and 2 group (among the remaining 5 species), and $P_5 = 5!$ choices for arranging species 3–7. So the number of arrangements with species 1 and 2 next to each other is $2! \cdot 6 \cdot 5! = 1440$. Therefore, the number of possible arrangements excluding adjacent placement of species 1 and 2 is $5040 - 1440 = 3600$.

Alternative Solution: There are 5 species that are easy to place since it does not matter who they are next to. For these 5 species the number of possible arrangements is just the number of permutations of the 5 species, that is, $P_5 = 5!$. The remaining two species must be separated from one another by at least one of the 5 "easy" species. This means that there are 6 possible openings in which each of these two species might be placed. These correspond to (i) the left-most position on the shelf, (ii) between the first and second aquaria, (iii) between the second and third aquaria, (iv) between the third and fourth aquaria, (v) between the fourth and fifth aquaria, and (vi) the rightmost position on the shelf. The number of ways to place the two species in one of these 6 openings can be obtained by noting that there are 6 possible choices of opening for the first of the two species, and then 5 possible choices of opening for the remaining species. This gives a total of $6 \cdot 5 = 30$ possibilities. From the Fundamental Principle of Counting, the total number of possible arrangements of all species is therefore $5! \cdot 30 = 3600$.

33. (a) The number of ways of choosing three coins to be heads from a total of five coins is $C_{5,3} = \dfrac{5!}{3!\,(5-3)!} = 10$. Similarly, the number of ways of choosing four and five coins to be heads from a total of five coins is $C_{5,4} = 5$ and $C_{5,5} = 1$ respectively. Therefore, the number of ways of choosing *at least* three coins to be heads is $10 + 5 + 1 = 16$.

(b) The number of ways of choosing zero coins to be tails from a total of five coins is $C_{5,0} = \dfrac{5!}{0!\,(5-0)!} = 1$. Similarly, the number of ways of choosing one and two coins to be tails from a total of five coins is $C_{5,1} = 5$ and $C_{5,2} = 10$ respectively. Therefore, the number of ways of choosing *at most* two coins to be tails is $1 + 5 + 10 = 16$. Note: Obtaining at most two tails is logically equivalent to obtaining at least three heads, so the answers to parts (a) and (b) are the same.

35. The number of ways of choosing 3 winners from 10 entries is $C_{10,3} = \dfrac{10!}{3!\,(10-3)!} = \dfrac{10 \cdot 9 \cdot 8}{3 \cdot 2 \cdot 1} = 120$.

Put another way, there are 10 possible choices for the first winner, and then 9 possible choices for the second winner and 8 for the third winner. This gives a total of $10 \cdot 9 \cdot 8$ possible choices if we account for their ordering. In this exercise, however, we care only about the identity of the individuals and not the order in which they were chosen. And for any particular choice of 3 winners there are 3! different orderings (that is, permutations) that contain the same individuals. Therefore, the number of ways to choose three winners is $\dfrac{10 \cdot 9 \cdot 8}{3!} = \dfrac{10!}{3!7!} = C_{10,3}$.

© 2016 Cengage Learning. All Rights Reserved. May not be scanned, copied or duplicated, or posted to a publicly accessible website, in whole or in part.

37. The number of combinations of a set of 10 individuals when taken 6 at a time is $C_{10,6} = \dfrac{10!}{6!\,(10-6)!} = \dfrac{10 \cdot 9 \cdot 8 \cdot 7}{4 \cdot 3 \cdot 2 \cdot 1} = 210$. Thus, there are 210 ways for six people to be HIV^+.

Put another way, there are 10 possible choices for the first HIV^+ case, and then 9 possible choices for the second, 8 possible choices for the third, 7 possible choices for the fourth, 6 possible choices for the fifth, and 5 possible choices for the sixth. This gives a total of $10 \cdot 9 \cdot 8 \cdot 7 \cdot 6 \cdot 5$ possible choices if we account for their ordering. In this exercise, however, we care only about the identity of the individuals and not the order in which they were chosen. And for any particular choice of 6 HIV^+ individuals there are 6! different orderings (that is, permutations) that contain the same individuals. Therefore, the number of ways to choose six HIV^+ individuals is $\dfrac{10 \cdot 9 \cdot 8 \cdot 7 \cdot 6 \cdot 5}{6!} = \dfrac{10!}{6!4!} = C_{10,6}$.

39. The number of ways of choosing 8 birds from a population of 20 is $C_{20,8} = \dfrac{20!}{8!\,(20-8)!} = 125{,}970$. Thus, there are 125,970 possible sets of 8 birds.

Put another way, there are 20 possible choices for the first bird, and then 19 possible choices for the second, 18 possible choices for the third, 17 possible choices for the fourth, 16 possible choices for the fifth, 15 possible choices for the sixth, 14 possible choices for the seventh, and 13 possible choices for the eighth. This gives a total of $20 \cdot 19 \cdot 18 \cdot 17 \cdot 16 \cdot 15 \cdot 14 \cdot 13$ possible choices if we account for their ordering. In this exercise, however, we care only about the identity of the birds and not the order in which they were chosen. And for any particular choice of 8 birds there are 8! different orderings (that is, permutations) that contain the same individuals. Therefore, the number of ways to choose eight birds is

$$\frac{20 \cdot 19 \cdot 18 \cdot 17 \cdot 16 \cdot 15 \cdot 14 \cdot 13}{8!} = \frac{20!}{8!12!} = C_{20,8}.$$

41. The product $(x+y)^n = (x+y)(x+y)\cdots(x+y)$ can be expanded using the distributive law. Each term in the resulting expansion has either an x or a y taken from every binomial in the product. The term obtained by taking k x's (leaving $n-k$ y's) is expressed as the product $x^k y^{n-k}$ and can be chosen in $C_{n,k} = \binom{n}{k}$ ways, since k x's are chosen from a set of n possibilities. This gives $\binom{n}{k} x^k y^{n-k}$ and after summing over all possible values of k from 0 to n we obtain the identity

$$(x+y)^n = \sum_{k=0}^{n} \binom{n}{k} x^k y^{n-k}.$$

43. The number of ways of choosing 2 white balls out of 4 possibilities is $C_{4,2} = \dfrac{4!}{2!\,(4-2)!} = 6$. For each of these combinations, there are $C_{8,2} = 28$ ways of choosing 2 black balls from a total of 8 and $C_{6,2} = 15$ ways of choosing 2 red balls from a total of 6. Therefore, from the Fundamental Principle of Counting, the number of ways of choosing 2 balls of each color is $6 \cdot 28 \cdot 15 = 2520$.

45. (a) Once 4 balls are chosen for Group A, the remaining 8 balls must belong to Group B. So finding the number of ways of forming the two groups is equivalent to finding the number of ways of forming Group A. The number of combinations of 12 balls taken 4 at a time is $C_{12,4} = \dfrac{12!}{4!\,(12-4)!} = \dfrac{12 \cdot 11 \cdot 10 \cdot 9}{4 \cdot 3 \cdot 2 \cdot 1} = 495$. Thus, there are 495 ways to form the two groups.

© 2016 Cengage Learning. All Rights Reserved. May not be scanned, copied or duplicated, or posted to a publicly accessible website, in whole or in part.

(b) Of the 4 balls in Group A, there are 3 red balls and 1 non-red ball. The number of ways of choosing the 3 reds from 5 possible red balls is $C_{5,3} = 10$ and the number of ways of choosing the 1 non-red from 7 possible non-red balls is $C_{7,1} = 7$. Therefore, by the Fundamental Principle of Counting, there are a total of $10 \cdot 7 = 70$ different ways to form the two groups (forming group A automatically determines the balls in Group B as well).

47. (a) The number of combinations of 18 bat species taken 14 at a time is $C_{18,14} = \dfrac{18!}{14!\,(18-14)!} = \dfrac{18 \cdot 17 \cdot 16 \cdot 15}{4 \cdot 3 \cdot 2 \cdot 1} = 3060$. So there are 3060 ways for 14 of the 18 bat species to be classified as fragile.

(b) Of the 14 fragile species, there are 9 that use caves and 5 that do not. The number of ways of choosing 9 species from 11 possible cave-roosting species is $C_{11,9} = 55$ and the number of ways of choosing 5 species from 7 possible non-cave-roosting species is $C_{7,5} = 21$. Therefore, by the Fundamental Principle of Counting, there are a total of $55 \cdot 21 = 1155$ different ways.

49. The number of ways of choosing 3 textbooks out of 5 possibilities is $C_{5,3} = 10$ and the number of ways of choosing 2 novels out of 6 possibilities is $C_{6,2} = 15$. Thus, the number of combinations of 3 textbooks and 2 novels is $10 \cdot 15 = 150$. Now, for each one of these combinations, there are $P_5 = 5! = 120$ ways of arranging the 5 books. Therefore, the total number of arrangements is $150 \cdot 120 = 18,000$.

12.2 What is Probability?

1. The sample space consists of all possible grade outcomes. Assuming one mark for each correct answer, this gives $\Omega = \{0, 1, 2, 3, \ldots, 20\}$.

3. If we identify the seat location using a numeral for the row followed by a letter, then the sample space is given by the entries in the table at right.

Seat\Row	1	2	3	$\cdots$	29	30
A	$1A$	$2A$	$3A$	$\cdots$	$29A$	$30A$
B	$1B$	$2B$	$3B$	$\cdots$	$29B$	$30B$
C	$1C$	$2C$	$3C$	$\cdots$	$29C$	$30C$
D	$1D$	$2D$	$3D$	$\cdots$	$29D$	$30D$

5. (a) $\Omega = \{1, 2, 3, 4, 5, 6\}$

(b) The event "getting an even number" consists of the outcomes $\{2, 4, 6\}$.

(c) The event "getting a number greater than 4" consists of the outcomes $\{5, 6\}$.

7. (a) There are 2 possible outcomes for each of the 3 students, so the FPC gives $n(\Omega) = 2 \cdot 2 \cdot 2 = 8$.

(b) Listing the outcomes of each student as an ordered set, the sample space is given by $\Omega = \{(P,P,P), (P,P,F), (P,F,P), (F,P,P), (P,F,F), (F,P,F), (F,F,P), (F,F,F)\}$.

(c) The event "exactly two students pass" consists of the outcomes $\{(P,P,F), (P,F,P), (F,P,P)\}$.

(d) The event "at most two students fail" consists of the same outcomes as the sample space excluding the outcome (F,F,F). Thus, the outcomes of this event are $\{(P,P,P), (P,P,F), (P,F,P), (F,P,P), (P,F,F), (F,P,F), (F,F,P)\}$.

(e) "All students have the same result" implies that all students receive the same mark. The outcomes of this event are $\{(P,P,P), (F,F,F)\}$.

© 2016 Cengage Learning. All Rights Reserved. May not be scanned, copied or duplicated, or posted to a publicly accessible website, in whole or in part.

9. (a) There are 2 possible outcomes for each of the 4 students, so the FPC gives $n(\Omega) = 2 \cdot 2 \cdot 2 \cdot 2 = 16$.

(b) We identify students infected with the flu by F and those not infected by N. Using an ordered list to identify the infection status of the 4 students, we can write the sample space as

$$\Omega = \{(F,F,F,F), (F,F,F,N), (F,F,N,F), (F,N,F,F), (N,F,F,F), (F,F,N,N), (F,N,F,N), (N,F,F,N), (N,F,N,F), (N,N,F,F), (F,N,N,F), (F,N,N,N), (N,F,N,N), (N,N,F,N), (N,N,N,F), (N,N,N,N)\}$$

(c) $\{(F,F,F,F), (F,F,F,N), (F,F,N,F), (F,N,F,F), (N,F,F,F)\}$

(d) $\{(F,N,N,N), (N,F,N,N), (N,N,F,N), (N,N,N,F), (N,N,N,N)\}$

(e) $\{(F,F,F,F)\}$

11. $A \cup C = \{2,4,5,6,8\} \cup \{1,3,5,7,9\} = \{1,2,3,4,5,6,7,8,9\} = \Omega$

13. $(A \cap B)^c = (\{2,4,5,6,8\} \cap \{1,2,5\})^c = \{2,5\}^c = \{1,3,4,6,7,8,9\}$

15. $(A^c \cap B)^c = (\{2,4,5,6,8\}^c \cap \{1,2,5\})^c = (\{1,3,7,9\} \cap \{1,2,5\})^c = \{1\}^c = \{2,3,4,5,6,7,8,9\}$

17. (a)

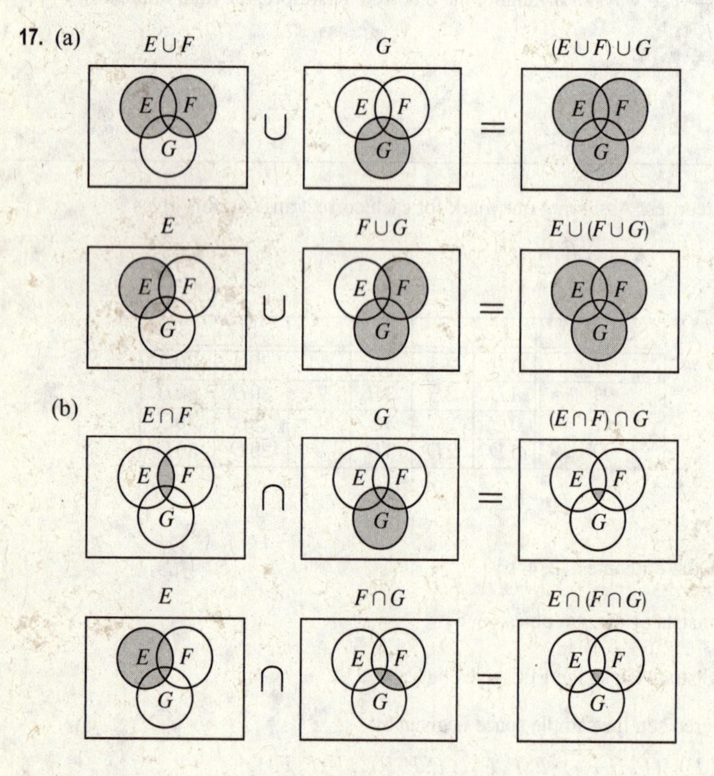

19. $(A \cup B) \cup C = (\{1,7,9\} \cup \{2,5,7\}) \cup \{1,3,5,8,9\} = \{1,2,5,7,9\} \cup \{1,3,5,8,9\} = \{1,2,3,5,7,8,9\}$

21. $C \cap (A \cap B) = \{1,3,5,8,9\} \cap (\{1,7,9\} \cap \{2,5,7\}) = \{1,3,5,8,9\} \cap \{7\} = \{\} = \emptyset$

23. $(A \cap B) \cup C = (\{1,7,9\} \cap \{2,5,7\}) \cup \{1,3,5,8,9\} = \{7\} \cup \{1,3,5,8,9\} = \{1,3,5,7,8,9\}$

© 2016 Cengage Learning. All Rights Reserved. May not be scanned, copied or duplicated, or posted to a publicly accessible website, in whole or in part.

25.

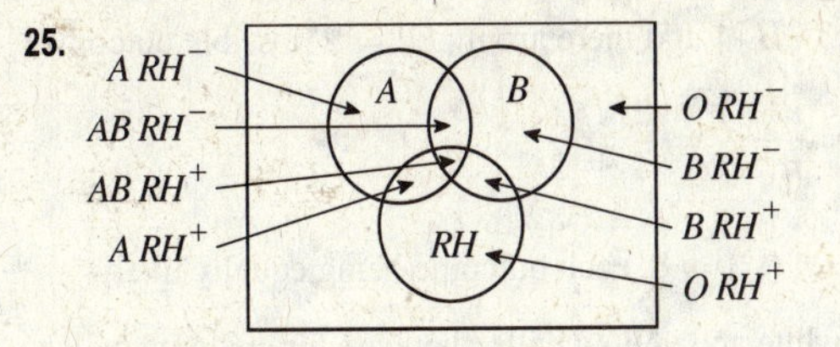

27. (a) $E_1 \cap E_2$ is the event in which both foxes and wolves are present.

(b) $E_1^c \cup E_2$ is the event in which either foxes are absent or wolves are present (or both).

(c) $(E_2 \cup E_3) \cap E_4$ is the event in which either wolves or chickadees are present and blue jays are present.

(d) E_3^c is the event in which no chickadees are present, that is, chickadees are absent.

29. The sample space for a six-sided die roll is $\Omega = \{1, 2, 3, 4, 5, 6\}$ and each outcome is equally likely.

(a) If the event E is rolling a two or a three, then $E = \{2, 3\}$ and there are $n(E) = 2$ possible outcomes. Therefore, the probability of rolling a two or a three is $P(E) = \dfrac{n(E)}{n(\Omega)} = \dfrac{2}{6} = \dfrac{1}{3}$.

(b) If the event E is rolling an odd number, then $E = \{1, 3, 5\}$ and there are $n(E) = 3$ possible outcomes. Therefore, the probability of rolling an odd number is $P(E) = \dfrac{n(E)}{n(\Omega)} = \dfrac{3}{6} = \dfrac{1}{2}$.

(c) If the event E is rolling a number divisible by 3, then $E = \{3, 6\}$ and there are $n(E) = 2$ possible outcomes. Therefore, the probability of rolling a number divisible by 3 is $P(E) = \dfrac{n(E)}{n(\Omega)} = \dfrac{2}{6} = \dfrac{1}{3}$.

31. The number of possible of outcomes when drawing a card from a standard deck is $n(\Omega) = 52$, each outcome being equally likely.

(a) Let event E be drawing a heart. There are 13 hearts in a deck so the number of outcomes is $n(E) = 13$. Therefore, the probability of drawing a heart is $P(E) = \dfrac{n(E)}{n(\Omega)} = \dfrac{13}{52} = \dfrac{1}{4}$.

(b) Let events E and F represent drawing a heart and drawing a spade respectively. There are 13 hearts and 13 spades in a deck so $n(E) = n(F) = 13$ and $P(E) = \dfrac{n(E)}{n(\Omega)} = \dfrac{13}{52} = \dfrac{1}{4} = P(F)$. Since events E and F are mutually exclusive, the probability of drawing either a heart or a spade is $P(E \cup F) = P(E) + P(F) = \dfrac{1}{4} + \dfrac{1}{4} = \dfrac{1}{2}$.

(c) Let event E be drawing a heart, a diamond, or a spade. There are 13 hearts, 13 diamonds, and 13 spades in a deck so the number of outcomes is $n(E) = 13 + 13 + 13 = 39$. Therefore, the probability of drawing a heart, a diamond, or a spade is $P(E) = \dfrac{n(E)}{n(\Omega)} = \dfrac{39}{52} = \dfrac{3}{4}$.

33. Each coin toss has two equally likely outcomes, either heads (H) or tails (T). Thus, the sample space for two coin tosses is $\Omega = \{(H, H), \{H, T\}, \{T, H\}, \{T, T\}\}$ and the total number of outcomes is $n(\Omega) = 4$.

(a) If the event E is getting heads at least one time, then $E = \{(H, H), \{H, T\}, \{T, H\}\}$ and there are $n(E) = 3$ possible outcomes. Therefore, the probability of getting heads at least one time is $P(E) = \dfrac{n(E)}{n(\Omega)} = \dfrac{3}{4}$.

© 2016 Cengage Learning. All Rights Reserved. May not be scanned, copied or duplicated, or posted to a publicly accessible website, in whole or in part.

(b) If the event E is getting heads exactly one time, then $E = \{\{H, T\}, \{T, H\}\}$ and there are $n(E) = 2$ possible outcomes.

Therefore, the probability of getting heads exactly one time is $P(E) = \dfrac{n(E)}{n(\Omega)} = \dfrac{2}{4} = \dfrac{1}{2}$.

35. There are a total of 8 balls in the jar, so the number of possible outcomes is $n(\Omega) = 8$, each outcome being equally likely.

(a) Let event E be drawing neither a white nor yellow ball. If we exclude white and yellow balls, there are 5 remaining balls to choose, so the number of outcomes is $n(E) = 5$. Therefore, the probability of drawing neither a white nor yellow ball is $P(E) = \dfrac{n(E)}{n(\Omega)} = \dfrac{5}{8}$.

Note: Drawing neither a white nor yellow ball is equivalent to drawing a red ball.

(b) Let event E be drawing a red, white, or yellow ball. There are 8 red, white, or yellow balls, so the number of outcomes is $n(E) = 8$. Therefore, the probability of drawing red, white, or yellow ball is $P(E) = \dfrac{n(E)}{n(\Omega)} = \dfrac{8}{8} = 1$.

(c) Let event E be drawing a non-white ball. There are 6 non-white balls in the jar so the number of outcomes is $n(E) = 6$.

Therefore, the probability of drawing a non-white ball is $P(E) = \dfrac{n(E)}{n(\Omega)} = \dfrac{6}{8} = \dfrac{3}{4}$.

37. (a)
$$\begin{aligned}\Omega = \{&(1,1),(1,2),(1,3),(1,4),(1,5),(1,6),\\ &(2,1),(2,2),(2,3),(2,4),(2,5),(2,6),\\ &(3,1),(3,2),(3,3),(3,4),(3,5),(3,6),\\ &(4,1),(4,2),(4,3),(4,4),(4,5),(4,6),\\ &(5,1),(5,2),(5,3),(5,4),(5,5),(5,6),\\ &(6,1),(6,2),(6,3),(6,4),(6,5),(6,6)\}\end{aligned}$$

The total number of outcomes is $n(\Omega) = 36$.

(b) If the event E is getting a sum of 7, then $E = \{(1,6),(2,5),(3,4),(4,3),(5,2),(6,1)\}$ and there are $n(E) = 6$ possible outcomes. Therefore, the probability of getting a sum of 7 is

$P(E) = \dfrac{n(E)}{n(\Omega)} = \dfrac{6}{36} = \dfrac{1}{6}$.

(c) If the event E is getting a sum of 9, then $E = \{(3,6),(4,5),(5,4),(6,3)\}$ and there are $n(E) = 4$ possible outcomes.

Therefore, the probability of getting a sum of 9 is $P(E) = \dfrac{n(E)}{n(\Omega)} = \dfrac{4}{36} = \dfrac{1}{9}$.

(d) If the event E is rolling doubles, then $E = \{(1,1),(2,2),(3,3),(4,4),(5,5),(6,6)\}$ and there are $n(E) = 6$ possible outcomes. Therefore, the probability of rolling doubles is $P(E) = \dfrac{n(E)}{n(\Omega)} = \dfrac{6}{36} = \dfrac{1}{6}$.

(e) If the event E is rolling doubles, then event E^c is not rolling doubles, that is, the two dice show different numbers. Using the Complement Rule and the result of part (d), the probability that the two dice show different numbers is $P(E^c) = 1 - P(E) = 1 - \dfrac{1}{6} = \dfrac{5}{6}$.

(f) If the event E is getting a sum of 9 or higher, then

$E = \{(3,6),(4,5),(4,6),(5,4),(5,5),(5,6),(6,3),(6,4),(6,5),(6,6)\}$ and there are $n(E) = 10$ possible outcomes. Therefore, the probability of getting a sum of 9 or higher is $P(E) = \dfrac{n(E)}{n(\Omega)} = \dfrac{10}{36} = \dfrac{5}{18}$.

© 2016 Cengage Learning. All Rights Reserved. May not be scanned, copied or duplicated, or posted to a publicly accessible website, in whole or in part.

39. The total number of 5 card hands is the number of combinations of 52 cards when taken 5 at a time, given by $n(\Omega) = \binom{52}{5} = 2,598,960$. Each of these outcomes is equally likely.

(a) There are a total of 13 hearts in a deck, so the number of ways of choosing 5 hearts is $\binom{13}{5} = 1287$. Therefore, the probability of drawing 5 hearts is $\dfrac{1287}{2,598,960} = \dfrac{33}{66,640}$.

(b) From Part (a), there are 1287 ways of drawing 5 hearts. Now, there are 4 different types of suits, so the number of ways of getting 5 cards of the same suit is $4 \cdot 1287 = 5148$. Therefore, the probability of this event is $\dfrac{5148}{2,598,960} = \dfrac{33}{16,660}$.

(c) Each of the 4 suits has 3 face cards, so there are a total of $3 \cdot 4 = 12$ face cards. The number of ways of choosing 5 face cards is $\binom{12}{5} = 792$. Therefore, the probability of this event is $\dfrac{792}{2,598,960} = \dfrac{33}{108,290}$.

(d) Since there are 4 suits, there are 4 ways of getting a royal flush. Therefore, the probability of obtaining a royal flush is $\dfrac{4}{2,598,960} = \dfrac{1}{649,740}$.

41. The possible offspring genotypes are $\Omega = \{\mathsf{Tt}, \mathsf{tt}\}$

(a) Only one outcome, Tt, will produce a tall offspring. Therefore, the probability of this event is $\dfrac{1}{2}$.

(b) Only one outcome, tt, will produce a short offspring. Therefore, the probability of this event is $\dfrac{1}{2}$.

43. As in Exercise 42, there are a total of $n(\Omega) = 8$ possible outcomes. Let event E represent having exactly one boy. The number of outcomes with exactly one boy is the number of combinations of the 3 children taken 1 at a time. So $n(E) = \binom{3}{1} = \dfrac{3!}{1!\,(3-1)!} = 3$ and the probability of having exactly one boy is therefore $P(E) = \dfrac{n(E)}{n(\Omega)} = \dfrac{3}{8}$.

45. The number of ways of picking a group of 6 moose from a population of 20 is given by $n(\Omega) = \binom{20}{6} = \dfrac{20!}{6!\,(20-6)!} = 38,760$. Now, the number of ways of selecting 3 tagged moose from 6 possibilities is $\binom{6}{3} = 20$, and the number of ways of selecting 3 non-tagged moose from 14 possibilities is $\binom{14}{3} = 364$. By the FPC, the number of ways of obtaining a group of 3 tagged and 3 non-tagged moose is $20 \cdot 364 = 7280$. Therefore, the probability of obtaining this outcome is $\dfrac{7280}{38,760} = \dfrac{182}{969}$.

47. The total number of different 5 cheetah samples selected from a population of 20 individuals is $\binom{20}{5} = 15,504$. Now, the number of ways of selecting 5 cheetahs all with gene A is $\binom{13}{5} = 1287$ and the number of ways of selecting 5 cheetahs all with gene B is $\binom{7}{5} = 21$. So the number of samples containing all the same gene is $1287 + 21 = 1308$, and the probability of this event is $\dfrac{1308}{15,504} = \dfrac{109}{1292}$.

© 2016 Cengage Learning. All Rights Reserved. May not be scanned, copied or duplicated, or posted to a publicly accessible website, in whole or in part.

49. If E is the event in which all genes in the sample are the same, then E^c is the event in which there are at least two different genes in the sample. Thus, we can compute $P(E)$ and then use the Complement Rule to determine $P(E^c)$. The total number of different 3 moose samples selected from a population of 25 individuals is $\binom{25}{3} = 2300$. Now, the number of ways of selecting 3 moose all with gene A, gene B, and gene C is $\binom{10}{3}$, $\binom{11}{3}$, and $\binom{4}{3}$ respectively. So the number of samples containing all the same gene is $\binom{10}{3} + \binom{11}{3} + \binom{4}{3} = 120 + 165 + 4 = 289$, and the probability of this event is $P(E) = \frac{289}{2300}$.

Therefore, the probability that there will be at least two different genes in the sample is $P(E^c) = 1 - P(E) = \frac{2011}{2300}$.

51. Assume $n(\Omega) \neq 0$, that is, the sample space contains at least one outcome. Dividing the inequality $0 \leq n(E) \leq n(\Omega)$ by $n(\Omega)$ gives $0 \leq \frac{n(E)}{n(\Omega)} \leq 1$. Substituting $P(E) = \frac{n(E)}{n(\Omega)}$ from Definition 2 gives $0 \leq P(E) \leq 1$, proving Property 3. To prove Property 4, we let $E = \Omega$ in Definition 2 which gives $P(\Omega) = \frac{n(\Omega)}{n(\Omega)} = 1$.

53. We can express the union of any two events E and F as $E \cup F = E \cup (E^c \cap F)$. Also, we can write $F = (E^c \cap F) \cup (E \cap F)$ (see Figure 6). Since $(E^c \cap F)$ and $(E \cap F)$ are mutually exclusive, the number of outcomes in their union is the sum of the number of outcomes in each event. So $n(F) = n(E^c \cap F) + n(E \cap F) \Leftrightarrow$ $n(E^c \cap F) = n(F) - n(E \cap F)$. Similarly, E and $(E^c \cap F)$ are mutually exclusive so $n(E \cup F) = n(E) + n(E^c \cap F) = n(E) + n(F) - n(E \cap F)$. Finally, dividing both sides by $n(\Omega)$, assuming $n(\Omega) \neq 0$, we get $\frac{n(E \cup F)}{n(\Omega)} = \frac{n(E)}{n(\Omega)} + \frac{n(F)}{n(\Omega)} - \frac{n(E \cap F)}{n(\Omega)}$ which after applying Definition 2 gives the result $P(E \cup F) = P(E) + P(F) - P(E \cap F)$ for equally likely outcomes.

55. (a) As in Exercise 12.1.45, the number of different ways of forming the two groups is given by $\binom{12}{4} = 495$. Also, the number of ways of getting 3 red balls and 1 non-red ball in group A is $\binom{5}{3} \cdot \binom{7}{1} = 10 \cdot 7 = 70$. Therefore, if the groups are created randomly so that each outcome is equally likely, then the probability of getting 3 of the red balls in group A is $\frac{70}{495} \approx 0.14$.

(b) The answer from Part (a) suggests there is a probability of about 0.14 of obtaining 3 red balls in group A if the groups were created randomly. This is almost three times bigger than the threshold value of 0.05, so we do not consider the probability of this outcome to be very small. Therefore, it seems possible that the balls were randomly assigned to the two groups.

57. (a) As in Exercise 12.1.47, the number of different ways for 14 of the 18 species to be fragile is $\binom{18}{14} = 3060$. Also, the number of different ways of having 9 cave-roosting species and 5 non-cave-roosting species is $\binom{11}{9} \cdot \binom{7}{5} = 55 \cdot 21 = 1155$. Therefore, if there is no relationship between cave use and endangerment status, then the probability of obtaining the given data is $\frac{1155}{3060} \approx 0.38$.

(b) Relative to a 0.05 threshold, the probability found in Part (a) is quite large, indicating that if there is no relationship between cave use and endangerment status, then there is about a 38% chance of obtaining the given data. Therefore, we cannot conclude that there is a relationship between cave use and endangerment status.

© 2016 Cengage Learning. All Rights Reserved. May not be scanned, copied or duplicated, or posted to a publicly accessible website, in whole or in part.

59. $\Omega = \{1,2,3,4,5,6\} \Rightarrow n(\Omega) = 6$

(a) The outcomes of each event are $E = \{4,5,6\}$ and $F = \{1,2,3,4\}$, so $E \cap F = \{4\}$ is nonempty which means E and F are not mutually exclusive. The probability of the event $E \cup F$ as given by Equation 8 is

$$P(E \cup F) = P(E) + P(F) - P(E \cap F) = \frac{n(E)}{n(\Omega)} + \frac{n(F)}{n(\Omega)} - \frac{n(E \cap F)}{n(\Omega)} = \frac{3}{6} + \frac{4}{6} - \frac{1}{6} = 1.$$

(b) The outcomes of each event are $E = \{3,6\}$ and $F = \{1,2\}$, so $E \cap F = \emptyset$ is empty which means E and F are mutually exclusive. The probability of the event $E \cup F$ as given by Equation 5 is

$$P(E \cup F) = P(E) + P(F) = \frac{n(E)}{n(\Omega)} + \frac{n(F)}{n(\Omega)} = \frac{2}{6} + \frac{2}{6} = \frac{2}{3}.$$

61. The number of outcomes in the sample space is $n(\Omega) = 52$.

(a) There are a total of $n(E) = 13$ clubs and $n(F) = 4$ kings in a deck. Also, there is $n(E \cap F) = 1$ card that is both a club and a king, so E and F are not mutually exclusive. The probability of the event $E \cup F$ as given by Equation 8 is

$$P(E \cup F) = P(E) + P(F) - P(E \cap F) = \frac{n(E)}{n(\Omega)} + \frac{n(F)}{n(\Omega)} - \frac{n(E \cap F)}{n(\Omega)} = \frac{13}{52} + \frac{4}{52} - \frac{1}{52} = \frac{16}{52} = \frac{4}{13}.$$

(b) There are a total of $n(E) = 4$ aces and $n(F) = 13$ spades in a deck. Also, there is $n(E \cap F) = 1$ card that is both an ace and a spade, so E and F are not mutually exclusive. The probability of the event $E \cup F$ as given by Equation 8 is

$$P(E \cup F) = P(E) + P(F) - P(E \cap F) = \frac{n(E)}{n(\Omega)} + \frac{n(F)}{n(\Omega)} - \frac{n(E \cap F)}{n(\Omega)} = \frac{4}{52} + \frac{13}{52} - \frac{1}{52} = \frac{16}{52} = \frac{4}{13}.$$

63. Let events H and T represent being infected with HIV and TB respectively. We can express each of the provided percentages as a fraction over 100 and since each individual in the population is equally likely to be selected, Definition (2) gives $P(H) = \frac{20}{100} = \frac{1}{5}$, $P(T) = \frac{40}{100} = \frac{2}{5}$, and $P(H \cap T) = \frac{15}{100} = \frac{3}{20}$. The probability of the individual having HIV or TB is $P(H \cup T) = P(H) + P(T) - P(H \cap T) = \frac{1}{5} + \frac{2}{5} - \frac{3}{20} = \frac{9}{20} = 0.45$.

65. Let event A represent Lab A producing a positive test and event B represent a positive test from Lab B. Then $P(A) = 0.1$, $P(B) = 0.05$, and $P(A \cap B) = 0.03$. Therefore, the probability that at least one lab produces a positive test is $P(A \cup B) = P(A) + P(B) - P(A \cap B) = 0.1 + 0.05 - 0.03 = 0.12$.

67. Consider an event A which contains all possible outcomes of an experiment, that is, $A = \Omega$. The complement event A^c includes all outcomes in Ω that are not in A, so $A^c = \emptyset$. The Complement Rule states that $P(A^c) = 1 - P(A) \Rightarrow P(\emptyset) = 1 - P(\Omega)$ and using Axiom 6(b), we have $P(\emptyset) = 1 - 1 = 0$.

69. Let E_i represent the ith outcome of a simple event in Ω, and since there are $n(\Omega)$ outcomes, we have $i = 1, \ldots, n(\Omega)$. Since the simple events are by definition pairwise disjoint, we can make use of the Exercise 68 result $P\left(\bigcup_{i=1}^{n(\Omega)} E_i\right) = P(E_1) + P(E_2) + \cdots + P\left(E_{n(\Omega)}\right)$. Now, all simple events are equally likely so $P(E_1) = P(E_2) = \cdots = P\left(E_{n(\Omega)}\right)$, and $\Omega = \bigcup_{i=1}^{n(\Omega)} E_i$. Thus, the probability expression from Exercise 68 becomes

$$P(\Omega) = n(\Omega)\,P(E_1) \Rightarrow P(E_1) = \frac{P(\Omega)}{n(\Omega)} = \frac{1}{n(\Omega)}$$ where we have used Axiom 6(b) in the final step. This indicates that a single outcome has probability $1/n(\Omega)$, so if event A consists of $n(A)$ of these outcomes, then the union rule gives

$$P(A) = \underbrace{\frac{1}{n(\Omega)} + \frac{1}{n(\Omega)} + \cdots + \frac{1}{n(\Omega)}}_{n(A)\text{ times}} = \frac{n(A)}{n(\Omega)}.$$

© 2016 Cengage Learning. All Rights Reserved. May not be scanned, copied or duplicated, or posted to a publicly accessible website, in whole or in part.

12.3 Conditional Probability

1. The sample space consists of the following equally likely outcomes $\Omega = \{1, 2, 3, 4, 5, 6\}$, so $n(\Omega) = 6$.

(a) Let F and T represent the events 'five' and 'greater than three'. Now, $F = \{5\}$, $T = \{4, 5, 6\}$ and $F \cap T = \{5\}$, so $P(T) = 3/6$ and $P(F \cap T) = 1/6$. Therefore, $P(F|T) = \dfrac{P(F \cap T)}{P(T)} = \dfrac{1/6}{3/6} = \dfrac{1}{3}$. Thus, the fraction of all outcomes greater than 3 that are also equal to 4 is $\frac{1}{3}$.

(b) Let T and O represent the events 'three' and 'odd'. Now, $T = \{3\}$, $O = \{1, 3, 5\}$ and $T \cap O = \{3\}$, so $P(O) = 3/6$ and $P(T \cap O) = 1/6$. Therefore, $P(T|O) = \dfrac{P(T \cap O)}{P(O)} = \dfrac{1/6}{3/6} = \dfrac{1}{3}$. Thus, the fraction of all odd outcomes that are also equal to 3 is $\frac{1}{3}$.

3. The sample space consists of the following outcomes $\Omega = \{(H, H), (H, T), (T, H), (T, T)\}$, so $n(\Omega) = 4$.

(a) Let A and B represent the events 'first toss heads' and 'two heads'. Then $A = \{(H, H), (H, T)\}$, $B = \{(H, H)\}$ and $B \cap A = \{(H, H)\}$, so $P(A) = 2/4$ and $P(B \cap A) = 1/4$. Therefore, $P(B|A) = \dfrac{P(B \cap A)}{P(A)} = \dfrac{1/4}{2/4} = \dfrac{1}{2}$. Thus, the fraction of 'first toss heads' outcomes that are also 'two heads' is $\frac{1}{2}$.

(b) Let A and B represent the events 'first toss tails' and 'both tosses same'. Then $A = \{(T, H), (T, T)\}$, $B = \{(H, H), (T, T)\}$ and $B \cap A = \{(T, T)\}$, so $P(A) = 2/4$ and $P(B \cap A) = 1/4$. Therefore, $P(B|A) = \dfrac{P(B \cap A)}{P(A)} = \dfrac{1/4}{2/4} = \dfrac{1}{2}$. Thus, the fraction of 'first toss tails' outcomes that also have 'both tosses the same' is $\frac{1}{2}$.

(c) Let A and B represent the events 'at least one head' and 'two heads'. Then $A = \{(H, H), (H, T), (T, H)\}$, $B = \{(H, H)\}$ and $B \cap A = \{(H, H)\}$, so $P(A) = 3/4$ and $P(B \cap A) = 1/4$. Therefore, $P(B|A) = \dfrac{P(B \cap A)}{P(A)} = \dfrac{1/4}{3/4} = \dfrac{1}{3}$. Thus, the fraction of 'at least one head' outcomes that also have two heads is $\frac{1}{3}$.

5. The sample space consists of the different possible color outcomes, either black (B) or white (W), for the two draws. This is given by $\Omega = \{(B, B), (B, W), (W, B), (W, W)\}$. Note that each outcome is not *equally* likely in this case.

(a) Let F and T represent the events 'first ball black' and 'both balls black'. Then $F = \{(B, B), (B, W)\}$, $B = \{(B, B)\}$ and $T \cap F = \{(B, B)\}$. There are 5 black balls out of a total of 12 balls, so the probability of obtaining a black on a single draw is $\frac{5}{12}$ and the probability of obtaining a white is $\frac{7}{12}$. Since the balls are drawn with replacement, the two draws are independent events. Thus, using the Multiplication and Union Rules, we find $P(F) = P(\{(B, B)\}) + P(\{(B, W)\}) = \frac{5}{12} \cdot \frac{5}{12} + \frac{5}{12} \cdot \frac{7}{12} = \frac{5}{12}$ and $P(T \cap F) = P(\{(B, B)\}) = \frac{5}{12} \cdot \frac{5}{12} = \frac{35}{144}$. Therefore, $P(T|F) = \dfrac{P(T \cap F)}{P(F)} = \dfrac{25/144}{5/12} = \dfrac{5}{12}$. Therefore, the fraction of 'first ball black' outcomes that also have a second black ball is $5/12$.

(b) Let A and T represent the events 'at least one white ball' and 'both balls white'. Then $A = \{(B, W), (W, B), (W, W)\}$, $T = \{(W, W)\}$ and $T \cap A = \{(W, W)\}$. As in part (a), the Multiplication and Union Rules are used to calculate the following probabilities: $P(A) = P(\{(B, W)\}) + P(\{(W, B)\}) + P(\{(W, W)\}) = \frac{5}{12} \cdot \frac{7}{12} + \frac{7}{12} \cdot \frac{5}{12} + \frac{7}{12} \cdot \frac{7}{12} = \frac{119}{144}$

© 2016 Cengage Learning. All Rights Reserved. May not be scanned, copied or duplicated, or posted to a publicly accessible website, in whole or in part.

and $P(T \cap A) = P(\{(W, W)\}) = \frac{7}{12} \cdot \frac{7}{12} = \frac{49}{144}$. Therefore, $P(T|A) = \dfrac{P(T \cap A)}{P(A)} = \dfrac{49/144}{119/144} = \dfrac{49}{119} = \dfrac{7}{17}$.
Thus, the fraction of 'at least one white ball' outcomes that also have two white balls is $7/17$.

7. (a) Let X^m represent a chromosome with an X-linked mutation. Then, the sample space consists of the following outcomes:

$$\Omega = \{\mathsf{XX}, \mathsf{XY}, \mathsf{X}^m\mathsf{X}, \mathsf{XX}^m, \mathsf{X}^m\mathsf{X}^m, \mathsf{X}^m\mathsf{Y}\}$$

(b) Let F and M represent the respective outcomes female and male. Female offspring receive two X chromosomes each having a probability μ of carrying the mutation. Thus, using the Fundamental Principle of Counting, we have $P(\mathsf{X}^m\mathsf{X}^m|F) = \mu \times \mu = \mu^2$, $P(\mathsf{XX}^m|F) = P(\mathsf{X}^m\mathsf{X}|F) = \mu(1-\mu)$ and $P(\mathsf{XX}|F) = (1-\mu)(1-\mu) = (1-\mu)^2$. Male offspring receive a single X chromosome, so the probabiliy of having the disease is $P(\mathsf{X}^m\mathsf{Y}|M) = \mu$ and the probability of not having the disease is $P(\mathsf{XY}|M) = (1-\mu)$. These results are displayed in the tree diagram at right. Genotypes that are shaded gray correspond to individuals with the disease.

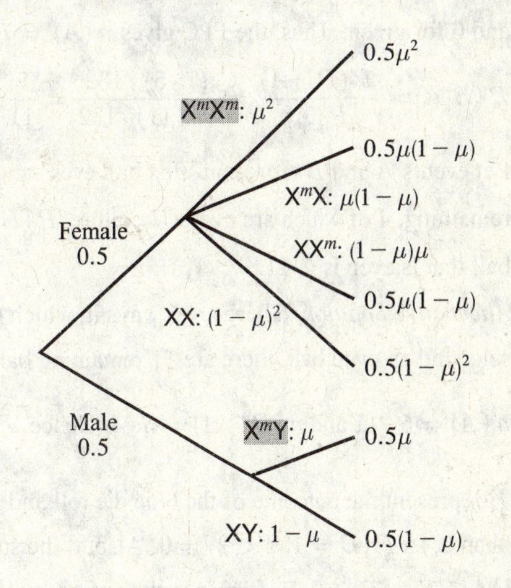

(c) Let D represent the outcome 'has disease'. Using the Law of Total Probability given in Equation 5, the probability of having the disease is $P(D) = 0.5\mu + 0.5\mu^2 = 0.5\mu(1+\mu)$.

(d) Referring to the tree diagram in part (b), we find

$$P(D|M) = \frac{P(D \cap M)}{P(M)} = \frac{0.5\mu}{0.5} = \mu$$

(e) Referring to the tree diagram in part (b), we find

$$P(D|F) = \frac{P(D \cap F)}{P(F)} = \frac{0.5\mu^2}{0.5} = \mu^2$$

9. The number of different outcomes when drawing 2 balls from a jar of 12 is $P_{12,2} = 132$ and each outcome is equally likely.

(a) Let events A and B represent 'first ball red' and 'second ball red'. If the first ball drawn is red, then there are 11 balls remaining, 4 of which are red. Therefore, $P(B|A) = 4/11$. The fraction of 'first ball red' outcomes that have a second red ball is $4/11$.

Alternative Solution: There are 5 ways in which to select a red ball first, after which there are 11 remaining balls to choose from, 4 of which are red. Thus, the FPC gives $n(A) = 5 \cdot 11$ and $n(B \cap A) = 5 \cdot 4$. Hence,

$$P(B|A) = \frac{P(B \cap A)}{P(A)} = \frac{(5 \cdot 4)/132}{(5 \cdot 11)/132} = \frac{4}{11}.$$

(b) Let events A and B represent 'first ball green' and 'second ball red'. If the first ball drawn is green, then there are 11 balls remaining, 5 of which are red. Therefore, $P(B|A) = 5/11$. The fraction of 'first ball green' outcomes that have a 'second ball red' is $5/11$.

Alternative Solution: There are 7 ways in which to select a green ball first, after which there are 11 remaining balls to choose from, 5 of which are red. Thus, the FPC gives $n(A) = 7 \cdot 11$ and $n(B \cap A) = 7 \cdot 5$. Hence,

$$P(B|A) = \frac{P(B \cap A)}{P(A)} = \frac{(7 \cdot 5)/132}{(7 \cdot 11)/132} = \frac{5}{11}.$$

© 2016 Cengage Learning. All Rights Reserved. May not be scanned, copied or duplicated, or posted to a publicly accessible website, in whole or in part.

(c) Let events A and B represent 'first ball odd' and 'second ball even'. If the first ball drawn is odd, then there are 11 balls remaining, 5 of which are even. Therefore, $P(B|A) = 5/11$. The fraction of 'first ball odd' outcomes that have a second ball that is even is $5/11$.

Alternative Solution: There are 7 ways in which to select an odd ball first: 1, 3, 5 for red and 1, 3, 5, 7 for green. After selecting an odd ball, there are 11 remaining balls to choose from, 5 of which are even: 2 and 4 for red and 2, 4, and 6 for green. Thus, the FPC gives $n(A) = 7 \cdot 11$ and $n(B \cap A) = 7 \cdot 5$. Hence,

$$P(B|A) = \frac{P(B \cap A)}{P(A)} = \frac{(7 \cdot 5)/132}{(7 \cdot 11)/132} = \frac{5}{11}.$$

(d) Let events A and B represent 'first ball even' and 'second ball even'. If the first ball drawn is even, then there are 11 balls remaining, 4 of which are even. Therefore, $P(B|A) = 4/11$. The fraction of 'first ball even' outcomes that have a second ball that is even is $4/11$.

Alternative Solution: There are 5 ways in which to select an even ball first: 2 and 4 for red and 2, 4, and 6 for green. After selecting an even ball, there are 11 remaining balls to choose from, now 4 of which are even. Thus, the FPC gives $n(A) = 5 \cdot 11$ and $n(B \cap A) = 5 \cdot 4$. Hence, $P(B|A) = \frac{P(B \cap A)}{P(A)} = \frac{(5 \cdot 4)/132}{(5 \cdot 11)/132} = \frac{4}{11}$.

11. Let B represent the outcome of the blue die roll and S represent the sum of the two dice rolled. It is impossible to roll a sum less than 2, so $P(B = 1|S < 2) = 0$. Also, if the sum rolled is greater than 7, then the blue die must have been at least 2, so $P(B = 1|S > 7) = 0$. Referring to the outcomes of a two die roll shown in Figure 3, observe that there is 1 outcome with $S = 2$, 2 outcomes with $S = 3$, 3 outcomes with $S = 4$, and so on, up to $S = 7$. Therefore, if $2 \le k \le 7$, then $n(S = k) = k - 1$ so that $P(B = 1|S = k) = \frac{1}{k-1}$. Combining the previous results we get the follow piecewise conditional probability function:

$$P(B = 1|S = k) = \begin{cases} 0 & \text{if } k < 2 \text{ or } k > 7 \\ \frac{1}{k-1} & \text{if } 2 \le k \le 7 \end{cases}$$

13. The possible offspring genotypes of a AT × AA cross are {AA, AT}, both of which have axial flowers and which are both equally likely. If X and G represent the events 'axial' and 'genotype AT', then $X = \{\text{AA}, \text{AT}\}$, $G = \{\text{AT}\}$ and $G \cap X = \{\text{AT}\}$. Therefore, $P(G|X) = \frac{P(G \cap X)}{P(X)} = \frac{1/2}{2/2} = \frac{1}{2}$. Hence, the fraction of axial offspring that have genotype AT is $1/2$.

15. $P(C|U) = \frac{P(C \cap U)}{P(U)} = \frac{0.32}{0.78} \approx 0.41$

17. $P(O|U) = \frac{P(O \cap U)}{P(U)} = \frac{0.09}{0.78} \approx 0.12$

19. $P(D|O) = \frac{P(D \cap O)}{P(O)} = \frac{0.01}{0.1} \approx 0.1$

21. If M and F represent the outcomes male and female, then the sample space consists of the following equally likely outcomes $\Omega = \{(M, M), (M, F), (F, M), (F, F)\} \Rightarrow n(\Omega) = 4$.

(a) Let B and A represent the events 'both chicks male' and 'at least 1 male chick'. Then $A = \{(M, M), (M, F), (F, M)\}$, $B = \{(M, M)\}$ and $B \cap A = \{(M, M)\}$, so that $n(A) = 3$ and $n(B \cap A) = 1$. Therefore, $P(B|A) = \frac{P(B \cap A)}{P(A)} = \frac{1/4}{3/4} = \frac{1}{3}$. Hence, the fraction of 'at least 1 male chick' outcomes that contain two male chicks is $1/3$.

© 2016 Cengage Learning. All Rights Reserved. May not be scanned, copied or duplicated, or posted to a publicly accessible website, in whole or in part.

(b) Let B and F represent the events 'both chicks female' and 'first chick female'. Then $F = \{(F, M), (F, F)\}$, $B = \{(F, F)\}$ and $B \cap F = \{(F, F)\}$, so that $n(F) = 2$ and $n(B \cap A) = 1$. Therefore, $P(B|F) = \dfrac{P(B \cap F)}{P(F)} = \dfrac{1/4}{2/4} = \dfrac{1}{2}$. Hence, the fraction of 'first chick female' outcomes that contain two female chicks is $1/2$.

23. If E is a subset of F, then $F \cap E = E$ as illustrated in the Venn diagram at right.

Therefore $P(F|E) = \dfrac{P(F \cap E)}{P(E)} = \dfrac{P(E)}{P(E)} = 1$.

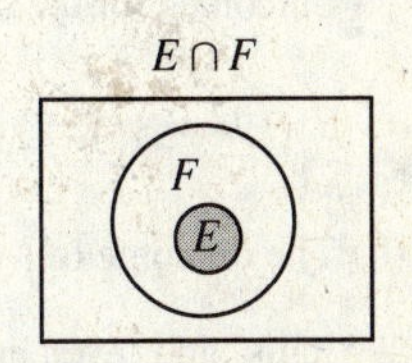

25. (a) Let F and S represent the events 'first ball is black' and 'second ball is white'. For the first draw, there are 7 black balls out of a total of 10 balls, so $P(F) = \frac{7}{10}$. If a black ball is selected in the first draw, then 3 of the 9 remaining balls are white, so $P(S|F) = \frac{3}{9} = \frac{1}{3}$. Therefore, the probability of drawing a black then a white ball is $P(F \cap S) = P(S|F)\,P(F) = \left(\frac{1}{3}\right)\left(\frac{7}{10}\right) = \frac{7}{30}$.

(b) Let F and S represent the events 'first ball is black' and 'second ball is black'. For the first draw, there are 7 black balls out of a total of 10 balls, so $P(F) = \frac{7}{10}$. If a black ball is selected in the first draw, then 6 of the 9 remaining balls are black, so $P(S|F) = \frac{6}{9} = \frac{2}{3}$. Therefore, the probability of drawing two black balls is $P(F \cap S) = P(S|F)\,P(F) = \left(\frac{2}{3}\right)\left(\frac{7}{10}\right) = \frac{7}{15}$.

27. (a) There are 6 possible outcomes for each die roll, so there are a total of $6^2 = 36$ different equally likely outcomes for the two die rolls and only 1 way to roll two sixes. Thus, $P(E \cap F) = \frac{1}{36}$.

(b) The outcome of the second die roll is unaffected by the outcome of the first and vice versa. Therefore, we expect that the events E and F will be independent. Verifying this mathematically, $P(E) = \frac{1}{6}$ and $P(F) = \frac{1}{6}$ $\Rightarrow$ $P(E)\,P(F) = \frac{1}{36} = P(E \cap F)$ which implies that E and F are independent by Definition 4.

29.

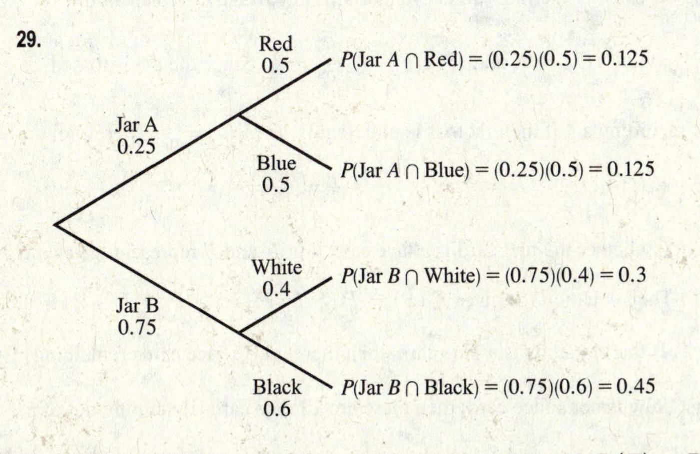

31. $P(E) = P(\{1\}) + P(\{2\}) = 0.16 + 0.04 = 0.2$ and $P(F) = P(\{3\}) + P(\{4\}) = 0.16 + 0.64 = 0.8$ so $P(E)\,P(F) = (0.2)(0.8) = 0.16$. Also, $E \cap F = \emptyset$ so $P(E \cap F) = 0 \neq P(E)\,P(F)$. Therefore, E and F are *not* independent events.

© 2016 Cengage Learning. All Rights Reserved. May not be scanned, copied or duplicated, or posted to a publicly accessible website, in whole or in part.

33. $P(E) = P(\{1\}) + P(\{2\}) = 0.16 + 0.04 = 0.2$ and $P(F) = P(\{2\}) + P(\{3\}) = 0.04 + 0.16 = 0.2$ so $P(E)P(F) = (0.2)(0.2) = 0.04$. Also, $E \cap F = \{2\}$ so $P(E \cap F) = P(\{2\}) = 0.04 = P(E)P(F)$. Therefore, E and F are independent events.

35. (a) Let T and E represent the events 'tails' and 'even number'. There are 2 possible outcomes for the coin flip and 6 possible outcomes for the die roll, so the number of outcomes in the sample space is $n(\Omega) = 2 \cdot 6 = 12$. Since $T \cap E = \{(T, 2), (T, 4), (T, 6)\}$, we have $P(T \cap E) = \dfrac{3}{12} = \dfrac{1}{4}$.

(b) The outcome of a die roll is unaffected by the outcome of a coin flip and vice versa. Therefore, we expect that the events 'tails' and 'even number' will be independent. Verifying this mathematically, we have $P(T) = \frac{1}{2}$ and $P(E) = \frac{1}{2}$ $\Rightarrow$ $P(T)P(E) = \frac{1}{4} = P(T \cap E)$ which implies that T and E are independent by Definition 4.

37. 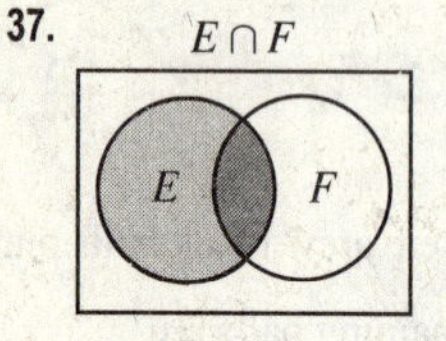

If events E and F are independent then $P(E \cap F) = P(E)P(F)$ $\Rightarrow$ $P(F) = P(E \cap F)/P(E)$. Thus, if events E and F are independent, the fraction of the area of E that overlaps with F is the same as the area of F in the Venn diagram.

39. Using M and F to denote the outcomes male and female, the sample space consists of the following outcomes $\Omega = \{(M, M), (M, F), (F, M), (F, F)\}$. Let event A represent having a boy then a girl and let event B represent having a girl then a boy. So $A = \{(M, F)\}$, $B = \{(F, M)\}$ and $A \cap B = \{\}$. Since $P(M) = P(F) = \frac{1}{2}$ and the genders of the first and second children are independent, we have $P(A) = P(M)P(F) = \frac{1}{2} \times \frac{1}{2} = \frac{1}{4}$ and $P(B) = P(F)P(M) = \frac{1}{2} \times \frac{1}{2} = \frac{1}{4}$. Therefore, using the Union Rule, the probability of having a daughter and a son is $P(A \cup B) = P(A) + P(B) - P(A \cap B) = \frac{1}{4} + \frac{1}{4} - 0 = \frac{1}{2}$.

41. We will use the Law of Total Probability, conditioning on whether the first roll is even or odd. Let E and O represent the events 'first roll even' and 'first roll odd'. Then $P(E) = \frac{1}{2}$, $P(O) = \frac{1}{2}$, and $P(4|E) = \dfrac{1/6}{1/2} = \frac{1}{3}$. Since the die is tossed again if the first roll is odd, there is a 1 in 6 chance of recording a 4 if the first toss is odd, that is, $P(4|O) = \frac{1}{6}$. Therefore, the probability of recording a 4 is $P(4) = P(4|E)P(E) + P(4|O)P(O) = \frac{1}{3} \times \frac{1}{2} + \frac{1}{6} \times \frac{1}{2} = \frac{3}{12} = \frac{1}{4}$.

43. We will use the Law of Total Probability, conditioning on whether the first card is a face card. Let F and S represent the events 'first draw is face card' and 'second draw is face card'. Then equation (5) gives $P(S) = P(S|F)P(F) + P(S|F^c)P(F^c)$. Now, $P(F) = \frac{12}{52} = \frac{3}{13}$ and $P(F^c) = 1 - \frac{3}{13} = \frac{10}{13}$. If the first draw is a face card, then there are 11 face cards remaining out of a total of 51 cards, so $P(S|F) = \frac{11}{51}$. If the first draw is not a face card, then there are 12 face cards remaining out of a total of 51 cards, so $P(S|F^c) = \frac{12}{51}$. Therefore, the probability that the second card is a face card is $P(S) = \left(\frac{11}{51}\right)\left(\frac{3}{13}\right) + \left(\frac{12}{51}\right)\left(\frac{10}{13}\right) = \frac{153}{663} = \frac{3}{13}$.

© 2016 Cengage Learning. All Rights Reserved. May not be scanned, copied or duplicated, or posted to a publicly accessible website, in whole or in part.

45. We define our sample space as the collection of all females that carry mutations and the events C, $B1$ and $B2$ as 'develops cancer', 'mutation in BCRCA1 gene' and 'mutation in BCRCA2 gene'. We are given $P(B1) = 1/3$, $P(B2) = 2/3$, $P(C|B1) = 3/5$, and $P(C|B2) = 1/5$. Using Equation (5), the probability that a female will develop breast cancer is $P(C) = P(C|B1)P(B1) + P(C|B2)P(B2) = \frac{3}{5} \times \frac{1}{3} + \frac{1}{5} \times \frac{2}{3} = \frac{1}{3}$.

47. Let F, M and H denote the events 'female', 'male', and 'has heart attack'. We can express each of the provided percentages as a fraction over 100 and since each individual in the population is equally likely to be selected, Definition 12.2.2 gives $P(F) = 0.51$, $P(H|F) = 0.002$, and $P(H|M) = 0.003$. Using the Complement Rule, we get $P(M) = 1 - 0.51 = 0.49$. Using Equation (5), the probability that an individual will have a heart attack is

$$P(H) = P(H|M)P(M) + P(H|F)P(F) = (0.003)(0.49) + (0.002)(0.51) = 0.00249 \approx 0.25\%.$$

49. A plant with terminal flowers must have genotype TT and a plant produced from a AA × AT cross may be AA or AT with equal probability. If the second parent is AA, then the cross between the two parents, namely TT × AA, will always produce an AT offspring. If instead the second parent is AT, then the cross between the two parents, namely TT × AT, can produce the following offsprings {AT, TT} with each outcome being equally likely. Using E and F to denote the events 'second parent is AA' and 'second parent is AT', we have $P(E) = P(F) = \frac{1}{2}$, so the probability of producing an axial offspring is

$$P(\text{axial}) = P(\text{axial}|E)P(E) + P(\text{axial}|F)P(F) = \frac{1}{1} \cdot \frac{1}{2} + \frac{1}{2} \cdot \frac{1}{2} = \frac{3}{4}$$

51. $P(F|E) = \dfrac{P(E|F)P(F)}{P(E)} = \dfrac{0.3 \times 0.1}{0.4} = 0.075$

53. $P(F|E) = \dfrac{P(E|F)P(F)}{P(E)} = \dfrac{0.4 \times 0.25}{0.4} = 0.25$

55. Let events A and B represent a mosquito carrying serotype A and B respectively. We can express each of the provided percentages as a fraction over 100 and since each mosquito in the population is equally likely to be selected, Definition (2) gives $P(B|A) = \frac{20}{100} = \frac{1}{5}$, $P(A) = \frac{10}{100} = \frac{1}{10}$, and $P(B) = \frac{35}{100} = \frac{7}{20}$. The probability of a mosquito carrying serotype A given that it carries serotype B is $P(A|B) = \dfrac{P(B|A)P(A)}{P(B)} = \dfrac{\frac{1}{5} \times \frac{1}{10}}{\frac{7}{20}} = \frac{2}{35} \approx 0.057$.

57. Let M and C denote the events 'carries mutant gene' and 'develops cancer' respectively. So $P(C|M) = 3/5 = 0.6$, $P(M) = 0.01$, and $P(C) = 0.1$. Then the probability that a randomly chosen cancer patient will carry the mutant BRCA1 gene is $P(M|C) = \dfrac{P(C|M)P(M)}{P(C)} = \dfrac{0.6 \times 0.01}{0.1} = 0.06$.

59. Let T^+, T^- and D denote the events 'a positive test result', 'a negative test result' and 'has diabetes' respectively, so that the complement event D^c is not having diabetes. We can express each of the provided percentages as a fraction over 100 and since each individual in the population is equally likely to be selected, Definition (2) gives $P(T^+|D) = 0.9$, $P(T^-|D^c) = 0.95$ and $P(D) = 0.083$. Using the Complement Rule, we find $P(D^c) = 1 - P(D) = 0.917$ and $P(T^-|D) = 1 - P(T^+|D) = 0.1$. Also, using equation (5) the probability of a negative test result is $P(T^-) = P(T^-|D)P(D) + P(T^-|D^c)P(D^c) = (0.1)(0.083) + (0.95)(0.917) = 0.87945$. Therefore, the probability of a randomly chosen US citizen having diabetes if their test is negative is

$$P(D|T^-) = \frac{P(T^-|D)P(D)}{P(T^-)} = \frac{0.1 \times 0.083}{0.87945} \approx 0.0094.$$

© 2016 Cengage Learning. All Rights Reserved. May not be scanned, copied or duplicated, or posted to a publicly accessible website, in whole or in part.

12.4 Discrete Random Variables

1. There are 5 possible values for X with each value being equally likely. Therefore, $p_i = 1/5$ for $i = 1, 2, \ldots, 5$.

(a)

i	PDF p_i
1	0.2
2	0.2
3	0.2
4	0.2
5	0.2

(b)

3. The sample space for three coin flips consists of the following outcomes $\Omega = \{HHH, HHT, HTH, THH, HTT, THT, TTH, TTT\} \Rightarrow n(\Omega) = 8$. If X is a random variable representing the number of tails, then the PDF is given by

$$P(X = 0) = n(0)/n(\Omega) = 1/8$$
$$P(X = 1) = n(1)/n(\Omega) = 3/8$$
$$P(X = 2) = n(2)/n(\Omega) = 3/8$$
$$P(X = 3) = n(3)/n(\Omega) = 1/8$$

Alternative Solution: X is a binomial random variable with $n = 3$ trials and a probability of success (tails) of $p = 1/2$. Therefore, $P(X = 0) = \binom{3}{0}\left(\frac{1}{2}\right)^0\left(\frac{1}{2}\right)^3 = \frac{1}{8}$, $P(X = 1) = \binom{3}{1}\left(\frac{1}{2}\right)^1\left(\frac{1}{2}\right)^2 = \frac{3}{8}$, $P(X = 2) = \binom{3}{2}\left(\frac{1}{2}\right)^2\left(\frac{1}{2}\right)^1 = \frac{3}{8}$, and $P(X = 3) = \binom{3}{3}\left(\frac{1}{2}\right)^3\left(\frac{1}{2}\right)^0 = \frac{1}{8}$.

5. The number of possible outcomes of rolling two dice is $n(\Omega) = 6^2 = 36$ and the sum of two dice must lie between 2 and 12 inclusive. Using the table of outcomes in Figure 12.3.3 to count the number of ways of achieving each sum, we find the PDF is given by

$$P(X = 2) = n(2)/n(\Omega) = 1/36$$
$$P(X = 3) = n(3)/n(\Omega) = 2/36 = 1/18$$
$$P(X = 4) = n(4)/n(\Omega) = 3/36 = 1/12$$
$$P(X = 5) = n(5)/n(\Omega) = 4/36 = 1/9$$
$$P(X = 6) = n(6)/n(\Omega) = 5/36$$
$$P(X = 7) = n(7)/n(\Omega) = 6/36 = 1/6$$
$$P(X = 8) = n(8)/n(\Omega) = 5/36$$
$$P(X = 9) = n(9)/n(\Omega) = 4/36 = 1/9$$
$$P(X = 10) = n(10)/n(\Omega) = 3/36 = 1/12$$
$$P(X = 11) = n(11)/n(\Omega) = 2/36 = 1/18$$
$$P(X = 12) = n(12)/n(\Omega) = 1/36$$

7. There are $n(\Omega) = 36$ different outcomes when two dice are rolled. We construct a table below for values of $X = i$ and calculate the corresponding probabilities.

i	Outcomes	$n(X = i)$	$P(X = i)$
0	$(1,1), (2,2), (3,3), (4,4), (5,5), (6,6)$	6	$6/36$
1	$(1,2), (2,1), (2,3), (3,2), (3,4), (4,3), (4,5), (5,4), (5,6), (6,5)$	10	$10/36$
2	$(1,3), (3,1), (2,4), (4,2), (3,5), (5,3), (4,6), (6,4)$	8	$8/36$
3	$(1,4), (4,1), (2,5), (5,2), (3,6), (6,3)$	6	$6/36$
4	$(1,5), (5,1), (2,6), (6,2)$	4	$4/36$
5	$(1,6), (6,1)$	2	$2/36$

Observe that the PDF for X can be expressed as the piecewise function $P(X = i) = \begin{cases} 1/6 & \text{if } i = 0 \\ (-2i + 12)/36 & \text{if } 1 \le i \le 5 \end{cases}$.

© 2016 Cengage Learning. All Rights Reserved. May not be scanned, copied or duplicated, or posted to a publicly accessible website, in whole or in part.

9. There are a total of 52 cards in a deck 12 of which are face cards. The number of different 3 card combinations is $\binom{52}{3} = 22,100$ and the number of ways of drawing no face cards is $\binom{12}{0}\binom{40}{3} = 9880$ where we have used the Multiplication Rule and the fact that 40 cards are not face cards. Therefore, the probability of obtaining no face cards is

$P(X=0) = \dfrac{9880}{22100} = \dfrac{38}{85}$. Similarly, $P(X=1) = \dfrac{\binom{12}{1}\binom{40}{2}}{22100} = \dfrac{36}{85}$, $P(X=2) = \dfrac{\binom{12}{2}\binom{40}{1}}{22100} = \dfrac{132}{1105}$, and

$P(X=3) = \dfrac{\binom{12}{3}\binom{40}{0}}{22100} = \dfrac{11}{1105}$.

11. There are a total of 9 balls, 2 of which are red. Therefore, $P(X=0) = 7/9$ and $P(X=1) = 2/9$.

13. The total number of 3 ball combinations chosen from a jar containing 9 balls is $\binom{9}{3} = 84$ with each outcome being equally likely. The number of ways of selecting a group of 3 white balls (or no red balls) from 7 possibilities is $\binom{7}{3} = 35$. The number of ways of selecting a group of 2 white ball and 1 red ball is $\binom{7}{2}\binom{2}{1} = 42$ where we have used the Multiplication Rule. Also, the number of ways of selecting a group of 1 white ball and 2 red ball is $\binom{7}{1}\binom{2}{2} = 7$. Note it is impossible to select a group of 3 red balls since there are only 2 red balls in the jar. Therefore, the probability of obtaining no red balls is $P(X=0) = \frac{35}{84} = \frac{5}{12}$. Similarly, the probability of obtaining 1, 2 or 3 red balls is $P(X=1) = \frac{42}{84} = \frac{1}{2}$, $P(X=2) = \frac{7}{84} = \frac{1}{12}$ and $P(X=3) = 0$ respectively.

15. (a) The PDF must satisfy $\sum\limits_i p_i = 1 \;\Rightarrow\; \sum\limits_{i=1}^{10} k\left(\frac{1}{4}\right)^i = 1 \;\Rightarrow\; k\left[\dfrac{\frac{1}{4}\left(1-(1/4)^{10}\right)}{1-1/4}\right] = 1$ [using Formula 2.1.5] $\Rightarrow$

$\dfrac{1}{3}\left(\dfrac{1,048,575}{1,048,576}\right)k = 1 \;\Rightarrow\; k = \dfrac{1,048,576}{349,525}$.

(b) The PDF must satisfy $\sum\limits_i p_i = 1 \;\Rightarrow\; \sum\limits_{i=1}^{n} k\left(\frac{1}{4}\right)^i = 1 \;\Rightarrow\; k\left[\dfrac{\frac{1}{4}\left(1-(1/4)^{n}\right)}{1-1/4}\right] = 1$ [using Formula 2.1.5] $\Rightarrow$

$\frac{1}{3}\left(1-\left(\frac{1}{4}\right)^n\right)k = 1 \;\Rightarrow\; k = \dfrac{3}{1-\left(\frac{1}{4}\right)^n}$.

17. (a)

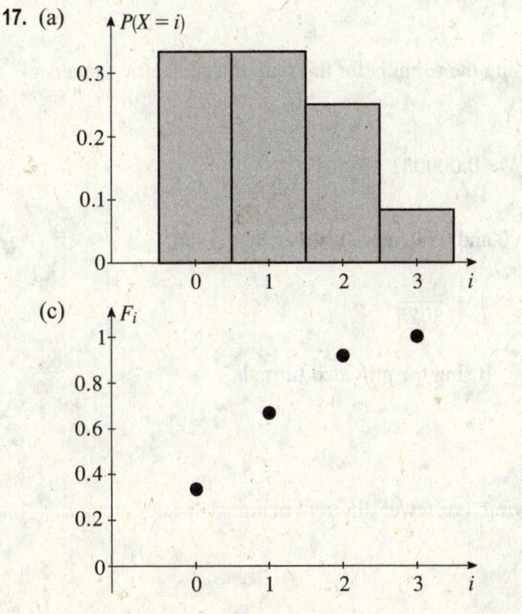

(b) We use Definition 5 along with the method of Example 6 to construct a table of CDF values. For example,

$F_0 = P(X \le 0) = \frac{1}{3}$

$F_1 = p_1 + F_0 = \frac{1}{3} + \frac{1}{3} = \frac{2}{3}$

$F_2 = p_2 + F_1 = \frac{1}{4} + \frac{2}{3} = \frac{11}{12}$

$F_3 = p_3 + F_2 = \frac{1}{12} + \frac{11}{12} = 1$

i	PDF p_i	CDF F_i
0	1/3	1/3
1	1/3	2/3
2	1/4	11/12
3	1/12	1

© 2016 Cengage Learning. All Rights Reserved. May not be scanned, copied or duplicated, or posted to a publicly accessible website, in whole or in part.

19. (a)

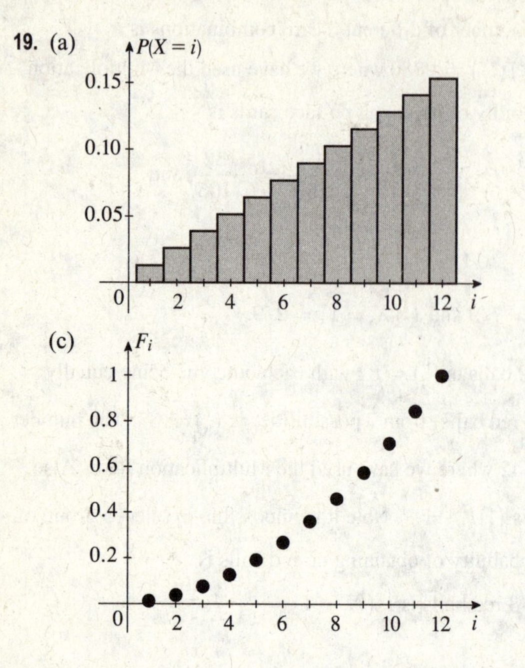

(b) $F_i = P(X \leq i) = \sum_{k=1}^{i} p_k$

$$= \frac{1}{78}\sum_{k=1}^{i} k = \frac{1}{78}\left(\frac{i\,(i+1)}{2}\right)$$

$$= \frac{i\,(i+1)}{156}$$

where in the second last equality, the summation was simplified using Formula 5.2.5.

21. We use Definition 5 along with the method of Example 6 to construct a table of PDF values. For example:

$$F_0 = P\,(X \leq 0) = p_0 = 0.5$$
$$p_1 = F_1 - F_0 = 0.5 - 0.5 = 0$$
$$p_2 = F_2 - F_1 = 0.5 - 0.5 = 0$$
$$p_3 = F_3 - F_2 = 1 - 0.5 = 0.5$$

i	CDF F_i	PDF p_i
0	0.5	0.5
1	0.5	0
2	0.5	0
3	1	0.5

23. The probability of the discrete random variable having a value between 10 and 20 inclusive is given by the sum

$$\sum_{i=10}^{20} p_i = \sum_{i=1}^{20} p_i - \sum_{i=1}^{9} p_i = 2\sum_{i=1}^{20}\left(\frac{1}{3}\right)^i - 2\sum_{i=1}^{9}\left(\frac{1}{3}\right)^i$$

$$= 2\left[\frac{1/3\,(1-(1/3)^{20})}{1-(1/3)} - \frac{1/3\,(1-(1/3)^{9})}{1-(1/3)}\right] \quad \text{[using the formula for the sum of a geometric sequence]}$$

$$= \frac{2}{3}\left[\frac{1-(1/3)^{20}}{2/3} - \frac{1-(1/3)^{9}}{2/3}\right] = (1/3)^9 - (1/3)^{20} \approx 0.000051$$

25. The probability of the discrete random variable having a value between 2 and 6 inclusive is given by the sum

$$\sum_{i=2}^{6} p_i = \sum_{i=1}^{6} p_i - p_1 = \sum_{i=1}^{6}\left(\frac{1}{3025}i^3\right) - \frac{1}{3025}$$

$$= \frac{1}{3025}\left(\frac{6^2\,(6+1)^2}{4}\right) - \frac{1}{3025} \quad \text{[using the provided formula]}$$

$$= \frac{441}{3025} - \frac{1}{3025} = \frac{8}{55}$$

27. (a) $F_3 = P\,(X \leq 3) = 0.71$. This represents the probability of observing 3 or fewer different mammal species along a transect through a forest.

© 2016 Cengage Learning. All Rights Reserved. May not be scanned, copied or duplicated, or posted to a publicly accessible website, in whole or in part.

(b) We use Definition 5 along with the method of Example 6 to construct a table of PDF values. For example,

$F_0 = P(X \leq 0) = p_0 = 0.01$, $p_1 = F_1 - F_0 = 0.16 - 0.01 = 0.15$, $p_2 = F_2 - F_1 = 0.36 - 0.16 = 0.20$, and so on.

The accompanying PDF histogram is plotted below.

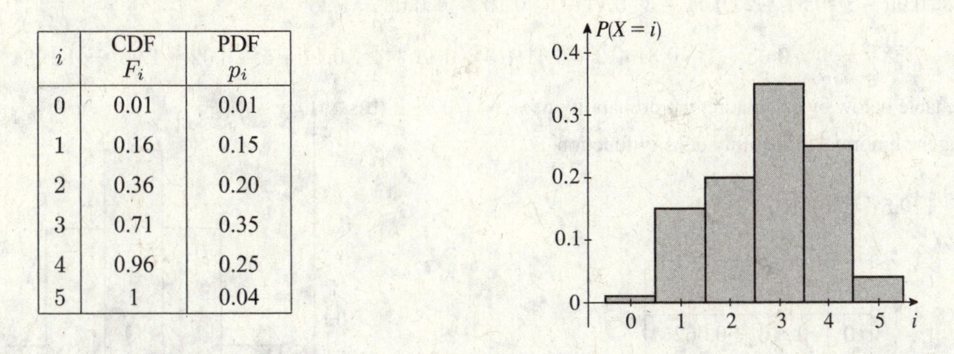

i	CDF F_i	PDF p_i
0	0.01	0.01
1	0.16	0.15
2	0.36	0.20
3	0.71	0.35
4	0.96	0.25
5	1	0.04

29-35. Let X be a random variable representing the dollar value of the winnings.

29. The probability of rolling a 6 is $1/6$ and the probability of not rolling a 6 is $5/6$. Therefore, $P(X = 10) = \frac{1}{6}$ and $P(X = -1) = \frac{5}{6}$.

(a) $E[X] = \sum_i ip_i = 10 \cdot \frac{1}{6} + (-1) \cdot \frac{5}{6} = \frac{5}{6} \approx \$0.83$

(b) $Var[X] = \sum_i i^2 p_i - E[X]^2 = 10^2 \cdot \frac{1}{6} + (-1)^2 \cdot \frac{5}{6} - \left(\frac{5}{6}\right)^2 = \frac{100}{6} + \frac{5}{6} - \frac{25}{36} = \frac{605}{36}$

31. Since a fair coin has an equal chance of landing heads or tails, we have $P(X = 3) = \frac{1}{2}$ and $P(X = 2) = \frac{1}{2}$.

(a) $E[X] = \sum_i ip_i = 3 \cdot \frac{1}{2} + 2 \cdot \frac{1}{2} = \frac{5}{2} = \$2.50$

(b) $Var[X] = \sum_i i^2 p_i - E[X]^2 = 3^2 \cdot \frac{1}{2} + 2^2 \cdot \frac{1}{2} - \left(\frac{5}{2}\right)^2 = \frac{9}{2} + 2 - \frac{25}{4} = \frac{1}{4}$

33. The probability of rolling an even number is $3/6$ and the probability of rolling an odd number is $3/6$. Therefore, $P(X = 2) = \frac{1}{2}$ and $P(X = -2) = \frac{1}{2}$.

(a) $E[X] = \sum_i ip_i = 2 \cdot \frac{1}{2} + (-2) \cdot \frac{1}{2} = \$0$

(b) $Var[X] = \sum_i i^2 p_i - E[X]^2 = 2^2 \cdot \frac{1}{2} + (-2)^2 \cdot \frac{1}{2} - 0^2 = 2 + 2 = 4$

35. Let's assume a silver dollar has a value of \$1 and a slug has a value of \$0. Since it costs \$0.50 to play, drawing a silver dollar has a net gain of $X = 1 - 0.5 = \$0.5$ and drawing a slug has a net gain of $X = -0.5$. Now, there is a $\frac{2}{10}$ chance of drawing a silver dollar and an $\frac{8}{10}$ chance of drawing a slug. Therefore, $P(X = 0.5) = \frac{1}{5}$ and $P(X = -0.5) = \frac{4}{5}$.

(a) $E[X] = \sum_i ip_i = 0.5 \cdot \frac{1}{5} + (-0.5) \cdot \frac{4}{5} = -\frac{3}{10} = -\$0.30$

(b) $Var[X] = \sum_i i^2 p_i - E[X]^2 = (0.5)^2 \cdot \frac{1}{5} + (-0.5)^2 \cdot \frac{4}{5} - \left(-\frac{3}{10}\right)^2 = \frac{4}{25} = 0.16$

37. The histogram is symmetric and can be balanced at the point $X = 5$. Thus, $E[X] = 5$.

39. We can view this histogram as a non-symmetric distribution of mass with the greatest concentration of mass occurring at $X = 2$ and $X = 3$. Since there is a tail of 'mass' to the right of the peak, we would balance the histogram at a point slightly larger than $X = 2.5$. Thus, we estimate $E[X] \approx 3$.

© 2016 Cengage Learning. All Rights Reserved. May not be scanned, copied or duplicated, or posted to a publicly accessible website, in whole or in part.

41. $E[X] = \sum_i ip_i = 1 \cdot p_1 + 2 \cdot p_2 + 3 \cdot p_3 = 1 \cdot 0.3 + 2 \cdot 0.55 + 3 \cdot 0.15 = 1.85$

$Var[X] = \sum_i i^2 p_i - E[X]^2 = 1^2 \cdot 0.3 + 2^2 \cdot 0.55 + 3^2 \cdot 0.15 - 1.85^2 = 0.4275$

43. $E[X] = \sum_i ip_i = 0 \cdot 0.05 + 1 \cdot 0.31 + 2 \cdot 0.41 + 3 \cdot 0.11 + 4 \cdot 0.10 + 5 \cdot 0.02 = 1.96$

$Var[X] = \sum_i i^2 p_i - E[X]^2 = 0^2 \cdot 0.05 + 1^2 \cdot 0.31 + 2^2 \cdot 0.41 + 3^2 \cdot 0.11 + 4^2 \cdot 0.10 + 5^2 \cdot 0.02 - 1.96^2 = 1.1984$

45. (a) We construct the table below by calculating the probability of i "successes" using the binomial probability density function

$$P(X = i) = \binom{4}{i} (0.5)^i (0.5)^{4-i}.$$

i	PDF p_i
0	$1/16 = 0.0625$
1	$1/4 = 0.25$
2	$3/8 = 0.375$
3	$1/4 = 0.25$
4	$1/16 = 0.0625$

(b)

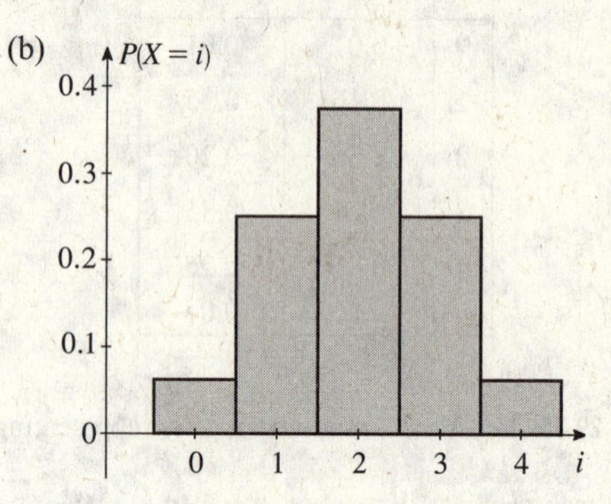

47. (a) We construct the table below by calculating the probability of i "successes" using the binomial probability density function

$$P(X = i) = \binom{7}{i} (0.2)^i (0.8)^{7-i}.$$

i	PDF p_i
0	0.2097152
1	0.3670016
2	0.2752512
3	0.1146880
4	0.0286720
5	0.0043008
6	0.0003584
7	0.0000128

(b)

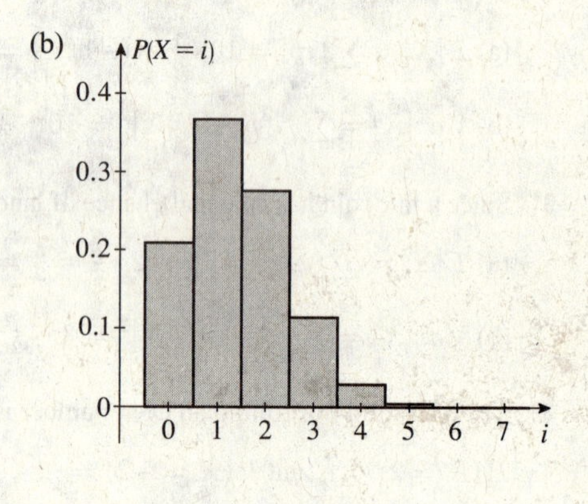

49. Treating '1' as a "success", we can use the binomial PDF in Definition 16 to calculate the probability of 3 successes in 3 trials with each success having probability $1/6$. This gives $P(X = 3) = \binom{3}{3} \left(\frac{1}{6}\right)^3 \left(\frac{5}{6}\right)^0 = \frac{1}{216} \approx 0.0046.$

51. Treating a 'hit' as a "success", we can use the binomial PDF in Definition 16 with $n = 7$ trials and $p = 0.8$ to calculate the probability of each event.

(a) $i = 0 \Rightarrow P(X = 0) = \binom{7}{0} (0.8)^0 (0.2)^7 = \frac{1}{78125} = 1.28 \times 10^{-5}$

(b) $i = 7 \Rightarrow P(X = 7) = \binom{7}{7} (0.8)^7 (0.2)^0 = \frac{16384}{78125} \approx 0.210$

© 2016 Cengage Learning. All Rights Reserved. May not be scanned, copied or duplicated, or posted to a publicly accessible website, in whole or in part.

(c) $P(X > 1) = 1 - P(X \le 1) = 1 - [P(X = 0) + P(X = 1)]$

$$= 1 - \left[\binom{7}{0}(0.8)^0(0.2)^7 + \binom{7}{1}(0.8)^1(0.2)^6\right] = 1 - \left[\frac{1}{78125} + \frac{28}{78125}\right]$$

$$= \frac{78096}{78125} \approx 0.9996$$

(d) $P(X \ge 5) = P(X = 5) + P(X = 6) + P(X = 7)$

$$= \binom{7}{5}(0.8)^5(0.2)^2 + \binom{7}{6}(0.8)^6(0.2)^1 + \binom{7}{7}(0.8)^7(0.2)^0 = \frac{21504}{78125} + \frac{28672}{78125} + \frac{16384}{78125}$$

$$= \frac{13312}{15625} \approx 0.8520$$

53. Assuming the outcomes for each child are independent, the probability of having i children with the condition is given by the binomial PDF $P(X = i) = \binom{4}{i}(0.25)^i(0.75)^{4-i}$.

(a) $P(X \ge 1) = 1 - P(X = 0) = 1 - \binom{4}{0}(0.25)^0(0.75)^4 = 1 - 0.75^4 \approx 0.684$

(b) $P(X \ge 3) = P(X = 3) + P(X = 4) = \binom{4}{3}(0.25)^3(0.75)^1 + \binom{4}{4}(0.25)^4(0.75)^0 = 0.05078$

55. The number of raccoons with rabies is a binomial random variable with $n = 4$ and $p = 0.1$ where 'has rabies' is treated as a "success". Thus, the probability of having at least one rabies carrier is $P(X \ge 1) = 1 - P(X = 0) = 1 - \binom{4}{0}(0.1)^0(0.9)^4 = 1 - 0.9^4 \approx 0.3439$.

57. (a) If we treat 'left-handed' as a "success", then the number of left-handed subjects is a binomial random variable with $n = 15$ and $p = 0.1$. Thus, the probability of having exactly 12 left-handed subjects is $P(X = 12) = \binom{15}{12}(0.1)^{12}(0.9)^3 = 455\,(0.1)^{12}(0.9)^3 \approx 3.32 \times 10^{-10}$.

(b) $P(X \ge 12) = P(X = 12) + P(X = 13) + P(X = 14) + P(X = 15)$

$$= \binom{15}{12}(0.1)^{12}(0.9)^3 + \binom{15}{13}(0.1)^{13}(0.9)^2 + \binom{15}{14}(0.1)^{14}(0.9)^1 + \binom{15}{15}(0.1)^{15}(0.9)^0$$

$$= 455\,(0.1)^{12}(0.9)^3 + 105\,(0.1)^{13}(0.9)^2 + 15\,(0.1)^{14}(0.9) + (0.1)^{15}$$

$$= 3.403 \times 10^{-10}$$

59. Let X be a binomial random variable in which 'boy' is treated as a "success". We are told that $p = 0.52$, and there are $n = 10$ children.

(a) $P(X = 10) = \binom{10}{10}(0.52)^{10}(0.48)^0 = (0.52)^{10} \approx 0.00145$

(b) If all children are girls, then there are no male offspring. The probability of this occurring is given by

$$P(X = 0) = \binom{10}{0}(0.52)^0(0.48)^{10} = (0.48)^{10} \approx 0.000649$$

(c) $P(X = 5) = \binom{10}{5}(0.52)^5(0.48)^5 = 252\,(0.52)^5(0.48)^5 \approx 0.244$

© 2016 Cengage Learning. All Rights Reserved. May not be scanned, copied or duplicated, or posted to a publicly accessible website, in whole or in part.

61. We can treat the number of NK cells X as a binomial random variable with $p = 0.07$ and $n = 10$ assuming that the outcome of each trial is independent.

(a) $P(X = 0) = \binom{10}{0}(0.07)^0(0.93)^{10} = (0.93)^{10} \approx 0.484$

(b) $P(X = 2) = \binom{10}{2}(0.07)^2(0.93)^8 = 45(0.07)^2(0.93)^8 \approx 0.123$

(c) $P(X \le 3) = P(0) + P(1) + P(2) + P(3)$

$$= \binom{10}{0}(0.07)^0(0.93)^{10} + \binom{10}{1}(0.07)^1(0.93)^9 + \binom{10}{2}(0.07)^2(0.93)^8 + \binom{10}{3}(0.07)^3(0.93)^7$$
$$= (0.93)^{10} + 10(0.07)(0.93)^9 + 45(0.07)^2(0.93)^8 + 120(0.07)^3(0.93)^7$$
$$\approx 0.996$$

63. We can treat the number of GC nucleotides X as a binomial random variable with $n = 12$ and $i = 3$ assuming that each GC nucleotide identity is independent of one another.

(a) $p = 0.5 \Rightarrow P(X = 3) = \binom{12}{3}(0.5)^3(0.5)^9 = 220(0.5)^{12} \approx 0.0537$

(b) $p = 0.3 \Rightarrow P(X = 3) = \binom{12}{3}(0.3)^3(0.7)^9 = 220(0.3)^3(0.7)^9 \approx 0.2397$

(c) If the GC content of the virus is p, then using Definition 16 with $i = 3$ and $n = 12$ gives

$$P(X = 3) = \binom{12}{3}p^3(1-p)^9 = 220p^3(1-p)^9$$

(d) The maximum value of $P(X = 3)$ must satisfy $\frac{d}{dp}\left[220p^3(1-p)^9\right] = 0 \Rightarrow 220\left[3p^2(1-p)^9 - 9p^3(1-p)^8\right] = 0$ $\Rightarrow p^2(1-p)^8[(1-p) - 3p] = 0 \Rightarrow p^2(1-p)^8[1-4p] = 0 \Rightarrow p = 0, 1, \frac{1}{4}$. Now, $P(X = 3) = 0$ when $p = 0$ and $p = 1$. Also, when $p = \frac{1}{4}$, $P(X = 3) = 220\left(\frac{1}{4}\right)^3\left(1 - \frac{1}{4}\right)^9 \approx 0.258$. Thus, by using the Closed Interval Method on the domain $[0, 1]$, we have found that the absolute maximum value of $P(X = 3)$ occurs when $p = 1/4$.

65. The PDF for a binomial random variable X is $p_i = \dfrac{n!}{i!(n-i)!}p^i(1-p)^{n-i}$. Using this to compute the expected value of X gives

$$E[X] = \sum_i ip_i = \sum_{i=0}^{n} i\frac{n!}{i!(n-i)!}p^i(1-p)^{n-i} = \sum_{i=1}^{n}\frac{n!}{(i-1)!(n-i)!}p^i(1-p)^{n-i}$$
$$= np\sum_{i=1}^{n}\frac{(n-1)!}{(i-1)!\,[(n-1)-(i-1)]!}p^{i-1}(1-p)^{(n-1)-(i-1)}$$
$$= np\sum_{i=1}^{n}\binom{n-1}{i-1}p^{i-1}(1-p)^{(n-1)-(i-1)}$$
$$= np\,(p + 1 - p)^n = np$$

The summation in the second last line above was simplified using the binomial theorem with x replaced by p, y by $1-p$, n by $n-1$, and k by $i-1$.

67. Drawing exactly k black balls also requires that exactly $n-k$ white balls are chosen. The number of combinations of b black balls when taken k at a time is $\binom{b}{k}$ and the number of combinations of w white balls when taken $n-k$ at a time is $\binom{w}{n-k}$. Thus, by the Fundamental Principle of Counting, the number of ways drawing exactly k black balls is $\binom{b}{k}\binom{w}{n-k}$. Also, the total number of different outcomes of drawing n balls from a collection of size N is $\binom{N}{n}$. Therefore, $p_k = \dfrac{\binom{b}{k}\binom{w}{n-k}}{\binom{N}{n}}$.

© 2016 Cengage Learning. All Rights Reserved. May not be scanned, copied or duplicated, or posted to a publicly accessible website, in whole or in part.

69. Since the prey are consumed without replacement, the probability of the event is given by the PDF of a hypergeometric distribution (see Exercise 67). The number of different combinations of 50 bivalves taken 30 at a time is $\binom{50}{30}$ and the number of different combinations of 50 chironomids taken 15 at a time is $\binom{50}{15}$. Therefore, using the Fundamental Principle of Counting, the number of different combinations of 30 bivalve and 15 chironomids is $\binom{50}{30}\binom{50}{15}$. Also, the total number of different outcomes of selecting 45 prey from a group of 100 is $\binom{100}{45}$. Since the fish has no food preference, all outcomes are equally likely and the probability of this choice of 45 food items is $\dfrac{\binom{50}{30}\binom{50}{15}}{\binom{100}{45}} \approx 0.001726$.

71. (a) If the first success occurs on the ith trial, then the first $i-1$ trials must be failures each of which occur with probability $1-p$. Therefore, the probability of obtaining these consecutive independent outcomes is $P(X=i)=(1-p)^{i-1}p$.

(b) $F_i = P(X \le i) = \displaystyle\sum_{k=1}^{i} P(X=k) = \sum_{k=1}^{i}(1-p)^{k-1}p = p\sum_{k=1}^{i}(1-p)^{k-1}$

$$= p\left[1+(1-p)+(1-p)^2+\ldots+(1-p)^{i-1}\right] = p\frac{\left[1-(1-p)^i\right]}{1-(1-p)} = 1-(1-p)^i$$

where we have used the formula for the sum of a finite geometric series given in (2.1.5).

73. The number of years after which the population goes extinct is given by a geometric random variable with $p=0.01$ (see Exercise 71). The geometric PDF is $P(X=i)=p(1-p)^{i-1} \Rightarrow$ $P(X=5)=(0.01)(1-0.01)^{5-1}=(0.01)(0.99)^4 \approx 0.0096$. Therefore, the probability that the population will go extinct after five years is approximately 0.96%.

12.5 Continuous Random Variables

1. (a) $\int_{30{,}000}^{40{,}000} f(x)\,dx$ is the probability that a randomly chosen tire will have a lifetime between 30,000 and 40,000 miles.

(b) $\int_{25{,}000}^{\infty} f(x)\,dx$ is the probability that a randomly chosen tire will have a lifetime of at least 25,000 miles.

3. (a) $\displaystyle\int_{25}^{40} f(t)\,dt$ is the probability that a randomly selected fruit fly will die between 25 and 40 days old.

(b) $\displaystyle\int_{0}^{14} f(t)\,dt$ is the probability that a randomly selected fruit fly will die before it is 14 days old.

5. (a) $e^{-x}>0$ for all $x\in\mathbb{R}$ so if $x\ge 0$ then $f(x)=xe^{-x}\ge 0$ which verifies condition (3). We verify condition (4) by checking that the area under the graph of $f(x)$ is 1 as follows:

$$\int_{-\infty}^{\infty} f(x)\,dx = \int_0^{\infty} xe^{-x}\,dx = \lim_{b\to\infty}\left[(-x-1)e^{-x}\right]_0^b = \lim_{b\to\infty}\left[(-b-1)e^{-b}-(-1)\right] = \lim_{b\to\infty}\left[\frac{-b-1}{e^b}\right]+1$$

$$\overset{\text{H}}{=} \lim_{b\to\infty}\left[\frac{-1}{e^b}\right]+1 = 1$$

(b) $P(1\le X\le 2) = \int_1^2 xe^{-x}\,dx = \left[(-x-1)e^{-x}\right]_1^2 = -3e^{-2}+2e^{-1} = 2/e-3/e^2$ $[\approx 0.33]$

© 2016 Cengage Learning. All Rights Reserved. May not be scanned, copied or duplicated, or posted to a publicly accessible website, in whole or in part.

7. (a) If $c \geq 0$, then $f(x) \geq 0$, so condition (3) is satisfied. For condition (4), we see that $\int_{-\infty}^{\infty} f(x)\,dx = \int_{-\infty}^{\infty} \frac{c}{1+x^2}\,dx$ and

$$\int_0^{\infty} \frac{c}{1+x^2}\,dx = \lim_{t\to\infty} \int_0^t \frac{c}{1+x^2}\,dx = c \lim_{t\to\infty} \left[\tan^{-1} x\right]_0^t = c \lim_{t\to\infty} \tan^{-1} t = c\left(\tfrac{\pi}{2}\right)$$

Similarly, $\int_{-\infty}^{0} \frac{c}{1+x^2}\,dx = c\left(\frac{\pi}{2}\right)$, so $\int_{-\infty}^{\infty} \frac{c}{1+x^2}\,dx = 2c\left(\frac{\pi}{2}\right) = c\pi$.

Since $c\pi$ must equal 1, we must have $c = 1/\pi$ so that f is a probability density function.

(b) $P(-1 < X < 1) = \int_{-1}^{1} \frac{1/\pi}{1+x^2}\,dx = \frac{2}{\pi}\int_0^1 \frac{1}{1+x^2}\,dx = \frac{2}{\pi}\left[\tan^{-1} x\right]_0^1 = \frac{2}{\pi}\left(\frac{\pi}{4} - 0\right) = \frac{1}{2}$

9. $f(x) = 1 > 0$ and $\int_{-\infty}^{\infty} f(x)\,dx = \int_0^1 (1)\,dx = 1$ so $f(x)$ satisfies conditions (3) and (4) for a PDF. The CDF is given by

$F(x) = \int_{-\infty}^{x} f(s)\,ds = \int_0^x (1)\,ds = x$ where $x \in [0,1]$. Thus, $P(\frac{1}{3} \leq X \leq \frac{2}{3}) = F(\frac{2}{3}) - F(\frac{1}{3}) = \frac{2}{3} - \frac{1}{3} = \frac{1}{3}$.

11. $f(x) = x^{-2} \geq 0$ and $\int_{-\infty}^{\infty} f(x)\,dx = \int_{1/2}^{1} x^{-2}\,dx = -\left[x^{-1}\right]_{1/2}^{1} = -[1-2] = 1$. So $f(x)$ satisfies conditions (3) and (4) for a PDF. The CDF is given by $F(x) = \int_{-\infty}^{x} f(s)\,ds = \int_{1/2}^{x} s^{-2}\,ds = -\left.s^{-1}\right]_{1/2}^{x} = 2 - x^{-1}$ where $\frac{1}{2} \leq x \leq 1$.

Thus, $P(2/3 \leq X \leq 3/4) = F(3/4) - F(2/3) = (2 - \frac{4}{3}) - (2 - \frac{3}{2}) = \frac{1}{6}$.

13. $f(t) = \frac{3}{2}\sqrt{t} \geq 0$ when $t \geq 0$ and $\int_{-\infty}^{\infty} f(t)\,dt = \int_0^1 \frac{3}{2}t^{1/2}\,dt = \left[t^{3/2}\right]_0^1 = 1$, so $f(x)$ satisfies conditions (3) and (4) for a PDF. The CDF is given by $F(t) = \int_{-\infty}^{t} f(s)\,ds = \int_0^t \frac{3}{2}s^{1/2}\,ds = \left.s^{3/2}\right]_0^t = t^{3/2}$ where $t \in [0,1]$.

Thus, $P(\frac{1}{4} \leq T \leq \frac{3}{4}) = F(\frac{3}{4}) - F(\frac{1}{4}) = (\frac{3}{4})^{3/2} - (\frac{1}{4})^{3/2} = \frac{3}{8}\sqrt{3} - \frac{1}{8} \approx 0.525$.

15. $f(x) = \frac{1}{2}e^{-|x|} \geq 0$ and

$$\int_{-\infty}^{\infty} f(x)\,dx = \int_{-\infty}^{0} \tfrac{1}{2}e^{x}\,dx + \int_0^{\infty} \tfrac{1}{2}e^{-x}\,dx = \tfrac{1}{2}\lim_{n\to\infty}\left\{\left[e^x\right]_{-n}^{0} - \left[e^{-x}\right]_0^n\right\} = \tfrac{1}{2}\lim_{n\to\infty}\left[1 - e^{-n} - (e^{-n} - 1)\right] = 1.$$

So $f(x)$ satisfies conditions (3) and (4) for a PDF.

When $x \leq 0$, the CDF is given by $F(x) = \int_{-\infty}^{x} f(s)\,ds = \int_{-\infty}^{x} \frac{1}{2}e^s\,ds = \frac{1}{2}\lim_{n\to\infty}\left[e^s\right]_{-n}^{x} = \frac{1}{2}\lim_{n\to\infty}\left[e^x - e^{-n}\right] = \frac{1}{2}e^x$.

When $x > 0$, the CDF is $F(x) = \int_{-\infty}^{x} f(s)\,ds = F(0) + \int_0^x \frac{1}{2}e^{-s}\,ds = \frac{1}{2} - \frac{1}{2}\left[e^{-s}\right]_0^x = \frac{1}{2} - \frac{1}{2}(e^{-x} - 1) = 1 - \frac{1}{2}e^{-x}$.

Therefore, $P(-5 \leq X \leq 10) = F(10) - F(-5) = (1 - \frac{1}{2}e^{-10}) - (\frac{1}{2}e^{-5}) \approx 0.9966$.

17. $f(x) = \dfrac{2}{\pi(1+x^2)} \geq 0$ and

$\int_{-\infty}^{\infty} f(x)\,dx = \lim_{b\to\infty}\int_0^b \frac{2}{\pi(1+x^2)}\,dx = \frac{2}{\pi}\lim_{b\to\infty}\left[\tan^{-1} x\right]_0^b = \frac{2}{\pi}\lim_{b\to\infty}\left[\tan^{-1} b - 0\right] = \frac{2}{\pi}\left(\frac{\pi}{2}\right) = 1$. So $f(x)$ satisfies conditions (3) and (4) for a PDF. The CDF is given by

$$F(x) = \int_{-\infty}^{x} f(s)\,ds = \frac{2}{\pi}\int_0^x \frac{1}{1+s^2}\,ds = \frac{2}{\pi}\left[\tan^{-1} s\right]_1^x = \frac{2}{\pi}\left(\tan^{-1} x - \tan^{-1} 1\right) = \frac{2}{\pi}\left(\tan^{-1} x - \frac{\pi}{4}\right)$$

$$= \frac{2}{\pi}\tan^{-1} x - \frac{1}{2}$$

where $x \geq 0$. Thus, $P(10 \leq X \leq 20) = F(20) - F(10) = \frac{2}{\pi}\tan^{-1} 20 - \frac{2}{\pi}\tan^{-1} 10 \approx 0.0316$.

© 2016 Cengage Learning. All Rights Reserved. May not be scanned, copied or duplicated, or posted to a publicly accessible website, in whole or in part.

19. (a) $P(T<48)=\int_0^{48} f(t)\,dt=\frac{1}{15676}\int_0^{48} t^2e^{-0.05t}\,dt$

$$=\frac{1}{15676}\left[\left[-20t^2e^{-0.05t}\right]_0^{48}+40\int_0^{48} te^{-0.05t}\,dt\right]\quad\left[\begin{array}{l}\text{by parts with}\\ u=t^2,\,dv=e^{-0.05t}dt\end{array}\right]$$

$$=\frac{1}{15676}\left[-46080e^{-2.4}+40\int_0^{48} te^{-0.05t}\,dt\right]$$

$$=\frac{1}{15676}\left[-46080e^{-2.4}+40\left(\left[-20te^{-0.05t}\right]_0^{48}+20\int_0^{48} e^{-0.05t}\,dt\right)\right]\quad\left[\begin{array}{l}\text{by parts with}\\ u=t,\,dv=e^{-0.05t}dt\end{array}\right]$$

$$=\frac{1}{15676}\left[-46080e^{-2.4}-38400e^{-2.4}-16000\left[e^{-0.05t}\right]_0^{48}\right]$$

$$=\frac{1}{15676}\left(-84480e^{-2.4}-16000e^{-2.4}+16000\right)=\frac{640}{15676}\left(25-157e^{-2.4}\right)\approx 0.439$$

(b) We calculate the probability that an infected patient will display symptoms within the first 36 hours using integration by parts twice as in Part (a).

$$P(T\le 36)=\int_0^{36} f(t)\,dt=\frac{1}{15676}\int_0^{36} t^2e^{-0.05t}\,dt$$

$$=\frac{1}{15676}\left[\left[-20t^2e^{-0.05t}\right]_0^{36}+40\int_0^{36} te^{-0.05t}\,dt\right]\quad\left[\begin{array}{l}\text{by parts with}\\ u=t^2,\,dv=e^{-0.05t}dt\end{array}\right]$$

$$=\frac{1}{15676}\left[-25920e^{-1.8}+40\int_0^{36} te^{-0.05t}\,dt\right]$$

$$=\frac{1}{15676}\left[-25920e^{-1.8}+40\left(\left[-20te^{-0.05t}\right]_0^{36}+20\int_0^{36} e^{-0.05t}\,dt\right)\right]\quad\left[\begin{array}{l}\text{by parts with}\\ u=t,\,dv=e^{-0.05t}dt\end{array}\right]$$

$$=\frac{1}{15676}\left[-25920e^{-1.8}-28800e^{-1.8}-16000\left[e^{-0.05t}\right]_0^{36}\right]$$

$$=\frac{1}{15676}\left(-54720e^{-1.8}-16000e^{-1.8}+16000\right)=\frac{320}{15676}\left(50-221e^{-1.8}\right)\approx 0.275$$

Therefore, the probability that an infected patient will not display symptoms until *after* 36 hours is $P(T>36)\approx 1-0.275=0.725$.

21. (a) $P(T<30)=\int_0^{30} f(t)\,dt=\int_0^{30}\frac{1}{1600}t\,dt=\frac{1}{3200}\left[t^2\right]_0^{30}=\frac{1}{3200}(900)=\frac{9}{32}=0.28125$

(b) The PDF changes at $t=40$, so we must break up the probability integral into two parts as follows.

$$P(30<T<60)=\int_{30}^{60} f(t)\,dt=\int_{30}^{40}\tfrac{1}{1600}t\,dt+\int_{40}^{60}\tfrac{1}{1600}(80-t)\,dt$$

$$=\left[\tfrac{1}{3200}t^2\right]_{30}^{40}+\tfrac{1}{1600}\left[80t-\tfrac{1}{2}t^2\right]_{40}^{60}$$

$$=\left(\tfrac{1}{2}-\tfrac{9}{32}\right)+\tfrac{1}{1600}(3000-2400)$$

$$=\tfrac{7}{32}+\tfrac{3}{8}=\tfrac{19}{32}=0.59375$$

23. $F(x)=x \;\Rightarrow\; f(x)=F'(x)=1$

25. $F(t)=\dfrac{t^2}{1+t^2} \;\Rightarrow\; f(t)=F'(t)=\dfrac{2t\left(1+t^2\right)-t^2(2t)}{\left(1+t^2\right)^2}=\dfrac{2t}{\left(1+t^2\right)^2}$

© 2016 Cengage Learning. All Rights Reserved. May not be scanned, copied or duplicated, or posted to a publicly accessible website, in whole or in part.

27. $F(t) = \dfrac{Ce^t}{1+Ce^t} \Rightarrow f(t) = F'(t) = \dfrac{Ce^t(1+Ce^t) - Ce^t(Ce^t)}{(1+Ce^t)^2} = \dfrac{Ce^t}{(1+Ce^t)^2}$

29. (a) $E[T] = \int_{-\infty}^{\infty} t\,f(t)\,dt = \int_0^1 t\,dt = \frac{1}{2}t^2\big]_0^1 = \frac{1}{2}$

(b) $Var[T] = \int_{-\infty}^{\infty} t^2 f(t)\,dt - E[T]^2 = \int_0^1 t^2\,dt - \left(\frac{1}{2}\right)^2 = \left[\frac{1}{3}t^3\right]_0^1 - \frac{1}{4} = \frac{1}{3} - \frac{1}{4} = \frac{1}{12}$

31. (a) $E[X] = \int_{-\infty}^{\infty} x\,f(x)\,dx = \int_{1/2}^1 x^{-1}\,dx = [\ln|x|]_{1/2}^1 = \ln 1 - \ln\frac{1}{2} = 0 + \ln 2 = \ln 2$

(b) $Var[X] = \int_{-\infty}^{\infty} x^2 f(x)\,dx - E[X]^2 = \int_{1/2}^1 1\,dx - (\ln 2)^2 = [x]_{1/2}^1 - (\ln 2)^2 = \left(1 - \frac{1}{2}\right) - (\ln 2)^2 = \frac{1}{2} - (\ln 2)^2$

33. (a) $E[T] = \int_{-\infty}^{\infty} t\,f(t)\,dt = \lim\limits_{x\to\infty} \int_0^x te^{-t}\,dt = \lim\limits_{x\to\infty}\left(-te^{-t}\big]_0^x + \int_0^x e^{-t}\,dt\right)$ [by parts with $u = t$, $dv = e^{-t}dt$]

$= \lim\limits_{x\to\infty}\left(-xe^{-x} - \left[e^{-t}\right]_0^x\right) = \lim\limits_{x\to\infty}\left[-xe^{-x} - (e^{-x} - 1)\right] = \lim\limits_{x\to\infty}\left[-\frac{x}{e^x} - e^{-x} + 1\right] \overset{H}{=} 0 + 0 + 1 = 1$

(b) $Var[T] = \int_{-\infty}^{\infty} t^2 f(t)\,dt - E[T]^2 = \lim\limits_{x\to\infty}\int_0^x t^2e^{-t}\,dt - (1)^2$

$= \lim\limits_{x\to\infty}\left(-2t^2e^{-t}\big]_0^x + 2\int_0^x te^{-t}\,dt\right) - 1$ [by parts with $u = t^2$, $dv = e^{-t}dt$]

$= \lim\limits_{x\to\infty}\left(-2x^2e^{-x} - \left[2te^{-t}\right]_0^x + 2\int_0^x e^{-t}\,dt\right) - 1$ [by parts with $u = t$, $dv = e^{-t}dt$]

$= \lim\limits_{x\to\infty}\left(-2x^2e^{-x} - 2xe^{-x} - 2\left[e^{-t}\right]_0^x\right) - 1 = \lim\limits_{x\to\infty}\left(-2\frac{x^2}{e^x} - 2\frac{x}{e^x} - 2\left[e^{-x} - 1\right]\right) - 1$

$\overset{H}{=} 0 - 0 - 2(0 - 1) - 1 = 1$

35. (a) $E[X] = \int_{-\infty}^{\infty} x\,f(x)\,dx = \frac{1}{2}\int_{-\infty}^{\infty} xe^{-|x|}\,dx = 0$ since $xe^{-|x|}$ is an odd function (see Equation 5.4.6).

(b) $Var[X] = \int_{-\infty}^{\infty} x^2 f(x)\,dx - E[X]^2 = \frac{1}{2}\int_{-\infty}^{\infty} x^2e^{-|x|}\,dx = \int_0^{\infty} x^2e^{-x}\,dx$ [since the integrand is even (see Equation 5.4.6)]

$= \lim\limits_{n\to\infty}\left(-x^2e^{-x}\big]_0^n + 2\int_0^n xe^{-x}\,dx\right)$ [by parts with $u = x^2$, $dv = e^{-x}dx$]

$= \lim\limits_{n\to\infty}\left(-n^2e^{-n} - \left[2xe^{-x}\right]_0^n + 2\int_0^n e^{-x}\,dx\right)$ [by parts with $u = x$, $dv = e^{-x}dx$]

$= \lim\limits_{n\to\infty}\left(-\frac{n^2}{e^n} - 2\frac{n}{e^n} - 2\left[e^{-x}\right]_0^n\right) = \lim\limits_{n\to\infty}\left[-\frac{n^2}{e^n} - 2\frac{n}{e^n} - 2\left(e^{-n} - 1\right)\right] \overset{H}{=} 0 - 0 - 2(0 - 1) = 2$

37. (a) Since the graph of the PDF is a horizontal line, a uniform random variable has a PDF given by $f(x) = C$ where C is a positive constant. The PDF must also satisfy $\int_{-\infty}^{\infty} f(x)\,dx = 1 \Rightarrow \int_a^b C\,dx = 1 \Rightarrow Cx]_a^b = C(b - a) = 1 \Rightarrow$ $C = 1/(b - a)$. Therefore, $f(x) = \dfrac{1}{b - a}$ where $x \in [a, b]$.

© 2016 Cengage Learning. All Rights Reserved. May not be scanned, copied or duplicated, or posted to a publicly accessible website, in whole or in part.

(b) $E[X] = \int_{-\infty}^{\infty} x\, f(x)\, dx = \frac{1}{b-a} \int_a^b x\, dx = \frac{1}{b-a} \left[\frac{1}{2} x^2\right]_a^b = \frac{1}{2(b-a)} (b^2 - a^2) = \frac{(b+a)(b-a)}{2(b-a)} = \frac{b+a}{2}$

(c) $Var[X] = \int_{-\infty}^{\infty} x^2 f(x)\, dx - E[X]^2 = \frac{1}{b-a} \int_a^b x^2\, dx - \left(\frac{b+a}{2}\right)^2 = \frac{1}{3(b-a)} \left[x^3\right]_a^b - \frac{(b+a)^2}{4}$

$= \frac{b^3 - a^3}{3(b-a)} - \frac{(b+a)^2}{4} = \frac{(b-a)(b^2 + ab + a^2)}{3(b-a)} - \frac{b^2 + 2ab + a^2}{4}$

$= \frac{4(b^2 + ab + a^2) - 3(b^2 + 2ab + a^2)}{12} = \frac{b^2 - 2ab + a^2}{12} = \frac{(b-a)^2}{12}$

39. If all locations along the chromosome are equally likely to contain the join point, then the PDF of X must be constant, that is, X is a uniform random variable. As found in Exercise 37, a uniform random variable on the interval $[0, 2]$ has an expected value $E[X] = \frac{1}{2}(0+2) = 1$ and a variance $Var[X] = \frac{1}{12}(2-0)^2 = \frac{1}{3}$. Therefore, the experimental findings agree with the hypothesis that X is a uniform random variable.

41. The PDF changes at $t = 40$, so we must break up the expectation and variance integrals into two parts.

(a) $E[T] = \int_{-\infty}^{\infty} t\, f(t)\, dt = \frac{1}{1600} \int_0^{40} t^2\, dt + \frac{1}{1600} \int_{40}^{80} (80t - t^2)\, dt = \frac{1}{1600} \left(\frac{1}{3} \left[t^3\right]_0^{40} + \left[40t^2 - \frac{1}{3}t^3\right]_{40}^{80}\right)$

$= \frac{1}{1600} \left[\frac{1}{3} 40^3 + \left(40(80)^2 - \frac{1}{3} 80^3\right) - \left(40(40)^2 - \frac{1}{3} 40^3\right)\right] = \frac{1}{1600} \left[\frac{2}{3} 40^3 + 40(80)^2 - \frac{1}{3} 80^3 - 40(40)^2\right]$

$= \frac{1}{1600}(64000) = 40$

(b) $Var[T] = \int_{-\infty}^{\infty} t^2 f(t)\, dt - E[T]^2 = \frac{1}{1600} \int_0^{40} t^3\, dt + \frac{1}{1600} \int_{40}^{80} (80t^2 - t^3)\, dt - 40^2$

$= \frac{1}{1600} \left[\frac{1}{4} t^4\right]_0^{40} + \frac{1}{1600} \left[\frac{80}{3} t^3 - \frac{1}{4} t^4\right]_{40}^{80} - 1600$

$= \frac{1}{1600} \left(\frac{1}{4} 40^4\right) + \frac{1}{1600} \left[\left(\frac{80}{3} 80^3 - \frac{1}{4} 80^4\right) - \left(\frac{80}{3} 40^3 - \frac{1}{4} 40^4\right)\right] - 1600$

$= 400 + \frac{4400}{3} - 1600 = \frac{800}{3}$

43. $\int_{-\infty}^{m} f(x)\, dx = \frac{1}{2} \Rightarrow \int_0^m dx = \frac{1}{2} \Rightarrow x]_0^m = \frac{1}{2} \Rightarrow m = \frac{1}{2}$

45. $\int_{-\infty}^{m} f(x)\, dx = \frac{1}{2} \Rightarrow \int_0^m 2(1+x)^{-2}\, dx = \frac{1}{2} \Rightarrow -2\left[(1+x)^{-1}\right]_0^m = \frac{1}{2} \Rightarrow -2\left[(1+m)^{-1} - 1\right] = \frac{1}{2} \Rightarrow$

$(1+m)^{-1} = \frac{3}{4} \Rightarrow 1 + m = \frac{4}{3} \Rightarrow m = \frac{1}{3}$

47. $\int_{-\infty}^{m} f(x)\, dx = \frac{1}{2} \Rightarrow \int_1^m \ln x\, dx = \frac{1}{2} \Rightarrow [x \ln x - x]_1^m = \frac{1}{2}$ [Using parts as in Example 5.5.2] $\Rightarrow m \ln m - m + 1 = \frac{1}{2} \Rightarrow$

$m \ln m - m + \frac{1}{2} = 0$ This equation has no algebraic solution, however, a calculator with root finding capabilities can be used to numerically solve the equation. Doing so, we find $m \approx 2.1555$.

49. $f(x) = x^{-2} \Rightarrow f'(x) = -2x^{-3} = -\frac{2}{x^3} \Rightarrow f'(x) \neq 0$ and is never undefined on its domain. Since $f'(x)$ is a decreasing function ($f'(x) < 0$) on $\left[\frac{1}{2}, 1\right]$, the maximum value of $f(x)$ occurs at the left-endpoint of its domain. Thus, the mode is $x = \frac{1}{2}$.

© 2016 Cengage Learning. All Rights Reserved. May not be scanned, copied or duplicated, or posted to a publicly accessible website, in whole or in part.

51. $f(x) = \frac{1}{2}e^{-|x|} = \begin{cases} \frac{1}{2}e^{-x} & \text{if } x \geq 0 \\ \frac{1}{2}e^{x} & \text{if } x < 0 \end{cases}$. $f(x)$ decreases when $x \geq 0$ (exponential decay) and increases when $x < 0$ (exponential growth). Therefore, there is a local maximum value at $x = 0$, which is also an absolute maximum since $\lim_{x\to\pm\infty} f(x) = 0$. Thus, the mode of X is 0.

53. $E[T] = \int_{-\infty}^{\infty} t\, f(t)\, dt = \frac{1}{2}\int_0^2 t^2\, dt = \frac{1}{6}\left[t^3\right]_0^2 = \frac{1}{6}(8) = \frac{4}{3}$

$Var[T] = \int_{-\infty}^{\infty} t^2 f(t)\, dt - E[T]^2 = \frac{1}{2}\int_0^2 t^3\, dt - \left(\frac{4}{3}\right)^2 = \frac{1}{8}\left[t^4\right]_0^2 - \frac{16}{9} = 2 - \frac{16}{9} = \frac{2}{9}$

Therefore, the standard deviation is $\sqrt{2/9} = \frac{1}{3}\sqrt{2}$.

55. As found in Exercise 34, $Var[X] = 3 - 4\ln 2 - (2\ln 2 - 1)^2$. Therefore, the standard deviation is $\sqrt{3 - 4\ln 2 - (2\ln 2 - 1)^2}$.

57. The PDF of an exponential random variable with mean $\bar{T} = 10$ min is $f(t) = \frac{1}{10}e^{-t/10}$ where $t \geq 0$.

$P(T > 20) = \int_{20}^{\infty} \frac{1}{10}e^{-t/10}\, dt = \lim_{x\to\infty}\left[-e^{-t/10}\right]_{20}^{x} = 0 + e^{-2} \approx 0.135$

59. We use an exponential density function with $\mu = 2.5$ min.

(a) $P(X > 4) = \int_4^{\infty} f(t)\, dt = \lim_{x\to\infty}\int_4^x \frac{1}{2.5}e^{-t/2.5}\, dt = \lim_{x\to\infty}\left[-e^{-t/2.5}\right]_4^x = 0 + e^{-4/2.5} \approx 0.202$

(b) $P(0 \leq X \leq 2) = \int_0^2 f(t)\, dt = \left[-e^{-t/2.5}\right]_0^2 = -e^{-2/2.5} + 1 \approx 0.551$

(c) We need to find a value a so that $P(X \geq a) = 0.02$, or, equivalently, $P(0 \leq X \leq a) = 0.98 \Leftrightarrow$ $\int_0^a f(t)\, dt = 0.98 \Leftrightarrow \left[-e^{-t/2.5}\right]_0^a = 0.98 \Leftrightarrow -e^{-a/2.5} + 1 = 0.98 \Leftrightarrow e^{-a/2.5} = 0.02 \Leftrightarrow$ $-a/2.5 = \ln 0.02 \Leftrightarrow a = -2.5\ln\frac{1}{50} = 2.5\ln 50 \approx 9.78 \text{ min} \approx 10 \text{ min}$. The ad should say that if you aren't served within 10 minutes, you get a free hamburger.

61. The PDF of an exponential random variable with mean $\bar{T} = 1$ year is $f(t) = e^{-t}$ where $t \geq 0$.

(a) $P(T > 10) = \int_{10}^{\infty} e^{-t}\, dt = \lim_{x\to\infty}\left[-e^{-t}\right]_{10}^{x} = 0 + e^{-10} \approx 4.54 \times 10^{-5}$

(b) $P(0 < T < 1) = \int_0^1 e^{-t}\, dt = \left[-e^{-t}\right]_0^1 = -e^{-1} + 1 = 1 - 1/e \approx 0.632$

(c) $Var[T] = \int_{-\infty}^{\infty} t^2 f(t)\, dt - E[T]^2 = \lim_{x\to\infty}\int_0^x t^2 e^{-t}\, dt - (1)^2 \overset{97}{=} \lim_{x\to\infty}\left(\left[-t^2e^{-t}\right]_0^x + 2\int_0^x te^{-t}\, dt\right) - 1$

$\overset{96}{=} \lim_{x\to\infty}\left(-x^2e^{-x} - 2\left[(t+1)e^{-t}\right]_0^x\right) - 1 = \lim_{x\to\infty}\left[-x^2e^{-x} - 2(x+1)e^{-x} + 2\right] - 1$

$= \lim_{x\to\infty}\left[-x^2e^{-x} - 2xe^{-x} - 2e^{-x}\right] + 1 \overset{\text{H}}{=} \lim_{x\to\infty}\left[0 - 0 - 0\right] + 1 = 1$

63. (a) Let's simplify notation by defining $c = \binom{n}{2}$ so that $f(t) = ce^{-ct}$, which we identify as the PDF of an exponential random variable. Thus, using the result of Example 8, we find $E[T] = 1/c = 1/\binom{n}{2}$.

© 2016 Cengage Learning. All Rights Reserved. May not be scanned, copied or duplicated, or posted to a publicly accessible website, in whole or in part.

(b) $Var[T] = \int_{-\infty}^{\infty} t^2\, f(t)\, dt - E[T]^2 = c \lim_{x\to\infty} \int_0^x t^2 e^{-ct}\, dt - (1/c)^2$

$\overset{97}{=} c \lim_{x\to\infty} \left(\left[-\frac{1}{c} t^2 e^{-ct} \right]_0^x + \frac{2}{c} \int_0^x t e^{-ct}\, dt \right) - \frac{1}{c^2} \overset{96}{=} c \lim_{x\to\infty} \left(-\frac{1}{c} x^2 e^{-cx} + \frac{2}{c} \left[\frac{1}{c^2} (-ct - 1) e^{-ct} \right]_0^x \right) - \frac{1}{c^2}$

$= c \lim_{x\to\infty} \left[-\frac{1}{c} x^2 e^{-cx} + \frac{2}{c^3} (-cxe^{-cx} - e^{-cx} + 1) \right] - \frac{1}{c^2}$

$= c \lim_{x\to\infty} \left[-\frac{x^2}{ce^{cx}} + \frac{2}{c^3} \left(-\frac{cx}{e^{cx}} - e^{-cx} + 1 \right) \right] - \frac{1}{c^2} \overset{H}{=} c \left[0 + \frac{2}{c^3} (0 - 0 + 1) \right] - \frac{1}{c^2} = \frac{2}{c^2} - \frac{1}{c^2}$

$= 1/c^2 = \left[\binom{n}{2} \right]^{-2}$

(c) $P(T < 20) = \int_0^{20} f(t)\, dt = \int_0^{20} \binom{n}{2} e^{-\binom{n}{2} t}\, dt = \left[-e^{-\binom{n}{2} t} \right]_0^{20} = 1 - e^{-20\binom{n}{2}}$

65. $E[X] = \int_{-\infty}^{\infty} x\, f(x)\, dx = \frac{1}{2} c \int_{-\infty}^{\infty} x e^{-c|x|}\, dx = 0$ since $xe^{-c|x|}$ is an odd function (see Equation 5.4.6).

67. The PDF of a Laplace random variable with mean $c = 1$ is $f(x) = \frac{1}{2} e^{-|x|}$ where $x \in \mathbb{R}$.

(a) $P(|X| > 2) = 1 - P(-2 \le X \le 2) = 1 - \frac{1}{2} \int_{-2}^{2} e^{-|x|}\, dx = 1 - \int_0^2 e^{-x}\, dx$ [since the integrand is even (see Equation 5.4.6)]

$= 1 - \left[-e^{-x} \right]_0^2 = 1 - (-e^{-2} + 1) = e^{-2} \approx 0.135$

(b) $P(-1 < X < 1) = \frac{1}{2} \int_{-1}^{1} e^{-|x|}\, dx = \int_0^1 e^{-x}\, dx = -e^{-x} \big]_0^1 = -e^{-1} + 1 = 1 - 1/e \approx 0.632$

69. (a) $P(0 \le X \le 100) = \int_0^{100} \frac{1}{8\sqrt{2\pi}} \exp\!\left(-\frac{(x - 112)^2}{2 \cdot 8^2} \right) dx \approx 0.0668$ (using a calculator or computer to estimate the integral), so there is about a 6.68% chance that a randomly chosen vehicle is traveling at a legal speed.

(b) $P(X \ge 125) = \int_{125}^{\infty} \frac{1}{8\sqrt{2\pi}} \exp\!\left(-\frac{(x - 112)^2}{2 \cdot 8^2} \right) dx = \int_{125}^{\infty} f(x)\, dx$. In this case, we could use a calculator or computer to estimate either $\int_{125}^{300} f(x)\, dx$ or $1 - \int_0^{125} f(x)\, dx$. Both are approximately 0.0521, so about 5.21% of the motorists are targeted.

71. (a) With $\mu = 69$ and $\sigma = 2.8$, we have $P(65 \le X \le 73) = \int_{65}^{73} \frac{1}{2.8\sqrt{2\pi}} \exp\left(-\frac{(x - 69)^2}{2 \cdot 2.8^2} \right) dx \approx 0.847$ (using a calculator or computer to estimate the integral).

(b) $P(X > 6 \text{ feet}) = P(X > 72 \text{ inches}) = 1 - P(0 \le X \le 72) \approx 1 - 0.858 = 0.142$, so 14.2% of the adult male population is more than 6 feet tall.

73. (a) With $\mu = 0$ and $\sigma = \sqrt{2}$, we have

$P(|X| > 2) = 1 - P(-2 \le X \le 2) = 1 - \int_{-2}^{2} \frac{1}{\sqrt{4\pi}} \exp\left(-\frac{x^2}{4} \right) dx \approx 1 - 0.843 = 0.157$ (using a calculator or computer to estimate the integral), so there is a 15.7% chance that an individual disperses more than 2 meters.

(b) $P(-1 \le X \le 1) = \int_{-1}^{1} \frac{1}{\sqrt{4\pi}} \exp\left(-\frac{x^2}{4} \right) dx \approx 0.520$, so there is a 52.0% chance that an individual disperses less than 1 meter.

© 2016 Cengage Learning. All Rights Reserved. May not be scanned, copied or duplicated, or posted to a publicly accessible website, in whole or in part.

75. $E[X] = \int_{-\infty}^{\infty} x\, f(x)\, dx = \int_{-\infty}^{\infty} \frac{1}{\sigma\sqrt{2\pi}} x e^{-(x-\mu)^2/(2\sigma^2)}\, dx$

$= \int_{-\infty}^{\infty} \frac{1}{\sigma\sqrt{2\pi}} (z+\mu)\, e^{-z^2/(2\sigma^2)}\, dz \qquad$ [substitute $z = x - \mu$, $dz = dx$]

$= \frac{1}{\sigma\sqrt{2\pi}} \int_{-\infty}^{\infty} z e^{-z^2/(2\sigma^2)}\, dz + \mu \int_{-\infty}^{\infty} \frac{1}{\sigma\sqrt{2\pi}} e^{-z^2/(2\sigma^2)}\, dz$

The first integral evaluates to 0 since the integrand is an odd function (see Equation 5.4.6). The second integral evaluates to 1 since the integrand is the PDF of a normal random variable with mean 0. Therefore, $E[X] = 0 + \mu(1) = \mu$.

77. $f(x) = \frac{1}{\sigma\sqrt{2\pi}} e^{-(x-\mu)^2/(2\sigma^2)} \;\Rightarrow\; f'(x) = \frac{1}{\sigma\sqrt{2\pi}} e^{-(x-\mu)^2/(2\sigma^2)} \frac{-2(x-\mu)}{2\sigma^2} = \frac{-1}{\sigma^3\sqrt{2\pi}} e^{-(x-\mu)^2/(2\sigma^2)}(x-\mu) \;\Rightarrow$

$f''(x) = \frac{-1}{\sigma^3\sqrt{2\pi}}\left[e^{-(x-\mu)^2/(2\sigma^2)} \cdot 1 + (x-\mu)e^{-(x-\mu)^2/(2\sigma^2)} \frac{-2(x-\mu)}{2\sigma^2}\right]$

$= \frac{-1}{\sigma^3\sqrt{2\pi}} e^{-(x-\mu)^2/(2\sigma^2)}\left[1 - \frac{(x-\mu)^2}{\sigma^2}\right] = \frac{1}{\sigma^5\sqrt{2\pi}} e^{-(x-\mu)^2/(2\sigma^2)}\left[(x-\mu)^2 - \sigma^2\right]$

$f''(x) < 0 \;\Rightarrow\; (x-\mu)^2 - \sigma^2 < 0 \;\Rightarrow\; |x-\mu| < \sigma \;\Rightarrow\; -\sigma < x - \mu < \sigma \;\Rightarrow\; \mu - \sigma < x < \mu + \sigma$ and similarly, $f''(x) > 0 \;\Rightarrow\; x < \mu - \sigma$ or $x > \mu + \sigma$. Thus, f changes concavity and has inflection points at $x = \mu \pm \sigma$.

12 Review

TRUE-FALSE QUIZ

1. True. A permutation of a set of objects is defined as an *ordered* arrangement of the objects without repetition (see Section 12.1).

3. True. By definition, two events E and F are disjoint or mutually exclusive if their intersection is empty. See Definition 12.2.1(d).

5. True. The Multiplication Rule states that $P(E \cap F) = P(E|F)\,P(F)$, which in the case of independent events, simplifies to $P(E \cap F) = P(E)\,P(F)$. See Equations (3) and (4) in Section 12.3 and the corresponding discussion on the Multiplication Rule and Independence.

7. False. The CDF for a continuous random variable can be obtained by *integrating* its PDF as in Equation 12.5.6.

9. False. For example, the PDF $f(x) = 2x$ defined on the interval $[0, 1]$ has mean

$E[X] = \int_{-\infty}^{\infty} x f(x)\, dx = 2\int_0^1 x^2\, dx = \frac{2}{3}\left[x^3\right]_0^1 = \frac{2}{3} \neq \frac{1}{2}.$

EXERCISES

1. For each of the 12 toppings, there are 2 possible choices: include it or exclude it. Using the Fundamental Principle of Counting with $n_1 = n_2 = \cdots = n_{12} = 2$, the number of possible pizzas is $n_1 \cdot n_2 \cdot \cdots \cdot n_{12} = 2 \cdot 2 \cdot \cdots \cdot 2 = 2^{12} = 4096$. Note, we assume that choosing no toppings corresponds to a plain pizza.

© 2016 Cengage Learning. All Rights Reserved. May not be scanned, copied or duplicated, or posted to a publicly accessible website, in whole or in part.

3. (a) There are 26 letter choices for each of the 3 letter selections, so there are $26 \cdot 26 \cdot 26 = 26^3$ possible letter arrangements. For each of these, the number of ways of arranging 10 numerals for the 3 numeral selections is $10 \cdot 10 \cdot 10 = 10^3$. Thus, using the Fundamental Principle of Counting, the number of different plates is $26^3 \cdot 10^3 = 17{,}576{,}000$.

(b) If repetition is not allowed, the number of permutations of 26 letters taken 3 at a time is $P_{26,3} = \dfrac{26!}{23!} = 15{,}600$ and the number of permutations of 10 numerals taken 3 at a time is $P_{10,3} = \dfrac{10!}{7!} = 720$. Thus, using the Fundamental Principle of Counting, the number of different plates is $15{,}600 \cdot 720 = 11{,}232{,}000$.

5. (a) The number of ways of arranging the 10 people in line is $P_{10} = 10! = 3{,}628{,}800$.

(b) The number of ways of arranging 6 people from a group of 10 is $P_{10,6} = \dfrac{10!}{4!} = 10 \cdot 9 \cdot 8 \cdot 7 \cdot 6 \cdot 5 = 151{,}200$.

7. We will count the total number of different tetraploid individuals by examining the number of gene variant repetitions that may occur. If all the gene variants are different, that is no gene variant is repeated, then we must choose 4 variants from 4 possibilities which gives $\binom{4}{4} = 1$ possible combination.

If there is one pair of repeated variants and two different variants, the 1 repeated pair can be chosen from 4 possibilities which gives $\binom{4}{1}$ combinations and the 2 different variants must be chosen from the remaining 3 variants which gives $\binom{3}{2}$ combinations. Thus, from the FPC, the number of individuals having a pair of repeated variants and two different variants is $\binom{4}{1}\binom{3}{2} = 4 \cdot 3 = 12$.

If there are two pairs of repeated variants, the identity of those 2 pairs can be chosen from 4 possibilities which gives $\binom{4}{2} = 6$ combinations.

If 3 of the variants are the same, the repeated variant can be chosen from 4 possibilities which gives $\binom{4}{1}$ combinations and the 1 different variant must be chosen from the remaining 3 variants which gives $\binom{3}{1}$ combinations. Thus, from the FPC, the number of individuals having exactly 3 identical variants is $\binom{4}{1}\binom{3}{1} = 4 \cdot 3 = 12$.

Lastly, if all 4 of the variants are identical, this repeated variant can be chosen from 4 possibilities which gives $\binom{4}{1} = 4$ possible combinations.

Since all the cases above are mutually exclusive, the number of different tetraploid individuals is $1 + 12 + 6 + 12 + 4 = 35$.

9. $$\binom{n}{r}\binom{n-r}{k} = \left(\frac{n!}{r!\,(n-r)!}\right)\left(\frac{(n-r)!}{k!\,(n-r-k)!}\right) = \frac{n!}{r!} \cdot \frac{1}{k!\,(n-r-k)!}$$
$$= \frac{n!}{k!\,(n-k)!} \cdot \frac{(n-k)!}{r!\,(n-k-r)!} = \binom{n}{k}\binom{n-k}{r}$$

11. $B \cup C^c = \{1,2,5,9\} \cup \{6\}^c = \{1,2,5,9\} \cup \{1,2,3,4,5,7,8,9\} = \{1,2,3,4,5,7,8,9\}$

13. $A \cup B \cup C = \{1,3,5,8\} \cup \{1,2,5,9\} \cup \{6\} = \{1,2,3,5,8,9\} \cup \{6\} = \{1,2,3,5,6,8,9\}$

© 2016 Cengage Learning. All Rights Reserved. May not be scanned, copied or duplicated, or posted to a publicly accessible website, in whole or in part.

15. (a)

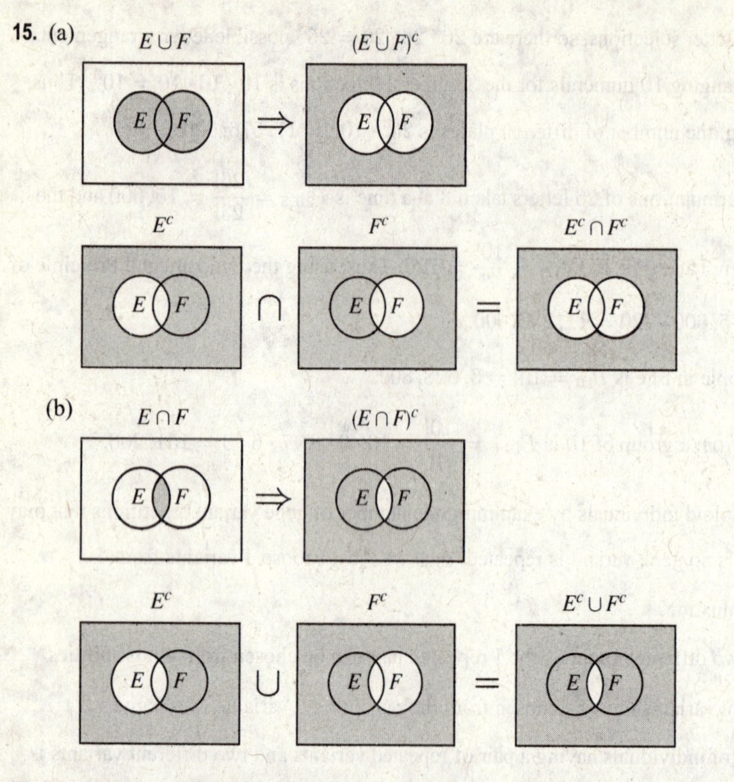

17. The number of possible of outcomes when drawing a card from a standard deck is $n(\Omega) = 52$, each outcome being equally likely.

(a) Let event E be drawing a numerical value between 3 and 5 inclusive. Since there are four cards for each of the three numerical values, the number of outcomes is $n(E) = 4 \cdot 3 = 12$. Therefore, the probability of drawing a 3, 4, or 5 is $P(E) = \dfrac{n(E)}{n(\Omega)} = \dfrac{12}{52} = \dfrac{3}{13}$.

(b) Let event E be drawing a card that is not a seven and not a heart. There are 12 numerical values that can be chosen and 3 possible suits for each, so the number of outcomes is $n(E) = 12 \cdot 3 = 36$. Therefore, the probability of drawing a card that is not a seven and not a heart is $P(E) = \dfrac{n(E)}{n(\Omega)} = \dfrac{36}{52} = \dfrac{9}{13}$.

(c) Let event E be drawing a card that is not a heart, not a diamond and not a spade. The only remaining suit is clubs for which there are $n(E) = 13$ cards in a deck. Therefore, the probability of event E is $P(E) = \dfrac{n(E)}{n(\Omega)} = \dfrac{13}{52} = \dfrac{1}{4}$.

19. There are a total of 13 balls in the jar, so the number of possible outcomes is $n(\Omega) = 13$, each outcome being equally likely.

(a) If neither a white nor a yellow ball is drawn, then a red ball must be drawn. Since there are 4 red balls, the probability of this event is $\frac{4}{13}$.

(b) The number of red, white and yellow balls is $4 + 3 + 6 = 13$, so the probability of drawing any one of these is $\frac{13}{13} = 1$.

(c) If a white ball is not drawn, then the ball must be red or yellow. Since there are $4 + 6 = 10$ red and yellow balls, the probability of not drawing a white ball is $\frac{10}{13}$.

© 2016 Cengage Learning. All Rights Reserved. May not be scanned, copied or duplicated, or posted to a publicly accessible website, in whole or in part.

21. (a) The total number of ways the customers can receive the cards is $P_5 = 5! = 120$. Now, there are 4 ways the first customer can receive the wrong card, after which there are 3 ways the second customer can receive the wrong card, after which there are 2 ways the second customer can receive the wrong card, and so on. Thus, the number of ways in which none of the customers get their own card back is $4! = 24$, and the probability of this event is $24/120 = 1/5$.

(b) If there are n people, the number of ways the customers can receive the cards is $P_n = n!$. Now, there are $n-1$ ways the first customer can receive the wrong card, after which there are $n-2$ ways the second customer can receive the wrong card, after which there are $n-3$ ways the second customer can receive the wrong card, and so on. Thus, the number of ways in which none of the customers get their own card back is $(n-1)\cdot(n-2)\cdot\dots\cdot 1 = (n-1)!$, so the probability of this event is $\dfrac{(n-1)!}{n!} = \dfrac{1}{n}$.

23. Of the 50 fish in the pond, 30 are marked and 20 are non-marked. The number of ways of selecting 4 marked fish from 30 possibilities is $\binom{30}{4}$, and the number of ways of selecting the remaining 6 non-marked fish from 20 possibilities is $\binom{20}{6}$. Thus, by the FPC, the number of ways of obtaining a group of 4 marked and 6 non-marked fish is $\binom{30}{4}\cdot\binom{20}{6}$. Also, the number of ways of picking a group of 10 fish from a population of 50 is given by $n(\Omega) = \binom{50}{10}$. Therefore, the probability of obtaining the described outcome is $\dfrac{\binom{30}{4}\cdot\binom{20}{6}}{\binom{50}{10}} = \dfrac{15,174,540}{146,746,831} \approx 0.1034$.

25. (a) $F - E$

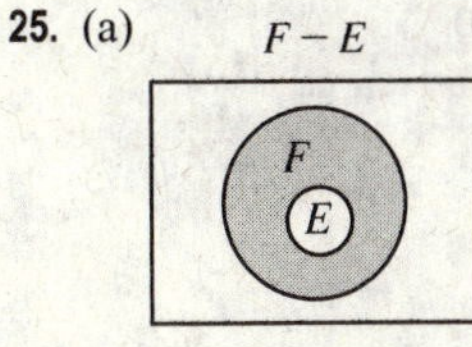

(b) As can be seen in the Venn diagram from part (a), the complement of event $F - E$ is $(F-E)^c = E \cup F^c \Leftrightarrow F - E = (E \cup F^c)^c \Rightarrow P(F-E) = P((E\cup F^c)^c) = 1 - P(E \cup F^c)$ [using the Complement Rule] $\Rightarrow P(F-E) = 1 - [P(E) + P(F^c) - P(E \cap F^c)]$ [using the Union Rule]. Now, $P(E \cap F^c) = 0$ because $E \subset F$, and $P(F^c) = 1 - P(F)$. Therefore, $P(F-E) = 1 - [P(E) + 1 - P(F)] = P(F) - P(E)$.

27. (a) If E and F are independent, then $P(E \cap F) = P(E)\,P(F) = \frac{1}{2}\times\frac{1}{2} = \frac{1}{4}$.

(b) If E and F are independent, then $P(E|F) = \dfrac{P(E\cap F)}{P(F)} = \dfrac{1/4}{1/2} = \dfrac{1}{2}$.

29. If the events HIV^+ and TB are *not* independent, then the probability of an individual being HIV^+ varies depending on whether the individual has symptoms of TB and vice versa. Biologically, this means that an HIV^+ individual may have a greater (or lesser) chance of displaying symptoms of TB than a randomly selected person. In fact, studies have shown that HIV^+ individuals are more likely to be infected with TB, so $P(TB|\text{HIV}^+) > P(TB)$.

31. $P(F|E) < P(F) \Leftrightarrow \dfrac{P(F\cap E)}{P(E)} < P(F) \Leftrightarrow \dfrac{P(F\cap E)}{P(F)} < P(E)$ [since probabilities are positive] $\Leftrightarrow$ $P(E|F) < P(E)$

© 2016 Cengage Learning. All Rights Reserved. May not be scanned, copied or duplicated, or posted to a publicly accessible website, in whole or in part.

33. Let X be a binomial random variable in which 'plant germinates' is treated as a success. We are told that $p = 0.75$, and there are $n = 4$ seeds.

(a) $P(X \geq 1) = 1 - P(X = 0) = 1 - \binom{4}{0}(0.75)^0(0.25)^4 = 1 - (0.25)^4 = \frac{255}{256} \approx 0.996$

(b) $P(X \geq 2) = P(X \geq 1) - P(X = 1) = \frac{255}{256} - \binom{4}{1}(0.75)^1(0.25)^3 = \frac{255}{256} - \frac{3}{64} = \frac{243}{256} \approx 0.949$

(c) $P(X = 4) = \binom{4}{4}(0.75)^4(0.25)^0 = (0.75)^4 = \frac{81}{256} \approx 0.316$

35. (a) $\int_0^1 f(x)\,dx = \int_0^1 \frac{3}{2}\sqrt{x}\,dx = \frac{3}{2}\int_0^1 x^{1/2}\,dx = \frac{3}{2}\left[\frac{2}{3}x^{3/2}\right]_0^1 = 1^{3/2} - 0^{3/2} = 1$

As required, the area under the graph of the PDF is 1.

(b) The PDF has domain $x \in [0, 1]$, so the probability that X is larger than $\frac{1}{2}$ is

$$P(X > \tfrac{1}{2}) = \textstyle\int_{1/2}^1 f(x)\,dx = \int_{1/2}^1 \frac{3}{2}x^{1/2}\,dx = \left[x^{3/2}\right]_{1/2}^1 = 1 - (1/2)^{3/2} \approx 0.646$$

(c) $P(\frac{1}{4} < X < \frac{3}{4}) = \int_{1/4}^{3/4} f(x)\,dx = \int_{1/4}^{3/4} \frac{3}{2}x^{1/2}\,dx = \left[x^{3/2}\right]_{1/4}^{3/4} = (3/4)^{3/2} - (1/4)^{3/2} \approx 0.525$

(d) $E[X] = \int_{-\infty}^{\infty} xf(x)\,dx = \frac{3}{2}\int_0^1 x^{3/2}\,dx = \frac{3}{2}\left[\frac{2}{5}x^{5/2}\right]_0^1 = \frac{3}{5}(1 - 0) = \frac{3}{5}$

(e) $Var[X] = \int_{-\infty}^{\infty} x^2 f(x)\,dx - E[X]^2 = \frac{3}{2}\int_0^1 x^{5/2}\,dx - \left(\frac{3}{5}\right)^2 = \frac{3}{2}\left[\frac{2}{7}x^{7/2}\right]_0^1 - \frac{9}{25} = \frac{3}{7}(1 - 0) - \frac{9}{25} = \frac{12}{175}$

37. The PDF of an exponential random variable with mean $\bar{T} = 8$ minutes is $f(t) = \frac{1}{8}e^{-t/8}$ where $t \geq 0$.

(a) $P(0 < T < 3) = \int_0^3 \frac{1}{8}e^{-t/8}\,dt = \left[-e^{-t/8}\right]_0^3 = -e^{-3/8} + 1 \approx 0.313$

(b) $P(T > 10) = \int_{10}^{\infty} \frac{1}{8}e^{-t/8}\,dt = \lim\limits_{x\to\infty}\left[-e^{-t/8}\right]_{10}^{x} = 0 + e^{-5/4} \approx 0.287$

(c) The median wait time must satisfy $\int_{-\infty}^{m} f(t)\,dt = \frac{1}{2} \Rightarrow \int_0^m \frac{1}{8}e^{-t/8}\,dt = \frac{1}{2} \Rightarrow$

$-e^{-t/8}\Big]_0^m = -e^{-m/8} + 1 = \frac{1}{2} \Rightarrow -m/8 = \ln\frac{1}{2} \Rightarrow m = 8\ln 2 \approx 5.55$ minutes.

39. (a) $\int_{-\infty}^{\infty} f(x)\,dx = 2\int_0^{\infty} \frac{1}{\pi(1+x^2)}\,dx$ [since $f(x)$ is even] $= \frac{2}{\pi}\lim\limits_{b\to\infty}\int_0^b \frac{1}{(1+x^2)}\,dx$

$$= \frac{2}{\pi}\lim_{b\to\infty}\left[\tan^{-1}x\right]_0^b = \frac{2}{\pi}\lim_{b\to\infty}\left(\tan^{-1}b - \tan^{-1}0\right) = \frac{2}{\pi}\left(\frac{\pi}{2} - 0\right)$$

$$= 1$$

As required, the area under the graph of the PDF is 1.

(b) $E[X] = \int_{-\infty}^{\infty} xf(x)\,dx = \int_{-\infty}^{\infty} \frac{x}{\pi(1+x^2)}\,dx = \int_{-\infty}^{0} \frac{x}{\pi(1+x^2)}\,dx + \int_0^{\infty} \frac{x}{\pi(1+x^2)}\,dx$ [by Definition 5.8.3]

Evaluating the first integral on the right side we have:

$$\int_{-\infty}^{0} \frac{x}{\pi(1+x^2)}\,dx = \lim_{a\to-\infty}\int_a^0 \frac{x}{\pi(1+x^2)}\,dx = \frac{1}{2}\lim_{a\to-\infty}\int_{1+a^2}^{1} \frac{1}{\pi u}\,du \quad [\text{substitute } u = 1 + x^2,\ du = 2x\,dx]$$

$$= \frac{1}{2\pi}\lim_{a\to-\infty}\left[\ln|u|\right]_{1+a^2}^{1} = \frac{1}{2\pi}\lim_{a\to-\infty}\left(\ln|1| - \ln\left|1 + a^2\right|\right) = -\frac{1}{2\pi}\lim_{a\to-\infty}\left(\ln\left|1 + a^2\right|\right) = -\infty$$

Since this integral is divergent, it follows that $\int_{-\infty}^{\infty} \frac{x}{\pi(1+x^2)}\,dx$ is divergent. Hence, $E[X]$ is undefined.

© 2016 Cengage Learning. All Rights Reserved. May not be scanned, copied or duplicated, or posted to a publicly accessible website, in whole or in part.

13 □ INFERENTIAL STATISTICS

13.1 The Sampling Distribution

1. (a) X_1 and X_2 are independent random variables, so Definition (2) states that

$$P(X_1 \le 10 \cap X_2 \le 8) = P(X_1 \le 10)\, P(X_2 \le 8) = F_1(10)\; F_2(8)$$

(b) $P(X_1 \le 5 \cap X_2 \ge 15) = P(X_1 \le 5)\, P(X_2 \ge 15) = F_1(5)\; [1 - F_2(15)]$

(c) $P(5 \le X_1 \le 10 \cap 3 \le X_2 \le 8) = P(5 \le X_1 \le 10)\, P(3 \le X_2 \le 8) = [F_1(10) - F_1(5)]\,[F_2(8) - F_2(3)]$

3. As discussed in Section 12.5, the PDF of an exponential random variable is given by $f(t) = \mu^{-1} e^{-t/\mu}$. Thus, the PDFs for T_1 and T_2 are $f(t_1) = e^{-t_1}$ and $f(t_2) = \frac{1}{2} e^{-t_2/2}$ respectively. Now, T_1 and T_2 are independent exponential random variables, so Definition (2) states that

$$P(T_1 \ge 2 \cap T_2 \ge 5) = P(T_1 \ge 2)\, P(T_2 \ge 5) = \left(\int_2^\infty e^{-t}\, dt\right)\left(\int_5^\infty \tfrac{1}{2} e^{-t/2}\, dt\right)$$

$$= \left(\lim_{x \to \infty} \int_2^x e^{-t}\, dt\right)\left(\lim_{y \to \infty} \int_5^y \tfrac{1}{2} e^{-t/2}\, dt\right) = \left(\lim_{x \to \infty} \left[-e^{-t}\right]_2^x\right)\left(\lim_{y \to \infty} \left[-e^{-t/2}\right]_5^y\right)$$

$$= \left(\lim_{x \to \infty} \left[-e^{-x} + e^{-2}\right]\right)\left(\lim_{y \to \infty} \left[-e^{-y/2} + e^{-5/2}\right]\right) = \left(e^{-2}\right)\left(e^{-5/2}\right) = e^{-9/2} \approx 0.011$$

5. The random variable of interest is the antibody level of a resident of the city. For the data set to constitute a set of i.i.d. random variables all of the antibody levels must come from a common distribution and be mutually independent. It is doubtful that the measured levels are mutually independent because people living in a common household probably have had similar influenza exposure, and therefore they will likely also have similar levels of antibodies.

7. The random variable of interest is the height of a tree in the specified forest. For the data set to constitute a set of i.i.d. random variables all of the heights must come from a common distribution and be mutually independent. The height of trees within a particular region might not be independent if, for example, high soil nutrients tend to be aggregated in certain regions. In this case, if a randomly chosen tree from a region was very tall then we might expect other trees from that same region to also be very tall.

9. The standard deviation is $\sigma = 7.6$ and there are 32 data points, so by Theorem 4, the standard deviation of the sample mean is $\dfrac{7.6}{\sqrt{32}} \approx 1.344$. The probability of obtaining a sample mean in the interval $[\mu - 2, \mu + 2]$ is then

$$\int_{\mu-2}^{\mu+2} \frac{1}{1.344\sqrt{2\pi}} e^{-(y-\mu)^2/(2(1.344)^2)}\, dy = \frac{1}{\sqrt{\pi}} \int_{\frac{-2}{1.344\sqrt{2}}}^{\frac{2}{1.344\sqrt{2}}} e^{-u^2}\, du \quad \left[\text{substitute } u = \frac{y-\mu}{1.344\sqrt{2}},\ dy = 1.344\sqrt{2}\, du\right]$$

$$\approx 0.863 \qquad \text{[using a numerical integrator]}$$

© 2016 Cengage Learning. All Rights Reserved. May not be scanned, copied or duplicated, or posted to a publicly accessible website, in whole or in part.

11. The standard deviation is $\sigma = 20$ and there are 20 data points, so by Theorem 4, the standard deviation of the sample mean is $\frac{20}{\sqrt{20}} \approx 4.472$. Using the Complement Rule, the probability of obtaining a sample mean in the interval $(\infty, \mu - 8) \cup (\mu + 8, \infty)$ is

$$1 - \int_{\mu-8}^{\mu+8} \frac{1}{4.472\sqrt{2\pi}} e^{-(y-\mu)^2/(2(4.472)^2)}\,dy = 1 - \frac{1}{\sqrt{\pi}} \int_{\frac{-8}{4.472\sqrt{2}}}^{\frac{8}{4.472\sqrt{2}}} e^{-u^2}\,du \quad \left[\text{substitute } u = \frac{y-\mu}{4.472\sqrt{2}}, dy = 4.472\sqrt{2}\,du\right]$$

$$\approx 0.0736 \qquad [\text{using a numerical integrator}]$$

13. The standard deviation is $\sigma = 0.12$ and the sample size of each study is 20, so by Theorem 4, the standard deviation of the sample mean is $\frac{0.12}{\sqrt{20}} \approx 0.0268$. Using the Complement Rule, the probability of obtaining a sample mean in the interval $(\infty, \mu - 0.0181) \cup (\mu + 0.0181, \infty)$ is

$$1 - \int_{\mu-0.0181}^{\mu+0.0181} \frac{1}{0.0268\sqrt{2\pi}} e^{-(y-\mu)^2/(2(0.0268)^2)}\,dy = 1 - \frac{1}{\sqrt{\pi}} \int_{\frac{-0.0181}{0.0268\sqrt{2}}}^{\frac{0.0181}{0.0268\sqrt{2}}} e^{-u^2}\,du \quad \left[\begin{array}{r}\text{substitute } u = \frac{y-\mu}{0.0268\sqrt{2}}, \\ dy = 0.0268\sqrt{2}\,du\end{array}\right]$$

$$\approx 0.500 \qquad [\text{using a numerical integrator}]$$

Now, $14\,(0.500) = 7$, so we expect roughly 7 studies will have a sample mean that falls further than 0.0181 from the true population mean.

15. The standard deviation is $\sigma = 0.9$ and the sample size of each study is 28, so by Theorem 4, the standard deviation of the sample mean is $\frac{0.9}{\sqrt{28}} \approx 0.170$. Using the Complement Rule, the probability of obtaining a sample mean in the interval $(\infty, \mu - 0.1) \cup (\mu + 0.1, \infty)$ is

$$1 - \int_{\mu-0.1}^{\mu+0.1} \frac{1}{0.17\sqrt{2\pi}} e^{-(y-\mu)^2/(2(0.17)^2)}\,dy = 1 - \frac{1}{\sqrt{\pi}} \int_{\frac{-0.1}{0.17\sqrt{2}}}^{\frac{0.1}{0.17\sqrt{2}}} e^{-u^2}\,du \quad \left[\text{substitute } u = \frac{y-\mu}{0.17\sqrt{2}}, dy = 0.17\sqrt{2}\,du\right]$$

$$\approx 0.557 \qquad [\text{using a numerical integrator}]$$

Now, the two studies are independent and each has a probability 0.557 of having a sample mean that deviates by more than 0.1 from the true population mean. Thus, by the definition of independence, the probability that the sample mean of *both* studies deviates by more than 0.1 from the true population mean is $0.557 \times 0.557 \approx 0.310$.

17. The standard deviation is $\sigma = 0.043$ and the sample size of each study is 10, so by Theorem 4, the standard deviation of the sample mean is $\frac{0.043}{\sqrt{10}} \approx 0.0136$. The probability of obtaining a sample mean in the interval $(\mu - 0.01, \mu - 0.01)$ is

$$\int_{\mu-0.01}^{\mu+0.01} \frac{1}{0.0136\sqrt{2\pi}} e^{-(y-\mu)^2/(2(0.0136)^2)}\,dy = \frac{1}{\sqrt{\pi}} \int_{\frac{-0.1}{0.0136\sqrt{2}}}^{\frac{0.1}{0.0136\sqrt{2}}} e^{-u^2}\,du \quad \left[\text{substitute } u = \frac{y-\mu}{0.0136\sqrt{2}}, dy = 0.0136\sqrt{2}\,du\right]$$

$$\approx 0.538 \qquad [\text{using a numerical integrator}]$$

Since the studies are all independent, the probability that the three studies fall within 0.01 of the population mean is $0.538^3 \approx 0.1557$. Now, using the Complement Rule, the probability that at least one of the sample means deviates by more than 0.01 from the population mean is $1 - 0.1557 \approx 0.844$.

© 2016 Cengage Learning. All Rights Reserved. May not be scanned, copied or duplicated, or posted to a publicly accessible website, in whole or in part.

19. $\bar{x} = \left[\begin{array}{l} 902 + 920 + 924 + 932 + 937 + 939 + 945 + 948 + 949 + 951 \\ +957 + 958 + 961 + 965 + 969 + 970 + 975 + 982 + 987 + 991 \end{array}\right] \Big/ 20 = \dfrac{19062}{20} = 953.1 \text{ lbs}$

$$s = \sqrt{\frac{1}{19}\left[\begin{array}{l} (902-953.1)^2 + (920-953.1)^2 + (924-953.1)^2 + (932-953.1)^2 + (937-953.1)^2 \\ +(939-953.1)^2 + (945-953.1)^2 + (948-953.1)^2 + (949-953.1)^2 + (951-953.1)^2 \\ +(957-953.1)^2 + (958-953.1)^2 + (961-953.1)^2 + (965-953.1)^2 + (969-953.1)^2 \\ +(970-953.1)^2 + (975-953.1)^2 + (982-953.1)^2 + (987-953.1)^2 + (991-953.1)^2 \end{array}\right]}$$

$= \sqrt{10251.8/19} \approx 23.2286 \text{ lbs}$

21. $\bar{x} = \left[\begin{array}{l} 6 + 10 + 5 + 12 + 5 + 3 + 41 + 12 + 6 + 15 \\ +20 + 4 + 3 + 7 + 13 + 16 + 18 + 29 + 23 + 17 \end{array}\right] \Big/ 20 = \dfrac{265}{20} = 13.25 \text{ days}$

$$s = \sqrt{\frac{1}{19}\left[\begin{array}{l} (6-13.25)^2 + (10-13.25)^2 + (5-13.25)^2 + (12-13.25)^2 + (5-13.25)^2 \\ +(3-13.25)^2 + (41-13.25)^2 + (12-13.25)^2 + (6-13.25)^2 + (15-13.25)^2 \\ +(20-13.25)^2 + (4-13.25)^2 + (3-13.25)^2 + (7-13.25)^2 + (13-13.25)^2 \\ +(16-13.25)^2 + (18-13.25)^2 + (29-13.25)^2 + (23-13.25)^2 + (17-13.25)^2 \end{array}\right]}$$

$= \sqrt{1795.75/19} \approx 9.721788 \text{ days}$

23. $\bar{x} = [71 + 70 + 69 + 68 + 67 + 66 + 64 + 62 + 58 + 57 + 55 + 53 + 52 + 48 + 45] / 15 = \dfrac{905}{15} = \dfrac{181}{3} \approx 60.33$

$$s = \sqrt{\frac{1}{14}\left[\begin{array}{l} \left(71-\frac{181}{3}\right)^2 + \left(70-\frac{181}{3}\right)^2 + \left(69-\frac{181}{3}\right)^2 + \left(68-\frac{181}{3}\right)^2 + \left(67-\frac{181}{3}\right)^2 \\ +\left(66-\frac{181}{3}\right)^2 + \left(64-\frac{181}{3}\right)^2 + \left(62-\frac{181}{3}\right)^2 + \left(58-\frac{181}{3}\right)^2 + \left(57-\frac{181}{3}\right)^2 \\ +\left(55-\frac{181}{3}\right)^2 + \left(53-\frac{181}{3}\right)^2 + \left(52-\frac{181}{3}\right)^2 + \left(48-\frac{181}{3}\right)^2 + \left(45-\frac{181}{3}\right)^2 \end{array}\right]}$$

$= \sqrt{(2968/3)/14} = \sqrt{212/3} \approx 8.406347$

25. (a) Since X_1 and X_2 are independent random variables, Definition (2) states that

$$P(X_1 \le x_1 \cap X_2 \le x_2) = P(X_1 \le x_1)\ P(X_2 \le x_2) = F(x_1)\ F(x_2)$$

(b) $G(x_1, x_2) = F(x_1)\ F(x_2) \;\Rightarrow\; \dfrac{\partial G}{\partial x_2} = F(x_1)\ F'(x_2) \;\Rightarrow\; \dfrac{\partial^2 G}{\partial x_1 \partial x_2} = F'(x_1)\ F'(x_2)$

From Equation 12.5.7, the derivative of the common CDF is $F'(x) = f(x) = \dfrac{1}{\sigma\sqrt{2\pi}}\, e^{-(x-\mu)^2/(2\sigma^2)}$. Therefore, the joint PDF is

$$g(x_1, x_2) = F'(x_1)\ F'(x_2) = \left[\frac{1}{\sigma\sqrt{2\pi}}\, e^{-(x_1-\mu)^2/(2\sigma^2)}\right]\left[\frac{1}{\sigma\sqrt{2\pi}}\, e^{-(x_2-\mu)^2/(2\sigma^2)}\right]$$

$$= \frac{1}{2\pi\sigma^2}\, e^{-(x_1-\mu)^2/(2\sigma^2)} e^{-(x_2-\mu)^2/(2\sigma^2)}$$

(c) $g(160, 170) = \dfrac{1}{2\pi\sigma^2}\, e^{-(160-\mu)^2/(2\sigma^2)}\, e^{-(170-\mu)^2/(2\sigma^2)} = \dfrac{1}{2\pi\sigma^2}\, e^{-(54500-660\mu+2\mu^2)/(2\sigma^2)}$

$$= \frac{1}{2\pi\sigma^2}\, e^{-(27250-330\mu+\mu^2)/(\sigma^2)}$$

© 2016 Cengage Learning. All Rights Reserved. May not be scanned, copied or duplicated, or posted to a publicly accessible website, in whole or in part.

(d) The likelihood function $L(\mu) = \dfrac{1}{2\pi\sigma^2}\, e^{-(27250-330\mu+\mu^2)/(\sigma^2)}$ is maximized when $L'(\mu) = 0 \;\Rightarrow$

$$\frac{1}{2\pi\sigma^2}\, e^{-(27250-330\mu+\mu^2)/(\sigma^2)}\left[(330-2\mu)/(\sigma^2)\right] = 0 \;\Rightarrow\; 330-2\mu = 0 \;\Rightarrow\; \mu = \frac{330}{2} = 165\,\text{cm}$$

Thus, the maximum likelihood estimate for the population mean is 165 cm, that is, the average of the two sampled heights.

13.2 Confidence Intervals

1. As calculated in the solution of Exercise 13.1.19, the sample mean of the Weddell seal data is $\bar{x} = 953.1$ lbs. The population standard deviation is known, so (3) gives $z_{95} = 1.96$. Thus the endpoints of the confidence interval are $953.1 \pm 1.96\dfrac{22.6}{\sqrt{20}}$ which gives an interval of approximately $(943.2, 963.0)$.

3. As calculated in the solution of Exercise 13.1.21, the sample mean of the "Hospital Stays" data is $\bar{x} = 13.25$ days. The population standard deviation is known, so (3) gives $z_{95} = 1.96$. Thus the endpoints of the confidence interval are $13.25 \pm 1.96\dfrac{9.5}{\sqrt{20}}$ which gives an interval of approximately $(9.1, 17.4)$.

5. (a) We construct a histogram containing 5 bins of width 0.1 starting at an initial value 0.05. Note that the choice of bin width is not unique and different bin widths will result in different histograms. The coral bleaching data does not appear to be normally distributed as it is skewed to the right.

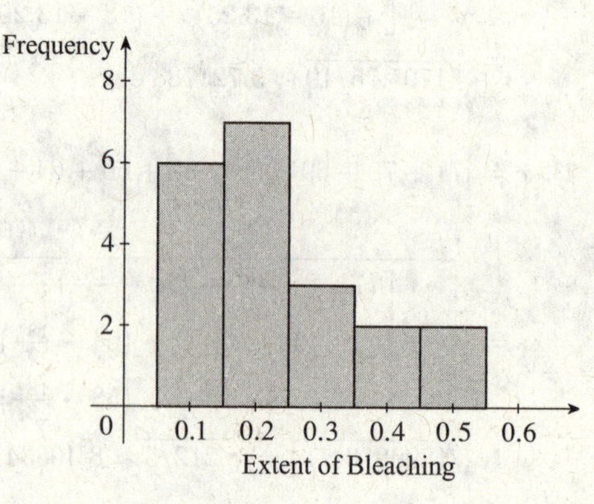

(b) The population standard deviation is unknown so we must carry out the steps outlined in (11).

$$\bar{x} = \frac{1}{20}\begin{bmatrix} 0.19+0.06+0.26+0.32+0.35+0.09+0.05+0.21+0.48+0.21 \\ +0.23+0.12+0.22+0.16+0.35+0.13+0.34+0.07+0.16+0.45 \end{bmatrix} = \frac{4.45}{20} = 0.2225$$

$$s = \sqrt{\frac{1}{19}\begin{bmatrix} (0.19-0.2225)^2+(0.06-0.2225)^2+(0.26-0.2225)^2+(0.32-0.2225)^2+(0.35-0.2225)^2 \\ +(0.09-0.2225)^2+(0.05-0.2225)^2+(0.21-0.2225)^2+(0.48-0.2225)^2+(0.21-0.2225)^2 \\ +(0.23-0.2225)^2+(0.12-0.2225)^2+(0.22-0.2225)^2+(0.16-0.2225)^2+(0.35-0.2225)^2 \\ +(0.13-0.2225)^2+(0.34-0.2225)^2+(0.07-0.2225)^2+(0.16-0.2225)^2+(0.45-0.2225)^2 \end{bmatrix}}$$

$$= \sqrt{0.300575/19} \approx 0.125777$$

Using either a calculator or Appendix H, the t value for a 90% confidence level and $20 - 1 = 19$ degrees of freedom is $t_{90,19} \approx 1.729$. Thus the endpoints of the confidence interval are $0.2225 \pm 1.729\dfrac{0.1258}{\sqrt{20}}$ which gives an interval of approximately $(0.17, 0.27)$. In determining this confidence interval, we found the t value that bounds 90% of Student's t distribution assuming the sample mean is normally distributed. However, we saw in part (a) that the coral bleaching data is not normally distributed so the calculations made using that t value may not produce the correct 90% confidence interval.

© 2016 Cengage Learning. All Rights Reserved. May not be scanned, copied or duplicated, or posted to a publicly accessible website, in whole or in part.

7. (a) We construct a histogram containing 6 bins of width 5 starting at an initial value of 45. (See the solution to Exercise 11.2.15 for a detailed construction of the frequency histogram.) Note that the choice of bin width is not unique and different bin widths will result in different histograms. The age at death does not appear to be normally distributed.

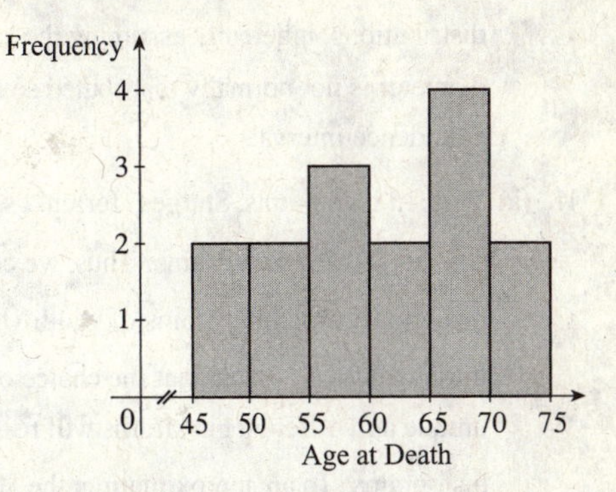

(b) The population standard deviation is known so we must carry out the steps outlined in (9). As calculated in the solution to Exercise 13.1.23, the sample mean is $\bar{x} = \frac{181}{3}$. From (3), a 90% confidence interval has $z_{90} = 1.64$. Thus the endpoints of the confidence interval are $\dfrac{181}{3} \pm 1.64\dfrac{8}{\sqrt{15}}$ which gives an interval of approximately $(56.9, 63.7)$. In calculating this confidence interval, we used the z value that bounds 90% of the area of the standard normal curve, inherently assuming the sample mean is normally distributed. However, we saw in part (a) that the age at death is not normally distributed so the calculations made using that z value may not produce the correct 90% confidence interval.

9. (a) We construct a histogram containing 5 bins of width 1.5 starting at an initial value of 1.5. (See the solution to Exercise 11.2.17 for a detailed construction of the frequency histogram.) Note that the choice of bin width is not unique and different bin widths will result in different histograms. The tumor diameter does not appear to be normally distributed.

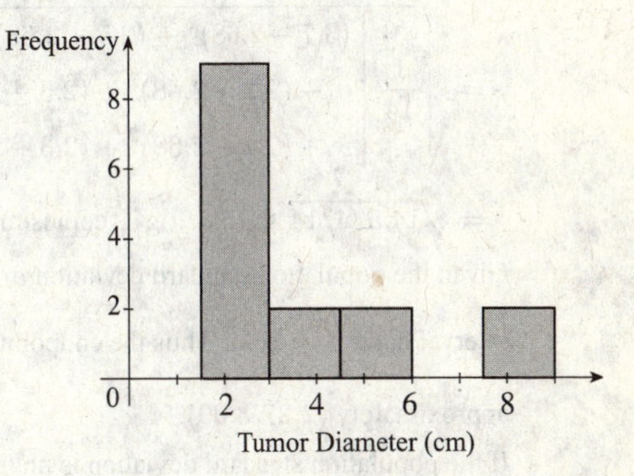

(b) $\bar{x} = \dfrac{1}{15}\left[1.5 + 1.5 + 1.6 + 1.6 + 1.7 + 2.0 + 2.1 + 2.2 + 2.6 + 3.2 + 4.0 + 4.9 + 5.5 + 7.5 + 7.6\right] = \dfrac{49.5}{15} = 3.3$

$$s = \sqrt{\frac{1}{14}\begin{bmatrix} (1.5-3.3)^2 + (1.5-3.3)^2 + (1.6-3.3)^2 + (1.6-3.3)^2 + (1.7-3.3)^2 \\ + (2.0-3.3)^2 + (2.1-3.3)^2 + (2.2-3.3)^2 + (2.6-3.3)^2 + (3.2-3.3)^2 \\ + (4.0-3.3)^2 + (4.9-3.3)^2 + (5.5-3.3)^2 + (7.5-3.3)^2 + (7.6-3.3)^2 \end{bmatrix}}$$

$$= \sqrt{63.68/14} \approx 2.133$$

Given the population standard deviation $\sigma = 2.1$, we can carry out the steps outlined in (9). From (3), a 90% confidence interval has $z_{90} = 1.64$. Thus the endpoints of the confidence interval are $3.3 \pm 1.64\dfrac{2.1}{\sqrt{15}}$ which gives an interval of approximately $(2.41, 4.19)$.

If the population standard deviation is unknown, we must carry out the steps outlined in (11). Using either a calculator or Appendix H, the t value for a 90% confidence level and $15 - 1 = 14$ degrees of freedom is $t_{90,14} \approx 1.761$. Thus the endpoints of the confidence interval are $3.3 \pm 1.761\dfrac{2.133}{\sqrt{15}}$ which gives an interval of approximately $(2.33, 4.27)$.

In calculating these confidence intervals, we used the z and t values that bound 90% of the area of their respective

[continued]

© 2016 Cengage Learning. All Rights Reserved. May not be scanned, copied or duplicated, or posted to a publicly accessible website, in whole or in part.

distributions, inherently assuming the sample mean is normally distributed. However, we saw in part (a) that the tumor diameter is not normally distributed so the calculations made using those z and t values may not produce the correct 90% confidence intervals.

11. (a) With 15 data points, Sturges' formula suggests $1 + \ln(15)/\ln 2 \approx 5$ bins. Thus, we construct a histogram containing 5 bins of width 1 starting at an initial value of 0. Note that the choice of bin width is not unique and different bin widths will result in different histograms. To an approximation, the HAART data appears to be normally distributed.

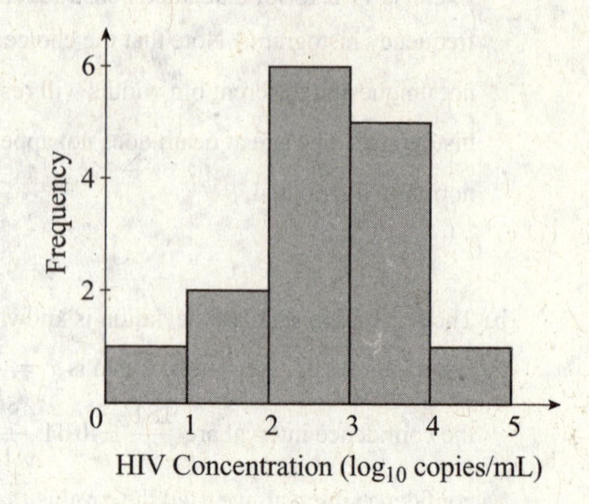

(b) $$\bar{x} = \left[\begin{array}{l} 3.7 + 2.5 + 3.5 + 3.6 + 0.7 + 2.2 + 2.5 \\ \quad +1.4 + 3.6 + 1.8 + 2.4 + 2.3 + 4.7 + 2.3 + 3.0 \end{array}\right] \Big/ 15 = \frac{40.2}{15} = 2.68 \ \log_{10} \text{ copies/mL}$$

$$s = \sqrt{\frac{1}{14}\left[\begin{array}{l} (3.7-2.68)^2 + (2.5-2.68)^2 + (3.5-2.68)^2 + (3.6-2.68)^2 + (0.7-2.68)^2 \\ +(2.2-2.68)^2 + (2.5-2.68)^2 + (1.4-2.68)^2 + (3.6-2.68)^2 + (1.8-2.68)^2 \\ +(2.4-2.68)^2 + (2.3-2.68)^2 + (4.7-2.68)^2 + (2.3-2.68)^2 + (3.0-2.68)^2 \end{array}\right]}$$

$$= \sqrt{14.584/14} \approx 1.02 \ \log_{10} \text{ copies/mL}$$

Given the population standard deviation $\sigma = 0.98$, we can carry out the steps outlined in (9). From (3), a 90% confidence interval has $z_{90} = 1.64$. Thus the endpoints of the confidence interval are $2.68 \pm 1.64\frac{0.98}{\sqrt{15}}$ which gives an interval of approximately $(2.27, 3.09)$.

If the population standard deviation is unknown, we must carry out the steps outlined in (11). Using either a calculator or Appendix H, the t value for a 90% confidence level and $15 - 1 = 14$ degrees of freedom is $t_{90,14} \approx 1.761$. Thus the endpoints of the confidence interval are $2.68 \pm 1.761\frac{1.02}{\sqrt{15}}$ which gives an interval of approximately $(2.22, 3.14)$.

The width of the confidence intervals calculated using the population and sample standard deviations are $3.09 - 2.27 = 0.82$ and $3.14 - 2.22 = 0.92$ respectively. The latter value is larger reflecting a greater uncertainty when using the sample data to estimate the population standard deviation

13. The population standard deviation is unknown in both samples so we must carry out the steps outlined in (11). The mean and standard deviation of the female data are $\bar{x}_f \approx 52.846$ and $s_f \approx 3.997$ and those of the male data are $\bar{x}_m = 39.824$ and $s_m \approx 9.146$. The female data has $13 - 1 = 12$ degrees of freedom while the male data has $17 - 1 = 16$. The corresponding t values for a 95% confidence level are $t_{95,12} = 2.179$ and $t_{95,16} = 2.120$ (found either from a calculator or Appendix H). Thus the endpoints of the confidence interval for females are $52.846 \pm 2.179\frac{3.997}{\sqrt{13}}$ which gives an interval of approximately $(50.4, 55.3)$. Similarly for the males the endpoints are $39.824 \pm 2.120\frac{9.146}{\sqrt{17}}$ which gives a confidence interval of approximately $(35.1, 44.5)$.

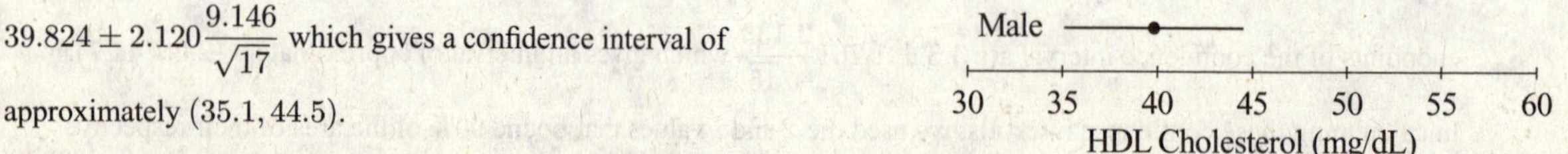

© 2016 Cengage Learning. All Rights Reserved. May not be scanned, copied or duplicated, or posted to a publicly accessible website, in whole or in part.

15. (a) $F_X(x) = P(X \le x) \Rightarrow F_X(\sigma z + \mu) = P(X \le \sigma z + \mu) = P(\sigma Z + \mu \le \sigma z + \mu) = P(Z \le z) = F_Z(z)$

(b) $F_Z(z) = F_X(\sigma z + \mu) \Rightarrow F'_Z(z) = F'_X(\sigma z + \mu)\dfrac{d}{dz}(\sigma z + \mu)$ [by the chain rule] $\Leftrightarrow f_Z(z) = \sigma f_X(\sigma z + \mu)$

Therefore, $f_Z(z) = \dfrac{\sigma}{\sigma\sqrt{2\pi}} e^{-\frac{1}{2}\left(\frac{\sigma z + \mu - \mu}{\sigma}\right)^2} = \dfrac{1}{\sqrt{2\pi}} e^{-\frac{1}{2}z^2}$. This is the PDF of a normal random variable having mean 0 and standard deviation 1. Thus, Z is a standard normal random variable.

17. Let Y_n be the sample mean from a sample of size n and define the standard normal variable $Z_n = \dfrac{Y_n - \mu}{\sigma/\sqrt{n}}$ where μ is the population mean and $\sigma/\sqrt{n}$ is the standard deviation of the mean. For a $C\%$ confidence interval, we require a value of z_c that satisfies $P(-z < Z_n < z) = \dfrac{C}{100} \Rightarrow P\left(-z < \dfrac{Y_n - \mu}{\sigma/\sqrt{n}} < z\right) = \dfrac{C}{100} \Rightarrow$

$$P\left(-z\frac{\sigma}{\sqrt{n}} < Y_n - \mu < z\frac{\sigma}{\sqrt{n}}\right) = \frac{C}{100} \Rightarrow P\left(-Y_n - z\frac{\sigma}{\sqrt{n}} < -\mu < -Y_n + z\frac{\sigma}{\sqrt{n}}\right) = \frac{C}{100} \Rightarrow$$

$$P\left(Y_n - z\frac{\sigma}{\sqrt{n}} < \mu < Y_n + z\frac{\sigma}{\sqrt{n}}\right) = \frac{C}{100}.$$

Therefore, the $C\%$ confidence interval for the mean has endpoints $\mu \pm z\dfrac{\sigma}{\sqrt{n}}$.

13.3 Hypothesis Testing

1. With 20 data points, Sturges' formula suggests $1 + \ln(20) / \ln 2 \approx 5$ bins. Thus, we construct a histogram containing 5 bins of width 20 starting at an initial value of 900. Note that the choice of bin width is not unique and different bin widths will result in different histograms. To an approximation, the Weddell seal weights appear to be normally distributed.

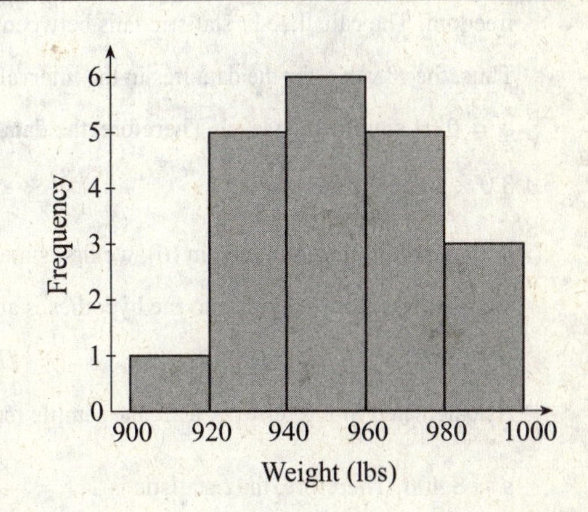

Following the steps outlined in (6), we first state the null and alternative hypotheses. We wish to test whether the population mean differs from 935 lbs, so the hypotheses are

$$H_0\colon \mu = 935 \qquad H_A\colon \mu \ne 935$$

We will use a significance level of $\alpha = 0.05$. The sample mean and sample standard deviation of the data are $Y_{20} = 953.1$ and $s = 23.229$. Therefore, the t statistic is $T_{19} = \dfrac{953.1 - 935}{23.229/\sqrt{20}} \approx 3.485$. We now consult the row of Table 2 in Appendix H that corresponds to $20 - 1 = 19$ degrees of freedom. The calculated t statistic falls between 2.861 and 3.883 with corresponding P values of 0.01 and 0.001 respectively. Thus, the P value for the data lies in the interval $0.001 < P < 0.01$. Since $P < 0.05$ we reject the null hypothesis. Therefore, the data provides evidence that the population mean weight differs from 935 pounds.

© 2016 Cengage Learning. All Rights Reserved. May not be scanned, copied or duplicated, or posted to a publicly accessible website, in whole or in part.

3. Following the steps outlined in (6), we first state the null and alternative hypotheses. We wish to test whether the population mean differs from 20 days, so the hypotheses are

$$H_0\colon \mu = 20 \qquad H_A\colon \mu \neq 20$$

We will use a significance level of $\alpha = 0.05$. As calculated in Exercise 13.1.21, the sample mean and sample standard deviation of the data are $Y_{20} = 13.25$ and $s \approx 9.722$. Therefore, the t statistic is $T_{19} = \dfrac{13.25 - 20}{9.722/\sqrt{20}} \approx -3.105$. We now consult the row of Table 2 in Appendix H that corresponds to $20 - 1 = 19$ degrees of freedom. The calculated t statistic falls between 2.861 and 3.883 with corresponding P values of 0.01 and 0.001 respectively. Thus, the P value for the data lies in the interval $0.001 < P < 0.01$. Since $P < 0.05$ we reject the null hypothesis. Therefore, the data provides evidence that the population has a mean hospital stay that differs from 20 days. In consulting Appendix H, we have inherently assumed that the calculated value T_{19} follows a t distribution with 19 degrees of freedom. This is only the case when the sample mean is normally distributed, which by Theorem 13.1.4, requires the population mean to be normally distributed.

5. Following the steps outlined in (6), we first state the null and alternative hypotheses. We wish to test whether the population mean differs from 3.0×10^6 cells per μL. Measuring RBC in units of (cells $\times 10^6)/\mu$L, the hypotheses are

$$H_0\colon \mu = 3.0 \qquad H_A\colon \mu \neq 3.0$$

The sample mean and sample standard deviation of the data are $Y_{18} \approx 3.744$ and $s \approx 1.395$. Therefore, the t statistic is $T_{17} = \dfrac{3.744 - 3.0}{1.395/\sqrt{18}} \approx 2.263$. We now consult the row of Table 2 in Appendix H that corresponds to $18 - 1 = 17$ degrees of freedom. The calculated t statistic falls between 2.110 and 2.898 with corresponding P values of 0.05 and 0.01 respectively. Thus, the P value for the data lies in the interval $0.01 < P < 0.05$. Since $P > 0.01$ we do *not* reject the null hypothesis at the $\alpha = 0.01$ significance level. Therefore, the data are consistent with the population having a mean RBC concentration of 3.0×10^6 cells per μL.

7. Following the steps outlined in (6), we first state the null and alternative hypotheses. We wish to test whether the population mean differs from 15 years, so the hypotheses are

$$H_0\colon \mu = 15 \qquad H_A\colon \mu \neq 15$$

As calculated in Exercise 13.1.23, the sample mean and sample standard deviation of the data are $Y_{15} = 181/3$ and $s \approx 8.406$. Therefore, the t statistic is $T_{14} = \dfrac{181/3 - 15}{8.406/\sqrt{15}} \approx 20.887$. We now consult the row of Table 2 in Appendix H that corresponds to $15 - 1 = 14$ degrees of freedom. The largest t statistic value is 4.140 which corresponds to a P value of 0.001. Observe that larger t statistic values have smaller P values in the table. Since the calculated t statistic 20.887 is larger than 4.140, we deduce that $P < 0.001$. (Alternatively, using a calculator or CAS, we find the more precise value $P \approx 6.0 \times 10^{-12}$). Thus, if the mean age of the population is 15, there is less than a 0.1% chance of having observed the given data. In consulting Appendix H, we have inherently assumed that the calculated value T_{14} follows a t distribution with 14 degrees of freedom. This is only the case when the sample mean is normally distributed, which by Theorem 13.1.4, requires the population mean to be normally distributed.

© 2016 Cengage Learning. All Rights Reserved. May not be scanned, copied or duplicated, or posted to a publicly accessible website, in whole or in part.

9. We construct a histogram containing 8 bins of width 0.5 starting at an initial value of 8.5. Note that the choice of bin width is not unique and different bin widths will result in different histograms. To an approximation, the wingspan data appears to be normally distributed.

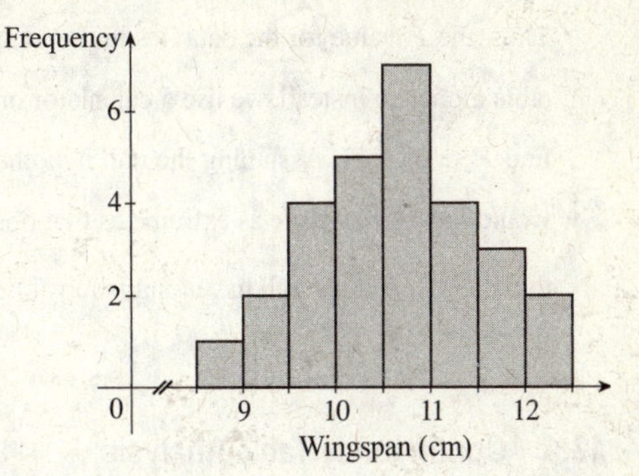

Following the steps outlined in (6), we first state the null and alternative hypotheses. We wish to test whether the population mean differs from 10 cm, so the hypotheses are

$$H_0\colon \mu = 10 \qquad H_A\colon \mu \neq 10$$

As calculated in Exercise 13.1.24, the sample mean and sample standard deviation of the data are $Y_{28} \approx 10.5857$ and $s \approx 0.8939$. Therefore, the t statistic is $T_{27} = \dfrac{10.5857 - 10}{0.8939/\sqrt{28}} \approx 3.467$. We now consult the row of Table 2 in Appendix H that corresponds to $28 - 1 = 27$ degrees of freedom. The calculated t statistic falls between 2.771 and 3.690 with corresponding P values of 0.01 and 0.001 respectively. Thus, the P value for the data lies in the interval $0.001 < P < 0.01$. Alternatively, using a calculator or CAS, we find the more precise value $P \approx 0.0018$. Thus, if the population mean wingspan is 10 cm, there is approximately a 0.18% chance of having observed the given data.

11. Following the steps outlined in (6), we first state the null and alternative hypotheses. We wish to test if HAART is effective, that is, whether the population mean change in HIV concentration after treatment differs from 0. So the hypotheses are

$$H_0\colon \mu = 0 \qquad H_A\colon \mu \neq 0$$

We will use a significance level of $\alpha = 0.05$. As calculated in Exercise 13.2.11, the sample mean and sample standard deviation of the data are $Y_{15} = 2.68$ and $s \approx 1.021$. Therefore, the t statistic is $T_{14} = \dfrac{2.68 - 0}{1.021/\sqrt{15}} \approx 10.17$. We now consult the row of Table 2 in Appendix H that corresponds to $15 - 1 = 14$ degrees of freedom. The largest t statistic value is 4.140 which corresponds to a P value of 0.001. Observe that larger t statistic values have smaller P values in the table. Since the calculated t statistic 10.17 is larger than 4.140, we deduce that $P < 0.001$. In fact using a calculator or CAS we find $P = 7.58 \times 10^{-8}$. Therefore, we reject the null hypothesis at the 0.05 significance level. In other words, the data provides evidence that the population has a mean change in HIV concentration that differs from 0. Thus, the HAART treatment appears to have an affect on HIV concentration.

13. We are told that the null hypothesis is $H_0\colon \mu = 50$ so the alternative hypothes is is $H_A\colon \mu \neq 50$. The sample mean and sample standard deviation of the data are $Y_{13} \approx 52.846$ and $s \approx 3.997$. Therefore, the t statistic is $T_{12} = \dfrac{52.846 - 50}{3.997/\sqrt{13}} \approx 2.567$. We now consult the row of Table 2 in Appendix H that corresponds to $13 - 1 = 12$ degrees of freedom. The calculated t statistic falls between 2.179 and 3.055 with corresponding P values of 0.05 and 0.01 respectively.

[continued]

© 2016 Cengage Learning. All Rights Reserved. May not be scanned, copied or duplicated, or posted to a publicly accessible website, in whole or in part.

Thus, the P value for the data lies in the interval $0.01 < P < 0.05$. We cannot determine a more precise P value from the table alone, so instead, we use a calculator or CAS to numerically compute the P value corresponding to $T_{12} \approx 11.768$ and find $P \approx 0.0247$. Assuming the null hypothesis is true, this suggests there is a probability 0.0247 that a sample of size 13 would have a t statistic as extreme as that observed. Therefore, from the Fundamental Principle of Counting, the probability that the two studies will have t statistic values that are more extreme than that for the given data is $(0.0247)^2 \approx 6.1 \times 10^{-4}$.

13.4 Contingency Table Analysis

1. Let S represent the event 'survived' so that the complement event, S^C, represents 'died'. Also, let's use V for the event 'vaccinated' and V^C for 'unvaccinated'. We wish to test for an association between vaccination status and survival. Thus, the goal is to determine whether the data suggest that $P(S|V)$ differs from $P(S|V^C)$. The null and alternative hypotheses are

$$H_0\text{: } P(S|V) = P(S|V^c) \qquad H_A\text{: } P(S|V) \neq P(S|V^c)$$

We will use a significance level of $\alpha = 0.05$. The observed results of the experiment for the two categories vaccination status and survival are recorded in the contingency table below.

	Survived (S)	Died (S^C)	Row Total
Vaccinated (V)	25 (12.5)	0 (12.5)	25
Unvaccinated (V^c)	0 (12.5)	25 (12.5)	25
Column Total	25	25	50

The expected numbers for each cell are calculated using Equation 8 as follows

$$E_{S,V} = \frac{25 \times 25}{50} = \frac{625}{50} = 12.5 \qquad E_{S^c,V} = \frac{25 \times 25}{50} = \frac{625}{50} = 12.5$$

$$E_{S,V^C} = \frac{25 \times 25}{50} = \frac{625}{50} = 12.5 \qquad E_{S^c,V^c} = \frac{25 \times 25}{50} = \frac{625}{50} = 12.5$$

These values appear in parentheses in the contingency table. Note that the expected values satisfy the two conditions in Step 7(c), so we are justified in using the chi-squared test. Calculating the test statistic using Equation 7(d) gives

$$\chi^2 = \sum_{i=1}^{4} \frac{(O_i - E_i)^2}{E_i} = \frac{(25 - 12.5)^2}{12.5} + \frac{(0 - 12.5)^2}{12.5} + \frac{(0 - 12.5)^2}{12.5} + \frac{(25 - 12.5)^2}{12.5} = 50$$

Now, the P value associated with this χ^2 value is determined by calculating the integral in Step 7(e) which gives

$$P = \int_{50}^{\infty} \frac{1}{\sqrt{2\pi}} x^{-1/2} e^{-x/2} dx \approx 1.537 \times 10^{-12} \qquad \text{[using a numerical integrator]}$$

Since $P < 0.05$, we reject the null hypothesis. Thus, the data provide evidence that vaccination status does affect survival.

© 2016 Cengage Learning. All Rights Reserved. May not be scanned, copied or duplicated, or posted to a publicly accessible website, in whole or in part.

3. Let M represent the event 'developed myopia' so that the complement event, M^C, represents 'did not develop myopia'. Also, let's use D for the event 'slept in darkness' and D^C for 'slept with light'. We wish to test for an association between night-time ambient light and myopia. Thus, the goal is to determine whether the data suggest that $P\left(M|D\right)$ differs from $P\left(M|D^C\right)$. The null and alternative hypotheses are

$$H_0\colon\ P(M|D) = P(M|D^c) \qquad H_A\colon\ P(M|D) \neq P(M|D^c)$$

We will use a significance level of $\alpha = 0.05$. The observed results of the study for the two categories level of ambient light and myopia are recorded in the contingency table below.

	Developed Myopia (M)	No Myopia (M^c)	Row Total
Darkness (D)	9 (17.03)	163 (154.97)	172
Night-light (D^c)	31 (22.97)	201 (209.03)	232
Column Total	40	364	404

The expected numbers for each cell are calculated using Equation 8 as follows

$$E_{M,D} = \frac{172 \times 40}{404} = \frac{6880}{404} \approx 17.03 \qquad E_{M^c,D} = \frac{172 \times 364}{404} = \frac{62608}{404} \approx 154.97$$

$$E_{M,D^C} = \frac{232 \times 40}{404} = \frac{9280}{404} \approx 22.97 \qquad E_{M^c,D^c} = \frac{232 \times 364}{404} = \frac{84448}{404} \approx 209.03$$

These values appear in parentheses in the contingency table. Note that the expected values satisfy the two conditions in Step 7(c), so we are justified in using the chi-squared test. Calculating the test statistic using Equation 7(d) gives

$$\chi^2 = \sum_{i=1}^{4} \frac{(O_i - E_i)^2}{E_i} = \frac{(9-17.03)^2}{17.03} + \frac{(163-154.97)^2}{154.97} + \frac{(31-22.97)^2}{22.97} + \frac{(201-209.03)^2}{209.03} \approx 7.318$$

Now, the P value associated with this χ^2 value is determined by calculating the integral in Step 7(e) which gives

$$P = \int_{7.318}^{\infty} \frac{1}{\sqrt{2\pi}} x^{-1/2} e^{-x/2} dx \approx 0.0068 \qquad \text{[using a numerical integrator]}$$

Since $P < 0.05$, we reject the null hypothesis. Thus, the data provide evidence that night-time ambient light and myopia are associated.

5. Let X represent the event 'species X present' so that the complement event, X^C, represents 'species X not present'. Also, let's use H for the event 'herbivore-dominated' and H^C for 'carnivore-dominated'. We wish to test for an association between community type and species presence. Thus, the goal is to determine whether the data suggest that $P\left(X|H\right)$ differs from $P\left(X|H^C\right)$. The null and alternative hypotheses are

$$H_0\colon\ P(X|H) = P(X|H^c) \qquad H_A\colon\ P(X|H) \neq P(X|H^c)$$

We will use a significance level of $\alpha = 0.05$. The observed results of the experiment for the two categories community type and species presence are recorded in the contingency table below.

	Species X present (X)	Species X not present (X^c)	Row Total
Herbivore-dominated (H)	9 (5.78)	3 (6.22)	12
Carnivore-dominated (H^c)	4 (7.22)	11 (7.78)	15
Column Total	13	14	27

[continued]

© 2016 Cengage Learning. All Rights Reserved. May not be scanned, copied or duplicated, or posted to a publicly accessible website, in whole or in part.

The expected numbers for each cell are calculated using Equation 8 as follows

$$E_{X,H} = \frac{12 \times 13}{27} = \frac{156}{27} \approx 5.78 \qquad E_{X^c,H} = \frac{12 \times 14}{27} = \frac{168}{27} \approx 6.22$$

$$E_{X,H^C} = \frac{15 \times 13}{27} = \frac{195}{27} \approx 7.22 \qquad E_{X^c,H^c} = \frac{15 \times 14}{27} = \frac{210}{27} \approx 7.78$$

These values appear in parentheses in the contingency table. Note that the expected values satisfy the two conditions in Step 7(c), so we are justified in using the chi-squared test. Calculating the test statistic using Equation 7(d) gives

$$\chi^2 = \sum_{i=1}^{4} \frac{(O_i - E_i)^2}{E_i} = \frac{(9-5.78)^2}{5.78} + \frac{(3-6.22)^2}{6.22} + \frac{(4-7.22)^2}{7.22} + \frac{(11-7.78)^2}{7.78} \approx 6.23$$

Now, the P value associated with this χ^2 value is determined by calculating the integral in Step 7(e) which gives

$$P = \int_{6.23}^{\infty} \frac{1}{\sqrt{2\pi}} x^{-1/2} e^{-x/2} dx \approx 0.0126 \qquad \text{[using a numerical integrator]}$$

Since $P < 0.05$, we reject the null hypothesis. Thus, the data provide evidence that community type and species X are associated, so species X is a good indicator species.

7. Let T represent the event 'positive test' so that the complement event, T^c, represents 'negative test'. Also, let's use C for the event 'cancer' and C^C for 'unvaccinated'. We wish to test for an association between cancer and PSA test result. Thus, the goal is to determine whether the data suggest that $P(T|C)$ differs from $P(T|C^C)$. The null and alternative hypotheses are

$$H_0\colon\ P(T|C) = P(T|C^c) \qquad H_A\colon\ P(T|C) \neq P(T|C^c)$$

We will use a significance level of $\alpha = 0.05$. There are $0.75\,(490) \approx 368$ men who had a positive PSA test and had prostate cancer, and there are $0.55\,(346) \approx 190$ men who had a positive PSA test but did not have cancer. The observed results of the experiment are recorded in the contingency table below.

	Positive Test (T)	Negative Test (T^c)	Row Total
Cancer (C)	368 (327.06)	122 (162.94)	490
No Cancer (C^c)	190 (230.94)	156 (115.06)	346
Column Total	558	278	836

The expected numbers for each cell are calculated using Equation 8 as follows

$$E_{S,C} = \frac{490 \times 558}{836} \approx 327.06 \qquad E_{S^c,C} = \frac{490 \times 278}{836} \approx 162.94$$

$$E_{S,C^C} = \frac{346 \times 558}{836} \approx 230.94 \qquad E_{S^c,C^c} = \frac{346 \times 278}{836} \approx 115.06$$

These values appear in parentheses in the contingency table. Note that the expected values satisfy the two conditions in Step 7(c), so we are justified in using the chi-squared test. Calculating the test statistic using Equation 7(d) gives

$$\chi^2 = \sum_{i=1}^{4} \frac{(O_i - E_i)^2}{E_i} = \frac{(368-327.06)^2}{327.06} + \frac{(122-162.94)^2}{162.94} + \frac{(190-230.94)^2}{230.94} + \frac{(156-115.06)^2}{115.06} \approx 37.2$$

Now, the P value associated with this χ^2 value is determined by calculating the integral in Step 7(e) which gives

$$P = \int_{37.2}^{\infty} \frac{1}{\sqrt{2\pi}} x^{-1/2} e^{-x/2} dx \approx 1.0 \times 10^{-9} \qquad \text{[using a numerical integrator]}$$

Since $P < 0.05$, we reject the null hypothesis. Thus, the data provide evidence that cancer and the PSA test result are associated. Noting that approximately $368/558 \approx 66\%$ of men who tested positive had cancer, we conclude that a positive test is indicative of cancer.

© 2016 Cengage Learning. All Rights Reserved. May not be scanned, copied or duplicated, or posted to a publicly accessible website, in whole or in part.

9. Suppose we wish to test for a relationship between two categorical variables each having two possible values. If the first variable has values E and E^c and the second variable has values F and F^c, then the null and alternative hypotheses are

$$H_0\colon P(E|F) = P(E|F^c) = q \qquad H_A\colon P(E|F) \neq P(E|F^c)$$

Now suppose a study or experiment produces the following contingency table

	E	E^C	Row Total
F	a	b	$a+b$
F^c	c	d	$c+d$
Column Total	$a+c$	$b+d$	$a+b+c+d$

where a, b, c and d are nonnegative integers. If the null hypothesis is true, then there is a probability q that the $a+b$ individuals in event F are also in E. So the number of individuals in E (who are also in F) is a binomial random variable with $n = a+b$ and a probability of success of q. Now, we can estimate q as the fraction of the total number of individuals in the study who are in E. This gives the estimate $\hat{q} = \dfrac{a+c}{a+b+c+d}$. Thus, the expected number of individuals in the top-left cell is $nq = \dfrac{(a+b)(a+c)}{a+b+c+d}$. Repeating this argument for the top-right cell, we have $\hat{q} = \dfrac{a+b}{a+b+c+d}$ and $n = b+d$ so that the expected number of individuals is $\dfrac{(a+b)(b+d)}{a+b+c+d}$. Similarly for the bottom-right and bottom-left cells, we find the expected number of individuals are $\dfrac{(c+d)(b+d)}{a+b+c+d}$ and $\dfrac{(c+d)(a+c)}{a+b+c+d}$. In general, we see that the expected numbers in each cell is given by $\dfrac{(\text{Row Total}) \times (\text{Column Total})}{(\text{Grand Total})}$.

13 Review

TRUE-FALSE QUIZ

1. True. See Theorem 13.1.4.

3. True, since the mean of the estimate $E[Y]$ is the same as the population mean μ. See Definition 13.1.5.

5. True. A 95% confidence interval will contain the population mean in 95% of all samples, whereas the 90% confidence interval will capture the population mean in 90% of samples. Since the 95% confidence interval has a greater chance of capturing the population mean it must be wider than the 90% confidence interval. See Figure 13.2.4

7. True. See Definition 13.3.4 and the discussion that follows.

9. True. Contingency table analyses involve performing a hypothesis test using the χ^2-statistic to determine whether the variables are associated. See Procedure 7 in Section 13.4.

© 2016 Cengage Learning. All Rights Reserved. May not be scanned, copied or duplicated, or posted to a publicly accessible website, in whole or in part.

EXERCISES

1. The standard deviation is $\sigma = 15$ and the sample size is 12, so by Theorem 13.1.4, the standard deviation of the sample mean is $\frac{15}{\sqrt{12}} \approx 4.330$. The probability of obtaining a sample mean in the interval $[\mu - 10, \mu + 10]$ is then

$$\int_{\mu-10}^{\mu+10} \frac{1}{4.33\sqrt{2\pi}} e^{-(y-\mu)^2/(2(4.33)^2)}\,dy = \frac{1}{\sqrt{\pi}} \int_{\frac{-10}{4.33\sqrt{2}}}^{\frac{10}{4.33\sqrt{2}}} e^{-u^2}\,du \quad \left[\text{substitute } u = \frac{y-\mu}{4.33\sqrt{2}},\ dy = 4.33\sqrt{2}\,du\right]$$

$$\approx 0.9791 \quad [\text{using a numerical integrator}]$$

3. The standard deviation is $\sigma = 10$ and the sample size is 14, so by Theorem 13.1.4, the standard deviation of the sample mean is $\frac{10}{\sqrt{14}} \approx 2.673$. Using the Complement Rule, the probability of obtaining a sample mean in the interval $(\infty, \mu - 5] \cup [\mu + 5, \infty)$ is

$$1 - \int_{\mu-5}^{\mu+5} \frac{1}{2.673\sqrt{2\pi}} e^{-(y-\mu)^2/(2(2.673)^2)}\,dy = 1 - \frac{1}{\sqrt{\pi}} \int_{\frac{-5}{2.673\sqrt{2}}}^{\frac{5}{2.673\sqrt{2}}} e^{-u^2}\,du \quad \left[\text{substitute } u = \frac{y-\mu}{2.673\sqrt{2}},\ dy = 2.673\sqrt{2}\,du\right]$$

$$\approx 0.0614 \quad [\text{using a numerical integrator}]$$

5. The population standard deviation is unknown so we must carry out the steps outlined in 13.2.11.

$$\bar{x} = \frac{1}{10}[8 + 15 + 9 + 22 + 18 + 18 + 16 + 21 + 20 + 18] = \frac{165}{10} = 16.5$$

$$s = \sqrt{\frac{1}{9}\left[\begin{array}{l}(8-16.5)^2 + (15-16.5)^2 + (9-16.5)^2 + (22-16.5)^2 + (18-16.5)^2 \\ +(18-16.5)^2 + (16-16.5)^2 + (21-16.5)^2 + (20-16.5)^2 + (18-16.5)^2\end{array}\right]}$$

$$= \sqrt{200.5/9} \approx 4.7199$$

Using either a calculator or Appendix H, the t value for a 95% confidence level and $10 - 1 = 9$ degrees of freedom is $t_{95,9} \approx 2.262$. Thus the endpoints of the confidence interval are $16.5 \pm 2.262\frac{4.7199}{\sqrt{10}}$ which gives an interval of approximately $(13.1, 19.9)$.

7. The population standard deviation is unknown so we must carry out the steps outlined in 13.2.11.

$$\bar{x} = [677 + 553 + 720 + 520 + 652 + 559 + 597 + 471 + 577 + 807 + 502 + 733]/\,12 = \frac{7368}{12} = 614\text{ mg}$$

$$s = \sqrt{\frac{1}{11}\left[\begin{array}{l}(677-614)^2 + (553-614)^2 + (720-614)^2 + (520-614)^2 + (652-614)^2 + (559-614)^2 \\ +(597-614)^2 + (471-614)^2 + (577-614)^2 + (807-614)^2 + (502-614)^2 + (733-614)^2\end{array}\right]}$$

$$= \sqrt{118292/11} \approx 103.7$$

Using either a calculator or Appendix H, the t value for a 95% confidence level and $12 - 1 = 11$ degrees of freedom is $t_{95,11} \approx 2.201$. Thus the endpoints of the confidence interval are $614 \pm 2.201\frac{103.7}{\sqrt{12}}$ which gives an interval of approximately $(548.1, 679.9)$.

© 2016 Cengage Learning. All Rights Reserved. May not be scanned, copied or duplicated, or posted to a publicly accessible website, in whole or in part.

9. The population standard deviation is unknown in both samples so we must carry out the steps outlined in 13.2.11. The mean and standard deviation of the 'crop rotation' data are $\bar{x}_r \approx 39.5$ and $s_r \approx 5.266$ and those of the 'no rotation' data are $\bar{x}_n \approx 33.833$ and $s_n \approx 4.469$. Both samples have 12 data points which gives $12 - 1 = 13$ degrees of freedom. Using either a calculator or Appendix H, the t value for a 95% confidence level and 13 degrees of freedom is $t_{95,14} \approx 2.201$.

Thus the endpoints of the confidence interval for the 'rotation' data are $39.5 \pm 2.201\frac{5.266}{\sqrt{12}}$ which gives an interval of approximately $(36.2, 42.8)$. Similarly for the 'no rotation' data the endpoints are $33.833 \pm 2.201\frac{4.469}{\sqrt{12}}$ which gives a confidence interval of approximately $(31.0, 36.7)$.

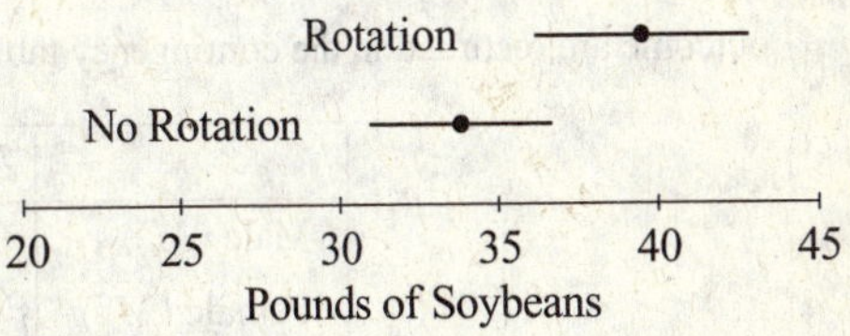

Rotation

No Rotation

20 25 30 35 40 45

Pounds of Soybeans

11. Following the steps outlined in Equation 13.3.6, we first state the null and alternative hypotheses. We wish to test whether the population mean differs from 15 mg, so the hypotheses are

$$H_0\colon\ \mu = 15 \qquad H_A\colon\ \mu \neq 15$$

We will use a significance level of $\alpha = 0.05$. As calculated in Exercise 5, the sample mean and sample standard deviation of the data are $Y_{10} = 16.5$ and $s \approx 4.7199$. Therefore, the t statistic is $T_9 = \dfrac{16.5 - 15}{4.7199/\sqrt{10}} \approx 1.005$. We now consult the row of Table 2 in Appendix H that corresponds to $10 - 1 = 9$ degrees of freedom. The smallest t statistic value is 1.383 which corresponds to a P value of 0.2. Observe that smaller t statistic values have larger P values in the table. Since the magnitude of the calculated t statistic 1.005 is less than 1.383, we deduce that $P > 0.2$ and hence also $P > 0.05$. Therefore, we do *not* reject the null hypothesis at the 0.05 significance level. In other words, the data are consistent with the population having a mean pollen mass of 15 mg.

13. Following the steps outlined in Equation 13.3.6, we first state the null and alternative hypotheses. We wish to test whether the population mean differs from 750 mg, so the hypotheses are

$$H_0\colon\ \mu = 750 \qquad H_A\colon\ \mu \neq 750$$

We will use a significance level of $\alpha = 0.05$. As calculated in Exercise 7, the sample mean and sample standard deviation of the data are $Y_{12} = 614$ and $s \approx 103.7$. Therefore, the t statistic is $T_{11} = \dfrac{614 - 750}{103.7/\sqrt{12}} \approx -4.543$. We now consult the row of Table 2 in Appendix H that corresponds to $12 - 1 = 11$ degrees of freedom. The largest t statistic value is 4.437 which corresponds to a P value of 0.001. Observe that larger t statistic values have smaller P values in the table. Since the magnitude of the calculated t statistic 4.543 is larger than 4.437, we deduce that $P < 0.001$ and hence also $P < 0.05$. Therefore, we reject the null hypothesis at the 0.05 significance level. In other words, the data provides evidence that the population has a mean mass that differs from 750 mg.

© 2016 Cengage Learning. All Rights Reserved. May not be scanned, copied or duplicated, or posted to a publicly accessible website, in whole or in part.

15. Let T represent the event 'positive test' so that the complement event, T^C, represents 'negative test'. Also, let's use M for the event 'male' and M^C for 'female'. We wish to test for an association between gender and test outcome. Thus, the goal is to determine whether the data suggest that $P(T|M)$ differs from $P\left(T|M^C\right)$. The null and alternative hypotheses are

$$H_0\colon\ P(T|M) = P(T|M^c) \qquad H_A\colon\ P(T|M) \neq P(T|M^c)$$

We will use a significance level of $\alpha = 0.05$. The observed results of the experiment for the two categories gender and test outcome are recorded in the contingency table below.

	Positive Test (T)	Negative Test (T^c)	Row Total
Male (M)	13 (13.42)	8 (7.58)	21
Female (M^c)	10 (9.58)	5 (5.42)	15
Column Total	23	13	36

The expected numbers for each cell are calculated using Equation 8 as follows

$$E_{T,M} = \frac{21 \times 23}{36} = \frac{483}{36} \approx 13.42 \qquad E_{T^c,M} = \frac{21 \times 13}{36} = \frac{273}{36} \approx 7.58$$

$$E_{T,M^C} = \frac{15 \times 23}{36} = \frac{345}{36} \approx 9.58 \qquad E_{T^c,M^c} = \frac{15 \times 13}{36} = \frac{195}{36} \approx 5.42$$

These values appear in parentheses in the contingency table. Note that the expected values satisfy the two conditions in Step 7(c), so we are justified in using the chi-squared test. Calculating the test statistic using Equation 7(d) gives

$$\chi^2 = \sum_{i=1}^{4} \frac{(O_i - E_i)^2}{E_i} = \frac{(13 - 13.42)^2}{13.42} + \frac{(8 - 7.58)^2}{7.58} + \frac{(5 - 5.42)^2}{5.42} + \frac{(10 - 9.58)^2}{9.58} \approx 0.087$$

Now, the P value associated with this χ^2 value is determined by calculating the integral in Step 7(e) which gives

$$P = \int_{0.087}^{\infty} \tfrac{1}{\sqrt{2\pi}} x^{-1/2} e^{-x/2}\,dx \approx 0.76 \qquad \text{[using a numerical integrator]}$$

Since $P > 0.05$, we do not reject the null hypothesis. Thus, the data are consistent with the hypothesis that there is no relationship between gender and test outcome.

© 2016 Cengage Learning. All Rights Reserved. May not be scanned, copied or duplicated, or posted to a publicly accessible website, in whole or in part.

APPENDIXES

A Intervals, Inequalities, and Absolute Values

1. $|5 - 23| = |-18| = 18$

3. $|\sqrt{5} - 5| = -(\sqrt{5} - 5) = 5 - \sqrt{5}$ because $\sqrt{5} - 5 < 0$.

5. If $x < 2$, $x - 2 < 0$, so $|x - 2| = -(x - 2) = 2 - x$.

7. $$|x+1| = \begin{cases} x+1 & \text{if } x+1 \geq 0 \\ -(x+1) & \text{if } x+1 < 0 \end{cases} = \begin{cases} x+1 & \text{if } x \geq -1 \\ -x-1 & \text{if } x < -1 \end{cases}$$

9. $|x^2 + 1| = x^2 + 1$ [since $x^2 + 1 \geq 0$ for all x].

11. $2x + 7 > 3 \;\Leftrightarrow\; 2x > -4 \;\Leftrightarrow\; x > -2$, so $x \in (-2, \infty)$.

13. $1 - x \leq 2 \;\Leftrightarrow\; -x \leq 1 \;\Leftrightarrow\; x \geq -1$, so $x \in [-1, \infty)$.

15. $0 \leq 1 - x < 1 \;\Leftrightarrow\; -1 \leq -x < 0 \;\Leftrightarrow\; 1 \geq x > 0$, so $x \in (0, 1]$.

17. $(x - 1)(x - 2) > 0$.

Case 1: (both factors are positive, so their product is positive) $x - 1 > 0 \;\Leftrightarrow\; x > 1$,

and $x - 2 > 0 \;\Leftrightarrow\; x > 2$, so $x \in (2, \infty)$.

Case 2: (both factors are negative, so their product is positive) $x - 1 < 0 \;\Leftrightarrow\; x < 1$,

and $x - 2 < 0 \;\Leftrightarrow\; x < 2$, so $x \in (-\infty, 1)$.

Thus, the solution set is $(-\infty, 1) \cup (2, \infty)$.

19. $x^2 < 3 \;\Leftrightarrow\; x^2 - 3 < 0 \;\Leftrightarrow\; (x - \sqrt{3})(x + \sqrt{3}) < 0$.

Case 1: $x > \sqrt{3}$ and $x < -\sqrt{3}$, which is impossible.

Case 2: $x < \sqrt{3}$ and $x > -\sqrt{3}$.

Thus, the solution set is $(-\sqrt{3}, \sqrt{3})$.

Another method: $x^2 < 3 \;\Leftrightarrow\; |x| < \sqrt{3} \;\Leftrightarrow\; -\sqrt{3} < x < \sqrt{3}$.

21. $x^3 - x^2 \leq 0 \;\Leftrightarrow\; x^2(x - 1) \leq 0$. Since $x^2 \geq 0$ for all x, the inequality is satisfied when $x - 1 \leq 0 \;\Leftrightarrow\; x \leq 1$.

Thus, the solution set is $(-\infty, 1]$.

© 2016 Cengage Learning. All Rights Reserved. May not be scanned, copied or duplicated, or posted to a publicly accessible website, in whole or in part.

23. $x^3 > x \iff x^3 - x > 0 \iff x(x^2-1) > 0 \iff x(x-1)(x+1) > 0$. Construct a chart:

Interval	x	$x-1$	$x+1$	$x(x-1)(x+1)$
$x < -1$	$-$	$-$	$-$	$-$
$-1 < x < 0$	$-$	$-$	$+$	$+$
$0 < x < 1$	$+$	$-$	$+$	$-$
$x > 1$	$+$	$+$	$+$	$+$

Since $x^3 > x$ when the last column is positive, the solution set is $(-1,0) \cup (1,\infty)$.

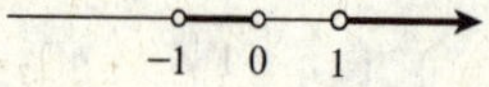

−1 0 1

25. $1/x < 4$. This is clearly true for $x < 0$. So suppose $x > 0$. then $1/x < 4 \iff$ $1 < 4x \iff \frac{1}{4} < x$. Thus, the solution set is $(-\infty, 0) \cup \left(\frac{1}{4}, \infty\right)$.

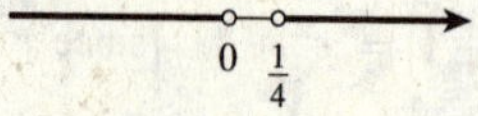

0 $\frac{1}{4}$

27. $C = \frac{5}{9}(F-32) \Rightarrow F = \frac{9}{5}C + 32$. So $50 \le F \le 95 \Rightarrow 50 \le \frac{9}{5}C + 32 \le 95 \Rightarrow 18 \le \frac{9}{5}C \le 63 \Rightarrow$ $10 \le C \le 35$. So the interval is $[10, 35]$.

29. (a) Let T represent the temperature in degrees Celsius and h the height in km. $T = 20$ when $h = 0$ and T decreases by 10°C for every km (1°C for each 100-m rise). Thus, $T = 20 - 10h$ when $0 \le h \le 12$.

(b) From part (a), $T = 20 - 10h \Rightarrow 10h = 20 - T \Rightarrow h = 2 - T/10$. So $0 \le h \le 5 \Rightarrow 0 \le 2 - T/10 \le 5 \Rightarrow$ $-2 \le -T/10 \le 3 \Rightarrow -20 \le -T \le 30 \Rightarrow 20 \ge T \ge -30 \Rightarrow -30 \le T \le 20$. Thus, the range of temperatures (in $^\circ$C) to be expected is $[-30, 20]$.

31. $|x+3| = |2x+1| \iff$ either $x + 3 = 2x + 1$ or $x + 3 = -(2x+1)$. In the first case, $x = 2$, and in the second case, $x + 3 = -2x - 1 \iff 3x = -4 \iff x = -\frac{4}{3}$. So the solutions are $-\frac{4}{3}$ and 2.

33. By Property 5 of absolute values, $|x| < 3 \iff -3 < x < 3$, so $x \in (-3, 3)$.

35. $|x-4| < 1 \iff -1 < x - 4 < 1 \iff 3 < x < 5$, so $x \in (3, 5)$.

37. $|x+5| \ge 2 \iff x + 5 \ge 2$ or $x + 5 \le -2 \iff x \ge -3$ or $x \le -7$, so $x \in (-\infty, -7] \cup [-3, \infty)$.

39. $|2x-3| \le 0.4 \iff -0.4 \le 2x - 3 \le 0.4 \iff 2.6 \le 2x \le 3.4 \iff 1.3 \le x \le 1.7$, so $x \in [1.3, 1.7]$.

41. $a(bx - c) \ge bc \iff bx - c \ge \dfrac{bc}{a} \iff bx \ge \dfrac{bc}{a} + c = \dfrac{bc + ac}{a} \iff x \ge \dfrac{bc + ac}{ab}$

43. $|ab| = \sqrt{(ab)^2} = \sqrt{a^2b^2} = \sqrt{a^2}\,\sqrt{b^2} = |a|\,|b|$

© 2016 Cengage Learning. All Rights Reserved. May not be scanned, copied or duplicated, or posted to a publicly accessible website, in whole or in part.

B Coordinate Geometry

1. Use the distance formula with $P_1(x_1, y_1) = (1, 1)$ and $P_2(x_2, y_2) = (4, 5)$ to get

$$|P_1P_2| = \sqrt{(4-1)^2 + (5-1)^2} = \sqrt{3^2 + 4^2} = \sqrt{25} = 5$$

3. The slope m of the line through $P(-3, 3)$ and $Q(-1, -6)$ is $m = \dfrac{-6-3}{-1-(-3)} = -\dfrac{9}{2}$.

5. Using $A(-2, 9)$, $B(4, 6)$, $C(1, 0)$, and $D(-5, 3)$, we have

$|AB| = \sqrt{[4-(-2)]^2 + (6-9)^2} = \sqrt{6^2 + (-3)^2} = \sqrt{45} = \sqrt{9}\,\sqrt{5} = 3\sqrt{5}$,

$|BC| = \sqrt{(1-4)^2 + (0-6)^2} = \sqrt{(-3)^2 + (-6)^2} = \sqrt{45} = \sqrt{9}\,\sqrt{5} = 3\sqrt{5}$,

$|CD| = \sqrt{(-5-1)^2 + (3-0)^2} = \sqrt{(-6)^2 + 3^2} = \sqrt{45} = \sqrt{9}\,\sqrt{5} = 3\sqrt{5}$, and

$|DA| = \sqrt{[-2-(-5)]^2 + (9-3)^2} = \sqrt{3^2 + 6^2} = \sqrt{45} = \sqrt{9}\,\sqrt{5} = 3\sqrt{5}$. So all sides are of equal length and we have a rhombus. Moreover, $m_{AB} = \dfrac{6-9}{4-(-2)} = -\dfrac{1}{2}$, $m_{BC} = \dfrac{0-6}{1-4} = 2$, $m_{CD} = \dfrac{3-0}{-5-1} = -\dfrac{1}{2}$, and

$m_{DA} = \dfrac{9-3}{-2-(-5)} = 2$, so the sides are perpendicular. Thus, A, B, C, and D are vertices of a square.

7. The graph of the equation $x = 3$ is a vertical line with x-intercept 3. The line does not have a slope.

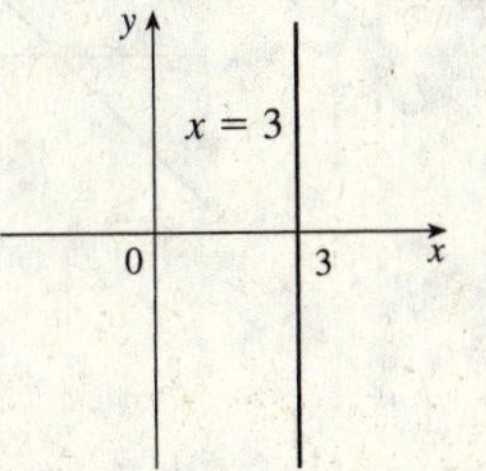

9. $xy = 0 \Leftrightarrow x = 0$ or $y = 0$. The graph consists of the coordinate axes.

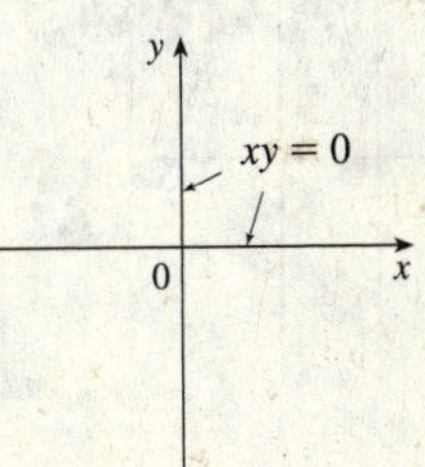

11. By the point-slope form of the equation of a line, an equation of the line through $(2, -3)$ with slope 6 is $y - (-3) = 6(x - 2)$ or $y = 6x - 15$.

13. The slope of the line through $(2, 1)$ and $(1, 6)$ is $m = \dfrac{6-1}{1-2} = -5$, so an equation of the line is $y - 1 = -5(x - 2)$ or $y = -5x + 11$.

15. By the slope-intercept form of the equation of a line, an equation of the line is $y = 3x - 2$.

17. Since the line passes through $(1, 0)$ and $(0, -3)$, its slope is $m = \dfrac{-3-0}{0-1} = 3$, so an equation is $y = 3x - 3$.

Another method: From Exercise 61, $\dfrac{x}{1} + \dfrac{y}{-3} = 1 \;\Rightarrow\; -3x + y = -3 \;\Rightarrow\; y = 3x - 3$.

19. The line is parallel to the x-axis, so it is horizontal and must have the form $y = k$. Since it goes through the point $(x, y) = (4, 5)$, the equation is $y = 5$.

© 2016 Cengage Learning. All Rights Reserved. May not be scanned, copied or duplicated, or posted to a publicly accessible website, in whole or in part.

21. Putting the line $x + 2y = 6$ into its slope-intercept form gives us $y = -\frac{1}{2}x + 3$, so we see that this line has slope $-\frac{1}{2}$. Thus, we want the line of slope $-\frac{1}{2}$ that passes through the point $(1, -6)$: $y - (-6) = -\frac{1}{2}(x - 1) \quad \Leftrightarrow \quad y = -\frac{1}{2}x - \frac{11}{2}$.

23. $2x + 5y + 8 = 0 \quad \Leftrightarrow \quad y = -\frac{2}{5}x - \frac{8}{5}$. Since this line has slope $-\frac{2}{5}$, a line perpendicular to it would have slope $\frac{5}{2}$, so the required line is $y - (-2) = \frac{5}{2}[x - (-1)] \quad \Leftrightarrow \quad y = \frac{5}{2}x + \frac{1}{2}$.

25. $x + 3y = 0 \quad \Leftrightarrow \quad y = -\frac{1}{3}x$, so the slope is $-\frac{1}{3}$ and the y-intercept is 0.

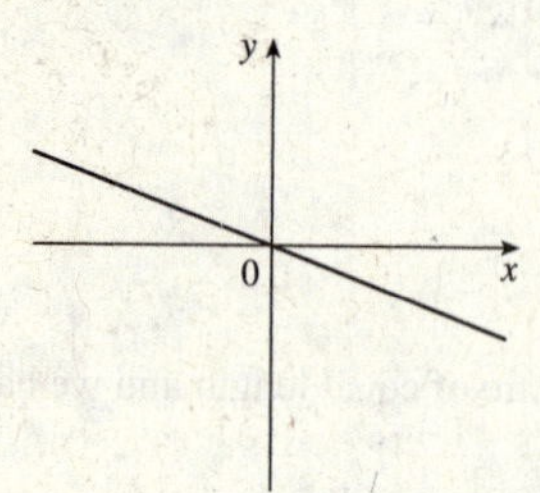

27. $3x - 4y = 12 \quad \Leftrightarrow \quad y = \frac{3}{4}x - 3$, so the slope is $\frac{3}{4}$ and the y-intercept is -3.

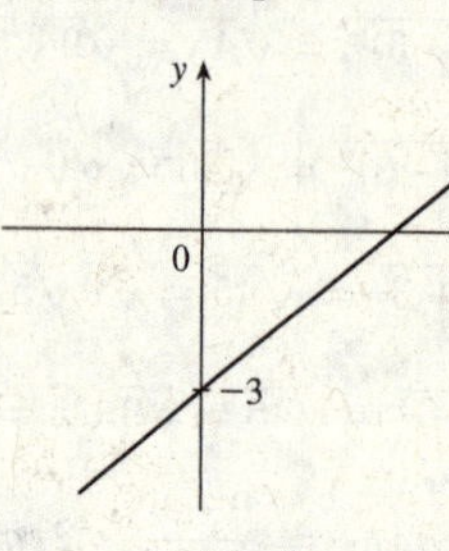

29. $\{(x, y) \mid x < 0\}$

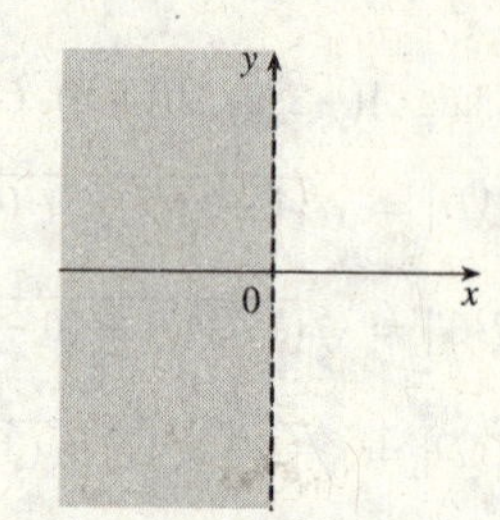

31. $\{(x, y) \mid |x| \leq 2\} = \{(x, y) \mid -2 \leq x \leq 2\}$

33. $\{(x, y) \mid 0 \leq y \leq 4, x \leq 2\}$

35. $\{(x, y) \mid 1 + x \leq y \leq 1 - 2x\}$

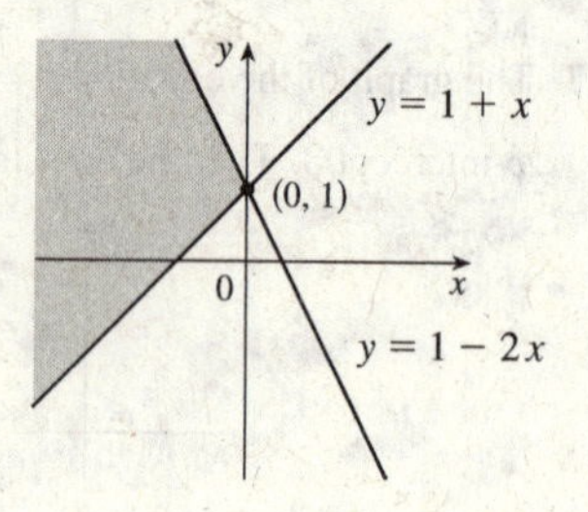

37. An equation of the circle with center $(3, -1)$ and radius 5 is $(x - 3)^2 + (y + 1)^2 = 5^2 = 25$.

39. $x^2 + y^2 - 4x + 10y + 13 = 0 \quad \Leftrightarrow \quad x^2 - 4x + y^2 + 10y = -13 \quad \Leftrightarrow$
$(x^2 - 4x + 4) + (y^2 + 10y + 25) = -13 + 4 + 25 = 16 \quad \Leftrightarrow \quad (x - 2)^2 + (y + 5)^2 = 4^2$. Thus, we have a circle with center $(2, -5)$ and radius 4.

41. $2x - y = 4 \quad \Leftrightarrow \quad y = 2x - 4 \quad \Rightarrow \quad m_1 = 2$ and $6x - 2y = 10 \quad \Leftrightarrow \quad 2y = 6x - 10 \quad \Leftrightarrow \quad y = 3x - 5 \quad \Rightarrow \quad m_2 = 3$.
Since $m_1 \neq m_2$, the two lines are not parallel. To find the point of intersection: $2x - 4 = 3x - 5 \quad \Leftrightarrow \quad x = 1 \quad \Rightarrow$
$y = -2$. Thus, the point of intersection is $(1, -2)$.

43. Let M be the point $\left(\frac{x_1 + x_2}{2}, \frac{y_1 + y_2}{2}\right)$. Then

$$|MP_1|^2 = \left(x_1 - \frac{x_1 + x_2}{2}\right)^2 + \left(y_1 - \frac{y_1 + y_2}{2}\right)^2 = \left(\frac{x_1 - x_2}{2}\right)^2 + \left(\frac{y_1 - y_2}{2}\right)^2$$

$$|MP_2|^2 = \left(x_2 - \frac{x_1 + x_2}{2}\right)^2 + \left(y_2 - \frac{y_1 + y_2}{2}\right)^2 = \left(\frac{x_2 - x_1}{2}\right)^2 + \left(\frac{y_2 - y_1}{2}\right)^2$$

Hence, $|MP_1| = |MP_2|$; that is, M is equidistant from P_1 and P_2.

© 2016 Cengage Learning. All Rights Reserved. May not be scanned, copied or duplicated, or posted to a publicly accessible website, in whole or in part.

45. With $A(1,4)$ and $B(7,-2)$, the slope of segment AB is $\frac{-2-4}{7-1} = -1$, so its perpendicular bisector has slope 1. The midpoint of AB is $\left(\frac{1+7}{2}, \frac{4+(-2)}{2}\right) = (4,1)$, so an equation of the perpendicular bisector is $y - 1 = 1(x-4)$ or $y = x - 3$.

47. If $P(x,y)$ is any point on the parabola, then the distance from P to the focus is $|PF| = \sqrt{x^2 + (y-p)^2}$ and the distance from P to the directrix is $|y+p|$. (Figure 14 in the text illustrates the case where $p > 0$.) The defining property of a parabola is that these distances are equal: $\sqrt{x^2 + (y-p)^2} = |y+p|$. We get an equivalent equation by squaring and simplifying: $x^2 + (y-p)^2 = |y+p|^2 = (y+p)^2 \quad \Leftrightarrow \quad x^2 + y^2 - 2py + p^2 = y^2 + 2py + p^2 \quad \Leftrightarrow \quad x^2 = 4py$. Thus, an equation of a parabola with focus $(0,p)$ and directrix $y = -p$ is $x^2 = 4py$.

49. See Figure 20 in the text. $P(x,y)$ is a point on the ellipse when $|PF_1| + |PF_2| = 2a$; that is,

$\sqrt{(x+c)^2 + y^2} + \sqrt{(x-c)^2 + y^2} = 2a$ or $\sqrt{(x-c)^2 + y^2} = 2a - \sqrt{(x+c)^2 + y^2}$. Squaring both sides, we have

$x^2 - 2cx + c^2 + y^2 = 4a^2 - 4a\sqrt{(x+c)^2 + y^2} + x^2 + 2cx + c^2 + y^2$, which simplifies to $a\sqrt{(x+c)^2 + y^2} = a^2 + cx$.

We square again: $a^2\left(x^2 + 2cx + c^2 + y^2\right) = a^4 + 2a^2cx + c^2x^2$, which becomes $(a^2 - c^2)x^2 + a^2y^2 = a^2\left(a^2 - c^2\right)$.

From triangle F_1F_2P in Figure 20, we see that $2c < 2a$, so $c < a$ and, therefore, $a^2 - c^2 > 0$. For convenience, let $b^2 = a^2 - c^2$. Then the equation of the ellipse becomes $b^2x^2 + a^2y^2 = a^2b^2$ or, if both sides are divided by a^2b^2,

$$\frac{x^2}{a^2} + \frac{y^2}{b^2} = 1.$$

51. From Figure 23 in the text, $|PF_1| - |PF_2| = \pm 2a \quad \Leftrightarrow \quad \sqrt{(x+c)^2 + y^2} - \sqrt{(x-c)^2 + y^2} = \pm 2a \quad \Leftrightarrow$

$\sqrt{(x+c)^2 + y^2} = \sqrt{(x-c)^2 + y^2} \pm 2a \quad \Leftrightarrow$

$(x+c)^2 + y^2 = (x-c)^2 + y^2 + 4a^2 \pm 4a\sqrt{(x-c)^2 + y^2} \quad \Leftrightarrow \quad 4cx - 4a^2 = \pm 4a\sqrt{(x-c)^2 + y^2} \quad \Leftrightarrow$

$c^2x^2 - 2a^2cx + a^4 = a^2\left(x^2 - 2cx + c^2 + y^2\right) \quad \Leftrightarrow \quad \left(c^2 - a^2\right)x^2 - a^2y^2 = a^2\left(c^2 - a^2\right) \quad \Leftrightarrow$

$b^2x^2 - a^2y^2 = a^2b^2$ [where $b^2 = c^2 - a^2$] $\quad \Leftrightarrow \quad \frac{x^2}{a^2} - \frac{y^2}{b^2} = 1$.

53.

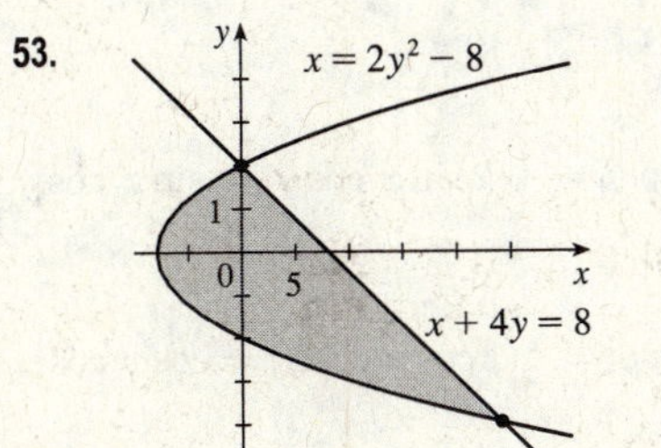

$x + 4y = 8$, $x = 2y^2 - 8$. Substitute x from the second equation into the first: $\left(2y^2 - 8\right) + 4y = 8 \quad \Leftrightarrow \quad 2y^2 + 4y - 16 = 0 \quad \Leftrightarrow$ $y^2 + 2y - 8 = 0 \quad \Leftrightarrow \quad (y+4)(y-2) = 0 \quad \Leftrightarrow \quad y = -4$ or 2. So the points of intersection are $(24, -4)$ and $(0, 2)$.

© 2016 Cengage Learning. All Rights Reserved. May not be scanned, copied or duplicated, or posted to a publicly accessible website, in whole or in part.

C Trigonometry

1. (a) $210^\circ = 210^\circ\left(\frac{\pi}{180^\circ}\right) = \frac{7\pi}{6}$ rad (b) $9^\circ = 9^\circ\left(\frac{\pi}{180^\circ}\right) = \frac{\pi}{20}$ rad

3. (a) 4π rad $= 4\pi\left(\frac{180^\circ}{\pi}\right) = 720^\circ$ (b) $-\frac{3\pi}{8}$ rad $= -\frac{3\pi}{8}\left(\frac{180^\circ}{\pi}\right) = -67.5^\circ$

5. Using Formula 3, $a = r\theta = 36 \cdot \frac{\pi}{12} = 3\pi$ cm.

7. Using Formula 3, $\theta = a/r = \frac{1}{1.5} = \frac{2}{3}$ rad $= \frac{2}{3}\left(\frac{180^\circ}{\pi}\right) = \left(\frac{120}{\pi}\right)^\circ \approx 38.2^\circ$.

9. (a)

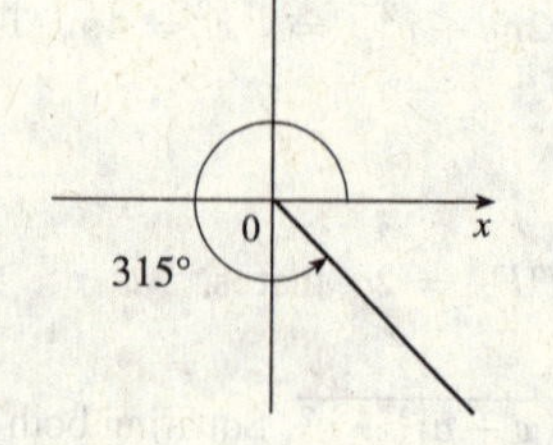

(b)

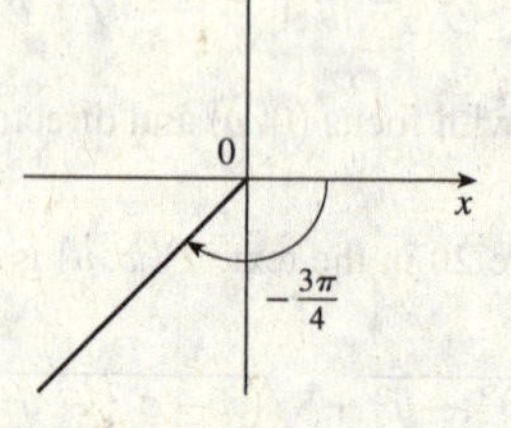

11.

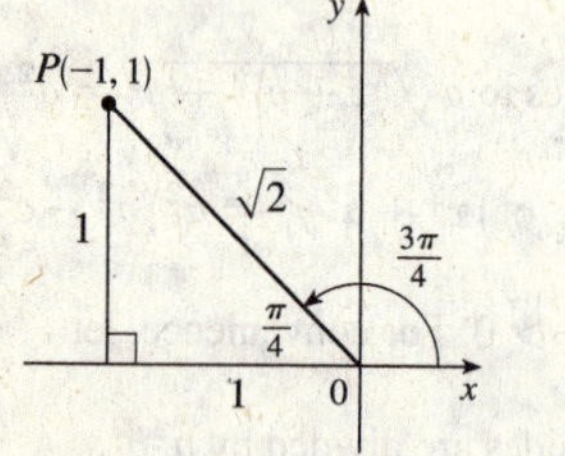

From the diagram we see that a point on the terminal side is $P(-1, 1)$. Therefore, taking $x = -1$, $y = 1$, $r = \sqrt{2}$ in the definitions of the trigonometric ratios, we have $\sin\frac{3\pi}{4} = \frac{1}{\sqrt{2}}$, $\cos\frac{3\pi}{4} = -\frac{1}{\sqrt{2}}$, $\tan\frac{3\pi}{4} = -1$, $\csc\frac{3\pi}{4} = \sqrt{2}$, $\sec\frac{3\pi}{4} = -\sqrt{2}$, and $\cot\frac{3\pi}{4} = -1$.

13. $\sin\theta = y/r = \frac{3}{5} \;\Rightarrow\; y = 3$, $r = 5$, and $x = \sqrt{r^2 - y^2} = 4$ (since $0 < \theta < \frac{\pi}{2}$). Therefore taking $x = 4$, $y = 3$, $r = 5$ in the definitions of the trigonometric ratios, we have $\cos\theta = \frac{4}{5}$, $\tan\theta = \frac{3}{4}$, $\csc\theta = \frac{5}{3}$, $\sec\theta = \frac{5}{4}$, and $\cot\theta = \frac{4}{3}$.

15. $\sin 35^\circ = \dfrac{x}{10} \;\Rightarrow\; x = 10\sin 35^\circ \approx 5.73576$ cm

17. $\tan\frac{2\pi}{5} = \dfrac{x}{8} \;\Rightarrow\; x = 8\tan\frac{2\pi}{5} \approx 24.62147$ cm

19.

y, P(b, a), c, a, θ, b, −θ, x, a, c, Q(b, −a)

(a) From the diagram we see that $\sin\theta = \dfrac{y}{r} = \dfrac{a}{c}$, and $\sin(-\theta) = \dfrac{-a}{c} = -\dfrac{a}{c} = -\sin\theta$.

(b) Again from the diagram we see that $\cos\theta = \dfrac{x}{r} = \dfrac{b}{c} = \cos(-\theta)$.

21. (a) Using (12a) and (13a), we have

$$\tfrac{1}{2}[\sin(x+y) + \sin(x-y)] = \tfrac{1}{2}[\sin x\,\cos y + \cos x\,\sin y + \sin x\,\cos y - \cos x\,\sin y] = \tfrac{1}{2}(2\sin x\,\cos y) = \sin x\,\cos y.$$

(b) This time, using (12b) and (13b), we have

$$\tfrac{1}{2}[\cos(x+y) + \cos(x-y)] = \tfrac{1}{2}[\cos x\,\cos y - \sin x\,\sin y + \cos x\,\cos y + \sin x\,\sin y] = \tfrac{1}{2}(2\cos x\,\cos y) = \cos x\,\cos y.$$

(c) Again using (12b) and (13b), we have

$$\begin{aligned}\tfrac{1}{2}[\cos(x-y) - \cos(x+y)] &= \tfrac{1}{2}[\cos x\,\cos y + \sin x\,\sin y - \cos x\,\cos y + \sin x\,\sin y]\\ &= \tfrac{1}{2}(2\sin x\,\sin y) = \sin x\,\sin y\end{aligned}$$

© 2016 Cengage Learning. All Rights Reserved. May not be scanned, copied or duplicated, or posted to a publicly accessible website, in whole or in part.

23. Using (12a), we have $\sin\left(\frac{\pi}{2} + x\right) = \sin\frac{\pi}{2}\cos x + \cos\frac{\pi}{2}\sin x = 1 \cdot \cos x + 0 \cdot \sin x = \cos x$.

25. Using (6), we have $\sin\theta\cot\theta = \sin\theta \cdot \dfrac{\cos\theta}{\sin\theta} = \cos\theta$.

27. Using (14a), we have $\tan 2\theta = \tan(\theta + \theta) = \dfrac{\tan\theta + \tan\theta}{1 - \tan\theta\,\tan\theta} = \dfrac{2\tan\theta}{1 - \tan^2\theta}$.

29. Since $\sin x = \frac{1}{3}$ we can label the opposite side as having length 1, the hypotenuse as having length 3, and use the Pythagorean Theorem to get that the adjacent side has length $\sqrt{8}$. Then, from the diagram, $\cos x = \frac{\sqrt{8}}{3}$. Similarly we have that $\sin y = \frac{3}{5}$. Now use (12a):

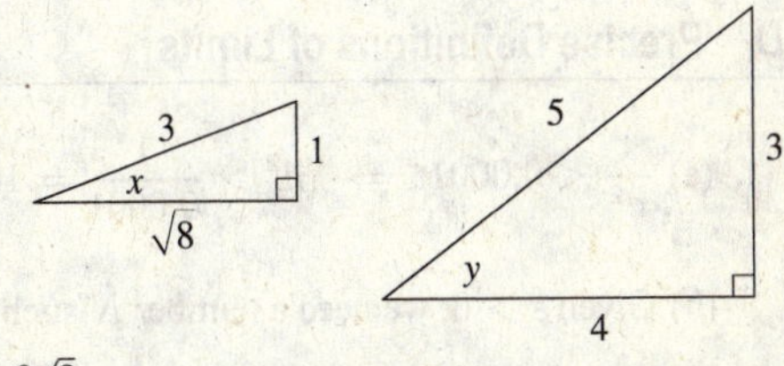

$\sin(x + y) = \sin x\,\cos y + \cos x\,\sin y = \frac{1}{3} \cdot \frac{4}{5} + \frac{\sqrt{8}}{3} \cdot \frac{3}{5} = \frac{4}{15} + \frac{3\sqrt{8}}{15} = \frac{4 + 6\sqrt{2}}{15}$.

31. $2\cos x - 1 = 0 \iff \cos x = \frac{1}{2} \Rightarrow x = \frac{\pi}{3}, \frac{5\pi}{3}$ for $x \in [0, 2\pi]$.

33. Using (15a), we have $\sin 2x = \cos x \iff 2\sin x\,\cos x - \cos x = 0 \iff \cos x(2\sin x - 1) = 0 \iff \cos x = 0$ or $2\sin x - 1 = 0 \Rightarrow x = \frac{\pi}{2}, \frac{3\pi}{2}$ or $\sin x = \frac{1}{2} \Rightarrow x = \frac{\pi}{6}$ or $\frac{5\pi}{6}$. Therefore, the solutions are $x = \frac{\pi}{6}, \frac{\pi}{2}, \frac{5\pi}{6}, \frac{3\pi}{2}$.

35. We know that $\sin x = \frac{1}{2}$ when $x = \frac{\pi}{6}$ or $\frac{5\pi}{6}$, and from Figure 13(a), we see that $\sin x \le \frac{1}{2} \Rightarrow 0 \le x \le \frac{\pi}{6}$ or $\frac{5\pi}{6} \le x \le 2\pi$ for $x \in [0, 2\pi]$.

37. $\tan x = -1$ when $x = \frac{3\pi}{4}, \frac{7\pi}{4}$, and $\tan x = 1$ when $x = \frac{\pi}{4}$ or $\frac{5\pi}{4}$. From Figure 14(a) we see that $-1 < \tan x < 1 \Rightarrow 0 \le x < \frac{\pi}{4}, \frac{3\pi}{4} < x < \frac{5\pi}{4}$, and $\frac{7\pi}{4} < x \le 2\pi$.

39. $y = \cos\left(x - \frac{\pi}{3}\right)$. We start with the graph of $y = \cos x$ and shift it $\frac{\pi}{3}$ units to the right.

41. $y = \frac{1}{3}\tan\left(x - \frac{\pi}{2}\right)$. We start with the graph of $y = \tan x$, shift it $\frac{\pi}{2}$ units to the right and compress it to $\frac{1}{3}$ of its original vertical size.

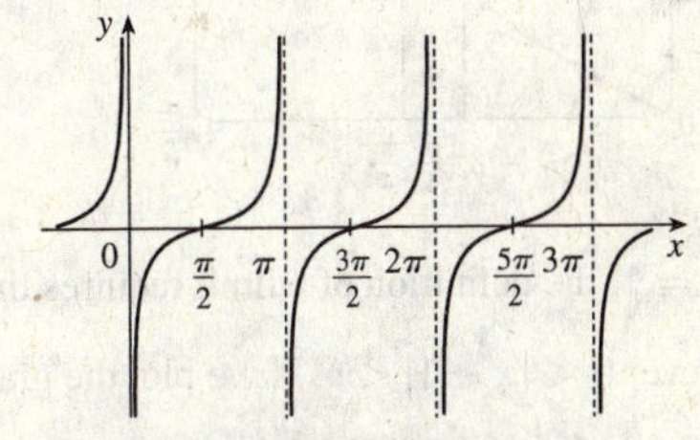

43. From the figure in the text, we see that $x = b\cos\theta$, $y = b\sin\theta$, and from the distance formula we have that the distance c from (x, y) to $(a, 0)$ is $c = \sqrt{(x - a)^2 + (y - 0)^2} \Rightarrow$

$$c^2 = (b\cos\theta - a)^2 + (b\sin\theta)^2 = b^2\cos^2\theta - 2ab\cos\theta + a^2 + b^2\sin^2\theta$$
$$= a^2 + b^2(\cos^2\theta + \sin^2\theta) - 2ab\cos\theta = a^2 + b^2 - 2ab\cos\theta \quad \text{[by (7)]}$$

45. Using the Law of Cosines, we have $c^2 = 1^2 + 1^2 - 2(1)(1)\cos(\alpha - \beta) = 2[1 - \cos(\alpha - \beta)]$. Now, using the distance formula, $c^2 = |AB|^2 = (\cos\alpha - \cos\beta)^2 + (\sin\alpha - \sin\beta)^2$. Equating these two expressions for c^2, we get $2[1 - \cos(\alpha - \beta)] = \cos^2\alpha + \sin^2\alpha + \cos^2\beta + \sin^2\beta - 2\cos\alpha\,\cos\beta - 2\sin\alpha\,\sin\beta \Rightarrow$ $1 - \cos(\alpha - \beta) = 1 - \cos\alpha\,\cos\beta - \sin\alpha\,\sin\beta \Rightarrow \cos(\alpha - \beta) = \cos\alpha\,\cos\beta + \sin\alpha\,\sin\beta$.

© 2016 Cengage Learning. All Rights Reserved. May not be scanned, copied or duplicated, or posted to a publicly accessible website, in whole or in part.

47. In Exercise 86 we used the subtraction formula for cosine to prove the addition formula for cosine. Using that formula with $x = \frac{\pi}{2} - \alpha$, $y = \beta$, we get $\cos\left[\left(\frac{\pi}{2} - \alpha\right) + \beta\right] = \cos\left(\frac{\pi}{2} - \alpha\right)\cos\beta - \sin\left(\frac{\pi}{2} - \alpha\right)\sin\beta \Rightarrow$ $\cos\left[\frac{\pi}{2} - (\alpha - \beta)\right] = \cos\left(\frac{\pi}{2} - \alpha\right)\cos\beta - \sin\left(\frac{\pi}{2} - \alpha\right)\sin\beta$. Now we use the identities given in the problem, $\cos\left(\frac{\pi}{2} - \theta\right) = \sin\theta$ and $\sin\left(\frac{\pi}{2} - \theta\right) = \cos\theta$, to get $\sin(\alpha - \beta) = \sin\alpha\cos\beta - \cos\alpha\sin\beta$.

D Precise Definitions of Limits

1. (a) $\frac{1}{n^2} < 0.0001 \quad\Rightarrow\quad n^2 > \frac{1}{0.0001} = 10,000 \quad\Rightarrow\quad n > 100$ (since n must be positive)

(b) Given $\varepsilon > 0$, we need a number N such that if $n > N$ then $\left|\frac{1}{n^2} - 0\right| < \varepsilon \;\Leftrightarrow\; \frac{1}{n^2} < \varepsilon \;\Leftrightarrow\; n > 1/\sqrt{\varepsilon}$ (since n is positive). Take $N = 1/\sqrt{\varepsilon}$. Now if $n > N = 1/\sqrt{\varepsilon}$ then $\left|\frac{1}{n^2} - 0\right| = \frac{1}{n^2} < \varepsilon$. Thus $\lim\limits_{n\to\infty} \frac{1}{n^2} = 0$ by Definition 1.

3. On the left side of $x = 2$, we need $|x - 2| < \left|\frac{10}{7} - 2\right| = \frac{4}{7}$. On the right side, we need $|x - 2| < \left|\frac{10}{3} - 2\right| = \frac{4}{3}$. For both of these conditions to be satisfied at once, we need the more restrictive of the two to hold, that is, $|x - 2| < \frac{4}{7}$. So we can choose $\delta = \frac{4}{7}$, or any smaller positive number.

5. The leftmost question mark is the solution of $\sqrt{x} = 1.6$ and the rightmost, $\sqrt{x} = 2.4$. So the values are $1.6^2 = 2.56$ and $2.4^2 = 5.76$. On the left side, we need $|x - 4| < |2.56 - 4| = 1.44$. On the right side, we need $|x - 4| < |5.76 - 4| = 1.76$. To satisfy both conditions, we need the more restrictive condition to hold—namely, $|x - 4| < 1.44$. Thus, we can choose $\delta = 1.44$, or any smaller positive number.

7.

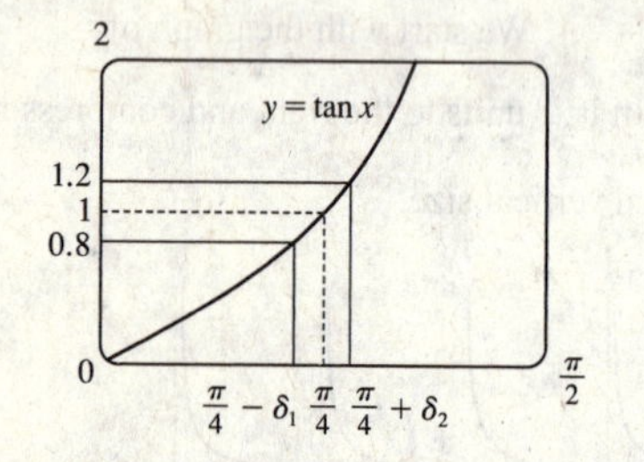

From the graph, we find that $y = \tan x = 0.8$ when $x \approx 0.675$, so $\frac{\pi}{4} - \delta_1 \approx 0.675 \quad\Rightarrow\quad \delta_1 \approx \frac{\pi}{4} - 0.675 \approx 0.1106$. Also, $y = \tan x = 1.2$ when $x \approx 0.876$, so $\frac{\pi}{4} + \delta_2 \approx 0.876 \quad\Rightarrow\quad \delta_2 = 0.876 - \frac{\pi}{4} \approx 0.0906$. Thus, we choose $\delta = 0.0906$ (or any smaller positive number) since this is the smaller of δ_1 and δ_2.

9. For $\varepsilon = 1$, the definition of a limit requires that we find δ such that $\left|(4 + x - 3x^3) - 2\right| < 1 \quad\Leftrightarrow\quad 1 < 4 + x - 3x^3 < 3$ whenever $0 < |x - 1| < \delta$. If we plot the graphs of $y = 1$, $y = 4 + x - 3x^3$ and $y = 3$ on the same screen, we see that we need $0.86 \le x \le 1.11$. So since $|1 - 0.86| = 0.14$ and $|1 - 1.11| = 0.11$, we choose $\delta = 0.11$ (or any smaller positive number). For $\varepsilon = 0.1$, we must find δ such that $\left|(4 + x - 3x^3) - 2\right| < 0.1 \quad\Leftrightarrow\quad 1.9 < 4 + x - 3x^3 < 2.1$ whenever $0 < |x - 1| < \delta$. From the graph, we see that we need $0.988 \le x \le 1.012$. So since $|1 - 0.988| = 0.012$ and $|1 - 1.012| = 0.012$, we choose $\delta = 0.012$ (or any smaller positive number) for the inequality to hold.

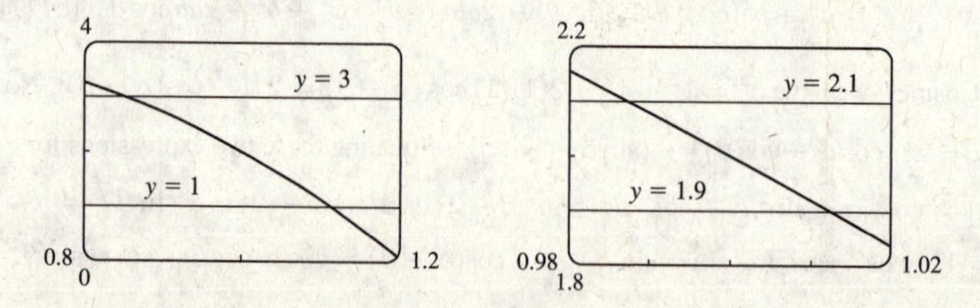

© 2016 Cengage Learning. All Rights Reserved. May not be scanned, copied or duplicated, or posted to a publicly accessible website, in whole or in part.

11. Given $\varepsilon > 0$, we need $\delta > 0$ such that if $|x| < \delta$ then $|x^3 - 0| < \varepsilon \;\Leftrightarrow\; |x|^3 < \varepsilon \;\Leftrightarrow\; |x| < \sqrt[3]{\varepsilon}$. Take $\delta = \sqrt[3]{\varepsilon}$.

Then $|x - 0| < \delta \;\Rightarrow\; |x^3 - 0| < \delta^3 = \varepsilon$. Thus, $\lim\limits_{x \to 0} x^3 = 0$ by the definition of a limit.

13. (a) $A = \pi r^2$ and $A = 1000 \text{ cm}^2 \;\Rightarrow\; \pi r^2 = 1000 \;\Rightarrow\; r^2 = \frac{1000}{\pi} \;\Rightarrow\; r = \sqrt{\frac{1000}{\pi}} \;\; (r > 0) \;\; \approx 17.8412$ cm.

(b) $|A - 1000| \le 5 \;\Rightarrow\; -5 \le \pi r^2 - 1000 \le 5 \;\Rightarrow\; 1000 - 5 \le \pi r^2 \le 1000 + 5 \;\Rightarrow$

$\sqrt{\frac{995}{\pi}} \le r \le \sqrt{\frac{1005}{\pi}} \;\Rightarrow\; 17.7966 \le r \le 17.8858$. $\sqrt{\frac{1000}{\pi}} - \sqrt{\frac{995}{\pi}} \approx 0.04466$ and $\sqrt{\frac{1005}{\pi}} - \sqrt{\frac{1000}{\pi}} \approx 0.04455$. So if the machinist gets the radius within 0.0445 cm of 17.8412, the area will be within 5 cm^2 of 1000.

(c) x is the radius, $f(x)$ is the area, a is the target radius given in part (a), L is the target area (1000), ε is the tolerance in the area (5), and δ is the tolerance in the radius given in part (b).

15. (a) $|4x - 8| = 4\,|x - 2| < 0.1 \;\Leftrightarrow\; |x - 2| < \frac{0.1}{4}$, so $\delta = \frac{0.1}{4} = 0.025$.

(b) $|4x - 8| = 4\,|x - 2| < 0.01 \;\Leftrightarrow\; |x - 2| < \frac{0.01}{4}$, so $\delta = \frac{0.01}{4} = 0.0025$.

17. Given $\varepsilon > 0$, we need $\delta > 0$ such that if $0 < |x - (-3)| < \delta$, then $|(1 - 4x) - 13| < \varepsilon$. But $|(1 - 4x) - 13| < \varepsilon \;\Leftrightarrow$ $|-4x - 12| < \varepsilon \;\Leftrightarrow\; |-4|\,|x + 3| < \varepsilon \;\Leftrightarrow\; |x - (-3)| < \varepsilon/4$. So if we choose $\delta = \varepsilon/4$, then $0 < |x - (-3)| < \delta \;\Rightarrow\; |(1 - 4x) - 13| < \varepsilon$.

Thus, $\lim\limits_{x \to -3} (1 - 4x) = 13$ by the definition of a limit.

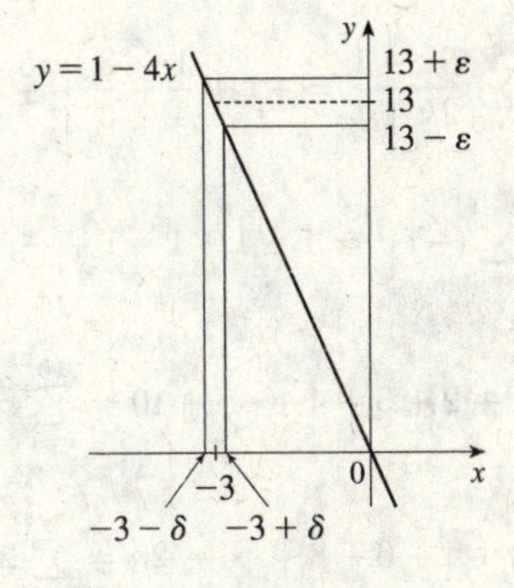

19. (a) Given $\varepsilon > 0$, we need a number N such that if $n > N$ then $\left|\frac{1}{2^n} - 0\right| < \varepsilon \;\Leftrightarrow\; \frac{1}{2^n} < \varepsilon \;\Leftrightarrow\; 2^n > 1/\varepsilon \;\Leftrightarrow$

$n > \log_2(1/\varepsilon)$. Take $N = \log_2(1/\varepsilon)$. Now if $n > N = \log_2(1/\varepsilon)$ then $\left|\frac{1}{2^n} - 0\right| = \frac{1}{2^n} < \varepsilon$. Thus $\lim\limits_{n \to \infty} \frac{1}{2^n} = 0$ by Definition 1.

(b) Given $\varepsilon > 0$, we need a number N such that if $n > N$ then $|r^n - 0| < \varepsilon \;\Leftrightarrow\; |r|^n < \varepsilon \;\Leftrightarrow\; n \ln|r| < \ln \varepsilon \;\Leftrightarrow$

$n > \frac{\ln \varepsilon}{\ln |r|}$ (since $\ln|r| < 0$ when $|r| < 1$). Take $N = \ln \varepsilon / \ln |r|$. Now if $n > N = \ln \varepsilon / \ln |r|$ then

$|r^n - 0| = |r|^n < \varepsilon$. Thus $\lim\limits_{n \to \infty} r^n = 0$ if $|r| < 1$ by Definition 1.

21. Let $\varepsilon > 0$. We want to find $\delta > 0$ such that

$$\left|\frac{xy}{\sqrt{x^2 + y^2}} - 0\right| < \varepsilon \qquad \text{whenever} \qquad 0 < \sqrt{x^2 + y^2} < \delta$$

that is,

$$\frac{|xy|}{\sqrt{x^2 + y^2}} < \varepsilon \qquad \text{whenever} \qquad 0 < \sqrt{x^2 + y^2} < \delta$$

[continued]

© 2016 Cengage Learning. All Rights Reserved. May not be scanned, copied or duplicated, or posted to a publicly accessible website, in whole or in part.

But $|x| = \sqrt{x^2} \le \sqrt{x^2+y^2}$ and $|y| = \sqrt{y^2} \le \sqrt{x^2+y^2}$, so

$$\frac{|xy|}{\sqrt{x^2+y^2}} \le \frac{\left(\sqrt{x^2+y^2}\right)^2}{\sqrt{x^2+y^2}} = \sqrt{x^2+y^2}$$

Thus, if we choose $\delta = \varepsilon$ and let $0 < \sqrt{x^2+y^2} < \delta$, then

$$\left| \frac{xy}{\sqrt{x^2+y^2}} - 0 \right| \le \sqrt{x^2+y^2} < \delta = \varepsilon$$

Hence, by Definition 3,

$$\lim_{(x,y)\to(0,0)} \frac{xy}{\sqrt{x^2+y^2}} = 0$$

F Sigma Notation

1. $\sum_{i=1}^{5} \sqrt{i} = \sqrt{1} + \sqrt{2} + \sqrt{3} + \sqrt{4} + \sqrt{5}$

3. $\sum_{i=4}^{6} 3^i = 3^4 + 3^5 + 3^6$

5. $\sum_{k=0}^{4} \frac{2k-1}{2k+1} = -1 + \frac{1}{3} + \frac{3}{5} + \frac{5}{7} + \frac{7}{9}$

7. $\sum_{i=1}^{n} i^{10} = 1^{10} + 2^{10} + 3^{10} + \cdots + n^{10}$

9. $\sum_{j=0}^{n-1} (-1)^j = 1 - 1 + 1 - 1 + \cdots + (-1)^{n-1}$

11. $1 + 2 + 3 + 4 + \cdots + 10 = \sum_{i=1}^{10} i$

13. $\frac{1}{2} + \frac{2}{3} + \frac{3}{4} + \frac{4}{5} + \cdots + \frac{19}{20} = \sum_{i=1}^{19} \frac{i}{i+1}$

15. $2 + 4 + 6 + 8 + \cdots + 2n = \sum_{i=1}^{n} 2i$

17. $1 + 2 + 4 + 8 + 16 + 32 = \sum_{i=0}^{5} 2^i$

19. $x + x^2 + x^3 + \cdots + x^n = \sum_{i=1}^{n} x^i$

21. $\sum_{i=4}^{8} (3i-2) = [3(4)-2] + [3(5)-2] + [3(6)-2] + [3(7)-2] + [3(8)-2] = 10 + 13 + 16 + 19 + 22 = 80$

23. $\sum_{j=1}^{6} 3^{j+1} = 3^2 + 3^3 + 3^4 + 3^5 + 3^6 + 3^7 = 9 + 27 + 81 + 243 + 729 + 2187 = 3276$

(For a more general method, see Exercise 47.)

25. $\sum_{n=1}^{20} (-1)^n = -1 + 1 - 1 + 1 - 1 + 1 - 1 + 1 - 1 + 1 - 1 + 1 - 1 + 1 - 1 + 1 - 1 + 1 - 1 + 1 = 0$

27. $\sum_{i=0}^{4} (2^i + i^2) = (1+0) + (2+1) + (4+4) + (8+9) + (16+16) = 61$

29. $\sum_{i=1}^{n} 2i = 2\sum_{i=1}^{n} i = 2 \cdot \frac{n(n+1)}{2}$ [by Theorem 3(c)] $= n(n+1)$

© 2016 Cengage Learning. All Rights Reserved. May not be scanned, copied or duplicated, or posted to a publicly accessible website, in whole or in part.

31. $\sum_{i=1}^{n}(i^2+3i+4)=\sum_{i=1}^{n}i^2+3\sum_{i=1}^{n}i+\sum_{i=1}^{n}4=\frac{n(n+1)(2n+1)}{6}+\frac{3n(n+1)}{2}+4n$

$=\frac{1}{6}[(2n^3+3n^2+n)+(9n^2+9n)+24n]=\frac{1}{6}(2n^3+12n^2+34n)=\frac{1}{3}n(n^2+6n+17)$

33. $\sum_{i=1}^{n}(i+1)(i+2)=\sum_{i=1}^{n}(i^2+3i+2)=\sum_{i=1}^{n}i^2+3\sum_{i=1}^{n}i+\sum_{i=1}^{n}2=\frac{n(n+1)(2n+1)}{6}+\frac{3n(n+1)}{2}+2n$

$=\frac{n(n+1)}{6}[(2n+1)+9]+2n=\frac{n(n+1)}{3}(n+5)+2n$

$=\frac{n}{3}[(n+1)(n+5)+6]=\frac{n}{3}(n^2+6n+11)$

35. $\sum_{i=1}^{n}(i^3-i-2)=\sum_{i=1}^{n}i^3-\sum_{i=1}^{n}i-\sum_{i=1}^{n}2=\left[\frac{n(n+1)}{2}\right]^2-\frac{n(n+1)}{2}-2n$

$=\frac{1}{4}n(n+1)[n(n+1)-2]-2n=\frac{1}{4}n(n+1)(n+2)(n-1)-2n$

$=\frac{1}{4}n[(n+1)(n-1)(n+2)-8]=\frac{1}{4}n[(n^2-1)(n+2)-8]=\frac{1}{4}n(n^3+2n^2-n-10)$

37. By Theorem 2(a) and Example 3, $\sum_{i=1}^{n}c=c\sum_{i=1}^{n}1=cn$.

39. $\sum_{i=1}^{n}[(i+1)^4-i^4]=(2^4-1^4)+(3^4-2^4)+(4^4-3^4)+\cdots+[(n+1)^4-n^4]$

$=(n+1)^4-1^4=n^4+4n^3+6n^2+4n$

On the other hand,

$\sum_{i=1}^{n}[(i+1)^4-i^4]=\sum_{i=1}^{n}(4i^3+6i^2+4i+1)=4\sum_{i=1}^{n}i^3+6\sum_{i=1}^{n}i^2+4\sum_{i=1}^{n}i+\sum_{i=1}^{n}1$

$=4S+n(n+1)(2n+1)+2n(n+1)+n \qquad \left[\text{where } S=\sum_{i=1}^{n}i^3\right]$

$=4S+2n^3+3n^2+n+2n^2+2n+n=4S+2n^3+5n^2+4n$

Thus, $n^4+4n^3+6n^2+4n=4S+2n^3+5n^2+4n$, from which it follows that

$4S=n^4+2n^3+n^2=n^2(n^2+2n+1)=n^2(n+1)^2$ and $S=\left[\frac{n(n+1)}{2}\right]^2$.

41. (a) $\sum_{i=1}^{n}\left[i^4-(i-1)^4\right]=(1^4-0^4)+(2^4-1^4)+(3^4-2^4)+\cdots+\left[n^4-(n-1)^4\right]=n^4-0=n^4$

(b) $\sum_{i=1}^{100}(5^i-5^{i-1})=(5^1-5^0)+(5^2-5^1)+(5^3-5^2)+\cdots+(5^{100}-5^{99})=5^{100}-5^0=5^{100}-1$

(c) $\sum_{i=3}^{99}\left(\frac{1}{i}-\frac{1}{i+1}\right)=\left(\frac{1}{3}-\frac{1}{4}\right)+\left(\frac{1}{4}-\frac{1}{5}\right)+\left(\frac{1}{5}-\frac{1}{6}\right)+\cdots+\left(\frac{1}{99}-\frac{1}{100}\right)=\frac{1}{3}-\frac{1}{100}=\frac{97}{300}$

(d) $\sum_{i=1}^{n}(a_i-a_{i-1})=(a_1-a_0)+(a_2-a_1)+(a_3-a_2)+\cdots+(a_n-a_{n-1})=a_n-a_0$

43. $\lim_{n\to\infty}\sum_{i=1}^{n}\frac{1}{n}\left(\frac{i}{n}\right)^2=\lim_{n\to\infty}\frac{1}{n^3}\sum_{i=1}^{n}i^2=\lim_{n\to\infty}\frac{1}{n^3}\frac{n(n+1)(2n+1)}{6}=\lim_{n\to\infty}\frac{1}{6}\left(1+\frac{1}{n}\right)\left(2+\frac{1}{n}\right)=\frac{1}{6}(1)(2)=\frac{1}{3}$

© 2016 Cengage Learning. All Rights Reserved. May not be scanned, copied or duplicated, or posted to a publicly accessible website, in whole or in part.

45. $\lim_{n\to\infty}\sum_{i=1}^{n}\frac{2}{n}\left[\left(\frac{2i}{n}\right)^3+5\left(\frac{2i}{n}\right)\right]=\lim_{n\to\infty}\sum_{i=1}^{n}\left[\frac{16}{n^4}i^3+\frac{20}{n^2}i\right]=\lim_{n\to\infty}\left[\frac{16}{n^4}\sum_{i=1}^{n}i^3+\frac{20}{n^2}\sum_{i=1}^{n}i\right]$

$$=\lim_{n\to\infty}\left[\frac{16}{n^4}\frac{n^2(n+1)^2}{4}+\frac{20}{n^2}\frac{n(n+1)}{2}\right]=\lim_{n\to\infty}\left[\frac{4(n+1)^2}{n^2}+\frac{10n(n+1)}{n^2}\right]$$

$$=\lim_{n\to\infty}\left[4\left(1+\frac{1}{n}\right)^2+10\left(1+\frac{1}{n}\right)\right]=4\cdot 1+10\cdot 1=14$$

47. Let $S=\sum_{i=1}^{n}ar^{i-1}=a+ar+ar^2+\cdots+ar^{n-1}$. Multiplying both sides by r gives us

$rS=ar+ar^2+\cdots+ar^{n-1}+ar^n$. Subtracting the first equation from the second, we find

$(r-1)S=ar^n-a=a(r^n-1)$, so $S=\dfrac{a(r^n-1)}{r-1}$ [since $r\neq 1$].

49. $\sum_{i=1}^{n}(2i+2^i)=2\sum_{i=1}^{n}i+\sum_{i=1}^{n}2\cdot 2^{i-1}=2\frac{n(n+1)}{2}+\frac{2(2^n-1)}{2-1}=2^{n+1}+n^2+n-2.$

For the first sum we have used Theorems 2(a) and 3(c), and for the second, Exercise 47 with $a=r=2$.

G Complex Numbers

1. $(5-6i)+(3+2i)=(5+3)+(-6+2)i=8+(-4)i=8-4i$

3. $(2+5i)(4-i)=2(4)+2(-i)+(5i)(4)+(5i)(-i)=8-2i+20i-5i^2=8+18i-5(-1)$

$$=8+18i+5=13+18i$$

5. $\overline{12+7i}=12-7i$

7. $\dfrac{1+4i}{3+2i}=\dfrac{1+4i}{3+2i}\cdot\dfrac{3-2i}{3-2i}=\dfrac{3-2i+12i-8(-1)}{3^2+2^2}=\dfrac{11+10i}{13}=\dfrac{11}{13}+\dfrac{10}{13}i$

9. $\dfrac{1}{1+i}=\dfrac{1}{1+i}\cdot\dfrac{1-i}{1-i}=\dfrac{1-i}{1-(-1)}=\dfrac{1-i}{2}=\dfrac{1}{2}-\dfrac{1}{2}i$

11. $i^3=i^2\cdot i=(-1)i=-i$

13. $\sqrt{-25}=\sqrt{25}\,i=5i$

15. $\overline{12-5i}=12+15i$ and $|12-15i|=\sqrt{12^2+(-5)^2}=\sqrt{144+25}=\sqrt{169}=13$

17. $\overline{-4i}=\overline{0-4i}=0+4i=4i$ and $|-4i|=\sqrt{0^2+(-4)^2}=\sqrt{16}=4$

19. $4x^2+9=0\;\Leftrightarrow\;4x^2=-9\;\Leftrightarrow\;x^2=-\frac{9}{4}\;\Leftrightarrow\;x=\pm\sqrt{-\frac{9}{4}}=\pm\sqrt{\frac{9}{4}}\,i=\pm\frac{3}{2}i.$

21. By the quadratic formula, $x^2+2x+5=0\;\Leftrightarrow\;x=\dfrac{-2\pm\sqrt{2^2-4(1)(5)}}{2(1)}=\dfrac{-2\pm\sqrt{-16}}{2}=\dfrac{-2\pm 4i}{2}=-1\pm 2i.$

23. By the quadratic formula, $z^2+z+2=0\;\Leftrightarrow\;z=\dfrac{-1\pm\sqrt{1^2-4(1)(2)}}{2(1)}=\dfrac{-1\pm\sqrt{-7}}{2}=-\dfrac{1}{2}\pm\dfrac{\sqrt{7}}{2}i.$

© 2016 Cengage Learning. All Rights Reserved. May not be scanned, copied or duplicated, or posted to a publicly accessible website, in whole or in part.

25. For $z = -3 + 3i$, $r = \sqrt{(-3)^2 + 3^2} = 3\sqrt{2}$ and $\tan\theta = \frac{3}{-3} = -1 \;\Rightarrow\; \theta = \frac{3\pi}{4}$ (since z lies in the second quadrant).

Therefore, $-3 + 3i = 3\sqrt{2}\left(\cos\frac{3\pi}{4} + i\sin\frac{3\pi}{4}\right)$.

27. For $z = 3 + 4i$, $r = \sqrt{3^2 + 4^2} = 5$ and $\tan\theta = \frac{4}{3} \;\Rightarrow\; \theta = \tan^{-1}\left(\frac{4}{3}\right)$ (since z lies in the first quadrant). Therefore,

$3 + 4i = 5\left\{\cos\left[\tan^{-1}\left(\frac{4}{3}\right)\right] + i\sin\left[\tan^{-1}\left(\frac{4}{3}\right)\right]\right\}$.

29. For $z = \sqrt{3} + i$, $r = \sqrt{\left(\sqrt{3}\right)^2 + 1^2} = 2$ and $\tan\theta = \frac{1}{\sqrt{3}} \;\Rightarrow\; \theta = \frac{\pi}{6} \;\Rightarrow\; z = 2\left(\cos\frac{\pi}{6} + i\sin\frac{\pi}{6}\right)$.

For $w = 1 + \sqrt{3}\,i$, $r = 2$ and $\tan\theta = \sqrt{3} \;\Rightarrow\; \theta = \frac{\pi}{3} \;\Rightarrow\; w = 2\left(\cos\frac{\pi}{3} + i\sin\frac{\pi}{3}\right)$.

Therefore, $zw = 2 \cdot 2\left[\cos\left(\frac{\pi}{6} + \frac{\pi}{3}\right) + i\sin\left(\frac{\pi}{6} + \frac{\pi}{3}\right)\right] = 4\left(\cos\frac{\pi}{2} + i\sin\frac{\pi}{2}\right)$,

$z/w = \frac{2}{2}\left[\cos\left(\frac{\pi}{6} - \frac{\pi}{3}\right) + i\sin\left(\frac{\pi}{6} - \frac{\pi}{3}\right)\right] = \cos\left(-\frac{\pi}{6}\right) + i\sin\left(-\frac{\pi}{6}\right)$, and $1 = 1 + 0i = 1(\cos 0 + i\sin 0) \;\Rightarrow$

$1/z = \frac{1}{2}\left[\cos\left(0 - \frac{\pi}{6}\right) + i\sin\left(0 - \frac{\pi}{6}\right)\right] = \frac{1}{2}\left[\cos\left(-\frac{\pi}{6}\right) + i\sin\left(-\frac{\pi}{6}\right)\right]$. For $1/z$, we could also use the formula that precedes Example 5 to obtain $1/z = \frac{1}{2}\left(\cos\frac{\pi}{6} - i\sin\frac{\pi}{6}\right)$.

31. For $z = 2\sqrt{3} - 2i$, $r = \sqrt{\left(2\sqrt{3}\right)^2 + (-2)^2} = 4$ and $\tan\theta = \frac{-2}{2\sqrt{3}} = -\frac{1}{\sqrt{3}} \;\Rightarrow\; \theta = -\frac{\pi}{6} \;\Rightarrow$

$z = 4\left[\cos\left(-\frac{\pi}{6}\right) + i\sin\left(-\frac{\pi}{6}\right)\right]$. For $w = -1 + i$, $r = \sqrt{2}$, $\tan\theta = \frac{1}{-1} = -1 \;\Rightarrow\; \theta = \frac{3\pi}{4} \;\Rightarrow$

$w = \sqrt{2}\left(\cos\frac{3\pi}{4} + i\sin\frac{3\pi}{4}\right)$. Therefore, $zw = 4\sqrt{2}\left[\cos\left(-\frac{\pi}{6} + \frac{3\pi}{4}\right) + i\sin\left(-\frac{\pi}{6} + \frac{3\pi}{4}\right)\right] = 4\sqrt{2}\left(\cos\frac{7\pi}{12} + i\sin\frac{7\pi}{12}\right)$,

$z/w = \frac{4}{\sqrt{2}}\left[\cos\left(-\frac{\pi}{6} - \frac{3\pi}{4}\right) + i\sin\left(-\frac{\pi}{6} - \frac{3\pi}{4}\right)\right] = \frac{4}{\sqrt{2}}\left[\cos\left(-\frac{11\pi}{12}\right) + i\sin\left(-\frac{11\pi}{12}\right)\right] = 2\sqrt{2}\left(\cos\frac{13\pi}{12} + i\sin\frac{13\pi}{12}\right)$, and

$1/z = \frac{1}{4}\left[\cos\left(-\frac{\pi}{6}\right) - i\sin\left(-\frac{\pi}{6}\right)\right] = \frac{1}{4}\left(\cos\frac{\pi}{6} + i\sin\frac{\pi}{6}\right)$.

33. For $z = 1 + i$, $r = \sqrt{2}$ and $\tan\theta = \frac{1}{1} = 1 \;\Rightarrow\; \theta = \frac{\pi}{4} \;\Rightarrow\; z = \sqrt{2}\left(\cos\frac{\pi}{4} + i\sin\frac{\pi}{4}\right)$. So by De Moivre's Theorem,

$$\begin{aligned}(1+i)^{20} &= \left[\sqrt{2}\left(\cos\tfrac{\pi}{4} + i\sin\tfrac{\pi}{4}\right)\right]^{20} = (2^{1/2})^{20}\left(\cos\tfrac{20\cdot\pi}{4} + i\sin\tfrac{20\cdot\pi}{4}\right) = 2^{10}(\cos 5\pi + i\sin 5\pi)\\ &= 2^{10}[-1 + i(0)] = -2^{10} = -1024\end{aligned}$$

35. For $z = 2\sqrt{3} + 2i$, $r = \sqrt{\left(2\sqrt{3}\right)^2 + 2^2} = \sqrt{16} = 4$ and $\tan\theta = \frac{2}{2\sqrt{3}} = \frac{1}{\sqrt{3}} \;\Rightarrow\; \theta = \frac{\pi}{6} \;\Rightarrow\; z = 4\left(\cos\frac{\pi}{6} + i\sin\frac{\pi}{6}\right)$.

So by De Moivre's Theorem,

$\left(2\sqrt{3} + 2i\right)^5 = \left[4\left(\cos\frac{\pi}{6} + i\sin\frac{\pi}{6}\right)\right]^5 = 4^5\left(\cos\frac{5\pi}{6} + i\sin\frac{5\pi}{6}\right) = 1024\left[-\frac{\sqrt{3}}{2} + \frac{1}{2}i\right] = -512\sqrt{3} + 512i$.

37. $1 = 1 + 0i = 1\,(\cos 0 + i\sin 0)$. Using Equation 3 with $r = 1$, $n = 8$, and $\theta = 0$, we have

$w_k = 1^{1/8}\left[\cos\left(\dfrac{0 + 2k\pi}{8}\right) + i\sin\left(\dfrac{0 + 2k\pi}{8}\right)\right] = \cos\dfrac{k\pi}{4} + i\sin\dfrac{k\pi}{4}$, where $k = 0, 1, 2, \ldots, 7$.

$w_0 = 1(\cos 0 + i\sin 0) = 1$, $w_1 = 1\left(\cos\frac{\pi}{4} + i\sin\frac{\pi}{4}\right) = \frac{1}{\sqrt{2}} + \frac{1}{\sqrt{2}}i$,

$w_2 = 1\left(\cos\frac{\pi}{2} + i\sin\frac{\pi}{2}\right) = i$, $w_3 = 1\left(\cos\frac{3\pi}{4} + i\sin\frac{3\pi}{4}\right) = -\frac{1}{\sqrt{2}} + \frac{1}{\sqrt{2}}i$,

$w_4 = 1(\cos\pi + i\sin\pi) = -1$, $w_5 = 1\left(\cos\frac{5\pi}{4} + i\sin\frac{5\pi}{4}\right) = -\frac{1}{\sqrt{2}} - \frac{1}{\sqrt{2}}i$,

$w_6 = 1\left(\cos\frac{3\pi}{2} + i\sin\frac{3\pi}{2}\right) = -i$, $w_7 = 1\left(\cos\frac{7\pi}{4} + i\sin\frac{7\pi}{4}\right) = \frac{1}{\sqrt{2}} - \frac{1}{\sqrt{2}}i$

Im
i
0
1
Re

© 2016 Cengage Learning. All Rights Reserved. May not be scanned, copied or duplicated, or posted to a publicly accessible website, in whole or in part.

39. $i = 0 + i = 1\left(\cos\frac{\pi}{2} + i\sin\frac{\pi}{2}\right)$. Using Equation 3 with $r = 1$, $n = 3$, and $\theta = \frac{\pi}{2}$, we have

$$w_k = 1^{1/3}\left[\cos\left(\frac{\frac{\pi}{2} + 2k\pi}{3}\right) + i\sin\left(\frac{\frac{\pi}{2} + 2k\pi}{3}\right)\right], \text{ where } k = 0, 1, 2.$$

$$w_0 = \left(\cos\tfrac{\pi}{6} + i\sin\tfrac{\pi}{6}\right) = \tfrac{\sqrt{3}}{2} + \tfrac{1}{2}i$$

$$w_1 = \left(\cos\tfrac{5\pi}{6} + i\sin\tfrac{5\pi}{6}\right) = -\tfrac{\sqrt{3}}{2} + \tfrac{1}{2}i$$

$$w_2 = \left(\cos\tfrac{9\pi}{6} + i\sin\tfrac{9\pi}{6}\right) = -i$$

Im

Re

0

$-i$

41. Using Euler's formula (6) with $y = \frac{\pi}{2}$, we have $e^{i\pi/2} = \cos\frac{\pi}{2} + i\sin\frac{\pi}{2} = 0 + 1i = i$.

43. Using Euler's formula (6) with $y = \dfrac{\pi}{3}$, we have $e^{i\pi/3} = \cos\dfrac{\pi}{3} + i\sin\dfrac{\pi}{3} = \dfrac{1}{2} + \dfrac{\sqrt{3}}{2}i$.

45. Using Equation 7 with $x = 2$ and $y = \pi$, we have $e^{2+i\pi} = e^2 e^{i\pi} = e^2(\cos\pi + i\sin\pi) = e^2(-1 + 0) = -e^2$.

47. Take $r = 1$ and $n = 3$ in De Moivre's Theorem to get

$$\begin{aligned}
[1(\cos\theta + i\sin\theta)]^3 &= 1^3(\cos 3\theta + i\sin 3\theta) \\
(\cos\theta + i\sin\theta)^3 &= \cos 3\theta + i\sin 3\theta \\
\cos^3\theta + 3(\cos^2\theta)(i\sin\theta) + 3(\cos\theta)(i\sin\theta)^2 + (i\sin\theta)^3 &= \cos 3\theta + i\sin 3\theta \\
\cos^3\theta + (3\cos^2\theta\,\sin\theta)i - 3\cos\theta\,\sin^2\theta - (\sin^3\theta)i &= \cos 3\theta + i\sin 3\theta \\
(\cos^3\theta - 3\sin^2\theta\,\cos\theta) + (3\sin\theta\,\cos^2\theta - \sin^3\theta)i &= \cos 3\theta + i\sin 3\theta
\end{aligned}$$

Equating real and imaginary parts gives $\cos 3\theta = \cos^3\theta - 3\sin^2\theta\,\cos\theta$ and $\sin 3\theta = 3\sin\theta\,\cos^2\theta - \sin^3\theta$.

49. $F(x) = e^{rx} = e^{(a+bi)x} = e^{ax+bxi} = e^{ax}(\cos bx + i\sin bx) = e^{ax}\cos bx + i(e^{ax}\sin bx) \Rightarrow$

$$\begin{aligned}
F'(x) &= (e^{ax}\cos bx)' + i(e^{ax}\sin bx)' \\
&= (ae^{ax}\cos bx - be^{ax}\sin bx) + i(ae^{ax}\sin bx + be^{ax}\cos bx) \\
&= a[e^{ax}(\cos bx + i\sin bx)] + b[e^{ax}(-\sin bx + i\cos bx)] \\
&= ae^{rx} + b[e^{ax}(i^2\sin bx + i\cos bx)] \\
&= ae^{rx} + bi[e^{ax}(\cos bx + i\sin bx)] = ae^{rx} + bie^{rx} = (a + bi)e^{rx} = re^{rx}
\end{aligned}$$

© 2016 Cengage Learning. All Rights Reserved. May not be scanned, copied or duplicated, or posted to a publicly accessible website, in whole or in part.